Fundamental Physical Constants

Avogadro	$N_0 = 6.022\ 136 \times 10^{23}$ l/mol
Boltzmann	$k = 1.380\ 658 \times 10^{-23}$ J/K
Planck	$h = 6.626\ 076 \times 10^{-34}$ Js
Gas Constant	$\bar{R} = N_0 k = 8.314\ 51$ J/mol K
Atomic Mass Unit	$m_0 = 1.660\ 540 \times 10^{-27}$ kg
Velocity of light	$c = 2.997\ 925 \times 10^{8}$ m/s
Electron Charge	$e = 1.602\ 177 \times 10^{-19}$ C
Electron Mass	$m_e = 9.109\ 389 \times 10^{-31}$ kg
Proton Mass	$m_p = 1.672\ 623 \times 10^{-27}$ kg
Gravitation (Std.)	$g = 9.806\ 65$ m/s^2
Stefan Boltzmann	$\sigma = 5.670\ 51 \times 10^{-8}$ W/m^2 K^4

Prefixes

10^{-1}	deci	d
10^{-2}	centi	c
10^{-3}	milli	m
10^{-6}	micro	μ
10^{-9}	nano	n
10^{-12}	pico	p
10^{-15}	femto	f
10^{1}	deka	da
10^{2}	hecto	h
10^{3}	kilo	k
10^{6}	mega	M
10^{9}	giga	G
10^{12}	tera	T
10^{15}	peta	P

Concentration

10^{-6} parts per million ppm

FUNDAMENTALS OF
THERMODYNAMICS

SIXTH EDITION

FUNDAMENTALS OF
THERMODYNAMICS

SIXTH EDITION

RICHARD E. SONNTAG
CLAUS BORGNAKKE
University of Michigan

GORDON J. VAN WYLEN
Hope College   (emeritus)

JOHN WILEY & SONS, INC.

Acquisitions Editor	Joseph Hayton
Production Editor	Sandra Russell
Marketing Manager	Katherine Hepburn
Senior Designer	Dawn L. Stanley
Production Management Services	UG / GGS Information Services, Inc.
Cover Photo	© Judy Dole / The Image Bank

This book was typeset in 10/12 Times New Roman By UG / GGS Information Services, Inc. and printed and bound by R. R. Donnelley and Sons (Willard). The cover was printed by Phoenix Color Corp.

The paper in this book was manufactured by a mill whose forest management programs include sustained yield harvesting of its timberlands. Sustained yield harvesting principles ensure that the number of trees cut each year does not exceed the amount of new growth.

This book is printed on acid-free paper.

Richard E. Sonntag, Claus Borgnakke, and Gordon J. Van Wylen
Fundamentals of Thermodynamics, sixth edition

ISBN 0-471-15232-3

Printed in the United States of America.

10 9 8 7 6 5 4 3 2 1

PREFACE

In this sixth edition we have retained the basic objective of the earlier editions:

- to present a comprehensive and rigorous treatment of classical thermodynamics while retaining an engineering perspective, and in doing so
- to lay the groundwork for subsequent studies in such fields as fluid mechanics, heat transfer, and statistical thermodynamics, and also
- to prepare the student to effectively use thermodynamics in the practice of engineering.

We have deliberately directed our presentation to students. New concepts and definitions are presented in the context where they are first relevant in a natural progression. The first thermodynamic properties to be defined (Chapter 2) are those that can be readily measured: pressure, specific volume, and temperature. In Chapter 3, tables of thermodynamic properties are introduced, but only in regard to these measurable properties. Internal energy and enthalpy are introduced in connection with the first law, entropy with the second law, and the Helmholtz and Gibbs functions in the chapter on thermodynamic relations. Many real world realistic examples have been included in the book to assist the student in gaining an understanding of thermodynamics, and the problems at the end of each chapter have been carefully sequenced to correlate with the subject matter and are grouped and identified as such. The early chapters in particular contain a much larger number of examples, illustrations, and problems than in previous editions, and throughout the book, chapter-end summaries are included, followed by a set of concept/study problems that should be of benefit to the students.

NEW FEATURES IN THIS EDITION

End-of-Chapter Summaries

The new end-of-chapter summaries provide a short review of the main concepts covered in the chapter, with highlighted key words. To further enhance the summary we have listed the set of skills that the student should have mastered after studying the chapter. These skills are among the outcomes that can be tested with the accompanying set of study-guide problems in addition to the main set of homework problems.

Main Concepts and Formulas

Main concepts and formulas are included at the end of each chapter for reference.

Study Guide Problems

We have made a set of study guide problems for each chapter as a quick check of the chapter material. These are selected to be short and directed toward a very specific concept. A student can answer all of these questions to assess their level of understanding, and determine if any of the subjects need to be studied further. These problems are also suitable to use together with the rest of the homework problems in assignments and included in the solution manual.

Homework Problems

The number of homework problems has been greatly expanded and now exceeds 2,400. A large number of introductory problems have been added to cover all aspects of the chapter material. We have furthermore separated the problems into sections according to subject for easy selection according to the particular coverage given. A number of more comprehensive problems have been retained and grouped in the end as review problems.

Tables

The tables of the substances have been expanded to include the specific internal energy in the superheated vapor region. The ideal gas tables have been printed on a mass basis as well as a mole basis, to reflect their use on mass basis early in the text, and mole basis for the combustion and chemical equilibrium chapters.

Revisions

In this edition we have incorporated a number of developments and approaches included in our recent textbook, *Introduction to Engineering Thermodynamics*, Richard E. Sonntag and Claus Borgnakke, John Wiley & Sons, Inc. (2001). In Chapter 3, we first introduce thermodynamic tables, and then note the behavior of superheated vapor at progressively lower densities, which leads to the definition of the ideal gas model, then the compressibility factor and equations of state. In Chapter 5, the result of ideal gas energy depending only on temperature follows the examination of steam table values at different temperatures and pressures. Second law presentation in Chapter 7 is streamlined, with better integration of the concepts of thermodynamic temperature and ideal gas temperature. Coverage of ideal gas and ideal gas mixtures focuses on unit mass basis, instead of mole basis, and is simpler. Development of availability and reversible work in Chapter 10 focuses on the steady-state process, and leads to the general expression for exergy. We have therefore included a new section on the general exergy balance to amplify the concept of transport and destruction of exergy. The chapter with property relations is slightly reorganized and streamlined to also focus on properties on a mass basis. Due to current technology developments we have expanded our discussion of the fuel cells and also updated the chapter with combustion.

Expanded Software Included

In this edition we have included the expanded software CATT2 that includes a number of additional substances besides those included in the printed tables in Appendix B. A number of hydrocarbon fuels, refrigerants, and cryogenic fluids are included and a printed version is available in the booklet *Thermodynamic and Transport Properties*, Claus Borgnakke and Richard E. Sonntag, John Wiley & Sons, Inc. (1997).

FLEXIBILITY IN COVERAGE AND SCOPE

We have attempted to cover fairly comprehensively the basic subject matter of classical thermodynamics, and believe that the book provides adequate preparation for study of the application of thermodynamics to the various professional fields as well as for study of more advanced topics in thermodynamics, such as those related to materials, surface phenomena, plasmas, and cryogenics. We also recognize that a number of colleges offer a single introductory course in thermodynamics for all departments, and we have tried to cover those topics that the various departments might wish to have included in such a course. However, since specific courses vary considerably in prerequisites, specific objectives, duration, and background of the students, we have arranged the material, particularly in the later chapters, so that there is considerable flexibility in the amount of material that may be covered.

Units

Our philosophy regarding units in this edition has been to organize the book so that the course or sequence can be taught entirely in SI units (Le Système International d'Unités). Thus, all the text examples are in SI units, as are the complete problem sets and the thermodynamic tables. In recognition, however, of the continuing need for engineering graduates to be familiar with English Engineering units, we have included an introduction to this system in Chapter 2. We have also repeated a sufficient number of examples, problems, and tables in these units, which should allow for suitable practice for those who wish to use these units. For dealing with English units, the force-mass conversion question between pound mass and pound force is treated simply as a units conversion, without using an explicit conversion constant. Throughout, symbols, units and sign conventions are treated as in previous editions.

Supplements and Additional Support

Additional support is made available through a companion website at **www.wiley.com/college/sonntag**. Tutorials and reviews of all the basic material as indicated in the main text by the ThermoNet icon are accessible through the companion website. The website allows students to go through a self-paced study developing the basic skill set associated with the various subjects usually covered in a first course in thermodynamics.

The chapter on compressible flow is also available at **www.wiley.com/college/sonntag** and revised with summary, study guide problems, and new homework problems. We recognize that in many cases this chapter is not included in the thermodynamics courses, but may instead be covered elsewhere in the curriculum.

We have tried to include material appropriate and sufficient for a two-semester course sequence, and to provide flexibility for choice of topic coverage. Instructors may want to visit the companion website at **www.wiley.com/college/sonntag** for information and suggestions on possible course structure and schedules, additional study problem material, and current errata for the book.

ACKNOWLEDGMENTS

We acknowledge with appreciation the suggestions, counsel, and encouragement of many colleagues, both at the University of Michigan and elsewhere. This assistance has been very helpful to us during the writing of this edition, as it was with the earlier editions of the book. Both undergraduate and graduate students have been of particular assistance, for their perceptive questions have often caused us to rewrite or rethink a given portion of the text, or to try to develop a better way of presenting the material in order to anticipate such questions or difficulties. Finally, for each of us, the encouragement and patience of our wives and families have been indispensable and have made this time of writing pleasant and enjoyable, in spite of the pressures of the project. A special thanks to a number of colleagues at other institutions, who have reviewed the book and provided input to the revisions. Some of the reviewers are

> Edward E. Anderson, Texas Tech University
> Haim H. Bau, University of Pennsylvania
> Pei-Feng Hsu, Florida Institute of Technology
> Gerald J. Micklow, University of Alabama
> Jayathi Y. Murthy, Carnegie Mellon University
> Anthony J. Wheeler, San Francisco State University

We also wish to thank our editor, Joseph Hayton, for his effort in the planning and the support and encouragement during the production of this edition.

Our hope is that this book will contribute to the effective teaching of thermodynamics to students who face very significant challenges and opportunities during their professional careers. Your comments, criticism, and suggestions will also be appreciated and you may channel that through Claus Borgnakke, claus@umich.edu.

RICHARD E. SONNTAG
CLAUS BORGNAKKE
GORDON J. VAN WYLEN
Ann Arbor, Michigan
April 2002

CONTENTS

SYMBOLS

a	acceleration
A	area
a, A	specific Helmholtz function and total Helmholtz function
AF	air-fuel ratio
B_S	adiabatic bulk modulus
B_T	isothermal bulk modulus
c	velocity of sound
c	mass fraction
C_D	coefficient of discharge
C_p	constant-pressure specific heat
C_v	constant-volume specific heat
C_{po}	zero-pressure constant-pressure specific heat
C_{vo}	zero-pressure constant-volume specific heat
e, E	specific energy and total energy
F	force
FA	fuel-air ratio
g	acceleration due to gravity
g, G	specific Gibbs function and total Gibbs function
h, H	specific enthalpy and total enthalpy
i	electrical current
I	irreversibility
J	proportionality factor to relate units of work to units of heat
k	specific heat ratio: C_p/C_v
K	equilibrium constant
KE	kinetic energy
L	length
m	mass
$\dot{m}$	mass flow rate
M	molecular weight
M	Mach number
n	number of moles
n	polytropic exponent
P	pressure
P_i	partial pressure of component i in a mixture
PE	potential energy
P_r	relative pressure as used in gas tables
q, Q	heat transfer per unit mass and total heat transfer
$\dot{Q}$	rate of heat transfer

Q_H, Q_L	heat transfer with high-temperature body and heat transfer with low-temperature body; sign determined from context
R	gas constant
$\overline{R}$	universal gas constant
s, S	specific entropy and total entropy
S_{gen}	entropy generation
$\dot{S}_{\text{gen}}$	rate of entropy generation
t	time
T	temperature
u, U	specific internal energy and total internal energy
v, V	specific volume and total volume
v_r	relative specific volume as used in gas tables
V	velocity
w, W	work per unit mass and total work
$\dot{W}$	rate of work, or power
w^{rev}	reversible work between two states
x	quality
y	gas-phase mole fraction
Z	elevation
Z	compressibility factor
Z	electrical charge

SCRIPT LETTERS

$\mathscr{E}$	electrical potential
$\mathscr{S}$	surface tension
$\mathscr{T}$	tension

GREEK LETTERS

α	residual volume
α_p	volume expansivity
β	coefficient of performance for a refrigerator
β'	coefficient of performance for a heat pump
β_S	adiabatic compressibility
β_T	isothermal compressibility
η	efficiency
μ	chemical potential
ν	stoichiometric coefficient
ρ	density
Φ	equivalence ratio
ϕ	relative humidity
ϕ, Φ	exergy or availability for a control mass
ψ	flow availability
ω	humidity ratio or specific humidity
ω	a centric factor

SUBSCRIPTS

c	property at the critical point
c.v.	control volume
e	state of a substance leaving a control volume
f	formation

f	property of saturated liquid
fg	difference in property for saturated vapor and saturated liquid
g	property of saturated vapor
i	state of a substance entering a control volume
i	property of saturated solid
if	difference in property for saturated liquid and saturated solid
ig	difference in property for saturated vapor and saturated solid
r	reduced property
s	isentropic process
0	property of the surroundings
0	stagnation property

SUPERSCRIPTS

—	bar over symbol denotes property on a molal basis (over V, H, S, U, A, G, the bar denotes partial molal property)
°	property at standard-state condition
*	ideal gas
*	property at the throat of a nozzle
rev	reversible

FUNDAMENTALS OF THERMODYNAMICS

SIXTH EDITION

SOME INTRODUCTORY COMMENTS 1

In the course of our study of thermodynamics, a number of the examples and problems presented refer to processes that occur in equipment such as a steam power plant, a fuel cell, a vapor-compression refrigerator, a thermoelectric cooler, a turbine or rocket engine, and an air separation plant. In this introductory chapter, a brief description of this equipment is given. There are at least two reasons for including such a chapter. First, many students have had limited contact with such equipment, and the solution of problems will be more meaningful when they have some familiarity with the actual processes and the equipment. Second, this chapter will provide an introduction to thermodynamics, including the use of certain terms (which will be more formally defined in later chapters), some of the problems to which thermodynamics can be applied, and some of the things that have been accomplished, at least in part, from the application of thermodynamics.

Thermodynamics is relevant to many other processes than those cited in this chapter. It is basic to the study of materials, chemical reactions, and plasmas. The student should bear in mind that this chapter is only a brief and necessarily very incomplete introduction to the subject of thermodynamics.

1.1 THE SIMPLE STEAM POWER PLANT

A schematic diagram of a recently installed steam power plant is shown in Fig. 1.1. High-pressure superheated steam leaves the steam drum at the top of the boiler, also referred to as a steam generator, and enters the turbine. The steam expands in the turbine and in doing so does work, which enables the turbine to drive the electric generator. The steam, now at low pressure, exits the turbine and enters the heat exchanger, where heat is transferred from the steam (causing it to condense) to the cooling water. Since large quantities of cooling water are required, power plants have traditionally been located near rivers or lakes, leading to thermal pollution of those water supplies. More recently, condenser cooling water is recycled by evaporating a fraction of the water in large cooling towers, thereby cooling the remainder of the water that remains as a liquid. In the power plant shown in Fig. 1.1, the plant is designed to recycle the condenser cooling water by using the heated water for district space heating.

The pressure of the condensate leaving the condenser is increased in the pump, enabling it to return to the steam generator for reuse. In many cases, an economizer or water preheater is used in the steam cycle, and in many power plants, the air that is used for combustion of the fuel may be preheated by the exhaust combustion-product gases. These exhaust gases must also be purified before being discharged to the atmosphere, such that there are many complications to the simple cycle.

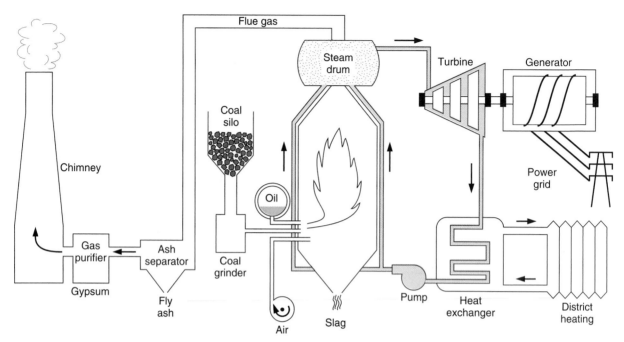

FIGURE 1.1 Schematic diagram of a steam power plant.

Figure 1.2 is a photograph of the power plant depicted in Fig. 1.1. The tall building shown at the left is the boiler house, next to which are buildings housing the turbine and other components. Also noted are the tall chimney or stack, and the coal supply ship at dock. This particular power plant is located in Denmark, and at the time of its installation, set a world record of efficiency, converting 45% of the 850 MW of coal combustion energy into electricity. Another 47% is reusable for district space heating, an amount that in older plants was simply thrown away to the environment with no benefit.

The steam power plant described utilizes coal as the combustion fuel. Other plants use natural gas, fuel oil, or biomass as the fuel. A number of power plants around the world operate on the heat released from nuclear reactions instead of fuel combustion. Figure 1.3 is a schematic diagram of a nuclear marine propulsion power plant. A secondary fluid circulates through the reactor, picking up heat generated by the nuclear reaction inside. This heat is then transferred to the water in the steam generator. The steam cycle processes are the same as in the previous example, but in this application the condenser cooling water is seawater, that is then returned at higher temperature to the sea.

1.2 FUEL CELLS

When a conventional power plant is viewed as a whole, as shown in Fig. 1.4, fuel and air enter the power plant and products of combustion leave the unit. There is also a transfer of heat to the cooling water, and work is done in the form of the electrical energy leaving the power plant. The overall objective of a power plant is to convert the availability (to do work) of the fuel into work (in the form of electrical energy) in the most efficient manner, taking into consideration cost, space, safety, and environmental concerns.

FIGURE 1.2 The Esbjerg, Denmark power station. (Courtesy Vestkraft 1996.)

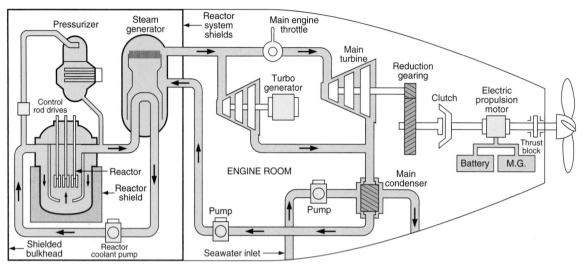

FIGURE 1.3 Schematic diagram of a shipboard nuclear propusion system. (Courtesy Babcock & Wilcox Co.)

We might well ask whether all the equipment in the power plant, such as the steam generator, the turbine, the condenser, and the pump, is necessary. Is it possible to produce electrical energy from the fuel in a more direct manner?

The fuel cell accomplishes this objective. Figure 1.5 shows a schematic arrangement of a fuel cell of the ion-exchange-membrane type. In this fuel cell, hydrogen and oxygen react to form water. Hydrogen gas enters at the anode side and is ionized at the surface of the ion-exchange membrane, as indicated in Fig. 1.5. The electrons flow through the external circuit to the cathode while the positive hydrogen ions migrate through the membrane to the cathode, where both react with oxygen to form water.

There is a potential difference between the anode and cathode, and thus there is a flow of electricity through a potential difference; this, in thermodynamic terms, is called work. There may also be a transfer of heat between the fuel cell and the surroundings.

At the present time the fuel used in fuel cells is usually either hydrogen or a mixture of gaseous hydrocarbons and hydrogen. The oxidizer is usually oxygen. However, current development is directed toward the production of fuel cells that use hydrocarbon fuels and

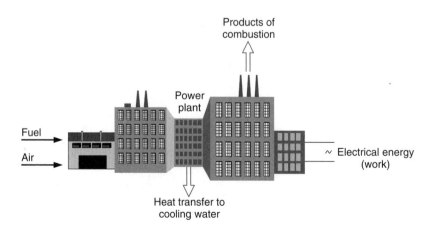

FIGURE 1.4
Schematic diagram of a
power plant.

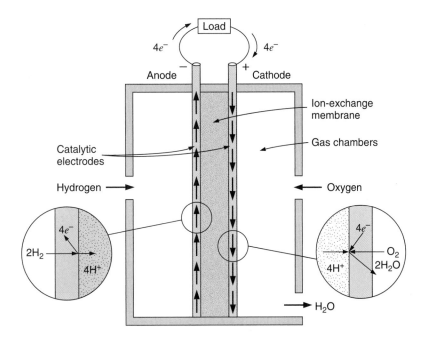

FIGURE 1.5
Schematic arrangement of an ion-exchange membrane type of fuel cell.

air. Although the conventional (or nuclear) steam power plant is still used in large-scale power-generating systems and conventional piston engines and gas turbines are still used in most transportation power systems, the fuel cell may eventually become a serious competitor. The fuel cell is already being used to produce power for space and other special applications.

Thermodynamics plays a vital role in the analysis, development, and design of all power-producing systems, including reciprocating internal-combustion engines and gas turbines. Considerations such as the increase of efficiency, improved design, optimum operating conditions, environmental pollution, and alternate methods of power generation involve, among other factors, the careful application of the fundamentals of thermodynamics.

1.3 THE VAPOR-COMPRESSION REFRIGERATION CYCLE

A simple vapor-compression refrigeration cycle is shown schematically in Fig. 1.6. The refrigerant enters the compressor as a slightly superheated vapor at a low pressure. It then leaves the compressor and enters the condenser as a vapor at some elevated pressure, where the refrigerant is condensed as heat is transferred to cooling water or to the surroundings. The refrigerant then leaves the condenser as a high-pressure liquid. The pressure of the liquid is decreased as it flows through the expansion valve, and as a result, some of the liquid flashes into cold vapor. The remaining liquid, now at a low pressure and temperature, is vaporized in the evaporator as heat is transferred from the refrigerated space. This vapor then reenters the compressor.

In a typical home refrigerator the compressor is located in the rear near the bottom of the unit. The compressors are usually hermetically sealed; that is, the motor and

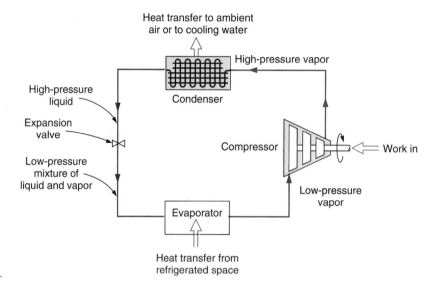

FIGURE 1.6
Schematic diagram of a
simple refrigeration cycle.

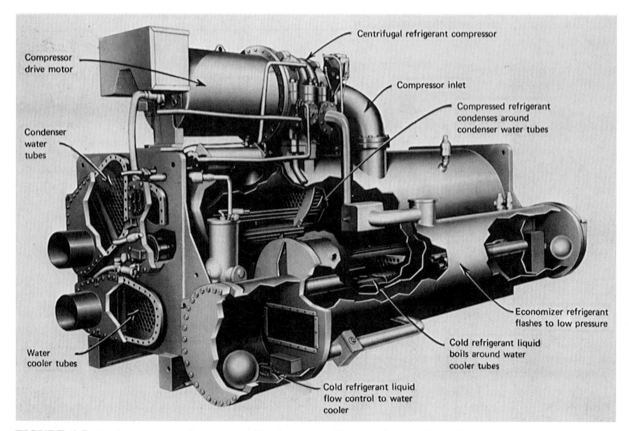

FIGURE 1.7 A refrigeration unit for an air-conditioning system. (Courtesy Carrier Air Conditioning Co.)

compressor are mounted in a sealed housing, and the electric leads for the motor pass through this housing. This seal prevents leakage of the refrigerant. The condenser is also located at the back of the refrigerator and is arranged so that the air in the room flows past the condenser by natural convection. The expansion valve takes the form of a long capillary tube, and the evaporator is located around the outside of the freezing compartment inside the refrigerator.

Figure 1.7 shows a large centrifugal unit that is used to provide refrigeration for an air-conditioning unit. In this unit, water is cooled and the circulated to provide cooling where needed.

1.4 THE THERMOELECTRIC REFRIGERATOR

We may well ask the same question about the vapor-compression refrigerator that we asked about the steam power plant—is it possible to accomplish our objective in a more direct manner? Is it possible, in the case of a refrigerator, to use the electrical energy (which goes to the electric motor that drives the compressor) to produce cooling in a more direct manner and thereby to avoid the cost of the compressor, condenser, evaporator, and all the related piping?

The thermoelectric refrigerator is such a device. This is shown schematically in Fig. 1.8a. The thermoelectric device, like the conventional thermocouple, uses two dissimilar materials. There are two junctions between these two materials in a thermoelectric refrigerator. One is located in the refrigerated space and the other in ambient surroundings. When a potential difference is applied, as indicated, the temperature of the junction located in the refrigerated space will decrease and the temperature of the other junction will

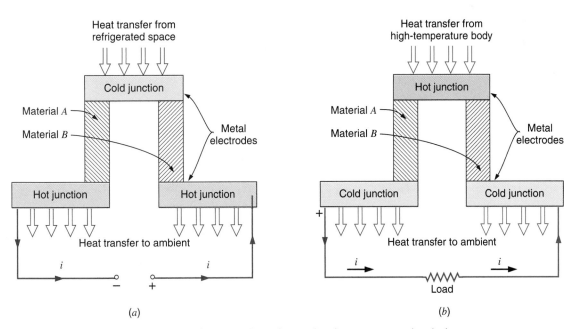

FIGURE 1.8 (a) A thermoelectric refrigerator. (b) A thermoelectric power generation device.

increase. Under steady-state operating conditions, heat will be transferred from the refrigerated space to the cold junction. The other junction will be at a temperature above the ambient, and heat will be transferred from the junction to the surroundings.

A thermoelectric device can also be used to generate power by replacing the refrigerated space with a body that is at a temperature above the ambient. Such a system is shown in Fig. 1.8*b*.

The thermoelectric refrigerator cannot yet compete economically with conventional vapor-compression units. However, in certain special applications, the thermoelectric refrigerator is already is use and, in view of research and development efforts under way in this field, it is quite possible that thermoelectric refrigerators will be much more extensively used in the future.

1.5 THE AIR SEPARATION PLANT

One process of great industrial significance is the air separation. In an air separation plant, air is separated into its various components. The oxygen, nitrogen, argon, and rare gases so produced are used extensively in various industrial, research, space, and consumer-goods applications. The air separation plant can be considered an example from two major fields, the chemical process industry and cryogenics. Cryogenics is a term applied to technology, processes, and research at very low temperatures (in general, below 150 K). In both chemical processing and cryogenics, thermodynamics is basic to an understanding of many phenomena that occur and to the design and development of processes and equipment.

A number of different designs of air separation plants have been developed. Consider Fig. 1.9, which shows a somewhat simplified sketch of a type of plant that is frequently used. Air from the atmosphere is compressed to a pressure of 2 to 3 MPa. It is then purified, particularly to remove carbon dioxide (which would plug the flow passages as it solidifies when the air is cooled to its liquefaction temperature). The air is then compressed to a pressure of 15 to 20 MPa, cooled to the ambient temperature in the aftercooler, and dried to remove the water vapor (which would also plug the flow passages as it freezes).

The basic refrigeration in the liquefaction process is provided by two different processes. In one process the air in the expansion engine expands. During this process the air does work and as a result the temperature of the air is reduced. In the other refrigeration process air passes through a throttle valve that is so designed and so located that there is a substantial drop in the pressure of the air and, associated with this, a substantial drop in the temperature of the air.

As shown in Fig. 1.9, the dry, high-pressure air enters a heat exchanger. The air temperature drops as it flows through the heat exchanger. At some intermediate point in the heat exchanger, part of the air is bled off and flows through the expansion engine. The remaining air flows through the rest of the heat exchanger and through the throttle valve. The two streams join (both are at the pressure of 0.5 to 1 MPa) and enter the bottom of the distillation column, which is referred to as the high-pressure column. The function of the distillation column is to separate the air into its various components, principally oxygen and nitrogen. Two streams of different composition flow from the high-pressure column through throttle valves to the upper column (also called the low-pressure column). One of these streams is an oxygen-rich liquid that flows from the bottom of the lower column,

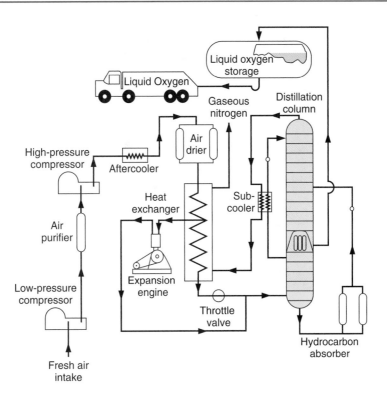

FIGURE 1.9 A simplified diagram of a liquid oxygen plant.

and the other is a nitrogen-rich stream that flows through the subcooler. The separation is completed in the upper column. Liquid oxygen leaves from the bottom of the upper column, and gaseous nitrogen leaves from the top of the column. The nitrogen gas flows through the subcooler and the main heat exchanger. It is the transfer of heat to this cold nitrogen gas that causes the high-pressure air entering the heat exchanger to become cooler.

Not only is a thermodynamic analysis essential to the design of the system as a whole, but essentially every component of such a system, including the compressors, the expansion engine, the purifiers and driers, and the distillation column, operates according to the principles of thermodynamics. In this separation process we are also concerned with the thermodynamic properties of mixtures and the principles and procedures by which these mixtures can be separated. This is the type of problem encountered in petroleum refining and many other chemical processes. It should also be noted that cryogenics is particularly relevant to many aspects of the space program, and a thorough knowledge of thermodynamics is essential for creative and effective work in cryogenics.

1.6 THE GAS TURBINE

The basic operation of a gas turbine is similar to that of the steam power plant, except that air is used instead of water. Fresh atmospheric air flows through a compressor that brings it to a high pressure. Energy is then added by spraying fuel into the air and igniting it so the combustion generates a high-temperature flow. This high-temperature, high-pressure gas enters a turbine, where it expands down to the exhaust pressure, producing a shaft

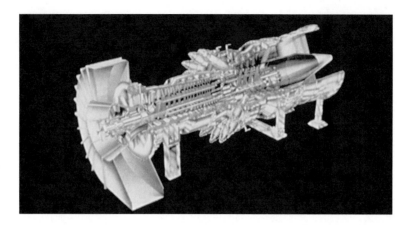

FIGURE 1.10 A 150-MW gas turbine. (Courtesy Westinghouse Electric Corporation.)

work output in the process. The turbine shaft work is used to drive the compressor and other devices, such as an electric generator that may be coupled to the shaft. The energy that is not used for shaft work comes out in the exhaust gases, so these have either a high temperature or a high velocity. The purpose of the gas turbine determines the design so that the most desirable energy form is maximized. An example of a large gas turbine for stationary power generation is shown in Fig. 1.10. The unit has sixteen stages of compression and four stages in the turbine and is rated at 150 MW. Notice that since the combustion of fuel uses the oxygen in the air, the exhaust gases cannot be recirculated as the water is in the steam power plant.

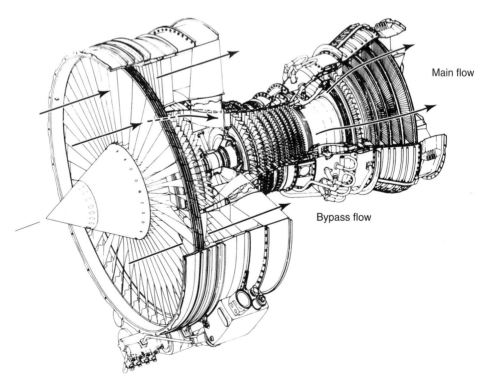

Main flow

Bypass flow

FIGURE 1.11 A turbofan jet engine. (Courtesy General Electric Aircraft Engines.)

A gas turbine is often the preferred power-generating device where a large amount of power is needed, but only a small physical size is possible. Examples are jet engines, turbofan jet engines, offshore oilrig power plants, ship engines, helicopter engines, smaller local power plants, or peak-load power generators in larger power plants. Since the gas turbine has relatively high exhaust temperatures, it can also be arranged so the exhaust gases are used to heat water that runs in a steam power plant before it exhausts to the atmosphere.

In the examples mentioned previously, the jet engine and turboprop applications utilize part of the power to discharge the gases at high velocity. This is what generates the thrust of the engine that moves the airplane forward. The gas turbines in these applications are therefore designed differently than for the stationary power plant, where the energy is taken out as shaft work to an electric generator. An example of a turbofan jet engine used in a commercial airplane is shown in Fig. 1.11. The large front-end fan also blows air past the engine, providing cooling and gives additional thrust.

1.7 THE CHEMICAL ROCKET ENGINE

The advent of missiles and satellites brought to prominence the use of the rocket engine as a propulsion power plant. Chemical rocket engines may be classified as either liquid propellant or solid propellant, according to the fuel used.

Figure 1.12 shows a simplified schematic diagram of a liquid-propellant rocket. The oxidizer and fuel are pumped through the injector plate into the combustion chamber where combustion takes place at high pressure. The high-pressure, high-temperature products of combustion expand as they flow through the nozzle, and as a result they leave the nozzle with a high velocity. The momentum change associated with this increase in velocity gives rise to the forward thrust on the vehicle.

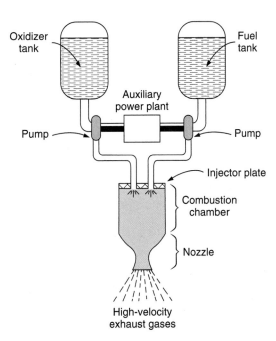

FIGURE 1.12
Simplified schematic diagram of a liquid-propellant rocket engine.

The oxidizer and fuel must be pumped into the combustion chamber, and some auxiliary power plant is necessary to drive the pumps. In a large rocket this auxiliary power plant must be very reliable and have a relatively high power output, yet it must be light in weight. The oxidizer and fuel tanks occupy the largest part of the volume of an actual rocket, and the range and payload of a rocket are determined largely by the amount of oxidizer and fuel that can be carried. Many different fuels and oxidizers have been considered and tested, and much effort has gone into the development of fuels and oxidizers that will give a higher thrust per unit mass rate of flow of reactants. Liquid oxygen is frequently used as the oxidizer in liquid-propellant rockets, and liquid hydrogen is frequently used as the fuel.

Much work has also been done on solid-propellant rockets. They have been very successfully used for jet-assisted takeoffs of airplanes, military missiles, and space vehicles. They are much simpler in both the basic equipment required for operation and the logistic problems involved in their use, but they are more difficult to control.

1.8 OTHER APPLICATIONS AND ENVIRONMENTAL ISSUES

There are many other applications in which thermodynamics is relevant. Many municipal landfill operations are now utilizing the heat produced by the decomposition of biomass waste to produce power, and they also capture the methane gas produced by these chemical reactions for use as a fuel. Geothermal sources of heat are also being utilized, as are solar- and windmill-produced electricity. Sources of fuel are being converted from one form to another, more usable or convenient form, such as in the gasification of coal or the conversion of biomass to liquid fuels. Hydroelectric plants have been in use for many years, as have other applications involving water power. Thermodynamics is also relevant to such processes as the curing of a poured concrete slab, which produces heat, the cooling of electronic equipment, in various applications in cryogenics (cryo-surgery, food fast-freezing), and many other diverse applications.

We must also be concerned with environmental issues related to these many devices and applications of thermodynamics. For example, the construction and operation of the steam power plant creates electricity, which is so deeply entrenched in our society that we take its ready availability for granted. In recent years, however, it has become increasingly apparent that we need to consider seriously the effects of such an operation on our environment. Combustion of hydrocarbon fuels releases carbon dioxide into the atmosphere, where its concentration is increasing. Carbon dioxide, as well as other gases, absorbs infrared radiation from the surface of the earth, holding it close to the planet and creating the "greenhouse effect," which in turn is believed to cause global warming and critical climatic changes around the earth. Power plant combustion, particularly of coal, releases sulfur dioxide, which is absorbed in clouds and later falls as acid rain in many areas. Combustion processes in power plants and gasoline and diesel engines also generate pollutants other than these two. Species such as carbon monoxide, nitric oxides, and partly burned fuels together with particulates all contribute to atmospheric pollution and are regulated by law for many applications. Catalytic converters on automobiles help to minimize the air pollution problem. In power plants, Fig. 1.1 indicates the fly ash cleanup and also the flue gas clean up processes that are now incorporated to address these problems. Thermal pollution associated with power plant cooling water requirements was discussed in Section 1.1.

Refrigeration and air-conditioning systems, as well as other industrial processes, have used certain chlorofluorocarbon fluids that eventually find their way to the upper atmosphere and destroy the protective ozone layer. Many countries have already banned the production of some of these compounds, and the search for improved replacement fluid continues.

These are only some of the many environmental problems caused by our efforts to produce goods and effects intended to improve our way of life. During our study of thermodynamics, which is the science of the conversion of energy from one form to another, we must continue to reflect on these issues. We must consider how we can eliminate or at least minimize damaging effects, as well as use our natural resources, efficiently and responsibly.

2 SOME CONCEPTS AND DEFINITIONS

One excellent definition of thermodynamics is that it is the science of energy and entropy. Since we have not yet defined these terms, an alternate definition in already familiar terms is: Thermodynamics is the science that deals with heat and work and those properties of substances that bear a relation to heat and work. Like all sciences, the basis of thermodynamics is experimental observation. In thermodynamics these findings have been formalized into certain basic laws, which are known as the first, second, and third laws of thermodynamics. In addition to these laws, the zeroth law of thermodynamics, which in the logical development of thermodynamics precedes the first law, has been set forth.

In the chapters that follow, we will present these laws and the thermodynamic properties related to these laws and apply them to a number of representative examples. The objective of the student should be to gain both a thorough understanding of the fundamentals and an ability to apply these fundamentals to thermodynamic problems. The examples and problems further this twofold objective. It is not necessary for the student to memorize numerous equations, for problems are best solved by the application of the definitions and laws of thermodynamics. In this chapter some concepts and definitions basic to thermodynamics are presented.

2.1 A THERMODYNAMIC SYSTEM AND THE CONTROL VOLUME

A thermodynamic system comprises a device or combination of devices containing a quantity of matter that is being studied. To define this more precisely, a control volume is chosen so that it contains the matter and devices inside a control surface. Everything external to the control volume is the surroundings, with the separation given by the control surface. The surface may be open or closed to mass flows, and it may have flows of energy in terms of heat transfer and work across it. The boundaries may be movable or stationary. In the case of a control surface that is closed to mass flow, so that no mass can escape or enter the control volume, it is called a control mass containing the same amount of matter at all times.

Selecting the gas in the cylinder of Fig. 2.1 as a control volume by placing a control surface around it, we recognize this as a control mass. If a Bunsen burner is placed under the cylinder, the temperature of the gas will increase and the piston will rise. As the piston rises, the boundary of the control mass moves. As we will see later, heat and work cross the boundary of the control mass during this process, but the matter that composes the control mass can always be identified and remains the same.

14

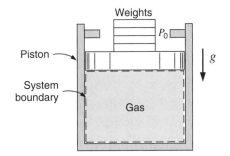

FIGURE 2.1 Example of a control mass.

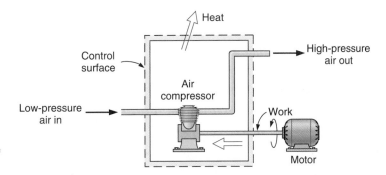

FIGURE 2.2 Example of a control volume.

An isolated system is one that is not influenced in any way by the surroundings. This means that no mass, heat, or work cross the boundary of the system. In many cases a thermodynamic analysis must be made of a device, such as an air compressor, which has a flow of mass into it, out of it, or both, as shown schematically in Fig. 2.2. The procedure followed in such an analysis is to specify a control volume that surrounds the device under consideration. The surface of this control volume is the control surface, which may have mass momentum, and also heat and work, cross it.

Thus the more general control surface defines a control volume, where mass may flow in or out, with a control mass as the special case of no mass flow in or out. Hence the control mass contains a fixed mass at all times, which explains its name. The difference in the formulation of the analysis is considered in detail in Chapter 6. The terms closed system (fixed mass) and open system (involving a flow of mass) are sometimes used to make this distinction. Here, we use the term system as a more general and loose description for a mass, device, or combination of devices that then is more precisely defined, when a control volume is selected. The procedure that will be followed in the presentation of the first and the second laws of thermodynamics is first to present these laws for a control mass and then to extend the analysis to the more general control volume.

2.2 MACROSCOPIC VERSUS MICROSCOPIC POINT OF VIEW

An investigation into the behavior of a system may be undertaken from either a microscopic or a macroscopic point of view. Let us briefly describe a system from a microscopic point of view. Consider a system consisting of a cube 25 mm on a side and containing a monatomic gas at atmospheric pressure and temperature. This volume contains approximately 10^{20} atoms. To describe the position of each atom, we need to specify three coordinates; to describe the velocity of each atom, we specify three velocity components.

Thus, to describe completely the behavior of this system from a microscopic point of view we must deal with at least 6×10^{20} equations. Even with a large digital computer, this is a quite hopeless computational task. However, there are two approaches to this problem that reduce the number of equations and variables to a few that can be computed relatively easily. One approach is the statistical approach, in which, on the basis of statistical considerations and probability theory, we deal with "average" values for all particles under consideration. This is usually done in connection with a model of the atom under consideration. This is the approach used in the disciplines known as kinetic theory and statistical mechanics.

The other approach to reducing the number of variables to a few that can be handled is the macroscopic point of view of classical thermodynamics. As the word macroscopic implies, we are concerned with the gross or average effects of many molecules. These effects can be perceived by our senses and measured by instruments. However, what we really perceive and measure is the time-averaged influence of many molecules. For example, consider the pressure a gas exerts on the walls of its container. This pressure results from the change in momentum of the molecules as they collide with the wall. From a macroscopic point of view, however, we are not concerned with the action of the individual molecules but with the time-averaged force on a given area, which can be measured by a pressure gauge. In fact, these macroscopic observations are completely independent of our assumptions regarding the nature of matter.

Although the theory and development in this book is presented from a macroscopic point of view, a few supplementary remarks regarding the significance of the microscopic perspective are included as an aid to the understanding of the physical processes involved. Another book in this series, *Introduction to Thermodynamics: Classical and Statistical*, by R. E. Sonntag and G. J. Van Wylen, includes thermodynamics from the microscopic and statistical point of view.

A few remarks should be made regarding the continuum. From the macroscopic view, we are always concerned with volumes that are very large compared to molecular dimensions and, therefore, with systems that contain many molecules. Because we are not concerned with the behavior of individual molecules, we can treat the substance as being continuous, disregarding the action of individual molecules. This continuum concept, of course, is only a convenient assumption that loses validity when the mean free path of the molecules approaches the order of magnitude of the dimensions of the vessel, as, for example, in high-vacuum technology. In much engineering work the assumption of a continuum is valid and convenient, going hand in hand with the macroscopic view.

2.3 PROPERTIES AND STATE OF A SUBSTANCE

If we consider a given mass of water, we recognize that this water can exist in various forms. If it is a liquid initially, it may become a vapor when it is heated or a solid when it is cooled. Thus, we speak of the different phases of a substance. A phase is defined as a quantity of matter that is homogeneous throughout. When more than one phase is present, the phases are separated from each other by the phase boundaries. In each phase the substance may exist at various pressures and temperatures or, to use the thermodynamic term, in various states. The state may be identified or described by certain observable, macroscopic properties; some familiar ones are temperature, pressure, and density. In later chapters other properties will be introduced. Each of the properties of a substance in a given state has only one definite value, and these properties always have the same value for a given state, regardless of how the substance arrived at the state. In fact, a property can be defined as any quantity that depends on the state of the system and is independent of the path (that is, the prior history) by which the system arrived at the given state. Conversely, the state is specified or described by the properties. Later we will consider the number of independent properties a substance can have, that is, the minimum number of properties that must be specified to fix the state of the substance.

Thermodynamic properties can be divided into two general classes, intensive and extensive properties. An intensive property is independent of the mass; the value of an extensive property varies directly with the mass. Thus, if a quantity of matter in a given state is divided into two equal parts, each part will have the same value of intensive properties as the original and half the value of the extensive properties. Pressure, temperature, and density are examples of intensive properties. Mass and total volume are examples of extensive properties. Extensive properties per unit mass, such as specific volume, are intensive properties.

Frequently we will refer not only to the properties of a substance but to the properties of a system. When we do so we necessarily imply that the value of the property has significance for the entire system, and this implies equilibrium. For example, if the gas that composes the system (control mass) in Fig. 2.1 is in thermal equilibrium, the temperature will be the same throughout the entire system, and we may speak of the temperature as a property of the system. We may also consider mechanical equilibrium, which is related to pressure. If a system is in mechanical equilibrium, there is no tendency for the pressure at any point to change with time as long as the system is isolated from the surroundings. There will be a variation in pressure with elevation because of the influence of gravitational forces, although under equilibrium conditions there will be no tendency for the pressure at any location to change. However, in many thermodynamic problems, this variation in pressure with elevation is so small that is can be neglected. Chemical equilibrium is also important and will be considered in Chapter 15. When a system is in equilibrium regarding all possible changes of state, we say that the system is in thermodynamic equilibrium.

2.4 PROCESSES AND CYCLES

Whenever one or more of the properties of a system change, we say that a change in state has occurred. For example, when one of the weights on the piston in Fig. 2.3 is removed, the piston rises and a change in state occurs, for the pressure decreases and the specific volume increases. The path of the succession of states through which the system passes is called the process.

Let us consider the equilibrium of a system as it undergoes a change in state. The moment the weight is removed from the piston in Fig. 2.3, mechanical equilibrium does not exist, and as a result the piston is moved upward until mechanical equilibrium is again restored. The question is this: Since the properties describe the state of a system only when it is in equilibrium, how can we describe the states of a system during a process if the actual process occurs only when equilibrium does not exist? One step in the answer to

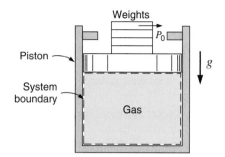

FIGURE 2.3 Example of a system that may undergo a quasi-equilibrium process.

this question concerns the definition of an ideal process, which we call a quasi-equilibrium process. A quasi-equilibrium process is one in which the deviation from thermodynamic equilibrium is infinitesimal, and all the states the system passes through during a quasi-equilibrium process may be considered equilibrium states. Many actual processes closely approach a quasi-equilibrium process and may be so treated with essentially no error. If the weights on the piston in Fig. 2.3 are small and are taken off one by one, the process could be considered quasi-equilibrium. However, if all the weights were removed at once, the piston would rise rapidly until it hit the stops. This would be a nonequilibrium process, and the system would not be in equilibrium at any time during this change of state.

For nonequilibrium processes, we are limited to a description of the system before the process occurs and after the process is completed and equilibrium is restored. We are unable to specify each state through which the system passes or the rate at which the process occurs. However, as we will see later, we are able to describe certain overall effects that occur during the process.

Several processes are described by the fact that one property remains constant. The prefix iso- is used to describe such a process. An isothermal process is a constant-temperature process, an isobaric (sometimes called isopiestic) process is a constant-pressure process, and an isochoric process is a constant-volume process.

When a system in a given initial state goes through a number of different changes of state or processes and finally returns to its initial state, the system has undergone a cycle. Therefore, at the conclusion of a cycle, all the properties have the same value they had at the beginning. Steam (water) that circulates through a steam power plant undergoes a cycle.

A distinction should be made between a thermodynamic cycle, which has just been described, and a mechanical cycle. A four-stroke-cycle internal-combustion engine goes through a mechanical cycle once every two revolutions. However, the working fluid does not go through a thermodynamic cycle in the engine, since air and fuel are burned and changed to products of combustion that are exhausted to the atmosphere. In this text the term cycle will refer to a thermodynamic "cycle" unless otherwise designated.

2.5 UNITS FOR MASS, LENGTH, TIME, AND FORCE

Since we are considering thermodynamic properties from a macroscopic perspective, we are dealing with quantities that can, either directly or indirectly, be measured and counted. Therefore, the matter of units becomes an important consideration. In the remaining sections of this chapter we will define certain thermodynamic properties and the basic units.

Because the relation between force and mass is often a difficult matter for students, it is considered in this section in some detail.

Force, mass, length, and time are related by Newton's second law of motion, which states that the force acting on a body is proportional to the product of the mass and the acceleration in the direction of the force:

$$F \propto ma$$

The concept of time is well established. The basic unit of time is the second (s), which in the past was defined in terms of the solar day, the time interval for one complete revolution of the earth relative to the sun. Since this period varies with the season of the year, an average value over a one-year period is called the mean solar day, and the mean solar second is 1/86 400 of the mean solar day. (The measurement of the earth's rotation is sometimes made relative to a fixed star, in which case the period is called a sidereal day.) In 1967, the General Conference of Weights and Measures (CGPM) adopted a definition of the second as the time required for a beam of cesium-133 atoms to resonate 9 192 631 770 cycles in a cesium resonator.

For periods of time less than a second, the prefixes milli, micro, nano, or pico, as listed in Table 2.1, are commonly used. For longer periods of time, the units minute (min), hour (h), or day (day) are frequently used. It should be pointed out that the prefixes in Table 2.1 are used with many other units as well.

The concept of length is also well established. The basic unit of length is the meter (m). For many years the accepted standard was the International Prototype Meter, the distance between two marks on a platinum–iridium bar under certain prescribed conditions. This bar in maintained at the International Bureau of Weights and Measures, in Sevres, France. In 1960, the CGPM adopted a definition of the meter as a length equal to 1 650 763.73 wavelengths in a vacuum of the orange-red line of krypton-86. Then in 1983, the CGPM adopted a more precise definition of the meter in terms of the speed of light (which is now a fixed constant): The meter is the length of the path traveled by light in a vacuum during a time interval of 1/299 792 458 of a second.

The fundamental unit of mass is the kilogram (kg). As adopted by the first CGPM in 1889 and restated in 1901, it is the mass of a certain platinum–iridium cylinder maintained under prescribed conditions at the International Bureau of Weights and Measures. A related unit that is used frequently in thermodynamics is the mole (mol), defined as an amount of substance containing as many elementary entities as there are atoms in 0.012 kg of carbon-12. These elementary entities must be specified; they may be atoms, molecules, electrons, ions, or other particles or specific groups. For example, one mole of diatomic oxygen, having a molecular weight of 32 (compared to 12 for carbon), has a mass of 0.032 kg. The mole is often termed a gram mole, since it is an amount of substance in

TABLE 2.1
Unit Prefixes

Factor	Prefix	Symbol	Factor	Prefix	Symbol
10^{12}	tera	T	10^{-3}	milli	m
10^{9}	giga	G	10^{-6}	micro	μ
10^{6}	mega	M	10^{-9}	nano	n
10^{3}	kilo	k	10^{-12}	pico	p

grams numerically equal to the molecular weight. In this text, when using the metric SI system we will find it preferable to use the kilomole (kmol), the amount of substance in kilograms numerically equal to the molecular weight, rather than the mole.

The system of units in use presently throughout most of the world is the metric International System, commonly referred to as SI units (from Le Système International d'Unités). In this system, the second, meter, and kilogram are the basic units for time, length, and mass, respectively, as just defined, and the unit of force is defined directly from Newton's second law.

Therefore, a proportionality constant is unnecessary, and we may write that law as an equality:

$$F = ma \qquad (2.1)$$

The unit of force is the newton (N), which by definition is the force required to accelerate a mass of one kilogram at the rate of one meter per second per second:

$$1 \text{ N} = 1 \text{ kg m/s}^2$$

It is worth noting that SI units derived from proper nouns use capital letters for symbols; others use the lowercase letters. The liter, with the symbol L, is an exception.

The traditional system of units used in the Untied States is the English Engineering System. In this system the unit of time is the second, which has been discussed earlier. The basic unit of length is the foot (ft), which at present is defined in terms of the meter as

$$1 \text{ ft} = 0.3048 \text{ m}$$

The inch (in.) is defined in terms of the foot

$$12 \text{ in.} = 1 \text{ ft}$$

The unit of mass in this system is the pound mass (lbm). It was originally the mass of a certain platinum cylinder kept in the Tower of London, but now it is defined in terms of the kilogram as

$$1 \text{ lbm} = 0.453 \ 592 \ 37 \text{ kg}$$

A related unit is the pound mole (lb mol), which is an amount of substance in pounds mass numerically equal to the molecular weight of that substance. It is important to distinguish between a pound mole and a mole (gram mole).

In the English Engineering System of Units, the unit of force is the pound force (lbf), defined as the force with which the standard pound mass is attracted to the earth under conditions of standard acceleration of gravity, which is that at 45° latitude and sea level elevation, 9.806 65 m/s² or 32.1740 ft/s². Thus, it follows from Newton's second law, that

$$1 \text{ lbf} = 32.174 \text{ lbm ft/s}^2$$

which is a necessary factor for the purpose of units conversion and consistency. Note that we must be careful to distinguish between a lbm and a lbf, and we do not use the term pound alone.

The term weight is often used with respect to a body and is sometimes confused with mass. Weight is really correctly used only as a force. When we say a body weighs so much, we mean that this is the force with which it is attracted to the earth (or some other body), that is, the product of its mass and the local gravitational acceleration. The mass of a substance remains constant with elevation, but its weight varies with elevation.

EXAMPLE 2.1   What is the weight of a one kg mass at an altitude where the local acceleration of gravity is 9.75 m/s²?

Solution

Weight is the force acting on the mass, which from Newton's second law is

$$F = mg = 1 \text{ kg} \times 9.75 \text{ m/s}^2 \times [1 \text{ N s}^2/\text{kg m}] = 9.75 \text{ N}$$

EXAMPLE 2.1E   What is the weight of a one lbm mass at an altitude where the local acceleration of gravity is 32.0 ft/s²?

Solution

Weight is the force acting on the mass, which from Newton's second law is

$$F = mg = 1 \text{ lbm} \times 32.0 \text{ ft/s}^2 \times [\text{lbf s}^2/32.174 \text{ lbm ft}] = 0.9946 \text{ lbf}$$

2.6 ENERGY

One of the very important concepts in a study of thermodynamics is that of energy. Energy is a fundamental concept, such as mass or force and, as is often the case with such concepts, is very difficult to define. Energy has been defined as the capability to produce an effect. Fortunately the word energy and the basic concept that this word represents are familiar to us in everyday usage, and a precise definition is not essential at this point.

Energy can be stored within a system and can be transferred (as heat, for example) from one system to another. In a study of statistical thermodynamics we would examine, from a molecular view, the ways in which energy can be stored. Because it is helpful in a study of classical thermodynamics to have some notion of how this energy is stored, a brief introduction is presented here.

Consider as a system a certain gas at a given pressure and temperature contained within a tank or pressure vessel. When considered from the molecular view, we identify three general forms of energy:

1. Intermolecular potential energy, which is associated with the forces between molecules.
2. Molecular kinetic energy, which is associated with the translational velocity of individual molecules.
3. Intramolecular energy (that within the individual molecules), which is associated with the molecular and atomic structure and related forces.

The first of these forms of energy, the intermolecular potential energy, depends on the magnitude of the intermolecular forces and the position the molecules have relative to each other at any instant of time. It is impossible to determine accurately the magnitude of this energy because we do not know either the exact configuration and orientation of the molecules at any time or the exact intermolecular potential function. However, there are two situations for which we can make good approximations. The first situation is at low or

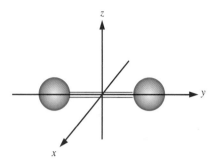

FIGURE 2.4 The coordinate system for a diatomic molecule.

moderate densities. In this case the molecules are relatively widely spaced, so that only two-molecule or two- and three-molecule interactions contribute to the potential energy. At these low and moderate densities, techniques are available for determining, with reasonable accuracy, the potential energy of a system composed of reasonably simple molecules. The second situation is at very low densities; under these conditions the average intermolecular distance between molecules is so large that the potential energy may be assumed to be zero. Consequently, we have in this case a system of independent particles (an ideal gas) and, therefore, from a statistical point of view, we are able to concentrate our efforts on evaluating the molecular translational and internal energies.

The translational energy, which depends only on the mass and velocities of the molecules, is determined by using the equations of mechanics—either quantum or classical.

The intramolecular internal energy is more difficult to evaluate because, in general, it may result from a number of contributions. Consider a simple monatomic gas such as helium. Each molecule consists of a helium atom. Such an atom possesses electronic energy as a result of both orbital angular momentum of the electrons about the nucleus and angular momentum of the electrons spinning on their axes. The electronic energy is commonly very small compared with the translational energies. (Atoms also possess nuclear energy, which, except in the case of nuclear reactions, is constant. We are not concerned with nuclear energy at this time.) When we consider more complex molecules, such as those composed of two or three atoms, additional factors must be considered. In addition to having electronic energy, a molecule can rotate about its center of gravity and thus have rotational energy. Furthermore, the atoms may vibrate with respect to each other and have vibrational energy. In some situations there may be an interaction between the rotational and vibrational modes of energy.

Consider a diatomic molecule, such as oxygen, as shown in Fig. 2.4. In addition to translation of the molecule as a solid body, the molecule can rotate about its center of mass in two normal directions, about the x axis and about the z axis (rotation about the y axis is negligible), and the two atoms can also vibrate, that is, stretch the bond joining the atoms along the y axis. A more rapid rotation increases the rotational energy, and a stronger vibration results in an increase of vibrational energy of the molecule.

More complex molecules, such as typical polyatomic molecules, are usually three-dimensional in structure and have multiple vibrational modes, each of which contributes to the energy storage of the molecule. The more complicated the molecule is, the larger the number of degrees of freedom that exist for energy storage. This subject of the modes of energy storage and their evaluation is discussed in some detail in Appendix C, for those interested in a further development of the quantitative effects from a molecular viewpoint.

This general discussion can be summarized by referring to Fig. 2.5. Let heat be transferred to the water. During this process the temperature of the liquid and vapor (steam) will increase, and eventually all the liquid will become vapor. From the macroscopic view we

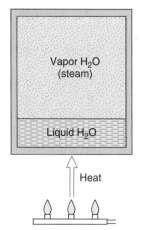

FIGURE 2.5 Heat transfer to water.

are concerned only with the energy that is transferred as heat, the change in properties, such as temperature and pressure, and the total amount of energy (relative to some base) that the H_2O contains at any instant. Thus, questions about how energy is stored in the H_2O do not concern us. From a microscopic viewpoint we are concerned about the way in which energy is stored in the molecules. We might be interested in developing a model of the molecule so that we could predict the amount of energy required to change the temperature a given amount. Although the focus in this book is on the macroscopic or classical viewpoint, it is helpful to keep in mind the microscopic or statistical perspective as well, as the relationship between the two helps us in understanding basic concepts such as energy.

2.7 SPECIFIC VOLUME AND DENSITY

The specific volume of a substance is defined as the volume per unit mass and is given the symbol v. The density of a substance is defined as the mass per unit volume, and it is therefore the reciprocal of the specific volume. Density is designated by the symbol ρ. Specific volume and density are intensive properties.

The specific volume of a system in a gravitational field may vary from point to point. For example, if the atmosphere is considered a system, the specific volume increases as the elevation increases. Therefore, the definition of specific volume involves the specific volume of a substance at a point in a system.

Consider a small volume δV of a system, and let the mass be designated δm. The specific volume is defined by the relation

$$v = \lim_{\delta V \to \delta V'} \frac{\delta V}{\delta m}$$

where $\delta V'$ is the smallest volume for which the mass can be considered a continuum. Volumes smaller than this will lead to the recognition that mass is not evenly distributed in space but is concentrated in particles as molecules, atoms, electrons, etc. This is tentatively indicated in Fig. 2.6, where in the limit of a zero volume the specific volume may be infinite (the volume does not contain any mass) or very small (the volume is part of a nucleus).

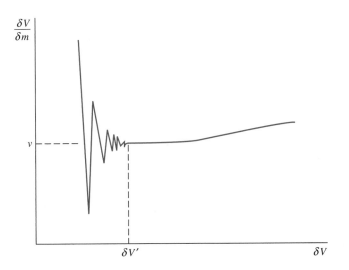

FIGURE 2.6 The continuum limit for the specific volume.

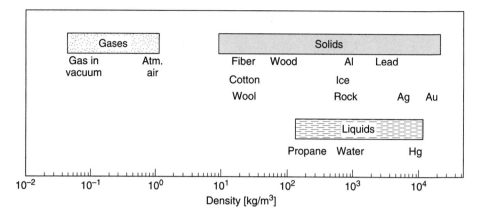

FIGURE 2.7 Density of common substances.

Thus, in a given system, we should speak of the specific volume or density at a point in the system, and recognize that this may vary with elevation. However, most of the systems that we consider are relatively small, and the change in specific volume with elevation is not significant. Therefore, we can speak of one value of specific volume or density for the entire system.

In this text, the specific volume and density will be given either on a mass or on a mole basis. A bar over the symbol (lowercase) will be used to designate the property on a mole basis. Thus, $\bar{v}$ will designate molal specific volume and $\bar{\rho}$ will designate the molal density. In SI units, those for specific volume are m^3/kg and m^3/mol (or $m^3/kmol$); for density the corresponding units are kg/m^3 and mol/m^3 (or $kmol/m^3$). In English units, those for specific volume are ft^3/lbm and ft^3/lb mol; the corresponding units for density are lbm/ft^3 and lb mol/ft^3.

Although the SI unit for volume is the cubic meter, a commonly used volume unit is the liter (L), which is a special name given to a volume of 0.001 cubic meters, that is, $1\ L = 10^{-3}\ m^3$. The general ranges of density for some common solids, liquids, and gases are shown in Fig. 2.7. Specific values for various solids, liquids and gases in SI units are listed in Tables A.3, A.4, and A.5, respectively, and in English units in Tables F.2, F.3, and F.4.

EXAMPLE 2.2

A 1-m^3 container, Fig. 2.8, is filled with 0.12 m^3 of granite, 0.15 m^3 of sand, 0.2 m^3 of liquid 25°C water, and the rest of the volume, 0.53 m^3, is air with a density of 1.15 kg/m^3. Find the overall (average) specific volume and density.

Solution

From the definition of specific volume and density we have:

$$v = V/m \quad \text{and} \quad \rho = m/V = 1/v$$

We need to find the total mass taking density from Tables A.3 and A.4

$$m_{granite} = \rho V_{granite} = 2750\ kg/m^3 \times 0.12\ m^3 = 330\ kg$$

$$m_{sand} = \rho_{sand}\ V_{sand} = 1500\ kg/m^3 \times 0.15\ m^3 = 225\ kg$$

$$m_{water} = \rho_{water}\ V_{water} = 997\ kg/m^3 \times 0.2\ m^3 = 199.4\ kg$$

$$m_{air} = \rho_{air}\ V_{air} = 1.15\ kg/m^3 \times 0.53\ m^3 = 0.61\ kg$$

Remark: It is misleading to include the numbers for ρ and V, as the air is separate from the rest of the mass.

FIGURE 2.8 Sketch for Example 2.2.

Now the total mass becomes

$$m_{\text{tot}} = m_{\text{granite}} + m_{\text{sand}} + m_{\text{water}} + m_{\text{air}} = 755 \text{ kg}$$

and the specific volume and density can be calculated

$$v = V_{\text{tot}}/m_{\text{tot}} = 1 \text{ m}^3/755 \text{ kg} = 0.001325 \text{ m}^3/\text{kg}$$

$$\rho = m_{\text{tot}}/V_{\text{tot}} = 755 \text{ kg}/1 \text{ m}^3 = 755 \text{ kg/m}^3$$

2.8 PRESSURE

When dealing with liquids and gases, we ordinarily speak of pressure; for solids we speak of stresses. The pressure in a fluid at rest at a given point is the same in all directions, and we define pressure as the normal component of force per unit area. More specifically, if δA is a small area, $\delta A'$ the smallest area over which we can consider the fluid a continuum, and δF_n the component of force normal to δA, we define pressure, P, as

$$P = \lim_{\delta A \to \delta A'} \frac{\delta F_n}{\delta A}$$

where the lower limit corresponds to sizes as mentioned for the specific volume, shown in Fig. 2.6. The pressure P at a point in a fluid in equilibrium is the same in all directions. In a viscous fluid in motion, the variation in the state of stress with orientation becomes an important consideration. These considerations are beyond the scope of this book, and we will consider pressure only in terms of a fluid in equilibrium.

The unit for pressure in the International System is the force of one newton acting on a square meter area, which is called the pascal (Pa). That is,

$$1 \text{ Pa} = 1 \text{ N/m}^2$$

Two other units, not part of the International System, continue to be widely used. These are the bar, where

$$1 \text{ bar} = 10^5 \text{ Pa} = 0.1 \text{ MPa}$$

and the standard atmosphere, where

$$1 \text{ atm} = 101\ 325 \text{ Pa} = 14.696 \text{ lbf/in}^2$$

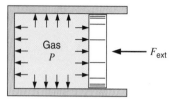

FIGURE 2.9 The balance of forces on a movable boundary relates to inside gas pressure.

which is slightly larger than the bar. In this text, we will normally use the SI unit, the pascal, and especially the multiples of kilopascal and megapascal. The bar will be utilized often in the examples and problems, but the atmosphere will not be used, except in specifying certain reference points.

Consider a gas contained in a cylinder fitted with a movable piston, as shown in Fig. 2.9. The pressure exerted by the gas on all its boundaries is the same, assuming that the gas is in an equilibrium state. This pressure is fixed by the external force acting on the piston, since there must be a balance of forces for the piston to remain stationary. Thus, the product of the pressure and the movable piston area must be equal to the external force. If the external force is now changed, in either direction, the gas pressure inside must accordingly adjust, with appropriate movement of the piston, to establish a force balance at a new equilibrium state. As another example, if the gas in the cylinder is heated by an outside body, which tends to increase the gas pressure, the piston will move instead, such that the pressure remains equal to whatever value is required by the external force.

EXAMPLE 2.3

The hydraulic piston/cylinder system shown in Fig. 2.10 has a cylinder diameter of $D = 0.1$ m with a piston and rod mass of 25 kg. The rod has a diameter of 0.01 m with an outside atmospheric pressure of 101 kPa. The inside hydraulic fluid pressure is 250 kPa. How large a force can the rod push within the upward direction?

Solution

We will assume a static balance of forces on the piston (positive upward) so

$$F_{net} = ma = 0$$

$$= P_{cyl} A_{cyl} - P_0 (A_{cyl} - A_{rod}) - F - m_p g$$

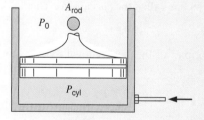

FIGURE 2.10 Sketch for Example 2.3.

solve for F

$$F = P_{cyl}A_{cyl} - P_0(A_{cyl} - A_{rod}) - m_p g$$

The areas are:

$$A_{cyl} = \pi r^2 = \pi D^2/4 = \frac{\pi}{4}\, 0.1^2\, m^2 = 0.007\,854\ m^2$$

$$A_{rod} = \pi r^2 = \pi D^2/4 = \frac{\pi}{4}\, 0.01^2\, m^2 = 0.000\,078\,54\ m^2$$

So the force becomes

$$F = [250 \times 0.007\,854 - 101(0.007\,854 - 0.000\,078\,54)]\,1000 - 25 \times 9.81$$

$$= 1963.5 - 785.32 - 245.25$$

$$= 932.9\ N$$

Note that we must convert kPa to Pa to get units of N.

In most thermodynamic investigations we are concerned with absolute pressure. Most pressure and vacuum gauges, however, read the difference between the absolute pressure and the atmospheric pressure existing at the gauge. This is referred to as gauge pressure. This is shown graphically in Fig. 2.11, and the following examples illustrate the principles. Pressures below atmospheric and slightly above atmospheric, and pressure differences (for example, across an orifice in a pipe), are frequently measured with a manometer, which contains water, mercury, alcohol, oil, or other fluids.

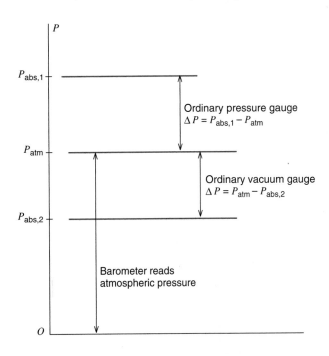

FIGURE 2.11
Illustration of terms used in pressure measurement.

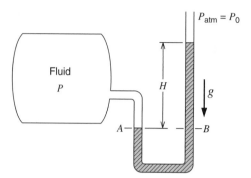

FIGURE 2.12
Example of pressure
measurement using a
column of fluid.

Consider the column of fluid of height H standing above point B in the manometer shown in Fig. 2.12. The force acting downward at the bottom of the column is

$$P_0A + mg = P_0A + \rho AgH$$

where m is the mass of the fluid column, A is its cross-sectional area, and ρ is its density. This force must be balanced by the upward force at the bottom of the column, which is P_BA. Therefore,

$$P_B - P_0 = \rho gH$$

Since points A and B are at the same elevation in columns of the same fluid, their pressures must be equal (the fluid being measured in the vessel has a much lower density, such that its pressure P is equal to P_A). Overall,

$$\Delta P = P - P_0 = \rho gH \tag{2.2}$$

For distinguishing between absolute and gauge pressure in this text, the term pascal will always refer to absolute pressure. Any gauge pressure will be indicated as such.

EXAMPLE 2.4 A mercury (Hg) manometer is used to measure the pressure in a vessel as shown in Fig. 2.12. The mercury has a density of 13 590 kg/m³, and the height difference between the two columns is measured to be 24 cm. We want to determine the pressure inside the vessel.

Solution

The manometer measures the gauge pressure as a pressure difference. From Eq. 2.2,

$$\Delta P = P_{gauge} = \rho gH = 13\ 590 \times 9.806\ 65 \times 0.24$$

$$= 31\ 985\ \frac{kg}{m^3}\frac{m}{s^2}\ m = 31\ 985\ Pa = 31.985\ kPa$$

$$= 0.316\ atm$$

To get the absolute pressure inside the vessel we have

$$P_A = P_{vessel} = P_B = \Delta P + P_{atm}$$

We need to know the atmospheric pressure measured by a barometer (absolute pressure). Assume this pressure is known as 750 mm Hg, being measured with a setup similar to the one in Fig. 2.12 with one side open to the atmosphere and the other side closed so there is mercury vapor with a very small pressure on top of the liquid column. The absolute pressure in the vessel becomes

$$P_{\text{vessel}} = \Delta P + P_{\text{atm}} = 31\,985 + 13\,590 \times 0.750 \times 9.806\,65$$

$$= 31\,985 + 99\,954 = 131\,940 \text{ Pa} = 1.302 \text{ atm}$$

EXAMPLE 2.4E A mercury (Hg) manometer is used to measure the pressure in a vessel as shown in Fig. 2.12. The mercury has a density of 848 lbm/ft³, and the height difference between the two columns is measured to be 9.5 in. We want to determine the pressure inside the vessel.

Solution

The manometer measures the gauge pressure as a pressure difference. From Eq. 2.2,

$$\Delta P = P_{\text{gauge}} = \rho g H$$

$$= 848 \frac{\text{lbm}}{\text{ft}^3} \times 32.174 \frac{\text{ft}}{\text{s}^2} \times 9.5 \text{ in.} \times \frac{1}{1728} \frac{\text{ft}^3}{\text{in}^3} \times \left[\frac{1 \text{ lbf s}^2}{32.174 \text{ lbm ft}} \right]$$

$$= 4.66 \text{ lbf/in}^2$$

To get the absolute pressure inside the vessel we have

$$P_A = P_{\text{vessel}} = P_0 = \Delta P + P_{\text{atm}}$$

We need to know the atmospheric pressure measured by a barometer (absolute pressure). Assume this pressure is known as 29.5 in. Hg, being measured with a setup similar to the one above with one side open to the atmosphere and the other side closed so there is mercury vapor with a very small pressure on top of the liquid column. The absolute pressure in the vessel becomes

$$P_{\text{vessel}} = \Delta P + P_{\text{atm}}$$

$$= 848 \times 32.174 \times 29.5 \times \frac{1}{1728} \times \left(\frac{1}{32.174} \right) + 4.66$$

$$= 19.14 \text{ lbf/in}^2$$

EXAMPLE 2.5 What is the pressure at the bottom of the 7.5-m-tall storage tank of fluid at 25°C shown in Fig. 2.13? Assume the fluid is gasoline with atmospheric pressure 101 kPa on the top surface. Repeat the question for liquid refrigerant R-134a when the top surface pressure is 1 MPa.

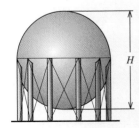

FIGURE 2.13 Sketch for Example 2.5.

Solution

The densities of the liquids are listed in Table A.4:

$$\rho_{\text{gasoline}} = 750 \text{ kg/m}^3; \qquad \rho_{\text{R-134a}} = 1206 \text{ kg/m}^3$$

The pressure difference due to the gravity is, from Eq. 2.2,

$$\Delta P = \rho g H$$

The total pressure is

$$P = P_{\text{top}} + \Delta P$$

For the gasoline we get

$$\Delta P = \rho g H = 750 \text{ kg/m}^3 \times 9.807 \text{ m/s}^2 \times 7.5 \text{ m} = 55\ 164 \text{ Pa}$$

Now convert all pressures to kPa

$$P = 101 + 55.164 = 156.2 \text{ kPa}$$

For the R-134a we get

$$\Delta P = \rho g H = 1206 \text{ kg/m}^3 \times 9.807 \text{ m/s}^2 \times 7.5 \text{ m} = 88\ 704 \text{ Pa}$$

Now convert all pressures to kPa

$$P = 1000 + 88.704 = 1089 \text{ kPa}$$

EXAMPLE 2.6

A piston/cylinder with cross-sectional area of 0.01 m² is connected with a hydraulic line to another piston cylinder of cross-sectional area of 0.05 m². Assume both chambers and the line are filled with hydraulic fluid of density 900 kg/m³ and the larger second piston/cylinder is 6 m higher up in elevation. The telescope arm and the buckets have hydraulic piston/cylinders moving them, as seen in Fig. 2.14. With an outside atmospheric pressure of 100 kPa and a net force of 25 kN on the smallest piston, what is the balancing force on the second larger piston?

Solution

When the fluid is stagnant and at the same elevation we have the same pressure throughout the fluid. The force balance on the smaller piston is then related to the pressure (we neglect the rod area) as

$$F_1 + P_0 A_1 = P_1 A_1$$

from which the fluid pressure is

$$P_1 = P_0 + F_1/A_1 = 100 \text{ kPa} + 25 \text{ kN}/0.01 \text{ m}^2 = 2600 \text{ kPa}$$

FIGURE 2.14 Sketch for Example 2.6.

The pressure at the higher elevation in piston/cylinder 2 is, from Eq. 2.2,

$$P_2 = P_1 - \rho g H = 2600 \text{ kPa} - 900 \text{ kg/m}^3 \times 9.81 \text{ m/s}^2 \times 6 \text{ m}/(1000 \text{ Pa/kPa})$$

$$= 2547 \text{ kPa}$$

where the second term is divided by 1000 to convert from Pa to kPa. Then the force balance on the second piston gives

$$F_2 + P_0 A_2 = P_2 A_2$$

$$F_2 = (P_2 - P_0)A_2 = (2547 - 100) \text{ kPa} \times 0.05 \text{ m}^2 = 122.4 \text{ kN}$$

2.9 EQUALITY OF TEMPERATURE

Although temperature is a familiar property, defining it exactly is difficult. We are aware of "temperature" first of all as a sense of hotness or coldness when we touch an object. We also learn early that when a hot body and a cold body are brought into contact, the hot body becomes cooler and the cold body becomes warmer. If these bodies remain in contact for some time, they usually appear to have the same hotness or coldness. However, we also realize that our sense of hotness or coldness is very unreliable. Sometimes very cold bodies may seem hot, and bodies of different materials that are at the same temperature appear to be at different temperatures.

Because of these difficulties in defining temperature, we define equality of temperature. Consider two blocks of copper, one hot and the other cold, each of which is in contact with a mercury-in-glass thermometer. If these two blocks of copper are brought into thermal communication, we observe that the electrical resistance of the hot block decreases with time and that of the cold block increases with time. After a period of time has elapsed, however, no further changes in resistance are observed. Similarly, when the blocks are first brought in thermal communication, the length of a side of the hot block decreases with time, but the length of a side of the cold block increases with time. After a period of time, no further change in length of either of the blocks is perceived. In addition, the mercury column of the thermometer in the hot block drops at first and that in the cold block rises, but after a period of time no further changes in height are observed. We may say, therefore, that two bodies have equality of temperature if, when they are in thermal communication, no change in any observable property occurs.

2.10 THE ZEROTH LAW OF THERMODYNAMICS

Now consider the same two blocks of copper and another thermometer. Let one block of copper be brought into contact with the thermometer until equality of temperature is established, and then remove it. Then let the second block of copper be brought into contact with the thermometer. Suppose that no change in the mercury level of the thermometer occurs during this operation with the second block. We then can say that both blocks are in thermal equilibrium with the given thermometer.

The zeroth law of thermodynamics states that when two bodies have equality of temperature with a third body, they in turn have equality of temperature with each other.

This seems obvious to us because we are so familiar with this experiment. Because the principle is not derivable from other laws, and because it precedes the first and second laws of thermodynamics in the logical presentation of thermodynamics, it is called the zeroth law of thermodynamics. This law is really the basis of temperature measurement. Every time a body has equality of temperature with the thermometer, we can say that the body has the temperature we read on the thermometer. The problem remains of how to relate temperatures that we might read on different mercury thermometers or obtain from different temperature-measuring devices, such as thermocouples and resistance thermometers. This observation suggests the need for a standard scale for temperature measurements.

2.11 TEMPERATURE SCALES

Two scales are commonly used for measuring temperature, namely the Fahrenheit (after Gabriel Fahrenheit, 1686–1736) and the Celsius. The Celsius scale was formerly called the centigrade scale but is now designated the Celsius scale after Anders Celsius (1701–1744), the Swedish astronomer who devised this scale.

The Fahrenheit temperature scale is used with the English Engineering system of units, and the Celsius scale with the SI unit system. Until 1954 both of these scales were based on two fixed, easily duplicated points—the ice point and the steam point. The temperature of the ice point is defined as the temperature of a mixture of ice and water that is in equilibrium with saturated air at a pressure of 1 atm. The temperature of the steam point is the temperature of water and steam, which are in equilibrium at a pressure of 1 atm. On the Fahrenheit scale these two points are assigned the numbers 32 and 212, respectively, and on the Celsius scale the points are 0 and 100, respectively. Why Fahrenheit chose these numbers is an interesting story. In searching for an easily reproducible point, Fahrenheit selected the temperature of the human body and assigned it the number 96. He assigned the number 0 to the temperature of a certain mixture of salt, ice, and salt solution. On this scale the ice point was approximately 32. When this scale was slightly revised and fixed in terms of the ice point and steam point, the normal temperature of the human body was found to be 98.6 F.

In this text the symbols F and °C will denote the Fahrenheit and Celsius scales, respectively. The symbol T will refer to temperature on all temperature scales.

At the tenth CGPM in 1954, the Celsius scale was redefined in terms of a single fixed point and the ideal-gas temperature scale. The single fixed point is the triple point of water (the state in which the solid, liquid, and vapor phases of water exist together in equilibrium). The magnitude of the degree is defined in terms of the ideal-gas temperature scale, which is discussed in Chapter 7. The essential features of this new scale are a single fixed point and a definition of the magnitude of the degree. The triple point of water is assigned the value of 0.01°C. On this scale the steam point is experimentally found to be 100.00°C. Thus, there is essential agreement between the old and new temperature scales.

We have not yet considered an absolute scale of temperature. The possibility of such a scale comes from the second law of thermodynamics and is discussed in Chapter 7. On the basis of the second law of thermodynamics, a temperature scale that is independent of any thermometric substance can be defined. This absolute scale is usually referred to as the thermodynamic scale of temperature. However, it is very complicated to use this scale directly, and therefore, a more practical scale, the International Temperature Scale, which closely represents the thermodynamic scale, has been adopted.

The absolute scale related to the Celsius scale is the Kelvin scale (after William Thomson, 1824–1907, who is also known as Lord Kelvin), and is designated K (without the degree symbol). The relation between these scales is

$$K = {}^\circ C + 273.15 \qquad (2.3)$$

In 1967, the CGPM defined the kelvin as $1/273.16$ of the temperature at the triple point of water. The Celsius scale is now defined by this equation instead of by its earlier definition.

The absolute scale related to the Fahrenheit scale is the Rankine scale and is designated R. The relation between these scales is

$$R = F + 459.67 \qquad (2.4)$$

A number of empirically based temperature scales, to standardize temperature measurement and calibration, have been in use over the last 70 years. The most recent of these is the International Temperature Scale of 1990, or ITS-90. It is based on a number of fixed and easily reproducible points that are assigned definite numerical values of temperature, and on specified formulas relating temperature to the readings on certain temperature-measuring instruments for the purpose of interpolation between the defining fixed points. Details of the ITS-90 are not considered further in this text. It is noted that this scale is a practical means for establishing measurements that conform closely to the absolute thermodynamic temperature scale.

SUMMARY

We introduce a thermodynamic system as a control volume, which for a fixed mass is a control mass. Such a system can be isolated, exchanging neither mass, momentum, or energy with its surroundings. A closed system versus an open system refers to the ability of mass exchange with the surroundings. If properties for a substance change, the state changes and a process occurs. When a substance has gone through several processes returning to the same initial state it has completed a cycle.

Basic units for thermodynamic and physical properties are mentioned and most are covered in Table A.1. Thermodynamic properties such as density ρ, specific volume v, pressure P, and temperature T are introduced together with units for these. Properties are classified as intensive, independent of mass (like v), or extensive, proportional to mass (like V). Students should already be familiar with other concepts from physics such as force F, velocity $\mathbf{V}$, and acceleration a. Application of Newton's law of motion leads to the variation of the static pressure in a column of fluid and the measurements of pressure (absolute and gauge) by barometers and manometers. The normal temperature scale and the absolute temperature scale are introduced.

You should have learned a number of skills and acquired abilities from studying this chapter that will allow you to

- Define (choose) a control volume (C.V.) around some matter; sketch the content and identify storage locations for mass; and identify mass and energy flows crossing the C.V. surface.
- Know properties P, T, v, and ρ and their units.
- Know how to look up conversion of units in Table A.1.
- Know that energy is stored as kinetic, potential, or internal (in molecules).

- Know that energy can be transferred.
- Know the difference between (v, ρ) and (V, m) intensive versus extensive.
- Apply a force balance to a given system and relate it to pressure P.
- Know the difference between a relative (gauge) and absolute pressure P.
- Understand the working of a manometer or a barometer and get ΔP or P from height H.
- Know the difference between a relative and absolute temperature T.
- Have an idea about magnitudes (v, ρ, P, T).

Most of these concepts will be repeated and reinforced in the following chapters such as properties in Chapter 3, energy transfer as heat and work in Chapter 4, and internal energy in Chapter 5 together with their applications.

KEY CONCEPTS AND FORMULAS

Control volume	everything inside a control surface
Pressure definition	$P = \dfrac{F}{A}$ (mathematical limit for small A)
Specific volume	$v = \dfrac{V}{m}$
Density	$\rho = \dfrac{m}{V}$ (Tables A.3, A.4, A.5, F.2, F.3, and F.4)
Static pressure variation	$\Delta P = \rho g H$ (depth H in fluid of density ρ)
Absolute temperature	$T[K] = T[°C] + 273.15$ $T[R] = T[F] + 459.67$
Units	Table A.1

Concepts from Physics

Newton's law of motion	$F = ma$
Acceleration	$a = \dfrac{d^2x}{dt^2} = \dfrac{d\mathbf{V}}{dt}$
Velocity	$\mathbf{V} = \dfrac{dx}{dt}$

CONCEPT-STUDY GUIDE PROBLEMS

2.1 Make a control volume (C.V.) around the turbine in the steam power plant in Fig. 1.1 and list the flows of mass and energy that are there.

2.2 Make a control volume around the whole power plant in Fig. 1.2 and with the help of Fig. 1.1 list what flows of mass and energy are in or out and any storage of energy. Make sure you know what is inside and what is outside your chosen C.V.

2.3 Make a control volume that includes the steam flow around in the main turbine loop in the nuclear

propulsion system in Fig. 1.3. Identify mass flows (hot or cold) and energy transfers that enter or leave the C.V.

2.4 Take a control volume around your kitchen refrigerator and indicate where the components shown in Fig. 1.6 are located and show all flows of energy transfer.

2.5 An electric dip heater is put into a cup of water and heats it from 20°C to 80°C. Show the energy flow(s) and storage and explain what changes.

2.6 Separate the list P, F, V, v, ρ, T, a, m, L, t, and $\mathbf{V}$ into intensive, extensive, and nonproperties.

2.7 An escalator brings four people of total mass 300 kg, 25 m up in a building. Explain what happens with respect to energy transfer and stored energy.

2.8 Water in nature exists in different phases—solid, liquid, and vapor (gas). Indicate the relative magnitude of density and specific volume for the three phases.

2.9 Is density a unique measure of mass distribution in a volume? Does it vary? If so, on what kind of scale (distance)?

2.10 Density of fibers, rock wool insulation, foams and cotton is fairly low. Why is that?

2.11 How much mass is there approximately in 1 L of mercury (Hg)? Atmospheric air?

2.12 Can you carry 1 m^3 of liquid water?

2.13 A manometer shows a pressure difference of 1 m of liquid mercury. Find ΔP in kPa.

2.14 You dive 5 m down in the ocean. What is the absolute pressure there?

2.15 What pressure difference does a 10-m column of atmospheric air show?

2.16 The pressure at the bottom of a swimming pool is evenly distributed. Suppose we look at a cast-iron plate of 7272 kg lying on the ground with an area of 100 m^2. What is the average pressure below that? Is it just as evenly distributed?

2.17 A laboratory room keeps a vacuum of 0.1 kPa. What net force does that put on the door of size 2 m by 1 m?

2.18 A tornado rips off a 100-m^2 roof with a mass of 1000 kg. What is the minimum vacuum pressure needed to do that if we neglect the anchoring forces?

2.19 What is a temperature of $-5°C$ in degrees Kelvin?

2.20 What is the smallest temperature in degrees Celsius you can have? Kelvin?

2.21 Density of liquid water is $\rho = 1008 - T/2$ [kg/m^3] with T in °C. If the temperature increases 10°C how much deeper does a 1-m layer of water become?

2.22 Convert the formula for water density in Problem 2.21 to be for T in Kelvin.

HOMEWORK PROBLEMS

Properties and Units

2.23 A steel cylinder of mass 2 kg contains 4 L of liquid water at 25°C at 200 kPa. Find the total mass and volume of the system. List two extensive and three intensive properties of the water.

2.24 An apple "weighs" 80 g and has a volume of 100 cm^3 in a refrigerator at 8°C. What is the apple density? List three intensive and two extensive properties of the apple.

2.25 One kilopond (1 kp) is the weight of 1 kg in the standard gravitational field. How many Newtons (N) is that?

2.26 A pressurized steel bottle is charged with 5 kg of oxygen gas and 7 kg of nitrogen gas. How many kmoles are in the bottle?

Force and Energy

2.27 The "standard" acceleration (at sea level and 45° latitude) due to gravity is 9.806 65 m/s^2. What is the force needed to hold a mass of 2 kg at rest in this gravitational field? How much mass can a force of 1 N support?

2.28 A force of 125 N is applied to a mass of 12 kg in addition to the standard gravitation. If the direction of the force is vertical up, find the acceleration of the mass.

2.29 A model car rolls down an incline with a slope such that the gravitational "pull" in the direction of motion is one-third of the standard gravitational force (see Problem 2.27). If the car has a mass of 0.45 kg, find the acceleration.

2.30 When you move up from the surface of the earth, the gravitation is reduced as $g = 9.807 - 3.32 \times 10^{-6}z$, with z as the elevation in meters. By how many percent is the weight of an airplane reduced when it cruises at 11 000 m?

2.31 A car is driven at 60 km/h and is brought to a full stop with constant deceleration in 5 s. If the total car and driver mass is 1075 kg, find the necessary force.

2.32 A car of mass 1775 kg travels with a velocity of 100 km/h. Find the kinetic energy. How high should the car be lifted in the standard gravitational field to have a potential energy that equals the kinetic energy?

2.33 A 1200 kg car moving at 20 km/h is accelerated at a constant rate of 4 m/s² up to a speed of 75 km/h. What are the force and total time required?

2.34 A steel plate of 950 kg accelerates from rest at 3 m/s² for a period of 10 s. What force is needed and what is the final velocity?

2.35 A 15-kg steel container has 1.75 kmole of liquid propane inside. A force of 2 kN now accelerates this system. What is the acceleration?

2.36 A bucket of concrete of total mass 200 kg is raised by a crane with an acceleration of 2 m/s² relative to the ground at a location where the local gravitational acceleration is 9.5 m/s². Find the required force.

2.37 On the moon the gravitational acceleration is approximately one-sixth that on the surface of the earth. A 5-kg mass is "weighed" with a beam balance on the surface of the moon. What is the expected reading? If this mass is weighed with a spring scale that reads correctly for standard gravity on earth (see Problem 2.27), what is the reading?

Specific Volume

2.38 A 5-m³ container is filled with 900 kg of granite (density of 2400 kg/m³) and the rest of the volume is air with density equal to 1.15 kg/m³. Find the mass of air and the overall (average) specific volume.

2.39 A tank has two rooms separated by a membrane. Room A has 1 kg of air and a volume of 0.5 m³; room B has 0.75 m³ of air with density 0.8 kg/m³. The membrane is broken and the air comes to a uniform state. Find the final density of the air.

2.40 A 1-m³ container is filled with 400 kg of granite stone, 200 kg of dry sand, and 0.2 m³ of liquid 25°C water. Use properties from Tables A.3 and A.4. Find the average specific volume and density of the masses when you exclude air mass and volume.

2.41 A 1-m³ container is filled with 400 kg of granite stone, 200 kg of dry sand, and 0.2 m³ of liquid 25°C water. Use properties from Tables A.3 and A.4 and use an air density of 1.1 kg/m³. Find the average specific volume and density of the 1-m³ volume.

2.42 One kilogram of diatomic oxygen (O₂, molecular weight of 32) is contained in a 500-L tank. Find the specific volume on both a mass and mole basis (v and $\bar{v}$).

2.43 A 15-kg steel gas tank holds 300 L of liquid gasoline, having a density of 800 kg/m³. If the system is decelerated at 6 m/s² what is the needed force?

Pressure

2.44 A hydraulic lift has a maximum fluid pressure of 500 kPa. What should the piston/cylinder diameter be so it can lift a mass of 850 kg?

2.45 A piston/cylinder with a cross-sectional area of 0.01 m² has a piston mass of 100 kg resting on the stops, as shown in Fig. P2.45. With an outside atmospheric pressure of 100 kPa, what should the water pressure be to lift the piston?

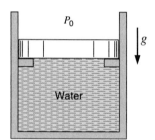

FIGURE P2.45

2.46 A vertical hydraulic cylinder has a 125-mm-diameter piston with hydraulic fluid inside the cylinder and an ambient pressure of 1 bar. Assuming standard gravity, find the piston mass that will create a pressure inside of 1500 kPa.

2.47 A valve in the cylinder shown in Fig. P2.47 has a cross-sectional area of 11 cm² with a pressure of 735 kPa inside the cylinder and 99 kPa outside. How large a force is needed to open the valve?

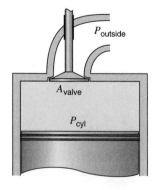

FIGURE P2.47

2.48 A cannnonball of 5 kg acts as a piston in a cylinder of 0.15 m diameter. As the gunpowder is burned, a pressure of 7 MPa is created in the gas behind the ball. What is the acceleration of the ball if the cylinder (cannon) is pointing horizontally?

2.49 Repeat the previous problem for a cylinder (cannon) pointing 40° up relative to the horizontal direction.

2.50 A large exhaust fan in a laboratory room keeps the pressure inside at 10 cm of water relative vacuum to the hallway. What is the net force on the door measuring 1.9 m by 1.1 m?

2.51 What is the pressure at the bottom of a 5-m-tall column of fluid with atmospheric pressure of 101 kPa on the top surface if the fluid is

a. water at 20°C

b. glycerine at 25°C

c. light oil

2.52 The hydraulic lift in an auto-repair shop has a cylinder diameter of 0.2 m. To what pressure should the hydraulic fluid be pumped to lift 40 kg of piston/arms and 700 kg of a car?

2.53 A 2.5-m-tall steel cylinder has a cross-sectional area of 1.5 m². At the bottom with a height of 0.5 m is liquid water on top of which is a 1-m-high layer of gasoline. This is shown in Fig. P2.53. The gasoline surface is exposed to atmospheric air at 101 kPa. What is the highest pressure in the water?

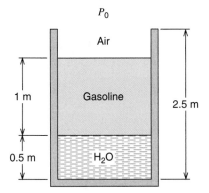

FIGURE P2.53

2.54 At the beach, atmospheric pressure is 1025 mbar. You dive 15 m down in the ocean and you later climb a hill up to 250 m in elevation. Assume the density of water is about 1000 kg/m³ and the density of air is 1.18 kg/m³. What pressure do you feel at each place?

2.55 A piston, $m_p = 5$ kg, is fitted in a cylinder, $A = 15$ cm², that contains a gas. The setup is in a centrifuge that creates an acceleration of 25 m/s² in the direction of piston motion toward the gas. Assuming standard atmospheric pressure outside the cylinder, find the gas pressure.

2.56 A steel tank of cross-sectional are a 3 m² and 16 m tall weighs 10 000 kg and is open at the top as shown in Fig. P2.56. We want to float it in the ocean so it sticks 10 m straight down by pouring concrete into the bottom of it. How much concrete should we put in?

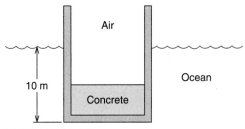

FIGURE P2.56

2.57 Liquid water with density ρ is filled on top of a thin piston in a cylinder with cross-sectional area A and total height H, as shown in Fig. P2.57. Air is let in under the piston so it pushes up, spilling the water over the edge. Deduce the formula for the air pressure as a function of the piston elevation from the bottom, h.

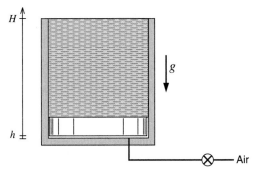

FIGURE P2.57

Manometers and Barometers

2.58 The density of atmospheric air is about 1.15 kg/m³, which we assume is constant. How large an absolute pressure will a pilot see when flying 1500 m above ground level where the pressure is 101 kPa?

2.59 A differential pressure gauge mounted on a vessel shows 1.25 MPa and a local barometer gives atmospheric pressure as 0.96 bar. Find the absolute pressure inside the vessel.

2.60 Two vertical cylindrical storage tanks are full of liquid water (density = 1000 kg/m³) with the top open to the atmosphere. One is 10 m tall and 2 m in diameter; the other is 2.5 m tall with diameter 4 m. What is the total force form the bottom of each tank to the water and what is the pressure at the bottom of each tank?

2.61 Blue manometer fluid of density 925 kg/m³ shows a column height difference of 3 cm vacuum with one end attached to a pipe and the other open to $P_0 = 101$ kPa. What is the absolute pressure in the pipe?

2.62 The absolute pressure in a tank is 85 kPa and the local ambient absolute pressure is 97 kPa. If a U-tube with mercury (density = 13 550 kg/m³) is attached to the tank to measure the vacuum, what column height difference would it show?

2.63 The pressure gauge on an air tank shows 75 kPa when the diver is 10 m down in the ocean. At what depth will the gauge pressure be zero? What does that mean?

2.64 A submarine maintains 101 kPa inside it and dives 240 m down in the ocean having an average density of 1030 kg/m³. What is the pressure difference between the inside and the outside of the submarine hull?

2.65 A barometer to measure absolute pressure shows a mercury column height of 725 mm. The temperature is such that the density of the mercury is 13 550 kg/m³. Find the ambient pressure.

2.66 An absolute pressure gauge attached to a steel cylinder shows 135 kPa. We want to attach a manometer using liquid water a day that $P_{atm} = 101$ kPa. How high a fluid level difference must we plan for?

2.67 The difference in height between the columns of a manometer is 200 mm with a fluid of density 900 kg/m³. What is the pressure difference? What is the height difference if the same pressure difference is measured using mercury (density = 13 600 kg/m³) as manometer fluid?

2.68 An exploration submarine should be able to go 4000 m down in the ocean. If the ocean density is 1020 kg/m³ what is the maximum pressure on the submarine hull?

2.69 Assume we use a pressure gauge to measure the air pressure at street level and at the roof of a tall building. If the pressure difference can be determined with an accuracy of 1 mbar (0.001 bar), what uncertainty in the height estimate does that correspond to?

2.70 A U-tube manometer filled with water (density = 1000 kg/m³) shows a height difference of 25 cm. What is the gauge pressure? If the right branch is tilted to make an angle of 30° with the horizontal, as shown in Fig. P2.70, what should the length of the column in the tilted tube be relative to the U-tube?

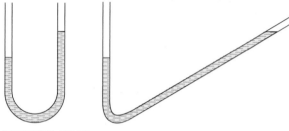

FIGURE P2.70

2.71 A barometer measures 760 mm Hg at street level and 735 mm Hg on top of a building. How tall is the building if we assume air density of 1.15 kg/m³?

2.72 A piece of experimental apparatus, Fig. P2.72, is located where $g = 9.5$ m/s² and the temperature is 5°C. An air flow inside the apparatus is determined by measuring the pressure drop across an orifice with a mercury manometer (see Problem 2.77 for density) showing a height difference of 200 mm. What is the pressure drop in kPa?

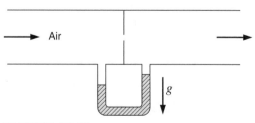

Air

g

FIGURE P2.72

2.73 Two piston/cylinder arrangements, *A* and *B*, have their gas chambers connected by a pipe, as shown in Fig. P2.73. Cross-sectional areas are $A_A = 75$ cm² and $A_B = 25$ cm², with the piston mass in *A* being $m_A = 25$ kg. Assume the outside pressure is 100 kPa and standard gravitation. Find the mass m_B so that none of the pistons have to rest on the bottom.

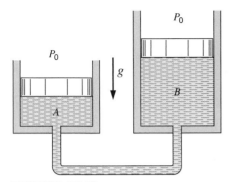

FIGURE P2.73

2.74 Two hydraulic piston/cylinders are of the same size and setup as in Problem 2.73, but with negligible piston masses. A single point force of 250 N presses down on piston A. Find the needed extra force on piston B so that none of the pistons have to move.

2.75 A pipe flowing light oil has a manometer attached as shown in Fig. P2.75. What is the absolute pressure in the pipe flow?

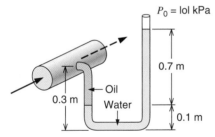

FIGURE P2.75

2.76 Two cylinders are filled with liquid water, $\rho \simeq$ 1000 kg/m³, and connected by a line with a closed valve, as shown in Fig. P2.76. A has 100 kg and B

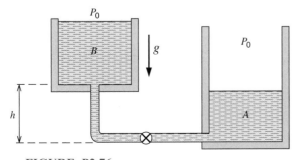

FIGURE P2.76

has 500 kg of water, their cross-sectional areas are $A_A = 0.1$ m² and $A_B = 0.25$ m², and the height h is 1 m. Find the pressure on each side of the valve. The valve is opened and water flows to an equilibrium. Find the final pressure at the valve location.

Temperature

2.77 The density of mercury changes approximately linearly with temperature as

$$\rho_{Hg} = 13\ 595 - 2.5\ T\ \text{kg/m}^3 \quad (T \text{ in Celsius})$$

so the same pressure difference will result in a manometer reading that is influenced by temperature. If a pressure difference of 100 kPa is measured in the summer at 35°C and in the winter at −15°C, what is the difference in column height between the two measurements?

2.78 A mercury thermometer measures temperature by measuring the volume expansion of a fixed mass of liquid Hg due to a change in the density (see Problem 2.77). Find the relative change (%) in volume for a change in temperature from 10°C to 20°C.

2.79 Using the freezing and boiling point temperatures for water in both Celsius and Fahrenheit scales, develop a conversion formula between the scales. Find the conversion formula between Kelvin and Rankine temperature scales.

2.80 The atmosphere becomes colder at higher elevation. As an average the standard atmospheric absolute temperature can be expressed as $T_{atm} = 288 - 6.5 \times 10^{-3}\ z$, where z is the elevation in meters. How cold is it outside an airplane cruising at 12 000 m expressed in Kelvin and in Celsius.

Review Problems

2.81 Repeat Problem 2.72 if the flow inside the apparatus is liquid water ($\rho \simeq 1000$ kg/m³) instead of air. Find the pressure difference between the two holes flush with the bottom of the channel. You cannot neglect the two unequal water columns.

2.82 The main waterline into a tall building has a pressure of 600 kPa at 5 m elevation below ground level. The building is shown in Fig. P2.82. How much extra pressure does a pump need to add to ensure a water line pressure of 200 kPa at the top floor 150 m above ground?

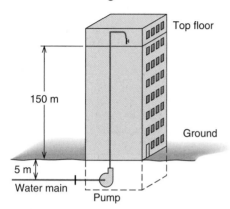

FIGURE P2.82

2.83 A 5-kg piston in a cylinder with diameter of 100 mm is loaded with a linear spring and the outside atmospheric pressure of 100 kPa as shown in Fig. P2.83. The spring exerts no force on the piston when it is at the bottom of the cylinder, and for the state shown, the pressure is 400 kPa with volume 0.4 L. The valve is opened to let some air in, causing the piston to rise 2 cm. Find the new pressure.

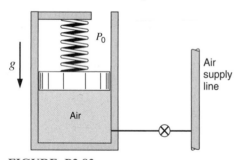

FIGURE P2.83

2.84 In the city water tower, water is pumped up to a level 25 m above ground in a pressurized tank

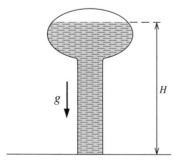

FIGURE P2.84

with air at 125 kPa over the water surface. This is illustrated in Fig. P2.84. Assuming the water density is 1000 kg/m^3 and standard gravity, find the pressure required to pump more water in at ground level.

2.85 Two cylinders are connected by a piston, as shown in Fig. P2.85. Cylinder A is used as a hydraulic lift and pumped up to 500 kPa. The piston mass is 25 kg and there is standard gravity. What is the gas pressure in cylinder B?

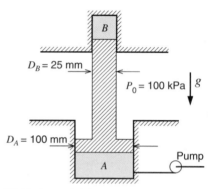

FIGURE P2.85

2.86 A dam retains a lake 6 m deep, shown in Fig. P2.86. To construct a gate in the dam we need to know the net horizontal force on a 5-m-wide and 6-m-tall port section that then replaces a 5-m section of the dam. Find the net horizontal force from the water on one side and air on the other side of the port.

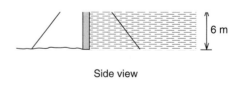

Side view

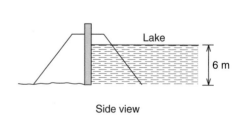

Side view

FIGURE P2.86

ENGLISH UNIT PROBLEMS

English Unit Concept Problems

2.87E A mass of 2 lbm has acceleration of 5 ft/s^2. What is the needed force in lbf?

2.88E How much mass is in 0.25 gal of liquid mercury (Hg)? Atmospheric air?

2.89E Can you easily carry a 1-gal bar of solid gold?

2.90E What is the temperature of −5 F in degrees Rankine?

2.91E What is the smallest temperature in degrees Fahrenheit you can have? Rankine?

English Unit Problems

2.92E An apple weighs 0.2 lbm and has a volume of 6 in^3 in a refrigerator at 38 F. What is the apple density? List three intensive and two extensive properties for the apple.

2.93E A 2500-lbm car moving at 15 mi/h is accelerated at a constant rate of 15 ft/s^2 up to a speed of 50 mi/h. What are the force and total time required?

2.94E Two pound moles of diatomic oxygen gas are enclosed in a 20-lbm steel container. A force of 2000 lbf now accelerates this system. What is the acceleration?

2.95E A valve in a cylinder has a cross-sectional area of 2 in^2 with a pressure of 100 psia inside the cylinder and 14.7 psia outside. How large a force is needed to open the valve?

2.96E One pound mass of diatomic oxygen (O_2 molecular weight 32) is contained in a 100-gal tank. Find the specific volume on both a mass and mole basis (v and $\bar{v}$).

2.97E A 30-lbm steel gas tank holds 10 ft^3 of liquid gasoline having a density of 50 lbm/ft^3. What force is needed to accelerate this combined system at a rate of 15 ft/s^2?

2.98E A laboratory room keeps a vacuum of 4 in. of water due to the exhaust fan. What is the net force on a door of size 6 ft by 3 ft?

2.99E A 7-ft-m tall steel cylinder has a cross-sectional area of 15 ft^2. At the bottom, with a height of 2 ft, is liquid water on top of which is a 4-ft-high layer of gasoline. The gasoline surface is exposed to atmospheric air at 14.7 psia. What is the highest pressure in the water?

2.100E A U-tube manometer filled with water, density 62.3 lbm/ft^3, shows a height difference of 10 in. What is the gauge pressure? If the right branch is tilted to make an angle of 30° with the horizontal, as shown in Fig. P2.70, what should the length of the column in the tilted tube be relative to the U-tube?

2.101E A piston/cylinder with cross-sectional area of 0.1 ft^2 has a piston mass of 200 lbm resting on the stops, as shown in Fig. P2.45. With an outside atmospheric pressure of 1 atm, what should the water pressure be to lift the piston?

2.102E The main waterline into a tall building has a pressure of 90 psia at 16 ft elevation below ground level. How much extra pressure does a pump need to add to ensure a waterline pressure of 30 psia at the top floor 450 ft above ground?

2.103E A piston, m_p = 10 lbm, is fitted in a cylinder, A = 2.5 in^2, that contains a gas. The setup is in a centrifuge that creates an acceleration of 75 ft/s^2. Assuming standard atmospheric pressure outside the cylinder, find the gas pressure.

2.104E The atmosphere becomes colder at higher elevation. As an average the standard atmospheric absolute temperature can be expressed as T_{atm} = $518 - 3.84 \times 10^{-3}\, z$, where z is the elevation in feet. How cold is it outside an airplane cruising at 32 000 ft expressed in Rankine and in Fahrenheit?

2.105E The density of mercury changes approximately linearly with temperature as

$$\rho_{Hg} = 851.5 - 0.086\, T \quad \text{lbm/ft}^3 \quad (T \text{ in degrees Fahrenheit})$$

so the same pressure difference will result in a manometer reading that is influenced by temperature. If a pressure difference of 14.7 lbf/in^2 is measured in the summer at 95 F and in the winter at 5 F, what is the difference in column height between the two measurements?

COMPUTER, DESIGN AND OPEN-ENDED PROBLEMS

2.106 Write a program to list corresponding temperatures in °C, K, F, and R from −50°C to 100°C in increments of 10 degrees.

2.107 Write a program that will input pressure in kPa or atm or lbf/in² and write the pressure out in kPa, atm, bar, and lbf/in².

2.108 Write a program to do the temperature correction on a mercury barometer reading (see Problem 2.62). Input reading and temperature and output corrected reading at 20°C and pressure in kPa.

2.109 Make a list of different weights and scales that are used to measure mass directly or indirectly. Investigate the ranges of mass and the accuracy that can be obtained.

2.110 Thermometers are based on several principles. Expansion of a liquid with a rise in temperature is used in many applications. Electrical resistance, thermistors, and thermocouples are common in instrumentation and remote probes. Investigate a variety of thermometers and make a list of their range, accuracy, advantages, and disadvantages.

2.111 Collect information for a resistance-, thermistor-, and thermocouple-based thermometer suitable for the range of temperatures form 0°C to 200°C. For each of the three types list the accuracy and response of the transducer (output per degree change). Is any calibration or corrections necessary when it is used in an instrument?

2.112 A thermistor is used as a temperature transducer. Its resistance changes with temperature approximately as

$$R = R_0 \exp[\alpha(1/T - 1/T_0)]$$

where it has the resistance R_0 at temperature T_0. Select the constants as $R_0 = 3000 \ \Omega$ and $T_0 = 298$ K, and compute α so it has the resistance of $200 \ \Omega$ at 100°C. Write a program to convert a measured resistance, R, into information about the temperature. Find information for actual thermistors and plot the calibration curves with the formula given in this problem and the recommended correction given by the manufacturer.

2.113 Investigate possible transducers for the measurement of temperature in a flame with temperatures near 1000 K. Are any available for a temperature of 2000 K?

2.114 Devices to measure pressure are available as differential or absolute pressure transducers. Make a list of 5 different differential pressure transducers to measure pressure differences in order of 100 kPa. Note their accuracy, response (linear or ?), and price.

2.115 A micromanometer uses a fluid with density 1000 kg/m³ and it is able to measure the height difference with an accuracy of ±0.5 mm. Its range is a maximum height difference of 0.5 m. Investigate if any transducers are available to replace the micromanometer.

2.116 An experiment involves the measurements of temperature and pressure of a gas flowing in a pipe at 300°C and 250 kPa. Write a report with a suggested set of transducers (at least two alternatives for each) and give the expected accuracy and cost.

PROPERTIES OF A PURE SUBSTANCE 3

In the previous chapter we considered three familiar properties of a substance—specific volume, pressure, and temperature. We now turn our attention to pure substances and consider some of the phases in which a pure substance may exist, the number of independent properties a pure substance may have, and methods of presenting thermodynamic properties.

Properties and the behavior of substances are very important for our studies of devices and thermodynamic systems. The steam power plant in Fig. 1.1 and the nuclear propulsion system in Fig. 1.3 have very similar processes, using water as the working substance. Water vapor (steam) is made by boiling at high pressure in the steam generator followed by an expansion in the turbine to a lower pressure, a cooling in the condenser, and a return to the boiler by a pump that raises the pressure. We must know the water properties to properly size the equipment such as the burners or heat exchangers, turbine, and pump for the desired transfer of energy and the flow of water. As the water is brought from liquid to vapor we need to know the temperature for the given pressure, and we must know the density or specific volume so that the piping can be properly dimensioned for the flow. If the pipes are too small, the expansion creates excessive velocities, leading to pressure losses and increased friction, and thus demanding a larger pump and reducing the turbine work output.

Another example is a refrigerator, shown in Fig. 1.6, where we need a substance that will boil from liquid to vapor at a low temperature, say $-20°C$. This absorbs energy from the cold space, keeping it cold. Inside the black grille in the back or at the bottom, the now hot substance is cooled by air flowing around the grille, so it condenses from vapor to liquid at a temperature slightly higher than room temperature. When such a system is designed, we need to know the pressures at which these processes take place and the amount of energy, covered in Chapter 5, that is involved. We also need to know how much volume the substance occupies, the specific volume, so that the piping diameters can be selected as mentioned for the steam power plant. The substance is selected so that the pressure is reasonable during these processes; it should not be too high, due to leakage and safety concerns, and not too low either, as air might leak into the system.

A final example of a situation where we need to know the substance properties is the gas turbine and a variation thereof, namely the jet engine shown in Fig. 1.11. In these systems, the working substance is a gas (very similar to air) and no phase change takes place. A combustion process burns fuel and air, freeing a large amount of energy, which heats the gas so that it expands. We need to know how hot the gas gets and how much the expansion is so that we can analyze the expansion process in the turbine and the exit nozzle of the jet engine. In this device, we do need large velocities inside the turbine section and for the exit of the jet engine. This high-velocity flow pushes on the blades in the turbine to create shaft work or pushes on the jet engine (something called thrust) to move the aircraft forward.

These are just a few examples of complete thermodynamic systems where a substance goes through several processes involving changes of its thermodynamic state and therefore its properties. As your studies progress, many other examples will be used to illustrate the general subjects.

3.1 THE PURE SUBSTANCE

A pure substance is one that has a homogeneous and invariable chemical composition. It may exist in more than one phase, but the chemical composition is the same in all phases. Thus, liquid water, a mixture of liquid water and water vapor (steam), and a mixture of ice and liquid water are all pure substances; every phase has the same chemical composition. In contrast, a mixture of liquid air and gaseous air is not a pure substance because the composition of the liquid phase is different from that of the vapor phase.

Sometimes a mixture of gases, such as air, is considered a pure substance as long as there is no change of phase. Strictly speaking, this is not true. As we will see later, we should say that a mixture of gases such as air exhibits some of the characteristics of a pure substance as long as there is no change of phase.

In this text the emphasis will be on simple compressible substances. This term designates substances whose surface effects, magnetic effects, and electrical effects are insignificant when dealing with the substances. But changes in volume, such as those associated with the expansion of a gas in a cylinder, are very important. Reference will be made, however, to other substances for which surface, magnetic, and electrical effects are important. We will refer to a system consisting of a simple compressible substance as a simple compressible system.

3.2 VAPOR–LIQUID–SOLID-PHASE EQUILIBRIUM IN A PURE SUBSTANCE

Consider as a system 1 kg of water contained in the piston/cylinder arrangement shown in Fig. 3.1a. Suppose that the piston and weight maintain a pressure of 0.1 MPa in the cylinder and that the initial temperature is 20°C. As heat is transferred to the water, the temperature increases appreciably, the specific volume increases slightly, and the pressure remains constant. When the temperature reaches 99.6°C, additional heat transfer results in a change of phase, as indicated in Fig. 3.1b. That is, some of the liquid becomes vapor, and during this process both the temperature and pressure remain constant, but the specific volume increases considerably. When the last drop of liquid has vaporized, further transfer of heat results in an increase in both temperature and specific volume of the vapor, as shown in Fig. 3.1c.

The term saturation temperature designates the temperature at which vaporization takes place at a given pressure. This pressure is called the saturation pressure for the given temperature. Thus, for water at 99.6°C the saturation pressure is 0.1 MPa, and for water at 0.1 MPa the saturation temperature is 99.6°C. For a pure substance there is a definite rela-

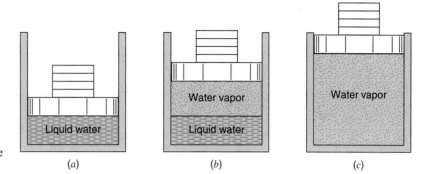

FIGURE 3.1
Constant-pressure change
from liquid to vapor phase
for a pure substance.

tion between saturation pressure and saturation temperature. A typical curve, called the vapor-pressure curve, is shown in Fig. 3.2.

If a substance exists as liquid at the saturation temperature and pressure, it is called saturated liquid. If the temperature of the liquid is lower than the saturation temperature for the existing pressure, it is called either a subcooled liquid (implying that the temperature is lower than the saturation temperature for the given pressure) or a compressed liquid (implying that the pressure is greater than the saturation pressure for the given temperature). Either term may be used, but the latter term will be used in this text.

When a substance exists as part liquid and part vapor at the saturation temperature, its quality is defined as the ratio of the mass of vapor to the total mass. Thus, in Fig. 3.1b, if the mass of the vapor is 0.2 kg and the mass of the liquid is 0.8 kg, the quality is 0.2 or 20%. The quality may be considered an intensive property and has the symbol x. Quality has meaning only when the substance is in a saturated state, that is, at saturation pressure and temperature.

If a substance exists as vapor at the saturation temperature, it is called saturated vapor. (Sometimes the term dry saturated vapor is used to emphasize that the quality is 100%.) When the vapor is at a temperature greater than the saturation temperature, it is said to exist as superheated vapor. The pressure and temperature of superheated vapor are independent properties, since the temperature may increase while the pressure remains constant. Actually, the substances we call gases are highly superheated vapors.

Consider Fig. 3.1 again. Let us plot on the temperature–volume diagram of Fig. 3.3 the constant-pressure line that represents the states through which the water passes as it is heated from the initial state of 0.1 MPa and 20°C. Let state A represent the initial state, B the saturated-liquid state (99.6°C), and line AB the process in which the liquid is heated from the initial temperature to the saturation temperature. Point C is the

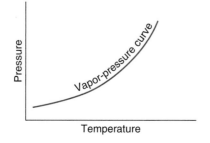

FIGURE 3.2 Vapor-pressure curve of a pure substance.

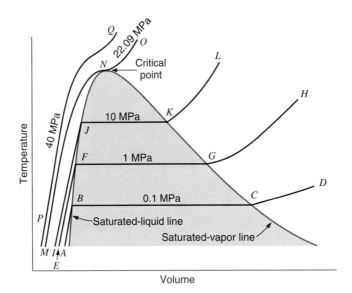

FIGURE 3.3
Temperature–volume
diagram for water
showing liquid and vapor
phases (not to scale).

saturated-vapor state, and line *BC* is the constant-temperature process in which the change of phase from liquid to vapor occurs. Line *CD* represents the process in which the steam is superheated at constant pressure. Temperature and volume both increase during this process.

Now let the process take place at a constant pressure of 1 MPa, starting from an initial temperature of 20°C. Point *E* represents the initial state, in which the specific volume is slightly less than that at 0.1 MPa and 20°C. Vaporization begins at point *F*, where the temperature is 179.9°C. Point *G* is the saturated-vapor state, and line *GH* is the constant-pressure process in which the steam in superheated.

In a similar manner, a constant pressure of 10 MPa is represented by line *IJKL*, for which the saturation temperature is 311.1°C.

At a pressure of 22.09 MPa, represented by line *MNO*, we find, however, that there is no constant-temperature vaporization process. Instead, point *N* is a point of inflection with a zero slope. This point is called the critical point. At the critical point the saturated-liquid and saturated-vapor states are identical. The temperature, pressure, and specific volume at the critical point are called the critical temperature, critical pressure, and critical volume. The critical-point data for some substances are given in Table 3.1. More extensive data are given in Table A.2 in the appendix.

TABLE 3.1
Some Critical-Point Data

	Critical Temperature, °C	Critical Pressure, MPa	Critical Volume, m³/kg
Water	374.14	22.09	0.003 155
Carbon dioxide	31.05	7.39	0.002 143
Oxygen	−118.35	5.08	0.002 438
Hydrogen	−239.85	1.30	0.032 192

A constant-pressure process at a pressure greater than the critical pressure is represented by line *PQ*. If water at 40 MPa and 20°C is heated in a constant-pressure process in a cylinder as shown in Fig. 3.1, there will never be two phases present and the state shown in Fig. 3.1*b* will never exist. Instead, there will be a continuous change in density and at all times there will be only one phase present. The question then is when do we have a liquid and when do we have a vapor? The answer is that this is not a valid question at supercritical pressures. We simply term the substance a fluid. However, rather arbitrarily, at temperatures below the critical temperature we usually refer to it as a compressed liquid and at temperatures above the critical temperature as a superheated vapor. It should be emphasized, however, that at pressures above the critical pressure we never have a liquid and vapor phase of a pure substance existing in equilibrium.

In Fig. 3.3, line *NJFB* represents the saturated-liquid line and line *NKGC* represents the saturated-vapor line.

By convention, the subscript *f* is used to designate a property of a saturated liquid and the subscript *g* a property of a saturated vapor (subscript *g* being used to denote saturation temperature and pressure). Thus, a saturation condition involving part liquid and part vapor such as in Fig. 3.1*b* can be shown on *T–v* coordinates as in Fig. 3.4. All of the liquid present is at state *f* with specific volume v_f and all of the vapor present is at state *g* with v_g. The total volume is the sum of the liquid volume and the vapor volume, or

$$V = V_{\text{liq}} + V_{\text{vap}} = m_{\text{liq}} v_f + m_{\text{vap}} v_g$$

The average specific volume of the system v is then

$$v = \frac{V}{m} = \frac{m_{\text{liq}}}{m} v_f + \frac{m_{\text{vap}}}{m} v_g = (1 - x) v_f + x v_g \qquad (3.1)$$

in terms of the definition of quality $x = m_{\text{vap}}/m$.

Using the definition

$$v_{fg} = v_g - v_f$$

Equation 3.1 can also be written as

$$v = v_f + x v_{fg} \qquad (3.2)$$

Now, the quality *x* can be viewed as the fraction $(v - v_f)/v_{fg}$ of the distance between saturated liquid and saturated vapor, as indicated in Fig. 3.4.

Let us now consider another experiment with the piston/cylinder arrangement. Suppose that the cylinder contains 1 kg of ice at −20°C, 100 kPa. When heat is transferred to the ice, the pressure remains constant, the specific volume increases slightly,

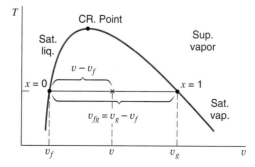

FIGURE 3.4 *T–v* diagram for the two-phase liquid–vapor region to show the quality specific volume relation.

TABLE 3.2
Some Solid–Liquid–Vapor Triple-Point Data

	Temperature, °C	Pressure, kPa
Hydrogen (normal)	−259	7.194
Oxygen	−219	0.15
Nitrogen	−210	12.53
Carbon dioxide	−56.4	520.8
Mercury	−39	0.000 000 13
Water	0.01	0.6113
Zinc	419	5.066
Silver	961	0.01
Copper	1083	0.000 079

and the temperature increases until it reaches 0°C, at which point the ice melts and the temperature remains constant. In this state the ice is called a saturated solid. For most substances the specific volume increases during this melting process, but for water the specific volume of the liquid is less than the specific volume of the solid. When all the ice has melted, a further heat transfer causes an increase in temperature of the liquid.

If the initial pressure of the ice at −20°C is 0.260 kPa, heat transfer to the ice results in an increase in temperature to −10°C. At this point, however, the ice passes directly from the solid phase to the vapor phase in the process known as sublimation. Further heat transfer results in superheating of the vapor.

Finally, consider an initial pressure of the ice of 0.6113 kPa and a temperature of −20°C. Through heat transfer let the temperature increase until it reaches 0.01°C. At this point, however, further heat transfer may cause some of the ice to become vapor and some to become liquid, for at this point it is possible to have the three phases in equilibrium. This point is called the triple point, which is defined as the state in which all three phases may be present in equilibrium. The pressure and temperature at the triple point for a number of substances are given in Table 3.2.

This whole matter is best summarized by the diagram of Fig. 3.5, which shows how the solid, liquid, and vapor phases may exist together in equilibrium. Along the sublimation line the solid and vapor phases are in equilibrium, along the fusion line the solid and liquid phases are in equilibrium, and along the vaporization line the liquid and vapor phases are in equilibrium. The only point at which all three phases may exist in equilibrium is the triple point. The vaporization line ends at the critical point because there is no distinct change from the liquid phase to the vapor phase above the critical point.

Consider a solid in state *A*, as shown in Fig. 3.5. When the temperature increases but the pressure (which is less than the triple-point pressure) is constant, the substance passes directly from the solid to the vapor phase. Along the constant-pressure line *EF*, the substance passes from the solid to the liquid phase at one temperature, and then from the liquid to the vapor phase at a higher temperature. Constant-pressure line *CD* passes through the triple point, and it is only at the triple point that the three phases may exist together in equilibrium. At a pressure above the critical pressure, such as *GH*, there is no sharp distinction between the liquid and vapor phases.

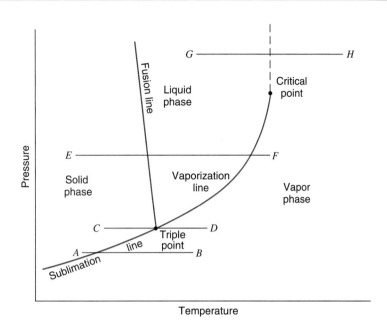

FIGURE 3.5

Pressure–temperature diagram for a substance such as water.

Although we have made these comments with rather specific reference to water (only because of our familiarity with water), all pure substances exhibit the same general behavior. However, the triple-point temperature and critical temperature vary greatly from one substance to another. For example, the critical temperature of helium, as given in Table A.2, is 5.3 K. Therefore, the absolute temperature of helium at ambient conditions is over 50 times greater than the critical temperature. In contrast, water has a critical temperature of 374.14°C (647.29 K), and at ambient conditions the temperature of water is less than half the critical temperature. Most metals have a much higher critical temperature than water. When we consider the behavior of a substance in a given state, it is often helpful to think of this state in relation to the critical state or triple point. For example, if the pressure is greater than the critical pressure, it is impossible to have a liquid phase and a vapor phase in equilibrium. Or, to consider another example, the states at which vacuum-melting a given metal is possible can be ascertained by a consideration of the properties at the triple point. Iron at a pressure just above 5 Pa (the triple-point pressure) would melt at a temperature of about 1535°C (the triple-point temperature).

Figure 3.6 shows the three-phase diagram for carbon dioxide, in which it is seen (see also Table 3.2) that the triple-point pressure is greater than normal atmospheric pressure, which is very unusual. Therefore, the commonly observed phase transition under conditions of atmospheric pressure of about 100 kPa is a sublimation from solid directly to vapor, without passing through a liquid phase, which is why solid carbon dioxide is commonly referred to as dry ice. We note from Fig. 3.6 that this phase transformation at 100 kPa occurs at a temperature below 200 K.

Finally, it should be pointed out that a pure substance can exist in a number of different solid phases. A transition from one solid phase to another is called an allotropic transformation. Figure 3.7 shows a number of solid phases for water. A pure substance can have a number of triple points, but only one triple point has a solid, liquid, and vapor equilibrium. Other triple points for a pure substance can have two solid phases and a liquid phase, two solid phases and a vapor phase, or three solid phases.

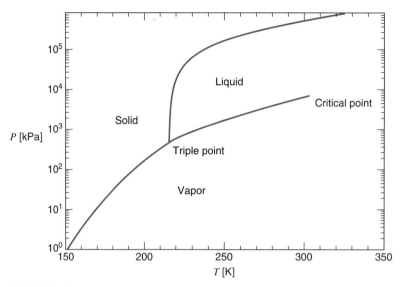

FIGURE 3.6 Carbon dioxide phase diagram.

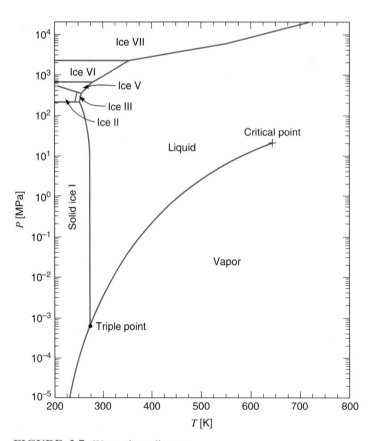

FIGURE 3.7 Water phase diagram.

3.3 INDEPENDENT PROPERTIES OF A PURE SUBSTANCE

One important reason for introducing the concept of a pure substance is that the state of a simple compressible pure substance (that is, a pure substance in the absence of motion, gravity, and surface, magnetic, or electrical effects) is defined by two independent properties. For example, if the specific volume and temperature of superheated steam are specified, the state of the steam is determined.

To understand the significance of the term independent property, consider the saturated-liquid and saturated-vapor states of a pure substance. These two states have the same pressure and the same temperature, but they are definitely not the same state. In a saturation state, therefore, pressure and temperature are not independent properties. Two independent properties such as pressure and specific volume or pressure and quality are required to specify a saturation state of a pure substance.

The reason for mentioning previously that a mixture of gases, such as air, has the same characteristics as a pure substance as long as only one phase is present, concerns precisely this point. The state of air, which is a mixture of gases of definite composition, is determined by specifying two properties as long as it remains in the gaseous phase. Air then can be treated as a pure substance.

3.4 TABLES OF THERMODYNAMIC PROPERTIES

Tables of thermodynamic properties of many substances are available, and in general, all these tables have the same form. In this section we will refer to the steam tables. The steam tables are selected both because they are a vehicle for presenting thermodynamic tables and because steam is used extensively in power plants and industrial processes. Once the steam tables are understood, other thermodynamic tables can be readily used.

Several different versions of steam tables have been published over the years. The set included in Appendix B, Table B.1, is a summary based on a complicated fit to the behavior of water. It is very similar to the *Steam Tables* by Keenan, Keyes, Hill, and Moore, published in 1969 and 1978. We will concentrate here on the three properties already discussed in Chapter 2 and in Section 3.2, namely T, P, and v, and note that the other properties listed in the set of Tables B.1, u, h, and s, will be introduced later.

The steam tables in Appendix B consist of five separate tables, as indicated in Fig. 3.8. The region of superheated vapor in Fig. 3.5 is given in Table B.1.3, and that of compressed liquid is given in Table B.1.4. The compressed-solid region shown in Fig. 3.5 is not listed in the appendix. The saturated-liquid and saturated-vapor region, as seen in the T and v diagram of Fig. 3.3 (and as the vaporization line in Fig. 3.5), is listed according to the values of T in Table B.1.1 and according to the values of P (T and P are not independent in the two-phase regions) in Table B.1.2. Similarly, the saturated-solid and saturated-vapor region is listed according to T in Table B.1.5, but the saturated-solid and saturated-liquid region, the third phase boundary line shown in Fig. 3.5 is not listed in the appendix.

In Table B.1.1, the first column after the temperature gives the corresponding saturation pressure in kilopascals. The next three columns give specific volume in cubic meters per kilogram. The first of these columns gives the specific volume of the saturated liquid, v_f; the third column gives the specific volume of the saturated vapor v_g; and the second column gives the difference between the two, v_{fg}, as defined in Section

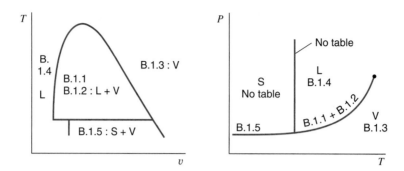

FIGURE 3.8 Listing of the steam tables.

3.2. Table B.1.2 lists the same information as Table B.1.1, but the data are listed according to pressure, as mentioned earlier.

As an example, let us calculate the specific volume of saturated steam at 200°C having a quality of 70%. Using Eq. 3.1 gives

$$v = 0.3(0.001\ 156) + 0.7(0.127\ 36)$$

$$= 0.0895\ \text{m}^3/\text{kg}.$$

Table B.1.3 gives the properties of superheated vapor. In the superheated region, pressure and temperature are independent properties; therefore, for each pressure a large number of temperatures are given, and for each temperature four thermodynamic properties are listed, the first one being specific volume. Thus, the specific volume of steam at a pressure of 0.5 MPa and 200°C is 0.4249 m³/kg.

Table B.1.4 gives the properties of the compressed liquid. To demonstrate the use of this table, consider a piston and a cylinder (as shown in Fig. 3.9) that contains 1 kg of saturated-liquid water at 100°C. Its properties are given in Table B.1.1, and we note that the pressure is 0.1013 MPa and the specific volume is 0.001 044 m³/kg. Suppose the pressure is increased to 10 MPa while the temperature is held constant at 100°C by the necessary transfer of heat, Q. Since water is slightly compressible, we would expect a slight decrease in specific volume during this process. Table B.1.4 gives this specific volume as 0.001 039 m³/kg. This is only a slight decrease, and only a small error would be made if one assumed that the volume of a compressed liquid is equal to the specific volume of the saturated liquid at the same temperature. In many situations this is the most convenient procedure, particularly when compressed-liquid data are not available. It is very important to note, however, that the specific volume of saturated liquid at the given pressure, 10

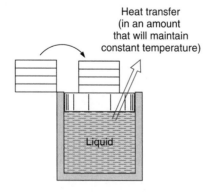

FIGURE 3.9
Illustration of compressed-liquid state.

MPa, does not give a good approximation. This value, from Table B.1.2, at a temperature of 311.1°C, is 0.001 452 m³/kg, which is in error by almost 40%.

Table B.1.5 of the steam tables gives the properties of saturated solid and saturated vapor that are in equilibrium. The first column gives the temperature, and the second column gives the corresponding saturation pressure. As would be expected, all these pressures are less than the triple-point pressure. The next two columns give the specific volume of the saturated solid and saturated vapor.

Appendix B also includes thermodynamic tables for several other substances; refrigerant fluids ammonia, R-12, R-22, and R-134a and the cryogenic fluids nitrogen and methane. In each case, only two tables are given—saturated liquid–vapor listed by temperature (equivalent to Table B.1.1 for water), and superheated vapor (equivalent to Table B.1.3).

Let us now consider a number of examples to illustrate the use of thermodynamic tables for water and also the other substances listed in Appendix B.

EXAMPLE 3.1 Determine the phase for each of the following water states using the Appendix B tables and indicate the relative position in the P–v, T–v, and P–T diagrams.

  a. 120°C, 500 kPa
  b. 120°C, 0.5 m³/kg

Solution

  a. Enter Table B.1.1 with 120°C. The saturation pressure is 198.5 kPa, so we have a compressed liquid, point a in Fig. 3.10. That is above the saturation line for 120°C. We could also have entered Table B.1.2 with 500 kPa and found the saturation temperature as 151.86°C, so we would say it is subcooled liquid. That is to the left of the saturation line for 500 kPa as seen in the P–T diagram.

  b. Enter Table B.1.1 with 120°C and notice

$$v_f = 0.00106 < v < v_g = 0.89186 \text{ m}^3/\text{kg}$$

so the state is a two phase mixture of liquid and vapor, point b in Fig. 3.10. The state is to the left of the saturated vapor state and to the right of the saturated liquid state both seen in the T–v diagram.

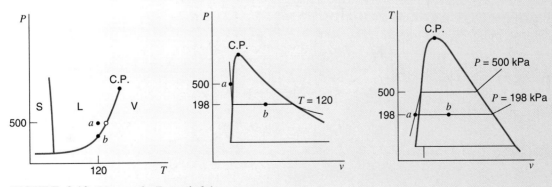

FIGURE 3.10 Diagram for Example 3.1.

EXAMPLE 3.2 Determine the phase for each of the following states using the Appendix B tables and indicate the relative position in the *P–v*, *T–v* and *P–T* diagrams, as in Figs. 3.11 and 3.12.

a. Ammonia 30°C, 1000 kPa
b. R-22 200 kPa, 0.15 m³/kg

Solution

a. Enter Table B.2.1 with 30°C. The saturation pressure is 1167 kPa. As we have a lower *P*, it is a superheated vapor state. We could also have entered with 1000 kPa and found a saturation temperature of slightly less than 25°C, so we have a state that is superheated about 5°C.
b. Enter Table B.4.1 with 200 kPa and notice

$$v > v_g \approx 0.1119 \text{ m}^3/\text{kg}$$

so from the *P–v* diagram the state is superheated vapor. We can find the state in Table B.4.2 between 40 and 50°C.

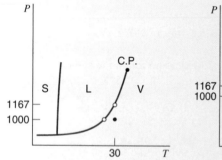
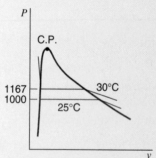
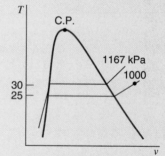

FIGURE 3.11 Diagram for Example 3.2*a*.

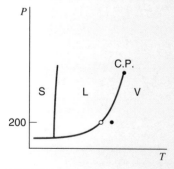
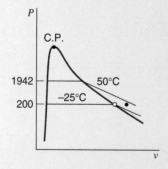
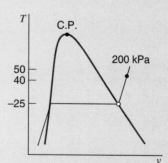

FIGURE 3.12 Diagram for Example 3.2*b*.

EXAMPLE 3.3 Determine the temperature and quality (if defined) for water at a pressure of 300 kPa and at each of these specific volumes:

a. 0.5 m³/kg
b. 1.0 m³/kg

Solution

For each state, it is necessary to determine what phase or phases are present, in order to know which table is the appropriate one to find the desired state information. That is, we must compare the given information with the appropriate phase boundary values. Consider a T–v diagram (or a P–v diagram) such as in Fig. 3.8. For the constant-pressure line of 300 kPa shown in Fig. 3.13, the values for v_f and v_g shown there are found from the saturation table, Table B.1.2.

a. By comparison with the values in Fig. 3.13, the state at which v is 0.5 m³/kg is seen to be in the liquid–vapor two-phase region, at which $T = 133.6°C$, and the quality x is found from Eq. 3.2 as

$$0.5 = 0.001\,073 + x\,0.604\,75, \qquad x = 0.825$$

Note that if we did not have Table B.1.2 (as would be the case with the other substances listed in Appendix B), we could have interpolated in Table B.1.1 between the 130°C and 135°C entries to get the v_f and v_g values for 300 kPa.

b. By comparison with the values in Fig. 3.13, the state at which v is 1.0 m³/kg is seen to be in the superheated vapor region, in which quality is undefined, and the temperature for which is found from Table B.1.3. In this case, T is found by linear interpolation between the 300 kPa specific-volume values at 300°C and 400°C, as shown in Fig. 3.14. This is an approximation for T, since the actual relation along the 300 kPa constant-pressure line is not exactly linear.

From the figure we have

$$\text{slope} = \frac{T - 300}{1.0 - 0.8753} = \frac{400 - 300}{1.0315 - 0.8753}$$

solving this gives $T = 379.8°C$.

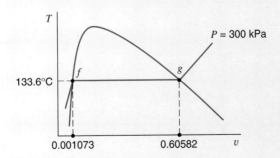

FIGURE 3.13 A T–v diagram for water at 300 kPa.

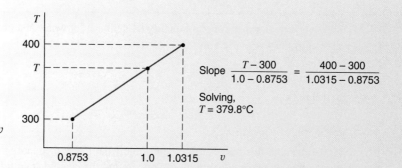

FIGURE 3.14 T and v values for superheated vapor water at 300 kPa.

EXAMPLE 3.4   A closed vessel contains 0.1 m³ of saturated liquid and 0.9 m³ of saturated vapor R-134a in equilibrium at 30°C. Determine the percent vapor on a mass basis.

Solution

Values of the saturation properties for R-134a are found from Table B.5.1. The mass–volume relations then give

$$V_{\text{liq}} = m_{\text{liq}} v_f, \qquad m_{\text{liq}} = \frac{0.1}{0.000\ 843} = 118.6 \text{ kg}$$

$$V_{\text{vap}} = m_{\text{vap}} v_g, \qquad m_{\text{vap}} = \frac{0.9}{0.026\ 71} = 33.7 \text{ kg}$$

$$m = 152.3 \text{ kg}$$

$$x = \frac{m_{\text{vap}}}{m} = \frac{33.7}{152.3} = 0.221$$

That is, the vessel contains 90% vapor by volume but only 22.1% vapor by mass.

EXAMPLE 3.4E   A closed vessel contains 0.1 ft³ of saturated liquid and 0.9 ft³ of saturated vapor R-134a in equilibrium at 90 F. Determine the percent vapor on a mass basis.

Solution

Values of the saturation properties for R-134a are found from Table F.10. The mass–volume relations then give

$$V_{\text{liq}} = m_{\text{liq}} v_f, \qquad m_{\text{liq}} = \frac{0.1}{0.0136} = 7.353 \text{ lbm}$$

$$V_{\text{vap}} = m_{\text{vap}} v_g, \qquad m_{\text{vap}} = \frac{0.9}{0.4009} = 2.245 \text{ lbm}$$

$$m = 9.598 \text{ lbm}$$

$$x = \frac{m_{\text{vap}}}{m} = \frac{2.245}{9.598} = 0.234$$

That is, the vessel contains 90% vapor by volume but only 23.4% vapor by mass.

EXAMPLE 3.5 A rigid vessel contains saturated ammonia vapor at 20°C. Heat is transferred to the system until the temperature reaches 40°C. What is the final pressure?

Solution

Since the volume does not change during this process, the specific volume also remains constant. From the ammonia tables, Table B.2.1, we have

$$v_1 = v_2 = 0.149\ 22\ \text{m}^3/\text{kg}$$

Since v_g at 40°C is less than 0.149 22 m³/kg, it is evident that in the final state the ammonia is superheated vapor. By interpolating between the 800- and 1000-kPa columns of Table B.2.2, we find that

$$P_2 = 945\ \text{kPa}$$

EXAMPLE 3.5E A rigid vessel contains saturated ammonia vapor at 70 F. Heat is transferred to the system until the temperature reaches 120 F. What is the final pressure?

Solution

Since the volume does not change during this process, the specific volume also remains constant. From the ammonia tables, Table F.8,

$$v_1 = v_2 = 2.311\ \text{ft}^3/\text{lbm}$$

Since v_g at 120 F is less than 2.311 ft³/lbm, it is evident that in the final state the ammonia is superheated vapor. By interpolating between the 125- and 150-lbf/in² columns of Table F.8, we find that

$$P_2 = 145\ \text{lbf/in}^2$$

EXAMPLE 3.6 Determine the missing property of P–v–T and x if applicable for the following states.

a. Nitrogen: $-53.2°C$, 600 kPa
b. Nitrogen: 100 K, 0.008 m³/kg

Solution

For nitrogen the properties are listed in Table B.6 with temperature in Kelvin.

a. Enter in Table B.6.1 with $T = 273.2 - 53.2 = 220$ K, which is higher than the critical T in the last entry. Then proceed to the superheated vapor tables. We would also have realized this by looking at the critical properties in Table A.2. From Table B.6.2 in the subsection for 600 kPa ($T_{\text{sat}} = 96.37$ K)

$$v = 0.10788\ \text{m}^3/\text{kg}$$

shown as point a in Fig. 3.15.

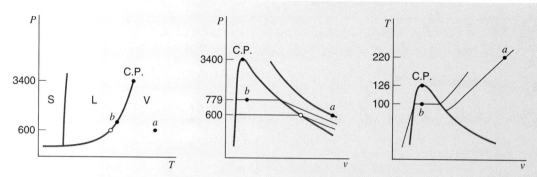

FIGURE 3.15 Diagram for Example 3.6.

b. Enter in Table B.6.1 with $T = 100$ K, and we see

$$v_f = 0.001\ 452 < v < v_g = 0.0312\ \text{m}^3/\text{kg}$$

so we have a two-phase state with a pressure as the saturation pressure, shown as b in Fig. 3.15

$$P_{\text{sat}} = 779.2\ \text{kPa}$$

and the quality from Eq. 3.2 becomes

$$x = (v - v_f)/v_{fg} = (0.008 - 0.001\ 452)/0.029\ 75 = 0.2201$$

EXAMPLE 3.7   Determine the pressure for water at 200°C with $v = 0.4\ \text{m}^3/\text{kg}$.

Solution

Start in Table B.1.1 with 200°C and note that $v > v_g = 0.127\ 36\ \text{m}^3/\text{kg}$ so we have superheated vapor. Proceed to Table B.1.3 at any subsection with 200°C; say we start at 200 kPa. There the $v = 1.080\ 34$, which is too large so the pressure must be higher. For

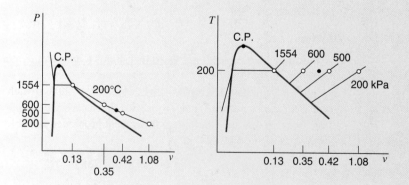

FIGURE 3.16
Diagram for Example 3.7.

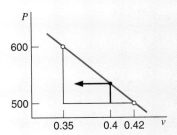

FIGURE 3.17 Linear interpolation for Example 3.7.

The real constant-T curve is slightly curved and not linear, but for manual interpolation we assume a linear variation.

500 kPa, $v = 0.424\ 92$, and for 600 kPa, $v = 0.352\ 02$, so it is bracketed. This is shown in Fig. 3.16.

A linear interpolation, Fig. 3.17, between the two pressures is done to get P at the desired v.

$$P = 500 + (600 - 500) \frac{0.4 - 0.424\ 92}{0.352\ 02 - 0.424\ 92} = 534.2 \text{ kPa}$$

3.5 THERMODYNAMIC SURFACES

The matter discussed to this point can be well summarized by a consideration of a pressure-specific volume–temperature surface. Two such surfaces are shown in Figs. 3.18 and 3.19. Figure 3.18 shows a substance such as water in which the specific volume increases during freezing. Figure 3.19 shows a substance in which the specific volume decreases during freezing.

In these diagrams the pressure, specific volume, and temperature are plotted on mutually perpendicular coordinates, and each possible equilibrium state is thus represented by a point on the surface. This follows directly from the fact that a pure substance has only two independent intensive properties. All points along a quasi-equilibrium process lie on the P–v–T surface, since such a process always passes through equilibrium states.

The regions of the surface that represent a single phase—the solid, liquid, and vapor phases—are indicated. These surfaces are curved. The two-phase regions—the solid–liquid, solid–vapor, and liquid–vapor regions—are ruled surfaces. By this we understand that they are made up of straight lines parallel to the specific-volume axis. This, of course, follows from the fact that in the two-phase region, lines of constant pressure are also lines of constant temperature, although the specific volume may change. The triple point actually appears as the triple line on the P–v–T surface, since the pressure and temperature of the triple point are fixed, but the specific volume may vary, depending on the proportion of each phase.

It is also of interest to note the pressure–temperature and pressure–volume projections of these surfaces. We have already considered the pressure–temperature diagram for a substance such as water. It is on this diagram that we observe the triple point. Various lines of constant temperature are shown on the pressure–volume diagram, and the

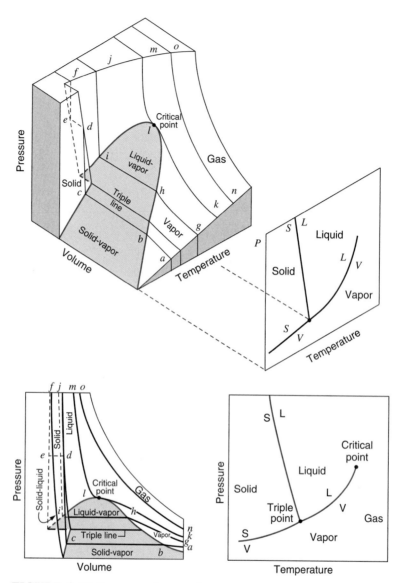

FIGURE 3.18 Pressure–volume–temperature surface for a substance that expands on freezing.

corresponding constant-temperature sections are lettered identically on the P–v–T surface. The critical isotherm has a point of inflection at the critical point.

One notices that for a substance such as water, which expands on freezing, the freezing temperature decreases with an increase in pressure. For a substance that contracts on freezing, the freezing temperature increases as the pressure increases. Thus, as the pressure of vapor is increased along the constant-temperature line *abcdef* in Fig. 3.18, a substance that expands on freezing first becomes solid and then liquid. For the substance that contracts on freezing, the corresponding constant-temperature line (Fig. 3.19) indicates that as the pressure on the vapor is increased, it first becomes liquid and then solid.

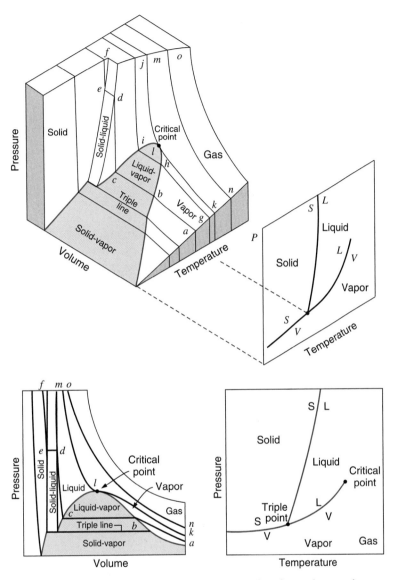

FIGURE 3.19 Pressure–volume–temperature surface for a substance that contracts on freezing.

3.6 THE *P–V–T* BEHAVIOR OF LOW- AND MODERATE-DENSITY GASES

One form of energy possession by a system discussed in Section 2.6 was intermolecular (IM) potential energy, that associated with the forces between molecules. It was stated there that at very low densities the average distances between molecules is so large that the IM potential energy may effectively be neglected. In such a case, the particles would be independent of one another, a situation referred to as an ideal gas. Under this approximation, it has been observed experimentally that, to a close degree, a very low density gas behaves according to the ideal gas equation of state

$$PV = n\overline{R}T, \qquad P\overline{v} = \overline{R}T \tag{3.3}$$

in which n is the number of kmol of gas, or

$$n = \frac{m}{M} = \frac{kg}{kg/kmol} \tag{3.4}$$

In Eq. 3.3, $\overline{R}$ is the universal gas constant, the value of which is, for any gas,

$$\overline{R} = 8.3145 \frac{kN\ m}{kmol\ K} = 8.3145 \frac{kJ}{kmol\ K}$$

and T is the absolute (ideal gas scale) temperature in kelvins (i.e., $T(K) = T(°C) + 273.15$). It is important to note that T must always be the absolute temperature whenever it is being used to multiply or divide in an equation. The ideal gas absolute temperature scale will be discussed in more detail in Chapter 7. In the English Engineering system,

$$\overline{R} = 1545 \frac{ft\ lbf}{lb\ mol\ R}$$

Substituting Eq. 3.4 into Eq. 3.3 and rearranging, we find that the ideal gas equation of state can be written conveniently in the form

$$PV = mRT, \qquad Pv = RT \tag{3.5}$$

where

$$R = \frac{\overline{R}}{M} \tag{3.6}$$

in which R is a different constant for each particular gas. The value of R for a number of substances is given in Table A.5 of Appendix A, and also in English units in Table F.4.

EXAMPLE 3.8 | What is the mass of air contained in a room 6 m × 10 m × 4 m if the pressure is 100 kPa and the temperature is 25°C?

Solution

Assume air to be an ideal gas. By using Eq. 3.5 and the value of R from Table A.5, we have

$$m = \frac{PV}{RT} = \frac{100\ kN/m^2 \times 240\ m^3}{0.287\ kN\ m/kg\ K \times 298.2\ K} = 280.5\ kg$$

EXAMPLE 3.9   A tank has a volume of 0.5 m³ and contains 10 kg of an ideal gas having a molecular weight of 24. The temperature is 25°C. What is the pressure?

Solution

The gas constant is determined first:

$$R = \frac{\overline{R}}{M} = \frac{8.3145 \text{ kN m/kmol K}}{24 \text{ kg/kmol}}$$

$$= 0.346\ 44 \text{ kN m/kg K}$$

We now solve for *P*:

$$P = \frac{mRT}{V} = \frac{10 \text{ kg} \times 0.346\ 44 \text{ kN m/kg K} \times 298.2 \text{ K}}{0.5 \text{ m}^3}$$

$$= 2066 \text{ kPa}$$

EXAMPLE 3.9E   A tank has a volume of 15 ft³ and contains 20 lbm of an ideal gas having a molecular weight of 24. The temperature is 80 F. What is the pressure?

Solution

The gas constant is determined first:

$$R = \frac{\overline{R}}{M} = \frac{1545 \text{ ft lbf/lb mol R}}{24 \text{ lbm/lb mol}} = 64.4 \text{ ft lbf/lbm R}$$

We now solve for *P*.

$$P = \frac{mRT}{V} = \frac{20 \text{ lbm} \times 64.4 \text{ ft lbf/lbm R} \times 540 \text{ R}}{144 \text{ in}^2/\text{ft}^2 \times 15 \text{ ft}^3} = 321 \text{ lbf/in}^2$$

EXAMPLE 3.10   A gas-bell is submerged in liquid water with its mass counterbalanced with rope and pulleys as shown in Fig. 3.20. The pressure inside is measured carefully to be 105 kPa, and the temperature is 21°C. A volume increase is measured to be 0.75 m³ over a period

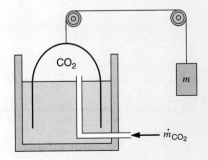

FIGURE 3.20 Sketch for Example 3.10.

of 185 s. What is the volume flow rate and the mass flow rate of the flow into the bell assuming it is carbon dioxide gas?

Solution

The volume flow rate is

$$\dot{V} = \frac{dV}{dt} = \frac{\Delta V}{\Delta t} = \frac{0.75}{185} = 0.040\ 54\ \text{m}^3/\text{s}$$

and the mass flow rate is $\dot{m} = \rho \dot{V} = \dot{V}/v$. At close to room conditions the carbon dioxide is an ideal gas, so $PV = mRT$ or $v = RT/P$, and from Table A.5 we have the ideal gas constant $R = 0.1889$ kJ/kg K. The mass flow rate becomes

$$\dot{m} = \frac{P\dot{V}}{RT} = \frac{105 \times 0.040\ 54}{0.1889\ (273.15 + 21)} \frac{\text{kPa m}^3/\text{s}}{\text{kJ/kg}} = 0.0766\ \text{kg/s}$$

Because of its simplicity, the ideal-gas equation of state is very convenient to use in thermodynamic calculations. However, two questions are now appropriate. The ideal-gas equation of state is a good approximation at low density. But what constitutes low density? Or, expressed in other words, over what range of density will the ideal-gas equation of state hold with accuracy? The second question is, how much does an actual gas at a given pressure and temperature deviate from ideal-gas behavior?

One specific example in response to these questions is shown in Fig. 3.21, a T–v diagram for water that indicates the error in assuming ideal gas for saturated vapor and for superheated vapor. As would be expected, at very low pressure or high temperature the error is small, but this becomes severe as the density increases. The same general trend would be the

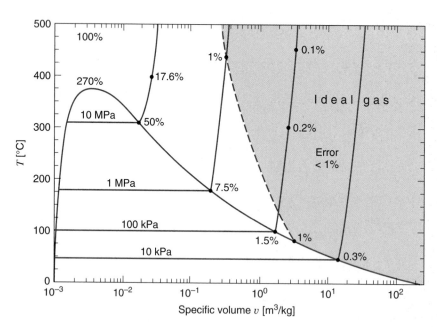

FIGURE 3.21
Temperature-specific
volume diagram for water.

case in referring to Fig. 3.18 or 3.19. As the state becomes further removed from the saturation region (i.e., high *T* or low *P*), the gas behavior becomes closer to the ideal-gas model.

A more quantitative study of the question of the ideal-gas approximation can be conducted by introducing the compressibility factor *Z*, defined as

$$Z = \frac{Pv}{RT}$$

or

$$Pv = ZRT \tag{3.7}$$

Note that for an ideal gas $Z = 1$, and the deviation of *Z* from unity is a measure of the deviation of the actual relation from the ideal-gas equation of state.

Figure 3.22 shows a skeleton compressibility chart for nitrogen. From this chart we make three observations. The first is that at all temperatures $Z \rightarrow 1$ as $P \rightarrow 0$. That is, as the pressure approaches zero, the *P–v–T* behavior closely approaches that predicted by the ideal-gas equation of state. Note also that at temperatures of 300 K and above (that is, room temperature and above) the compressibility factor is near unity up to pressure of about 10 MPa. This means that the ideal-gas equation of state can be used for nitrogen (and, as it happens, air) over this range with considerable accuracy.

We further note that at lower temperatures or at very high pressures, the compressibility factor deviates significantly from the ideal-gas value. Moderate-density forces of attraction tend to pull molecules together, resulting in a value of $Z < 1$, whereas very high density forces of repulsion tend to have the opposite effect.

If we examine compressibility diagrams for other pure substances, we find that the diagrams are all similar in the characteristics described above for nitrogen, at least in a qualitative sense. Quantitatively the diagrams are all different, since the critical temperatures and pressures of different substances vary over wide ranges, as evidenced from the values listed

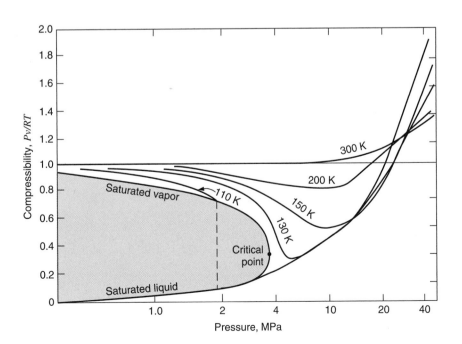

FIGURE 3.22
Compressibility of nitrogen.

in Table A.2. Is there a way in which we can put all of these substances on a common basis? To do so, we "reduce" the properties with respect to the values at the critical point. The reduced properties are defined as

$$\text{reduced pressure} = P_r = \frac{P}{P_c}, \qquad P_c = \text{critical pressure}$$

$$\text{reduced temperature} = T_r = \frac{T}{T_c}, \qquad T_c = \text{critical temperature} \qquad (3.8)$$

These equations state that the reduced property for a given state is the value of this property in this state divided by the value of this same property at the critical point.

If lines of constant T_r are plotted on a Z versus P_r diagram, a plot such as that in Fig. D.1 is obtained. The striking fact is that when such Z versus P_r diagrams are prepared for a number of different substances, all of them very nearly coincide, especially when the substances have simple, essentially spherical molecules. Correlations for substances with more complicated molecules are reasonably close, except near or at saturation or at high density. Thus, Fig. D.1 is actually a generalized diagram for simple molecules, which means that it represents the average behavior for a number of different simple substances. When such a diagram is used for a particular substance, the results will generally be somewhat in error. However, if P–v–T information is required for a substance in a region where no experimental measurements have been made, this generalized compressibility diagram will give reasonably accurate results. We need know only the critical pressure and critical temperature to use this basic generalized chart.

In our study of thermodynamics, we will use Fig. D.1 primarily to help us decide whether, in a given circumstance, it is reasonable to assume ideal-gas behavior as a model. For example, we note from the chart that if the pressure is very low (that is, $\ll P_c$), the ideal-gas model can be assumed with good accuracy, regardless of the temperature. Furthermore, at high temperatures (that is, greater than about twice T_c), the ideal-gas model can be assumed with good accuracy to pressures as high as four or five times P_c. When the temperature is less than about twice the critical temperature and the pressure is not extremely low, we are in a region, commonly termed superheated vapor, in which the deviation from ideal-gas behavior may be considerable. In this region it is preferable to use tables of thermodynamic properties or charts for a particular substance, as discussed in Section 3.4.

EXAMPLE 3.11   Is it reasonable to assume ideal-gas behavior at each of the given states?

a. Nitrogen at 20°C, 1.0 MPa

b. Carbon dioxide at 20°C, 1.0 MPa

c. Ammonia at 20°C, 1.0 MPa

Solution

In each case it is first necessary to check phase boundary and critical state data.

a. For nitrogen, the critical properties are, from Table A.2, 126.2 K, 3.39 MPa. Since the given temperature, 293.2 K is more than twice T_c and the reduced pressure is less than 0.3, ideal gas behavior is a very good assumption.

b. For carbon dioxide, the critical properties are 304.1 K, 7.38 MPa. Therefore, the reduced properties are 0.96 and 0.136. From Appendix Fig D.1, CO_2 is a gas (although $T < T_c$) with a Z of about 0.95, so the ideal-gas model is accurate to within about 5% in this case.

c. The ammonia tables, Appendix B.2, give the most accurate information. From Table B.2.1 at 20°C, P_g = 858 kPa. Since the given pressure of 1 MPa is greater than P_g, this state is a compressed liquid, and not a gas.

EXAMPLE 3.12 Determine the specific volume for R-134a at 100°C, 3.0 MPa, for the following models:

a. The R-134a tables, Table B.5
b. Ideal gas
c. The generalized chart, Fig. D.1

Solution

a. From Table B.5.2 at 100°C, 3 MPa,

$$v = 0.006\ 65 \text{ m}^3/\text{kg} \quad \text{(most accurate value)}$$

b. Assuming ideal gas, we have

$$R = \frac{\overline{R}}{M} = \frac{8.3145}{102.03} = 0.081\ 49\ \frac{\text{kJ}}{\text{kg K}}$$

$$v = \frac{RT}{P} = \frac{0.081\ 49 \times 373.2}{3000} = 0.010\ 14 \text{ m}^3/\text{kg}$$

which is more than 50% too large.

c. Using the generalized chart, Fig. D.1, we obtain

$$T_r = \frac{373.2}{374.2} = 1.0, \qquad P_r = \frac{3}{4.06} = 0.74, \qquad Z = 0.67$$

$$v = Z \times \frac{RT}{P} = 0.67 \times 0.010\ 14 = 0.006\ 79 \text{ m}^3/\text{kg}$$

which is only 2% too large.

EXAMPLE 3.13 Propane in a steel bottle of volume 0.1 m³ has a quality of 10% at a temperature of 15°C. Use the generalized compressibility chart to estimate the total propane mass and to find the pressure.

Solution

To use the generalized chart we need the reduced pressure and temperature. From Table A.2 for propane, P_c = 4250 kPa and T_c = 369.8 K. The reduced temperature is, from Eq. 3.8,

$$T_r = \frac{T}{T_c} = \frac{273.15 + 15}{369.8} = 0.779\ 2 = 0.78$$

From Figure D.1, shown in Fig. 3.23, we can read for the saturated states

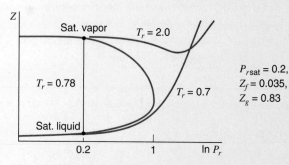

$P_{r\,\text{sat}} = 0.2,$
$Z_f = 0.035,$
$Z_g = 0.83$

FIGURE 3.23 Diagram for Example 3.13.

For the two-phase state the pressure is the saturated pressure

$$P = P_{r\,\text{sat}} \times P_c = 0.2 \times 4250 \text{ kPa} = 850 \text{ kPa}$$

The overall compressibility factor becomes, as Eq. 3.1 for v

$$Z = (1 - x)Z_f + xZ_g = 0.9 \times 0.035 + 0.1 \times 0.83 = 0.1145$$

The gas constant from Table A.5 is $R = 0.1886$ kJ/kg K, so the gas law is Eq. 3.7.

$$PV = mZRT$$

$$m = \frac{PV}{ZRT} = \frac{850 \times 0.1}{0.1145 \times 0.1886 \times 288.15} \frac{\text{kPa m}^3}{\text{kJ/kg}} = 13.66 \text{ kg}$$

Instead of the ideal-gas model to represent gas behavior, or even the generalized compressibility chart, which is approximate, it is desirable to have an equation of state that accurately represents the P–v–T behavior for a particular gas over the entire superheated vapor region. Such an equation is necessarily more complicated and consequently more difficult to use. Many such equations have been proposed and used to correlate the observed behavior of gases. To illustrate the nature and complexity of these equations, we present one of the best known, the Benedict–Webb–Rubin equation of state:

$$P = \frac{RT}{v} + \frac{RTB_0 - A_0 - C_0/T^2}{v^2} + \frac{RTb - a}{v^3} + \frac{a\alpha}{v^6} + \frac{c}{v^3 T^2}\left(1 + \frac{\gamma}{v^2}\right)e^{-\gamma/v^2} \quad (3.9)$$

This equation contains eight empirical constants and is accurate to densities of about twice the critical density. The empirical constants for the Benedict–Webb–Rubin equation for a number of substances are given in Appendix D.

An equation of state that accurately describes the relation among pressure, temperature, and specific volume is rather cumbersome and obtaining the solution requires considerable time. When we use a digital computer, it is often most convenient to determine the thermodynamic properties in a given state from such equations. However, in hand calculations, it is much more convenient to tabulate values of pressure, temperature, specific volume, and other thermodynamic properties for various substances. Such tables have been presented in Appendix B.

3.7 COMPUTERIZED TABLES

Most of the tables in the appendix are supplied in a computer program on the disk accompanying this book. The main program operates with a visual interface in the Windows environment on a PC-type computer and is generally self-explanatory.

The main program covers the full set of tables for water, refrigerants, and cryogenic fluids, as in Tables B.1 to B.7 including the compressed liquid region, which is only printed for water. For these substances a small graph with the $P-v$ diagram shows the region around the critical point down toward the triple line covering the compressed liquid, two-phase liquid–vapor, dense fluid, and superheated vapor regions. As a state is selected and the properties computed, a thin crosshair set of lines indicates the state in the diagram so this can be seen with a visual impression of the state's location.

Ideal gases are covered corresponding to the Tables A.7 for air and A.8 or A.9 for other ideal gases. You are able to select the substance and the units to work in for all the various table sections giving a wider choice than the printed tables. Metric units (SI) or standard English units for the properties can be used as well as a mass basis (kg or lbm) or a mole basis, satisfying the need for the most common applications.

The generalized chart, Fig. D.1, with the compressibility factor, is included to allow a more accurate value of Z to be obtained than can be read from the graph. This is particularly useful for the case of a two-phase mixture where the saturated liquid and saturated vapor values are needed. Besides the compressibility factor, this part of the program includes correction terms beyond ideal-gas approximations for changes in the other thermodynamic properties.

The only mixture application that is included with the program is moist air.

EXAMPLE 3.14   Find the states in Examples 3.1 and 3.2 with the computer-aided thermodynamics tables, CATT, and list the missing property of $P-v-T$ and x if applicable.

Solution

Water states from Example 3.1: Click Water, click Calculator and then select Case 1 (T, P). Input $(T, P) = (120, 0.5)$. The result is as shown in Fig. 3.24.

$\Rightarrow$ Compressed liquid $v = 0.0106$ m^3/kg (same as in Table B.1.4)

Click Calculator and then select Case 2 (T, v). Input $(T, v) = (120, 0.5)$

$\Rightarrow$ Two-phase $x = 0.5601, P = 198.5$ kPa

Ammonia state from Example 3.2: Click Cryogenics; check that it is ammonia. Otherwise select Ammonia, click Calculator, and then select Case 1 (T, P). Input $(T, P) = (30, 1)$

$\Rightarrow$ Superheated vapor $v = 0.1321$ m^3/kg (same as in Table B.2.2)

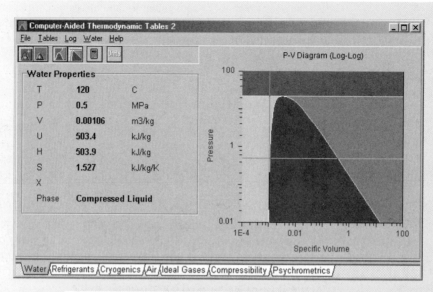

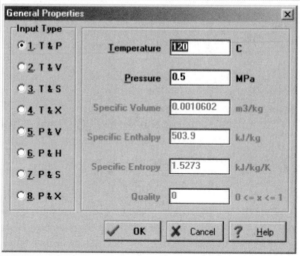

FIGURE 3.24 CATT
Result for Example 3.1.

R-22 state from Example 3.2: Click Refrigerants; check that it is R-22. Otherwise select
R-22 (Alt-R) click Calculator and then select Case 5 (P, v). Input $(P, v) = (0.2, 0.15)$

$$\Rightarrow \text{Superheated vapor} \qquad T = 46.26°C$$

SUMMARY   Thermodynamic properties of a pure substance and the phase boundaries for solid, liquid,
and vapor states are discussed. Phase equilibrium for vaporization (boiling liquid to
vapor), with the opposite direction being condensation (vapor to liquid); sublimation
(solid to vapor) or the opposite solidification (vapor to solid); and melting (solid to liquid)
or the opposite solidifying (liquid to solid) should be recognized. The three-dimensional
$P–v–T$ surface and the two-dimensional representations in the (P, T), (T, v) and (P, v) dia-

grams, and the vaporization, sublimation, and fusion lines are related to the printed tables in Appendix B. Properties from printed and computer tables covering a number of substances are introduced, including two-phase mixtures, for which we use the mass fraction of vapor (quality). The ideal-gas law approximates the limiting behavior for low density. An extension of the ideal-gas law is shown with the compressibility factor Z, and other more complicated equations of state are mentioned.

You should have learned a number of skills and acquired abilities from studying this chapter that will allow you to

- Know phases and the nomenclature used for states and interphases.
- Identify a phase given a state (T, P).
- Locate states relative to the critical point and know Tables A.2 (F.1) and 3.2.
- Recognize phase diagrams and interphase locations.
- Locate states in the Appendix B tables with any entry: (T, P), (T, v) or (P, v).
- Recognize how the tables show parts of the (T, P), (T, v) or (P, v) diagrams.
- Find properties in the two-phase regions; use quality x.
- Locate states using any combination of (T, P, v, x) including linear interpolation.
- Know when you have a liquid or solid and the properties in Tables A.3, A.4 (F.2, F.3).
- Know when a vapor is an ideal gas (or how to find out).
- Know the ideal-gas law and Table A.5 (F.4).
- Know the compressibility factor Z and the compressibility chart Fig. D.1.
- Know the existence of more general equations of state.
- Know how to get properties from the computer program.

KEY CONCEPTS AND FORMULAS

Phases	Solid, liquid, and vapor (gas)
Phase equilibrium	T_{sat}, P_{sat}, v_f, v_g, v_i
Multiphase boundaries	Vaporization, sublimation, and fusion lines:
	Figs. 3.5 (general), 3.6 (CO_2) and 3.7 (water)
	Critical point: Table 3.1, Table A.2 (F.1)
	Triple point: Table 3.2
Equilibrium state	Two independent properties (#1, #2)
Quality	$x = m_{vap}/m$ (vapor mass fraction)
	$1 - x = m_{liq}/m$ (liquid mass fraction)
Average specific volume	$v = (1 - x)v_f + xv_g$ (only two-phase mixture)
Equilibrium surface	$P–v–T$ Tables or equation of state
Ideal-gas law	$Pv = RT$ $PV = mRT = n\bar{R}T$
Universal gas constant	$\bar{R} = 8.3145$ kJ/kmol K
Gas constant	$R = \bar{R}/M$ kJ/kg K, Table A.5 or M from Table A.2
	ft lbf/lbm R, Table F.4 or M from Table F.1
Compressibility factor Z	$Pv = ZRT$ Chart for Z in Fig. D.1
Reduced properties	$P_r = \dfrac{P}{P_c}$ $T_r = \dfrac{T}{T_c}$ Entry to compressibility chart
Equations of state	Cubic, pressure explicit: Appendix D, Table D.1
	B–W–R: Eq. 3.7 and Table D.2 for various substances
	Lee Kesler: Appendix D, Table D.3, and Fig. D.1

CONCEPT-STUDY GUIDE PROBLEMS

3.1 What is the lowest temperature (approximately) at which water can be liquid?

3.2 What is the percent change in volume as liquid water freezes? Mention some effects in nature and for our households the volume change can have.

3.3 When you skate on ice a thin liquid film forms under the skate; how can that be?

3.4 An external water tap has the valve activated by a long spindle so that the closing mechanism is located well inside the wall. Why is that?

3.5 Some tools should be cleaned in water at least 150°C. How high a P is needed?

3.6 Are the pressures in the tables absolute or gauge pressures?

3.7 If I have 1 L of ammonia at room pressure and temperature (100 kPa, 20°C) how much mass is that?

3.8 How much is the change in liquid specific volume for water at 20°C as you move up from state i toward state j in Fig. 3.18, reaching 15 000 kPa?

3.9 For water at 100 kPa with a quality of 10% find the volume fraction of vapor.

3.10 Sketch two constant-pressure curves (500 kPa and 30 000 kPa) in a T–v diagram and indicate on the curves where in the water tables you see the properties.

3.11 Locate the state of ammonia at 200 kPa, −10°C. Indicate in both the P–v and the T–v diagrams the

location of the nearest states listed in the printed Table B.2.

3.12 Why are most of the compressed liquid or solid regions not included in the printed tables?

3.13 Water at 120°C with a quality of 25% has its temperature raised 20°C in a constant volume process. What is the new quality and pressure?

3.14 Water at 200 kPa with a quality of 25% has its temperature raised 20°C in a constant pressure process. What is the new quality and volume?

3.15 Why is it not typical to find tables for Ar, He, Ne or air like an Appendix B table?

3.16 What is the relative (%) change in P if we double the absolute temperature of an ideal gas keeping mass and volume constant? Repeat if we double V having m and T constant.

3.17 Calculate the ideal gas constant for argon and hydrogen based on Table A.2 and verify the value with Table A.5.

3.18 How close to ideal gas behavior (find Z) is ammonia at saturated vapor, 100 kPa? How about saturated vapor at 2000 kPa?

3.19 Find the volume of 2 kg of ethylene at 270 K, 2500 kPa using Z from Fig. D.1.

3.20 With $T_r = 0.85$ and a quality of 0.6 find the compressibility factor using Fig. D.1.

HOMEWORK PROBLEMS

Phase Diagrams; Triple and Critical Points

3.21 Modern extraction techniques can be based on dissolving material in supercritical fluids such as carbon dioxide. How high are the pressure and density of carbon dioxide when the pressure and temperature are around the critical point? Repeat for ethyl alcohol.

3.22 Find the lowest temperature at which it is possible to have water in the liquid phase. At what pressure must the liquid exist?

3.23 Water at 27°C can exist in different phases dependent on the pressure. Give the approximate pressure range in kPa for water being in each one of the three phases, vapor, liquid, or solid.

3.24 What is the lowest temperature in Kelvins for which you can see metal as a liquid if the metal is a. silver or b. copper?

3.25 If density of ice is 920 kg/m³, find the pressure at the bottom of a 1000-m-thick ice cap on the North Pole. What is the melting temperature at that pressure?

3.26 Dry ice is the name of solid carbon dioxide. How cold must it be at atmospheric (100 kPa) pressure? If it is heated at 100 kPa what eventually happens?

3.27 A substance is at 2 MPa and 17°C in a rigid tank. Using only the critical properties, can the phase of the mass be determined if the substance is nitrogen, water, or propane?

3.28 Give the phase for the following states:
  a. CO_2 at $T = 267°C$ and $P = 0.5$ MPa
  b. Air at $T = 20°C$ and $P = 200$ kPa
  c. NH_3 at $T = 170°C$ and $P = 600$ kPa

General Tables

3.29 Determine the phase of the substance at the given state using Appendix B tables.
  a. Water: 100°C, 500 kPa
  b. Ammonia: −10°C, 150 kPa
  c. R-12: 0°C, 350 kPa

3.30 Determine whether water at each of the following states is a compressed liquid, a superheated vapor, or a mixture of saturated liquid and vapor.
  a. 10 MPa, 0.003 m³/kg   c. 200°C, 0.1 m³/kg
  b. 1 MPa, 190°C   d. 10 kPa, 10°C

3.31 Give the phase for the following states:
  a. H_2O at $T = 275°C$ and $P = 5$ MPa
  b. H_2O at $T = -2°C$ and $P = 100$ kPa

3.32 Determine whether refrigerant R-22 in each of the following states is a compressed liquid, a superheated vapor, or a mixture of saturated liquid and vapor.
  a. 50°C, 0.05 m³/kg   c. 0.1 MPa, 0.1 m³/kg
  b. 1.0 MPa, 20°C   d. −20°C, 200 kPa

3.33 Fill out the following table for substance water:

	P[kPa]	T[°C]	v[m³/kg]	x
a.	500	20		
b.	500		0.20	
c.	1400	200		
d.		300		0.8

3.34 Place the four states a–d listed in Problem 3.33 as labeled dots in a sketch of the P–v and T–v diagrams.

3.35 Determine the phase and the specific volume for ammonia at these states using the Appendix B table.
  a. −10°C, 150 kPa
  b. 20°C, 100 kPa
  c. 60°C, quality 25%

3.36 Give the phase and the specific volume for the following:
  a. R-22 at $T = -25°C$ and $P = 100$ kPa
  b. R-22 at $T = -25°C$ and $P = 300$ kPa
  c. R-12 at $T = 5°C$ and $P = 200$ kPa

3.37 Fill out the following table for substance ammonia:

	P[kPa]	T[°C]	v[m³/kg]	x
a.		50	0.1185	
b.		50		0.5

3.38 Place the two states a–b listed in Problem 3.37 as labeled dots in a sketch of the P–v and T–v diagrams.

3.39 Calculate the following specific volumes:
  a. R-134a: 50°C, 80% quality
  b. Water: 4 MPa, 90% quality
  c. Nitrogen: 120 K, 60% quality

3.40 Give the phase and the missing property of P, T, v, and x.
  a. R-134a, $T = -20°C$, $P = 150$ kPa
  b. R-134a, $P = 300$ kPa, $v = 0.072$ m³/kg
  c. CH_4, $T = 155$ K, $v = 0.04$ m³/kg
  d. CH_4, $T = 350$ K, $v = 0.25$ m³/kg

3.41 A sealed rigid vessel has volume of 1 m³ and contains 2 kg of water at 100°C. The vessel is now heated. If a safety pressure valve is installed, at what pressure should the valve be set to have a maximum temperature of 200°C?

3.42 Saturated liquid water at 60°C is put under pressure to decrease the volume by 1% while keeping the temperature constant. To what pressure should it be compressed?

3.43 Saturated water vapor at 200 kPa is in a constant-pressure piston/cylinder assembly. At this state the piston is 0.1 m from the cylinder bottom. How much is this distance if the temperature is changed to
  a. 200°C
  b. 100°C.

3.44 You want a pot of water to boil at 105°C. How heavy a lid should you put on the 15-cm-diameter pot when $P_{atm} = 101$ kPa?

3.45 In your refrigerator the working substance evaporates from liquid to vapor at −20°C inside a pipe around the cold section. Outside (on the back or below) is a black grille, inside of which the working substance condenses form vapor to liquid at +40°C. For each location find the pressure and the change in specific volume (v) if

a. the substance is R-12

b. the substance is ammonia

3.46 Repeat the previous problem with the substances

a. R-134a

b. R-22

3.47 A water storage tank contains liquid and vapor in equilibrium at 110°C. The distance from the bottom of the tank to the liquid level is 8 m. What is the absolute pressure at the bottom of the tank?

3.48 Saturated water vapor at 200 kPa is in a constant-pressure piston/cylinder assembly. At this state the piston is 0.1 m from the cylinder bottom. How much is this distance and what is the temperature if the water is cooled to occupy half the original volume?

3.49 Two tanks are connected as shown in Fig. P3.49, both containing water. Tank A is at 200 kPa, $v = 0.5$ m³/kg, $V_A = 1$ m³, and tank B contains 3.5 kg at 0.5 MPa and 400°C. The valve is now opened and the two come to a uniform state. Find the final specific volume.

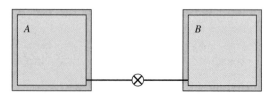

FIGURE P3.49

3.50 Determine the mass of methane gas stored in a 2 m³ tank at −30°C, 3 MPa. Estimate the percent error in the mass determination if the ideal gas model is used.

3.51 Saturated water vapor at 60°C has its pressure decreased to increase the volume by 10% while keeping the temperature constant. To what pressure should it be expanded?

3.52 Saturated water vapor at 200 kPa is in a constant-pressure piston/cylinder device. At this state the piston is 0.1 m from the cylinder bottom. How much is this distance and what is the temperature if the water is heated to occupy twice the original volume?

3.53 A boiler feed pump delivers 0.05 m³/s of water at 240°C, 20 MPa. What is the mass flow rate (kg/s)? What would be the percent error if the properties of saturated liquid at 240°C were used in the calculation? What if the properties of saturated liquid at 20 MPa were used?

3.54 Saturated vapor R-134a at 50°C changes volume at constant temperature. Find the new pressure, and quality if saturated, if the volume doubles. Repeat the question for the case where the volume is reduced to half the original volume.

3.55 A storage tank holds methane at 120 K, with a quality of 25%, and it warms up by 5°C per hour due to a failure in the refrigeration system. How much time will it take before the methane becomes single phase and what is the pressure then?

3.56 A glass jar is filled with saturated water at 500 kPa of quality 25%, and a tight lid is put on. Now it is cooled to −10°C. What is the mass fraction of solid at this temperature?

3.57 Saturated (liquid + vapor) ammonia at 60°C is contained in a rigid steel tank. It is used in an experiment, where it should pass through the critical point when the system is heated. What should the initial mass fraction of liquid be?

3.58 A steel tank contains 6 kg of propane (liquid + vapor) at 20°C with a volume of 0.015 m³. The tank is now slowly heated. Will the liquid level inside eventually rise to the top or drop to the bottom of the tank? What if the initial mass is 1 kg instead of 6 kg?

3.59 A 400-m³ storage tank is being constructed to hold LNG, liquified natural gas, which may be assumed to be essentially pure methane. If the tank is to contain 90% liquid and 10% vapor, by volume, at 100 kPa, what mass of LNG (kg) will the tank hold? What is the quality in the tank?

3.60 A sealed rigid vessel of 2 m³ contains a saturated mixture of liquid and vapor R-134a at 10°C. If it is heated to 50°C, the liquid phase disappears. Find the pressure at 50°C and the initial mass of the liquid.

3.61 A pressure cooker (closed tank) contains water at 100°C with the liquid volume being 1/10 of the vapor volume. It is heated until the pressure reaches 2.0 MPa. Find the final temperature. Has the final state more or less vapor than the initial state?

3.62 A pressure cooker has the lid screwed on tight. A small opening with $A = 5$ mm² is covered with a petcock that can be lifted to let steam escape. How much mass should the petcock have to allow boiling at 120°C with an outside atmosphere at 101.3 kPa?

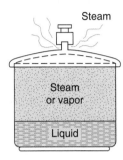

Steam

Steam or vapor

Liquid

FIGURE P3.62

3.63 Ammonia at 10°C with a mass of 10 kg is in a piston/cylinder assembly with an initial volume of 1 m³. The piston initially resting on the stops has a mass such that a pressure of 900 kPa will float it. Now the ammonia is slowly heated to 50°C. Find the final pressure and volume.

Ideal Gas

3.64 A cylinder fitted with a frictionless piston contains butane at 25°C, 500 kPa. Can the butane reasonably be assumed to behave as an ideal gas at this state?

3.65 A spherical helium balloon 10 m in diameter is at ambient T and P, 15°C and 100 kPa. How much helium does it contain? It can lift a total mass that equals the mass of displaced atmospheric air. How much mass of the balloon fabric and cage can then be lifted?

3.66 Is it reasonable to assume that at the given states the substance behaves as an ideal gas?
   a. Oxygen at 30°C, 3 MPa
   b. Methane at 30°C, 3 MPa
   c. Water at 30°C, 3 MPa
   d. R-134a at 30°C, 3 MPa
   e. R-134a at 30°C, 100 kPa

3.67 A 1-m³ tank is filled with a gas at room temperature (20°C) and pressure (100 kPa). How much mass is there if the gas is a. air, b. neon, or c. propane?

3.68 A rigid tank of 1 m³ contains nitrogen gas at 600 kPa, 400 K. By mistake someone lets 0.5 kg flow

out. If the final temperature is 375 K, what is then the final pressure?

3.69 A cylindrical gas tank 1 m long, with inside diameter of 20 cm, is evacuated and then filled with carbon dioxide gas at 25°C. To what pressure should it be charged if there should be 1.2 kg of carbon dioxide?

3.70 A glass is cleaned in 45°C hot water and placed on the table bottom up. The room air at 20°C that was trapped in the glass gets heated up to 40°C and some of it leaks out so that the net resulting pressure inside is 2 kPa above the ambient pressure of 101 kPa. Now the glass and the air inside cool down to room temperature. What is the pressure inside the glass?

3.71 A hollow metal sphere of 150 mm inside diameter is weighed on a precision beam balance when evacuated and again after being filled to 875 kPa with an unknown gas. The difference in mass is 0.0025 kg, and the temperature is 25°C. What is the gas, assuming it is a pure substance listed in Table A.5?

3.72 A vacuum pump is used to evacuate a chamber where some specimens are dried at 50°C. The pump rate of volume displacement is 0.5 m³/s with an inlet pressure of 0.1 kPa and temperature 50°C. How much water vapor has been removed over a 30-min period?

3.73 A 1-m³ rigid tank has propane at 100 kPa, 300 K and connected by a valve to another tank of 0.5 m³ with propane at 250 kPa, 400 K. The valve is opened and the two tanks come to a uniform state at 325 K. What is the final pressure?

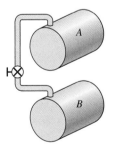

FIGURE P3.73

3.74 Verify the accuracy of the ideal-gas model when it is used to calculate specific volume for saturated water vapor as shown in Fig. 3.21. Do the calculation for 10 kPa and 1 MPa.

3.75 Assume we have three states of saturated vapor R-134a at +40°C, 0°C, and −40°C. Calculate the specific volume at the set of temperatures and corresponding saturated pressure assuming ideal-gas behavior. Find the percent relative error = $100(v - v_g)/v_g$ with v_g from the saturated R-134a table.

3.76 Do Problem 3.75, but for the substance R-12.

3.77 Do Problem 3.75, but for the substance ammonia.

3.78 Air in an automobile tire is initially at $-10°C$ and 190 kPa. After the automobile is driven awhile, the temperature gets up to $10°C$. Find the new pressure. You must make one assumption on your own.

→ Air

FIGURE P3.78

3.79 An initially deflated and flat balloon is connected by a valve to a 12-m^3 storage tank containing helium gas at 2 MPa and ambient temperature, $20°C$. The valve is opened and the balloon is inflated at constant pressure, $P_0 = 100$ kPa, equal to ambient pressure, until it becomes spherical at $D_1 = 1$ m. If the balloon is larger than this, the balloon material is stretched giving a pressure inside as

$$P = P_0 + C\left(1 - \frac{D_1}{D}\right)\frac{D_1}{D}$$

Compressibility Factor

3.80 Argon is kept in a rigid 5-m^3 tank at $-30°C$ and 3 MPa. Determine the mass using the compressibility factor. What is the error (%) if the ideal-gas model is used?

3.81 What is the percent error in specific volume if the ideal-gas model is used to represent the behavior of superheated ammonia at $40°C$ and 500 kPa? What if the generalized compressibility chart, Fig. D.1, is used instead?

3.82 A new refrigerant R-125 is stored as a liquid at $-20°C$ with a small amount of vapor. For a total of 1.5 kg R-125 find the pressure and the volume.

3.83 Many substances that normally do not mix well do so easily under supercritical pressures. A mass of

125 kg ethylene at 7.5 MPa and 296.5 K is stored for such a process. How much volume does it occupy?

3.84 Carbon dioxide at 330 K is pumped at a very high pressure, 10 MPa, into an oil well. As it penetrates the rock/oil, the oil viscosity is lowered so it flows out easily. For this process we need to know the density of the carbon dioxide being pumped.

3.85 To plan a commercial refrigeration system using R-123 we would like to know how much more volume saturated vapor R-123 occupies per kg at $-30°C$ compared to the saturated liquid state.

3.86 A bottle with a volume of 0.1 m^3 contains butane with a quality of 75% and a temperature of 300 K. Estimate the total butane mass in the bottle using the generalized compressibility chart.

3.87 Refrigerant R-32 is at $-10°C$ with a quality of 15%. Find the pressure and specific volume.

3.88 A mass of 2 kg of acetylene is in a 0.045-m^3 rigid container at a pressure of 4.3 MPa. Use the compressibility chart to estimate the temperature by trial and error.

3.89 A substance is at 2 MPa and $17°C$ in a 0.25 m^3 rigid tank. Estimate the mass using the compressibility factor if the substance is a. air, b. butane, or c. propane.

Review Problems

3.90 Determine the quality (if saturated) or temperature (if superheated) of the following substances at the given two states:
a. Water at
1: $120°C$, 1 m^3/kg; 2: 10 MPa, 0.01 m^3/kg
b. Nitrogen at
1: 1 MPa, 0.03 m^3/kg 2: 100 K, 0.03 m^3/kg

3.91 Fill out the following table for substance ammonia:

	P[kPa]	T[°C]	v[m^3/kg]	x
a.	400	-10		
b.			0.15	1.0

3.92 Find the phase, quality x if applicable, and the missing property P or T.
a. H_2O at $T = 120°C$ with $v = 0.5$ m^3/kg
b. H_2O at $P = 100$ kPa with $v = 1.8$ m^3/kg
c. H_2O at $T = 263$ K with $v = 200$ m^3/kg

3.93 Find the phase, quality x if applicable, and the missing property P or T.

  a. NH_3 at $P = 800$ kPa with $v = 0.2$ m³/kg

  b. NH_3 at $T = 20°C$ with $v = 0.1$ m³/kg

3.94 Give the phase and the missing properties of P, T, v, and x.

  a. R-22 at $T = 10°C$ with $v = 0.01$ m³/kg

  b. H_2O at $T = 350°C$ with $v = 0.2$ m³/kg

  c. R-12 at $T = -5°C$ and $P = 200$ kPa

  d. R-134a at 294 kPa and $v = 0.05$ m³/kg

3.95 Give the phase and the missing properties of P, T, v, and x. These may be a little more difficult if the appendix tables are used instead of the software.

  a. R-22, $T = 10°C$, $v = 0.036$ m³/kg

  b. H_2O, $v = 0.2$ m³/kg, $x = 0.5$

  c. H_2O, $T = 60°C$, $v = 0.001016$ m³/kg

  d. NH_3, $T = 30°C$, $P = 60$ kPa

  e. R-134a, $v = 0.005$ m³/kg, $x = 0.5$

3.96 A 5-m-long vertical tube of cross-sectional area 200 cm² is placed in a water fountain. It is filled with 15°C water; the bottom is closed and the top is open to the 100 kPa atmosphere.

  a. How much water is in the tube?

  b. What is the pressure at the bottom of the tube?

3.97 Consider two tanks, A and B, connected by a valve, as shown in Fig. P3.97. Each has a volume of 200 L, and tank A has R-12 at 25°C, 10% liquid and 90% vapor by volume, while tank B is evacuated. The valve is now opened and saturated vapor flows from A to B until the pressure in B has reached that in A, at which point the valve is closed. This process occurs slowly such that all temperatures stay at 25°C throughout the process. How much has the quality changed in tank A during the process?

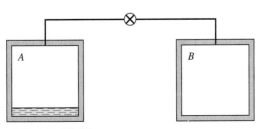

FIGURE P3.97

3.98 A spring-loaded piston/cylinder assembly contains water at 500°C and 3 MPa. The setup is such that pressure is proportional to volume, $P = CV$. It is now cooled until the water becomes saturated vapor. Sketch the P–v diagram and find the final pressure.

3.99 A 1-m³ rigid tank has air at 1500 kPa and ambient 300 K connected by a valve to a piston/cylinder, Fig. P3.99. The piston of area 0.1 m² requires 250 kPa below it to float. The valve is opened and the piston moves slowly 2 m up and the valve is closed. During the process air temperature remains at 300 K. What is the final pressure in the tank?

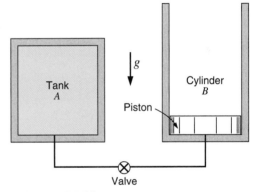

FIGURE P3.99

3.100 A tank contains 2 kg of nitrogen at 100 K with a quality of 50%. Through a volume flowmeter and valve, 0.5 kg is now removed while the temperature remains constant. Find the final state inside the tank and the volume of nitrogen removed if the valve/meter is located at

  a. The top of the tank

  b. The bottom of the tank

3.101 A piston/cylinder arrangement is loaded with a linear spring and the outside atmosphere. It contains water at 5 MPa, 400°C with the volume being 0.1 m³, as shown in Fig. P3.101. If the piston is at the

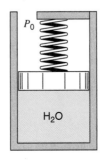

FIGURE P3.101

bottom, the spring exerts a force such that $P_{\text{lift}} = 200$ kPa. The system now cools until the pressure reaches 1200 kPa. Find the mass of water, the final state (T_2, v_2) and plot the P–v diagram for the process.

3.102 Water in a piston/cylinder is at 90°C, 100 kPa, and the piston loading is such that pressure is proportional to volume, $P = CV$. Heat is now added until the temperature reaches 200°C. Find the final pressure and also the quality if in the two-phase region.

3.103 A container with liquid nitrogen at 100 K has a cross-sectional area of 0.5 m² as shown in Fig. P3.103. Due to heat transfer, some of the liquid evaporates and in one hour the liquid level drops 30 mm. The vapor leaving the container passes through a valve and a heater and exits at 500 kPa, 260 K. Calculate the volume rate of flow of nitrogen gas exiting the heater.

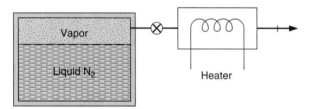

FIGURE P3.103

3.104 A cylinder containing ammonia is fitted with a piston restrained by an external force that is proportional to cylinder volume squared. Initial conditions are 10°C, 90% quality, and a volume of 5 L. A valve on the cylinder is opened and additional ammonia flows into the cylinder until the mass inside has doubled. If at this point the pressure is 1.2 MPa, what is the final temperature?

3.105 A cylinder/piston arrangement contains water at 105°C, 85% quality, with a volume of 1 L. The

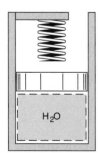

FIGURE P3.105

system is heated, causing the piston to rise and encounter a linear spring, as shown in Fig. P3.105. At this point the volume is 1.5 L, the piston diameter is 150 mm, and the spring constant is 100 N/mm. The heating continues, so the piston compresses the spring. What is the cylinder temperature when the pressure reaches 200 kPa?

3.106 Refrigerant-12 in a piston/cylinder arrangement is initially at 50°C with $x = 1$. It is then expanded in a process so that $P = Cv^{-1}$ to a pressure of 100 kPa. Find the final temperature and specific volume.

3.107 A 1-m³ rigid tank with air at 1 MPa and 400 K is connected to an air line as shown in Fig. P3.107. The valve is opened and air flows into the tank until the pressure reaches 5 MPa, at which point the valve is closed and the temperature inside is 450 K.

a. What is the mass of air in the tank before and after the process?

b. The tank eventually cools to room temperature, 300 K. What is the pressure inside the tank then?

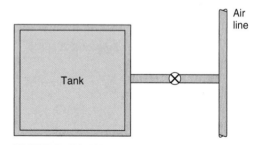

FIGURE P3.107

3.108 Ammonia in a piston/cylinder arrangement is at 700 kPa and 80°C. It is now cooled at constant pressure to saturated vapor (state 2) at which point the piston is locked with a pin. The cooling continues to −10°C (state 3). Show the processes 1 to 2 and 2 to 3 on both a P–v and a T–v diagram.

3.109 A cylinder has a thick piston initially held by a pin as shown in Fig. P.3.109. The cylinder contains carbon dioxide at 200 kPa and ambient temperature of 290 K. The metal piston has a density of 8000 kg/m³ and the atmospheric pressure is

101 kPa. The pin is now removed, allowing the piston to move and after a while the gas returns to ambient temperature. Is the piston against the stops?

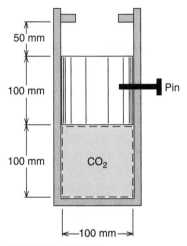

FIGURE P3.109

3.110 For a certain experiment, R-22 vapor is contained in a sealed glass tube at 20°C. It is desired to know the pressure at this condition, but there is no means of measuring it, since the tube is sealed. However, if the tube is cooled to −20°C small droplets of liquid are observed on the glass walls. What is the initial pressure?

3.111 A piston/cylinder arrangement, shown in Fig. P3.111, contains air at 250 kPa and 300°C. The 50-kg piston has a diameter of 0.1 m and initially pushes against the stops. The atmosphere is at 100 kPa and 20°C. The cylinder now cools as heat is transferred to the ambient surroundings.

a. At what temperature does the piston begin to move down?

b. How far has the piston dropped when the temperature reaches ambient?

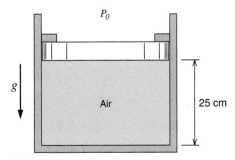

FIGURE P3.111

3.112 Air in a tank is at 1 MPa and room temperature of 20°C. It is used to fill an initially empty balloon to a pressure of 200 kPa, at which point the diameter is 2 m and the temperature is 20°C. Assume the pressure in the balloon is linearly proportional to its diameter and that the air in the tank also remains at 20°C throughout the process. Find the mass of air in the balloon and the minimum required volume of the tank.

3.113 A cylinder is fitted with a 10-cm-diameter piston that is restrained by a linear spring (force proportional to distance), as shown in Fig. P3.113. The spring force is 80 kN/m and the piston initially rests on the stops, with a cylinder volume of 1 L. The valve to the air line is opened and the piston begins to rise when the cylinder pressure is 150 kPa. When the valve is closed, the cylinder volume is 1.5 L and the temperature is 80°C. What mass of air is inside the cylinder?

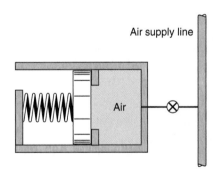

FIGURE P3.113

3.114 A 500-L tank stores 100 kg of nitrogen gas at 150 K. To design the tank the pressure must be estimated and three different methods are suggested. Which is the most accurate, and how different in percent are the other two?

a. Nitrogen tables, Table B.6

b. Ideal gas

c. Generalized compressibility chart, Fig. D.1

3.115 What is the percent error in pressure if the ideal-gas model is used to represent the behavior of superheated vapor R-22 at 50°C, 0.03082 m^3/kg? What if the generalized compressibility chart, Fig. D.1, is used instead? (Note that iterations are needed.)

Linear Interpolation

3.116 Find the pressure and temperature for saturated vapor R-12 with $v = 0.1$ m^3/kg

3.117 Use a linear interpolation to estimate properties of ammonia to fill out the table below.

	P [kPa]	T [°C]	v [m³/kg]	x
a.	550			0.75
b.	80	20		
c.		10	0.4	

3.118 Use a linear interpolation to estimate T_{sat} at 900 kPa for nitrogen. Sketch by hand the curve $P_{sat}(T)$ by using a few table entries around 900 kPa from Table B.6.1. Is your linear interpolation above or below the actual curve?

3.119 Use a double linear interpolation to find the pressure for superheated R-134a at 13°C with $v = 0.3$ m^3/kg.

3.120 Find the specific volume of ammonia at 140 kPa and 0°C.

3.121 Find the pressure of water at 200°C and specific volume of 1.5 m^3/kg.

Computer Tables

3.122 Use the computer software to find the properties for water at the four states in Problem 3.33.

3.123 Use the computer software to find the properties for ammonia at the four states listed in Problem 3.37.

3.124 Use the computer software to find the properties for ammonia at the three states listed in Problem 3.117.

3.125 Find the value of the saturated temperature for nitrogen by linear interpolation in Table B.6.1 for a pressure of 900 kPa. Compare this to the value given by the computer software.

3.126 Write a computer program that lists the states P, T, and v along the process curves in Problem 3.111.

3.127 Use the computer software to sketch the variation of pressure with temperature in Problem 3.41. Extend the curve a little into the single-phase region.

ENGLISH UNIT PROBLEMS

English Unit Concept Problems

3.128E Cabbage needs to be cooked (boiled) at 250 F. What pressure should the pressure cooker be set for?

3.129E If I have 1 ft^3 of ammonia at 15 psia, 60 F how much mass is that?

3.130E For water at 1 atm with a quality of 10% find the volume fraction of vapor.

3.131E Locate the state of R-134a at 30 psia, 20 F. Indicate in both the P–v and the T–v diagrams the location of the nearest states listed in the printed Table F.8.

3.132E Calculate the ideal gas constant for argon and hydrogen based on Table F.1 and verify the value with Table F.4.

English Unit Problems

3.133E Water at 80 F can exist in different phases dependent on the pressure. Give the approximate pressure range in lbf/in^2 for water being in each one of the three phases, vapor, liquid, or solid.

3.134E A substance is at 300 lbf/in^2, 65 F in a rigid tank. Using only the critical properties, can the phase of the mass be determined if the substance is nitrogen, water, or propane?

3.135E Determine whether water at each of the following states is a compressed liquid, a superheated vapor, or a mixture of saturated liquid and vapor.
   a. 1800 lbf/in^2, 0.03 ft^3/lbm
   b. 150 lbf/in^2, 320 F
   c. 380 F, 3 ft^3/lbm

3.136E Determine whether water at each of the following states is a compressed liquid, a superheated vapor, or a mixture of saturated liquid and vapor.
   a. 2 lbf/in^2, 50 F
   b. 270 F, 30 lbf/in^2
   c. 160 F, 10 ft^3/lbm

3.137E Give the phase and the missing property of P, T, v and x.
   a. R-134a, $T = -10$ F, $P = 18$ psia
   b. R-134a, $P = 50$ psia, $v = 1.3$ ft^3/lbm
   c. NH_3, $T = 120$ F, $v = 0.9$ ft^3/lbm
   d. NH_3, $T = 200$ F, $v = 11$ ft^3/lbm

3.138E Give the phase and the specific volume.
    a. R-22, $T = -10$ F, $P = 30$ lbf/in^2
    b. R-22, $T = -10$ F, $P = 40$ lbf/in^2
    c. H$_2$O, $T = 280$ F, $P = 35$ lbf/in^2
    d. NH$_3$, $T = 60$ F, $P = 15$ lbf/in^2

3.139E A water storage tank contains liquid and vapor in equilibrium at 220 F. The distance from the bottom of the tank to the liquid level is 25 ft. What is the absolute pressure at the bottom of the tank?

3.140E A sealed rigid vessel has volume of 35 ft^3 and contains 2 lbm of water at 200 F. The vessel is now heated. If a safety pressure valve is installed, at what pressure should the valve be set to have a maximum temperature of 400 F?

3.141E You want a pot of water to boil at 220 F. How heavy a lid should you put on the 6-in.-diameter pot when $P_{atm} = 14.7$ psia?

3.142E Saturated water vapor at 200 F has its pressure decreased to increase the volume by 10%, keeping the temperature constant. To what pressure should it be expanded?

3.143E A boiler feed pump delivers 100 ft^3/min of water at 400 F, 3000 lbf/in^2. What is the mass flowrate (lbm/s)? What would be the percent error if the properties of saturated liquid at 400 F were used in the calculation? What if the properties of saturated liquid at 3000 lbf/in^2 were used?

3.144E A pressure cooker has the lid screwed on tight. A small opening with $A = 0.0075$ in^2 is covered with a petcock that can be lifted to let steam escape. How much mass should the petcock have to allow boiling at 250 F with an outside atmosphere at 15 psia?

3.145E A steel tank contains 14 lbm of propane (liquid + vapor) at 70 F with a volume of 0.25 ft^3. The tank is now slowly heated. Will the liquid level inside eventually rise to the top or drop to the bottom of the tank? What if the initial mass is 2 lbm instead of 14 lbm?

3.146E A cylindrical gas tank 3 ft long, inside diameter of 8 in., is evacuated and then filled with carbon dioxide gas at 77 F. To what pressure should it be charged if there should be 2.6 lbm of carbon dioxide?

3.147E A spherical helium balloon of 30 ft in diameter is at ambient T and P, 60 F and 14.69 psia. How much helium does it contain? It can lift a total mass that equals the mass of displaced atmospheric air. How much mass of the balloon fabric and cage can then be lifted?

3.148E Give the phase and the specific volume.
    a. CO$_2$, $T = 510$ F, $P = 75$ lbf/in^2
    b. Air, $T = 68$ F, $P = 2$ atm
    c. Ar, $T = 300$ F, $P = 30$ lbf/in^2

3.149E What is the percent error in specific volume if the ideal-gas model is used to represent the behavior of superheated ammonia at 100 F, 80 lbf/in^2? What if the generalized compressibility chart, Fig. D.1, is used instead?

3.150E A cylinder is fitted with a 4-in.-diameter piston that is restrained by a linear spring (force proportional to distance) as shown in Fig. P3.113. The spring force constant is 400 lbf/in. and the piston initially rests on the stops, with a cylinder volume of 60 in^3. The valve to the air line is opened and the piston begins to rise when the cylinder pressure is 22 lbf/in^2. When the valve is closed, the cylinder volume is 90 in^3 and the temperature is 180 F. What mass of air is inside the cylinder?

3.151E A 35-ft^3 rigid tank has propane at 15 psia, 540 R and connected by a valve to another tank of 20 ft^3 with propane at 40 psia, 720 R. The valve is opened and the two tanks come to a uniform state at 600 R. What is the final pressure?

3.152E Two tanks are connected together as shown in Fig. P3.49, both containing water. Tank A is at 30 lbf/in^2, $v = 8$ ft^3/lbm, $V = 40$ ft^3, and tank B contains 8 lbm at 80 lbf/in^2, 750 F. The valve is now opened and the two come to a uniform state. Find the final specific volume.

3.153E A 35-ft^3 rigid tank has air at 225 psia and ambient 600 R connected by a valve to a piston/cylinder. The piston of area 1 ft^2 requires 40 psia below it to float (see Fig. P3.99). The valve is opened and the piston moves slowly 7 ft up and the valve is closed. During the process air temperature remains at 600 R. What is the final pressure in the tank?

3.154E Give the phase and the missing properties of P, T, v, and x. These may be a little more difficult if the appendix tables are used instead of the software.

    a. R-22, $T = 50$ F, $v = 0.6$ ft³/lbm
    b. H_2O, $v = 2$ ft³/lbm, $x = 0.5$
    c. H_2O, $T = 150$ F, $v = 0.01632$ ft³/lbm
    d. NH_3, $T = 80$ F, $P = 13$ lbf/in²
    e. R-134a, $v = 0.08$ ft³/lbm, $x = 0.5$

3.155E A pressure cooker (closed tank) contains water at 200 F with the liquid volume being 1/10 of the vapor volume. It is heated until the pressure reaches 300 lbf/in². Find the final temperature.

Has the final state more or less vapor than the initial state?

3.156E Refrigerant-22 in a piston/cylinder arrangement is initially at 120 F, $x = 1$. It is then expanded in a process so that $P = Cv^{-1}$ to a pressure of 30 lbf/in². Find the final temperature and specific volume.

3.157E A substance is at 70 F, 300 lbf/in² in a 10-ft³ tank. Estimate the mass from the compressibility chart if the substance is (a) air, (b) butane, or (c) propane.

3.158E Determine the mass of an ethane gas stored in a 25-ft³ tank at 250 F, 440 lbf/in² using the compressibility chart. Estimate the error (%) if the ideal gas model is used.

COMPUTER, DESIGN AND OPEN-ENDED PROBLEMS

3.159 Make a spreadsheet that will tabulate and plot the saturated pressure versus temperature for ammonia starting with $T = -40°C$ ending at the critical point in steps of 10°C.

3.160 Make a spreadsheet that will tabulate and plot values of P and T along a constant specific volume line for water. Starting state is 100 kPa, quality of 50% and the ending state is 800 kPa.

3.161 Write a computer program that lists the states P, T, and v along with the process curves in Problem 3.111.

3.162 Use the computer software to sketch the variation of pressure with temperature in Problem 3.55. Extend the curve a little into the single phase region.

3.163 By the use of the computer software find a few of the states between the beginning and end states and show the variation of pressure and temperature as a function of volume for Problem 3.102.

3.164 In Problem 3.106 we wish to follow the path of the process for the R-12 for any state between the initial and final states inside the cylinder.

3.165 For any specified substance in Tables B.1–B.7, fit a polynomial equation of degree n to tabular data for pressure as a function of density along any given isotherm in the superheated vapor region.

3.166 The refrigerant fluid in a household refrigerator changes phase from liquid to vapor at the low temperature in the refrigerator. It changes phase

from vapor to liquid at the higher temperature in the heat exchanger that gives the energy to the room air. Measure or otherwise estimate these temperatures. Based on these temperatures make a table with the refrigerant pressures for the refrigerants for which tables are available in Appendix B. Discuss the results and the requirements for a substance to be a potential refrigerant.

3.167 Repeat the previous problem for refrigerants that are listed in Table A.2 and use the compressibility chart Fig. D.1 to estimate the pressures.

3.168 The saturated pressure as a function of temperature follows the correlation developed by Wagner as

$$\ln P_r = [w_1\tau + w_2\tau^{1.5} + w_3\tau^3 + w_4\tau^6]/T_r$$

where the reduced pressure and temperature are $P_r = P/P_c$ and $T_r = T/T_c$. The temperature variable is $\tau = 1 - T_r$. The parameters are found for R-12 and R-134a as

	w_1	w_2	w_3	w_4
R-12	−6.91826	1.49560	−2.65015	−0.63170
R-134a	−7.59884	1.48886	−3.79873	1.81379

Compare these correlations to the tables in Appendix B.

3.169 Find the constants in the curve fit for the saturation pressure using Wagner's correlation as shown in the previous problem for water and methane. Find other correlations in the literature and compare them to the tables and give the maximum deviation.

3.170 The specific volume of saturated liquid can be approximated by the Rackett equation as

$$v_f = \frac{\overline{R}T_c}{MP_c} Z_c^n; \; n = 1 + (1 - T_r)^{2/7}$$

with the reduced temperature, $T_r = T/T_c$, and the compressibility factor, $Z_c = P_c v_c / RT_c$. Using values from Table A.2 with the critical constants, compare the formula to the tables for substances where the saturated specific volume is available.

4 WORK AND HEAT

In this chapter we consider work and heat. It is essential for the student of thermodynamics to understand clearly the definitions of both work and heat, because the correct analysis of many thermodynamic problems depends on distinguishing between them.

Work and heat are energy in transfer from one system to another and thus play a crucial role in most thermodynamic systems or devices. As we want to analyze such systems, we need to model the heat and work as functions of properties and parameters characteristic of the system or how it functions. An understanding of the physics involved allows us to construct a model for the heat and work and use the result in our analysis of the energy transfers and changes, which we will do with the first law of thermodynamics in Chapter 5.

To facilitate an understanding of the basic concepts we present a number of physical arrangements that will enable us to express the work done from changes in the system during a process. We will also examine work that is the result of a given process without going into details about how the process physically can be made to occur. This is done because such a description will be too complex and involve concepts that are not covered so far, but at least we can examine the result of the process.

A general description of heat transfer in different situations is a subject that usually is studied separately. However, a very simple introduction is beneficial so that the heat transfer does not become too abstract and we can relate it to the processes we examine. Heat transfer by conduction, convection (flow), and radiation is presented in terms of very simple models, emphasizing that it is driven by a temperature difference.

4.1 DEFINITION OF WORK

Work is usually defined as a force F acting through a displacement x, where the displacement is in the direction of the force. That is,

$$W = \int_1^2 F \, dx \tag{4.1}$$

This is a very useful relationship because it enables us to find the work required to raise a weight, to stretch a wire, or to move a charged particle through a magnetic field.

However, when treating thermodynamics from a macroscopic point of view, it is advantageous to tie in the definition of work with the concepts of systems, properties, and processes. We therefore define work as follows: Work is done by a system if the sole effect on the surroundings (everything external to the system) could be the raising of a weight. Notice that the raising of a weight is in effect a force acting through a distance. Notice also that our definition does not state that a weight was actually raised or that a

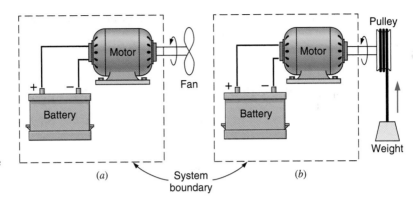

FIGURE 4.1 Example of work crossing the boundary of a system.

(a)   System boundary   (b)

force actually acted through a given distance, but that the sole effect external to the system could be the raising of a weight. Work done *by* a system is considered positive and work done *on* a system is considered negative. The symbol W designates the work done by a system.

In general, work is a form of energy in transit, that is, energy being transferred across a system boundary. The concept of energy and energy storage or possession has been discussed in some detail in Section 2.6. Work is the form of energy that fulfills the definition given in the preceding paragraph.

Let us illustrate this definition of work with a few examples. Consider as a system the battery and motor of Fig. 4.1a and let the motor drive a fan. Does work cross the boundary of the system? To answer this question using the definition of work given earlier, replace the fan with the pulley and weight arrangement shown in Fig. 4.1b. As the motor turns, the weight is raised, and the sole effect external to the system is the raising of a weight. Thus, for our original system of Fig. 4.1a, we conclude that work is crossing the boundary of the system, since the sole effect external to the system could be the raising of a weight.

Let the boundaries of the system be changed now to include only the battery shown in Fig. 4.2. Again we ask the question, does work cross the boundary of the system? To answer this question, we need to ask a more general question: Does the flow of electrical energy across the boundary of a system constitute work?

The only limiting factor in having the sole external effect be the raising of a weight is the inefficiency of the motor. However, as we design a more efficient motor, with lower bearing and electrical losses, we recognize that we can approach a certain limit that meets the requirement of having the only external effect be the raising of a weight. Therefore,

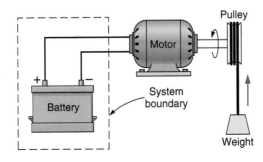

FIGURE 4.2 Example of work crossing the boundary of a system because of a flow of an electric current across the system boundary.

we can conclude that when there is a flow of electricity across the boundary of a system, as in Fig. 4.2, it is work.

4.2 UNITS FOR WORK

As already noted, work done *by* a system, such as that done by a gas expanding against a piston, is positive, and work done *on* a system, such as that done by a piston compressing a gas, is negative. Thus, positive work means that energy leaves the system, and negative work means that energy is added to the system.

Our definition of work involves raising of a weight, that is, the product of a unit force (one newton) acting through a unit distance (one meter). This unit for work in SI units is called the joule (J).

$$1 \, J = 1 \, N \, m$$

Power is the time rate of doing work and is designated by the symbol $\dot{W}$:

$$\dot{W} \equiv \frac{\delta W}{dt}$$

The unit for power is a rate of work of one joule per second, which is a watt (W):

$$1 \, W = 1 \, J/s$$

A familiar unit for power in English units is the horsepower (hp), where

$$1 \, hp = 550 \, ft \, lbf/s$$

Note that the work crossing the boundary of the system in Fig. 4.1 is that associated with a rotating shaft. To get the expression for power we use the differential work from Eq. 4.1 as

$$\delta W = F \, dx = Fr \, d\theta = T \, d\theta$$

that is, force acting through a distance dx or a torque ($T = Fr$) acting through an angle of rotation as shown in Fig. 4.3. Now the power becomes

$$\dot{W} = \frac{\delta W}{dt} = F \frac{dx}{dt} = F \mathbf{V} = Fr \frac{d\theta}{dt} = T\omega \tag{4.2}$$

that is, a force times rate of displacement (velocity) or a torque times angular velocity.

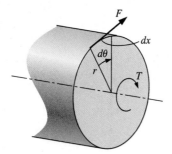

FIGURE 4.3 Force acting at radius r gives a torque $T = Fr$.

It is often convenient to speak of the work per unit mass of the system, often termed "specific work." This quantity is designated w and is defined

$$w \equiv \frac{W}{m}$$

4.3 WORK DONE AT THE MOVING BOUNDARY OF A SIMPLE COMPRESSIBLE SYSTEM

We have already noted that there are a variety of ways in which work can be done on or by a system. These include work done by a rotating shaft, electrical work, and the work done by the movement of the system boundary, such as the work done in moving the piston in a cylinder. In this section we will consider in some detail the work done at the moving boundary of a simple compressible system during a quasi-equilibrium process.

Consider as a system the gas contained in a cylinder and piston, as in Fig. 4.4. Let one of the small weights be removed from the piston, which will cause the piston to move upward a distance dL. We can consider this quasi-equilibrium process and calculate the amount of work W done by the system during this process. The total force on the piston is PA, where P is the pressure of the gas and A is the area of the piston. Therefore, the work δW is

$$\delta W = PA \, dL$$

But $A \, dL = dV$, the change in volume of the gas. Therefore,

$$\delta W = P \, dV \tag{4.3}$$

The work done at the moving boundary during a given quasi-equilibrium process can be found by integrating Eq. 4.3. However, this integration can be performed only if we know the relationship between P and V during this process. This relationship may be expressed in the form of an equation, or it may be shown in the form of a graph.

Let us consider a graphical solution first. We use as an example a compression process such as occurs during the compression of air in a cylinder, Fig. 4.5. At the beginning of the process the piston is at position 1, and the pressure is relatively low. This state is represented on a pressure–volume diagram (usually referred to as a P–V diagram). At the conclusion of the process the piston is in position 2, and the corresponding state of the gas is shown at point 2 on the P–V diagram. Let us assume that this compression was a quasi-equilibrium process and that during the process the system passed through the states shown by the line connecting states 1 and 2 on the P–V diagram. The assumption of a quasi-equilibrium process is essential here because each point on line 1–2 represents a definite state, and these states will correspond to the actual state of the system only if the deviation from equilibrium is infinitesimal. The work done on the air during this compression process can be found by integrating Eq. 4.3:

$$_1W_2 = \int_1^2 \delta W = \int_1^2 P \, dV \tag{4.4}$$

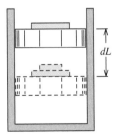

FIGURE 4.4 Example of work done at the moving boundary of a system in a quasi-equilibrium process.

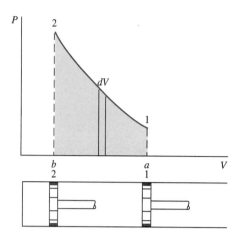

FIGURE 4.5 Use of pressure–volume diagram to show work done at the moving boundary of a system in a quasi-equilibrium process.

The symbol $_1W_2$ is to be interpreted as the work done during the process from state 1 to state 2. It is clear from examining the P–V diagram that the work done during this process,

$$\int_1^2 P\,dV$$

is represented by the area under the curve 1–2, area a–1–2–b–a. In this example the volume decreased, and the area a–1–2–b–a represents work done on the system. If the process had proceeded from state 2 to state 1 along the same path, the same area would represent work done by the system.

Further consideration of a P–V diagram, such as Fig. 4.6, leads to another important conclusion. It is possible to go from state 1 to state 2 along many different quasi-equilibrium paths, such as A, B, or C. Since the area underneath each curve represents the work for each process, the amount of work done during each process not only is a function of the end states of the process but depends on the path that is followed in going from one state to another. For this reason work is called a path function or, in mathematical parlance, δW is an inexact differential.

This concept leads to a brief consideration of point and path functions or, to use another term, exact and inexact differentials. Thermodynamic properties are point functions, a name that comes from the fact that for a given point on a diagram (such as Fig. 4.6) or surface (such as Fig. 3.18), the state is fixed, and thus there is a definite value of each

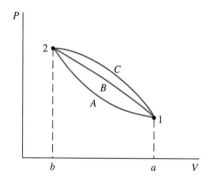

FIGURE 4.6 Various quasi-equilibrium processes between two given states, indicating that work is a path function.

property corresponding to this point. The differentials of point functions are exact differentials, and the integration is simply

$$\int_1^2 dV = V_2 - V_1$$

Thus, we can speak of the volume in state 2 and the volume in state 1, and the change in volume depends only on the initial and final states.

Work, however, is a path function, for, as has been indicated, the work done in a quasi-equilibrium process between two given states depends on the path followed. The differentials of path functions are inexact differentials, and the symbol δ will be used in this text to designate inexact differentials (in contrast to d for exact differentials). Thus, for work, we write

$$\int_1^2 \delta W = {}_1W_2$$

It would be more precise to use the notation ${}_1W_{2A}$, which would indicate the work done during the change from state 1 to state 2 along path A. However, it is implied in the notation ${}_1W_2$ that the process between states 1 and 2 has been specified. It should be noted that we never speak about the work in the system in state 1 or state 2, and thus we would never write $W_2 - W_1$.

In evaluating the integral of Eq. 4.4, we should always keep in mind that we wish to determine the area under the curve in Fig. 4.6. In connection with this point, we identify the following two classes of problems.

1. The relationship between P and V is given in terms of experimental data or in graphical form (as, for example, the trace on an oscilloscope). Therefore, we may evaluate the integral, Eq. 4.4, by graphical or numerical integration.
2. The relationship between P and V makes it possible to fit an analytical relationship between them. We may then integrate directly.

One common example of this second type of functional relationship is a process called a polytropic process, one in which

$$PV^n = \text{constant}$$

throughout the process. The exponent n may possibly be any value from $-\infty$ to $+\infty$, depending on the particular process. For this type of process, we can integrate Eq. 4.4 as follows:

$$PV^n = \text{constant} = P_1 V_1^n = P_2 V_2^n$$

$$P = \frac{\text{constant}}{V^n} = \frac{P_1 V_1^n}{V^n} = \frac{P_2 V_2^n}{V^n}$$

$$\int_1^2 P \, dV = \text{constant} \int_1^2 \frac{dV}{V^n} = \text{constant} \left(\frac{V^{-n+1}}{-n+1} \right) \Bigg|_1^2$$

$$\int_1^2 P \, dV = \frac{\text{constant}}{1-n} (V_2^{1-n} - V_1^{1-n}) = \frac{P_2 V_2^n V_2^{1-n} - P_1 V_1^n V_1^{1-n}}{1-n}$$

$$= \frac{P_2 V_2 - P_1 V_1}{1-n} \tag{4.5}$$

Note that the resulting Eq. 4.5 is valid for any exponent n, except $n = 1$. Where $n = 1$,

$$PV = \text{constant} = P_1 V_1 = P_2 V_2$$

and

$$\int_1^2 P \, dV = P_1 V_1 \int_1^2 \frac{dV}{V} = P_1 V_1 \ln \frac{V_2}{V_1} \qquad (4.6)$$

Note that in Eqs. 4.5 and 4.6 we did not say that the work is equal to the expressions given in these equations. These expressions give us the value of a certain integral, that is, a mathematical result. Whether or not that integral equals the work in a particular process depends on the result of a thermodynamic analysis of that process. It is important to keep the mathematical result separate from the thermodynamic analysis, for there are many situations in which work is not given by Eq. 4.4.

The polytropic process as described demonstrates one special functional relationship between P and V during a process. There are many other possible relations, some of which will be examined in the problems at the end of this chapter.

EXAMPLE 4.1 Consider as a system the gas in the cylinder shown in Fig. 4.7; the cylinder is fitted with a piston on which a number of small weights are placed. The initial pressure is 200 kPa, and the initial volume of the gas is 0.04 m³.

a. Let a Bunsen burner be placed under the cylinder, and let the volume of the gas increase to 0.1 m³ while the pressure remains constant. Calculate the work done by the system during this process.

$$_1W_2 = \int_1^2 P \, dV$$

Since the pressure is constant, we conclude from Eq. 4.4 that

$$_1W_2 = P \int_1^2 dV = P(V_2 - V_1)$$

$$_1W_2 = 200 \text{ kPa} \times (0.1 - 0.04)\text{m}^3 = 12.0 \text{ kJ}$$

b. Consider the same system and initial conditions, but at the same time as the Bunsen burner is under the cylinder and the piston is rising, let weights be removed from the piston at such a rate that, during the process, the temperature of the gas remains constant.

If we assume that the ideal-gas model is valid, then, from Eq. 3.5,

$$PV = mRT$$

We note that this is a polytropic process with exponent $n = 1$. From our analysis, we conclude that the work is given by Eq. 4.4 and that the integral in this equation is given by Eq. 4.6. Therefore,

$$_1W_2 = \int_1^2 P \, dV = P_1 V_1 \ln \frac{V_2}{V_1}$$

$$= 200 \text{ kPa} \times 0.04 \text{ m}^3 \times \ln \frac{0.10}{0.04} = 7.33 \text{ kJ}$$

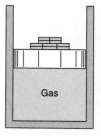

FIGURE 4.7 Sketch for Example 4.1.

Gas

c. Consider the same system, but during the heat transfer let the weights be removed at such a rate that the expression $PV^{1.3}$ = constant describes the relation between pressure and volume during the process. Again the final volume is 0.1 m³. Calculate the work.

This is a polytropic process in which $n = 1.3$. Analyzing the process, we conclude again that the work is given by Eq. 4.4 and that the integral is given by Eq. 4.5. Therefore,

$$P_2 = 200 \left(\frac{0.04}{0.10}\right)^{1.3} = 60.77 \text{ kPa}$$

$$_1W_2 = \int_1^2 P \, dV = \frac{P_2V_2 - P_1V_1}{1 - 1.3} = \frac{60.77 \times 0.1 - 200 \times 0.04}{1 - 1.3} \text{ kPa m}^3$$

$$= 6.41 \text{ kJ}$$

d. Consider the system and initial state given in the first three examples, but let the piston be held by a pin so that the volume remains constant. In addition, let heat be transferred from the system until the pressure drops to 100 kPa. Calculate the work.

Since $\delta W = P \, dV$ for a quasi-equilibrium process, the work is zero, because there is no change in volume.

The process for each of the four examples is shown on the P–V diagram of Fig. 4.8. Process 1–2a is a constant-pressure process, and area 1–2a–f–e–1 represents the work. Similarly, line 1–2b represents the process in which PV = constant, line 1–2c the process in which $PV^{1.3}$ = constant, and line 1–2d the constant-volume process. The student should compare the relative areas under each curve with the numerical results obtained for the amounts of work done.

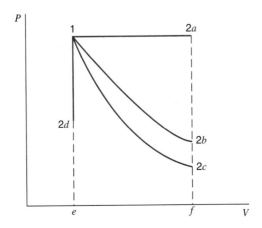

FIGURE 4.8
Pressure–volume diagram showing work done in the various processes of Example 4.1.

EXAMPLE 4.2

Consider a slightly different piston/cylinder arrangement as shown in Fig. 4.9. In this example the piston is loaded with a mass, m_p, the outside atmosphere P_0, a linear spring, and a single point force F_1. The piston traps the gas inside with a pressure P. A force balance on the piston in the direction of motion yields

$$m_p a \cong 0 = \sum F_\uparrow - \sum F_\downarrow$$

with a zero acceleration in a quasi-equilibrium process. The forces, when the spring is in contact with the piston, are

$$\sum F_\uparrow = PA, \qquad \sum F_\downarrow = m_p g + P_0 A + k_s(x - x_0) + F_1$$

with the linear spring constant, k_s. The piston position for a relaxed spring is x_0, which depends on how the spring is installed. The force balance then gives the gas pressure by division with the area A as

$$P = P_0 + [m_p g + F_1 + k_s(x - x_0)]/A$$

To illustrate the process in a P–V diagram, the distance x is converted to volume by division and multiplication with A:

$$P = P_0 + \frac{m_p g}{A} + \frac{F_1}{A} + \frac{k_s}{A^2}(V - V_0) = C_1 + C_2 V$$

This relation gives the pressure as a linear function of the volume, with the line having a slope of $C_2 = k_s/A^2$. Possible values of P and V are as shown in Fig. 4.10 for an expansion. Regardless of what substance is inside, any process must proceed along the line in the P–V diagram. The work term in a quasi-equilibrium process then follows as

$$_1W_2 = \int_1^2 P\, dV = \text{area under the process curve}$$

$$_1W_2 = \frac{1}{2}(P_1 + P_2)(V_2 - V_1)$$

For a contraction instead of expansion, the process would proceed in the opposite direction from the initial point 1 along a line of the same slope shown in Fig. 4.10.

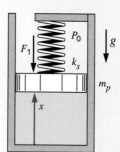

FIGURE 4.9 Sketch of physical system for Example 4.2.

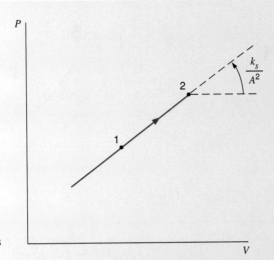

FIGURE 4.10 The process curve showing possible P–V combinations for Example 4.2.

EXAMPLE 4.3   The cylinder/piston setup of Example 4.2 contains 0.5 kg ammonia at $-20°C$ with a quality of 25%. The ammonia is now heated to $+20°C$, at which state the volume is observed to be 1.41 times larger. Find the final pressure and the work the ammonia produced.

Solution

The forces acting on the piston, gravitation constant, external atmosphere at constant pressure and the linear spring give a linear relation between P and $v(V)$.

State 1: (T_1, x_1) from Table B.2.1

$P_1 = P_{sat} = 190.2$ kPa

$v_1 = v_f + x_1 v_{fg} = 0.001\ 504 + 0.25 \times 0.621\ 84 = 0.156\ 96$ m³/kg

State 2: $(T_2, v_2 = 1.41\ v_1 = 1.41 \times 0.156\ 96 = 0.2213$ m³/kg)
Table B.2.2 state very close to $P_2 = 600$ kPa.

Process: $P = C_1 + C_2 v$

The work term can now be integrated knowing P versus v and can be seen as the area in the P–v diagram, shown in Fig. 4.11.

$$_1W_2 = \int_1^2 P\, DV = \int_1^2 Pm\, dv = \text{area} = m\frac{1}{2}(P_1 + P_2)(v_2 - v_1)$$

$$= 0.5 \text{ kg} \frac{1}{2}(190.2 + 600) \text{ kPa } (0.2213 - 0.156\ 96) \text{ m³/kg}$$

$$= 12.71 \text{ kJ}$$

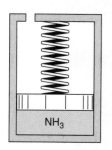
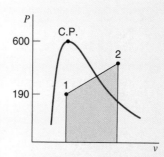
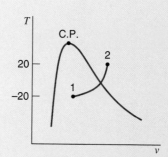

FIGURE 4.11 Diagrams for Example 4.3.

EXAMPLE 4.4 The piston/cylinder setup shown in Figure 4.12 contains 0.1 kg of water at 1000 kPa, 500°C. The water is now cooled with a constant force on the piston until it reaches half the initial volume. After this it cools to 25°C while the piston is against the stops. Find the final water pressure and the work in the overall process, and show the process in a P–v diagram.

Solution

We recognize this is a two-step process, one of constant P and one of constant V. This behavior is dictated by the construction of the device.

$$\begin{aligned}
\textit{State 1:} &\quad (P, T) \quad \text{From Table B.1.3; } v_1 = 0.354\,11 \text{ m}^3/\text{kg.}\\
\textit{Process 1-1a:} &\quad P = \text{constant} = F/A\\
1a\text{-}2: &\quad v = \text{constant} = v_{1a} = v_2 = v_1/2\\
\textit{State 2:} &\quad (T, v_2 = v_1/2 = 0.177\,06 \text{ m}^3/\text{kg})
\end{aligned}$$

From Table B.1.1, $v_2 < v_g$, so the process is two phase and $P_2 = P_{\text{sat}} = 3.169$ kPa.

$$_1W_2 = \int_1^2 P\, dV = m\int_1^2 P\, dv = mP_1(v_{1a} - v_1) + 0$$

$$= 0.1 \text{ kg} \times 1000 \text{ kPa } (0.177\,06 - 0.354\,11) \text{ m}^3/\text{kg} = -17.7 \text{ kJ}$$

Note that the work done from $1a$ to 2 is zero (no change in volume) as shown in Fig. 4.13.

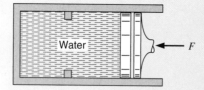

FIGURE 4.12 Sketch for Example 4.4.

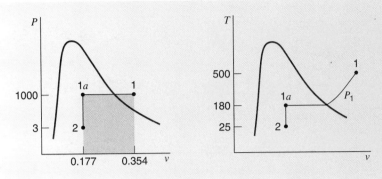

FIGURE 4.13
Diagrams for Example 4.4.

In this section we have discussed boundary movement work in a quasi-equilibrium process. We should also realize that there may very well be boundary movement work in a nonequilibrium process. Then the total force exerted on the piston by the gas inside the cylinder, PA, does not equal the external force, F_{ext}, and the work is not given by Eq. 4.3. The work can, however, be evaluated in terms of F_{ext} or, dividing by area, an equivalent external pressure, P_{ext}. The work done at the moving boundary in this case is

$$\delta W = F_{ext}\, dL = P_{ext}\, dV \tag{4.7}$$

Evaluation of Eq. 4.7 in any particular instance requires a knowledge of how the external force or pressure changes during the process.

EXAMPLE 4.5   Consider the system shown in Fig. 4.14 in which the piston of mass m_p is initially held in place by a pin. The gas inside the cylinder is initially at pressure P_1 and volume V_1. When the pin is now released, the external force per unit area acting on the system (gas) boundary is comprised of two parts:

$$P_{ext} = F_{ext}/A = P_0 + m_p g/A$$

Calculate the work done by the system when the piston has come to rest.

After the piston is released, the system is exposed to the boundary pressure equal to P_{ext}, which dictates the pressure inside the system, as discussed in Section 2.8 in connection with Fig. 2.9. We further note that neither of the two components of this external force will change with a boundary movement, since the cylinder is vertical (gravitational force) and the top is open to the ambient surroundings (movement upward merely pushes the air out of the way). If the initial pressure P_1 is greater than that resisting the boundary, the piston will move upward at a finite rate, that is, in a nonequilibrium process, with the cylinder pressure eventually coming to equilibrium at the value P_{ext}. If we were able to trace the average cylinder pressure as a function of time, it would typically behave as shown in Fig. 4.15. However, the work done by the system during this process is done against the force resisting the boundary movement and is therefore given by Eq. 4.7. Also, since the external force is constant during this process, the result is

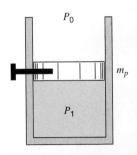

FIGURE 4.14
Example of a nonequilibrium process.

$$_1W_2 = \int_1^2 P_{ext}\, dV = P_{ext}(V_2 - V_1)$$

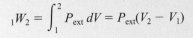

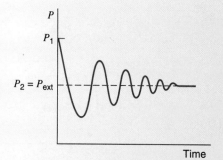

FIGURE 4.15
Cylinder pressure as a
function of time.

where V_2 is greater than V_1, and the work by the system is positive. If the initial pressure had been less than the boundary pressure, the piston would have moved downward, compressing the gas, with the system eventually coming to equilibrium at P_{ext}, at a volume less than the initial volume, and the work would be negative, that is, done on the system by its surroundings.

4.4 OTHER SYSTEMS THAT INVOLVE WORK

In the preceding section we considered the work done at the moving boundary of a simple compressible system during a quasi-equilibrium process and also during a nonequilibrium process. There are other types of systems in which work is done at a moving boundary. In this section we briefly consider three such systems, a stretched wire, a surface film, and electrical work.

Consider as a system a stretched wire that is under a given tension $\mathcal{T}$. When the length of the wire changes by the amount dL, the work done by the system is

$$\delta W = -\mathcal{T}\, dL \tag{4.8}$$

The minus sign is necessary because work is done by the system when dL is negative. This equation can be integrated to have

$$_1W_2 = -\int_1^2 \mathcal{T}\, dL \tag{4.9}$$

The integration can be performed either graphically or analytically if the relation between $\mathcal{T}$ and L is known. The stretched wire is a simple example of the type of problem in solid-body mechanics that involves the calculation of work.

EXAMPLE 4.6   A metallic wire of initial length L_0 is stretched. Assuming elastic behavior, determine the work done in terms of the modulus of elasticity and the strain.

Let σ = stress, e = strain, and E = the modulus of elasticity.

$$\sigma = \frac{\mathcal{T}}{A} = Ee$$

Therefore,

$$\mathcal{T} = AEe$$

From the definition of strain,

$$de = \frac{dL}{L_0}$$

Therefore,

$$\delta W = -\mathcal{T}\, dL = -AEeL_0\, de$$

$$W = -AEL_0 \int_{e=0}^{e} e\, de = -\frac{A\, EL_0}{2}\, (e)^2$$

Now consider a system that consists of a liquid film having a surface tension $\mathcal{S}$. A schematic arrangement of such a film, maintained on a wire frame, one side of which can be moved, is shown in Fig. 4.16. When the area of the film is changed, for example, by sliding the movable wire along the frame, work is done on or by the film. When the area changes by an amount dA, the work done by the system is

$$\delta W = -\mathcal{S}\, dA \qquad (4.10)$$

For finite changes,

$$_1W_2 = -\int_1^2 \mathcal{S}\, dA \qquad (4.11)$$

We have already noted that electrical energy flowing across the boundary of a system is work. We can gain further insight into such a process by considering a system in which the only work mode is electrical. As examples of such a system, we can think of a charged condenser, an electrolytic cell, and the type of fuel cell described in Chapter 1. Consider a quasi-equilibrium process for such a system, and during this process let the potential difference be $\mathcal{E}$ and the amount of electrical charge that flows into the system be dZ. For this quasi-equilibrium process the work is given by the relation

$$\delta W = -\mathcal{E}\, dZ \qquad (4.12)$$

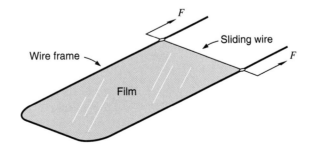

FIGURE 4.16
Schematic arrangement showing work done on a surface film.

Since the current, i, equals dZ/dt (where $t =$ time), we can also write

$$\delta W = -\mathcal{E}i\, dt \qquad (4.12)$$

$$_1W_2 = -\int_1^2 \mathcal{E}i\, dt \qquad (4.13)$$

Equation 4.13 may also be written as a rate equation for work (the power).

$$\dot{W} = \frac{\delta W}{dt} = -\mathcal{E}i \qquad (4.14)$$

Since the ampere (electric current) is one of the fundamental units in the International System, and the watt has been defined previously, this relation serves as the definition of the unit for electric potential, the volt (V), which is one watt divided by one ampere.

4.5 CONCLUDING REMARKS REGARDING WORK

The similarity of the expressions for work in the three processes discussed in Section 4.4 and in the processes in which work is done at a moving boundary should be noted. In each of these quasi-equilibrium processes, the work is given by the integral of the product of an intensive property and the change of an extensive property. The following is a summary list of these processes and their work expressions

Simple compressible system $\qquad\qquad _1W_2 = \int_1^2 P\, dV$

Stretched wire $\qquad\qquad\qquad\qquad _1W_2 = -\int_1^2 \mathcal{T}\, dL$

Surface film $\qquad\qquad\qquad\qquad\quad _1W_2 = -\int_1^2 \mathcal{S}\, dA$

System in which the work is completely electrical $\quad _1W_2 = -\int_1^2 \mathcal{E}\, dZ \quad (4.15)$

Although we will deal primarily with systems in which there is only one mode of work, it is quite possible to have more than one work mode in a given process. Thus, we could write

$$\delta W = P\, dV - \mathcal{T}\, dL - \mathcal{S}\, dA - \mathcal{E}\, dZ + \cdots \qquad (4.16)$$

where the dots represent other products of an intensive property and the derivative of a related extensive property. In each term the intensive property can be viewed as the driving force that causes a change to occur in the related extensive property, which is often termed the displacement. Just as we could derive the expression for power for the single point force in Eq. 4.2, the rate form of Eq. 4.16 expresses the power as

$$\dot{W} = \frac{dW}{dt} = P\dot{V} - \mathcal{T}\mathbf{V} - \mathcal{S}\dot{A} - \mathcal{E}\dot{Z} + \cdots \qquad (4.17)$$

It should also be noted that many other forms of work can be identified in processes that are not quasi-equilibrium processes. For example, there is the work done by shearing forces in the friction in a viscous fluid or the work done by a rotating shaft that crosses the system boundary.

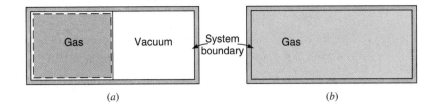

FIGURE 4.17
Example of process involving a change of volume for which the work is zero.

The identification of work is an important aspect of many thermodynamic problems. We have already noted that work can be identified only at the boundaries of the system. For example, consider Fig. 4.17, which shows a gas separated from the vacuum by a membrane. Let the membrane rupture and the gas fill the entire volume. Neglecting any work associated with the rupturing of the membrane, we can ask whether work is done in the process. If we take as our system the gas and the vacuum space, we readily conclude that no work is done because no work can be identified at the system boundary. If we take the gas as a system, we do have a change of volume, and we might be tempted to calculate the work from the integral

$$\int_1^2 P\,dV$$

However, this is not a quasi-equilibrium process, and therefore the work cannot be calculated from this relation. Because there is no resistance at the system boundary as the volume increases, we conclude that for this system no work is done in this process of filling the vacuum.

Another example can be cited with the aid of Fig. 4.18. In Fig. 4.18*a* the system consists of the container plus the gas. Work crosses the boundary of the system at the point where the system boundary intersects the shaft, and this work can be associated with the shearing forces in the rotating shaft. In Fig. 4.18*b* the system includes the shaft and weight as well as the gas and the container. Therefore, no work crosses the system boundary as the weight moves downward. As we will see in the next chapter, we can identify a change of potential energy within the system, but this should not be confused with work crossing the system boundary.

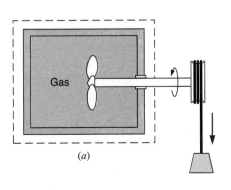

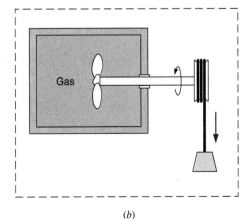

FIGURE 4.18
Example showing how selection of the system determines whether work is involved in a process.

4.6 DEFINITION OF HEAT

The thermodynamic definition of heat is somewhat different from the everyday under-standing of the word. It is essential to understand clearly the definition of heat given here, because it plays a part in so many thermodynamic problems.

If a block of hot copper is placed in a beaker of cold water, we know from experi-ence that the block of copper cools down and the water warms up until the copper and water reach the same temperature. What causes this decrease in the temperature of the copper and the increase in the temperature of the water? We say that it is the result of the transfer of energy from the copper block to the water. It is out of such a transfer of energy that we arrive at a definition of heat.

Heat is defined as the form of energy that is transferred across the boundary of a system at a given temperature to another system (or the surroundings) at a lower tem-perature by virtue of the temperature difference between the two systems. That is, heat is transferred from the system at the higher temperature to the system at the lower tem-perature, and the heat transfer occurs solely because of the temperature difference be-tween the two systems. Another aspect of this definition of heat is that a body never contains heat. Rather, heat can be identified only as it crosses the boundary. Thus, heat is a transient phenomenon. If we consider the hot block of copper as one system and the cold water in the beaker as another system, we recognize that originally neither system contains any heat (they do contain energy, of course). When the copper block is placed in the water and the two are in thermal communication, heat is transferred from the cop-per to the water until equilibrium of temperature is established. At this point we no longer have heat transfer, because there is no temperature difference. Neither system contains heat at the conclusion of the process. It also follows that heat is identified at the boundary of the system, for heat is defined as energy being transferred across the system boundary.

Heat, like work, is a form of energy transfer to or from a system. Therefore, the units for heat, and to be more general, for any other form of energy as well, are the same as the units for work, or are at least directly proportional to them. In the International Sys-tem the unit for heat (energy) is the joule. Similarly, in the English System, the foot pound force is an appropriate unit for heat. However, another unit came to be used naturally over the years, the result of an association with the process of heating water, such as that used in connection with defining heat in the previous section. Consider as a system 1 lbm of water at 59.5 F. Let a block of hot copper of appropriate mass and temperature be placed in the water so that when thermal equilibrium is established the temperature of the water is 60.5 F. This unit amount of heat transferred from the copper to the water in this process is called the British thermal unit (Btu). More specifically, it is called the 60-degree Btu, defined as the amount of heat required to raise 1 lbm of water from 59.5 F to 60.5 F. (The Btu as used today is actually defined in terms of the standard SI units.) It is worth noting here that a unit of heat in metric units, the calorie, originated naturally in a manner similar to the origin of the Btu in the English system. The calorie is defined as the amount of heat required to raise 1 g of water from 14.5°C to 15.5°C.

Heat transferred *to* a system is considered positive, and heat transferred *from* a system is negative. Thus, positive heat represents energy transferred to a system, and negative heat represents energy transferred from a system. The symbol Q represents heat. A process in which there is no heat transfer ($Q = 0$) is called an adiabatic process.

From a mathematical perspective, heat, like work, is a path function and is recog-nized as an inexact differential. That is, the amount of heat transferred when a system undergoes a change from state 1 to state 2 depends on the path that the system follows during the change of state. Since heat is an inexact differential, the differential is written δQ. On integrating, we write

$$\int_1^2 \delta Q = {}_1Q_2$$

In words, ${}_1Q_2$ is the heat transferred during the given process between states 1 and 2.

The rate at which heat is transferred to a system is designated by symbol $\dot{Q}$.

$$\dot{Q} \equiv \frac{\delta Q}{dt}$$

It is also convenient to speak of the heat transfer per unit mass of the system, q, often termed specific heat transfer, which is defined as

$$q \equiv \frac{Q}{m}$$

4.7 HEAT TRANSFER MODES

Heat transfer is the transport of energy due to a temperature difference between different amounts of matter. We know that an ice cube taken out of the freezer will melt as it is placed in a warmer environment such as a glass of liquid water or on a plate with room air around it. From the discussion about energy in Section 2.6 we realize that molecules of matter have translational (kinetic), rotational, and vibrational energy. Energy in these modes can be transmitted to the nearby molecules by interactions (collisions) or by exchange of molecules such that energy is given out by molecules that have more in the average (higher temperature) to those that have less in the average (lower temperature). This energy exchange between molecules is heat transfer by conduction, and it increases with the temperature difference and the ability of the substance to make the transfer. This is expressed in Fourier's law of conduction

$$\dot{Q} = -kA\frac{dT}{dx} \quad \text{(W)} \tag{4.18}$$

giving the rate of heat transfer as proportional to the conductivity, k, the total area, A, and the temperature gradient. The minus sign gives a direction of the heat transfer from a higher temperature to a lower temperature region. Often the gradient is evaluated as a temperature difference divided by a distance when an estimate has to be done if a mathematical or numerical solution is not available.

Values of the conductivity, k, range from the order of 100 W/m K for metals, 1 to 10 for nonmetallic solids as glass, ice and rock, from 0.1 to 10 for liquids, around 0.1 for insulation materials, and from 0.1 down to less than 0.01 for gases.

A different mode of heat transfer takes place when a medium is flowing, called convective heat transfer. In this mode the bulk motion of a substance moves matter with a certain energy level over or near a surface with a different temperature. Now the heat transfer by conduction is dominated by the manner in which the bulk motion brings the two substances in contact or close proximity. Examples of this are the wind blowing over a building or flow through heat exchangers, which can be air flowing over/through a radiator with water flowing inside the radiator piping. The overall heat transfer is typically correlated with Newton's law of cooling as

$$\dot{Q} = Ah\,\Delta T \qquad (4.19)$$

where the transfer properties are lumped into the heat transfer coefficient, h, which then becomes a function of the media properties, the flow and geometry. A more detailed study of fluid mechanics and heat transfer aspects of the overall process is necessary to evaluate the heat transfer coefficient for a given situation.

Typical values for the convection coefficient (all in W/mK^2) are

Natural convection	$h = 5$–25, gas	$h = 50$–1000, liquid
Forced convection	$h = 25$–250, gas	$h = 50$–$20\,000$, liquid
Boiling phase change	$h = 2500$–$100\,000$	

The final mode of heat transfer is radiation, which transmits energy as electromagnetic waves in space. The transfer can happen in empty space and does not require any matter, but the emission (generation) of the radiation and the absorption does require a substance to be present. Surface emission is usually written as a fraction, emissivity ε, of a perfect black body emission as

$$\dot{Q} = \varepsilon\sigma A T_s^4 \quad \text{(W)} \qquad (4.20)$$

with the surface temperature, T_s, and the Stefan-Boltzmann constant, σ. Typical values of the emissivity range from 0.92 for nonmetallic surfaces to 0.6 to 0.9 for nonpolished metallic surfaces, to less than 0.1 for highly polished metal surfaces. Radiation is distributed over a range of wavelengths and it is emitted and absorbed differently for different surfaces, but such a description is beyond the scope of the present text.

EXAMPLE 4.7   Consider the constant transfer of energy from a warm room at 20°C inside a house to the colder ambient at −10°C through a single-pane window as shown in Fig. 4.19. The temperature variation with distance from the outside glass surface is shown with an outside convection heat transfer layer, but no such layer is inside the room (as a simplification). The glass pane has a thickness of 5 mm (0.005 m) with a conductivity of 1.4 W/m K and a total surface area of 0.5 m^2. The outside wind is blowing so that the convective heat transfer coefficient is 100 W/m^2 K. With an outer glass sur-

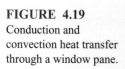

FIGURE 4.19
Conduction and convection heat transfer through a window pane.

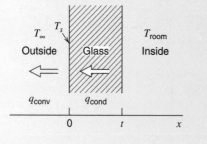

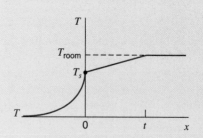

face temperature of 12.1°C we would like to know the rate of heat transfer in the glass and the convective layer.

For the conduction through the glass we have

$$\dot{Q} = -kA \frac{dT}{dx} = -kA \frac{\Delta T}{\Delta x} = -1.4 \frac{W}{m\,K} \times 0.5\,m^2 \frac{20 - 12.1}{0.005} \frac{K}{m} = -1106\,W$$

and the negative sign shows that energy is leaving the room. For the outside convection layer we have

$$\dot{Q} = hA\,\Delta T = 100 \frac{W}{m^2\,K} \times 0.5\,m^2\,[12.1 - (-10)]\,K = 1105\,W$$

with a direction from the higher to the lower temperature, i.e., toward the outside.

4.8 COMPARISON OF HEAT AND WORK

At this point it is evident that there are many similarities between heat and work.

1. Heat and work are both transient phenomena. Systems never possess heat or work, but either or both cross the system boundary when a system undergoes a change of state.

2. Both heat and work are boundary phenomena. Both are observed only at the boundaries of the system, and both represent energy crossing the boundary of the system.

3. Both heat and work are path functions and inexact differentials.

It should also be noted that in our sign convention, $+Q$ represents heat transferred *to* the system and thus is energy added to the system, and $+W$ represents work done *by* the system and thus represents energy leaving the system.

A final illustration may help explain the difference between heat and work. Figure 4.20 shows a gas contained in a rigid vessel. Resistance coils are wound around the outside of the vessel. When current flows through the resistance coils, the temperature of the gas increases. Which crosses the boundary of the system, heat or work?

In Fig. 4.20*a* we consider only the gas as the system. The energy crosses the boundary of the system because the temperature of the walls is higher than the temperature of the gas. Therefore, we recognize that heat crosses the boundary of the system.

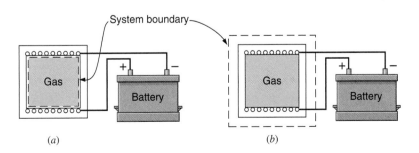

FIGURE 4.20 An example showing the difference between heat and work.

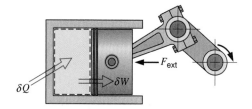

FIGURE 4.21 The effects of heat addition to a control volume that also can give out work.

In Fig. 4.20*b* the system includes the vessel and the resistance heater. Electricity crosses the boundary of the system and, as indicated earlier, this is work.

Consider a gas in a cylinder fitted with a movable piston, as shown in Fig. 4.21. There is a positive heat transfer to the gas, which tends to make the temperature increase. It also tends to increase the gas pressure. However, the pressure is dictated by the external force acting on its movable boundary, as discussed in Section 2.8. If this remains constant, then the volume increases instead. There are also the opposite tendencies for a negative heat transfer, that is, one out of the gas. Consider again the positive heat transfer, except that in this case the external force simultaneously decreases. This causes the gas pressure to decrease, such that the temperature tends to go down. In this case, there are simultaneous tendencies for temperature change in the opposite direction, which effectively decouples directions of heat transfer and temperature change.

Often when we want to evaluate a finite amount of energy transferred as either work or heat we must integrate the instantaneous rate over time.

$$_1W_2 = \int_1^2 \dot{W}\, dt, \qquad _1Q_2 = \int_1^2 \dot{Q}\, dt$$

In order to perform the integration we must know how the rate varies with time. For time periods where the rate does not change significantly, a simple average may be of sufficient accuracy to allow us to write

$$_1W_2 = \int_1^2 \dot{W}\, dt = \dot{W}_{avg}\, \Delta t \tag{4.21}$$

which is similar to the information given on your electric utility bill as kilowatt-hours.

SUMMARY

Work and heat are energy transfers between a control volume and its surroundings. Work is energy that can be transferred mechanically (or electrically, or chemically) from one system to another and must cross the control surface either as a transient phenomenon or as a steady rate of work, which is power. Work is a function of the process path as well as the beginning state and end state. The displacement work is equal to the area below the process curve drawn in a *P–V* diagram if we have an equilibrium process. A number of ordinary processes can be expressed as polytropic processes having a particular simple mathematical form for the *P–V* relation. Work involved by the action of surface tension, single-point forces, or electrical systems should be recognized and treated separately. Any nonequilibrium processes (say, dynamic forces, which are important due to accelerations) should be identified so that only equilibrium force or pressure is used to evaluate the work term.

Heat transfer is energy transferred due to a temperature difference, and the conduction, convection, and radiation modes are discussed.

You should have learned a number of skills and acquired abilities from studying this chapter that will allow you to

- Recognize force and displacement in a system.
- Know power as rate of work (force × velocity, angular torque × velocity)
- Know work is a function of the end states and the path followed in process
- Calculate the work term knowing the P–V or the F–x relationship
- Evaluate the work involved in a polytropic process between two states
- Know work is the area under the process curve in a P–V diagram
- Apply a force balance on a mass and determine work in a process from it
- Distinguish between an equilibrium process and a nonequilibrium process
- Recognize the three modes of heat transfer: conduction, convection and radiation
- Be familiar with Fourier's law of conduction and its use in simple applications
- Know the simple models for convection and radiation heat transfer
- Understand the difference between the rates ($\dot{W}$, $\dot{Q}$) and the amounts ($_1W_2$, $_1Q_2$).

KEY CONCEPTS AND FORMULAS

Work	Energy in transfer—mechanical, electrical, and chemical
Heat	Energy in transfer, caused by a ΔT
Displacement work	$W = \int_1^2 F\, dx = \int_1^2 P\, dV = \int_1^2 \mathcal{S}\, dA = \int_1^2 T\, d\theta$
Specific work	$w = W/m$ (work per unit mass)
Power, rate of work	$\dot{W} = F\mathbf{V} = P\dot{V} = T\omega$ ($\dot{V}$ displacement rate)
	Velocity $\mathbf{V} = r\omega$, torque $T = Fr$, angular velocity = ω
Polytropic process	$PV^n = $ constant or $Pv^n = $ constant
Polytropic process work	$_1W_2 = \dfrac{1}{1-n}(P_2V_2 - P_1V_1)$ (if $n \neq 1$)
	$_1W_2 = P_1V_1 \ln\dfrac{V_2}{V_1}$ (if $n = 1$)
Conduction heat transfer	$\dot{Q} = -kA\dfrac{dT}{dx}$
Conductivity	k (W/m K)
Convection heat transfer	$\dot{Q} = hA\,\Delta T$
Convection coefficient	h (W/m² K)
Radiation heat transfer (net to ambient)	$\dot{Q} = \epsilon\sigma A(T_s^4 - T_{amb}^4)$ ($\sigma = 5.67 \times 10^{-8}$ W/m² K⁴)
Rate integration	$_1Q_2 = \int \dot{Q}\, dt \approx \dot{Q}_{avg}\,\Delta t$

CONCEPT-STUDY GUIDE PROBLEMS

4.1 The electric company charges the customers per kilowatt-hour. What is that is SI units?

4.2 A car engine is rated at 160 hp. What is the power in SI units?

4.3 A 1200-hp dragster engine has a drive shaft rotating at 2000 RPM. How much torque is on the shaft?

4.4 A 1200-hp dragster engine drives the car with a speed of 100 km/h. How much force is between the tires and the road?

4.5 Two hydraulic piston/cylinders are connected through a hydraulic line so that they have roughly the same pressure. If they have diameters of D_1 and

$D_2 = 2D_1$ respectively, what can you say about the piston forces F_1 and F_2?

4.6 Normally pistons have a flat head, but in diesel engines pistons can have bowls in them and protruding ridges. Does this geometry influence the work term?

4.7 What is roughly the relative magnitude of the work in the process 1–2c versus the process 1–2a shown in Fig. 4.8?

4.8 A hydraulic cylinder of area 0.01 m^2 must push a 1000-kg arm and shovel 0.5 m straight up. What pressure is needed and how much work is done?

4.9 A work of 2.5 kJ must be delivered on a rod from a pneumatic piston/cylinder where the air pressure is limited to 500 kPa. What diameter cylinder should I have to restrict the rod motion to maximum 0.5 m?

4.10 Helium gas expands from 125 kPa, 350 K, and 0.25 m^3 to 100 kPa in a polytropic process with $n = 1.667$. Is the work positive, negative, or zero?

4.11 An ideal gas goes through an expansion process whereby the volume doubles. Which process will lead to the larger work output, an isothermal process or a polytropic process with $n = 1.25$?

4.12 Show how the polytropic exponent n can be evaluated if you know the end state properties, (P_1, V_1) and (P_2, V_2).

4.13 A drag force on an object moving through a medium (like a car through air or a submarine through water) is $F_d = 0.225\, A\rho \mathbf{V}^2$. Verify that the unit become newtons.

4.14 A force of 1.2 kN moves a truck with 60 km/h up a hill. What is the power?

4.15 Electric power is volts times amperes ($P = Vi$). When a car battery at 12 V is charged with 6 A for 3 h, how much energy is delivered?

4.16 Torque and energy and work have the same units (N m). Explain the difference.

4.17 Find the rate of conduction heat transfer through a 1.5-cm-thick hardwood board, $k = 0.16$ W/m K, with a temperature difference between the two sides of 20°C.

4.18 A 2-m^2 window has a surface temperature of 15°C, and the outside wind is blowing air at 2°C across it with a convection heat transfer coefficient of $h = 125$ W/m^2 K. What is the total heat transfer loss?

4.19 A radiant heating lamp has a surface temperature of 1000 K with $\varepsilon = 0.8$. How large a surface area is needed to provide 250 W of radiation heat transfer?

HOMEWORK PROBLEMS

Force Displacement Work

4.20 A piston of mass 2 kg is lowered 0.5 m in the standard gravitational field. Find the required force and the work involved in the process.

4.21 An escalator raises a 100-kg bucket of sand 10 m in 1 min. Determine the total amount of work done during the process.

4.22 A bulldozer pushes 500 kg of dirt 100 m with a force of 1500 N. It then lifts the dirt 3 m up to put it in a dump truck. How much work did it do in each situation?

4.23 A hydraulic cylinder has a piston of cross-sectional area 25 cm^2 and a fluid pressure of 2 MPa. If the piston is moved 0.25 m, how much work is done?

4.24 Two hydraulic cylinders maintain a pressure of 1200 kPa. One has a cross-sectional area of 0.01 m^2, the other one of 0.03 m^2. To deliver 1 kJ of work to the piston, how large a displacement (V) and piston motion H are needed for each cylinder? Neglect P_{atm}.

4.25 A linear spring, $F = k_s(x - x_0)$ with spring constant $k_s = 500$ N/m is stretched until it is 100 mm longer. Find the required force and the work input.

4.26 A nonlinear spring has a force versus the displacement relation of $F = k_s(x - x_0)^n$. If the spring end is moved to x_1 from the relaxed state, determine the formula for the required work.

4.27 The rolling resistance of a car depends on its weight as $F = 0.006$ mg. How long will a car of 1400 kg drive for a work input of 25 kJ?

4.28 A car drives for half an hour at constant speed and uses 30 MJ over a distance of 40 km. What was the traction force to the road and its speed?

4.29 The air drag force on a car is $0.225\, A\rho \mathbf{V}^2$. Assume air at 290 K, 100 kPa and a car frontal area of 4 m^2

driving at 90 km/h. How much energy is used to overcome the air drag driving for 30 min?

4.30 Two hydraulic piston/cylinders are connected with a line. The master cylinder has an area of 5 cm², creating a pressure of 1000 kPa. The slave cylinder has an area of 3 cm². If 25 J is the work input to the master cylinder, what is the force and displacement of each piston and the work output of the slave cylinder piston?

Boundary Work: Simple One-Step Process

4.31 A constant-pressure piston/cylinder assembly contains 0.2 kg of water as saturated vapor at 400 kPa. It is now cooled so that the water occupies half the original volume. Find the work done in the process.

4.32 A steam radiator in a room at 25°C has saturated water vapor at 110 kPa flowing through it when the inlet and exit valves are closed. What are the pressure and the quality of the water when it has cooled to 25°C? How much work is done?

4.33 A 400-L tank, *A* (see Fig. P4.33), contains argon gas at 250 kPa and 30°C. Cylinder *B*, having a frictionless piston of such mass that a pressure of 150 kPa will float it, is initially empty. The valve is opened and argon flows into *B* and eventually reaches a uniform state of 150 kPa and 30°C throughout. What is the work done by the argon?

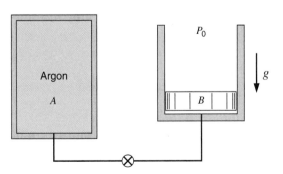

FIGURE P4.33

4.34 A piston/cylinder contains air at 600 kPa, 290 K and a volume of 0.01 m³. A constant-pressure process gives 54 kJ of work out. Find the final volume and temperature of the air.

4.35 Saturated water vapor at 200 kPa is in a constant-pressure piston/cylinder. In this state, the piston is

0.1 m from the cylinder bottom and the cylinder area is 0.25 m². The temperature is then changed to 200°C. Find the work in the process.

4.36 A cylinder fitted with a frictionless piston contains 5 kg of superheated refrigerant R-134a vapor at 1000 kPa and 140°C. The setup is cooled at constant pressure until the R-134a reaches a quality of 25%. Calculate the work done in the process.

4.37 Find the specific work in Problem 3.54 for the case where the volume is reduced.

4.38 A piston/cylinder has 5 m of liquid 20°C water on top of the piston (*m* = 0) with cross-sectional area of 0.1 m², see Fig. P2.57. Air is let in under the piston that rises and pushes the water out over the top edge. Find the necessary work to push all the water out and plot the process in a *P–V* diagram.

4.39 Air in a spring-loaded piston/cylinder setup has a pressure that is linear with volume, *P* = *A* + *BV*. With an initial state of *P* = 150 kPa, *V* = 1 L and a final state of 800 kPa and volume 1.5 L, it is similar to the setup in Problem 3.113. Find the work done by the air.

4.40 Find the specific work in Problem 3.43.

4.41 A piston/cylinder contains 1 kg of water at 20°C with volume 0.1 m³. By mistake someone locks the piston, preventing it from moving while we heat the water to saturated vapor. Find the final temperature and volume, and the process work.

4.42 A piston/cylinder assembly contains 1 kg of liquid water at 20°C and 300 kPa, as shown in Fig. P4.42. There is a linear spring mounted on the piston such that when the water is heated the pressure reaches 3 MPa with a volume of 0.1 m³.

 a. Find the final temperature.

 b. Plot the process in a *P–v* diagram.

 c. Find the work in the process.

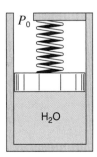

FIGURE P4.42

4.43 A piston/cylinder assembly contains 3 kg of air at 20°C and 300 kPa. It is now heated in a constant pressure process to 600 K.

a. Find the final volume.

b. Plot the process path in a *P—v* diagram.

c. Find the work in the process.

4.44 A piston/cylinder assembly contains 0.5 kg of air at 500 kPa and 500 K. The air expands in a process such that *P* is linearly decreasing with volume to a final state of 100 kPa, 300 K. Find the work in the process.

4.45 Consider the nonequilibrium process described in Problem 3.109. Determine the work done by the carbon dioxide in the cylinder during the process.

4.46 Consider the problem of inflating the helium balloon, as described in Problem 3.79. For a control volume that consists of the helium inside the balloon determine the work done during the filling process when the diameter changes from 1 m to 4 m.

Polytropic Process

4.47 Consider a mass going through a polytropic process where pressure is directly proportional to volume ($n = -1$). The process starts with $P = 0$, $V = 0$ and ends with $P = 600$ kPa, $V = 0.01$ m^3. Find the boundary work done by the mass.

4.48 The piston/cylinder arrangement shown in Fig. P4.48 contains carbon dioxide at 300 KPa and 100°C with a volume of 0.2 m^3. Weights are added to the piston such that the gas compresses according to the relation $PV^{1.2} = $ constant to a final temperature of 200°C. Determine the work done during the process.

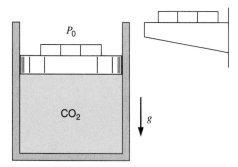

FIGURE P4.48

4.49 A gas initially at 1 MPa and 500°C is contained in a piston and cylinder arrangement with an initial

volume of 0.1 m^3. The gas is then slowly expanded according to the relation $PV = $ constant until a final pressure of 100 kPa is reached. Determine the work for this process.

4.50 Helium gas expands from 125 kPa, 350 K, and 0.25 m^3 to 100 kPa in a polytropic process with $n = 1.667$. How much work does it give out?

4.51 Air goes through a polytropic process from 125 kPa and 325 K to 300 kPa and 500 K. Find the polytropic exponent n and the specific work in the process.

4.52 A piston/cylinder device contains 0.1 kg of air at 100 kPa and 400 K that goes through a polytropic compression process with $n = 1.3$ to a pressure of 300 kPa. How much work has the air done in the process?

4.53 A balloon behaves so the pressure is $P = C_2 V^{1/3}$, $C_2 = 100$ kPa/m. The balloon is blown up with air from a starting volume of 1 m^3 to a volume of 3 m^3. Find the final mass of the air, assuming it is at 25°C, and the work done by the air.

4.54 A balloon behaves such that the pressure inside is proportional to the diameter squared. It contains 2 kg of ammonia at 0°C, with 60% quality. The balloon and ammonia are now heated so that a final pressure of 600 kPa is reached. Considering the ammonia as a control mass, find the amount of work done in the process.

4.55 Consider a piston/cylinder setup with 0.5 kg of R-134a as saturated vapor at -10°C. It is now compressed to a pressure of 500 kPa in a polytropic process with $n = 1.5$. Find the final volume and temperature, and determine the work done during the process.

4.56 Consider the process described in Problem 3.98. With 1 kg of water as a control mass, determine the boundary work during the process.

4.57 Find the work done in Problem 3.106.

4.58 A piston/cylinder contains water at 500°C, 3 MPa. It is cooled in a polytropic process to 200°C, 1 MPa. Find the polytropic exponent and the specific work in the process.

Boundary Work: Multistep Process

4.59 Consider a two-part process with an expansion from 0.1 to 0.2 m^3 at a constant pressure of 150 kPa followed by an expansion from 0.2 to 0.4 m^3 with a linearly rising pressure from 150 kPa ending at 300 kPa. Show the process in a *P–V* diagram and find the boundary work.

4.60 A cylinder containing 1 kg of ammonia has an externally loaded piston. Initially the ammonia is at 2 MPa and 180°C. It is now cooled to saturated vapor at 40°C and then further cooled to 20°C, at which point the quality is 50%. Find the total work for the process, assuming a piecewise linear variation of *P* versus *V*.

4.61 A piston/cylinder arrangement shown in Fig. P4.61 initially contains air at 150 kPa and 400°C. The setup is allowed to cool to the ambient temperature of 20°C.

a. Is the piston resting on the stops in the final state? What is the final pressure in the cylinder?

b. What is the specific work done by the air during the process?

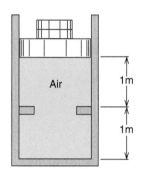

FIGURE P4.61

4.62 A piston/cylinder has 1.5 kg of air at 300 K and 150 kPa. It is now heated up in a two step process. First constant volume to 1000 K (state 2) then followed by a constant pressure process to 1500 K, state 3. Find the final volume and the work in the process.

4.63 A piston/cylinder assembly (Fig. P4.63) has 1 kg of R-134a at state 1 with 110°C, 600 kPa. It is then

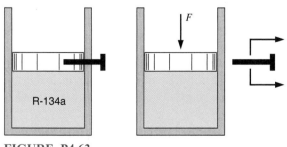

FIGURE P4.63

brought to saturated vapor, state 2, by cooling while the piston is locked with a pin. Now the piston is balanced with an additional constant force and the pin is removed. The cooling continues to a state 3, where the R-134a is saturated liquid. Show the processes in a *P–V* diagram and find the work in each of the two steps, 1 to 2 and 2 to 3.

4.64 The refrigerant R-22 is contained in a piston/cylinder as shown in Fig. P4.64, where the volume is 11 L when the piston hits the stops. The initial state is −30°C, 150 kPa, with a volume of 10 L. This system is brought indoors and warms up to 15°C.

a. Is the piston at the stops in the final state?

b. Find the work done by the R-22 during this process.

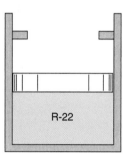

FIGURE P4.64

4.65 A piston/cylinder assembly contains 50 kg of water at 200 kPa with a volume of 0.1 m³. Stops in the cylinder restrict the enclosed volume to 0.5 m³, similar to the setup in Problem 4.64. The water is now heated to 200°C. Find the final pressure, volume, and work done by the water.

4.66 Find the work in Problem 3.108.

4.67 A piston/cylinder assembly contains 1 kg of liquid water at 20°C and 300 kPa. Initially the piston floats, similar to the setup in Problem 4.64, with a maximum enclosed volume of 0.002 m³ if the piston touches the stops. Now heat is added so that a final pressure of 600 kPa is reached. Find the final volume and the work in the process.

4.68 Ten kilograms of water in a piston/cylinder arrangement exist as saturated liquid/vapor at 100 kPa, with a quality of 50%. It is now heated so the volume triples. The mass of the piston is such that a cylinder pressure of 200 kPa will float it (see Fig. P4.68).

a. Find the final temperature and volume of the water.

b. Find the work given out by the water.

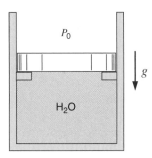

FIGURE P4.68

4.69 Find the work in Problem 3.63.

4.70 A piston/cylinder setup similar to Problem 4.68 contains 0.1 kg saturated liquid and vapor water at 100 kPa with quality 25%. The mass of the piston is such that a pressure of 500 kPa will float it. The water is heated to 300°C. Find the final pressure, volume, and work, $_1W_2$.

Other Types of Work and General Concepts

4.71 A 0.5-m-long steel rod with a 1-cm diameter is stretched in a tensile test. What is the work required to obtain a relative strain of 0.1%? The modulus of elasticity of steel is 2×10^8 kPa.

4.72 A copper wire of diameter 2 mm is 10 m long and stretched out between two posts. The normal stress (pressure) $\sigma = E(L - L_0)/L_0$, depends on the length L versus the unstretched length L_0 and Young's modulus $E = 1.1 \times 10^6$ kPa. The force is $F = A\sigma$ and measured to be 110 N. How much longer is the wire and how much work was put in?

4.73 A film of ethanol at 20°C has a surface tension of 22.3 mN/m and is maintained on a wire frame as shown in Fig. P4.73. Consider the film with two surfaces as a control mass and find the work done when the wire is moved 10 mm to make the film 20×40 mm.

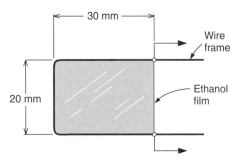

FIGURE P4.73

4.74 Assume a balloon material with a constant surface tension of $\mathscr{S} = 2$ N/m. What is the work required to stretch a special balloon up to a radius of $r = 0.5$ m? Neglect any effect from atmospheric pressure.

4.75 A soap bubble has a surface tension of $\mathscr{S} = 3 \times 10^{-4}$ N/cm as it sits flat on a rigid ring of diameter 5 cm. You now blow on the film to create a half-sphere surface of diameter 5 cm. How much work was done?

4.76 Assume we fill a spherical balloon from a bottle of helium gas. The helium gas provides work $\int P \, dV$ that stretches the balloon material $\int S \, dA$ and pushes back the atmosphere $\int P_0 \, dV$. Write the incremental balance for $dW_{\text{helium}} = dW_{\text{stretch}} + dW_{\text{atm}}$ to establish the connection between the helium pressure, the surface tension S, and P_0 as a function of radius.

4.77 A sheet of rubber is stretched out over a ring of radius 0.25 m. I pour liquid water at 20°C on it, as in Fig. P4.77, so that the rubber forms a half sphere (cup). Neglect the rubber mass and find the surface tension near the ring.

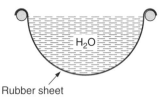

Rubber sheet　　　　　　　　　　**FIGURE P4.77**

4.78 Consider a window-mounted air-conditioning unit used in the summer to cool incoming air. Examine the system boundaries for rates of work and heat transfer, including signs.

4.79 Consider a hot-air heating system for a home. Examine the following systems for heat transfer.

　a. The combustion chamber and combustion gas side of the heat transfer area

　b. The furnace as a whole, including the hot- and cold-air ducts and chimney

4.80 Consider a household refrigerator that has just been filled up with room-temperature food. Define a control volume (mass) and examine its boundaries for rates of work and heat transfer, including sign.

　a. Immediately after the food is placed in the refrigerator

　b. After a long period of time has elapsed and the food is cold

4.81 A room is heated with an electric space heater on a winter day. Examine the following control volumes, regarding heat transfer and work, including sign:

a. The space heater

b. Room

c. The space heater and the room together

Rates of Work

4.82 An escalator raises a 100-kg bucket 10 m in 1 min. Determine the rate of work in the process.

4.83 A car uses 25 hp to drive at a horizontal level at a constant speed of 100 km/h. What is the traction force between the tires and the road?

4.84 A piston/cylinder of cross-sectional area 0.01 m^2 maintains constant pressure. It contains 1 kg of water with a quality of 5% at 150°C. If we heat so that 1 g/s of liquid turns into vapor, what is the rate of work out?

4.85 A crane lifts a bucket of cement with a total mass of 450 kg vertically upward with a constant velocity of 2 m/s. Find the rate of work needed to do this.

4.86 Consider the car with the rolling resistance as in Problem 4.27. How fast can it drive using 30 hp?

4.87 Consider the car with the air drag force as in Problem 4.29. How fast can it drive using 30 hp?

4.88 Consider a 1400-kg car having rolling resistance as in Problem 4.27 and air resistance as in Problem 4.29. How fast can it drive using 30 hp?

4.89 A battery is well insulated while being charged by 12.3 V at a current of 6 A. Take the battery as a control mass and find the instantaneous rate of work and the total work done over 4 h.

4.90 A current of 10 A runs through a resistor with a resistance of 15 Ω. Find the rate of work that heats the resistor up.

4.91 A pressure of 650 kPa pushes a piston of diameter 0.25 m with **V** = 5 m/s. What is the volume displacement rate, the force, and the transmitted power?

4.92 Assume the process in Problem 4.59 takes place with a constant rate of change in volume over 2 minutes. Show the power (rate of work) as a function of time.

4.93 Air at a constant pressure in a piston/cylinder is at 300 kPa, 300 K and has a volume of 0.1 m^3. It is heated to 600 K over 30 s in a process with con-

stant piston velocity. Find the power delivered to the piston.

4.94 A torque of 650 N m rotates a shaft of diameter 0.25 m with ω = 50 rad/s. What is the shaft surface speed and the transmitted power?

Heat Transfer Rates

4.95 The sun shines on a 150-m^2 road surface so that it is at 45°C. Below the 5-cm-thick asphalt, with average conductivity of 0.06 W/m K, is a layer of compacted rubble at a temperature of 15°C. Find the rate of heat transfer to the rubble.

4.96 A steel pot, with conductivity of 50 W/m K and a 5-mm-thick bottom, is filled with 15°C liquid water. The pot has a diameter of 20 cm and is now placed on an electric stove that delivers 250 W as heat transfer. Find the temperature on the outer pot bottom surface assuming the inner surface is at 15°C.

4.97 A water heater is covered up with insulation boards over a total surface area of 3 m^2. The inside board surface is at 75°C, the outside surface is at 20°C, and the board material has a conductivity of 0.08 W/m K. How thick should the board be to limit the heat transfer loss to 200 W?

4.98 You drive a car on a winter day with the atmospheric air at -15°C, and you keep the outside front windshield surface temperature at $+2$°C by blowing hot air on the inside surface. If the windshield is 0.5 m^2 and the outside convection coefficient is 250 W/m^2 K, find the rate of energy loss through the front windshield. For that heat transfer rate and a 5-mm-thick glass with k = 1.25 W/m K, what is then the inside windshield surface temperature?

4.99 A large condenser (heat exchanger) in a power plant must transfer a total of 100 MW from steam running in a pipe to seawater being pumped through the heat exchanger. Assume the wall separating the steam and seawater is 4 mm of steel, with conductivity of 15 W/m K, and that a maximum of 5°C difference between the two fluids is allowed in the design. Find the required minimum area for the heat transfer, neglecting any convective heat transfer in the flows.

4.100 The black grille on the back of a refrigerator has a surface temperature of 35°C with a total surface area of 1 m^2. Heat transfer to the room air at 20°C takes place with an average convective heat transfer coefficient of 15 W/m^2 K. How much energy can be removed during 15 minutes of operation?

4.101 Owing to a faulty door contact, the small light bulb (25 W) inside a refrigerator is kept on and limited insulation lets 50 W of energy from the outside seep into the refrigerated space. How much of a temperature difference to the ambient surroundings at 20°C must the refrigerator have in its heat exchanger with an area of 1 m^2 and an average heat transfer coefficient of 15 W/m^2 K to reject the leaks of energy?

4.102 The brake shoe and steel drum of a car continuously absorb 25 W as the car slows down. Assume a total outside surface area of 0.1 m^2 with a convective heat transfer coefficient of 10 W/m^2 K to the air at 20°C. How hot does the outside brake and drum surface become when steady conditions are reached?

4.103 A wall surface on a house is 30°C with an emissivity of $\varepsilon = 0.7$. The surrounding ambient air is at 15°C with an average emissivity of 0.9. Find the rate of radiation energy from each of those surfaces per unit area.

4.104 A log of burning wood in the fireplace has a surface temperature of 450°C. Assume the emissivity is 1 (perfect black body) and find the radiant emission of energy per unit surface area.

4.105 A radiant heat lamp is a rod, 0.5 m long and 0.5 cm in diameter, through which 400 W of electric energy is deposited. Assume the surface has an emissivity of 0.9 and neglect incoming radiation. What will the rod surface temperature be?

Review Problems

4.106 A vertical cylinder (Fig. P4.106) has a 61.18 kg piston locked with a pin, trapping 10 L of R-22 at 10°C with 90% quality inside. Atmospheric pressure is 100 kPa, and the cylinder cross-sectional area is 0.006 m^2. The pin is removed, allowing the piston to move and come to rest with a final temperature of 10°C for the R-22. Find the final pressure, final volume, and work done by the R-22.

4.107 A piston/cylinder assembly contains butane, C_4H_{10}, at 300°C and 100 kPa with a volume of 0.02 m^3. The gas is now compressed slowly in an isothermal process to 300 kPa.

  a. Show that it is reasonable to assume that butane behaves as an ideal gas during this process.

  b. Determine the work done by the butane during the process.

4.108 A cylinder fitted with a piston contains propane gas at 100 kPa and 300 K with a volume of 0.2 m^3. The gas is now slowly compressed according to the relation $PV^{1.1}$ = constant to a final temperature of 340 K. Justify the use of the ideal-gas model. Find the final pressure and the work done during the process.

4.109 The gas space above the water in a closed storage tank contains nitrogen at 25°C and 100 kPa. Total tank volume is 4 m^3, and there is 500 kg of water at 25°C. An additional 500 kg of water is now forced into the tank. Assuming constant temperature throughout, find the final pressure of the nitrogen and the work done on the nitrogen in this process.

4.110 Two kilograms of water are contained in a piston/cylinder (Fig. P4.110) with a massless piston loaded with a linear spring and the outside atmosphere. Initially the spring force is zero and $P_1 = P_0 = 100$ kPa with a volume of 0.2 m^3. If the piston just hits the upper stops, the volume is 0.8 m^3 and $T = 600$°C. Heat is now added until the pressure reaches 1.2 MPa. Find the final temperature, show the P–V diagram and find the work done during the process.

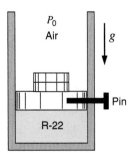

FIGURE P4.106

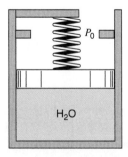

FIGURE P4.110

4.111 A cylinder having an initial volume of 3 m³ contains 0.1 kg of water at 40°C. The water is then compressed in an isothermal quasi-equilibrium process until it has a quality of 50%. Calculate the work done by splitting the process into two steps. Assume the water vapor is an ideal gas during the fist step of the process.

4.112 Air at 200 kPa, 30°C is contained in a cylinder/piston arrangement with initial volume 0.1 m³. The inside pressure balances ambient pressure of 100 kPa plus an externally imposed force that is proportional to $V^{0.5}$. Now heat is transferred to the system to a final pressure of 225 kPa. Find the final temperature and the work done in the process.

4.113 A spring-loaded piston/cylinder arrangement contains R-134a at 20°C, 24% quality with a volume 50 L. The setup is heated and thus expands, moving the piston. It is noted that when the last drop of liquid disappears the temperature is 40°C. The heating is stopped when $T = 130$°C. Verify that the final pressure is about 1200 kPa by iteration and find the work done in the process.

4.114 A piston/cylinder setup (Fig. P4.68) contains 1 kg of water at 20°C with a volume of 0.1 m³. Initially, the piston rests on some stops with the top surface open to the atmosphere, P_0, and a mass such that a water pressure of 400 kPa will lift it.

To what temperature should the water be heated to lift the piston? If it is heated to saturated vapor, find the final temperature, volume, and work, $_1W_2$.

4.115 Two springs with the same spring constant are installed in a massless piston/cylinder arrangement with the outside air at 100 kPa. If the piston is at the bottom, both springs are relaxed, and the second spring comes in contact with the piston at $V = 2$ m³. The cylinder (Fig. P4.115) contains ammonia initially at -2°C, $x = 0.13$, $V = 1$ m³, which is then heated until the pressure finally reaches 1200 kPa. At what pressure will the piston touch the second spring? Find the final temperature and the total work done by the ammonia.

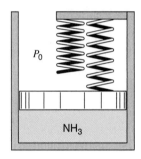

FIGURE P4.115

4.116 Find the work in the process described in Problem 3.101.

ENGLISH UNIT PROBLEMS

English Unit Concept Problems

4.117E The electric company charges the customers per kilowatt-hour. What is that in English units?

4.118E Work as $F \Delta x$ has units of lbf/ft. What is that in Btu?

4.119E Work of 2.5 Btu must be delivered on a rod from a pneumatic piston/cylinder where the air pressure is limited to 75 psia. What diameter cylinder should I have to restrict the rod motion to maximum 2 ft?

4.120E A force of 300 lbf moves a truck at 40 mi/h up a hill. What is the power?

4.121E A 1200-hp dragster engine drives a car with a speed of 65 mi/h. How much force is between the tires and the road?

4.122E A 1200-hp dragster engine has a drive shaft rotating at 2000 RPM. How much torque is on the shaft?

English Unit Problems

4.123E A bulldozer pushes 1000 lbm of dirt 300 ft with a force of 400 lbf. It then lifts the dirt 10 ft up to put it in a dump truck. How much work did it do in each situation?

4.124E A steam radiator in a room at 75 F has saturated water vapor at 16 lbf/in² flowing through it, when the inlet and exit valves are closed. What is the pressure and the quality of the water when it has cooled to 75 F? How much work is done?

4.125E A linear spring, $F = k_s(x - x_0)$, with spring constant $k_s = 35$ lbf/ft, is stretched until it is

2.5 in. longer. Find the required force and work input.

4.126E Two hydraulic cylinders maintain a pressure of 175 psia. One has a cross-sectional area of 0.1 ft², the other one of 0.3 ft². To deliver 1 Btu of work to the piston, how large a displacement (V) and piston motion H are needed for each cylinder? Neglect P_{atm}.

4.127E A piston/cylinder has 15 ft of liquid 70 F water on top of the piston ($m = 0$) with cross-sectional area of 1 ft² (see Fig. P2.57). Air is let in under the piston, which rises and pushes the water out over the top edge. Find the necessary work to push all the water out, and plot the process in a P–V diagram.

4.128E A cylinder fitted with a frictionless piston contains 10 lbm of superheated refrigerant R-134a vapor at 100 lbf/in², 300 F. The setup is cooled at constant pressure until the R-134a reaches a quality of 25%. Calculate the work done in the process.

4.129E The gas space above the water in a closed storage tank contains nitrogen at 80 F, 15 lbf/in². Total tank volume is 150 ft³, and there is 1000 lbm of water at 80 F. An additional 1000 lbm of water is now forced into the tank. Assuming constant temperature throughout, find the final pressure of the nitrogen and the work done on the nitrogen in this process.

4.130E A cylinder having an initial volume of 100 ft³ contains 0.2 lbm of water at 100 F. The water is then compressed in an isothermal quasi-equilibrium process until it has a quality of 50%. Calculate the work done in the process assuming water vapor is an ideal gas.

4.131E Helium gas expands from 20 psia, 600 R, and 9 ft³ to 15 psia in a polytropic process with $n = 1.667$. How much work does it give out?

4.132E Consider a mass going through a polytropic process where pressure is directly proportional to volume ($n = -1$). The process starts with $P = 0$, $V = 0$ and ends with $P = 90$ lbf/in², $V = 0.4$ ft³. The physical setup could be as in Problem 2.83. Find the boundary work done by the mass.

4.133E The piston/cylinder shown in Fig. P4.48 contains carbon dioxide at 50 lbf/in², 200 F with a volume of 5 ft³. Mass is added at such a rate that the gas compresses according to the relation $PV^{1.2} = $ constant to a final temperature of 350 F. Determine the work done during the process.

4.134E Find the specific work for Problem 3.156E.

4.135E Consider a two-part process with an expansion from 3 to 6 ft³ at a constant pressure of 20 lbf/in² followed by an expansion from 6 to 12 ft³ with a linearly rising pressure from 20 lbf/in² ending at 40 lbf/in². Show the process in a P–V diagram and find the boundary work.

4.136E A piston/cylinder has 2 lbm of R-134a at state 1 with 200 F, 90 lbf/in², and is then brought to saturated vapor, state 2, by cooling while the piston is locked with a pin. Now the piston is balanced with an additional constant force and the pin is removed. The cooling continues to state 3, where the R-134a is saturated liquid. Show the processes in a P–V diagram and find the work in each of the two steps, 1 to 2 and 2 to 3.

4.137E A cylinder containing 2 lbm of ammonia has an externally loaded piston. Initially the ammonia is at 280 lbf/in², 360 F. It is now cooled to saturated vapor at 105 F, and then further cooled to 65 F, at which point the quality is 50%. Find the total work for the process, assuming a piecewise linear variation of P versus V.

4.138E A 1-ft-long steel rod with a 0.5-in. diameter is stretched in a tensile test. What is the required work to obtain a relative strain of 0.1%? The modulus of elasticity of steel is 30×10^6 lbf/in².

4.139E An escalator raises a 200-lbm bucket of sand 30 ft in 1 min. Determine the total amount of work done and the instantaneous rate of work during the process.

4.140E A piston/cylinder of diameter 10 in. moves a piston with a velocity of 18 ft/s. The instantaneous pressure if 100 psia. What is the volume displacement rate, the force and the transmitted power?

4.141E The sun shines on a 1500-ft² road surface so it is at 115 F. Below the 2-in.-thick asphalt, average conductivity of 0.035 Btu/h ft F, is a layer of compacted rubble at a temperature of 60 F. Find the rate of heat transfer to the rubble.

4.142E A water heater is covered up with insulation boards over a total surface area of 30 ft². The inside board surface is at 175 F, the outside surface is at 70 F, and the board material has a conductivity of 0.05 Btu/h ft F. How thick should the board be to limit the heat transfer loss to 720 Btu/h?

4.143E The black grille on the back of a refrigerator has a surface temperature of 95 F with a total surface area of 10 ft². Heat transfer to the room air at 70 F takes place with an average convective heat transfer coefficient of 3 Btu/h ft² R. How much energy can be removed during 15 min of operation?

COMPUTER, DESIGN, AND OPEN-ENDED PROBLEMS

4.144 In Problem 4.48, determine the work done by the carbon dioxide at any point during the process.

4.145 In Problem 4.112, determine the work done by the air at any point during the process.

4.146 A piston/cylinder arrangement of initial volume 0.025 m³ contains saturated water vapor at 200°C. The steam now expands in a quasi-equilibrium isothermal process to a final pressure of 200 kPa while it does work against the piston. Determine the work done in this process by a numerical integration (summation) of the area below the *P–V* process curve. Compute about 10 points along the curve by using the computerized software to get the volume at 200°C and the various pressures. How different is the work calculated if ideal gas is assumed?

4.147 Reconsider the process in Problem 4.60 in which three states were specified. Solve the problem by fitting a single smooth curve (*P* versus *v*) through the three points. Map out the path followed (including temperature and quality) during the process.

4.148 Write a computer program to determine the boundary movement work for a specified substance undergoing a process for a given set of data (values of pressure and corresponding volume during the process).

4.149 Ammonia vapor is compressed inside a cylinder by an external force acting on the piston. The ammonia is initially at 30°C, 500 kPa, and the final pressure is 1400 kPa. The following data have been measured for the process:

Pressure, kPa	500	653	802	945	1100	1248	1400
Volume, L	1.25	1.08	0.96	0.84	0.72	0.60	0.50

Determine the work done by the ammonia by summing the area below the *P–V* process curve. As you plot it, *P* is the height and the change in volume is the base of a number of rectangles.

4.150 A substance is brought from a state of P_1, v_1 to a state of P_2, v_2 in a piston/cylinder arrangement. Assume that the process can be approximated as a polytropic process. Write a program that will find the polytropic exponent, *n*, and the boundary work per unit mass. The four state properties are input variables. Check the program with cases that you can easily hand calculate.

4.151 Assume that you have a plate of $A = 1$ m² with thickness $L = 0.02$ m over which there is a temperature difference of 20°C. Find the conductivity, k, from the literature and compare the heat transfer rates if the plate substance is a metal like aluminum or steel, or wood, foam insulation, air, argon, or liquid water. Assume the average substance temperature is 25°C.

4.152 Make a list of household appliances such as refrigerators, electric heaters, vacuum cleaners, hair dryers, TVs, stereo sets, and any others you may think of. For each, list its energy consumption and explain where you have energy transfer as work and where there is heat transfer.

5 THE FIRST LAW OF THERMODYNAMICS

Having completed our consideration of basic definitions and concepts, we are ready to proceed to a discussion of the first law of thermodynamics. This law is often called the conservation of energy law and, as we will see later, this is essentially true. Our procedure will be to state this law for a system (control mass) undergoing a cycle and then for a change of state of a system.

After the energy equation is formulated we will use it to relate change of state inside a control volume to the amount of energy that is transferred in a process as work or heat transfer. When a car engine has transferred some work to the car, the car's speed is increased, so we can relate the kinetic energy increase to the work, or if a stove provides a certain amount of heat transfer to a pot with water we can relate the water temperature increase to the heat transfer. More complicated processes can also occur, such as the expansion of very hot gases in a piston cylinder, as in a car engine, in which work is given out and at the same time heat is transferred to the colder walls. In other applications we can also see a change in the state without any work or heat transfer, such as a falling object that changes kinetic energy at the same time it is changing elevation. The energy equation then relates the two forms of energy of the object.

5.1 THE FIRST LAW OF THERMODYNAMICS FOR A CONTROL MASS UNDERGOING A CYCLE

The first law of thermodynamics states that during any cycle a system (control mass) undergoes, the cyclic integral of the heat is proportional to the cyclic integral of the work.

To illustrate this law, consider as a control mass the gas in the container shown in Fig. 5.1. Let this system go through a cycle that is made up of two processes. In the first process work is done on the system by the paddle that turns as the weight is lowered. Let the system then return to its initial state by transferring heat from the system until the cycle has been completed.

Historically, work was measured in mechanical units of force times distance, such as foot pounds force or joules, and heat was measured in thermal units, such as the British thermal unit or the calorie. Measurements of work and heat were made during a cycle for a wide variety of systems and for various amounts of work and heat. When the amounts of work and heat were compared, it was found that they were always proportional. Such

116

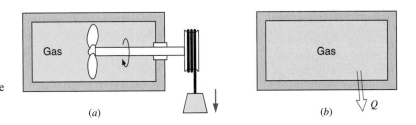

FIGURE 5.1 Example of a control mass undergoing a cycle.

(a) (b)

observations led to the formulation of the first law of thermodynamics, which in equation form is written

$$J \oint \delta Q = \oint \delta W \qquad (5.1)$$

The symbol $\oint \delta Q$, which is called the cyclic integral of the heat transfer, represents the net heat transfer during the cycle, and $\oint \delta W$, the cyclic integral of the work, represents the net work during the cycle. Here, J is a proportionality factor that depends on the units used for work and heat.

The basis of every law of nature is experimental evidence, and this is true also of the first law of thermodynamics. Many different experiments have been conducted on the first law, and every one thus far has verified it either directly or indirectly. The first law has never been disproved.

As was discussed in Chapter 4, the units for work and heat or for any other form of energy either are the same or are directly proportional. In SI units, the joule is used as the unit for both work and heat and for any other energy unit. In English units, the basic unit for work is the foot pound force, and the basic unit for heat is the British thermal unit (Btu). James P. Joule (1818–1889) did the first accurate work in the 1840s on measurement of the proportionality factor J, which relates these units. Today, the Btu is defined in terms of the basic SI metric units,

1 Btu = 778.17 ft lbf

This unit is termed the International British thermal unit. For much engineering work, the accuracy of other data does not warrant more accuracy than the relation 1 Btu = 778 ft lbf, which is the value used with English units in the problems in this text. Because these units are equivalent, it is not necessary to include the factor J explicitly in Eq. 5.1, but simply to recognize that for any system of units, each equation must have consistent units throughout. Therefore, we may write Eq. 5.1 as

$$\oint \delta Q = \oint \delta W \qquad (5.2)$$

which can be considered the basic statement of the first law of thermodynamics.

5.2 THE FIRST LAW OF THERMODYNAMICS FOR A CHANGE IN STATE OF A CONTROL MASS

Equation 5.2 states the first law of thermodynamics for a control mass during a cycle. Many times, however, we are concerned with a process rather than a cycle. We now consider the first law of thermodynamics for a control mass that undergoes a change of state.

We begin by introducing a new property, the energy, which is given the symbol E. Consider a system that undergoes a cycle in which it changes from state 1 to state 2 by process A and returns from state 2 to state 1 by process B. This cycle is shown in Fig. 5.2 on a pressure (or other intensive property)–volume (or other extensive property) diagram. From the first law of thermodynamics, Eq. 5.2, we have

$$\oint \delta Q = \oint \delta W$$

Considering the two separate processes, we have

$$\int_1^2 \delta Q_A + \int_2^1 \delta Q_B = \int_1^2 \delta W_A + \int_2^1 \delta W_B$$

Now consider another cycle in which the control mass changes from state 1 to state 2 by process C and returns to state 1 by process B, as before. For this cycle we can write

$$\int_1^2 \delta Q_C + \int_2^1 \delta Q_B = \int_1^2 \delta W_C + \int_2^1 \delta W_B$$

Subtracting the second of these equations from the first, we obtain

$$\int_1^2 \delta Q_A - \int_1^2 \delta Q_C = \int_1^2 \delta W_A - \int_1^2 \delta W_C$$

or, by rearranging,

$$\int_1^2 (\delta Q - \delta W)_A = \int_1^2 (\delta Q - \delta W)_C \qquad (5.3)$$

Since A and C represent arbitrary processes between states 1 and 2, the quantity $\delta Q - \delta W$ is the same for all processes between states 1 and 2. Therefore, $\delta Q - \delta W$ depends only on the initial and final states and not on the path followed between the two states. We conclude that this is a point function, and therefore it is the differential of a property of the mass. This property is the energy of the mass and is given the symbol E. Thus we can write

$$dE = \delta Q - \delta W \qquad (5.4)$$

Because E is a property, its derivative is written dE. When Eq. 5.4 is integrated from an initial state 1 to a final state 2, we have

$$E_2 - E_1 = {}_1Q_2 - {}_1W_2 \qquad (5.5)$$

where E_1 and E_2 are the initial and final values of the energy E of the control mass, ${}_1Q_2$ is the heat transferred to the control mass during the process from state 1 to state 2, and ${}_1W_2$ is the work done by the control mass during the process.

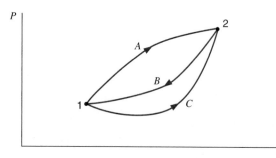

FIGURE 5.2

Demonstration of the existence of thermodynamic property E.

Note that a control mass may be made up of several different subsystems, as shown in Fig. 5.3. In this case, each part must be analyzed and included separately in applying the first law, Eq. 5.5. We further note that Eq. 5.5 is an expression of the general form

$$\Delta \text{ Energy} = + \text{ in} - \text{out}$$

in terms of the standard sign conventions for heat and work.

The physical significance of the property E is that it represents all the energy of the system in the given state. This energy might be present in a variety of forms, such as the kinetic or potential energy of the system as a whole with respect to the chosen coordinate frame, energy associated with the motion and position of the molecules, energy associated with the structure of the atom, chemical energy present in a storage battery, energy present in a charged condenser, or any of a number of other forms.

In the study of thermodynamics, it is convenient to consider the bulk kinetic and potential energy separately and then to consider all the other energy of the control mass in a single property that we call the internal energy and to which we give the symbol U. Thus, we would write

$$E = \text{Internal energy} + \text{kinetic energy} + \text{potential energy}$$

or

$$E = U + \text{KE} + \text{PE}$$

The kinetic and potential energy of the control mass are associated with the coordinate frame that we select and can be specified by the macroscopic parameters of mass, velocity, and elevation. The internal energy U includes all other forms of energy of the control mass and is associated with the thermodynamic state of the system.

Since the terms comprising E are point functions, we can write

$$dE = dU + d(\text{KE}) + d(\text{PE}) \tag{5.6}$$

The first law of thermodynamics for a change of state may therefore be written

$$dE = dU + d(\text{KE}) + d(\text{PE}) = \delta Q - \delta W \tag{5.7}$$

In words this equation states that as a control mass undergoes a change of state, energy may cross the boundary as either heat or work, and each may be positive or negative. The net change in the energy of the system will be exactly equal to the net energy that

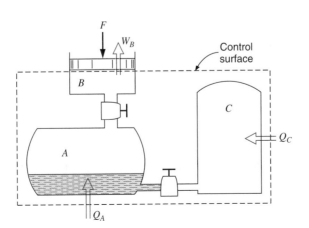

FIGURE 5.3 A control mass with several different subsystems.

crosses the boundary of the system. The energy of the system may change in any of three ways—by a change in internal energy, in kinetic energy, or in potential energy.

This section concludes by deriving an expression for the kinetic and potential energy of a control mass. Consider a mass that is initially at rest relative to the earth, which is taken as the coordinate frame. Let this system be acted on by an external horizontal force F that moves the mass a distance dx in the direction of the force. Thus, there is no change in potential energy. Let there be no heat transfer and no change in internal energy. Then from the first law, Eq. 5.7, we have

$$\delta W = -F\,dx = -d\mathrm{KE}$$

But

$$F = ma = m\frac{d\mathbf{V}}{dt} = m\frac{dx}{dt}\frac{d\mathbf{V}}{dx} = m\mathbf{V}\frac{d\mathbf{V}}{dx}$$

Then

$$d\mathrm{KE} = F\,dx = m\mathbf{V}\,d\mathbf{V}$$

Integrating, we obtain

$$\int_{\mathrm{KE}=0}^{\mathrm{KE}} d\mathrm{KE} = \int_{\mathbf{V}=0}^{\mathbf{V}} m\mathbf{V}\,d\mathbf{V}$$

$$\mathrm{KE} = \frac{1}{2}m\mathbf{V}^2 \tag{5.8}$$

A similar expression for potential energy can be found. Consider a control mass that is initially at rest and at the elevation of some reference level. Let this mass be acted on by a vertical force F of such magnitude that it raises (in elevation) the mass with constant velocity an amount dZ. Let the acceleration due to gravity at this point be g. From the first law, Eq. 5.7, we have

$$\delta W = -F\,dZ = -d\,\mathrm{PE}$$
$$F = ma = mg$$

Then

$$d\,\mathrm{PE} = F\,dZ = mg\,dZ$$

Integrating gives

$$\int_{\mathrm{PE}_1}^{\mathrm{PE}_2} d\,\mathrm{PE} = m\int_{Z_1}^{Z_2} g\,dZ$$

Assuming that g does not vary with Z (which is a very reasonable assumption for moderate changes in elevation), we obtain

$$\mathrm{PE}_2 - \mathrm{PE}_1 = mg(Z_2 - Z_1) \tag{5.9}$$

EXAMPLE 5.1 A car of mass 1100 kg drives with a velocity such that it has a kinetic energy of 400 kJ (see Fig. 5.4). Find the velocity. If the car is raised with a crane how high should it be lifted in the standard gravitational field to have a potential energy that equals the kinetic energy?

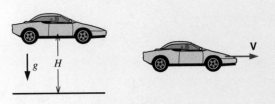

FIGURE 5.4 Sketch for Example 5.1.

Solution

The standard kinetic energy of the mass is

$$\text{KE} = \frac{1}{2} m \mathbf{V}^2 = 400 \text{ kJ}$$

From this we can solve for the velocity

$$\mathbf{V} = \sqrt{\frac{2 \text{ KE}}{m}} = \sqrt{\frac{2 \times 400 \text{ kJ}}{1100 \text{ kg}}}$$

$$= \sqrt{\frac{800 \times 1000 \text{ N m}}{1100 \text{ kg}}} = \sqrt{\frac{8000 \text{ kg m s}^{-2} \text{ m}}{11 \text{ kg}}} = 27 \text{ m/s}$$

Standard potential energy is

$$\text{PE} = mgH$$

so when this is equal to the kinetic energy we get

$$H = \frac{\text{KE}}{mg} = \frac{400\,000 \text{ N m}}{1100 \text{ kg} \times 9.807 \text{ m s}^{-2}} = 37.1 \text{ m}$$

Notice the necessity of converting the kJ to J in both calculations.

EXAMPLE 5.1E   A car of mass 2400 lbm drives with a velocity such that it has a kinetic energy of 400 Btu. Find the velocity. If the car is raised with a crane, how high should it be lifted in the standard gravitational field to have a potential energy that equals the kinetic energy?

Solution

The standard kinetic energy of the mass is

$$\text{KE} = \frac{1}{2} m \mathbf{V}^2 = 400 \text{ Btu}$$

From this we can solve for the velocity

$$\mathbf{V} = \sqrt{\frac{2 \text{ KE}}{m}} = \sqrt{\frac{2 \times 400 \text{ Btu} \times 778.17 \frac{\text{ft lbf}}{\text{Btu}} \times 32.174 \frac{\text{lbm ft}}{\text{lbf s}^2}}{2400 \text{ lbm}}}$$

$$= 91.4 \text{ ft/s}$$

Standard potential energy is

$$PE = mgH$$

so when this is equal to the kinetic energy KE we get

$$H = \frac{KE}{mg} = \frac{400 \text{ Btu} \times 778.17 \frac{\text{ft lbf}}{\text{Btu}} \times 32.174 \frac{\text{lbm ft}}{\text{lbf s}^2}}{2400 \text{ lbm} \times 32.174 \frac{\text{ft}}{\text{s}^2}}$$

$$= 129.7 \text{ ft}$$

Note the necessity of using the conversion constant $32.174 \frac{\text{lbm ft}}{\text{lbf s}^2}$ in both calculations.

Now, substituting the expressions for kinetic and potential energy into Eq. 5.6, we have

$$dE = dU + m\mathbf{V} \, d\mathbf{V} + mg \, dZ$$

Integrating for a change of state from state 1 to state 2 with constant g, we get

$$E_2 - E_1 = U_2 - U_1 + \frac{m\mathbf{V}_2^2}{2} - \frac{m\mathbf{V}_1^2}{2} + mgZ_2 - mgZ_1$$

Similarly, substituting these expressions for kinetic and potential energy into Eq. 5.7, we have

$$dE = dU + \frac{d(m\mathbf{V}^2)}{2} + d(mgZ) = \delta Q - \delta W \tag{5.10}$$

Assuming g is a constant, in the integrated form of this equation,

$$U_2 - U_1 + \frac{m(\mathbf{V}_2^2 - \mathbf{V}_1^2)}{2} + mg(Z_2 - Z_1) = {}_1Q_2 - {}_1W_2 \tag{5.11}$$

Three observations should be made regarding this equation. The first observation is that the property E, the energy of the control mass, was found to exist, and we were able to write the first law for a change of state using Eq. 5.5. However, rather than deal with this property E, we find it more convenient to consider the internal energy and the kinetic and potential energies of the mass. In general, this procedure will be followed in the rest of this book.

The second observation is that Eqs. 5.10 and 5.11 are in effect a statement of the conservation of energy. The net change of the energy of the control mass is always equal to the net transfer of energy across the boundary as heat and work. This is somewhat analogous to a joint checking account shared by a husband and wife. There are two ways in which deposits and withdrawals can be made—either by the husband or by the wife—and the balance will always reflect the net amount of the transaction. Similarly, there are two ways in which energy can cross the boundary of a control mass—either as heat or as work—and the energy of the mass will change by the exact amount of the net energy crossing the boundary. The concept of energy and the law of the conservation of energy are basic to thermodynamics.

The third observation is that Eqs. 5.10 and 5.11 can give only changes in internal energy, kinetic energy, and potential energy. We can learn nothing about absolute values of these quantities from these equations. If we wish to assign values to internal energy, kinetic energy, and potential energy, we must assume reference states and assign a value to the quantity in this reference state. The kinetic energy of a body with zero velocity relative to the earth is assumed to be zero. Similarly, the value of the potential energy is assumed to be zero when the body is at some reference elevation. With internal energy, therefore, we must also have a reference state if we wish to assign values of this property. This matter is considered in the following section.

EXAMPLE 5.2   A tank containing a fluid is stirred by a paddle wheel. The work input to the paddle wheel is 5090 kJ. The heat transfer from the tank is 1500 kJ. Consider the tank and the fluid inside a control surface and determine the change in internal energy of this control mass.

The first law of thermodynamics is (Eq. 5.11)

$$U_2 - U_1 + \frac{1}{2} m(\mathbf{V}_2^2 - \mathbf{V}_1^2) + mg(Z_2 - Z_1) = {}_1Q_2 - {}_1W_2$$

Since there is no change in kinetic and potential energy, this reduces to

$$U_2 - U_1 = {}_1Q_2 - {}_1W_2$$
$$U_2 - U_1 = -1500 - (-5090) = 3590 \text{ kJ}$$

EXAMPLE 5.3   Consider a stone having a mass of 10 kg and a bucket containing 100 kg of liquid water. Initially the stone is 10.2 m above the water, and the stone and the water are at the same temperature, state 1. The stone then falls into the water.

Determine ΔU, ΔKE, ΔPE, Q, and W for the following changes of state, assuming standard gravitational acceleration of 9.806 65 m/s^2.

a. The stone is about to enter the water, state 2.

b. The stone has just come to rest in the bucket, state 3.

c. Heat has been transferred to the surroundings in such an amount that the stone and water are at the same temperature, T_1, state 4.

Analysis and Solution

The first law for any of the steps is

$$Q = \Delta U + \Delta KE + \Delta PE + W$$

and each term can be identified for each of the changes of state.

a. The stone has fallen from Z_1 to Z_2, and we assume no heat transfer as it falls. The water has not changed state; thus

$$\Delta U = 0, \qquad {}_1Q_2 = 0, \qquad {}_1W_2 = 0$$

and the first law reduces to

$$\Delta KE + \Delta PE = 0$$

$$\Delta KE = -\Delta PE = -mg(Z_2 - Z_1)$$

$$= -10 \text{ kg} \times 9.806\ 65 \text{ m/s}^2 \times (-10.2 \text{ m})$$

$$= 1000 \text{ J} = 1 \text{ kJ}$$

That is, for the process from state 1 to state 2,

$$\Delta KE = 1 \text{ kJ} \quad \text{and} \quad \Delta PE = -1 \text{ kJ}$$

b. For the process from state 2 to state 3 with zero kinetic energy, we have

$$\Delta PE = 0, \qquad {}_2Q_3 = 0, \qquad {}_2W_3 = 0$$

Then

$$\Delta U + \Delta KE = 0$$

$$\Delta U = -\Delta KE = 1 \text{ kJ}$$

c. In the final state, there is no kinetic, nor potential energy, and the internal energy is the same as in state 1.

$$\Delta U = -1 \text{ kJ}, \qquad \Delta KE = 0, \qquad \Delta PE = 0, \qquad {}_3W_4 = 0$$

$$_3Q_4 = \Delta U = -1 \text{ kJ}$$

5.3 INTERNAL ENERGY—A THERMODYNAMIC PROPERTY

Internal energy is an extensive property because it depends on the mass of the system. Similarly, kinetic and potential energies are extensive properties.

The symbol U designates the internal energy of a given mass of a substance. Following the convention used with other extensive properties, the symbol u designates the internal energy per unit mass. We could speak of u as the specific internal energy, as we do with specific volume. However, because the context will usually make it clear whether u or U is referred to, we will simply use the term internal energy to refer to both internal energy per unit mass and the total internal energy.

In Chapter 3 we noted that in the absence of motion, gravity, surface effects, electricity, or other effects, the state of a pure substance is specified by two independent properties. It is very significant that, with these restrictions, the internal energy may be one of the independent properties of a pure substance. This means, for example, that if we specify the pressure and internal energy (with reference to an arbitrary base) of superheated steam, the temperature is also specified.

Thus, in a table of thermodynamic properties such as the steam tables, the value of internal energy can be tabulated along with other thermodynamic properties. Tables 1 and 2 of the steam tables (Tables B.1.1 and B.1.2) list the internal energy for saturated states. Included are the internal energy of saturated liquid u_f, the internal energy of saturated vapor u_g, and the difference between the internal energy of saturated liquid and saturated vapor u_{fg}. The values are given in relation to an arbitrarily assumed reference state, which, for water in the steam tables, is taken as zero for saturated liquid at the triple-point temperature, 0.01°C. All values of internal energy in the steam tables are then calculated relative to this reference (note that the reference state cancels out when finding a difference in u between any two states). Values for internal energy are found in the steam tables in the same manner as for specific volume. In the liquid–vapor saturation region,

$$U = U_{\text{liq}} + U_{\text{vap}}$$

or

$$mu = m_{\text{liq}}u_f + m_{\text{vap}}u_g$$

Dividing by m and introducing the quality x gives

$$u = (1 - x)u_f + xu_g$$

$$u = u_f + xu_{fg}$$

As an example, the specific internal energy of saturated seam having a pressure of 0.6 MPa and a quality of 95% can be calculated as

$$u = u_f + xu_{fg} = 669.9 + 0.95(1897.5) = 2472.5 \text{ kJ/kg}$$

Values for u in the superheated vapor region are tabulated in Table B.1.3, for compressed liquid in Table B.1.4, and for solid–vapor in Table B.1.5.

EXAMPLE 5.4 Determine the missing property $(P, T \text{ or } x)$ and also v for water at each of the following states:

a. $T = 300°C$, $u = 2780$ kJ/kg
b. $P = 2000$ kPa, $u = 2000$ kJ/kg

For each case, the two properties given are independent properties and therefore fix the state. For each, we must first determine the phase by comparison of the given information with phase boundary values.

a. At 300°C, from Table B.1.1, $u_g = 2563.0$ kJ/kg. The given $u > u_g$, so the state is in the superheated vapor region at some P less than P_g, which is 8581 kPa. Searching through Table B.1.3 at 300°C, we find that the value $u = 2780$ is between given values of u at 1600 kPa (2781.0) and 1800 kPa (2776.8). Interpolating linearly, we obtain

$$P = 1648 \text{ kPa.}$$

Note that quality is undefined in the superheated vapor region. At this pressure, by linear interpolation, we have $v = 0.1542$ m³/kg.

b. At $P = 2000$ kPa, from Table B.1.2, the given u of 2000 kJ/kg is greater than u_f (906.4) but less than u_g (2600.3). Therefore, this state is in the two-phase region with $T = T_g = 212.4°C$, and

$$u = 2000 = 906.4 + x1693.8, \qquad x = 0.6456$$

Then,

$$v = 0.001\ 177 + 0.6456 \times 0.098\ 45 = 0.064\ 74 \text{ m}^3/\text{kg}.$$

5.4 PROBLEM ANALYSIS AND SOLUTION TECHNIQUE

At this point in our study of thermodynamics, we have progressed sufficiently far (that is, we have accumulated sufficient tools with which to work) that it is worthwhile to develop a somewhat formal technique or procedure for analyzing and solving thermodynamic problems. For the time being it may not seem entirely necessary to use such a rigorous procedure for many of our problems, but we should keep in mind that as we acquire more analytical tools the problems that we are capable of dealing with will become much more complicated. Thus, it is appropriate that we begin to practice this technique now in anticipation of these future problems.

Our problem analysis and solution technique is contained within the framework of the following set of questions that must be answered in the process of an orderly solution of a thermodynamic problem.

1. What is the control mass or control volume? Is it useful, or necessary, to choose more than one? It may be helpful to draw a sketch of the system at this point, illustrating all heat and work flows, and indicating forces such as external pressures and gravitation.
2. What do we know about the initial state (i.e., which properties are known)?
3. What do we know about the final state?
4. What do we know about the process that takes place? Is anything constant or zero? Is there some known functional relation between two properties?
5. Is it helpful to draw a diagram of the information in steps 2 to 4 (for example, a T–v or P–v diagram)?
6. What is our thermodynamic model for the behavior of the substance (for example, steam tables, ideal gas, and so on)?
7. What is our analysis of the problem (i.e., do we examine control surfaces for various work modes or use the first law or conservation of mass)?
8. What is our solution technique? In other words, from what we have done so far in steps 1–7, how do we proceed to find whatever it is that is desired? Is a trial-and-error solution necessary?

It is not always necessary to write out all these steps, and in the majority of the examples throughout this text we will not do so. However, when faced with a new and unfamiliar problem, the student should always at least think through this set of questions to develop the ability to solve more challenging problems. In solving the following example, we will use this technique in detail.

EXAMPLE 5.5   A vessel having a volume of 5 m³ contains 0.05 m³ of saturated liquid water and 4.95 m³ of saturated water vapor at 0.1 MPa. Heat is transferred until the vessel is filled with saturated vapor. Determine the heat transfer for this process.

Control mass: All the water inside the vessel.
Sketch: Fig. 5.5.
Initial state: Pressure, volume of liquid, volume of vapor; therefore, state 1 is fixed.
Final state: Somewhere along the saturated-vapor curve; the water was heated, so $P_2 > P_1$.
Process: Constant volume and mass; therefore, constant specific volume.
Diagram: Fig. 5.6.
Model: Steam tables.

Analysis

From the first law we have

$$_1Q_2 = U_2 - U_1 + m\,\frac{\mathbf{V}_2^2 - \mathbf{V}_1^2}{2} + mg(Z_2 - Z_1) + {}_1W_2$$

From examining the control surface for various work modes, we conclude that the work for this process is zero. Furthermore, the system is not moving, so there is no change in kinetic energy. There is a small change in the center of mass of the system but we will assume that the corresponding change in potential energy is negligible (in kilojoules). Therefore,

$$_1Q_2 = U_2 - U_1$$

Solution

The heat transfer will be found from the first law. State 1 is known, so U_1 can be calculated. The specific volume at state 2 is also known (from state 1 and the process). Since

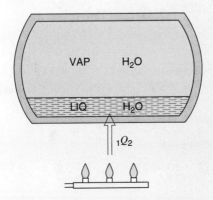

FIGURE 5.5 Sketch for Example 5.5.

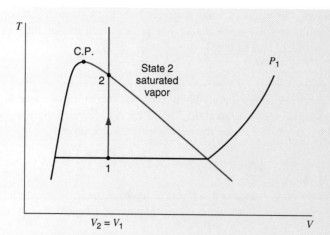

FIGURE 5.6 Diagram for Example 5.5.

state 2 is saturated vapor, state 2 is fixed, as is seen from Fig. 5.6. Therefore, U_2 can also be found.

The solution proceeds as follows:

$$m_{1\,\text{liq}} = \frac{V_{\text{liq}}}{v_f} = \frac{0.05}{0.001\,043} = 47.94 \text{ kg}$$

$$m_{1\,\text{vap}} = \frac{V_{\text{vap}}}{v_g} = \frac{4.95}{1.6940} = 2.92 \text{ kg}$$

Then

$$U_1 = m_{1\,\text{liq}}u_{1\,\text{liq}} + m_{1\,\text{vap}}u_{1\,\text{vap}}$$

$$= 47.94(417.36) + 2.92(2506.1) = 27\,326 \text{ kJ}$$

To determine u_2 we need to know two thermodynamic properties, since this determines the final state. The properties we know are the quality, $x = 100\%$, and v_2, the final specific volume, which can readily be determined.

$$m = m_{1\,\text{liq}} + m_{1\,\text{vap}} = 47.94 + 2.92 = 50.86 \text{ kg}$$

$$v_2 = \frac{V}{m} = \frac{5.0}{50.86} = 0.098\,31 \text{ m}^3/\text{kg}$$

In Table B.1.2 we find, by interpolation, that at a pressure of 2.03 MPa, $v_g = 0.098\,31$ m^3/kg. The final pressure of the steam is therefore 2.03 MPa. Then

$$u_2 = 2600.5 \text{ kJ/kg}$$

$$U_2 = mu_2 = 50.86(2600.5) = 132\,261 \text{ kJ}$$

$${}_1Q_2 = U_2 - U_1 = 132\,261 - 27\,326 = 104\,935 \text{ kJ}$$

EXAMPLE 5.5E

A vessel having a volume of 100 ft³ contains 1 ft³ of saturated liquid water and 99 ft³ of saturated water vapor at 14.7 lbf/in². Heat is transferred until the vessel is filled with saturated vapor. Determine the heat transfer for this process.

Control mass: All the water inside the vessel.

Sketch: Fig. 5.5.

Initial state: Pressure, volume of liquid, volume of vapor; therefore, state 1 is fixed.

Final state: Somewhere along the saturated-vapor curve; the water was heated, so $P_2 > P_1$.

Process: Constant volume and mass; therefore, constant specific volume.

Diagram: Fig. 5.6.

Model: Steam tables.

Analysis

First law:
$$_1Q_2 = U_2 - U_1 + m\frac{(\mathbf{V}_2^2 - \mathbf{V}_1^2)}{2} + mg(Z_2 - Z_1) + {_1W_2}$$

From examining the control surface for various work modes, we conclude that the work for this process is zero. Furthermore, the system is not moving, so there is no change in kinetic energy. There is a small change in center of mass of the system, but we will assume that the corresponding change in potential energy is negligible (compared to other terms). Therefore,

$$_1Q_2 = U_2 - U_1$$

Solution

The heat transfer will be found from the first law. State 1 is known, so U_1 can be calculated. Also, the specific volume at state 2 is known (from state 1 and the process). Since state 2 is saturated vapor, state 2 is fixed, as is seen from Fig. 5.6. Therefore, U_2 can also be found.

The solution proceeds as follows:

$$m_{1\,\text{liq}} = \frac{V_{\text{liq}}}{v_f} = \frac{1}{0.016\,72} = 59.81 \text{ lbm}$$

$$m_{1\,\text{vap}} = \frac{V_{\text{vap}}}{v_g} = \frac{99}{26.80} = 3.69 \text{ lbm}$$

Then,

$$U_1 = m_{1\,\text{liq}}u_{1\,\text{liq}} + m_{1\,\text{vap}}u_{1\,\text{vap}}$$
$$= 59.81(180.1) + 3.69(1077.6) = 14\,748 \text{ Btu}$$

To determine u_2 we need to know two thermodynamic properties, since this determines the final state. The properties we know are the quality, $x = 100\%$, and v_2, the final specific volume, which can readily be determined.

$$m = m_{1\,liq} + m_{1\,vap} = 59.81 + 3.69 = 63.50 \text{ lbm}$$

$$v_2 = \frac{V}{m} = \frac{100}{63.50} = 1.575 \text{ ft}^3/\text{lbm}$$

In Table F7.1 of the steam tables we find, by interpolation, that at a pressure of 294 lbf/in², $v_g = 1.575$ ft³/lbm. The final pressure of the steam is therefore 294 lbf/in². Then,

$$u_2 = 1117.0 \text{ Btu/lbm}$$

$$U_2 = mu_2 = 63.50(1117.0) = 70\,930 \text{ Btu}$$

$${}_1Q_2 = U_2 - U_1 = 70\,930 - 14\,748 = 56\,182 \text{ Btu}$$

5.5 THE THERMODYNAMIC PROPERTY ENTHALPY

In analyzing specific types of processes, we frequently encounter certain combinations of thermodynamic properties, which are therefore also properties of the substance undergoing the change of state. To demonstrate one such situation, let us consider a control mass undergoing a quasi-equilibrium constant-pressure process, as shown in Fig. 5.7. Assume that there are no changes in kinetic or potential energy and that the only work done during the process is that associated with the boundary movement. Taking the gas as our control mass and applying the first law, Eq. 5.11, we have, in terms of Q,

$${}_1Q_2 = U_2 - U_1 + {}_1W_2$$

The work done can be calculated from the relation

$${}_1W_2 = \int_1^2 P\,dV$$

Since the pressure is constant,

$${}_1W_2 = P\int_1^2 dV = P(V_2 - V_1)$$

Therefore,

$${}_1Q_2 = U_2 - U_1 + P_2V_2 - P_1V_1$$
$$= (U_2 + P_2V_2) - (U_1 + P_1V_1)$$

We find that, in this very restricted case, the heat transfer during the process is given in terms of the change in the quantity $U + PV$ between the initial and final states. Because all these quantities are thermodynamic properties, that is, functions only of the state of the system, their combination must also have these same characteristics. Therefore, we find it convenient to define a new extensive property, the enthalpy,

$$H \equiv U + PV \tag{5.12}$$

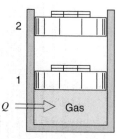

FIGURE 5.7 The constant-pressure quasi-equilibrium process.

or, per unit mass,

$$h \equiv u + Pv \tag{5.13}$$

As for internal energy, we could speak of specific enthalpy, h, and total enthalpy, H. However, we will refer to both as enthalpy, since the context will make it clear which is being discussed.

The heat transfer in a constant-pressure quasi-equilibrium process is equal to the change in enthalpy, which includes both the change in internal energy and the work for this particular process. This is by no means a general result. It is valid for this special case only because the work done during the process is equal to the difference in the PV product for the final and initial states. This would not be true if the pressure had not remained constant during the process.

The significance and use of enthalpy is not restricted to the special process just described. Other cases in which this same combination of properties $u + Pv$ appear will be developed later, notably in Chapter 6 in which we discuss control volume analyses. Our reason for introducing enthalpy at this time is that although the tables in Appendix B list values for internal energy, many other tables and charts of thermodynamic properties give values for enthalpy but not for the internal energy. Therefore, it is necessary to calculate the internal energy at a state using the tabulated values and Eq. 5.13:

$$u = h - Pv$$

Students often become confused about the validity of this calculation when analyzing system processes that do not occur at constant pressure, for which enthalpy has no physical significance. We must keep in mind that enthalpy, being a property, is a state or point function, and its use in calculating internal energy at the same state is not related to, or dependent on, any process that may be taking place.

Tabular values of internal energy and enthalpy, such as those included in Tables B.1 through B.7, are all relative to some arbitrarily selected base. In the steam tables, the internal energy of saturated liquid at 0.01°C is the reference state and is given a value of zero. For refrigerants, such as ammonia and chlorofluorocarbons R-12 and R-22, the reference state is arbitrarily taken as saturated liquid at $-40°C$. The enthalpy in this reference state is assigned the value of zero. Cryogenic fluids, such as nitrogen, have other arbitrary reference states chosen for enthalpy values listed in their tables. Because each of these reference states is arbitrarily selected, it is always possible to have negative values for enthalpy, as for saturated-solid water in Table B.1.5. When enthalpy and internal energy are given values relative to the same reference state, as they are in essentially all thermodynamic tables, the difference between internal energy and enthalpy at the reference state is equal to Pv. Since the specific volume of the liquid is very small, this product is negligible as far as the significant figures of the tables are concerned, but the principle should be kept in mind, for in certain cases it is significant.

In many thermodynamic tables, values of the specific internal energy u are not given. As mentioned earlier, these values can be readily calculated from the relation $u = h - Pv$, though it is important to keep the units in mind. As an example, let us calculate the internal energy u of superheated R-134a at 0.4 MPa, 70°C.

$$u = h - Pv$$

$$= 460.545 - 400 \times 0.066\,484$$

$$= 433.951 \text{ kJ/kg}$$

The enthalpy of a substance in a saturation state and with a given quality is found in the same way as the specific volume and internal energy. The enthalpy of saturated liquid has the symbol h_f, saturated vapor h_g, and the increase in enthalpy during vaporization h_{fg}. For a saturation state, the enthalpy can be calculated by one of the following relations:

$$h = (1 - x)h_f + xh_g$$

$$h = h_f + xh_{fg}$$

The enthalpy of compressed liquid water may be found from Table B.1.4. For substances for which compressed-liquid tables are not available, the enthalpy is taken as that of saturated liquid at the same temperature.

EXAMPLE 5.6   A cylinder fitted with a piston has a volume of 0.1 m³ and contains 0.5 kg of steam at 0.4 MPa. Heat is transferred to the steam until the temperature is 300°C, while the pressure remains constant.

Determine the heat transfer and the work for this process.

Control mass:	Water inside cylinder.
Initial state:	P_1, V_1, m; therefore v_1, is known, state 1 is fixed (at P_1, v_1, check steam tables—two-phase region).
Final state:	P_2, T_2; therefore state 2 is fixed (superheated).
Process:	Constant pressure.
Diagram:	Fig. 5.8.
Model:	Steam tables.

Analysis

There is no change in kinetic energy or change in potential energy. Work is done by movement at the boundary. Assume the process to be quasi-equilibrium. Since the pressure is constant, we have

$$_1W_2 = \int_1^2 P\,dV = P\int_1^2 dV = P(V_2 - V_1) = m(P_2v_2 - P_1v_1)$$

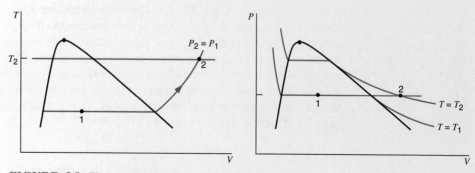

FIGURE 5.8 The constant-pressure quasi-equilibrium process.

Therefore, the first law is, in terms of Q,

$$_1Q_2 = m(u_2 - u_1) + {_1W_2}$$
$$= m(u_2 - u_1) + m(P_2v_2 - P_1v_1) = m(h_2 - h_1)$$

Solution

There is a choice of procedures to follow. State 1 is known, so v_1 and h_1 (or u_1) can be found. State 2 is also known, so v_2 and h_2 (or u_2) can be found. Using the first law and the work equation, we can calculate the heat transfer and work. Using the enthalpies, we have

$$v_1 = \frac{V_1}{m} = \frac{0.1}{0.5} = 0.2 = 0.001\,084 + x_1 0.4614$$

$$x_1 = \frac{0.1989}{0.4614} = 0.4311$$

$$h_1 = h_f + x_1 h_{fg}$$
$$= 604.74 + 0.4311 \times 2133.8 = 1524.7 \text{ kJ/kg}$$

$$h_2 = 3066.8 \text{ kJ/kg}$$

$$_1Q_2 = 0.5(3066.8 - 1524.7) = 771.1 \text{ kJ}$$

$$_1W_2 = mP(v_2 - v_1) = 0.5 \times 400(0.6548 - 0.2) = 91.0 \text{ kJ}$$

Therefore,

$$U_2 - U_1 = {_1Q_2} - {_1W_2} = 771.1 - 91.0 = 680.1 \text{ kJ}$$

The heat transfer could also have been found from u_1 and u_2:

$$u_1 = u_f + x_1 u_{fg}$$
$$= 604.31 + 0.4311 \times 1949.3 = 1444.7 \text{ kJ/kg}$$

$$u_2 = 2804.8 \text{ kJ/kg}$$

and

$$_1Q_2 = U_2 - U_1 + {_1W_2}$$
$$= 0.5(2804.8 - 1444.7) + 91.0 = 771.1 \text{ kJ}$$

5.6 THE CONSTANT-VOLUME AND CONSTANT-PRESSURE SPECIFIC HEATS

In this section we will consider a homogeneous phase of a substance of constant composition. This phase may be a solid, a liquid, or a gas, but no change of phase will occur. We will then define a variable termed the specific heat, the amount of heat required per unit mass to raise the temperature by one degree. Since it would be of interest to examine the relation between the specific heat and other thermodynamic variables, we note first that the heat transfer is given by Eq. 5.10. Neglecting changes in kinetic and potential energies, and assuming a simple compressible substance and quasi-equilibrium process, for which the work in Eq. 5.10 is given by Eq. 4.2, we have

$$\delta Q = dU + \delta W = dU + P\, dV$$

We find that this expression can be evaluated for two separate special cases.

1. Constant volume, for which the work term ($P\, dV$) is zero, so that the specific heat (at constant volume) is

$$C_v = \frac{1}{m}\left(\frac{\delta Q}{\delta T}\right)_v = \frac{1}{m}\left(\frac{\partial U}{\partial T}\right)_v = \left(\frac{\partial u}{\partial T}\right)_v \qquad (5.14)$$

2. Constant pressure, for which the work term can be integrated and the resulting PV terms at the initial and final states can be associated with the internal energy terms, as in Section 5.5, thereby leading to the conclusion that the heat transfer can be expressed in terms of the enthalpy change. The corresponding specific heat (at constant pressure) is

$$C_p = \frac{1}{m}\left(\frac{\delta Q}{\delta T}\right)_p = \frac{1}{m}\left(\frac{\partial H}{\partial T}\right)_p = \left(\frac{\partial h}{\partial T}\right)_p \qquad (5.15)$$

Note that in each of these special cases, the resulting expression, Eq. 5.14 or 5.15, contains only thermodynamic properties, from which we conclude that the constant-volume and constant-pressure specific heats must themselves be thermodynamic properties. This means that, although we began this discussion by considering the amount of heat transfer required to cause a unit temperature change and then proceeded through a very specific development leading to Eq. 5.14 (or 5.15), the result ultimately expresses a relation among a set of thermodynamic properties and therefore constitutes a definition that is independent of the particular process leading to it (in the same sense that the definition of enthalpy in the previous section is independent of the process used to illustrate one situation in which the property is useful in a thermodynamic analysis). As an example, consider the two identical fluid masses shown in Fig. 5.9. In the first system 100 kJ of heat is transferred to it, and in the second system 100 kJ of work is done on it. Thus, the change of internal energy is the same for each, and therefore the final state and the final temperature are the same in each. In accordance with Eq. 5.14, therefore, exactly the same value for the average constant-volume specific heat would be found for this substance for the two processes, even though the two processes are very different as far as heat transfer is concerned.

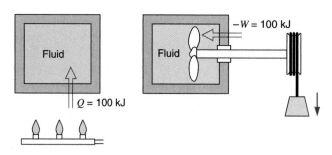

FIGURE 5.9 Sketch showing two ways in which a given ΔU may be achieved.

Solids and Liquids

As a special case, consider either a solid or a liquid. Since both of these phases are nearly incompressible,

$$dh = du + d(Pv) \approx du + v\, dP \qquad (5.16)$$

Also, for both of these phases, the specific volume is very small, such that in many cases

$$dh \approx du \approx C\, dT \qquad (5.17)$$

where C is either the constant-volume or the constant-pressure specific heat, as the two would be nearly the same. In many processes involving a solid or a liquid, we might further assume that the specific heat in Eq. 5.17 is constant (unless the process occurs at low temperature or over a wide range of temperatures). Equation 5.17 can then be integrated to

$$h_2 - h_1 \simeq u_2 - u_1 \simeq C(T_2 - T_1) \qquad (5.18)$$

Specific heats for various solids and liquids are listed in Tables A.3 and A.4.

In other processes for which it is not possible to assume constant specific heat, there may be a known relation for C as a function of temperature. Equation 5.17 could then also be integrated.

5.7 THE INTERNAL ENERGY, ENTHALPY, AND SPECIFIC HEAT OF IDEAL GASES

In general, for any substance the internal energy u depends on the two independent properties specifying the state. For a low-density gas, however, u depends primarily on T and much less on the second property, P or v. For example, consider several values for superheated vapor steam from Table B.1.3, shown in Table 5.1. From these values, it is evident that u depends strongly on T, but not much on P. Also, we note that the dependence of u on P is less at low pressure and is much less at high temperature; that is, as the density decreases, so does dependence of u on P (or v). It is therefore reasonable to extrapolate this behavior to very low density and to assume that as gas density becomes so low that the ideal-gas model is appropriate, internal energy does not depend on pressure at all but is a function only of temperature. That is, for an ideal gas,

$$Pv = RT \quad \text{and} \quad u = f(T) \text{ only} \qquad (5.19)$$

TABLE 5.1
Internal Energy for Superheated Vapor Steam

T, °C	P, kPa			
	10	100	500	1000
200	2661.3	2658.1	2642.9	2621.9
700	3479.6	3479.2	3477.5	3475.4
1200	4467.9	4467.7	4466.8	4465.6

The relation between the internal energy u and the temperature can be established by using the definition of constant-volume specific heat given by Eq. 5.14:

$$C_v = \left(\frac{\partial u}{\partial T}\right)_v$$

Because the internal energy of an ideal gas is not a function of specific volume, for an ideal gas we can write

$$C_{v0} = \frac{du}{dT}$$

$$du = C_{v0}\, dT \tag{5.20}$$

where the subscript 0 denotes the specific heat of an ideal gas. For a given mass m,

$$dU = mC_{v0}\, dT \tag{5.21}$$

From the definition of enthalpy and the equation of state of an ideal gas, it follows that

$$h = u + Pv = u + RT \tag{5.22}$$

Since R is a constant and u is a function of temperature only, it follows that the enthalpy, h, of an ideal gas is also a function of temperature only. That is,

$$h = f(T) \tag{5.23}$$

The relation between enthalpy and temperature is found from the constant-pressure specific heat as defined by Eq. 5.15:

$$C_p = \left(\frac{\partial h}{\partial T}\right)_p$$

Since the enthalpy of an ideal gas is a function of the temperature only and is independent of the pressure, it follows that

$$C_{p0} = \frac{dh}{dT}$$

$$dh = C_{p0}\, dT \tag{5.24}$$

For a given mass m,

$$dH = mC_{p0}\, dT \tag{5.25}$$

The consequences of Eqs. 5.20 and 5.24 are demonstrated in Fig. 5.10, which shows two lines of constant temperature. Since internal energy and enthalpy are functions of temperature only, these lines of constant temperature are also lines of constant internal energy and constant enthalpy. From state 1 the high temperature can be reached by a variety of paths, and in each case the final state is different. However, regardless of the path, the change in internal energy is the same, as is the change in enthalpy, for lines of constant temperature are also lines of constant u and constant h.

Because the internal energy and enthalpy of an ideal gas are functions of temperature only, it also follows that the constant-volume and constant-pressure specific heats are also functions of temperature only. That is,

$$C_{v0} = f(T), \qquad C_{p0} = f(T) \tag{5.26}$$

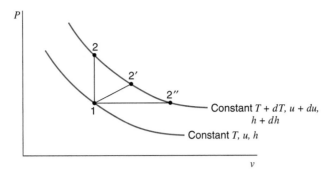

FIGURE 5.10
Pressure–volume diagram
for an ideal gas.

Because all gases approach ideal-gas behavior as the pressure approaches zero, the ideal-gas specific heat for a given substance is often called the zero-pressure specific heat, and the zero-pressure, constant-pressure specific heat is given the symbol C_{p0}. The zero-pressure, constant-volume specific heat is given the symbol C_{v0}. Figure 5.11 shows C_{p0} as a function of temperature for a number of different substances. These values are determined by the techniques of statistical thermodynamics and will not be discussed here. A brief summary presentation of this subject is given in Appendix C. It is noted there that the principal factor causing specific heat to vary with temperature is molecular vibration. More complex molecules have multiple vibrational modes and therefore show a greater temperature dependency, as is seen in Fig. 5.11. This is an important consideration when deciding whether or not to account for specific heat variation with temperature in any particular application.

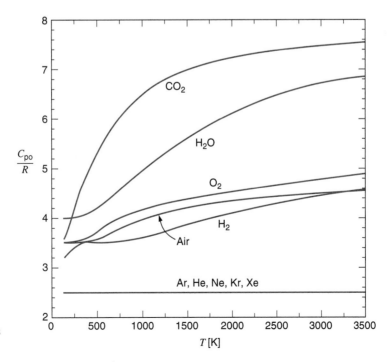

FIGURE 5.11 Heat
capacity for some gases as
function of temperature.

A very important relation between the constant-pressure and constant-volume specific heats of an ideal gas may be developed from the definition of enthalpy:

$$h = u + Pv = u + RT$$

Differentiating and substituting Eqs. 5.20 and 5.24, we have

$$dh = du + R\,dT$$

$$C_{p0}\,dT = C_{v0}\,dT + R\,dT$$

Therefore,

$$C_{p0} - C_{v0} = R \tag{5.27}$$

On a mole basis this equation is written

$$\overline{C}_{p0} - \overline{C}_{v0} = \overline{R} \tag{5.27}$$

This tells us that the difference between the constant-pressure and constant-volume specific heats of an ideal gas is always constant, though both are functions of temperature. Thus, we need examine only the temperature dependency of one, and the other is given by Eq. 5.27.

Let us consider the specific heat C_{p0}. There are three possibilities to examine. The situation is simplest if we assume constant specific heat, that is, no temperature dependence. Then it is possible to integrate Eq. 5.24 directly to

$$h_2 - h_1 = C_{p0}(T_2 - T_1) \tag{5.29}$$

We note from Fig. 5.11 the circumstances under which this will be an accurate model. It should be added, however, that it may be a reasonable approximation under other conditions, especially if an average specific heat in the particular temperature range is used in Eq. 5.29. Values of specific heat at room temperature and gas constants for various gases are given in Table A.5.

The second possibility for the specific heat is to use an analytical equation for C_{p0} as a function of temperature. Because the results of specific-heat calculations from statistical thermodynamics do not lend themselves to convenient mathematical forms, these results have been approximated empirically. The equations for C_{p0} as a function of temperature are listed in Table A.6 for a number of gases.

The third possibility is to integrate the results of the calculations of statistical thermodynamics from an arbitrary reference temperature to any other temperature T and to define a function

$$h_T = \int_{T_0}^{T} C_{p0}\,dT$$

This function can then be tabulated in a single-entry (temperature) table. Then, between any two states 1 and 2,

$$h_2 - h_1 = \int_{T_0}^{T_2} C_{p0}\,dT - \int_{T_0}^{T_1} C_{p0}\,dT = h_{T_2} - h_{T_1} \tag{5.30}$$

and it is seen that the reference temperature cancels out. This function h_T (and a similar function $u_T = h_T - RT$) is listed for air in Table A.7. These functions are listed for other gases in Table A.8.

To summarize the three possibilities, we note that using the ideal-gas tables, Tables A.7 and A.8, gives us the most accurate answer, but that the equations in Table A.6 would give a close empirical approximation. Constant specific heat would be less accurate, except for monatomic gases and gases below room temperature. It should be remembered that all these results are a part of the ideal-gas model, which in many of our problems is not a valid assumption for the behavior of the substance.

EXAMPLE 5.7 Calculate the change of enthalpy as 1 kg of oxygen is heated from 300 to 1500 K. Assume ideal-gas behavior.

Solution

For an ideal gas, the enthalpy change is given by Eq. 5.24. However, we also need to make an assumption about the dependence of specific heat on temperature. Let us solve this problem in several ways and compare the answers.

Our most accurate answer for the ideal-gas enthalpy change for oxygen between 300 and 1500 K would be from the ideal-gas tables, Table A.8. This result is, using Eq. 5.30,

$$h_2 - h_1 = 1540.2 - 273.2 = 1267.0 \text{ kJ/kg}$$

The empirical equation from Table A.6 should give a good approximation to this result. Integrating Eq. 5.24, we have

$$h_2 - h_1 = \int_{T_1}^{T_2} C_{p0} \, dT = \int_{\theta_1}^{\theta_2} C_{p0}(\theta) \times 1000 \, d\theta$$

$$= 1000 \left[0.88\theta - \frac{0.0001}{2} \theta^2 + \frac{0.54}{3} \theta^3 - \frac{0.33}{4} \theta^4 \right]_{\theta_1 = 0.3}^{\theta_2 = 1.5}$$

$$= 1241.5 \text{ kJ/kg}$$

which is lower than the first result by 2.0%.

If we assume constant specific heat, we must be concerned about what value we are going to use. If we use the value at 300 K from Table A.5, we find, from Eq. 5.29, that

$$h_2 - h_1 = C_{p0}(T_2 - T_1) = 0.922 \times 1200 = 1106.4 \text{ kJ/kg}$$

which is low by 12.7%. However, suppose we assume that the specific heat is constant at its value at 900 K, the average temperature. Substituting 900 K into the equation for specific heat from Table A.6, we have

$$C_{p0} = 0.88 - 0.0001(0.9) + 0.54(0.9)^2 - 0.33(0.9)^3$$

$$= 1.0767 \text{ kJ/kg K}$$

Substituting this value into Eq. 5.29 gives the result

$$h_2 - h_1 = 1.0767 \times 1200 = 1292.1 \text{ kJ/kg}$$

which is high by about 2.0%, a much closer result than the one using the room temperature specific heat. It should be kept in mind that part of the model involving ideal gas with constant specific heat also involves a choice of what value is to be used.

EXAMPLE 5.8

A cylinder fitted with a piston has an initial volume of 0.1 m³ and contains nitrogen at 150 kPa, 25°C. The piston is moved, compressing the nitrogen until the pressure is 1 MPa and the temperature is 150°C. During this compression process heat is transferred from the nitrogen, and the work done on the nitrogen is 20 kJ. Determine the amount of this heat transfer.

Control mass: Nitrogen.
Initial state: P_1, T_1, V_1; state 1 fixed.
Final state: P_2, T_2; state 2 fixed.
Process: Work input known.
Model: Ideal gas, constant specific heat with value at 300 K, Table A.5.

Analysis

From the first law we have

$$_1Q_2 = m(u_2 - u_1) + {}_1W_2$$

Solution

The mass of nitrogen is found from the equation of state with the value of R from Table A.5:

$$m = \frac{PV}{RT} = \frac{150 \text{ kPa} \times 0.1 \text{ m}^3}{0.2968 \frac{\text{kJ}}{\text{kg K}} \times 298.15 \text{ K}} = 0.1695 \text{ kg}$$

Assuming constant specific heat as given in Table A.5, we have

$$_1Q_2 = mC_{v0}(T_2 - T_1) + {}_1W_2$$

$$= 0.1695 \text{ kg} \times 0.745 \frac{\text{kJ}}{\text{kg K}} \times (150 - 25) \text{ K} - 20.0$$

$$= 15.8 - 20.0 = -4.2 \text{ kJ}$$

It would, of course, be somewhat more accurate to use Table A.8 than to assume constant specific heat (room temperature value), but often the slight increase in accuracy does not warrant the added difficulties of manually interpolating the tables.

EXAMPLE 5.8E

A cylinder fitted with a piston has an initial volume of 2 ft³ and contains nitrogen at 20 lbf/in², 80 F. The piston is moved, compressing the nitrogen until the pressure is 160 lbf/in² and the temperature is 300 F. During this compression process heat is transferred from the nitrogen, and the work done on the nitrogen is 9.15 Btu. Determine the amount of this heat transfer.

Control mass: Nitrogen.
Initial state: P_1, T_1, V_1; state 1 fixed.
Final state: P_2, T_2; state 2 fixed.
Process: Work input known.
Model: Ideal gas, constant specific heat with value at 540 R, Table F.4.

Analysis

First law:
$$_1Q_2 = m(u_2 - u_1) + {}_1W_2$$

Solution

The mass of nitrogen is found from the equation of state with the value of R from Table F.4.

$$m = \frac{PV}{RT} = \frac{20\,\frac{\text{lbf}}{\text{in}^2} \times 144 \times \frac{\text{in}^2}{\text{ft}^2}\,2\,\text{ft}^3}{55.15\,\frac{\text{ft lbf}}{\text{lbm R}} \times 540\,R} = 0.1934\,\text{lbm}$$

Assuming constant specific heat as given in Table F.4,

$$_1Q_2 = mC_{v0}(T_2 - T_1) + {}_1W_2$$

$$= 0.1934\,\text{lbm} \times 0.177\,\frac{\text{Btu}}{\text{lbm}\,R} \times (300 - 80)\,R - 9.15$$

$$= 7.53 - 9.15 = -1.62\,\text{Btu}$$

It would, of course, be somewhat more accurate to use Table F.6 than to assume constant specific heat (room temperature value), but often the slight increase in accuracy does not warrant the added difficulties of manually interpolating the tables.

5.8 THE FIRST LAW AS A RATE EQUATION

We frequently find it desirable to use the first law as a rate equation that expresses either the instantaneous or average rate at which energy crosses the control surface as heat and work and the rate at which the energy of the control mass changes. In so doing we are departing from a strictly classical point of view, because basically classical thermodynamics deals with systems that are in equilibrium, and time is not a relevant parameter for systems that are in equilibrium. However, since these rate equations are developed from the concepts of classical thermodynamics and are used in many applications of thermodynamics, they are included in this book. This rate form of the first law will be used in the development of the first law for the control volume in Section 6.2, and in this form the first law finds extensive applications in thermodynamics, fluid mechanics, and heat transfer.

Consider a time interval δt during which an amount of heat δQ crosses the control surface, an amount of work δW is done by the control mass, the internal energy change is ΔU, the kinetic energy change is ΔKE, and the potential energy change is ΔPE. From the first law we can write

$$\Delta U + \Delta KE + \Delta PE = \delta Q - \delta W$$

Dividing by δt we have the average rate of energy transfer as heat work and increase of the energy of the control mass:

$$\frac{\Delta U}{\delta t} + \frac{\Delta KE}{\delta t} + \frac{\Delta PE}{\delta t} = \frac{\delta Q}{\delta t} - \frac{\delta W}{\delta t}$$

Taking the limit for each of these quantities as δt approaches zero, we have

$$\lim_{\delta t \to 0} \frac{\Delta U}{\delta t} = \frac{dU}{dt}, \quad \lim_{\delta t \to 0} \frac{\Delta(KE)}{\delta t} = \frac{d(KE)}{dt}, \quad \lim_{\delta t \to 0} \frac{\Delta(PE)}{\delta t} = \frac{d(PE)}{dt}$$

$$\lim_{\delta t \to 0} \frac{\delta Q}{\delta t} = \dot{Q} \quad \text{(the heat transfer rate)}$$

$$\lim_{\delta t \to 0} \frac{\delta W}{\delta t} = \dot{W} \quad \text{(the power)}$$

Therefore, the rate equation form of the first law is

$$\frac{dU}{dt} + \frac{d(KE)}{dt} + \frac{d(PE)}{dt} = \dot{Q} - \dot{W} \tag{5.31}$$

We could also write this in the form

$$\frac{dE}{dt} = \dot{Q} - \dot{W} \tag{5.32}$$

EXAMPLE 5.9

During the charging of a storage battery, the current i is 20 A and the voltage $\mathcal{E}$ is 12.8 V. The rate of heat transfer from the battery is 10 W. At what rate is the internal energy increasing?

Solution

Since changes in kinetic and potential energy are insignificant, the first law can be written as a rate equation in the form of Eq. 5.31:

$$\frac{dU}{dt} = \dot{Q} - \dot{W}$$

$$\dot{W} = \mathcal{E}i = -12.8 \times 20 = -256 \text{ W} = -256 \text{ J/s}$$

Therefore,

$$\frac{dU}{dt} = \dot{Q} - \dot{W} = -10 - (-256) = 246 \text{ J/s}$$

EXAMPLE 5.10

A 25-kg cast-iron wood-burning stove, shown in Fig. 5.12, contains 5 kg of soft pine wood and 1 kg of air. All the masses are at room temperature, 20°C, and pressure, 101 kPa. The wood now burns and heats all the mass uniformly, releasing 1500 watts. Neglect any air flow and changes in mass and heat losses. Find the rate of change of the temperature (dT/dt) and estimate the time it will take to reach a temperature of 75°C.

Solution

C.V.: The iron, wood and air.
This is a control mass.

Energy equation rate form: $\qquad\qquad \dot{E} = \dot{Q} - \dot{W}$

FIGURE 5.12 Sketch for Example 5.10.

We have no changes in kinetic or potential energy and no change in mass, so

$$U = m_{air}u_{air} + m_{wood}u_{wood} + m_{iron}u_{iron}$$

$$\dot{E} = \dot{U} = m_{air}\dot{u}_{air} + m_{wood}\dot{u}_{wood} + m_{iron}\dot{u}_{iron}$$

$$= (m_{air}C_{Vair} + m_{wood}C_{wood} + m_{iron}C_{iron})\frac{dT}{dt}$$

Now the energy equation has zero work, an energy release of $\dot{Q}$, and becomes

$$(m_{air}C_{Vair} + m_{wood}C_{wood} + m_{iron}C_{iron})\frac{dT}{dt} = \dot{Q} - 0$$

$$\frac{dT}{dt} = \frac{\dot{Q}}{(m_{air}C_{Vair} + m_{wood}C_{wood} + m_{iron}C_{iron})}$$

$$= \frac{1500}{1 \times 0.717 + 5 \times 1.38 + 25 \times 0.42}\frac{W}{kg\ (kJ/kg)} = 0.0828\ K/s$$

Assuming the rate of temperature rise is constant, we can find the elapsed time as

$$\Delta T = \int \frac{dT}{dt}\,dt = \frac{dT}{dt}\Delta t$$

$$\Rightarrow \Delta t = \frac{\Delta T}{\frac{dT}{dt}} = \frac{75 - 20}{0.0828} = 664\ s = 11\ min$$

5.9 CONSERVATION OF MASS

In the previous sections we considered the first law of thermodynamics for a control mass undergoing a change of state. A control mass is defined as a fixed quantity of mass. The question now is whether the mass of such a system changes when its energy changes. If it does, our definition of a control mass as a fixed quantity of mass is no longer valid when the energy changes.

We know from relativistic considerations that mass and energy are related by the well-known equation

$$E = mc^2 \qquad (5.33)$$

where c = velocity of light and E = energy. We conclude from this equation that the mass of a control mass does change when its energy changes. Let us calculate the magnitude of this change of mass for a typical problem and determine whether this change in mass is significant.

Consider a rigid vessel that contains a 1-kg stoichiometric mixture of a hydrocarbon fuel (such as gasoline) and air. From our knowledge of combustion, we know that after combustion takes place it will be necessary to transfer about 2900 kJ from the system to restore it to its initial temperature. From the first law

$$_1Q_2 = U_2 - U_1 + {}_1W_2$$

we conclude that since $_1W_2 = 0$ and $_1Q_2 = -2900$ kJ, the internal energy of this system decreases by 2900 kJ during the heat transfer process. Let us now calculate the decrease in mass during this process using Eq. 5.33.

The velocity of light, c, is 2.9979×10^8 m/s. Therefore,

$$2900 \text{ kJ} = 2\,900\,000 \text{ J} = m \text{ (kg)} \times (2.9979 \times 10^8 \text{ m/s})^2$$

and so

$$m = 3.23 \times 10^{-11} \text{ kg}$$

Thus, when the energy of the control mass decreases by 2900 kJ, the decrease in mass is 3.23×10^{-11} kg.

A change in mass of this magnitude cannot be detected by even our most accurate chemical balance. Certainly, a fractional change in mass of this magnitude is beyond the accuracy required in essentially all engineering calculations. Therefore, if we use the laws of conservation of mass and conservation of energy as separate laws, we will not introduce significant error into most thermodynamic problems and our definition of a control mass as having a fixed mass can be used even though the energy changes.

SUMMARY

Conservation of energy is expressed for a cycle, and changes of total energy are then written for a control mass. Kinetic and potential energy can be changed through the work of a force acting on the control mass, and they are part of the total energy.

The internal energy and the enthalpy are introduced as substance properties with the specific heats (heat capacity) as derivatives of these with temperature. Property variations for limited cases are presented for incompressible states of a substance such as liquids and solids, and for a highly compressible state as an ideal gas. The specific heat for solids and liquids changes little with temperature, whereas the specific heat for a gas can change substantially with temperature.

The energy equation is also shown in a rate form to cover transient processes.

You should have learned a number of skills and acquired abilities from studying this chapter that will allow you to

- Recognize the components of total energy stored in a control mass
- Write the energy equation for a single uniform control mass
- Find the properties u and h for a given state in the Appendix B tables
- Locate a state in the tables with an entry such as (P, h)
- Find changes in u and h for liquid or solid states using Tables A.3 and A.4 or F.2 and F.3
- Find changes in u and h for ideal-gas states using Table A.5 or F.4
- Find changes in u and h for ideal-gas states using Tables A.7 and A.8 or F.5 and F.6
- Recognize that forms for C_p in Table A.6 are approximations to what is shown in Fig. 5.11 and the more accurate tabulations in Tables A.7, A.8, F.5, and F.6
- Formulate the conservation of mass and energy for a control mass that goes through a process involving work and heat transfers and different states
- Formulate the conservation of mass and energy for a more complex control mass where there are different masses with different states

- Use the energy equation in a rate form
- Know the difference between the general laws as the conservation of mass (continuity equation), conservation of energy (first law) and the specific laws that describes a device behavior or process

KEY CONCEPTS AND FORMULAS		
Total energy		$E = U + KE + KE = mu + \frac{1}{2}m\mathbf{V}^2 + mgZ$
Kinetic energy		$KE = \frac{1}{2}m\mathbf{V}^2$
Potential energy		$KE = mgZ$
Specific energy		$e = u + \frac{1}{2}\mathbf{V}^2 + gZ$
Enthalpy		$h \equiv u + Pv$
Two-phase mass average		$u = u_f + xu_{fg} = (1-x)u_f + xu_g$
		$h = h_f + xh_{fg} = (1-x)h_f + xh_g$
Specific heat, heat capacity		$C_v = \left(\frac{\partial u}{\partial T}\right)_v ; C_p = \left(\frac{\partial h}{\partial T}\right)_p$
Solids and liquids		Incompressible, so $v = $ constant $\cong v_f$ and v very small
		$C = C_v = C_p$ [Tables A.3 and A.4 (F.2 and F.3)]
		$u_2 - u_1 = C(T_2 - T_1)$
		$h_2 - h_1 = u_2 - u_1 + v(P_2 - P_1)$ (Often the second term is small.)
		$h = h_f + v_f(P - P_{sat}); u \cong u_f$ (saturated at same T)
Ideal gas		$h = u + Pv = u + RT$ (only functions of T)
		$C_v = \frac{du}{dT}; C_p = \frac{dh}{dT} = C_v + R$
		$u_2 - u_1 = \int C_v dT \cong C_v(T_2 - T_1)$
		$h_2 - h_1 = \int C_p dT \cong C_p(T_2 - T_1)$
		Left-hand side from Table A.7 or A.8, middle from Table A.6 and right-hand side from Table A.6 at a T_{avg} or from Table A.5 at 25°C
		Left-hand side from Table F.5 or F.6, right-hand side from Table F.4 at 77 F
Energy equation rate form		$\dot{E} = \dot{Q} - \dot{W}$ (rate = + in − out)
Energy equation integrated		$E_2 - E_1 = {}_1Q_2 - {}_1W_2$ (change = + in − out)
		$m(e_2 - e_1) = m(u_2 - u_1) + \frac{1}{2}m(\mathbf{V}_2^2 - \mathbf{V}_1^2) + mg(Z_2 - Z_1)$
Multiple masses, states		$E = m_A e_A + m_B e_B + m_C e_C + \cdots$

CONCEPT-STUDY GUIDE PROBLEMS

5.1 What is 1 cal in SI units and what is the name given to 1 N m?

5.2 In a complete cycle what is the net change in energy and in volume?

5.3 Why do we write ΔE or $E_2 - E_1$ whereas we write ${}_1Q_2$ and ${}_1W_2$?

5.4 When you wind a spring up in a toy or stretch a rubber band, what happens in terms of work, energy, and heat transfer? Later, when they are released, what happens then?

5.5 Explain in words what happens with the energy terms for the stone in Example 5.2. What would happen if it were a bouncing ball falling to a hard surface?

5.6 Make a list of at least five systems that store energy, explaining which form of energy.

5.7 A 1200-kg car is accelerated from 30 to 50 km/h in 5 s. How much work is that? If you continue from 50 to 70 km/h in 5 s, is that the same?

5.8 A crane used 2 kW to raise a 100-kg box 20 m. How much time did it take?

5.9 Saturated water vapor has a maximum for u and h at around 235°C. Is this similar to other substances?

5.10 A pot of water is boiling on a stove supplying 325 W to the water. What is the rate of mass (kg/s) vaporizing, assuming a constant pressure process?

5.11 A constant mass goes through a process whereby 100 W of heat transfer comes in and 100 W of work leaves. Does the mass change state?

5.12 I have 2 kg of liquid water at 20°C, 100 kPa. I now add 20 kJ of energy at a constant pressure. How hot does the water get if it is heated? How fast does it

move if it is pushed by a constant horizontal force? How high does it go if it is raised straight up?

5.13 Water is heated from 100 kPa, 20°C to 1000 kPa, 200°C. In one case pressure is raised at $T = C$, then T is raised at $P = C$. In a second case the opposite order is followed. Does that make a difference for $_1Q_2$ and $_1W_2$?

5.14 Two kilograms of water at 120°C with a quality of 25% has its temperature raised 20°C in a constant-volume process. What are the new quality and specific internal energy?

5.15 Two kilograms of water at 200 kPa with a quality 25% has its temperature raised 20°C in a constant-pressure process. What is the change in enthalpy?

5.16 You heat a gas 10 K at $P = C$. Which one in Table A.5 requires the most energy? Why?

5.17 Air is heated from 300 to 350 K at $V = C$. Find $_1q_2$. What if the air is heated from 1300 to 1350 K?

5.18 A mass of 3 kg of nitrogen gas at 2000 K, $V = C$, cools with 500 W. What is dT/dt?

5.19 A drag force on a car, with frontal area $A = 2$ m², driving at 80 km/h in air at 20°C, is $F_d = 0.225$ $A\rho_{air}\mathbf{V}^2$. How much power is needed and what is the traction force?

HOMEWORK PROBLEMS

Kinetic and Potential Energy

5.20 A hydraulic hoist raises a 1750-kg car 1.8 m in an auto repair shop. The hydraulic pump has a constant pressure of 800 kPa on its piston. What is the increase in potential energy of the car and how much volume should the pump displace to deliver that amount of work?

5.21 A piston motion moves a 25-kg hammerhead vertically down 1 m from rest to a velocity of 50 m/s in a stamping machine. What is the change in total energy of the hammerhead?

5.22 Airplane takeoff from an aircraft carrier is assisted by a steam-driven piston/cylinder with an average pressure of 1250 kPa. A 17 500-kg airplane should be accelerated from zero to a speed of 30 m/s with 30% of the energy coming from the steam piston. Find the needed piston displacement volume.

5.23 Solve Problem 5.22, but assume the steam pressure in the cylinder starts at 1000 kPa, dropping linearly

with volume to reach 100 kPa at the end of the process.

5.24 A 1200-kg car accelerates from zero to 100 km/h over a distance of 400 m. The road at the end of the 400 m is at 10 m higher elevation. What is the total increase in the car kinetic and potential energy?

5.25 A 25 kg piston is above a gas in a long vertical cylinder. Now the piston is released from rest and accelerates up in the cylinder reaching the end 5 m higher at a velocity of 25 m/s. The gas pressure drops during the process, so the average is 600 kPa with an outside atmosphere at 100 kPa. Neglect the change in gas kinetic and potential energy, and find the needed change in the gas volume.

5.26 The rolling resistance of a car depends on its weight as $F = 0.006$ mg. How far will a car of 1200 kg roll if the gear is put in neutral when it drives at 90 km/h on a level road without air resistance?

5.27 A mass of 5 kg is tied to an elastic cord 5 m long and dropped from a tall bridge. Assume the cord, once straight, acts as a spring with $k = 100$ N/m. Find the velocity of the mass when the cord is straight (5 m down). At what level does the mass come to rest after bouncing up and down?

Properties (u, h) from General Tables

5.28 Find the missing properties.

a. H_2O, $\quad T = 250°C$, $\quad P = ? u = ?$
$\quad v = 0.02$ m³/kg,

b. N_2, $\quad T = 120$ K, $\quad x = ? h = ?$
$\quad P = 0.8$ MPa,

c. H_2O, $\quad T = -2°C$, $\quad u = ? v = ?$
$\quad P = 100$ kPa,

d. R-134a, $\quad P = 200$ kPa, $\quad u = ? T = ?$
$\quad v = 0.12$ m³/kg,

5.29 Find the missing properties of T, P, v, u, h, and x if applicable and plot the location of the three states as points in the T–v and the P–v diagrams.

a. Water at 5000 kPa, $u = 800$ kJ/kg
b. Water at 5000 kPa, $v = 0.06$ m³/kg
c. R-134a at 35°C, $v = 0.01$ m³/kg

5.30 Find the missing properties and give the phase of the substance.

a. NH_3, $\quad T = 65°C$, $\quad u = ? v = ?$
$\quad P = 600$ kPa,

b. NH_3, $\quad T = 20°C$, $\quad u = ? v = ?$
$\quad P = 100$ kPa, $\quad x = ?$

c. Ammonia, $\quad T = 50°C$, $\quad P = ? u = ?$
$\quad v = 0.1185$ m³/kg, $\quad x = ?$

5.31 i. Find the phase and missing properties of P, T, v, u, and x.

a. H_2O at $P = 5000$ kPa, $u = 1000$ kJ/kg (steam table reference)
b. R-134a at $T = 20°C$, $u = 300$ kJ/kg
c. N_2 at 250 K, $P = 200$ kPa

ii. Show the three states as labeled dots in a T–v diagram with correct position relative to the two-phase region.

5.32 Find the missing properties and give the phase of the substance.

a. H_2O, $\quad T = 120°C$, $\quad u = ? P = ? x = ?$
$\quad v = 0.5$ m³/kg,

b. H_2O, $\quad T = 100°C$, $\quad u = ? x = ? v = ?$
$\quad P = 10$ MPa,

c. N_2, $\quad T = 200$ K, $\quad v = ? u = ?$
$\quad P = 200$ kPa,

d. NH_3, $\quad T = 100°C$, $\quad P = ? x = ?$
$\quad v = 0.1$ m³/kg,

e. N_2, $\quad T = 100$ K, $\quad v = ? u = ?$
$\quad x = 0.75$,

5.33 Find the missing properties among T, P, v, u, h, and x (if applicable), give the phase of the substance, and indicate the states relative to the two-phase region in both a T–v and a P–v diagram.

a. R-12, $\quad P = 500$ kPa, $h = 230$ kJ/kg
b. R-22, $\quad T = 10°C$, $u = 200$ kJ/kg
c. R-134a, $\quad T = 40°C$, $h = 400$ kJ/kg

5.34 Saturated liquid water at 20°C is compressed to a higher pressure with constant temperature. Find the changes in u and h from the initial state when the final pressure is

a. 500 kPa
b. 2000 kPa
c. 20 000 kPa

Energy Equation: Simple Process

5.35 A 100-L rigid tank contains nitrogen (N_2) at 900 K and 3 MPa. The tank is now cooled to 100 K. What are the work and heat transfer for the process?

5.36 A rigid container has 0.75 kg of water at 300°C, 1200 kPa. The water is now cooled to a final pressure of 300 kPa. Find the final temperature, the work, and the heat transfer in the process.

5.37 A cylinder fitted with a frictionless piston contains 2 kg of superheated refrigerant R-134a vapor at 350 kPa, 100°C. The cylinder is now cooled so that the R-134a remains at constant pressure until it reaches a quality of 75%. Calculate the heat transfer in the process.

5.38 Ammonia at 0°C with a quality of 60% is contained in a rigid 200-L tank. The tank and ammonia are now heated to a final pressure of 1 MPa. Determine the heat transfer for the process.

5.39 Water in a 150-L closed, rigid tank is at 100°C and 90% quality. The tank is then cooled to $-10°C$. Calculate the heat transfer for the process.

5.40 A piston/cylinder contains 1 kg of water at 20°C with volume 0.1 m³. By mistake someone locks the piston, preventing it from moving while we heat the water to saturated vapor. Find the final temperature and the amount of heat transfer in the process.

5.41 A test cylinder with constant volume of 0.1 L contains water at the critical point. It now cools down to room temperature of 20°C. Calculate the heat transfer from the water.

5.42 A 10-L rigid tank contains R-22 at −10°C with a quality of 80%. A 10-A electric current (from a 6-V battery) is passed through a resistor inside the tank for 10 min, after which the R-22 temperature is 40°C. What was the heat transfer to or from the tank during this process?

5.43 A piston/cylinder device contains 50 kg of water at 200 kPa with a volume of 0.1 m³. Stops in the cylinder are placed to restrict the enclosed volume to a maximum of 0.5 m³. The water is now heated until the piston reaches the stops. Find the necessary heat transfer.

5.44 A constant-pressure piston/cylinder assembly contains 0.2 kg of water as saturated vapor at 400 kPa. It is now cooled so that the water occupies half the original volume. Find the heat transfer in the process.

5.45 Two kilograms of water at 120°C with a quality of 25% has its temperature raised 20°C in a constant-volume process as in Fig. P5.45. What are the heat transfer and work in the process?

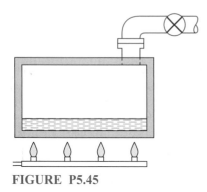

FIGURE P5.45

5.46 A 25-kg mass moves at 25 m/s. Now a brake system brings the mass to a complete stop with a constant deceleration over a period of 5 s. The brake energy is absorbed by 0.5 kg of water initially at 20°C and 100 kPa. Assume the mass is at constant *P* and *T*. Find the energy the brake removes from the mass and the temperature increase of the water, assuming its pressure is constant.

5.47 An insulated cylinder fitted with a piston contains R-12 at 25°C with a quality of 90% and a volume of 45 L. The piston is allowed to move, and the R-12 expands until it exists as saturated vapor. During this process the R-12 does 7.0 kJ of work against the piston. Determine the final temperature, assuming the process is adiabatic.

5.48 A water-filled reactor with volume of 1 m³ is at 20 MPa and 360°C and placed inside a containment room, as shown in Fig. P5.48. The room is well insulated and initially evacuated. Due to a failure, the reactor ruptures and the water fills the containment

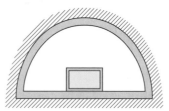

FIGURE P5.48

room. Find the minimum room volume so that the final pressure does not exceed 200 kPa.

5.49 A piston/cylinder arrangement contains water of quality $x = 0.7$ in the initial volume of 0.1 m³, where the piston applies a constant pressure of 200 kPa. The system is now heated to a final temperature of 200°C. Determine the work and the heat transfer in the process.

5.50 A piston/cylinder arrangement has the piston loaded with outside atmospheric pressure and the piston mass to a pressure of 150 kPa, as shown in Fig. P5.50. It contains water at −2°C, which is then heated until the water becomes saturated vapor. Find the final temperature and specific work and heat transfer for the process.

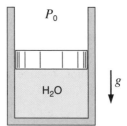

FIGURE P5.50

5.51 A piston/cylinder assembly contains 1 kg of liquid water at 20°C and 300 kPa. There is a linear spring mounted on the piston such that when the water is heated the pressure reaches 1 MPa with a volume of 0.1 m³. Find the final temperature and the heat transfer in the process.

5.52 A closed steel bottle contains ammonia at $-20°C$, $x = 20\%$ and the volume is 0.05 m³. It has a safety valve that opens at a pressure of 1.4 MPa. By accident, the bottle is heated until the safety valve opens. Find the temperature and heat transfer when the valve first opens.

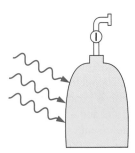

FIGURE P5.52

5.53 Two kilograms of water at 200 kPa with a quality of 25% has its temperature raised 20°C in a constant-pressure process. What are the heat transfer and work in the process?

5.54 Two kilograms of nitrogen at 100 K, $x = 0.5$ are heated in a constant pressure process to 300 K in a piston/cylinder arrangement. Find the initial and final volumes and the total heat transfer required.

5.55 A 1-L capsule of water at 700 kPa and 150°C is placed in a larger insulated and otherwise evacuated vessel. The capsule breaks and its contents fill the entire volume. If the final pressure should not exceed 125 kPa, what should the vessel volume be?

5.56 Superheated refrigerant R-134a at 20°C and 0.5 MPa is cooled in a piston/cylinder arrangement at constant temperature to a final two-phase state with quality of 50%. The refrigerant mass is 5 kg, and during this process 500 kJ of heat is removed. Find the initial and final volumes and the necessary work.

5.57 A cylinder having a piston restrained by a linear spring (of spring constant 15 kN/m) contains 0.5 kg of saturated vapor water at 120°C, as shown in Fig. P5.57. Heat is transferred to the water, causing the piston to rise. If the piston cross-sectional area is 0.05 m² and the pressure varies linearly with volume until a final pressure of 500 kPa is reached, find the final temperature in the cylinder and the heat transfer for the process.

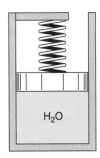

FIGURE P5.57

5.58 A rigid tank is divided into two rooms, both containing water, by a membrane, as shown in Fig. P5.58. Room A is at 200 kPa, $v = 0.5$ m³/kg, $V_A = 1$ m³, and room B contains 3.5 kg at 0.5 MPa, 400°C. The membrane now ruptures and heat transfer takes place so the water comes to a uniform state at 100°C. Find the heat transfer during the process.

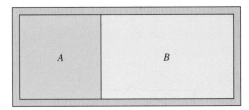

FIGURE P5.58

5.59 A 10-m-high open cylinder, with $A_{cyl} = 0.1$ m², contains 20°C water above and 2 kg of 20°C water below a 198.5-kg thin insulated floating piston, as shown in Fig. P5.59. Assume standard g, P_0. Now heat is added to the water below the piston so that it expands, pushing the piston up, causing the water on top to spill over the edge. This process continues until the piston reaches the top of the cylinder. Find the final state of the water below the piston (T, P, v) and the heat added during the process.

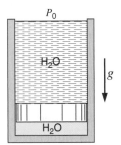

FIGURE P5.59

5.60 Assume the same setup as in Problem 5.48, but the room has a volume of 100 m³. Show that the final state is two phase and find the final pressure by trial and error.

Energy Equation: Multistep Solution

5.61 Ten kilograms of water in a piston/cylinder arrangement exist as saturated liquid/vapor at 100 kPa, with a quality of 50%. The system is now heated so that the volume triples. The mass of the piston is such that a cylinder pressure of 200 kPa will float it, as in Fig. P4.68. Find the final temperature and the heat transfer in the process.

5.62 Two tanks, each with a volume of 1 m³, are connected by a valve and line, as shown in Fig. P5.62. Tank A is filled with R-134a at 20°C with a quality of 15%. Tank B is evacuated. The valve is opened and saturated vapor flows from A into B until the pressures become equal. The process occurs slowly enough that all temperatures stay at 20°C during the process. Find the total heat transfer to the R-134a during the process.

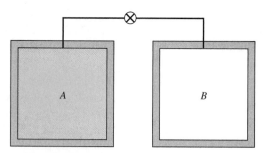

FIGURE P5.62

5.63 Consider the same system as in the previous problem. Let the valve be opened and transfer enough heat to both tanks so that all the liquid disappears. Find the necessary heat transfer.

5.64 A vertical cylinder fitted with a piston contains 5 kg of R-22 at 10°C, as shown in Fig. P5.64. Heat is transferred to the system, causing the piston to rise until it reaches a set of stops, at which point the volume has doubled. Additional heat is transferred until the temperature inside reaches 50°C, at which point the pressure inside the cylinder is 1.3 MPa.

a. What is the quality at the initial state?

b. Calculate the heat transfer for the overall process.

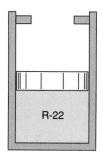

FIGURE P5.64

5.65 Find the heat transfer in Problem 4.67.

5.66 Refrigerant-12 is contained in a piston/cylinder arrangement at 2 MPa and 150°C with a massless piston against the stops, at which point $V = 0.5$ m³. The side above the piston is connected by an open valve to an air line at 10°C and 450 kPa, as shown in Fig. P5.66. The whole setup now cools to the surrounding temperature of 10°C. Find the heat transfer and show the process in a P–v diagram.

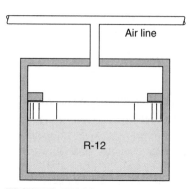

FIGURE P5.66

5.67 Find the heat transfer in Problem 4.114.

5.68 A rigid container has two rooms filled with water, each of 1 m³, separated by a wall (see Fig. P5.58). Room A has $P = 200$ kPa with a quality of $x = 0.80$. Room B has $P = 2$ MPa and $T = 400$°C. The partition wall is removed, and because of heat transfer the water comes to a uniform state with a temperature of 200°C. Find the final pressure and the heat transfer in the process.

5.69 The cylinder volume below the constant loaded piston has two compartments A and B filled with water, as shown in Fig. P5.69. A has 0.5 kg at 200 kPa and 150°C and B has 400 kPa with a quality of 50% and a volume of 0.1 m³. The valve is opened and heat is transferred so that the water comes to a uniform state with a total volume of 1.006 m³. Find the total mass of water and the total initial volume. Find the work and the heat transfer in the process.

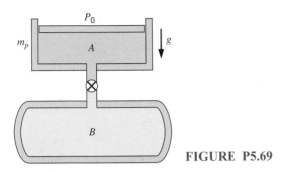

FIGURE P5.69

5.70 A rigid tank A of volume 0.6 m³ contains 3 kg of water at 120°C, and rigid tank B is 0.4 m³ with water at 600 kPa, 200°C. They are connected to a piston/cylinder initially empty with closed valves as shown in Fig. P5.70. The pressure in the cylinder should be 800 kPa to float the piston. Now the valves are slowly opened and heat is transferred so the water reaches a uniform state at 250°C with the valves open. Find the final volume and pressure, and the work and heat transfer in the process.

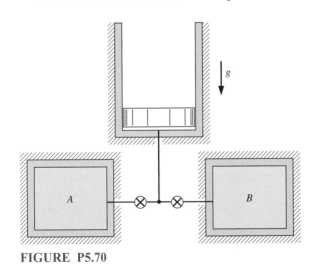

FIGURE P5.70

5.71 Calculate the heat transfer for the process described in problem 4.60.

5.72 Calculate the heat transfer for the process described in Problem 4.70.

5.73 A cylinder/piston arrangement contains 5 kg of water at 100°C with $x = 20\%$ and the piston, of $m_p = 75$ kg, resting on some stops, similar to Fig. P5.73. The outside pressure is 100 kPa, and the cylinder area is $A = 24.5$ cm². Heat is now added until the water reaches a saturated vapor state. Find the initial volume, final pressure, work, and heat transfer terms and show the P–v diagram.

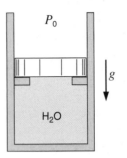

FIGURE P5.73

Energy Equation: Solids and Liquids

5.74 Because a hot water supply must also heat some pipe mass as it is turned on, the water does not come out hot right away. Assume 80°C liquid water at 100 kPa is cooled to 45°C as it heats 15 kg of copper pipe from 20 to 45°C. How much mass (kg) of water is needed?

5.75 A house is being designed to use a thick concrete floor mass as thermal storage material for solar energy heating. The concrete is 30 cm thick and the area exposed to the sun during the daytime is 4 m × 6 m. It is expected that this mass will undergo an average temperature rise of about 3°C during the day. How much energy will be available for heating during the nighttime hours?

5.76 A copper block of volume 1 L is heat treated at 500°C and now cooled in a 200-L oil bath initially at 20°C, as shown in Fig. P5.76. Assuming no heat transfer with the surroundings, what is the final temperature?

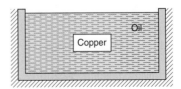

FIGURE P5.76

5.77 A 1-kg steel pot contains 1 kg of liquid water, both at 15°C. It is now put on the stove, where it is heated to the boiling point of the water. Neglect any air being heated and find the total amount of energy needed.

5.78 A car with mass 1275 kg is driven at 60 km/h when the brakes are applied quickly to decrease its speed to 20 km/h. Assume that the brake pads have a 0.5-kg mass with a heat capacity of 1.1 kJ/kg K and that the brake disks/drums are 4.0 kg of steel. Further assume that both masses are heated uniformly. Find the temperature increase in the brake assembly.

5.79 Saturated, $x = 1\%$, water at 25°C is contained in a hollow spherical aluminum vessel with inside diameter of 0.5 m and a 1-cm-thick wall. The vessel is heated until the water inside becomes saturated vapor. Considering the vessel and water together as a control mass, calculate the heat transfer for the process.

5.80 A 25-kg steel tank initially at −10°C is filled up with 100 kg of milk (assumed to have the same properties as water) at 30°C. The milk and the steel come to a uniform temperature of +5°C in a storage room. How much heat transfer is needed for this process?

5.81 An engine, shown in Fig. P5.81, consists of a 100-kg cast iron block with a 20-kg aluminum head, 20 kg of steel parts, 5 kg of engine oil, and 6 kg of glycerine (antifreeze). Everything begins at 5°C, and as the engine starts we want to know how hot it becomes if it absorbs a net of 7000 kJ before it reaches a steady uniform temperature.

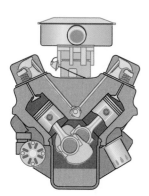

Automobile engine   **FIGURE P5.81**

Properties (u, h, C_v, and C_p), Ideal Gas

5.82 Use the ideal-gas air Table A.7 to evaluate the heat capacity C_p at 300 K as a slope of the curve $h(T)$ by

$\Delta h / \Delta T$. How much larger is it at 1000 K and at 1500 K?

5.83 We want to find the change in u for carbon dioxide between 600 K and 1200 K.
   a. Find it from a constant C_{v0} from Table A.5.
   b. Find it from a C_{v0} evaluated from the equation in Table A.6 at the average T.
   c. Find it from the values of u listed in Table A.8.

5.84 Do Problem 5.83 for oxygen gas.

5.85 Water at 20°C and 100 kPa is brought to 100 kPa and 1500°C. Find the change in the specific internal energy, using the water tables and ideal gas tables.

5.86 We want to find the increase in temperature of nitrogen gas at 1200 K when the specific internal energy is increased with 40 kJ/kg.
   a. Find it from a constant C_{v0} from Table A.5.
   b. Find it from a C_{v0} evaluated from the equation in Table A.6 at 1200 K.
   c. Find it from the values of u listed in Table A.8.

5.87 For a particular application the change in enthalpy of carbon dioxide from 30 to 1500°C at 100 kPa is needed. Consider the following methods and indicate the most accurate one.
   a. Using a constant specific heat and reading the value from Table A.5.
   b. Using a constant specific heat and obtaining the value at average temperature from the equation in Table A.6.
   c. Using a variable specific heat and integrating the equation in Table A.6.
   d. Reading the enthalpy from ideal gas tables in Table A.8.

5.88 An ideal gas is heated from 500 to 1500 K. Find the change in enthalpy using constant specific heat from Table A.5 (room temperature value) and discuss the accuracy of the result if the gas is
   a. Argon
   b. Oxygen
   c. Carbon dioxide

Energy Equation: Ideal Gas

5.89 A 250-L rigid tank contains methane at 500 K, 1500 kPa. It is now cooled down to 300 K. Find the mass of methane and the heat transfer using (a) the ideal-gas and (b) the methane tables.

5.90 A rigid insulated tank is separated into two rooms by a stiff plate. Room A of 0.5 m³ contains air at 250 kPa and 300 K and room B of 1 m³ has air at 150 kPa and 1000 K. The plate is removed and the air comes to a uniform state without any heat transfer. Find the final pressure and temperature.

5.91 A rigid container has 2 kg of carbon dioxide gas at 100 kPa and 1200 K that is heated to 1400 K. Solve for the heat transfer using (a) the heat capacity from Table A.5 and (b) properties from Table A.8.

5.92 Do the previous problem for nitrogen, N_2, gas.

5.93 A 10-m-high cylinder, with a cross-sectional area of 0.1 m², has a massless piston at the bottom with water at 20°C on top of it, as shown in Fig. P5.93. Air at 300 K, with a volume of 0.3 m³, under the piston is heated so that the piston moves up, spilling the water out over the side. Find the total heat transfer to the air when all the water has been pushed out.

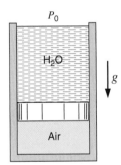

FIGURE P5.93

5.94 Find the heat transfer in Problem 4.43.

5.95 An insulated cylinder is divided into two parts of 1 m³ each by an initially locked piston, as shown in Fig. P5.95. Side A has air at 200 kPa, 300 K, and side B has air at 1.0 MPa, 1000 K. The piston is now unlocked so that it is free to move, and it

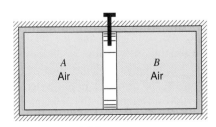

FIGURE P5.95

conducts heat so that the air comes to a uniform temperature $T_A = T_B$. Find the mass in both A and B and the final T and P.

5.96 A piston/cylinder contains air at 600 kPa, 290 K and a volume of 0.01 m³. A constant-pressure process gives 54 kJ of work out. Find the final temperature of the air and the heat transfer input.

5.97 A cylinder with a piston restrained by a linear spring contains 2 kg of carbon dioxide at 500 kPa and 400°C. It is cooled to 40°C, at which point the pressure is 300 kPa. Calculate the heat transfer for the process.

5.98 Water at 100 kPa and 400 K is heated electrically adding 700 kJ/kg in a constant pressure process. Find the final temperature using

a. The water Table B.1

b. The ideal-gas Table A.8

c. Constant specific heat from Table A.5

5.99 A piston/cylinder has 0.5 kg of air at 2000 kPa, 1000 K as shown. The cylinder has stops, so $V_{min} = 0.03$ m³. The air now cools to 400 K by heat transfer to the ambient. Find the final volume and pressure of the air (does it hit the stops?) and the work and heat transfer in the process.

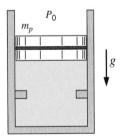

FIGURE P5.99

5.100 A spring-loaded piston/cylinder contains 1.5 kg of air at 27°C and 160 kPa. It is now heated to 900 K in a process where the pressure is linear in volume to a final volume of twice the initial volume. Plot the process in a $P–v$ diagram and find the work and heat transfer.

5.101 Air in a piston/cylinder assembly at 200 kPa and 600 K is expanded in a constant-pressure process to twice the initial volume, state 2, as shown in Fig. P5.101. The piston is then locked with a pin and heat is transferred to a final temperature of 600 K. Find P, T, and h for states 2 and 3, and find the work and heat transfer in both processes.

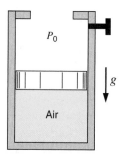

FIGURE P5.101

5.102 A vertical piston/cylinder setup has a linear spring mounted on the piston so that at zero cylinder volume a balancing pressure inside is zero. The cylinder contains 0.25 kg of air at 500 kPa and 27°C. Heat is now added so that the volume doubles.

a. Show the process path in a $P–V$ diagram.

b. Find the final pressure and temperature.

c. Find the work and heat transfer.

Energy Equation: Polytropic Process

5.103 A piston/cylinder device contains 0.1 kg of air at 300 K and 100 kPa. The air is now slowly compressed in an isothermal (T = constant) process to a final pressure of 250 kPa. Show the process in a $P–V$ diagram and find both the work and heat transfer in the process.

5.104 Oxygen at 300 kPa and 100°C is in a piston/cylinder arrangement with a volume of 0.1 m³. It is now compressed in a polytropic process with exponent n = 1.2 to a final temperature of 200°C. Calculate the heat transfer for the process.

5.105 A piston/cylinder setup contains 0.001 m³ air at 300 K and 150 kPa. The air is now compressed in a process in which $PV^{1.25} = C$ to a final pressure of 600 kPa. Find the work performed by the air and the heat transfer.

5.106 Helium gas expands from 125 kPa, 350 K and 0.25 m³ to 100 kPa in a polytropic process with n = 1.667. How much heat transfer is involved?

5.107 A piston/cylinder assembly in a car contains 0.2 L of air at 90 kPa and 20°C, as shown in Fig. P5.107. The air is compressed in a quasi-equilibrium polytropic process with polytropic exponent n = 1.25

to a final volume six times smaller. Determine the final pressure and temperature, and the heat transfer for the process.

FIGURE P5.107

5.108 A piston/cylinder has nitrogen gas at 750 K and 1500 kPa shown in Fig. P5.108. Now it is expanded in a polytropic process with n = 1.2 to P = 750 kPa. Find the final temperature, the specific work and specific heat transfer in the process.

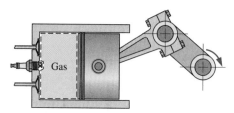

FIGURE P5.108

5.109 A piston/cylinder arrangement of initial volume 0.025 m³ contains saturated water vapor at 180°C. The steam now expands in a polytropic process with exponent n = 1 to a final pressure of 200 kPa while it does work against the piston. Determine the heat transfer for this process.

5.110 Air is expanded from 400 kPa and 600 K in a polytropic process to 150 kPa and 400 K in a piston/cylinder arrangement. Find the polytropic exponent n and the work and heat transfer per kg of air using constant heat capacity from Table A.5.

5.111 A piston/cylinder assembly has 1 kg of propane gas at 700 kPa and 40°C. The piston cross-sectional area is 0.5 m², and the total external force restraining the piston is directly proportional to the cylinder volume squared. Heat is transferred to the propane until its temperature reaches 700°C. Determine the final pressure inside the cylinder, the work done by the propane, and the heat transfer during the process.

5.112 An air pistol contains compressed air in a small cylinder, as shown in Fig. P5.112. Assume that the volume is 1 cm³, the pressure is 1 MPa, and the temperature is 27°C when armed. A bullet, with $m = 15$ g, acts as a piston initially held by a pin (trigger); when released, the air expands in an isothermal process ($T =$ constant). If the air pressure is 0.1 MPa in the cylinder as the bullet leaves the gun, find

a. the final volume and the mass of air

b. the work done by the air and work done on the atmosphere

c. the work done to the bullet and the bullet exit velocity

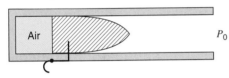

FIGURE P5.112

5.113 A spherical balloon contains 2 kg of R-22 at 0°C with a quality of 30%. This system is heated until the pressure in the balloon reaches 600 kPa. For this process, it can be assumed that the pressure in the balloon is directly proportional to the balloon diameter. How does pressure vary with volume and what is the heat transfer for the process?

5.114 Calculate the heat transfer for the process in Problem 4.55.

5.115 A piston/cylinder setup contains argon gas at 140 kPa and 10°C, and the volume is 100 L. The gas is compressed in a polytropic process to 700 kPa and 280°C. Calculate the polytropic exponent and the heat transfer during the process.

Energy Equation in Rate Form

5.116 A crane lifts a load of 450 kg vertically upward with a power input of 1 kW. How fast can the crane lift the load?

5.117 A computer in a closed room of volume 200 m³ dissipates energy at a rate of 10 kW. The room has 50 kg of wood, 25 kg of steel, and air, with all material at 300 K and 100 kPa. Assuming all the mass heats up uniformly, how long will it take to increase the temperature 10°C?

5.118 The rate of heat transfer to the surroundings from a person at rest is about 400 kJ/h. Suppose that the ventilation system fails in an auditorium containing 100 people. Assume the energy goes into the air of volume 1500 m³ initially at 300 K and 101 kPa. Find the rate (degrees per minute) of the air temperature change.

5.119 A piston/cylinder of cross-sectional area 0.01 m² maintains constant pressure. It contains 1 kg of water with a quality of 5% at 150°C. If we heat so that 1 g/s liquid turns into vapor, what is the rate of heat transfer needed?

5.120 The heaters in a spacecraft suddenly fail. Heat is lost by radiation at the rate of 100 kJ/h, and the electric instruments generate 75 kJ/h. Initially, the air is at 100 kPa and 25°C with a volume of 10 m³. How long will it take to reach an air temperature of −20°C?

5.121 A steam-generating unit heats saturated liquid water at constant pressure of 200 kPa in a piston/cylinder device. If 1.5 kW of power is added by heat transfer, find the rate (kg/s) at which saturated vapor is made.

5.122 A small elevator is being designed for a construction site. It is expected to carry four 75-kg workers to the top of a 100-m-tall building in less than 2 min. The elevator cage will have a counterweight to balance its mass. What is the smallest size (power) electric motor that can drive this unit?

5.123 As fresh poured concrete hardens, the chemical transformation releases energy at a rate of 2 W/kg. Assume the center of a poured layer does not have any heat loss and that it has an average heat capacity of 0.9 kJ/kg K. Find the temperature rise during 1 h of the hardening (curing) process.

5.124 A 100-W heater is used to melt 2 kg of solid ice at −10°C to liquid at +5°C at a constant pressure of 150 kPa.

a. Find the change in the total volume of the water.

b. Find the energy the heater must provide to the water.

c. Find the time the process will take assuming uniform T in the water.

5.125 Water is in a piston/cylinder maintaining constant P at 700 kPa, quality 90% with a volume of 0.1

m^3. A heater is turned on, heating the water with 2.5 kW. How long does it take to vaporize all the liquid?

Review Problems

5.126 Ten kilograms of water in a piston/cylinder setup with constant pressure are at 450°C and occupy a volume of 0.633 m^3. The system is now cooled to 20°C. Show the P–v diagram and find the work and heat transfer for the process.

5.127 Consider the system shown in Fig. P5.127. Tank A has a volume of 100 L and contains saturated vapor R-134a at 30°C. When the valve is cracked open, R-134a flows slowly into cylinder B. The piston requires a pressure of 200 kPa in cylinder B to raise it. The process ends when the pressure in tank A has fallen to 200 kPa. During this process heat is exchanged with the surroundings such that the R-134a always remains at 30°C. Calculate the heat transfer for the process.

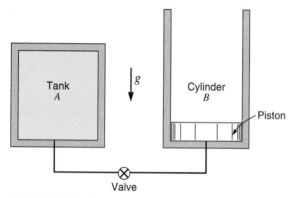

FIGURE P5.127

5.128 Ammonia, NH_3, is contained in a sealed rigid tank at 0°C, $x = 50\%$ and is then heated to 100°C. Find the final state P_2, u_2 and the specific work and heat transfer.

5.129 A piston/cylinder setup contains 1 kg of ammonia at 20°C with a volume of 0.1 m^3, as shown in Fig. P5.129. Initially the piston rests on some stops with the top surface open to the atmosphere, P_0, so that a pressure of 1400 kPa is required to lift it. To what temperature should the ammonia be heated to lift the piston? If it is heated to saturated vapor find the final temperature, volume, and heat transfer, $_1Q_2$.

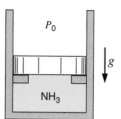

FIGURE P5.129

5.130 A piston held by a pin in an insulated cylinder, shown in Fig. P5.130, contains 2 kg of water at 100°C, with a quality of 98%. The piston has a mass of 102 kg, with cross-sectional area of 100 cm^2, and the ambient pressure is 100 kPa. The pin is released, which allows the piston to move. Determine the final state of the water, assuming the process to be adiabatic.

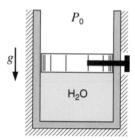

FIGURE P5.130

5.131 A piston/cylinder arrangement has a linear spring and the outside atmosphere acting on the piston shown in Fig. P5.131. It contains water at 3 MPa and 400°C with a volume of 0.1 m^3. If the piston is at the bottom, the spring exerts a force such that a pressure of 200 kPa inside is required to balance the forces. The system now cools until the pressure reaches 1 MPa. Find the heat transfer for the process.

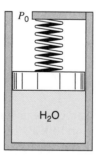

FIGURE P5.131

5.132 Consider the piston/cylinder arrangement shown in Fig. P5.132. A frictionless piston is free to move between two sets of stops. When the piston rests on the lower stops, the enclosed volume is 400 L. When the piston reaches the upper stops, the volume is 600 L. The cylinder initially contains water at 100 kPa, with 20% quality. It is heated until the water eventually exists as saturated vapor. The mass of the piston requires 300 kPa pressure to move it against the outside ambient pressure. Determine the final pressure in the cylinder, the heat transfer, and the work for the overall process.

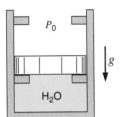

FIGURE P5.132

5.133 A piston/cylinder setup, shown in Fig. P5.133, contains R-12 at $-30°C$, $x = 20\%$. The volume is 0.2 m^3. It is known that $V_{stop} = 0.4 \text{ m}^3$, and if the piston sits at the bottom, the spring force balances the other loads on the piston. The system is now heated up to 20°C. Find the mass of the fluid and show the P–v diagram. Find the work and heat transfer.

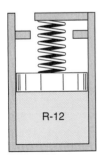

FIGURE P5.133

5.134 A piston/cylinder arrangement B is connected to a 1-m^3 tank A by a line and valve, shown in Fig. P5.134. Initially both contain water, with A at 100 kPa, saturated vapor and B at 400°C, 300 kPa, 1 m^3. The valve is now opened, and the water in both A and B comes to a uniform state.

a. Find the initial mass in A and B.

b. If the process results in $T_2 = 200°C$, find the heat transfer and the work.

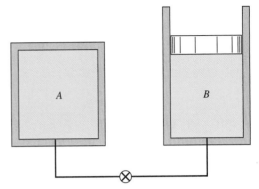

FIGURE P5.134

5.135 A small flexible bag contains 0.1 kg of ammonia at $-10°C$ and 300 kPa. The bag material is such that the pressure inside varies linear with volume. The bag is left in the sun with an incident radiation of 75 W, losing energy with an average 25 W to the ambient ground and air. After a while the bag is heated to 30°C at which time the pressure is 1000 kPa. Find the work and heat transfer in the process and the elapsed time.

5.136 Water at 150°C, 50% quality is contained in a cylinder/piston arrangement with initial volume 0.05 m^3. The loading of the piston is such that the inside pressure is linear with the square root of volume as $P = 100 + CV^{0.5}$ kPa. Now heat is transferred to the cylinder to a final pressure of 600 kPa. Find the heat transfer in the process.

5.137 A 1-m^3 tank containing air at 25°C and 500 kPa is connected through a valve to another tank containing 4 kg of air at 60°C and 200 kPa. Now the valve is opened and the entire system reaches thermal equilibrium with the surroundings at 20°C. Assume constant specific heat at 25°C and determine the final pressure and the heat transfer.

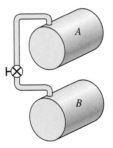

FIGURE P5.137

5.138 A closed cylinder is divided into two rooms by a frictionless piston held in place by a pin, as shown

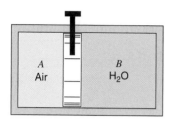

FIGURE P5.138

in Fig. P5.138. Room *A* has 10 L of air at 100 kPa, 30°C, and room *B* has 300 L of saturated water vapor at 30°C. The pin is pulled, releasing the piston, and both rooms come to equilibrium at 30°C, and as the water is compressed it becomes two-phase. Considering a control mass of the air and water, determine the work done by the system and the heat transfer to the cylinder.

ENGLISH UNIT PROBLEMS

English Unit Concept Problems

5.139E What is 1 cal in English units? What is 1 Btu in ft lbf?

5.140E Work as $F \, \Delta x$ has units of lbf ft. What is that in Btu?

5.141E A 2500-lbm car is accelerated from 25 mi/h to 40 mi/h. How much work is that?

5.142E A crane uses 7000 Btu/h to raise a 200-lbm box 60 ft. How much time does it take?

5.143E I have 4 lbm of liquid water at 70 F, 15 psia. I now add 20 Btu of energy at a constant pressure. How hot does it get if it is heated? How fast does it move if it is pushed by a constant horizontal force? How high does it go if it is raised straight up?

5.144E Air is heated from 540 R to 640 R at $V = C$. Find $_1q_2$. What if the air is heated from 2400 to 2500 R?

English Unit Problems

5.145E Airplane takeoff from an aircraft carrier is assisted by a steam-driven piston/cylinder with an average pressure of 200 psia. A 38 500-lbm airplane should be accelerated from zero to a speed of 100 ft/s with 30% of the energy coming from the steam piston. Find the needed piston displacement volume.

5.146E A hydraulic hoist raises a 3650-lbm car 6 ft in an auto repair shop. The hydraulic pump has a constant pressure of 100 lbf/in^2 on its piston. What is the increase in potential energy of the car and how much volume should the pump displace to deliver that amount of work?

5.147E A piston motion moves a 50-lbm hammerhead vertically down 3 ft from rest to a velocity of 150 ft/s in a stamping machine. What is the change in total energy of the hammerhead?

5.148E Find the missing properties and give the phase of the substance.

a. H_2O, $u = 1000$ Btu/lbm, $h = ? v = ?$
        $T = 270$ F, $x = ?$
b. H_2O, $u = 450$ Btu/lbm, $T = ? x = ?$
        $P = 1500$ lbf/in^2, $v = ?$
c. R-22, $T = 30$ F, $h = ? x = ?$
        $P = 75$ lbf/in^2,

5.149E Find the missing properties among (P, T, v, u, h) together with x, if applicable, and give the phase of the substance.

a. R-22, $T = 50$ F, $u = 85$ Btu/lbm
b. H_2O, $T = 600$ F, $h = 1322$ Btu/lbm
c. R-22, $P = 150$ lbf/in^2, $h = 115.5$ Btu/lbm

5.150E Find the missing properties among (P, T, v, u, h) together with x, if applicable, and give the phase of the substance.

a. R-134a, $T = 140$ F, $h = 185$ Btu/lbm
b. NH_3, $T = 170$ F, $P = 60$ lbf/in^2
c. R-134a, $T = 100$ F, $u = 175$ Btu/lbm

5.151E A cylinder fitted with a frictionless piston contains 4 lbm of superheated refrigerant R-134a vapor at 400 lbf/in^2, 200 F. The cylinder is now cooled so that the R-134a remains at constant pressure until it reaches a quality of 75%. Calculate the heat transfer in the process.

5.152E Ammonia at 30 F, quality 60% is contained in a rigid 8-ft^3 tank. The tank and ammonia are now heated to a final pressure of 140 lbf/in^2. Determine the heat transfer for the process.

5.153E Water in a 6-ft³ closed, rigid tank is at 200 F, 90% quality. The tank is then cooled to 20 F. Calculate the heat transfer during the process.

5.154E A constant-pressure piston/cylinder has 2 lbm of water at 1100 F and 2.26 ft³. It is now cooled to occupy 1/10 of the original volume. Find the heat transfer in the process.

5.155E A piston/cylinder arrangement has the piston loaded with outside atmospheric pressure and the piston mass to a pressure of 20 lbf/in², shown in Fig. P5.50. It contains water at 25 F, which is then heated until the water becomes saturated vapor. Find the final temperature and specific work and heat transfer for the process.

5.156E A water-filled reactor with volume of 50 ft³ is at 2000 lbf/in², 560 F and placed inside a containment room, as shown in Fig. P5.48. The room is well insulated and initially evacuated. Due to a failure, the reactor ruptures and the water fills the containment room. Find the minimum room volume so the final pressure does not exceed 30 lbf/in².

5.157E A piston/cylinder contains 2 lbm of liquid water at 70 F, and 30 lbf/in². There is a linear spring mounted on the piston such that when the water is heated the pressure reaches 300 lbf/in² with a volume of 4 ft³. Find the final temperature and plot the P–v diagram for the process. Calculate the work and the heat transfer for the process.

5.158E A twenty-pound-mass of water in a piston/cylinder with constant pressure is at 1100 F and a volume of 22.6 ft³. It is now cooled to 100 F. Show the P–v diagram and find the work and heat transfer for the process.

5.159E A vertical cylinder fitted with a piston contains 10 lbm of R-22 at 50 F, shown in Fig. P5.64. Heat is transferred to the system, causing the piston to rise until it reaches a set of stops at which point the volume has doubled. Additional heat is transferred until the temperature inside reaches 120 F, at which point the pressure inside the cylinder is 200 lbf/in².

a. What is the quality at the initial state?

b. Calculate the heat transfer for the overall process.

5.160E A piston/cylinder contains 2 lbm of water at 70 F with a volume of 0.1 ft³, shown in Fig. P5.129. Initially the piston rests on some stops with the top surface open to the atmosphere, P_0, so a pressure of 40 lbf/in² is required to lift it. To what temperature should the water be heated to lift the piston? If it is heated to saturated vapor, find the final temperature, volume, and the heat transfer.

5.161E Two tanks are connected by a valve and line, as shown in Fig. P5.62. The volumes are both 35 ft³ with R-134a at 70 F, quality 25% in A, and tank B is evacuated. The valve is opened, and saturated vapor flows from A into B until the pressures become equal. The process occurs slowly enough that all temperatures stay at 70 F during the process. Find the total heat transfer to the R-134a during the process.

5.162E Ammonia, NH_3, is contained in a sealed rigid tank at 30 F, $x = 50\%$ and is then heated to 200 F. Find the final state P_2, u_2 and the specific work and heat transfer.

5.163E Water at 70 F, 15 lbf/in², is brought to 30 lbf/in², 2700 F. Find the change in the specific internal energy, using the water tables and ideal-gas table.

5.164E A car with mass 3250 lbm is driven at 60 mi/h when the brakes are applied to quickly decrease its speed to 20 mi/h. Assume the brake pads are 1 lbm/in with a heat capacity of 0.2 Btu/lbm R, the brake disks/drums are 8 lbm of steel, and both masses are heated uniformly. Find the temperature increase in the brake assembly.

5.165E A 2-lbm steel pot contains 2 lbm of liquid water at 60 F. It is now put on the stove, where it is heated to the boiling point of the water. Neglect any air being heated and find the total amount of energy needed.

5.166E A copper block of volume 60 in³ is heat treated at 900 F and now cooled in a 3-ft³ oil bath initially at 70 F. Assuming no heat transfer with the surroundings, what is the final temperature?

5.167E An engine, shown in Fig. P5.81, consists of a 200-lbm cast iron block with a 40-lbm aluminum head, 40 lbm of steel parts, 10 lbm of engine oil and 12 lbm of glycerine (antifreeze). Everything begins at 40 F, and as the engine starts it absorbs a net of 7000 Btu before it reaches a steady uniform temperature. We want to know how hot it becomes.

5.168E A cylinder with a piston restrained by a linear spring contains 4 lbm of carbon dioxide at 70 lbf/in^2, 750 F. It is cooled to 75 F, at which point the pressure is 45 lbf/in^2. Calculate the heat transfer for the process.

5.169E An insulated cylinder is divided into two parts of 10 ft^3 each by an initially locked piston. Side A has air at 2 atm, 600 R, and side B has air at 10 atm, 2000 R, as shown in Fig. P5.95. The piston is now unlocked so it is free to move, and it conducts heat so the air comes to a uniform temperature $T_A = T_B$. Find the mass in both A and B and also the final T and P.

5.170E A 65-gal rigid tank contains methane gas at 900 R, 200 psia. It is now cooled down to 540 R. Assume an ideal gas and find the needed heat transfer.

5.171E Air in a piston/cylinder at 30 lbf/in^2, 1080 R is shown in Fig. P5.101. It is expanded in a constant-pressure process to twice the initial volume (state 2). The piston is then locked with a pin, and heat is transferred to a final temperature of 1080 R. Find P, T, and h for states 2 and 3, and find the work and heat transfer in both processes.

5.172E A 30-ft-high cylinder, cross-sectional area 1 ft^2, has a massless piston at the bottom with water at 70 F on top of it, as shown in Fig. P5.93. Air at 540 R, volume 10 ft^3 under the piston is heated so that the piston moves up, spilling the water out over the side. Find the total heat transfer to the air when all the water has been pushed out.

5.173E An air pistol contains compressed air in a small cylinder, as shown in Fig. P5.112. Assume that the volume is 1 in^3, pressure is 10 atm, and the temperature is 80 F when armed. A bullet, $m = 0.04$ lbm, acts as a piston initially held by a pin (trigger); when released, the air expands in an isothermal process ($T = $ constant). If the air pressure is 1 atm in the cylinder as the bullet leaves the gun, find

a. the final volume and the mass of air.

b. the work done by the air and work done on the atmosphere.

c. the work to the bullet and the bullet exit velocity.

5.174E A piston/cylinder in a car contains 12 in^3 of air at 13 lbf/in^2, 68 F, shown in Fig. P5.107. The air is compressed in a quasi-equilibrium polytropic process with polytropic exponent $n = 1.25$ to a final volume six times smaller. Determine the final pressure, temperature, and heat transfer for the process.

5.175E Oxygen at 50 lbf/in^2, 200 F is in a piston/cylinder arrangement with a volume of 4 ft^3. It is now compressed in a polytropic process with exponent, $n = 1.2$, to a final temperature of 400 F. Calculate the heat transfer for the process.

5.176E Helium gas expands from 20 psia, 600 R and 9 ft^3 to 15 psia in a polytropic process with $n = 1.667$. How much heat transfer is involved?

5.177E A cylinder fitted with a frictionless piston contains R-134a at 100 F, 80% quality, at which point the volume is 3 gal. The external force on the piston is now varied in such a manner that the R-134a slowly expands in a polytropic process to 50 lbf/in^2, 80 F. Calculate the work and the heat transfer for this process.

5.178E A piston/cylinder contains argon at 20 lbf/in^2, 60 F, and the volume is 4 ft^3. The gas is compressed in a polytropic process to 100 lbf/in^2, 550 F. Calculate the heat transfer during the process.

5.179E A small elevator is being designed for a construction site. It is expected to carry four 150-lbm workers to the top of a 300-ft-tall building in less than 2 min. The elevator cage will have a counterweight to balance its mass. What is the smallest size (power) electric motor that can drive this unit?

5.180E Water is in a piston/cylinder maintaining constant P at 330 F, quality 90%, with a volume of 4 ft^3. A heater is turned on heating the water with 10 000 Btu/h. What is the elapsed time to vaporize all the liquid?

5.181E A computer in a closed room of volume 5000 ft^3 dissipates energy at a rate of 10 hp. The room has 100 lbm of wood, 50 lbm of steel, and air, with all material at 540 R, 1 atm. Assuming all the mass heats up uniformly, how much time will it take to increase the temperature by 20 F?

5.182E A closed cylinder is divided into two rooms by a frictionless piston held in place by a pin, as shown in Fig. P5.138. Room *A* has 0.3 ft^3 air at 14.7 lbf/in^2, 90 F, and room *B* has 10 ft^3 saturated water vapor at 90 F. The pin is pulled, releasing the piston, and both rooms come to equilibrium at 90 F. Considering a control mass of the air and water, determine the work done by the system and the heat transfer to the cylinder.

COMPUTER, DESIGN, AND OPEN-ENDED PROBLEMS

5.183 Use the supplied software to track the process in Problem 5.37 in steps of 10°C until the two-phase region is reached, after that step with jumps of 5% in the quality. At each step write out *T*, *x*, and the heat transfer to reach that state from the initial state.

5.184 For one of the substances in Table A.6, compare the enthalpy change between any two temperatures, T_1 and T_2, as calculated by integrating the specific heat equation; by assuming constant specific heat at the average temperature; and by assuming constant specific heat at temperature T_1.

5.185 Track the process described in Problem 5.54 so that you can sketch the amount of heat transfer added and the work given out as a function of the volume.

5.186 Using states with given (*P*, v) and properties from the supplied software, track the process in Problem 5.57. Select five pressures away from the initial toward the final pressure so that you can plot the temperature, the heat added, and the work given out as a function of the volume.

5.187 Examine the sensitivity of the final pressure to the containment room volume in Problem 5.48. Solve for the volume for a range of final pressures, 100–250 kPa, and sketch the pressure versus volume curve.

5.188 Write a program to solve Problem 5.78 for a range of initial velocities. Let the car mass and final velocity be input variables.

5.189 Write a program for Problem 5.107, where the initial state, the volume ratio, and the polytropic exponent are input variables. To simplify the formulation, use constant specific heat.

5.190 Consider a general version of Problem 5.89 with a substance listed in Table A.6. Write a program where the initial temperature and pressure, and the final temperature are program inputs.

5.191 Examine a process where air at 300 K, 100 kPa is compressed in a piston/cylinder arrangement to 600 kPa. Assume the process is polytropic with exponents in the 1.2–1.6 range. Find the work and heat transfer per unit mass of air. Discuss the different cases and how they may be accomplished by insulating the cylinder or by providing heating or cooling.

5.192 A cylindrical tank of height 2 m with a cross-sectional area of 0.5 m^2 contains hot water at 80°C, 125 kPa. It is in a room with temperature $T_0 = 20$°C, so it slowly loses energy to the room air proportional to the temperature difference as

$$\dot{Q}_{loss} = CA(T - T_0)$$

with the tank surface area, *A*, and *C* is a constant. For different values of the constant *C*, estimate the time it takes to bring the water to 50°C. Make enough simplifying assumptions so that you can solve the problem mathematically, that is find a formula for *T*(*t*).

6 FIRST-LAW ANALYSIS FOR A CONTROL VOLUME

In the preceding chapter we developed the first-law analysis (energy balance) for a control mass going through a process. Many applications in thermodynamics do not readily lend themselves to a control mass approach but are conveniently handled by the more general control volume technique, as discussed in Chapter 2. The present chapter is concerned with development of the control volume forms of the conservation of mass and energy in situations where there are flows of substance present.

6.1 CONSERVATION OF MASS AND THE CONTROL VOLUME

A control volume is a volume in space in which one has interest for a particular study or analysis. The surface of this control volume is referred to as a control surface and always consists of a closed surface. The size and shape of the control volume are completely arbitrary and are so defined as to best suit the analysis to be made. The surface may be fixed, or it may move so that it expands or contracts. However, the surface must be defined relative to some coordinate system. In some analyses it may be desirable to consider a rotating or moving coordinate system and to describe the position of the control surface relative to such a coordinate system.

Mass as well as heat and work can cross the control surface, and the mass in the control volume, as well as the properties of this mass, can change with time. Figure 6.1 shows a schematic diagram of a control volume that includes heat transfer, shaft work, moving boundary work, accumulation of mass within the control volume, and several mass flows. It is important to identify and label each flow of mass and energy and the parts of the control volume that can store (accumulate) mass.

Let us consider the conservation of mass law as it relates to the control volume. The physical law concerning mass, recalling Section 5.9, says that we cannot create or destroy mass. We will express this law in a mathematical statement about the mass in the control volume. To do this we must consider all the mass flows into and out of the control volume and the net increase of mass within the control volume. As a somewhat simpler control volume we consider a tank with a cylinder and piston and two pipes attached as shown in Fig. 6.2. The rate of change of mass inside the control volume can be different from zero if we add or take a flow of mass out as

$$\text{Rate of change} = +\text{in} - \text{out}$$

162

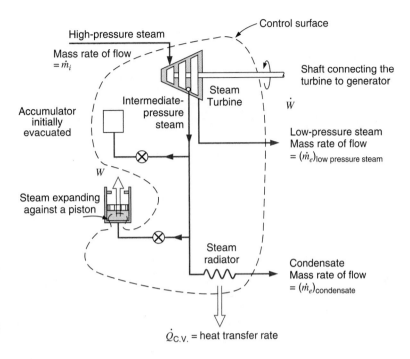

FIGURE 6.1
Schematic diagram of a
control volume showing
mass and energy transfers
and accumulation.

With several possible flows this is written as

$$\frac{dm_{\text{C.V.}}}{dt} = \sum \dot{m}_i - \sum \dot{m}_e \tag{6.1}$$

stating that if the mass inside the control volume changes with time it is because we add
some mass or take some mass out. There are no other means by which the mass inside the
control volume could change. Equation 6.1 expressing the conservation of mass is commonly termed the continuity equation. While this form of the equation is sufficient for the
majority of applications in thermodynamics, it is frequently rewritten in terms of the local
fluid properties in the study of fluid mechanics and heat transfer. In this text we are

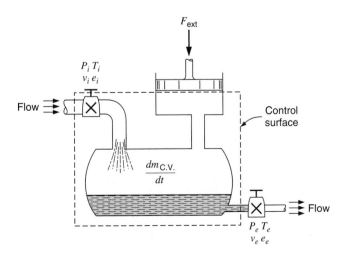

FIGURE 6.2
Schematic diagram of a
control volume for the
analysis of the continuity
equation.

FIGURE 6.3 The flow across a control volume surface with a flow cross-sectional area of A. Left of valve shown as an average velocity and to the right of valve shown as a distributed flow across area.

mainly concerned with the overall mass balance and thus consider Eq. 6.1 as the general expression for the continuity equation.

Since Eq. 6.1 is written for the total mass (lumped form) inside the control volume we may have to consider several contributions to the mass as

$$m_{C.V.} = \int \rho \, dV = \int (1/v)dV = m_A + m_B + m_C + \cdots$$

Such a summation is needed when the control volume has several accumulation units with different states of the mass.

Let us now consider the mass flow rates across the control volume surface in a little more detail. For simplicity we assume the fluid is flowing in a pipe or duct as illustrated in Fig. 6.3. We wish to relate the total flow rate that appears in Eq. 6.1 to the local properties of the fluid state. The flow across the control volume surface can be indicated with an average velocity shown to the left of the valve or with a distributed velocity over the cross section as shown to the right of the valve.

The volume flow rate is

$$\dot{V} = \mathbf{V}A = \int \mathbf{V}_{local} \, dA \tag{6.2}$$

so the mass flow rate becomes

$$\dot{m} = \rho_{avg} \dot{V} = \dot{V}/v = \int (\mathbf{V}_{local}/v)dA = \mathbf{V}A/v \tag{6.3}$$

where often the average velocity is used. It should be noted that this result, Eq. 6.3, has been developed for a stationary control surface and we tacitly assumed the flow was normal to the surface. This expression for the mass flow rate applies to any of the various flow streams entering or leaving the control volume, subject to the assumptions mentioned.

EXAMPLE 6.1   Air is flowing in a 0.2-m-diameter pipe at a uniform velocity of 0.1 m/s. The temperature is 25°C and the pressure 150 kPa. Determine the mass flow rate.

Solution

From Eq. 6.3 the mass flow rate is

$$\dot{m} = \mathbf{V}A/v$$

For air, using R from Table A.5, we have

$$v = \frac{RT}{P} = \frac{0.287 \text{ kJ/kg K} \times 298.2 \text{ K}}{150 \text{ kPa}} = 0.5705 \text{ m}^3/\text{kg}$$

The cross-sectional area is

$$A = \frac{\pi}{4}(0.2)^2 = 0.0314 \text{ m}^2$$

Therefore,

$$\dot{m} = \mathbf{V}A/v = 0.1 \text{ m/s} \times 0.0314 \text{ m}^2/0.5705 \text{ m}^3/\text{kg} = 0.0055 \text{ kg/s}$$

6.2 THE FIRST LAW OF THERMODYNAMICS FOR A CONTROL VOLUME

We have already considered the first law of thermodynamics for a control mass, which consists of a fixed quantity of mass, and noted, Eq. 5.5, that it may be written

$$E_2 - E_1 = {}_1Q_2 - {}_1W_2$$

We have also noted that this may be written as an instantaneous rate equation as

$$\frac{dE_{\text{C.M.}}}{dt} = \dot{Q} - \dot{W} \tag{6.4}$$

To write the first law as a rate equation for a control volume, we proceed in a manner analogous to that used in developing a rate equation for the law of conservation of mass. For this purpose a control volume is shown in Fig. 6.4 that involves rate of heat transfer, rates of work, and mass flows. The fundamental physical law states that we cannot create or destroy energy such that any rate of change of energy must be caused by rates of energy into or out of the control volume. We have already included rates of heat transfer and work in Eq. 6.4, so the additional explanations we need are associated with the mass flow rates.

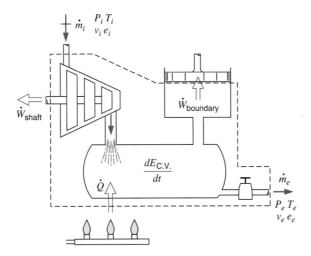

FIGURE 6.4
Schematic diagram to illustrate terms in the energy equation for a general control volume.

The fluid flowing across the control surface enters or leaves with an amount of energy per unit mass as

$$e = u + \frac{1}{2}\mathbf{V}^2 + gZ$$

relating to the state and position of the fluid. Whenever a fluid mass enters a control volume at state i, or exits at state e, there is a boundary movement work associated with that process.

To explain this in more detail consider an amount of mass flowing into the control volume. As this mass flows in there is a pressure at *its* back surface, so as this mass moves into the control volume *it* is being pushed by the mass behind *it*, which is the surroundings. The net effect is that after the mass has entered the control volume the surroundings have pushed it in against the local pressure with a velocity giving it a rate of work in the process. Similarly a fluid exiting the control volume at state e must push the surrounding fluid ahead of it, doing work on it, which is work leaving the control volume. The velocity and the area correspond to a certain volume per unit time entering the control volume, enabling us to relate that to the mass flow rate and the specific volume at the state of the mass going in. Now we are able to express the rate of flow work as

$$\dot{W}_{flow} = F\mathbf{V} = \int P\mathbf{V}\, dA = P\dot{V} = Pv\dot{m} \tag{6.5}$$

For the flow that leaves the control volume work is being done by the control volume, $P_e v_e \dot{m}_e$, and for the mass that enters, the surroundings do the rate of work, $P_i v_i \dot{m}_i$. The flow work per unit mass is then Pv, and the total energy associated with the flow of mass is

$$e + Pv = u + Pv + \frac{1}{2}\mathbf{V}^2 + gZ = h + \frac{1}{2}\mathbf{V}^2 + gZ \tag{6.6}$$

In this equation we have used the definition of the thermodynamic property enthalpy, and it is the appearance of the combination $(u + Pv)$ for the energy in connection with a mass flow that is the primary reason for the definition of the property enthalpy. Its introduction earlier in conjunction with the constant-pressure process was to facilitate use of the tables of thermodynamic properties at that time.

EXAMPLE 6.2   Assume we are standing next to the local city's main water line. The liquid water inside flows at a pressure of say 600 kPa (6 atm) with a temperature of about 10°C. We want to add a smaller amount, 1 kg, of liquid to the line through a side pipe and valve mounted on the main line. How much work will be involved in this process?

If the 1 kg of liquid water is in a bucket and we open the valve to the water main trying to pour it down into the pipe opening, we will realize that the water flows the other way. The water will flow from a higher to a lower pressure, that is, from inside the main line to the atmosphere (from 600 kPa to 101 kPa).

We must take the 1 kg of liquid water and put it into a piston cylinder (like a handheld pump) and attach the cylinder to the water pipe. Now we can press on the piston until the water pressure inside is 600 kPa and then open the valve to the main line and slowly squeeze the 1 kg of water in. The work done at the piston surface to the water is

$$W = \int P\, dV = P_{water}\, mv = 600 \text{ kPa} \times 1 \text{ kg} \times 0.001 \text{ m}^3/\text{kg} = 0.6 \text{ kJ}$$

and this is the necessary flow work for adding the 1 kg of liquid.

The extension of the first law of thermodynamics from Eq. 6.4 becomes

$$\frac{dE_{\text{C.V.}}}{dt} = \dot{Q}_{\text{C.V.}} - \dot{W}_{\text{C.V.}} + \dot{m}_i e_i - \dot{m}_e e_e + \dot{W}_{\text{flow in}} - \dot{W}_{\text{flow out}}$$

and the substitution of Eq. 6.5 gives

$$\frac{dE_{\text{C.V.}}}{dt} = \dot{Q}_{\text{C.V.}} - \dot{W}_{\text{C.V.}} + \dot{m}_i(e_i + P_i v_i) - \dot{m}_e(e_e + P_e v_e)$$

$$= \dot{Q}_{\text{C.V.}} - \dot{W}_{\text{C.V.}} + \dot{m}_i\left(h_i + \frac{1}{2}\mathbf{V}_i^2 + gZ_i\right) - \dot{m}_e\left(h_e + \frac{1}{2}\mathbf{V}_e^2 + gZ_e\right)$$

In this form of the energy equation the rate of work term is the sum of all shaft work terms and boundary work terms and any other types of work given out by the control volume; however, the flow work is now listed separately and included with the mass flow rate terms.

For the general control volume we may have several entering or leaving mass flow rates, so a summation over those terms is often needed. The final form of the first law of thermodynamics then becomes

$$\frac{dE_{\text{C.V.}}}{dt} = \dot{Q}_{\text{C.V.}} - \dot{W}_{\text{C.V.}} + \sum \dot{m}_i\left(h_i + \frac{1}{2}\mathbf{V}_i^2 + gZ_i\right) - \sum \dot{m}_e\left(h_e + \frac{1}{2}\mathbf{V}_e^2 + gZ_e\right) \tag{6.7}$$

expressing that the rate of change of energy inside the control volume is due to a net rate of heat transfer, a net rate of work (measured positive out), and the summation of energy fluxes due to mass flows into and out of the control volume. As with the conservation of mass, this equation can be written for the total control volume and can therefore be put in the lumped or integral form where

$$E_{\text{C.V.}} = \int \rho e \, dV = me = m_A e_A + m_B e_B + m_C e_C + \cdots$$

As the kinetic and potential energy terms per unit mass appear together with the enthalpy in all the flow terms, a shorter notation is often used

$$h_{\text{tot}} = h + \frac{1}{2}\mathbf{V}^2 + gZ$$

$$h_{\text{stag}} = h + \frac{1}{2}\mathbf{V}^2$$

defining the total enthalpy and the stagnation enthalpy (used in fluid mechanics). The shorter equation then becomes

$$\frac{dE_{\text{C.V.}}}{dt} = \dot{Q}_{\text{C.V.}} - \dot{W}_{\text{C.V.}} + \sum \dot{m}_i h_{\text{tot},i} - \sum \dot{m}_e h_{\text{tot},e} \tag{6.8}$$

giving the general energy equation on a rate form. All applications of the energy equation start with the form in Eq. 6.8, and for special cases this will result in a slightly simpler form as shown in the subsequent sections.

6.3 THE STEADY-STATE PROCESS

Our first application of the control volume equations will be to develop a suitable analytical model for the long-term steady operation of devices such as turbines, compressors, nozzles, boilers, condensers—a very large class of problems of interest in thermodynamic analysis. This model will not include the short-term transient start-up or shutdown of such devices, but only the steady operating period of time.

Let us consider a certain set of assumptions (beyond those leading to Eqs. 6.1 and 6.7) that lead to a reasonable model for this type of process, which we refer to as the steady-state process.

1. The control volume does not move relative to the coordinate frame.
2. The state of the mass at each point in the control volume does not vary with time.
3. As for the mass that flows across the control surface, the mass flux and the state of this mass at each discrete area of flow on the control surface do not vary with time. The rates at which heat and work cross the control surface remain constant.

As an example of a steady-state process consider a centrifugal air compressor that operates with constant mass rate of flow into and out of the compressor, constant properties at each point across the inlet and exit ducts, a constant rate of heat transfer to the surroundings, and a constant power input. At each point in the compressor the properties are constant with time, even though the properties of a given elemental mass of air vary as it flows through the compressor. Often, such a process is referred to as a steady-flow process, since we are concerned primarily with the properties of the fluid entering and leaving the control volume. However, in the analysis of certain heat transfer problems in which the same assumptions apply, we are primarily interested in the spatial distribution of properties, particularly temperature, and such a process is referred to as a steady-state process. Since this is an introductory book we will use the term steady-state process for both. The student should realize that the terms steady-state process and steady-flow process are both used extensively in the literature.

Let us now consider the significance of each of these assumptions for the steady-state process.

1. The assumption that the control volume does not move relative to the coordinate frame means that all velocities measured relative to the coordinate frame are also velocities relative to the control surface, and there is no work associated with the acceleration of the control volume.
2. The assumption that the state of the mass at each point in the control volume does not vary with time requires that

$$\frac{dm_{C.V.}}{dt} = 0$$

and also

$$\frac{dE_{C.V.}}{dt} = 0$$

Therefore, we conclude that for the steady-state process we can write, from Eqs. 6.1 and 6.7,

$$\text{Continuity equation: } \sum \dot{m}_i = \sum \dot{m}_e \tag{6.9}$$

$$\text{First law: } \dot{Q}_{C.V.} + \sum \dot{m}_i \left(h_i + \frac{\mathbf{V}_i^2}{2} + gZ_i \right) = \sum \dot{m}_e \left(h_e + \frac{\mathbf{V}_e^2}{2} + gZ_e \right) + \dot{W}_{C.V.} \tag{6.10}$$

3. The assumption that the various mass flows, states, and rates at which heat and work cross the control surface remain constant requires that every quantity in Eqs.

6.9 and 6.10 be steady with time. This means that application of Eqs. 6.9 and 6.10 to the operation of some device is independent of time.

Many of the applications of the steady-state model are such that there is only one flow stream entering and one leaving the control volume. For this type of process, we can write

$$\text{Continuity equation: } \dot{m}_i = \dot{m}_e = \dot{m} \tag{6.11}$$

$$\text{First law: } \dot{Q}_{\text{C.V.}} + \dot{m}\left(h_i + \frac{\mathbf{V}_i^2}{2} + gZ_i\right) = \dot{m}\left(h_e + \frac{\mathbf{V}_e^2}{2} + gZ_e\right) + \dot{W}_{\text{C.V.}} \tag{6.12}$$

Rearranging this equation, we have

$$q + h_i + \frac{\mathbf{V}_i^2}{2} + gZ_i = h_e + \frac{\mathbf{V}_e^2}{2} + gZ_e + w \tag{6.13}$$

where, by definition,

$$q = \frac{\dot{Q}_{\text{C.V.}}}{\dot{m}} \quad \text{and} \quad w = \frac{\dot{W}_{\text{C.V.}}}{\dot{m}} \tag{6.14}$$

Note that the units for q and w are kJ/kg. From their definition, q and w can be thought of as the heat transfer and work (other than flow work) per unit mass flowing into and out of the control volume for this particular steady-state process.

The symbols q and w are also used for the heat transfer and work per unit mass of a control mass. However, since it is always evident from the context whether it is a control mass (fixed mass) or control volume (involving a flow of mass) with which we are concerned, the significance of the symbols q and w will also be readily evident in each situation.

The steady-state process is often used in the analysis of reciprocating machines, such as reciprocating compressors or engines. In this case the rate of flow, which may actually be pulsating, is considered to be the average rate of flow for an integral number of cycles. A similar assumption is made regarding the properties of the fluid flowing across the control surface and the heat transfer and work crossing the control surface. It is also assumed that for an integral number of cycles the reciprocating device undergoes, the energy and mass within the control volume do not change.

A number of examples are now given to illustrate the analysis of steady-state processes.

6.4 EXAMPLES OF STEADY-STATE PROCESSES

In this section, we consider a number of examples of steady-state processes in which there is one fluid stream entering and one leaving the control volume, such that the first law can be written in the form of Eq. 6.13. Some may instead utilize control volumes that include more than one fluid stream, such that it is necessary to write the first law in the more general form of Eq. 6.10.

Heat Exchanger

A steady-state heat exchanger is a simple fluid flow through a pipe or system of pipes, where heat is transferred to or from the fluid. The fluid may be heated or cooled, and may or may not boil, liquid to vapor, or condense, vapor to liquid. One such example is the

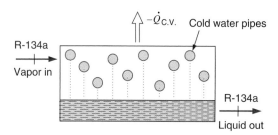

FIGURE 6.5 A refrigeration system condenser.

condenser in an R-134a refrigeration system, as shown in Fig. 6.5. Superheated vapor enters the condenser, and liquid exits. The process tends to occur at constant pressure, since a fluid flowing in a pipe usually undergoes only a small pressure drop, because of fluid friction at the walls. The pressure drop may or may not be taken into account in a particular analysis. There is no means for doing any work (shaft work, electrical work, etc.), and changes in kinetic and potential energies are commonly negligibly small. (One exception may be a boiler tube in which liquid enters and vapor exits at a much larger specific volume. In such a case, it may be necessary to check the exit velocity using Eq. 6.3.) The heat transfer in most heat exchangers is then found from Eq. 6.13 as the change in enthalpy of the fluid. In the condenser shown in Fig. 6.5, the heat transfer out of the condenser then goes to whatever is receiving it, perhaps a stream of air or of cooling water. It is often simpler to write the first law around the entire heat exchanger, including both flow streams, in which case there is little or no heat transfer with the surroundings. Such a situation is the subject of the following example.

EXAMPLE 6.3   Consider a water-cooled condenser in a large refrigeration system in which R-134a is the refrigerant fluid. The refrigerant enters the condenser at 1.0 MPa and 60°C, at the rate of 0.2 kg/s, and exits as a liquid at 0.95 MPa and 35°C. Cooling water enters the condenser at 10°C and exits at 20°C. Determine the rate at which cooling water flows through the condenser.

Control volume:	Condenser.
Sketch:	Fig. 6.6
Inlet states:	R-134a—fixed; water—fixed.
Exit states:	R134a—fixed; water—fixed.
Process:	Steady-state.
Model:	R-134a tables; steam tables.

Analysis

With this control volume we have two fluid streams, the R-134a and the water, entering and leaving the control volume. It is reasonable to assume that both kinetic and potential energy changes are negligible. We note that the work is zero, and we make the other reasonable assumption that there is no heat transfer across the control surface. Therefore, the first law, Eq. 6.10, reduces to

$$\sum \dot{m}_i h_i = \sum \dot{m}_e h_e$$

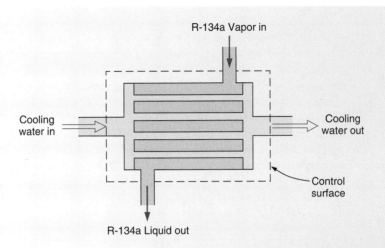

FIGURE 6.6
Schematic diagram of an
R-134a condenser.

Using the subscript r for refrigerant and w for water, we write

$$\dot{m}_r(h_i)_r + \dot{m}_w(h_i)_w = \dot{m}_r(h_e)_r + \dot{m}_w(h_e)_w$$

Solution

From the R-134a and steam tables, we have

$$(h_i)_r = 441.89 \text{ kJ/kg}, \quad (h_i)_w = 42.00 \text{ kJ/kg}$$

$$(h_e)_r = 249.10 \text{ kJ/kg}, \quad (h_e)_w = 83.95 \text{ kJ/kg}$$

Solving the above equation for $\dot{m}_w$, the rate of flow of water, we obtain

$$\dot{m}_w = \dot{m}_r \frac{(h_i - h_e)_r}{(h_e - h_i)_w} = 0.2 \text{ kg/s} \frac{(441.89 - 249.10) \text{ kJ/kg}}{(83.95 - 42.00) \text{ kJ/kg}} = 0.919 \text{ kg/s}$$

This problem can also be solved by considering two separate control volumes, one having the flow of R-134a across its control surface and the other having the flow of water across its control surface. Further, there is heat transfer from one control volume to the other.

The heat transfer for the control volume involving R-134a is calculated first. In this case the steady-state energy equation, Eq. 6.10, reduces to

$$\dot{Q}_{C.V.} = \dot{m}_r(h_e - h_i)_r$$

$$= 0.2 \text{ kg/s} \times (249.10 - 441.89) \text{ kJ/kg} = -38.558 \text{ kW}$$

This is also the heat transfer to the other control volume, for which $\dot{Q}_{C.V.} = +38.558$ kW.

$$\dot{Q}_{C.V.} = \dot{m}_w(h_e - h_i)_w$$

$$\dot{m}_w = \frac{38.558 \text{ kW}}{(83.95 - 42.00) \text{ kJ/kg}} = 0.919 \text{ kg/s}$$

Nozzle

A nozzle is a steady-state device whose purpose is to create a high-velocity fluid stream at the expense of the fluid's pressure. It is contoured in an appropriate manner to expand a flowing fluid smoothly to a lower pressure, thereby increasing its velocity. There is no means to do any work—there are no moving parts. There is little or no change in potential energy and usually little or no heat transfer. An exception is the large nozzle on a liquid-propellant rocket, such as was described in Section 1.7, in which the cold propellant is commonly circulated around the outside of the nozzle walls before going to the combustion chamber, in order to keep the nozzle from melting. This case, a nozzle with significant heat transfer, is the exception and would be noted in such an application. In addition, the kinetic energy of the fluid at the nozzle inlet is usually small and would be neglected if its value is not known.

EXAMPLE 6.4   Steam at 0.6 MPa and 200°C enters an insulated nozzle with a velocity of 50 m/s. It leaves at a pressure of 0.15 MPa and a velocity of 600 m/s. Determine the final temperature if the steam is superheated in the final state and the quality if it is saturated.

    Control volume: Nozzle.
    Inlet state: Fixed (see Fig. 6.7).
    Exit state: P_e known.
    Process: Steady-state.
    Model: Steam tables.

Analysis

We have

$$\dot{Q}_{C.V.} = 0 \quad \text{(nozzle insulated)}$$
$$\dot{W}_{C.V.} = 0$$
$$PE_i \approx PE_e$$

The first law (Eq. 6.13) yields

$$h_i + \frac{\mathbf{V}_i^2}{2} = h_e + \frac{\mathbf{V}_e^2}{2}$$

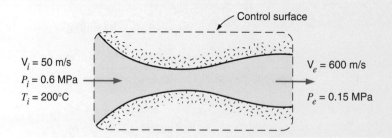

FIGURE 6.7
Illustration for Example 6.4.

Solution

Solving for h_e we obtain

$$h_e = 2850.1 + \left[\frac{(50)^2}{2 \times 1000} - \frac{(600)^2}{2 \times 1000} \right] \frac{m^2/s^2}{J/kJ} = 2671.4 \text{ kJ/kg}$$

The two properties of the fluid leaving that we now know are pressure and enthalpy, and therefore the state of this fluid is determined. Since h_e is less than h_g at 0.15 MPa, the quality is calculated.

$$h = h_f + x h_{fg}$$

$$2671.4 = 467.1 + x_e 2226.5$$

$$x_e = 0.99$$

EXAMPLE 6.4E Steam at 100 lbf/in², 400 F, enters an insulated nozzle with a velocity of 200 ft/s. It leaves at a pressure of 20.8 lbf/in² and a velocity of 2000 ft/s. Determine the final temperature if the steam is superheated in the final state, and the quality if it is saturated.

$$
\begin{array}{rl}
\textit{Control volume}: & \text{Nozzle.} \\
\textit{Inlet state}: & \text{Fixed (see Fig. 6.7E).} \\
\textit{Exit state}: & P_e \text{ known.} \\
\textit{Process}: & \text{Steady-state.} \\
\textit{Model}: & \text{Steam tables.}
\end{array}
$$

Analysis

$$\dot{Q}_{\text{C.V.}} = 0 \quad \text{(nozzle insulated)}$$

$$\dot{W}_{\text{C.V.}} = 0, \qquad \text{PE}_i = \text{PE}_e$$

First law (Eq. 6.13):

$$h_i + \frac{\mathbf{V}_i^2}{2} = h_e + \frac{\mathbf{V}_e^2}{2}$$

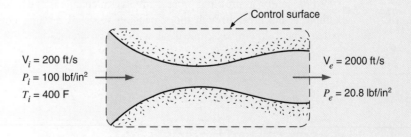

FIGURE 6.7E
Illustration for Example 6.4E.

Solution

$$h_e = 1227.5 + \frac{(200)^2}{2 \times 32.17 \times 778} - \frac{(2000)^2}{2 \times 32.17 \times 778} = 1148.3 \text{ Btu/lbm}$$

The two properties of the fluid leaving that we now know are pressure and enthalpy, and therefore the state of this fluid is determined. Since h_e is less than h_g at 20.8 lbf/in², the quality is calculated.

$$h = h_f + xh_{fg}$$

$$1148.3 = 198.31 + x_e 958.81$$

$$x_e = 0.99$$

Diffuser

A steady-state diffuser is a device constructed to decelerate a high-velocity fluid in a manner that results in an increase in pressure of the fluid. In essence, it is the exact opposite of a nozzle, and it may be thought of as a fluid flowing in the opposite direction through a nozzle, with the opposite effects. The assumptions are similar to those for a nozzle, with a large kinetic energy at the diffuser inlet and a small, but usually not negligible, kinetic energy at the exit being the only terms besides the enthalpies remaining in the first law, Eq. 6.13.

Throttle

A throttling process occurs when a fluid flowing in a line suddenly encounters a restriction in the flow passage. This may be a plate with a small hole in it, as shown in Fig. 6.8, it may be a partially closed valve protruding into the flow passage, or it may be a change to a much smaller diameter tube, called a capillary tube, which is normally found on a refrigerator. The result of this restriction is an abrupt pressure drop in the fluid, as it is forced to find its way through a suddenly smaller passageway. This process is drastically unlike the smoothly contoured nozzle expansion and area change, which results in a significant velocity increase. There is typically some increase in velocity in a throttle, but both inlet and exit kinetic energies are usually small enough to be neglected. There is no means for doing work and little or no change in potential energy. Usually, there is neither time nor opportunity for an appreciable heat transfer, such that the only terms left in the first law, Eq. 6.13, are the inlet and exit enthalpies. We conclude that a steady-state throttling process is approximately a pressure drop at constant enthalpy, and we will assume this to be the case unless otherwise noted.

Frequently, a throttling process involves a change in the phase of the fluid. A typical example is the flow through the expansion valve of a vapor-compression refrigeration system. The following example deals with this problem.

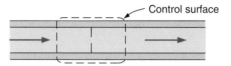

FIGURE 6.8 The throttling process.

EXAMPLE 6.5 Consider the throttling process across the expansion valve or through the capillary tube in a vapor-compression refrigeration cycle. In this process the pressure of the refrigerant drops from the high pressure in the condenser to the low pressure in the evaporator, and during this process some of the liquid flashes into vapor. If we consider this process to be adiabatic, the quality of the refrigerant entering the evaporator can be calculated.

Consider the following process, in which ammonia is the refrigerant. The ammonia enters the expansion valve at a pressure of 1.50 MPa and a temperature of 35°C. Its pressure on leaving the expansion valve is 291 kPa. Calculate the quality of the ammonia leaving the expansion valve.

Control volume: Expansion valve or capillary tube.
Inlet state: P_i, T_i known; state fixed.
Exit state: P_e known.
Process: Steady-state.
Model: Ammonia tables.

Analysis

We can use standard throttling process analysis and assumptions. The first law reduces to

$$h_i = h_e$$

Solution

From the ammonia tables we get

$$h_i = 346.8 \text{ kJ/kg}$$

(The enthalpy of a slightly compressed liquid is essentially equal to the enthalpy of saturated liquid at the same temperature.)

$$h_e = h_i = 346.8 = 134.4 + x_e(1296.4)$$

$$x_e = 0.1638 = 16.38\%$$

Turbine

A turbine is a rotary steady-state machine whose purpose is to produce shaft work (power, on a rate basis) at the expense of the pressure of the working fluid. Two general classes of turbines are steam (or other working fluid) turbines, in which the steam exiting the turbine passes to a condenser, where it is condensed to liquid, and gas turbines, in which the gas usually exhausts to the atmosphere from the turbine. In either type, the turbine exit pressure is fixed by the environment into which the working fluid exhausts, and the turbine inlet pressure has been reached by previously pumping or compressing the working fluid in another process. Inside the turbine, there are two distinct processes. In the first, the working fluid passes through a set of nozzles, or the equivalent—fixed blade passages contoured to expand the fluid to a lower pressure and to a high velocity. In the second process inside the turbine, this high-velocity fluid stream is directed onto a set of moving (rotating) blades, in which the velocity is reduced before being discharged from the passage. This directed velocity decrease produces a torque on the rotating shaft, resulting in a shaft work output. The low-velocity, low-pressure fluid then exhausts from the turbine.

The first law for this process is either Eq. 6.10 or 6.13. Usually, changes in potential energy are negligible, as is the inlet kinetic energy. Often, the exit kinetic energy is neglected, and any heat rejection from the turbine is undesirable and is commonly small. We therefore normally assume that a turbine process is adiabatic, and the work output in this case reduces to the decrease in enthalpy form the inlet to exit states. In the following example, however, we include all the terms in the first law and study their relative importance.

EXAMPLE 6.6   The mass rate of flow into a steam turbine is 1.5 kg/s, and the heat transfer from the turbine is 8.5 kW. The following data are known for the steam entering and leaving the turbine.

	Inlet Conditions	Exit Conditions
Pressure	2.0 MPa	0.1 MPa
Temperature	350°C	
Quality		100%
Velocity	50 m/s	100 m/s
Elevation above reference plane $g = 9.8066 \text{ m/s}^2$	6 m	3 m

Determine the power output of the turbine.

$$
\begin{aligned}
\textit{Control volume}: &\quad \text{Turbine (Fig. 6.9).} \\
\textit{Inlet state}: &\quad \text{Fixed (above).} \\
\textit{Exit state}: &\quad \text{Fixed (above).} \\
\textit{Process}: &\quad \text{Steady-state.} \\
\textit{Model}: &\quad \text{Steam tables.}
\end{aligned}
$$

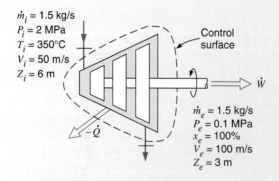

FIGURE 6.9
Illustration for Example 6.6.

Analysis

From the first law (Eq. 6.12) we have

$$\dot{Q}_{\text{C.V.}} + \dot{m}\left(h_i + \frac{\mathbf{V}_i^2}{2} + gZ_i\right) = \dot{m}\left(h_e + \frac{\mathbf{V}_e^2}{2} + gZ_e\right) + \dot{W}_{\text{C.V.}}$$

with

$$\dot{Q}_{\text{C.V.}} = -8.5 \text{ kW}$$

Solution

From the steam tables, $h_i = 3137.0$ kJ/kg. Substituting inlet conditions gives

$$\frac{\mathbf{V}_i^2}{2} = \frac{50 \times 50}{2 \times 1000} = 1.25 \text{ kJ/kg}$$

$$gZ_i = \frac{6 \times 9.8066}{1000} = 0.059 \text{ kJ/kg}$$

Similarly, for the exit $h_e = 2675.5$ kJ/kg and

$$\frac{\mathbf{V}_e^2}{2} = \frac{100 \times 100}{2 \times 1000} = 5.0 \text{ kJ/kg}$$

$$gZ_e = \frac{3 \times 9.8066}{1000} = 0.029 \text{ kJ/kg}$$

Therefore, substituting into Eq. 6.12, we obtain

$$-8.5 + 1.5(3137 + 1.25 + 0.059) = 1.5(2675.5 + 5.0 + 0.029) + \dot{W}_{\text{C.V.}}$$

$$\dot{W}_{\text{C.V.}} = -8.5 + 4707.5 - 4020.8 = 678.2 \text{ kW}$$

If Eq. 6.13 is used, the work per kilogram of fluid flowing is found first.

$$q + h_i + \frac{\mathbf{V}_i^2}{2} + gZ_i = h_e + \frac{\mathbf{V}_e^2}{2} + gZ_e + w$$

$$q = \frac{-8.5}{1.5} = -5.667 \text{ kJ/kg}$$

Therefore, substituting into Eq. 6.13, we get

$$-5.667 + 3137 + 1.25 + 0.059 = 2675.5 + 5.0 + 0.029 + w$$

$$w = 452.11 \text{ kJ/kg}$$

$$\dot{W}_{\text{C.V.}} = 1.5 \text{ kg/s} \times 452.11 \text{ kJ/kg} = 678.2 \text{ kW}$$

Two further observations can be made by referring to this example. First, in many engineering problems, potential energy changes are insignificant when compared with the other energy quantities. In the above example the potential energy change did not affect any of the significant figures. In most problems where the change in elevation is small the potential energy terms may be neglected.

Second, if velocities are small—say, under 20 m/s—in many cases the kinetic energy is insignificant compared with other energy quantities. Furthermore, when the velocities entering and leaving the system are essentially the same, the change in kinetic energy is small. Since it is the change in kinetic energy that is important in the steady-state energy equation, the kinetic energy terms can usually be neglected when there is no significant difference between the velocity of the fluid entering and that leaving the control volume. Thus, in many thermodynamic problems, one must make judgments as to which quantities may be negligible for a given analysis.

The preceding discussion and example concerned the turbine, which is a rotary work-producing device. There are other nonrotary devices that produce work, which can be called expanders as a general name. In such devices, the first-law analysis and assumptions are generally the same as for turbines, except that in a piston/cylinder type expander, there would in most cases be a larger heat loss or rejection during the process.

Compressor and Pump

The purpose of a steady-state compressor (gas) or pump (liquid) is the same: to increase the pressure of a fluid by putting in shaft work (power, on a rate basis). There are two fundamentally different classes of compressors. The most common is a rotary-type compressor (either axial flow or radial/centrifugal flow), in which the internal processes are essentially the opposite of the two processes occurring inside a turbine. The working fluid enters the compressor at low pressure, moving into a set of rotation blades, from which it exits at high velocity, a result of the shaft work input to the fluid. The fluid then passes through a diffuser section, in which it is decelerated in a manner that results in a pressure increase. The fluid then exits the compressor at high pressure.

The first law for the compressor is either Eq. 6.10 or 6.13. Usually, changes in potential energy are negligible, as is the inlet kinetic energy. Often the exit kinetic energy is neglected, as well. Heat rejection from the working fluid during compression would be desirable, but it is usually small in a rotary compressor, which is a high-volume flow-rate machine, and there is not sufficient time to transfer much heat from the working fluid. We therefore normally assume that a rotary compressor process is adiabatic, and the work input in this case reduces to the change in enthalpy from the inlet to exit states.

In a piston/cylinder-type compressor, the cylinder usually contains fins to promote heat rejection during compression (or the cylinder may be water-jacketed in a large compressor for even greater cooling rates). In this type of compressor, the heat transfer from the working fluid is significant and is not neglected in the first law. As a general rule, in any example or problem in this text, we will assume that a compressor is adiabatic unless otherwise noted.

EXAMPLE 6.7　The compressor in a plant (see Fig. 6.10) receives carbon dioxide at 100 kPa, 280 K, with a low velocity. At the compressor discharge, the carbon dioxide exits at 1100 kPa, 500 K, with velocity of 25 m/s and then flows into a constant-pressure aftercooler (heat exchanger) where it is cooled down to 350 K. The power input to the compressor is 50 kW. Determine the heat transfer rate in the aftercooler.

Solution

C.V. compressor, steady state, single inlet and exit flow.

Energy Eq. 6.13: $\quad q + h_1 + \frac{1}{2}\mathbf{V}_1^2 = h_2 + \frac{1}{2}\mathbf{V}_2^2 + w$

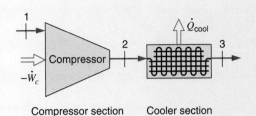

FIGURE 6.10 Sketch for Example 6.7.

Compressor section   Cooler section

Here we assume $q \cong 0$ and $\mathbf{V}_1 \cong 0$, so, getting h from Table A.8,

$$-w = h_2 - h_1 + \frac{1}{2}\mathbf{V}_2^2 = 401.52 - 198 + \frac{(25)^2}{2 \times 1000} = 203.5 + 0.3 = 203.8 \text{ kJ/kg}$$

Remember here to convert kinetic energy J/kg to kJ/kg by division by 1000.

$$\dot{m} = \frac{\dot{W}_c}{w} = \frac{-50}{-203.8} = 0.245 \text{ kg/s}$$

C.V. aftercooler, steady state, single inlet and exit flow, and no work.

Energy Eq. 6.13: $\qquad q + h_2 + \frac{1}{2}\mathbf{V}_2^2 = h_3 + \frac{1}{2}\mathbf{V}_3^2$

Here we assume no significant change in kinetic energy (notice how unimportant it was) and again we look for h in Table A.8

$$q = h_3 - h_2 = 257.9 - 401.5 = -143.6 \text{ kJ/kg}$$

$$\dot{Q}_{cool} = -\dot{Q}_{C.V.} = -\dot{m}q = 0.245 \text{ kg/s} \times 143.6 \text{ kJ/kg} = 35.2 \text{ kW}$$

EXAMPLE 6.8 A small liquid water pump is located 15 m down in a well (see Fig. 6.11), taking water in at 10°C, 90 kPa at a rate of 1.5 kg/s. The exit line is a pipe of diameter 0.04 m that goes up to a receiver tank maintaining a gauge pressure of 400 kPa. Assume the process is adiabatic with the same inlet and exit velocities and that the water stays at 10°C. Find the required pump work.

C.V. pump + pipe. Steady state, 1 inlet, 1 exit flow. Assume same velocity in and out and no heat transfer.

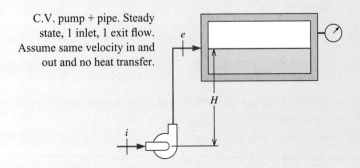

FIGURE 6.11 Sketch for Example 6.8.

Solution

Continuity equation: $\dot{m}_{in} = \dot{m}_{ex} = \dot{m}$

Energy Eq. 6.12: $\dot{m}\left(h_{in} + \frac{1}{2}\mathbf{V}^2_{in} + gZ_{in}\right) = \dot{m}\left(h_{ex} + \frac{1}{2}\mathbf{V}^2_{ex} + gZ_{ex}\right) + \dot{W}$

States: $h_{ex} = h_{in} + (P_{ex} - P_{in})v$ (v is constant and u is constant.)

From the energy equation

$$\dot{W} = \dot{m}(h_{in} + gZ_{in} - h_{ex} - gZ_{ex}) = \dot{m}[g(Z_{in} - Z_{ex}) - (P_{ex} - P_{in})v]$$

$$= 1.5\,\frac{kg}{s} \times \left[9.807\,\frac{m}{s^2} \times \frac{-15 - 0}{1000}\,m - (400 + 101.3 - 90)\,kPa\,0.001\,001\,\frac{m^3}{kg}\right]$$

$$= 1.5 \times (-0.147 - 0.412) = -0.84\,kW)$$

That is, the pump requires a power input of 840 W.

Power Plant and Refrigerator

The following examples illustrate the incorporation of several of the devices and machines already discussed in this section into a complete thermodynamic system, which is built for a specific purpose.

EXAMPLE 6.9   Consider the simple steam power plant, as shown in Fig. 6.12. The following data are for such a power plant.

Location	Pressure	Temperature or Quality
Leaving boiler	2.0 MPa	300°C
Entering turbine	1.9 MPa	290°C
Leaving turbine, entering condenser	15 kPa	90%
Leaving condenser, entering pump	14 kPa	45°C
Pump work = 4 kJ/kg		

Determine the following quantities per kilogram flowing through the unit:

a. Heat transfer in line between boiler and turbine.

b. Turbine work.

c. Heat transfer in condenser.

d. Heat transfer in boiler.

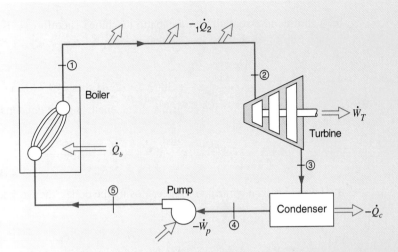

FIGURE 6.12 Simple steam power plant.

There is a certain advantage in assigning a number to various points in the cycle. For this reason the subscripts i and e in the steady-state energy equation are often replaced by appropriate numbers.

Since there are several control volumes to be considered in the solution to this problem, let us consolidate our solution procedure somewhat in this example. Using the notation of Fig. 6.12, we have:

$\quad$ *All processes*: $\quad$ Steady-state.

$\qquad\qquad$ *Model*: $\quad$ Steam tables.

From the steam tables:

$$h_1 = 3023.5 \text{ kJ/kg}$$

$$h_2 = 3002.5 \text{ kJ/kg}$$

$$h_3 = 226.0 + 0.9(2373.1) = 2361.8 \text{ kJ/kg}$$

$$h_4 = 188.5 \text{ kJ/kg}$$

$\quad$ *All analyses*: $\quad$ No changes in kinetic or potential energy will be considered in the solution. In each case, the first law is given by Eq. 6.13.

Now, we proceed to answer the specific questions raised in the problem statement.

a. For the control volume for the pipe line between the boiler and the turbine, the first law and solution are

$$_1q_2 + h_1 = h_2$$

$$_1q_2 = h_2 - h_1 = 3002.5 - 3023.5 = -21.0 \text{ kJ/kg}$$

b. A turbine is essentially an adiabatic machine. Therefore, it is reasonable to neglect heat transfer in the first law, so that

$$h_2 = h_3 + {_2}w_3$$

$${_2}w_3 = 3002.5 - 2361.8 = 640.7 \text{ kJ/kg}$$

c. There is no work for the control volume enclosing the condenser. Therefore, the first law and solution are

$${_3}q_4 + h_3 = h_4$$

$${_3}q_4 = 188.5 - 2361.8 = -2173.3 \text{ kJ/kg}$$

d. If we consider a control volume enclosing the boiler, the work is equal to zero, so that the first law becomes

$${_5}q_1 + h_5 = h_1$$

A solution requires a value for h_5, which can be found by taking a control volume around the pump:

$$h_4 = h_5 + {_4}w_5$$

$$h_5 = 188.5 - (-4) = 192.5 \text{ kJ/kg}$$

Therefore, for the boiler,

$${_5}q_1 + h_5 = h_1$$

$${_5}q_1 = 3023.5 - 192.5 = 2831 \text{ kJ/kg}$$

EXAMPLE 6.10   The refrigerator shown in Fig. 6.13 uses R-134a as the working fluid. The mass flow rate through each component is 0.1 kg/s, and the power input to the compressor is 5.0 kW. The following state data are known, using the state notation of Fig. 6.13.

$$P_1 = 100 \text{ kPa}, \qquad T_1 = -20°C$$

$$P_2 = 800 \text{ kPa}, \qquad T_2 = 50°C$$

$$T_3 = 30°C, \qquad x_3 = 0.0$$

$$T_4 = -25°C$$

Determine the following:

a. The quality at the evaporator inlet.
b. The rate of heat transfer to the evaporator.
c. The rate of heat transfer from the compressor.

All processes:	Steady-state.
Model:	R-134a tables.
All analyses:	No changes in kinetic or potential energy. The first law in each case is given by Eq. 6.10.

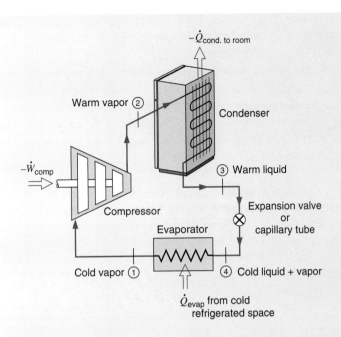

FIGURE 6.13
Refrigerator.

Solution

a. For a control volume enclosing the throttle, the first law gives

$$h_4 = h_3 = 241.8 \text{ kJ/kg}$$
$$h_4 = 241.8 = h_{f4} + x_4 h_{fg4} = 167.4 + x_4 \times 215.6$$
$$x_4 = 0.345$$

b. For a control volume enclosing the evaporator, the first law gives

$$\dot{Q}_{evap} = \dot{m}(h_1 - h_4)$$
$$= 0.1(387.2 - 241.8) = 14.54 \text{ kW}$$

c. And for the compressor, the first law gives

$$\dot{Q}_{comp} = \dot{m}(h_2 - h_1) + \dot{W}_{comp}$$
$$= 0.1(435.1 - 387.2) - 5.0 = -0.21 \text{ kW}$$

6.5 THE TRANSIENT PROCESS

In Sections 6.3 and 6.4 we considered the steady-state process and several examples of its application. Many processes of interest in thermodynamics involve unsteady flow and do not fit into this category. A certain group of these—for example, filling closed tanks with a gas or liquid, or discharge from closed vessels—can be reasonably represented to a first approximation by another simplified model. We call this process the transient process, for convenience, recognizing that our model includes specific assumptions that are not always valid. Our transient model assumptions are as follows:

1. The control volume remains constant relative to the coordinate frame.
2. The state of the mass within the control volume may change with time, but at any instant of time the state is uniform throughout the entire control volume (or over several identifiable regions that make up the entire control volume).
3. The state of the mass crossing each of the areas of flow on the control surface is constant with time although the mass flow rates may be time varying.

Let us examine the consequence of these assumptions and derive an expression for the first law that applies to this process. The assumption that the control volume remains stationary relative to the coordinate frame has already been discussed in Section 6.3. The remaining assumptions lead to the following simplifications for the continuity equation and the first law.

The overall process occurs during time t. At any instant of time during the process, the continuity equation is

$$\frac{dm_{\text{C.V.}}}{dt} + \sum \dot{m}_e - \sum \dot{m}_i = 0$$

where the summation is over all areas on the control surface through which flow occurs. Integrating over time t gives the change of mass in the control volume during the overall process:

$$\int_0^t \left(\frac{dm_{\text{C.V.}}}{dt} \right) dt = (m_2 - m_1)_{\text{C.V.}}$$

The total mass leaving the control volume during time t is

$$\int_0^t \left(\sum \dot{m}_e \right) dt = \sum m_e$$

and the total mass entering the control volume during time t is

$$\int_0^t \left(\sum \dot{m}_i \right) dt = \sum m_i$$

Therefore, for this period of time t, we can write the continuity equation for the transient process as

$$(m_2 - m_1)_{\text{C.V.}} + \sum m_e - \sum m_i = 0 \tag{6.15}$$

In writing the first law of the transient process we consider Eq. 6.7, which applies at any instant of time during the process:

$$\dot{Q}_{\text{C.V.}} + \sum \dot{m}_i \left(h_i + \frac{\mathbf{V}_i^2}{2} + gZ_i \right) = \frac{dE_{\text{C.V.}}}{dt} + \sum \dot{m}_e \left(h_e + \frac{\mathbf{V}_e^2}{2} + gZ_e \right) + \dot{W}_{\text{C.V.}}$$

Since at any instant of time the state within the control volume is uniform, the first law for the transient process becomes

$$\dot{Q}_{\text{C.V.}} + \sum \dot{m}_i \left(h_i + \frac{\mathbf{V}_i^2}{2} + gZ_i \right) = \sum \dot{m}_e \left(h_e + \frac{\mathbf{V}_e^2}{2} + gZ_e \right)$$
$$+ \frac{d}{dt} \left[m \left(u + \frac{\mathbf{V}^2}{2} + gZ \right) \right]_{\text{C.V.}} + \dot{W}_{\text{C.V.}}$$

Let us now integrate this equation over time t, during which time we have

$$\int_0^t \dot{Q}_{\text{C.V.}}\, dt = Q_{\text{C.V.}}$$

$$\int_0^t \left[\sum \dot{m}_i \left(h_i + \frac{\mathbf{V}_i^2}{2} + gZ_i\right)\right] dt = \sum m_i \left(h_i + \frac{\mathbf{V}_i^2}{2} + gZ_i\right)$$

$$\int_0^t \left[\sum \dot{m}_e \left(h_e + \frac{\mathbf{V}_e^2}{2} + gZ_e\right)\right] dt = \sum m_e \left(h_e + \frac{\mathbf{V}_e^2}{2} + gZ_e\right)$$

$$\int_0^t \dot{W}_{\text{C.V.}}\, dt = W_{\text{C.V.}}$$

$$\int_0^t \frac{d}{dt}\left[m\left(u + \frac{\mathbf{V}^2}{2} + gZ\right)\right]_{\text{C.V.}} dt = \left[m_2\left(u_2 + \frac{\mathbf{V}_2^2}{2} + gZ_2\right) - m_1\left(u_1 + \frac{\mathbf{V}_1^2}{2} + gZ_1\right)\right]_{\text{C.V.}}$$

Therefore, for this period of time t, we can write the first law for the transient process as

$$Q_{\text{C.V.}} + \sum m_i \left(h_i + \frac{\mathbf{V}_i^2}{2} + gZ_i\right)$$

$$= \sum m_e \left(h_e + \frac{\mathbf{V}_e^2}{2} + gZ_e\right)$$

$$+ \left[m_2\left(u_2 + \frac{\mathbf{V}_2^2}{2} + gZ_2\right) - m_1\left(u_1 + \frac{\mathbf{V}_1^2}{2} + gZ_1\right)\right]_{\text{C.V.}} + W_{\text{C.V.}} \qquad (6.16)$$

As an example of the type of problem for which these assumptions are valid and Eq. 6.16 is appropriate, let us consider the classic problem of flow into an evacuated vessel. This is the subject of Example 6.11.

EXAMPLE 6.11   Steam at a pressure of 1.4 MPa and temperature of 300°C is flowing in a pipe (Fig. 6.14). Connected to this pipe through a valve is an evacuated tank. The valve is opened and the tank fills with steam until the pressure is 1.4 MPa, and then the valve is closed. The process takes place adiabatically and kinetic energies and potential energies are negligible. Determine the final temperature of the steam.

Control volume:	Tank, as shown in Fig. 6.14.
Initial state (in tank):	Evacuated, mass $m_1 = 0$.
Final state:	P_2 known.
Inlet state:	P_i, T_i (in line) known.
Process:	Transient.
Model:	Steam tables.

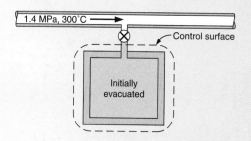

FIGURE 6.14 Flow into an evacuated vessel—control volume analysis.

Analysis

From the first law, Eq. 6.16, we have

$$\dot{Q}_{C.V.} + \sum m_i \left(h_i + \frac{\mathbf{V}_i^2}{2} + gZ_i \right)$$

$$= \sum m_e \left(h_e + \frac{\mathbf{V}_e^2}{2} + gZ_e \right)$$

$$+ \left[m_2 \left(u_2 + \frac{\mathbf{V}_2^2}{2} + gZ_2 \right) - m_1 \left(u_1 + \frac{\mathbf{V}_1^2}{2} + gZ_1 \right) \right]_{C.V.} + W_{C.V.}$$

We note that $Q_{C.V.} = 0$, $W_{C.V.} = 0$, $m_e = 0$, and $(m_1)_{C.V.} = 0$. We further assume that changes in kinetic and potential energy are negligible. Therefore, the statement of the first law for this process reduces to

$$m_i h_i = m_2 u_2$$

From the continuity equation for this process, Eq. 6.15, we conclude that

$$m_2 = m_i$$

Therefore, combining the continuity equation with the first law, we have

$$h_i = u_2$$

That is, the final internal energy of the steam in the tank is equal to the enthalpy of the steam entering the tank.

Solution

From the steam tables we obtain

$$h_i = u_2 = 3040.4 \text{ kJ/kg}$$

Since the final pressure is given as 1.4 MPa, we know two properties at the final state and therefore the final state is determined. The temperature corresponding to a pressure of 1.4 MPa and an internal energy of 3040.4 kJ/kg is found to be 452°C.

This problem can also be solved by considering the steam that enters the tank and the evacuated space as a control mass, as indicated in Fig. 6.15.

The process is adiabatic, but we must examine the boundaries for work. If we visualize a piston between the steam that is included in the control mass and the steam that flows behind, we readily recognize that the boundaries move and that the

THE TRANSIENT PROCESS ■ 187

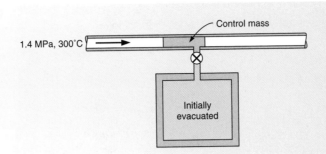

FIGURE 6.15 Flow into an evacuated vessel—control mass.

steam in the pipe does work on the steam that comprises the control mass. The amount of this work is

$$-W = P_1 V_1 = m P_1 v_1$$

Writing the first law for the control mass, Eq. 5.11, and noting that kinetic and potential energies can be neglected, we have

$$_1Q_2 = U_2 - U_1 + _1W_2$$

$$0 = U_2 - U_1 - P_1 V_1$$

$$0 = m u_2 - m u_1 - m P_1 v_1 = m u_2 - m h_1$$

Therefore,

$$u_2 = h_1$$

which is the same conclusion that was reached using a control volume analysis.

The two other examples that follow illustrate further the transient process.

EXAMPLE 6.12   Let the tank of the previous example have a volume of 0.4 m3 and initially contain saturated vapor at 350 kPa. The valve is then opened and steam from the line at 1.4 MPa and 3007C flows into the tank until the pressure is 1.4 MPa.

Calculate the mass of steam that flows into the tank.

     Control volume: Tank, as in Fig. 6.14.

        Initial state: P_1, saturated vapor; state fixed.

         Final state: P_2.

          Inlet state: P_i, T_i; state fixed.

            Process: Transient.

               Model: Steam tables.

Analysis

The situation is the same as in Example 6.10, except that the tank is not evacuated initially. Again we note that $Q_{C.V.} = 0$, $W_{C.V.} = 0$, and $m_e = 0$, and we assume that changes in kinetic and potential energy are zero. The statement of the first law for this process, Eq. 6.16, reduces to

$$m_i h_i = m_2 u_2 - m_1 u_1$$

The continuity equation, Eq. 6.15, reduces to

$$m_2 - m_1 = m_i$$

Therefore, combining the continuity equation with the first law, we have

$$(m_2 - m_1)h_i = m_2u_2 - m_1u_1$$

$$m_2(h_i - u_2) = m_1(h_i - u_1) \qquad \text{(a)}$$

There are two unknowns in this equation—m_2 and u_2. However, we have one additional equation:

$$m_2v_2 = V = 0.4 \ m^3 \qquad \text{(b)}$$

Substituting (b) into (a) and rearranging, we have

$$\frac{V}{v_2}(h_i - u_2) - m_1(h_i - u_1) = 0 \qquad \text{(c)}$$

in which the only unknowns are v_2 and u_2, both functions of T_2 and P_2. Since T_2 is unknown, it means that there is only one value of T_2 for which Eq. (c) will be satisfied, and we must find it by trial and error.

Solution

We have

$$v_1 = 0.5243 \ m^3/kg, \qquad m_1 = \frac{0.4}{0.5243} = 0.763 \ kg$$

$$h_i = 3040.4 \ kJ/kg, \qquad u_1 = 2548.9 \ kJ/kg$$

Assume that

$$T_2 = 300°C$$

For this temperature and the known value of P_2, we get

$$v_2 = 0.1823 \ m^3/kg, \qquad u_2 = 2785.2 \ kJ/kg$$

Substituting into (c), we obtain

$$\frac{0.4}{0.1823}(3040.4 - 2785.2) - 0.763(3040.4 - 2548.9) = +185.0 \ kJ$$

Now assume instead that

$$T_2 = 350°C$$

For this temperature and the known P_2, we get

$$v_2 = 0.2003 \ m^3/kg, \qquad u_2 = 2869.1 \ kJ/kg$$

Substituting these values into (c), we obtain

$$\frac{0.4}{0.2003}(3040.4 - 2869.1) - 0.763(3040.4 - 2548.9) = -32.9 \text{ kJ}$$

and we find that the actual T_2 must be between these two assumed values, in order that (c) be equal to zero. By interpolation,

$$T_2 = 342°C \quad \text{and} \quad v_2 = 0.1974 \text{ m}^3/\text{kg}$$

The final mass inside the tank is

$$m_2 = \frac{0.4}{0.1974} = 2.026 \text{ kg}$$

and the mass of steam that flows into the tank is

$$m_i = m_2 - m_1 = 2.026 - 0.763 = 1.263 \text{ kg}$$

EXAMPLE 6.13   A tank of 2 m³ volume contains saturated ammonia at a temperature of 40°C. Initially the tank contains 60% liquid and 50% vapor by volume. Vapor is withdrawn from the top of the tank until the temperature is 10°C. Assuming that only vapor (i.e., no liquid) leaves and that the process is adiabatic, calculate the mass of ammonia that is withdrawn.

Control volume: Tank.
Initial state: T_1, V_{liq}, V_{vap} ; state fixed.
Final state: T_2.
Exit state: Saturated vapor (temperature changing).
Process: Transient.
Model: Ammonia tables.

Analysis

In the first law, Eq. 6.16, we note that $Q_{\text{C.V.}} = 0$, $W_{\text{C.V.}} = 0$, and $m_i = 0$, and we assume that changes in kinetic and potential energy are negligible. However, the enthalpy of saturated vapor varies with temperature, and therefore we cannot simply assume that the enthalpy of the vapor leaving the tank remains constant. However, we note that at 40°C, $h_g = 1470.2$ kJ/kg and at 10°C, $h_g = 1452.0$ kJ/kg. Since the change in h_g during this process is small, we may accurately assume that h_e is the average of the two values given above. Therefore,

$$(h_e)_{\text{av}} = 1461.1 \text{ kJ/kg}$$

and the first law reduces to

$$m_e h_e + m_2 u_2 - m_1 u_1 = 0$$

and the continuity equation (from Eq. 6.15) becomes

$$(m_2 - m_1)_{\text{C.V.}} + m_e = 0$$

Combining these two equations we have

$$m_2(h_e - u_2) = m_1 h_e - m_1 u_1$$

Solution

The following values are from the ammonia tables:

$$v_{f1} = 0.001\ 725\ \text{m}^3/\text{kg}, \qquad v_{g1} = 0.083\ 13\ \text{m}^3/\text{kg}$$

$$v_{f2} = 0.001\ 60, \qquad v_{fg2} = 0.203\ 81$$

$$u_{f1} = 368.7\ \text{kJ/kg}, \qquad u_{g1} = 1341.0$$

$$u_{f2} = 226.0, \qquad u_{fg2} = 1099.7$$

Calculating first the initial mass, m_1, in the tank, we find that the mass of the liquid initially present, m_{f1}, is

$$m_{f1} = \frac{V_f}{v_{f1}} = \frac{1.0}{0.001\ 725} = 579.7\ \text{kg}$$

Similarly, the initial mass of vapor, m_{g1}, is

$$m_{g1} = \frac{V_g}{v_{g1}} = \frac{1.0}{0.083\ 13} = 12.0\ \text{kg}$$

$$m_1 = m_{f1} + m_{g1} = 579.7 + 12.0 = 591.7\ \text{kg}$$

$$m_1 h_e = 591.7 \times 1461.1 = 864\ 533\ \text{kJ}$$

$$m_1 u_1 = (mu)_{f1} + (mu)_{g1} = 579.7 \times 368.7 + 12.0 \times 1341.0$$

$$= 229\ 827\ \text{kJ}$$

Substituting these into the first law, we obtain

$$m_2(h_e - u_2) = m_1 h_e - m_1 u_1 = 864\ 533 - 229\ 827 = 634\ 706$$

There are two unknowns, m_2 and u_2, in this equation. However,

$$m_2 = \frac{V}{v_2} = \frac{2.0}{0.001\ 60 + x_2(0.203\ 81)}$$

and

$$u_2 = 226.0 + x_2(1099.7)$$

and thus both are functions only of x_2, the quality at the final state. Consequently,

$$\frac{2.0(1461.1 - 226.0 - 1099.7x_2)}{0.001\ 60 + 0.203\ 81x_2} = 634\ 706$$

Solving for x_2 we get

$$x_2 = 0.011\ 057$$

Therefore,

$$v_2 = 0.001\ 60 + 0.011\ 057 \times 0.203\ 81 = 0.003\ 853\ 5\ \text{m}^3/\text{kg}$$

$$m_2 = \frac{V}{v_2} = \frac{2}{0.003\ 853\ 5} = 519\ \text{kg}$$

and the mass of ammonia withdrawn, m_e, is

$$m_e = m_1 - m_2 = 591.7 - 519 = 72.7\ \text{kg}$$

EXAMPLE 6.13E A tank of 50 ft³ volume contains saturated ammonia at a temperature of 100 F. Initially the tank contains 50% liquid and 50% vapor by volume. Vapor is withdrawn from the top of the tank until the temperature is 50 F. Assuming that only vapor (i.e., no liquid) leaves and that the process is adiabatic, calculate the mass of ammonia that is withdrawn.

> *Control volume*:   Tank.
> *Initial state*: T_1, V_{liq}, V_{vap}; state fixed.
> *Final state*: T_2.
> *Exit state*:   Saturated vapor (temperature changing).
> *Process*:   Transient.
> *Model*:   Ammonia tables.

Analysis

In the first law, Eq. 6.16, we note that $Q_{\text{C.V.}} = 0$, $W_{\text{C.V.}} = 0$, and $m_i = 0$, and we assume that changes in kinetic and potential energy are negligible. However, the enthalpy of saturated vapor varies with temperature, and therefore we cannot simply assume that the enthalpy of the vapor leaving the tank remains constant. We note that at 100 F, $h_g = 631.8$ Btu/lbm and at 50 F, $h_g = 624.26$ Btu/lbm. Since the change in h_g during this process is small, we may accurately assume that h_e is the average of the two values given above. Therefore

$$(h_e)_{\text{avg}} = 628\ \text{Btu/lbm}$$

and the first law reduces to

$$m_e h_e + m_2 u_2 - m_1 u_1 = 0$$

and the continuity equation (from Eq. 6.15) is

$$(m_2 - m_1)_{\text{C.V.}} + m_e = 0$$

Combining these two equations we have

$$m_2(h_e - u_2) = m_1 h_e - m_1 u_1$$

The following values are from the ammonia tables:

$$v_{f1} = 0.0274\ 7\ \text{ft}^3/\text{lbm}, \qquad v_{g1} = 1.4168\ \text{ft}^3/\text{lbm}$$

$$v_{f2} = 0.0256\ 4\ \text{ft}^3/\text{lbm} \qquad v_{fg2} = 3.264\ 7\ \text{ft}^3/\text{lbm}$$

$$u_{f1} = 153.89\ \text{Btu/lbm}, \qquad u_{g1} = 576.23\ \text{Btu/lbm}$$

$$u_{f2} = 97.16\ \text{Btu/lbm}, \qquad u_{fg2} = 472.78\ \text{Btu/lbm}$$

Calculating first the initial mass, m_1, in the tank, the mass of the liquid initially present, m_{f1}, is

$$m_{f1} = \frac{V_f}{v_{f1}} = \frac{25}{0.027\ 47} = 910.08\ \text{lbm}$$

Similarly, the initial mass of vapor, m_{g1}, is

$$m_{g1} = \frac{V_g}{v_{g1}} = \frac{25}{1.416\ 8} = 17.65\ \text{lbm}$$

$$m_1 = m_{f1} + m_{g1} = 910.08 + 17.65 = 927.73\ \text{lbm}$$

$$m_1 h_e = 927.73 \times 628 = 582\ 614\ \text{Btu}$$

$$m_1 u_1 = (mu)_{f1} + (mu)_{g1} = 915 \times 149.9 + 16.6 + 577.0 \times 146\ 700\ \text{Btu}$$

Substituting these into the first law,

$$m_2(h_e - u_2) = m_1 h_e - m_1 u_1 = 582\ 614 - 150\ 223 = 432\ 391\ \text{Btu}$$

There are two unknowns, m_2 and u_2, in this equation. However,

$$m_2 = \frac{V}{v_2} = \frac{50}{0.025\ 64 + x_2(3.264\ 7)}$$

and

$$u_2 = 97.16 + x_2(472.78)$$

both functions only of x_2, the quality at the final state. Consequently,

$$\frac{50(628 - 97.16 - x_2\ 472.78)}{0.025\ 64 + 3.2647\ x_2} = 432\ 391$$

Solving, $x_2 = 0.010\ 768$

Therefore,

$$v_2 = 0.025\ 64 + 0.010\ 768 \times 3.264\ 7 = 0.060\ 794 \text{ ft}^3/\text{lbm}$$

$$m_2 = \frac{V}{v_2} = \frac{50}{0.060\ 794} = 822.4 \text{ lbm}$$

and the mass of ammonia withdrawn, m_e, is

$$m_e = m_1 - m_2 = 927.73 - 822.4 = 105.3 \text{ lbm}$$

SUMMARY   Conservation of mass is expressed as a rate of change of total mass due to mass flows in or out of the control volume. The control mass energy equation is extended to include mass flows that also carry energy (internal, kinetic, and potential) and the flow work needed to push the flow in or out of the control volume against the prevailing pressure. The conservation of mass (continuity equation) and the conservation of energy (first law) are applied to a number of standard devices.

A steady-state device has no storage effects, with all properties constant with time, and constitutes the majority of all flow-type devices. A combination of several devices forms a complete system built for a specific purpose, such as a power plant, jet engine, or refrigerator.

A transient process with a change in mass (storage) such as filling or emptying of a container is considered based on an average description. It is also realized that the start-up or shutdown of a steady-state device leads to a transient process.

You should have learned a number of skills and acquired abilities from studying this chapter that will allow you to

- Understand the physical meaning of the conservation equations. Rate $= +$in $-$ out
- Understand the concepts of mass flow rate, volume flow rate, and local velocity
- Recognize the flow and nonflow terms in the energy equation
- Know how the most typical devices work and if they have heat or work transfers
- Have a sense about devices where kinetic and potential energies are important
- Analyze steady-state single-flow devices such as nozzles, throttles, turbines, or pumps
- Extend the application to a multiple-flow device such as a heat exchanger, mixing chamber, or turbine, given the specific setup
- Apply the conservation equations to complete systems as a whole or to the individual devices and recognize their connections and interactions
- Recognize and use the proper form of the equations for transient problems
- Be able to assume a proper average value for any flow term in a transient
- Recognize the difference between storage of energy (dE/dt) and flow ($\dot{m}h$)

A number of steady-flow devices are listed in Table 6.1 with a very short statement of the device's purpose, known facts about work and heat transfer, and a common assumption if appropriate. This list is not complete with respect to the number of devices nor with respect to the facts listed but is meant as a short list of typical devices, some of which may be unfamiliar to many readers.

TABLE 6.1
Typical Steady-Flow Devices

Device	Purpose	Given	Assumption
Aftercooler	Cool a flow after a compressor	$w = 0$	P = constant
Boiler	Bring substance to a vapor state	$w = 0$	P = constant
Condenser	Take q out to bring substance to liquid state	$w = 0$	P = constant
Combustor	Burn fuel; acts like heat transfer in	$w = 0$	P = constant
Compressor	Bring a substance to higher pressure	w in	$q = 0$
Deaerator	Remove gases dissolved in liquids	$w = 0$	P = constant
Dehumidifier	Remove water from air		P = constant
Desuperheater	Add liquid water to superheated vapor steam to make it saturated vapor	$w = 0$	P = constant
Diffuser	Convert KE energy to higher P	$w = 0$	$q = 0$
Economizer	Low-T, low-P heat exchanger	$w = 0$	P = constant
Evaporator	Bring a substance to vapor state	$w = 0$	P = constant
Expander	Similar to a turbine, but may have a q		
Fan/blower	Move a substance, typically air	w in, KE up	$P = C$, $q = 0$
Feedwater heater	Heat liquid water with another flow	$w = 0$	P = constant
Flash evaporator	Generate vapor by expansion (throttling)	$w = 0$	$q = 0$
Heat engine	A device that converts heat into work	q in, w out	
Heat exchanger	Transfer heat from one median to another	$w = 0$	P = constant
Heat pump	A device moving a Q from T_{low} to T_{high}, requires a work input, refrigerator	w in	
Heater	Heat a substance	$w = 0$	P = constant
Humidifier	Add water to air–water mixture	$w = 0$	P = constant
Intercooler	Heat exchanger between compressor stages	$w = 0$	P = constant
Nozzle	Create KE; P drops Measure flow rate	$w = 0$	$q = 0$
Mixing chamber	Mix two or more flows	$w = 0$	$q = 0$
Pump	Same as compressor, but handles liquid	w in, P up	$q = 0$
Reactor	Allow reaction between two or more substances	$w = 0$	$q = 0$, $P = C$
Regenerator	Usually a heat exchanger to recover energy	$w = 0$	P = constant
Steam generator	Same as boiler, heat liquid water to superheat vapor	$w = 0$	P = constant
Supercharger	A compressor driven by engine shaft work to drive air into an automotive engine	w in	
Superheater	A heat exchanger that brings T up over T_{sat}	$w = 0$	P = constant
Turbine	Creat shaft work from high P flow	w out	$q = 0$
Turbocharger	A compressor driven by an exhaust flow turbine to charge air into an engine	$\dot{W}_{\text{turbine}} = \dot{W}_{\text{C.V.}}$	
Throttle	Same as valve		
Valve	Control flow by restriction; P drops	$w = 0$	$q = 0$

KEY CONCEPTS AND FORMULAS

Volume flow rate	$\dot{V} = \int \mathbf{V}\, dA = A\mathbf{V}$	(using average velocity)
Mass flow rate	$\dot{m} = \int \rho\, \mathbf{V}\, dA = \rho A\mathbf{V} = A\mathbf{V}/v$	(using average values)
Flow work rate	$\dot{W}_{flow} = P\dot{V} = \dot{m}Pv$	
Flow direction	From higher P to lower P unless significant KE or PE.	

Instantaneous Process

Continuity equation
$$\dot{m}_{C.V.} = \sum \dot{m}_i - \sum \dot{m}_e$$

Energy equation
$$\dot{E}_{C.V.} = \dot{Q}_{C.V.} - \dot{W}_{C.V.} + \sum \dot{m}_i h_{tot\,i} - \sum \dot{m}_e h_{tot\,e}$$

Total enthalpy
$$h_{tot} = h + \frac{1}{2}\mathbf{V}^2 + gZ = h_{stagnation} + gZ$$

Steady State

No storage: $\dot{m}_{C.V.} = 0;\qquad \dot{E}_{C.V.} = 0$

Continuity equation
$$\sum \dot{m}_i - \sum \dot{m}_e \quad (\text{in} = \text{out})$$

Energy equation
$$\dot{Q}_{C.V.} + \sum \dot{m}_i h_{tot\,i} = \dot{W}_{C.V.} + \sum \dot{m}_e h_{tot\,e} \quad (\text{in} = \text{out})$$

Specific heat transfer
$$q = \dot{Q}_{C.V.}/\dot{m} \quad (\text{steady state only})$$

Specific work
$$w = \dot{W}_{C.V.}/\dot{m} \quad (\text{steady state only})$$

Steady-state single flow energy equation
$$q + h_{tot\,i} = w + h_{tot\,e} \quad (\text{in} = \text{out})$$

Transient Process

Continuity equation
$$m_2 - m_1 = \sum m_i - \sum m_e$$

Energy equation
$$E_2 - E_1 = {}_1Q_2 - {}_1W_2 + \sum m_i h_{tot\,i} - \sum m_e h_{tot\,e}$$

$$E_2 - E_1 = m_2(u_2 + \frac{1}{2}\mathbf{V}_2^2 + gZ_2) - m_1(u_1 + \frac{1}{2}\mathbf{V}_1^2 + gZ_1)$$

$$h_{tot\,e} = h_{tot\,exit\,average} \approx \frac{1}{2}(h_{tot\,e1} + h_{tot\,e2})$$

CONCEPT-STUDY GUIDE PROBLEMS

6.1 A mass flow rate into a control volume requires a normal velocity component. Why?

6.2 A temperature difference drives a heat transfer. Does a similar concept apply to $\dot{m}$?

6.3 Can a steady-state device have boundary work?

6.4 Can you say something about changes in $\dot{m}$ and $\dot{V}$ through a steady-flow device?

6.5 How does a nozzle or sprayhead generate kinetic energy?

6.6 Liquid water at 15°C flows out of a nozzle straight up 15 m. What is the nozzle $\mathbf{V}_{exit}$?

6.7 What is the difference between a nozzle flow and a throttle process?

6.8 If you throttle a saturated liquid, what happens to the fluid state? What if this is done to an ideal gas?

6.9 R-134a at 30°C, 800 kPa is throttled so that it becomes cold at −10°C. What is the exit P?

6.10 Air at 500 K, 500 kPa is expanded to 100 kPa in two steady-flow cases. Case 1 is a throttle and case 2 is a turbine. Which has the highest exit T? Why?

6.11 A turbine at the bottom of a dam has a flow of liquid water through it. How does that produce power? Which terms in the energy equation are important?

6.12 A windmill takes a fraction of the wind kinetic energy out as power on a shaft. In what manner does the temperature and wind velocity influence the power? Hint: Write the power as mass flow rate times specific work.

6.13 If you compress air the temperature goes up. Why? When the hot air, high P, flows in long pipes, it eventually cools to ambient T. How does that change the flow?

6.14 In a boiler you vaporize some liquid water at 100 kPa flowing at 1 m/s. What is the velocity of the saturated vapor at 100 kPa if the pipe size is the same? Can the flow then be constant P?

6.15 A mixing chamber has all flows at the same P, neglecting losses. A heat exchanger has separate flows exchanging energy, but they do not mix. Why have both kinds?

6.16 In a coflowing (same direction) heat exchanger, 1 kg/s of air at 500 K flows into one channel and 2 kg/s of air flows into the neighboring channel at 300 K. If the heat exchanger is infinitely long, what

is the exit temperature? Sketch the variation of T in the two flows.

6.17 Air at 600 K flows with 3 kg/s into a heat exchanger and out at 100°C. How much (kg/s) water coming in at 100 kPa, 20°C can the air heat to the boiling point?

6.18 Steam at 500 kPa, 300°C is used to heat cold water at 15°C to 75°C for a domestic hot water supply. How much steam per kilogram of liquid water is needed if the steam should not condense?

6.19 Air at 20 m/s, 260 K, 75 kPa flows at 5 kg/s into a jet engine and flows out at 500 m/s, 800 K, 75 kPa. What is the change (power) in flow of kinetic energy?

6.20 An initially empty cylinder is filled with air from 20°C, 100 kPa until it is full. Assuming no heat transfer, is the final temperature larger, equal to, or smaller than 20°C? Does the final T depend on the size of the cylinder?

6.21 A cylinder has 0.1 kg of air at 25°C, 200 kPa with a 5-kg piston on top. A valve at the bottom is opened to let the air out, and the piston drops 0.25 m toward the bottom. What is the work involved in this process? What happens to the energy?

Homework Problems

Continuity Equation and Flow Rates

6.22 Air at 35°C and 105 kPa flows in a 100-mm × 150-mm rectangular duct in a heating system. The volumetric flow rate is 0.015 m³/s. What is the velocity of the air flowing in the duct and what is the mass flow rate?

6.23 A boiler receives a constant flow of 5000 kg/h liquid water at 5 MPa and 20°C, and it heats the flow such that the exit state is 450°C with a pressure of 4.5 MPa. Determine the necessary minimum pipe flow area in both the inlet and exit pipe(s) if there should be no velocities larger than 20 m/s.

6.24 An empty bath tub has its drain closed and is being filled with water from the faucet at a rate of 10 kg/min. After 10 min the drain is opened and 4 kg/min flows out, and at the same time the inlet flow is reduced to 2 kg/min. Plot the mass of the

water in the bathtub versus time and determine the time from the very beginning when the tub will be empty.

6.25 Nitrogen gas flowing in a 50-mm-diameter pipe at 15°C and 200 kPa, at the rate of 0.05 kg/s, encounters a partially closed valve. If there is a pressure drop of 30 kPa across the valve and essentially no temperature change, what are the velocities upstream and downstream of the valve?

6.26 Saturated vapor R-134a leaves the evaporator in a heat pump system at 10°C, with a steady mass flow rate of 0.1 kg/s. What is the smallest diameter tubing that can be used at this location if the velocity of the refrigerant is not to exceed 7 m/s?

6.27 A hot-air home heating system takes 0.25 m³/s air at 100 kPa, 17°C into a furnace, heats it to 52°C, and delivers the flow to a square duct 0.2 m by 0.2 m at 110 kPa (see Fig. P6.27). What is the velocity in the duct?

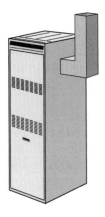

FIGURE P6.27

6.28 Steam at 3 MPa and 400°C enters a turbine with a volume flow rate of 5 m³/s. An extraction of 15% of the inlet mass flow rate exits at 600 kPa and 200°C. The rest exits the turbine at 20 kPa with a quality of 90% and a velocity of 20 m/s. Determine the volume flow rate of the extraction flow and the diameter of the final exit pipe.

6.29 A household fan of diameter 0.75 m takes air in at 98 kPa, 22°C and delivers it at 105 kPa, 23°C with a velocity of 1.5 m/s (see Fig. P6.29). What are the mass flow rate (kg/s), the inlet velocity, and the outgoing volume flow rate in m³/s?

FIGURE P6.29

Single-Flow, Single-Device Processes

Nozzles, Diffusers

6.30 Nitrogen gas flows into a convergent nozzle at 200 kPa and 400 K and very low velocity. It flows out of the nozzle at 100 kPa and 330 K. If the nozzle is insulated, find the exit velocity.

6.31 A nozzle receives 0.1 kg/s of steam at 1 MPa and 400°C with negligible kinetic energy. The exit is at

500 kPa and 350°C, and the flow is adiabatic. Find the nozzle exit velocity and the exit area.

6.32 Superheated vapor ammonia enters an insulated nozzle at 20°C and 800 kPa, as shown in Fig. P6.32, with a low velocity and at a steady rate of 0.01 kg/s. The ammonia exits at 300 kPa with a velocity of 450 m/s. Determine the temperature (or quality, if saturated) and the exit area of the nozzle.

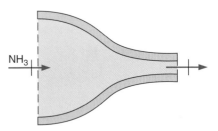

FIGURE P6.32

6.33 In a jet engine a flow of air at 1000 K, 200 kPa, and 30 m/s enters a nozzle, as shown in Fig. P6.33, where the air exits at 850 K, 90 kPa. What is the exit velocity assuming no heat loss?

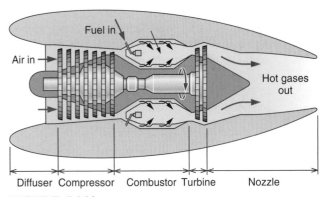

FIGURE P6.33

6.34 In a jet engine a flow of air at 1000 K, 200 kPa, and 40 m/s enters a nozzle, where the air exits at 500 m/s, 90 kPa. What is the exit temperature, assuming no heat loss?

6.35 A sluice gate dams water up 5 m. A 1-cm-diameter hole at the bottom of the gate allows liquid water at 20°C to come out. Neglect any changes in internal energy and find the exit velocity and mass flow rate.

6.36 A diffuser, shown in Fig. P6.36, has air entering at 100 kPa and 300 K with a velocity of 200 m/s. The inlet cross-sectional area of the diffuser is 100 mm^2. At the exit, the area is 860 mm^2, and the exit velocity is 20 m/s. Determine the exit pressure and temperature of the air.

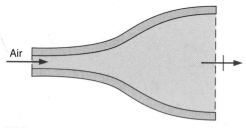

FIGURE P6.36

6.37 A diffuser receives an ideal-gas flow at 100 kPa and 300 K with a velocity of 250 m/s, and the exit velocity is 25 m/s. Determine the exit temperature if the gas is argon, helium, or nitrogen.

6.38 Air flows into a diffuser at 300 m/s, 300 K, and 100 kPa. At the exit the velocity is very small but the pressure is high. Find the exit temperature assuming zero heat transfer.

6.39 The front of a jet engine acts as a diffuser, receiving air at 900 km/h, −5°C, and 50 kPa, bringing it to 80 m/s relative to the engine before entering the compressor (see Fig. P6.39). If the flow area is reduced to 80% of the inlet area, find the temperature and pressure in the compressor inlet.

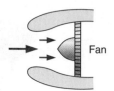

FIGURE P6.39

Throttle Flow

6.40 Helium is throttled from 1.2 MPa and 20°C to a pressure of 100 kPa. The diameter of the exit pipe is so much larger than that of the inlet pipe that the inlet and exit velocities are equal. Find the exit temperature of the helium and the ratio of the pipe diameters.

6.41 Saturated vapor R-134a at 500 kPa is throttled to 200 kPa in a steady flow through a valve. The ki-

netic energy in the inlet and exit flow is the same. What is the exit temperature?

6.42 Saturated liquid R-12 at 25°C is throttled to 150.9 kPa in your refrigerator. What is the exit temperature? Find the percent increase in the volume flow rate.

6.43 Water is flowing in a line at 400 kPa, and saturated vapor is taken out through a valve to 100 kPa. What is the temperature as it leaves the valve, assuming no changes in kinetic energy and no heat transfer?

6.44 Liquid water at 180°C and 2000 kPa is throttled into a flash evaporator chamber having a pressure of 500 kPa. Neglect any change in the kinetic energy. What is the fraction of liquid and vapor in the chamber?

6.45 Water at 1.5 MPa and 150°C is throttled adiabatically through a valve to 200 kPa. The inlet velocity is 5 m/s, and the inlet and exit pipe diameters are the same. Determine the state (neglecting kinetic energy in the energy equation) and the velocity of the water at the exit.

6.46 R-134a is throttled in a line flowing at 25°C and 750 kPa with negligible kinetic energy to a pressure of 165 kPa. Find the exit temperature and the ratio of exit pipe diameter to that of the inlet pipe (D_{ex}/D_{in}) so that the velocity stays constant.

6.47 Methane at 3 MPa, 300 K is throttled through a valve to 100 kPa. Calculate the exit temperature assuming no changes in the kinetic energy and ideal-gas behavior. Repeat the answer for real-gas behavior.

Turbines, Expanders

6.48 A steam turbine has an inlet of 2 kg/s water at 1000 kPa and 350°C with velocity of 15 m/s. The exit is at 100 kPa, $x = 1$, and very low velocity. Find the specific work and the power produced.

6.49 A small, high-speed turbine operating on compressed air produces a power output of 100 W. The inlet state is 400 kPa, 50°C, and the exit state is 150 kPa, −30°C. Assuming the velocities to be low and the process to be adiabatic, find the required mass flow rate of air through the turbine.

6.50 A liquid water turbine receives 2 kg/s water at 2000 kPa and 20°C with a velocity of 15 m/s. The exit is at 100 kPa, 20°C, and very low velocity. Find the specific work and the power produced.

6.51 Hoover Dam across the Colorado River dams up Lake Mead 200 m higher than the river downstream (see Fig. P6.51). The electric generators driven by water-powered turbines deliver 1300 MW of power. If the water is 17.5°C, find the minimum amount of water running through the turbines.

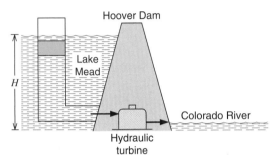

FIGURE P6.51

6.52 A windmill with rotor diameter of 30 m takes 40% of the kinetic energy out as shaft work on a day with 20°C and wind speed of 30 km/h. What power is produced?

6.53 A small turbine, shown in Fig. P6.53, is operated at part load by throttling a 0.25-kg/s steam supply at 1.4 MPa and 250°C down to 1.1 MPa before it enters the turbine, and the exhaust is at 10 kPa. If the turbine produces 110 kW, find the exhaust temperature (and quality if saturated).

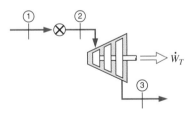

FIGURE P6.53

6.54 A small expander (a turbine with heat transfer) has 0.05 kg/s helium entering at 1000 kPa, 550 K and it leaves at 250 kPa, 300 K. The power output on the shaft is measured to 55 kW. Find the rate of heat transfer neglecting kinetic energies.

Compressors, Fans

6.55 A compressor in a commercial refrigerator receives R-22 at −25°C and $x = 1$. The exit is at 800 kPa and 40°C. Neglect kinetic energies and find the specific work.

6.56 The compressor of a large gas turbine receives air from the ambient surroundings at 95 kPa and 20°C with a low velocity. At the compressor discharge, air exits at 1.52 MPa and 430°C with velocity of 90 m/s. The power input to the compressor is 5000 kW. Determine the mass flow rate of air through the unit.

6.57 A compressor brings R-134a from 150 kPa, −10°C to 1200 kPa, 50°C. It is water cooled with a heat loss estimated as 40 kW, and the shaft work input is measured to be 150 kW. How much is the mass flow rate through the compressor?

6.58 An ordinary portable fan blows 0.2 kg/s of room air with a velocity of 18 m/s (see Fig. P6.29). What is the minimum power electric motor that can drive it? Hint: Are there any changes in P or T?

6.59 An air compressor takes in air at 100 kPa and 17°C, and delivers it at 1 MPa and 600 K to a constant-pressure cooler, which the air exits at 300 K (see Fig. P6.59). Find the specific compressor work and the specific heat transfer in the cooler.

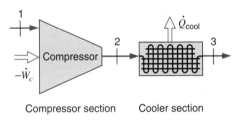

FIGURE P6.59

6.60 A 4-kg/s steady flow of ammonia runs through a device where it goes through a polytropic process. The inlet state is 150 kPa, −20°C and the exit state is 400 kPa, 80°C, where all kinetic and potential energies can be neglected. The specific work input has been found to be given as $(\frac{n}{n-1}) \Delta(Pv)$

a. Find the polytropic exponent n.

b. Find the specific work and the specific heat transfer.

6.61 An exhaust fan in a building should be able to move 2.5 kg/s of air at 98 kPa and 20°C through a 0.4-m-diameter vent hole. How high a velocity must it generate and how much power is required to do that?

6.62 How much power is needed to run the fan in Problem 6.29?

Heaters, Coolers

6.63 Carbon dioxide enters a steady-state, steady-flow heater at 300 kPa and 15°C, and exits at 275 kPa and 1200°C, as shown in Fig. P6.63. Changes in kinetic and potential energies are negligible. Calculate the required heat transfer per kilogram of carbon dioxide flowing through the heater.

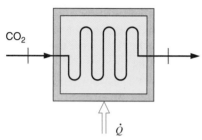

CO_2

$\dot{Q}$

FIGURE P6.63

6.64 A condenser (cooler) receives 0.05 kg/s of R-22 at 800 kPa and 40°C, and cools it to 15°C. There is a small pressure drop so that the exit state is saturated liquid. What cooling capacity (kW) must the condenser have?

6.65. A chiller cools liquid water for air-conditioning purposes. Assume 2.5 kg/s water at 20°C and 100 kPa is cooled to 5°C in a chiller. How much heat transfer (kW) is needed?

6.66 Saturated liquid nitrogen at 500 kPa enters a boiler at a rate of 0.005 kg/s and exits as saturated vapor (see Fig. P6.66). It then flows into a superheater, also at 500 kPa, where it exits at 500 kPa and 275 K. Find the rate of heat transfer in the boiler and in the superheater.

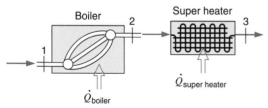

Boiler
2
Super heater
3
1
$\dot{Q}_{super\ heater}$
$\dot{Q}_{boiler}$

FIGURE P6.66

6.67 In a steam generator, compressed liquid water at 10 MPa and 30°C enters a 30-mm-diameter tube at a rate of 3 L/s. Steam at 9 MPa and 400°C exits the tube. Find the rate of heat transfer to the water.

6.68 The air conditioner in a house or a car has a cooler that brings atmospheric air from 30°C to 10°C with both states at 101 kPa. If the flow rate is 0.5 kg/s, find the rate of heat transfer.

6.69 A flow of liquid glycerine flows around an engine, cooling it as it absorbs energy. The glycerine enters the engine at 60°C and receives 9 kW of heat transfer. What is the required mass flow rate if the glycerine should come out at maximum 95°C?

6.70 Liquid nitrogen at 90 K and 400 kPa flows into a probe used in cryogenic survey. In the return line the nitrogen is then at 160 K and 400 kPa. Find the specific heat transfer to the nitrogen. If the return line has a cross-sectional area 100 times larger than that of the inlet line, what is the ratio of the return velocity to the inlet velocity?

Pumps, Pipe and Channel Flows

6.71 A small stream with 20°C water runs out over a cliff, creating a 100-m-tall waterfall. Estimate the downstream temperature when you neglect the horizontal flow velocities upstream and downstream from the waterfall. How fast was the water dropping just before it splashed into the pool at the bottom of the waterfall?

6.72 A small water pump is used in an irrigation system. The pump takes water in from a river at 10°C and 100 kPa at a rate of 5 kg/s. The exit line enters a pipe that goes up to an elevation 20 m above the pump and river, where the water runs into an open channel. Assume the process is adiabatic and that the water stays at 10°C. Find the required pump work.

6.73 A steam pipe for a 300-m-tall building receives superheated steam at 200 kPa at ground level. At the top floor the pressure is 125 kPa, and the heat loss in the pipe is 110 kJ/kg. What should the inlet temperature be so that no water will condense inside the pipe?

6.74 The main water line into a tall building has a pressure of 600 kPa at 5 m below ground level, as shown in Fig. P6.74. A pump brings the pressure up so the water can be delivered at 200 kPa at the top floor 150 m above ground level. Assume a flow rate of 10 kg/s liquid water at 10°C and neglect any difference in kinetic energy and internal energy u. Find the pump work.

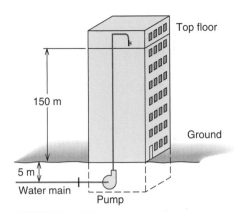

FIGURE P6.74

6.75 Consider a water pump that receives liquid water at 15°C and 100 kPa, and delivers it to a same diameter short pipe having a nozzle with exit diameter of 1 cm (0.01 m) to the atmosphere at 100 kPa (see Fig. P6.75). Neglect the kinetic energy in the pipes and assume constant u for the water. Find the exit velocity and the mass flow rate if the pump draws 1 kW of power.

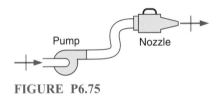

FIGURE P6.75

6.76 A cutting tool uses a nozzle that generates a high-speed jet of liquid water. Assume an exit velocity of 1000 m/s of 20°C liquid water with a jet diameter of 2 mm (0.002 m). How much mass flow rate is this? What size (power) pump is needed to generate this from a steady supply of 20°C liquid water at 200 kPa?

6.77 A pipe flows water at 15°C from one building to another. In the winter time the pipe loses an estimated 500 W of heat transfer. What is the minimum required mass flow rate that will ensure that the water does not freeze (i.e., reach 0°C)?

Multiple-Flow, Single-Device Processes

Turbines, Compressors, Expanders

6.78 A steam turbine receives water at 15 MPa and 600°C at a rate of 100 kg/s, as shown in Fig. P6.78. In the middle section 20 kg/s is withdrawn at 2 MPa and 350°C, and the rest exits the turbine at 75 kPa, with 95% quality. Assuming no heat transfer and no changes in kinetic energy, find the total turbine power output.

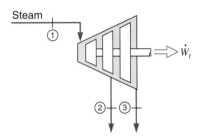

FIGURE P6.78

6.79 A steam turbine receives steam from two boilers (see Fig. P6.79). One flow is 5 kg/s at 3 MPa and 700°C, and the other flow is 15 kg/s at 800 kPa and 500°C. The exit state is 10 kPa, with a quality of 96%. Find the total power out of the adiabatic turbine.

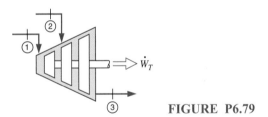

FIGURE P6.79

6.80 Two steady flows of air enter a control volume, shown in Fig. P6.80. One is 0.025 kg/s of flow at 350 kPa, 250°C (state 1), and the other enters at 450 kPa, 15°C (state 2). A single flow exits at 100 kPa, −40°C (state 3). The control volume rejects 1 kW of heat to the surroundings and produces 4 kW of power output. Neglect kinetic energies and determine the mass flow rate at state 2.

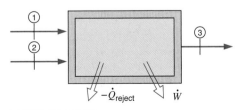

FIGURE P6.80

6.81 A large steady expansion engine has two low-velocity flows of water entering. High-pressure steam enters at point 1 with 2.0 kg/s at 2 MPa and 500°C, and 0.5 kg/s of cooling water at 120 kPa and 30°C centers at point 2. A single flow exits at point 3, with 150 kPa and 80% quality, through a 0.15-m-diameter exhaust pipe. There is a heat loss of 300 kW. Find the exhaust velocity and the power output of the engine.

6.82 Cogeneration is often used where a steam supply is needed for industrial process energy. Assume a supply of 5 kg/s steam at 0.5 MPa is needed. Rather than generating this from a pump and boiler, the setup in Fig. P6.82 is used to extract the supply from the high-pressure turbine. Find the power the turbine now cogenerates in this process.

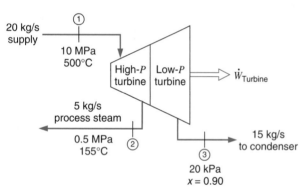

FIGURE P6.82

6.83 A compressor receives 0.1 kg/s of R-134a at 150 kPa, −10°C and delivers it at 1000 kPa, 40°C. The power input is measured to be 3 kW. The compressor has heat transfer to air at 100 kPa coming in at 20°C and leaving at 25°C. How much is the mass flow rate of air?

Heat Exchangers

6.84 A condenser (heat exchanger) brings 1 kg/s water flow at 10 kPa from 300°C to saturated liquid at 10 kPa, as shown in Fig. P6.84. The cooling is done by lake water at 20°C that returns to the lake at 30°C. For an insulated condenser, find the flow rate of cooling water.

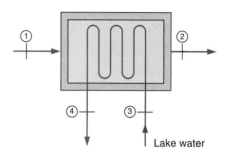

FIGURE P6.84

6.85 A cooler in an air conditioner brings 0.5 kg/s of air at 35°C to 5°C, both at 101 kPa. It then mixes the output with a flow of 0.25 kg/s air at 20°C and 101 kPa, sending the combined flow into a duct. Find the total heat transfer in the cooler and the temperature in the duct flow.

6.86 A heat exchanger, shown in Fig. P6.86, is used to cool an air flow from 800 to 360 K, with both states at 1 MPa. The coolant is a water flow at 15°C and 0.1 MPa. If the water leaves as saturated vapor, find the ratio of the flow rates $\dot{m}_{water}/\dot{m}_{air}$.

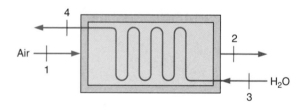

FIGURE P6.86

6.87 A superheater brings 2.5 kg/s of saturated water vapor at 2 MPa to 450°C. The energy is provided by hot air at 1200 K flowing outside the steam tube in the opposite direction as the water, a setup known as a counterflowing heat exchanger (similar to Fig. P6.86). Find the smallest possible mass flow rate of the air to ensure that its exit temperature is 20°C larger than the incoming water temperature.

6.88 An automotive radiator has glycerine at 95°C enter and return at 55°C as shown in Fig. P6.88. Air flows in at 20°C and leaves at 25°C. If the radiator should transfer 25 kW, what is the mass flow rate of the glycerine and what is the volume flow rate of air in at 100 kPa?

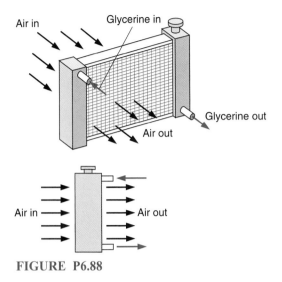

FIGURE P6.88

6.89 A two-fluid heat exchanger has 2 kg/s of liquid ammonia at 20°C, 1003 kPa entering at state 3 and exiting at state 4. It is heated by a flow of 1 kg/s nitrogen at 1500 K, state 1, and leaving at 600 K, state 2, similar to Fig. P6.86. Find the total rate of heat transfer inside the heat exchanger. Sketch the temperature versus distance for the ammonia and find state 4 (T, v) of the ammonia.

6.90 A copper wire has been heat treated to 1000 K and is now pulled into a cooling chamber that has 1.5 kg/s of air coming in at 20°C; the air leaves the other end at 60°C. If the wire moves 0.25 kg/s of copper, how hot is the copper as it comes out?

Mixing Processes

6.91 An open feedwater heater in a power plant heats 4 kg/s water at 45°C and 100 kPa by mixing it with steam from the turbine at 100 kPa and 250°C as in Fig. P6.91. Assume the exit flow is saturated liquid at the given pressure and find the mass flow rate from the turbine.

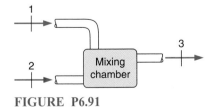

FIGURE P6.91

6.92 A desuperheater mixes superheated water vapor with liquid water in a ratio that produces saturated water vapor as output without any external heat transfer. A flow of 0.5 kg/s superheated vapor at 5 MPa and 400°C, and a flow of liquid water at 5 MPa and 40°C enters a desuperheater. If saturated water vapor at 4.5 MPa is produced, determine the flow rate of the liquid water.

6.93 Two air flows are combined to a single flow. One flow is 1 m³/s at 20°C and the other is 2 m³/s at 200°C, both at 100 kPa as in Fig. P6.93. They mix without any heat transfer to produce an exit flow at 100 kPa. Neglect kinetic energies and find the exit temperature and volume flow rate.

FIGURE P6.93

6.94 A mixing chamber with heat transfer receives 2 kg/s of R-22 at 1 MPa and 40°C in one line and 1 kg/s of R-22 at 30°C with a quality of 50% in a line with a valve. The outgoing flow is at 1 MPa and 60°C. Find the rate of heat transfer to the mixing chamber.

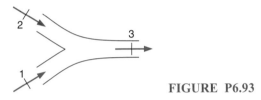

FIGURE P6.94

6.95 Two flows are mixed to form a single flow. Flow at state 1 is 1.5 kg/s of water at 400 kPa and 200°C, and flow at state 2 is at 500 kPa and 100°C. Which mass flow rate at state 2 will produce an exit $T_3 = 150$°C if the exit pressure is kept at 300 kPa?

6.96 An insulated mixing chamber receives 2 kg/s of R-134a at 1 MPa and 100°C in a line with low velocity. Another line with R-134a as saturated liquid at 60°C flows through a valve to the mixing chamber at 1 MPa after the valve, as shown in Fig. P6.94. The exit flow is saturated vapor at 1 MPa flowing at 20 m/s. Find the flow rate for the second line.

6.97 To keep a jet engine cool, some intake air bypasses the combustion chamber. Assume 2 kg/s of hot air at 2000 K and 500 kPa is mixed with 1.5 kg/s air at 500 K, 500 kPa without any external heat transfer as in Fig. P6.97. Find the exit temperature using constant heat capacity from Table A.5.

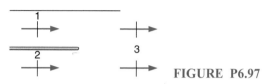

FIGURE P6.97

6.98 Solve the previous problem using values from Table A.7.

Multiple Devices, Cycle Processes

6.99 The following data are for a simple steam power plant as shown in Fig. P6.99. State 6 has $x_6 = 0.92$ and velocity of 200 m/s. The rate of steam flow is 25 kg/s, with 300 kW of power input to the pump. Piping diameters are 200 mm from the steam generator to the turbine and 75 mm from the condenser to the economizer and steam generator. Determine the velocity at state 5 and the power output of the turbine.

State	1	2	3	4	5	6	7
P, kPa	6200	6100	5900	5700	5500	10	9
T, °C		45	175	500	490		40
h, kJ/kg		194	744	3426	3404		168

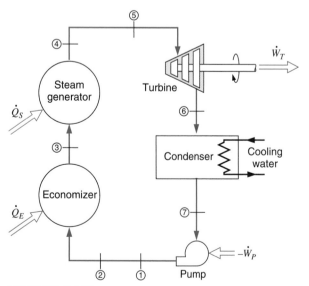

FIGURE P6.99

6.100 For the steam power plant shown in Problem 6.99, assume the cooling water comes from a lake at 15°C and is returned at 25°C. Determine the rate of heat transfer in the condenser and the mass flow rate of cooling water from the lake.

6.101 For the steam power plant shown in Problem 6.99, determine the rate of heat transfer in the economizer, which is a low-temperature heat exchanger. Also find the rate of heat transfer needed in the steam generator.

6.102 A somewhat simplified flow diagram for a nuclear power plant is given in Fig. P6.102. Mass flow rates and the various states in the cycle are shown in the accompanying table.

Point	ṁ, kg/s	P, kPa	T, °C	h, kJ/kg
1	75.6	7240	sat vap	
2	75.6	6900		2765
3	62.874	345		2517
4		310		
5		7		2279
6	75.6	7	33	138
7		415		140
8	2.772	35		2459
9	4.662	310		558
10		35	34	142
11	75.6	380	68	285
12	8.064	345		2517
13	75.6	330		
14				349
15	4.662	965	139	584
16	75.6	7930		565
17	4.662	965		2593
18	75.6	7580		688
19	1386	7240	277	1220
20	1386	7410		1221
21	1386	7310		

The cycle includes a number of heaters in which heat is transferred from steam, taken out of the turbine at some intermediate pressure, to liquid water pumped from the condenser on its way to the steam drum. The heat exchanger in the reactor supplies 157 MW, and it may be assumed that there is no heat transfer in the turbines.

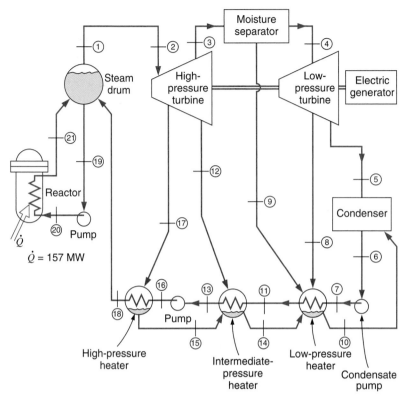

FIGURE P6.102

a. Assuming the moisture separator has no heat transfer between the two turbine sections, determine the enthalpy and quality (h_4, x_4).

b. Determine the power output of the low-pressure turbine.

c. Determine the power output of the high-pressure turbine.

d. Find the ratio of the total power output of the two turbines to the total power delivered by the reactor.

6.103 Consider the power plant as described in the previous problem.

a. Determine the quality of the steam leaving the reactor.

b. What is the power to the pump that feeds water to the reactor?

6.104 A gas turbine set-up to produce power during peak demand is shown in Fig. P6.104. The turbine provides power to the air compressor and the electric generator. If the electric generator should provide 5 MW what is the needed air flow at state 1 and the combustion heat transfer between states 2 and 3? The states are

1. 90 kPa, 290 K
2. 900 kPa, 560 K
3. 900 kPa, 1400 K
4. 100 kPa, 850 K

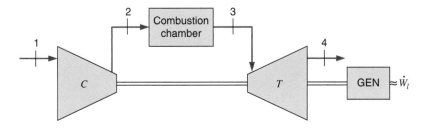

FIGURE P6.104

6.105 A proposal is made to use a geothermal supply of hot water to operate a steam turbine, as shown in Fig. P6.105. The high-pressure water at 1.5 MPa and 180°C is throttled into a flash evaporator chamber, which forms liquid and vapor at a lower pressure of 40 kPa. The liquid is discarded while the saturated vapor feeds the turbine and exits at 10 kPa with a 90% quality. If the turbine should produce 1 MW, find the required mass flow rate of hot geothermal water in kilograms per hour.

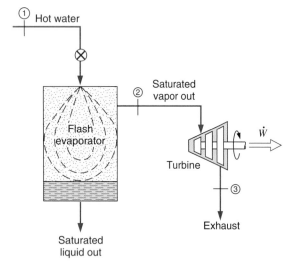

FIGURE P6.105

6.106 An R-12 heat pump cycle shown in Fig. P6.106 has an R-12 flow rate of 0.05 kg/s with 4 kW into the compressor. The following data are given:

State	1	2	3	4	5	6
P, kPa	1250	1230	1200	320	300	290
T, °C	120	110	45		0	5
h, kJ/kg	260	253	79.7		188	191

Calculate the heat transfer from the compressor, the heat transfer from the R-12 in the condenser, and the heat transfer to the R-12 in the evaporator.

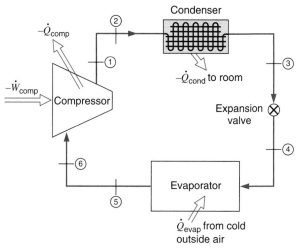

FIGURE P6.106

6.107 A modern jet engine has a temperature after combustion of about 1500 K at 3200 kPa as it enters the turbine section (see state 3, Fig. P6.107). The compressor inlet is at 80 kPa, 260 K (state 1) and the outlet (state 2) is at 3300 kPa, 780 K; the turbine outlet (state 4) into the nozzle is at 400 kPa, 900 K and the nozzle exit (state 5) is at 80 kPa, 640 K. Neglect any heat transfer and neglect kinetic energy except in and out of the nozzle. Find the compressor and turbine specific work terms and the nozzle exit velocity.

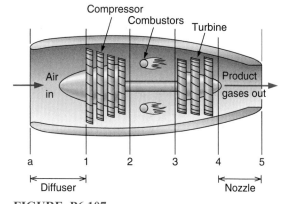

FIGURE P6.107

Transient Processes

6.108 A 1-m³, 40-kg rigid steel tank contains air at 500 kPa, and both tank and air are at 20°C. The tank is connected to a line flowing air at 2 MPa and 20°C. The valve is opened, allowing air to flow into the tank until the pressure reaches 1.5 MPa, and is then closed. Assume the air and tank are always at the same temperature and the final temperature is 35°C. Find the final air mass and the heat transfer.

6.109 An evacuated 150-L tank is connected to a line flowing air at room temperature, 25°C, and 8 MPa pressure. The valve is opened, allowing air to flow into the tank until the pressure inside is 6 MPa. At this point the valve is closed. This filling process occurs rapidly and is essentially adiabatic. The tank is then placed in storage where it eventually returns to room temperature. What is the final pressure?

6.110 An initially empty bottle is filled with water from a line at 0.8 MPa and 350°C. Assume no heat transfer and that the bottle is closed when the pressure reaches the line pressure. If the final mass is 0.75 kg, find the final temperature and the volume of the bottle.

6.111 A 25-L tank, shown in Fig. P6.111, that is initially evacuated is connected by a valve to an air supply line flowing air at 20°C and 800 kPa. The valve is opened, and air flows into the tank until the pressure reaches 600 kPa. Determine the final temperature and mass inside the tank, assuming the process is adiabatic. Develop an expression for the relation between the line temperature and the final temperature using constant specific heats.

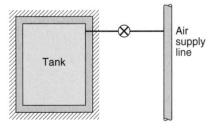

FIGURE P6.111

6.112 Helium in a steel tank is at 250 kPa, 300 K with a volume of 0.1 m³. It is used to fill a balloon. When the tank pressure drops to 150 kPa, the flow of helium stops by itself. If all the helium still is at 300 K how big a balloon did I get? Assume the

pressure in the balloon varies linearly with volume from 100 kPa ($V = 0$) to the final 150 kPa. How much heat transfer took place?

6.113 A rigid 100-L tank contains air at 1 MPa and 200°C. A valve on the tank is now opened, and air flows out until the pressure drops to 100 kPa. During this process, heat is transferred from a heat source at 200°C, such that when the valve is closed, the temperature inside the tank is 50°C. What is the heat transfer?

6.114 A 1-m³ tank contains ammonia at 150 kPa and 25°C. The tank is attached to a line flowing ammonia at 1200 kPa and 60°C. The valve is opened, and mass flows in until the tank is half full of liquid (by volume) at 25°C. Calculate the heat transferred from the tank during this process.

6.115 An empty canister of volume 1 L is filled with R-134a from a line flowing saturated liquid R-134a at 0°C. The filling is done quickly, so it is adiabatic. How much mass of R-134a is there after filling? The canister is placed on a storage shelf, where it slowly heats up to room temperature 20°C. What is the final pressure?

6.116 A piston/cyclinder assembly contains 1 kg of water at 20°C with a constant load on the piston such that the pressure is 250 kPa. A nozzle in a line to the cylinder is opened to enable a flow to the outside atmosphere at 100 kPa. The process continues until half the mass has flowed out and there is no heat transfer. Assume constant water temperature and find the exit velocity and total work done in the process.

6.117 A 200-L tank (see Fig. P6.117) initially contains water at 100 kPa and a quality of 1%. Heat is

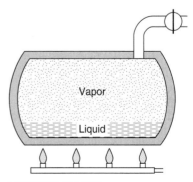

FIGURE P6.117

transferred to the water, thereby raising its pressure and temperature. At a pressure of 2 MPa a safety valve opens and saturated vapor at 2 MPa flows out. The process continues, maintaining 2 MPa inside until the quality in the tank is 90%, then stops. Determine the total mass of water that flowed out and the total heat transfer.

6.118 A 100-L rigid tank contains carbon dioxide gas at 1 MPa and 300 K. A valve is cracked open, and carbon dioxide escapes slowly until the tank pressure has dropped to 500 kPa. At this point the valve is closed. The gas remaining inside the tank may be assumed to have undergone a polytropic expansion, with polytropic exponent $n = 1.15$. Find the final mass inside and the heat transferred to the tank during the process.

6.119 A nitrogen line, at 300 K and 0.5 MPa, shown in Fig. P6.119, is connected to a turbine that exhausts to a closed initially empty tank of 50 m^3. The turbine operates to a tank pressure of 0.5 MPa, at which point the temperature is 250 K. Assuming the entire process is adiabatic, determine the turbine work.

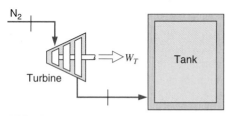

FIGURE P6.119

6.120 A 2-m-tall cyclinder has a small hole in the bottom as in Fig. P6.120. It is filled with liquid water 1 m high, on top of which is a 1-m-high air column at atmospheric pressure of 100 kPa. As the liquid water near the hole has a higher P than

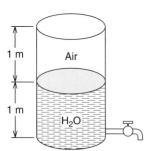

FIGURE P6.120

100 kPa, it runs out. Assume a slow process with constant T. Will the flow ever stop? When?

6.121 A 2-m^3 insulated vessel, shown in Fig. P6.121, contains saturated vapor steam at 4 MPa. A valve on the top of the tank is opened, and steam is allowed to escape. During the process any liquid formed collects at the bottom of the vessel, so only saturated vapor exits. Calculate the total mass that has escaped when the pressure inside reaches 1 MPa.

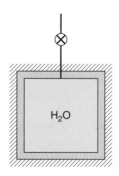

FIGURE P6.121

6.122 A 750-L rigid tank, shown in Fig. P6.122, initially contains water at 250°C, which is 50% liquid and 50% vapor, by volume. A valve at the bottom of the tank is opened, and liquid is slowly withdrawn. Heat transfer takes place such that the temperature remains constant. Find the amount of heat transfer required to reach the state where half the initial mass is withdrawn.

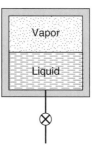

FIGURE P6.122

6.123 Consider the previous problem but let the line and valve be located in the top of the tank. Now saturated vapor is slowly withdrawn while heat transfer keeps the temperature inside constant. Find the heat transfer required to reach a state where half the original mass is withdrawn.

Review Problems

6.124 A flow of 2 kg/s of water at 500 kPa and 20°C is heated in a constant-pressure process to 1700°C. Find the best estimate for the rate of heat transfer needed.

6.125 In a glass factory a 2-m-wide sheet of glass at 1500 K comes out of the final rollers, which fix the thickness at 5 mm with a speed of 0.5 m/s (see Fig. P6.125). Cooling air in the amount of 20 kg/s comes in at 17°C from a slot 2 m wide and flows parallel with the glass. Suppose this setup is very long so the glass and air comes to nearly the same temperature (a coflowing heat exchanger) what is the exit temperature?

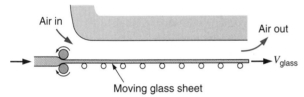

FIGURE P6.125

6.126 Assume a setup similar to the previous problem but with the air flowing in the opposite direction as the glass—it comes in where the glass goes out. How much air flow at 17°C is required to cool the glass to 450 K assuming the air must be at least 120 K cooler than the glass at any location?

6.127 Three air flows, all at 200 kPa, are connected to the same exit duct and mix without external heat transfer. Flow 1 has 1 kg/s at 400 K, flow 2 has 3 kg/s at 290 K, and flow 3 has 2 kg/s at 700 K. Neglect kinetic energies and find the volume flow rate in the exit flow.

6.128 Consider the power plant as described in Problem 6.102.

  a. Determine the temperature of the water leaving the intermediate pressure heater, T_{13}, assuming no heat transfer to the surroundings.

  b. Determine the pump work between states 13 and 16.

6.129 Consider the power plant as described in Problem 6.102.

  a. Find the power removed in the condenser by the cooling water (not shown).

  b. Find the power to the condensate pump.

  c. Do the energy terms balance for the low-pressure heater or is there a heat transfer not shown?

6.130 A 500-L insulated tank contains air at 40°C and 2 MPa. A valve on the tank is opened, and air escapes until half the original mass is gone, at which point the valve is closed. What is the pressure inside then?

6.131 A steam engine based on a turbine is shown in Fig. P6.131. The boiler tank has a volume of 100 L and initially contains saturated liquid with a very small amount of vapor at 100 kPa. Heat is now added by the burner. The pressure regulator, which keeps the pressure constant, does not open before the boiler pressure reaches 700 kPa. The saturated vapor enters the turbine at 700 kPa and is discharged to the atmosphere as saturated vapor at 100 kPa. The burner is turned off when no more liquid is present in the boiler. Find the total turbine work and the total heat transfer to the boiler for this process.

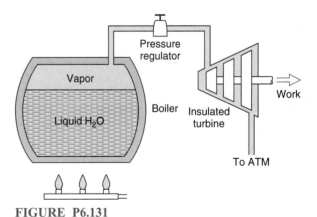

FIGURE P6.131

6.132 An insulated spring-loaded piston/cylinder device, shown in Fig. P6.132, is connected to an air

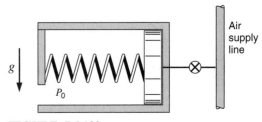

FIGURE P6.132

line flowing air at 600 kPa and 700 K by a valve. Initially the cylinder is empty and the spring force is zero. The valve is then opened until the cylinder pressure reaches 300 kPa. By noting that $u_2 = u_{line} + C_v(T_2 - T_{line})$ and $h_{line} - u_{line} = RT_{line}$, find an expression for T_2 as a function of P_2, P_0, and T_{line}. With $P_0 = 100$ kPa, find T_2.

6.133 A mass-loaded piston/cylinder shown in Fig. P6.133, containing air is at 300 kPa, 17°C with a volume of 0.25 m³, while at the stops $V = 1$ m³. An air line, 500 kPa, 600 K, is connected by a valve that is then opened until a final inside pressure of 400 kPa is reached, at which point $T = 350$ K. Find the air mass that enters, the work, and the heat transfer.

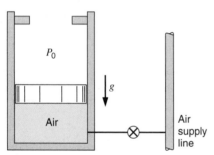

FIGURE P6.133

6.134 A 2-m³ storage tank contains 95% liquid and 5% vapor by volume of liquified natural gas (LNG) at 160 K, as shown in Fig. P6.134. It may be assumed that LNG has the same properties as pure methane. Heat is transferred to the tank and saturated vapor at 160 K flows into the steady flow

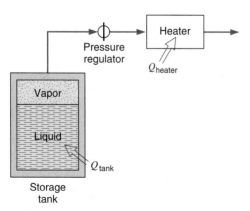

FIGURE P6.134

heater, which it leaves at 300 K. The process continues until all the liquid in the storage tank is gone. Calculate the total amount of heat transfer to the tank and the total amount of heat transferred to the heater.

Heat Transfer Problems

6.135 Liquid water at 80°C flows with 0.2 kg/s inside a square duct 2 cm on a side, insulated with a 1-cm-thick layer of foam, $k = 0.1$ W/m K. If the outside foam surface is at 25°C, how much has the water temperature dropped for 10 m length of duct? Neglect the duct material and any corner effects ($A = 4\,sL$).

6.136 A counterflowing heat exchanger conserves energy by heating cold outside fresh air at 10°C with the outgoing combustion gas (air) at 100°C as in Fig. P6.136. Assume both flows are 1 kg/s and the temperature difference between the flows at any point is 50°C. What is the incoming fresh air temperature after the heat exchanger? What is the equivalent (single) convective heat transfer coefficient between the flows if the interface area is 2 m²?

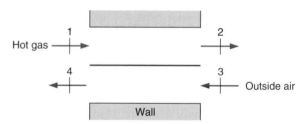

FIGURE P6.136

6.137 Saturated liquid water at 1000 kPa flows at 2 kg/s inside a 10-cm-outer-diameter steel pipe, and outside of the pipe is a flow of hot gases at 1000 K with a convection coefficient of $h = 150$ W/m² K. Neglect any $\triangle T$ in the steel and any inside convection h, and find the length of pipe needed to bring the water to saturated vapor.

6.138 A flow of 1000 K, 100 kPa air with 0.5 kg/s in a furnace flows over a steel plate of surface temperature 400 K. The flow is such that the convective heat transfer coefficient is $h = 125$ W/m² K. How much of a surface area does the air have to flow over to exit with a temperature of 800 K? How about 600 K?

ENGLISH UNIT PROBLEMS

Concept Problems

6.139E Liquid water at 60 F flows out of a nozzle straight up 40 ft. What is the nozzle $\mathbf{V}_{exit}$?

6.140E R-134a at 90 F, 125 psia is throttled so that it becomes cold at 10 F. What is the exit P?

6.141E In a boiler you vaporize some liquid water at 103 psia flowing at 3 ft/s. What is the velocity of the saturated vapor at 103 psia if the pipe size is the same? Can the flow then be constant P?

6.142E Air at 60 ft/s, 480 R, 11 psia flows at 10 lbm/s into a jet engine and flows out at 1500 ft/s, 1440 R, 11 psia. What is the change (power) in flow of kinetic energy?

6.143E An initially empty cylinder is filled with air from 70 F, 15 psia until it is full. Assuming no heat transfer, is the final temperature larger, equal to or smaller than 70 F? Does the final T depend on the size of the cyclinder?

6.144E Air at 95 F, 16 lbf/in^2 flows in a 4-in. × 6-in. rectangular duct in a heating system. The volumetric flow rate is 30 cfm (ft^3/min). What is the velocity of the air flowing in the duct?

6.145E A hot-air home heating system takes 500 ft^3/min (cfm) air at 14.7 psia, 65 F into a furnace and heats it to 130 F and delivers the flow to a square duct 0.5 ft by 0.5 ft at 15 psia. What is the velocity in the duct?

6.146E Saturated vapor R-134a leaves the evaporator in a heat pump at 50 F, with a steady mass flow rate of 0.2 lbm/s. What is the smallest diameter tubing that can be used at this location if the velocity of the refrigerant is not to exceed 20 ft/s?

6.147E A pump takes 40 F liquid water from a river at 14 lbf/in^2 and pumps it up to an irrigation canal 60 ft higher than the river surface. All pipes have a diameter of 4 in., and the flow rate is 35 lbm/s. Assume the pump exit pressure is just enough to carry a water column of the 60 ft height with 15 lbf/in^2 at the top. Find the flow work into and out of the pump and the kinetic energy in the flow.

6.148E In a jet engine a flow of air at 1800 R, 30 psia, and 90 ft/s enters a nozzle, where the air exits at 1500 R, 13 psia, as shown in Fig. P6.33. What is the exit velocity assuming no heat loss?

6.149E Nitrogen gas flows into a convergent nozzle at 30 lbf/in^2, 600 R and very low velocity. It flows out of the nozzle at 15 lbf/in^2, 500 R. If the nozzle is insulated, find the exit velocity.

6.150E A diffuser, shown in Fig. P6.36, has air entering at 14.7 lbf/in^2, 540 R, with a velocity of 600 ft/s. The inlet cross-sectional air of the diffuser is 0.2 in^2. At the exit, the area is 1.75 in^2, and the exit velocity is 60 ft/s. Determine the exit pressure and temperature of the air.

6.151E Helium is throttled from 175 lbf/in^2, 70 F, to a pressure of 15 lbf/in^2. The diameter of the exit pipe is so much larger than the inlet pipe that the inlet and exit velocities are equal. Find the exit temperature of the helium and the ratio of the pipe diameters.

6.152E Water flowing in a line at 60 lbf/in^2, saturated vapor is taken out through a valve to 14.7 lbf/in^2. What is the temperature as it leaves the valve assuming no changes in kinetic energy and no heat transfer?

6.153E A small, high-speed turbine operating on compressed air produces a power output of 0.1 hp. The inlet state is 60 lbf/in^2, 120 F, and the exit state is 14.7 lbf/in^2, −20 F. Assuming the velocities to be low and the process to be adiabatic, find the required mass flow rate of air through the turbine.

6.154E Hoover Dam, across the Colorado River, dams up Lake Mead 600 ft higher than the river downstream, as shown in Fig. P6.51. The electric generators driven by water-powered turbines deliver 1.2×10^6 Btu/s. If the water is 65 F, find the minimum amount of water running through the turbines.

6.155E A small expander (a turbine with heat transfer) has 0.1 lbm/s of helium entering at 160 psia, 1000 R and leaving at 40 psia, 540 R. The power output on the shaft is measured as 55 Btu/s. Find the rate of heat transfer, neglecting kinetic energies.

6.156E An exhaust fan in a building should be able to move 5 lbm/s of air at 14.4 psia, 68 F through a 1.25-ft-diameter vent hole. How high a velocity must the fan generate and how much power is required to do that?

6.157E In a steam generator, compressed liquid water at 1500 lbf/in^2, 100 F enters a 1-in-diameter tube at the rate of 5 ft^3/min. Steam at 1250 lbf/in^2, 750 F exits the tube. Find the rate of heat transfer to the water.

6.158E Carbon dioxide gas enters a steady-state, steady-flow heater at 45 lbf/in^2, 60 F and exits at 40 lbf/in^2, 1800 F. It is shown in Fig. P6.63, where changes in kinetic and potential energies are negligible. Calculate the required heat transfer per lbm of carbon dioxide flowing through the heater.

6.159E A flow of liquid glycerine flows around an engine, cooling it as it absorbs energy. The glycerine enters the engine at 140 F and receives 13 hp of heat transfer. What is the required mass flow rate if the glycerine should come out at a maximum 200 F?

6.160E A small water pump is used in an irrigation system. The pump takes water in from a river at 50 F, 1 atm at a rate of 10 lbm/s. The exit line enters a pipe that goes up to an elevation 60 ft above the pump and river, where the water runs into an open channel. Assume that the process is adiabatic and that the water stays at 50 F. Find the required pump work.

6.161E A steam turbine receives water at 2000 lbf/in^2, 1200 F at a rate of 200 lbm/s, as shown in Fig. P6.78. In the middle section 40 lbm/s is withdrawn at 300 lbf/in^2, 650 F and the rest exits the turbine at 10 lbf/in^2, 95% quality. Assuming no heat transfer and no changes in kinetic energy, find the total turbine power output.

6.162E A condenser, as the heat exchanger shown in Fig. P6.84, brings 1 lbm/s water flow at 1 lbf/in^2 from 500 F to saturated liquid at 1 lbf/in^2. The cooling is done by lake water at 70 F that returns to the lake at 90 F. For an insulated condenser, find the flow rate of cooling water.

6.163E A heat exchanger is used to cool an air flow from 1400 to 680 R, both states at 150 lbf/in^2. The coolant is a water flow at 60 F, 15 lbf/in^2, and it is shown in Fig. P6.86. If the water leaves as saturated vapor, find the ratio of the flow rates $\dot{m}_{water}/\dot{m}_{air}$.

6.164E An automotive radiator has glycerine at 200 F enter and return at 130 F as shown in Fig. P6.88. Air flows in at 68 F and leaves at 77 F. If the radiator should transfer 33 hp, what is the mass flow rate of the glycerine and what is the volume flow rate of air in at 15 psia?

6.165E An insulated mixing chamber as shown in Fig. P6.94 receives 4 lbm/s of R-134a at 150 lbf/in^2, 220 F in a line with low velocity. Another line with R-134a of saturated liquid at 130 F flows through a valve to the mixing chamber at 150 lbf/in^2 after the valve. The exit flow is saturated vapor at 150 lbf/in^2 flowing at 60 ft/s. Find the mass flow rate for the second line.

6.166E An air compressor as shown in Fig. P6.59 takes in air at 14 lbf/in^2, 60 F and delivers it at 140 lbf/in^2, 1080 R to a constant-pressure cooler, which the air exits at 560 R. Find the specific compressor work and the specific heat transfer.

6.167E The following data are for a simple steam power plant as shown in Fig. P6.99.

State	1	2	3	4	5	6	7
P lbf/in^2	900	890	860	830	800	1.5	1.4
T F		115	350	920	900		110
h, Btu/lbm		85.3	323	1468	1456	1029	78

State 6 has $x_6 = 0.92$ and velocity of 600 ft/s. The rate of steam flow is 200 000 lbm/h, with 400 hp input to the pump. Piping diameters are 8 in. from the steam generator to the turbine and 3 in. from the condenser to the steam generator. Determine the power output of the turbine and the heat transfer rate in the condenser.

6.168E For the same steam power plant as shown in Fig. P6.99 and Problem 6.167 determine the rate of heat transfer in the economizer, which is a low-temperature heat exchanger, and the steam generator. Determine also the flow rate of cooling water through the condenser, if the cooling water increases from 55 to 75 F in the condenser.

6.169E A geothermal supply of hot water operates a steam turbine, as shown in Fig. P6.105. The high-pressure water at 200 lbf/in^2, 350 F is throttled into a flash evaporator chamber, which forms liquid and vapor at a lower pressure of 60 lbf/in^2. The liquid is discarded while the saturated vapor feeds the turbine and exits at 1 lbf/in^2, 90% quality. If the turbine should produce 1000 hp, find the required mass flow rate of hot geothermal water.

6.170E A 1-ft^3 tank, shown in Fig. P6.111, that is initially evacuated is connected by a valve to an air supply line flowing air at 70 F, 120 lbf/in^2. The valve is opened, and air flows into the tank until the pressure reaches 90 lbf/in^2. Determine the final temperature and mass inside the tank, assuming the process is adiabatic. Develop an expression for the relation between the line temperature and the final temperature using constant specific heats.

6.171E Helium in a steel tank is at 40 psia, 540 R with a volume of 4 ft^3. It is used to fill a balloon. When the tank pressure drops to 24 psia, the flow of helium stops by itself. If all the helium still is at 540 R, how big a balloon did I get? Assume the pressure in the balloon varies linearly with volume from 14.7 psia ($V = 0$) to the final 24 psia. How much heat transfer took place?

6.172E A 20-ft^3 tank contains ammonia at 20 lbf/in^2, 80 F. The tank is attached to a line flowing ammonia at 180 lbf/in^2, 140 F. The valve is opened, and mass flows in until the tank is half full of liquid, by volume at 80 F. Calculate the heat transferred from the tank during this process.

6.173E An initially empty bottle, $V = 10$ ft^3, is filled with water from a line at 120 lbf/in^2, 500 F. Assume no heat transfer and that the bottle is closed when the pressure reaches line pressure. Find the final temperature and mass in the bottle.

6.174E A nitrogen line, 540 R, 75 lbf/in^2 is connected to a turbine that exhausts to a closed initially empty tank of 2000 ft^3, as shown in Fig. P6.119. The turbine operates to a tank pressure of 75 lbf/in^2, at which point the temperature is 450 R. Assuming the entire process is adiabatic, determine the turbine work.

6.175E A mass-loaded piston/cylinder containing air is at 45 lbf/in^2, 60 F with a volume of 9 ft^3, while at the stops $V = 36$ ft^3. An air line, 75 lbf./in^2, 1100 R is connected by a valve, as shown in Fig. P6.133. The valve is then opened until a final inside pressure of 60 lbf/in^2 is reached, at which point $T = 630$ R. Find the air mass that enters, work done, and heat transfer.

COMPUTER, DESIGN, AND OPEN-ENDED PROBLEMS

6.176 Fit a polynomial expression of degree n in the temperature for ideal-gas specific heat. Use the ideal-gas enthalpy values for one of the substances listed in Table A.8 as data. The accuracy of the correlation should be studied as a function of the temperature range of the fit as well as of the polynominal degree n.

6.177 An insulated tank of volume V contains a specified ideal gas (with constant specific heat) as P_1, T_1. A valve is opened, allowing the gas to flow out until the pressure inside drops to P_2. Determine T_2 and m_2 using a stepwise solution in increments of pressure between P_1 and P_2; the number of increments is variable.

6.178 We wish to solve Problem 6.121, using a stepwise solution, whereby the process is subdivided into several parts to minimize the effects of a linear average enthalpy approximation. Divide the process into two or three steps so that you can get a better estimate for the mass times enthalpy leaving the tank.

6.179 The air–water counterflowing heat exchanger given in Problem 6.86 has an air exit temperature of 360 K. Suppose the air exit temperature is listed as 300 K, then a ratio of the mass flow rates is found from the energy equation to be 5. Show that this is an impossible process by looking at air and water temperatures at several locations inside the heat exchanger. Discuss how this puts a limit on the energy that can be extracted from the air.

6.180 A coflowing heat exchanger receives air at 800 K, 1 MPa and liquid water at 15°C, 100 kPa, as shown in Fig. P6.180. The air line heats the water so that at the exit the air temperature is 20°C above the water temperature. Investigate the limits for the air and water exit temperatures as a function of the ratio of the two mass flow rates. Plot the temperatures of the air and water inside the heat exchanger along the flow path.

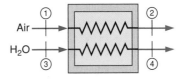

FIGURE P6.180

7 THE SECOND LAW OF THERMODYNAMICS

The first law of thermodynamics states that during any cycle that a system undergoes, the cyclic integral of the heat is equal to the cyclic integral of the work. The first law, however, places no restrictions on the direction of flow of heat and work. A cycle in which a given amount of heat is transferred from the system and an equal amount of work is done on the system satisfies the first law just as well as a cycle in which the flows of heat and work are reversed. However, we know from our experience that because a proposed cycle does not violate the first law does not ensure that the cycle will actually occur. It is this kind of experimental evidence that led to the formulation of the second law of thermodynamics. Thus, a cycle will occur only if both the first and second laws of thermodynamics are satisfied.

In its broader significance, the second law acknowledges that processes proceed in a certain direction but not in the opposite direction. A hot cup of coffee cools by virtue of heat transfer to the surroundings, but heat will not flow from the cooler surroundings to the hotter cup of coffee. Gasoline is used as a car drives up a hill, but the fuel level in the gasoline tank cannot be restored to its original level when the car coasts down the hill. Such familiar observations as these, and a host of others, are evidence of the validity of the second law of thermodynamics.

In this chapter, we consider the second law for a system undergoing a cycle, and in the next two chapters we extend the principles to a system undergoing a change of state and then to a control volume.

7.1 HEAT ENGINES AND REFRIGERATORS

Consider the system and the surroundings previously cited in the development of the first law, as shown in Fig. 7.1. Let the gas constitute the system and, as in our discussion of the first law, let this system undergo a cycle in which work is first done on the system by the paddle wheel as the weight is lowered. Then let the cycle be completed by transferring heat to the surroundings.

We know from our experience that we cannot reverse this cycle. That is, if we transfer heat to the gas, as shown by the dotted arrow, the temperature of the gas will increase, but the paddle wheel will not turn and raise the weight. With the given surroundings (the container, the paddle wheel, and the weight), this system can operate in a cycle in which the heat transfer and work are both negative, but it cannot operate in a cycle in which both the heat transfer and work are positive, even though this would not violate the first law.

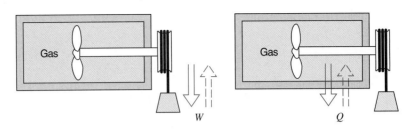

FIGURE 7.1 A system that undergoes a cycle involving work and heat.

Consider another cycle, known from our experience to be impossible actually to complete. Let two systems, one at a high temperature and the other at a low temperature, undergo a process in which a quantity of heat is transferred from the high-temperature system to the low-temperature system. We know that this process can take place. We also know that the reverse process, in which heat is transferred from the low-temperature system to the high-temperature system, does not occur, and that it is impossible to complete the cycle by heat transfer only. This impossibility is illustrated in Fig. 7.2.

These two examples lead us to a consideration of the heat engine and the refrigerator, which is also referred to as a heat pump. With the heat engine we can have a system that operates in a cycle and performs a net positive work and a net positive heat transfer. With the heat pump we can have a system that operates in a cycle and has heat transferred to it from a low-temperature body and heat transferred from it to a high-temperature body, though work is required to do this. Three simple heat engines and two simple refrigerators will be considered.

The first heat engine is shown in Fig. 7.3. It consists of a cylinder fitted with appropriate stops and a piston. Let the gas in the cylinder constitute the system. Initially the piston rests on the lower stops, with a weight on the platform. Let the system now undergo a process in which heat is transferred from some high-temperature body to the gas, causing it to expand and raise the piston to the upper stops. At this point the weight is removed. Now let the system be restored to its initial state by transferring heat from the gas to a low-temperature body, thus completing the cycle. Since the weight was raised during the cycle, it is evident that work was done by the gas during the cycle. From the first law we conclude that the net heat transfer was positive and equal to the work done during the cycle.

Such a device is called a heat engine, and the substance to which and from which heat is transferred is called the working substance or working fluid. A heat engine may be defined as a device that operates in a thermodynamic cycle and does a certain amount of net positive work through the transfer of heat from a high-temperature body to a low-temperature body. Often the term heat engine is used in a broader sense to include all devices that produce work, either through heat transfer or through combustion, even though the device does not operate in a thermodynamic cycle. The internal combustion engine

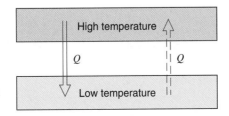

FIGURE 7.2 An example showing the impossibility of completing a cycle by transferring heat from a low-temperature body to a high-temperature body.

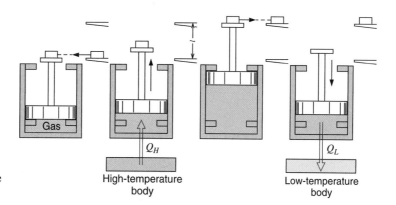

FIGURE 7.3 A simple heat engine.

and the gas turbine are examples of such devices, and calling them heat engines is an acceptable use of the term. In this chapter, however, we are concerned with the more restricted form of heat engine, as just defined, one that operates on a thermodynamic cycle.

A simple steam power plant is an example of a heat engine in this restricted sense. Each component in this plant may be analyzed individually as a steady-state, steady-flow process, but as a whole it may be considered a heat engine (Fig. 7.4) in which water (steam) is the working fluid. An amount of heat, Q_H, is transferred from a high-temperature body, which may be the products of combustion in a furnace, a reactor, or a secondary fluid that in turn has been heated in a reactor. In Fig. 7.4 the turbine is shown schematically as driving the pump. What is significant, however, is the net work that is delivered during the cycle. The quantity of heat Q_L is rejected to a low-temperature body, which is usually the cooling water in a condenser. Thus, the simple steam power plant is a heat engine in the restricted sense, for it has a working fluid, to which and from which heat is transferred, and which does a certain amount of work as it undergoes a cycle.

Another example of a heat engine is the thermoelectric power generation device that was discussed in Chapter 1 and shown schematically in Fig. 1.8b. Heat is transferred from a high-temperature body to the hot junction (Q_H), and heat is transferred from the cold junction to the surroundings (Q_L). Work is done in the form of electrical energy. Since there is no working fluid, we do not usually think of this as a device that operates in a

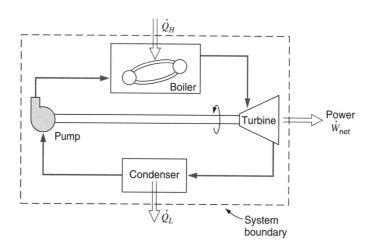

FIGURE 7.4 A heat engine involving steady-state processes.

cycle. However, if we adopt a microscopic point of view, we could regard a cycle as the flow of electrons. Furthermore, as with the steam power plant, the state at each point in the thermoelectric power generator does not change with time under steady-state conditions.

Thus, by means of a heat engine, we are able to have a system operate in a cycle and have both the net work and the net heat transfer positive, which we were not able to do with the system and surroundings of Fig. 7.1.

We note that in using the symbols Q_H and Q_L, we have departed from our sign connotation for heat, because for a heat engine Q_L is negative when the working fluid is considered as the system. In this chapter it will be advantageous to use the symbol Q_H to represent the heat transfer to or from the high-temperature body, and Q_L to represent the heat transfer to or from the low-temperature body. The direction of the heat transfer will be evident from the context.

At this point, it is appropriate to introduce the concept of thermal efficiency of a heat engine. In general, we say that efficiency is the ratio of output, the energy sought, to input, the energy that costs, but the output and input must be clearly defined. At the risk of oversimplification, we may say that in a heat engine the energy sought is the work, and the energy that costs money is the heat from the high-temperature source (indirectly, the cost of the fuel). Thermal efficiency is defined as

$$\eta_{\text{thermal}} = \frac{W(\text{energy sought})}{Q_H(\text{energy that costs})} = \frac{Q_H - Q_L}{Q_H} = 1 - \frac{Q_L}{Q_H} \qquad (7.1)$$

Heat engines vary greatly in size and shape, from large steam engines, gas turbines, or jet engines, to gasoline engines for cars and diesel engines for trucks or cars, to much smaller engines for lawn mowers or hand-held devices such as chain saws or trimmers. Typical values for the thermal efficiency of real engines are about 35–50% for large power plants, 30–35% for gasoline engines, and 35–40% for diesel engines. Smaller utility-type engines may have only about 20% efficiency, owing to their simple carburetion and controls and to the fact that some losses scale differently with size and therefore represent a larger fraction for smaller machines.

EXAMPLE 7.1

An automobile engine produces 136 hp on the output shaft with a thermal efficiency of 30%. The fuel it burns gives 35 000 kJ/kg as energy release. Find the total rate of energy rejected to the ambient and the rate of fuel consumption in kg/s.

Solution

From the definition of a heat engine efficiency, Eq. 7.1, and the conversion of hp from Table A.1 we have:

$$\dot{W} = \eta_{\text{eng}}\dot{Q}_H = 136 \text{ hp} \times 0.7355 \text{ kW/hp} = 100 \text{ kW}$$

$$\dot{Q}_H = \dot{W}/\eta_{\text{eng}} = 100/0.3 = 333 \text{ kW}$$

The energy equation for the overall engine gives:

$$\dot{Q}_L = \dot{Q}_H - \dot{W} = (1 - 0.3)\dot{Q}_H = 233 \text{ kW}$$

From the energy release in the burning we have: $\dot{Q}_H = \dot{m}q_H$, so

$$\dot{m} = \dot{Q}_H/q_H = \frac{333 \text{ kW}}{35\ 000 \text{ kJ/kg}} = 0.0095 \text{ kg/s}$$

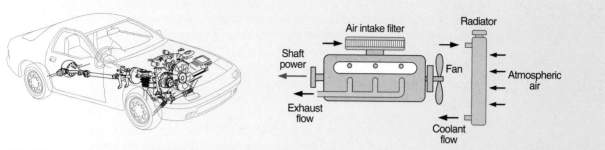

FIGURE 7.5 Sketch for Example 7.1.

An actual engine shown in Fig. 7.5 rejects energy to the ambient through the radiator cooled by atmospheric air, as heat transfer from the exhaust system and the exhaust flow of hot gases.

The second cycle that we were not able to complete was the one indicating the impossibility of transferring heat directly from a low-temperature body to a high-temperature body. This can of course be done with a refrigerator or heat pump. A vapor-compression refrigerator cycle, which was introduced in Chapter 1 and shown in Fig. 1.7, is shown again in Fig. 7.6. The working fluid is the refrigerant, such as R-134a or ammonia, which goes through a thermodynamic cycle. Heat is transferred to the refrigerant in the evaporator, where its pressure and temperature are low. Work is done on the refrigerant in the compressor, and heat is transferred from it in the condenser, where its pressure and temperature are high. The pressure drops as the refrigerant flows through the throttle valve or capillary tube.

Thus, in a refrigerator or heat pump, we have a device that operates in a cycle, that requires work, and that accomplishes the objective of transferring heat from a low-temperature body to a high-temperature body.

The thermoelectric refrigerator, which was discussed in Chapter 1 and shown schematically in Fig. 1.8a, is another example of a device that meets our definition of a refrigerator. The work input to the thermoelectric refrigerator is in the form of electrical energy, and heat is transferred from the refrigerated space to the cold junction (Q_L) and from the hot junction to the surroundings (Q_H).

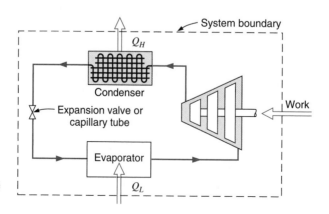

FIGURE 7.6 A simple refrigeration cycle.

The "efficiency" of a refrigerator is expressed in terms of the coefficient of performance, which we designate with the symbol β. For a refrigerator the objective, that is, the energy sought, is Q_L, the heat transferred from the refrigerated space. The energy that costs is the work W. Thus, the coefficient of performance, β,[1] is

$$\beta = \frac{Q_L \text{ (energy sought)}}{W \text{(energy that costs)}} = \frac{Q_L}{Q_H - Q_L} = \frac{1}{Q_H/Q_L - 1} \qquad (7.2)$$

A household refrigerator may have a coefficient of performance (often referred to as COP) of about 2.5, whereas a deep freeze unit will be closer to 1.0. Lower cold temperature space of higher warm temperature space will result in lower values of COP, as will be found in Section 7.6. For a heat pump operating over a moderate temperature range, a value of its COP can be around 4, with this value decreasing sharply as the heat pump's operating temperature range is broadened.

EXAMPLE 7.2   The refrigerator in a kitchen shown in Fig. 7.7 receives an electrical input power of 150 W to drive the system, and it rejects 400 W to the kitchen air. Find the rate of energy taken out of the cold space and the coefficient of performance of the refrigerator.

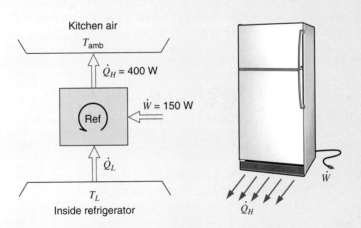

FIGURE 7.7 Sketch for Example 7.2.

---

[1]It should be noted that a refrigeration or heat pump cycle can be used with either of two objectives. It can be used as a refrigerator, in which case the primary objective is Q_L, the heat transferred to the refrigerant from the refrigerated space. It can also be used as a heating system (in which case it is usually referred to as a heat pump), the objective being Q_H, the heat transferred from the refrigerant to the high-temperature body, which is the space to be heated. Q_L is transferred to the refrigerant from the ground, the atmospheric air, or well water. The coefficient of performance for this case, β', is

$$\beta' = \frac{Q_H \text{(energy sought)}}{W \text{ (energy that costs)}} = \frac{Q_H}{Q_H - Q_L} = \frac{1}{1 - Q_L/Q_H}$$

It also follows that for a given cycle.

$$\beta' - \beta = 1$$

Unless otherwise specified, the term coefficient of performance will always refer to a refrigerator as defined by Eq. 7.2.

Solution

C.V. refrigerator. Assume steady state so there is no storage of energy. The information provided is $\dot{W} = 150$ W, and the heat rejected is $\dot{Q}_H = 400$ W.

The energy equation gives:

$$\dot{Q}_L = \dot{Q}_H - \dot{W} = 400 - 150 = 250 \text{ W}$$

This is also the rate of energy transfer into the cold space from the warmer kitchen due to heat transfer and exchange of cold air inside with warm air when you open the door.

From the definition of the coefficient of performance, Eq. 7.2

$$\beta_{\text{REFRIG}} = \frac{\dot{Q}_L}{\dot{W}} = \frac{250}{150} = 1.67$$

Before we state the second law, the concept of a thermal reservoir should be introduced. A thermal reservoir is a body to which and from which heat can be transferred indefinitely without change in the temperature of the reservoir. Thus, a thermal reservoir always remains at constant temperature. The ocean and the atmosphere approach this definition very closely. Frequently it will be useful to designate a high-temperature reservoir and a low-temperature reservoir. Sometimes a reservoir from which heat is transferred is called a source, and a reservoir to which heat is transferred is called as sink.

7.2 THE SECOND LAW OF THERMODYNAMICS

On the basis of the matter considered in the previous section, we are now ready to state the second law of thermodynamics. There are two classical statements of the second law, known as the Kelvin–Planck statement and the Clausius statement.

> *The Kelvin–Planck statement*: It is impossible to construct a device that will operate in a cycle and produce no effect other than the raising of a weight and the exchange of heat with a single reservoir. See Fig. 7.8.

This statement ties in with our discussion of the heat engine. In effect, it states that it is impossible to construct a heat engine that operates in a cycle, receives a given amount of heat from a high-temperature body, and does an equal amount of work. The only alter-

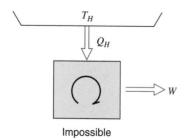

FIGURE 7.8 The Kelvin–Planck statement.

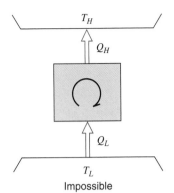

T_H

Q_H

Q_L

T_L

Impossible

FIGURE 7.9 The Clausius statement.

native is that some heat must be transferred from the working fluid at a lower temperature to a low-temperature body. Thus, work can be done by the transfer of heat only if there are two temperature levels, and heat is transferred from the high-temperature body to the heat engine and also from the heat engine to the low-temperature body. This implies that it is impossible to build a heat engine that has a thermal efficiency of 100%.

The Clausius statement: It is impossible to construct a device that operates in a cycle and produces no effect other than the transfer of heat from a cooler body to a hotter body. See Fig. 7.9.

This statement is related to the refrigerator or heat pump. In effect, it states that it is impossible to construct a refrigerator that operates without an input of work. This also implies that the coefficient of performance is always less than infinity.

Three observations should be made about these two statements. The first observation is that both are negative statements. It is of course impossible to "prove" these negative statements. However, we can say that the second law of thermodynamics (like every other law of nature) rests on experimental evidence. Every relevant experiment that has been conducted either directly or indirectly verifies the second law, and no experiment has ever been conducted that contradicts the second law. The basis of the second law is therefore experimental evidence.

A second observation is that these two statements of the second law are equivalent. Two statements are equivalent if the truth of each statement implies the truth of the other, or if the violation of each statement implies the violation of the other. That a violation of the Clausius statement implies a violation of the Kelvin–Planck statement may be shown. The device at the left in Fig. 7.10 is a refrigerator that requires no work and thus violates the Clausius statement. Let an amount of heat Q_L be transferred from the low-temperature reservoir to this refrigerator, and let the same amount of heat Q_L be transferred to the high-temperature reservoir. Let an amount of heat Q_H that is greater than Q_L be transferred from the high-temperature reservoir to the heat engine, and let the engine reject the amount of heat Q_L as it does an amount of work W, which equals $Q_H - Q_L$. Because there is no net heat transfer to the low-temperature reservoir, the low-temperature reservoir, along with the heat engine and the refrigerator, can be considered together as a device that operates in a cycle and produces no effect other than the raising of a weight (work) and the exchange of heat with a single reservoir. Thus, a violation of the Clausius statement implies a violation of the Kelvin–Planck statement. The complete equivalence

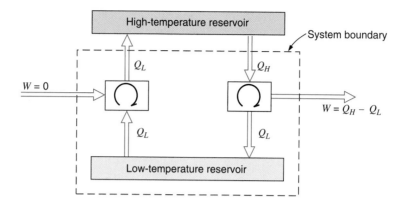

FIGURE 7.10
Demonstration of the equivalence of the two statements of the second law.

of these two statements is established when it is also shown that a violation of the Kelvin–Planck statement implies a violation of the Clausius statement. This is left as an exercise for the student.

The third observation is that frequently the second law of thermodynamics has been stated as the impossibility of constructing a perpetual-motion machine of the second kind. A perpetual-motion machine of the first kind would create work from nothing or create mass or energy, thus violating the first law. A perpetual-motion machine of the second kind would extract heat from a source and then convert this heat completely into other forms of energy, thus violating the second law. A perpetual-motion machine of the third kind would have no friction, and thus would run indefinitely but produce no work.

A heat engine that violated the second law could be made into a perpetual-motion machine of the second kind by taking the following steps. Consider Fig. 7.11, which might be the power plant of a ship. An amount of heat Q_L is transferred from the ocean to a high-temperature body by means of a heat pump. The work required is W', and the heat transferred to the high-temperature body is Q_H. Let the same amount of heat be transferred to a heat engine that violates the Kelvin–Planck statement of the second law and does an amount of work $W = Q_H$. Of this work an amount $Q_H - Q_L$ is required to drive the heat pump, leaving the net work ($W_{net} = Q_L$) available for driving the ship. Thus, we

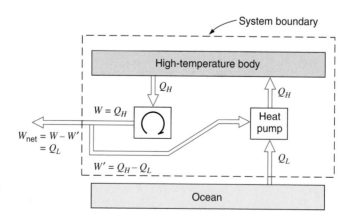

FIGURE 7.11 A perpetual-motion machine of the second kind.

have a perpetual-motion machine in the sense that work is done by utilizing freely available sources of energy such as the ocean or atmosphere.

7.3 THE REVERSIBLE PROCESS

The question that can now logically be posed is this: If it is impossible to have a heat engine of 100% efficiency, what is the maximum efficiency one can have? The first step in the answer to this question is to define an ideal process, which is called a reversible process.

A reversible process for a system is defined as a process that once having taken place can be reversed and in so doing leave no change in either system or surroundings.

Let us illustrate the significance of this definition for a gas contained in a cylinder that is fitted with a piston. Consider first Fig. 7.12, in which a gas, which we define as the system, is restrained at high pressure by a piston that is secured by a pin. When the pin is removed, the piston is raised and forced abruptly against the stops. Some work is done by the system, since the piston has been raised a certain amount. Suppose we wish to restore the system to its initial state. One way of doing this would be to exert a force on the piston and thus compress the gas until the pin can be reinserted in the piston. Since the pressure on the face of the piston is greater on the return stroke than on the initial stroke, the work done on the gas in this reverse process is greater than the work done by the gas in the initial process. An amount of heat must be transferred from the gas during the reverse stroke so that the system has the same internal energy as it had originally. Thus, the system is restored to its initial state, but the surroundings have changed by virtue of the fact that work was required to force the piston down and heat was transferred to the surroundings. The initial process therefore is an irreversible one because it could not be reversed without leaving a change in the surroundings.

In Fig. 7.13 let the gas in the cylinder comprise the system, and let the piston be loaded with a number of weights. Let the weights be slid off horizontally one at a time, allowing the gas to expand and do work in raising the weights that remain on the piston. As the size of the weights is made smaller and their number is increased, we approach a process that can be reversed, for at each level of the piston during the reverse process there will be a small weight that is exactly at the level of the platform and thus can be placed on the platform without requiring work. In the limit, therefore, as the weights become very small, the reverse process can be accomplished in such a manner that both the

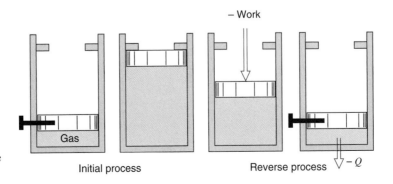

FIGURE 7.12 An example of an irreversible process.

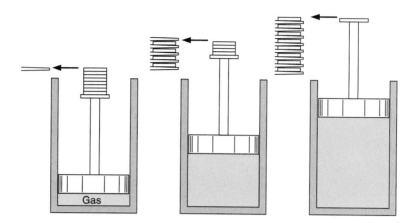

FIGURE 7.13 An example of a process that approaches being reversible.

system and surroundings are in exactly the same state they were initially. Such a process is a reversible process.

7.4 FACTORS THAT RENDER PROCESSES IRREVERSIBLE

There are many factors that make processes irreversible. Four of those factors—friction, unrestrained expansion, heat transfer through a finite temperature difference, and mixing of two different substances—are considered in this section.

Friction

It is readily evident that friction makes a process irreversible, but a brief illustration may amplify the point. Let a block and an inclined plane make up a system, as in Fig. 7.14, and let the block be pulled up the inclined plane by weights that are lowered. A certain amount of work is needed to do this. Some of this work is required to overcome the friction between the block and the plane, and some is required to increase the potential energy of the block. The block can be restored to its initial position by removing some of the weights and thus allowing the block to slide back down the plane. Some heat transfer from the system to the surroundings will no doubt be required to restore the block to its initial temperature. Since the surroundings are not restored to their initial state at the conclusion of the reserve process, we conclude that friction has rendered the process irre-

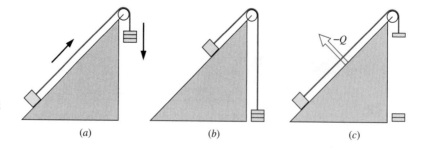

FIGURE 7.14 Demonstration of the fact that friction makes processes irreversible.

(a) (b) (c)

versible. Another type of frictional effect is that associated with the flow of viscous fluids in pipes and passages and in the movement of bodies through viscous fluids.

Unrestrained Expansion

The classic example of an unrestrained expansion, as shown in Fig. 7.15, is a gas separated from a vacuum by a membrane. Consider what happens when the membrane breaks and the gas fills the entire vessel. It can be shown that this is an irreversible process by considering what would be necessary to restore the system to its original state. The gas would have to be compressed and heat transferred from the gas until its initial state is reached. Since the work and heat transfer involve a change in the surroundings, the surroundings are not restored to their initial state, indicating that the unrestrained expansion was an irreversible process. The process described in Fig. 7.12 is also an example of an unrestrained expansion.

In the reversible expansion of a gas, there must be only an infinitesimal difference between the force exerted by the gas and the restraining force, so that the rate at which the boundary moves will be infinitesimal. In accordance with our previous definition, this is a quasi-equilibrium process. However, actual systems have a finite difference in forces, which causes a finite rate of movement of the boundary, and thus the processes are irreversible in some degree.

Heat Transfer through a Finite Temperature Difference

Consider as a system a high-temperature body and a low-temperature body, and let heat be transferred from the high-temperature body to the low-temperature body. The only way in which the system can be restored to its initial state is to provide refrigeration, which requires work from the surroundings, and some heat transfer to the surroundings will also be necessary. Because of the heat transfer and the work, the surroundings are not restored to their original state, indicating that the process was irreversible.

An interesting question is now posed. Heat is defined as energy that is transferred through a temperature difference. We have just shown that heat transfer through a temperature difference is an irreversible process. Therefore, how can we have a reversible heat-transfer process? A heat-transfer process approaches a reversible process as the temperature difference between the two bodies approaches zero. Therefore, we define a reversible heat-transfer process as one in which the heat is transferred through an infinitesimal temperature difference. We realize of course that to transfer a finite amount of heat through an infinitesimal temperature difference would require an infinite amount of time or infinite area. Therefore, all actual heat transfers are through a finite temperature difference and hence are irreversible, with the greater the temperature difference, the greater the irreversibility. We will find, however, that the concept of reversible heat transfer is very useful in describing ideal processes.

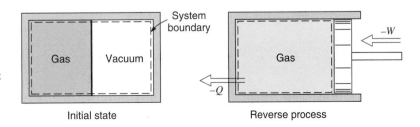

FIGURE 7.15
Demonstration of the fact that unrestrained expansion makes processes irreversible.

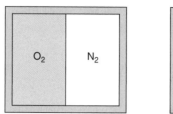
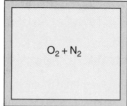

FIGURE 7.16
Demonstration of the fact that the mixing of two different substances is an irreversible process.

Mixing of Two Different Substances

Figure 7.16 illustrates the process of mixing two different gases separated by a membrane. When the membrane is broken, a homogeneous mixture of oxygen and nitrogen fills the entire volume, This process will be considered in some detail in Chapter 12. We can say here that this may be considered a special case of an unrestrained expansion, for each gas undergoes an unrestrained expansion as it fills the entire volume. A certain amount of work is necessary to separate these gases. Thus, an air separation plant such as described in Chapter 1 requires an input of work to accomplish the separation.

Other Factors

A number of other factors make processes irreversible, but they will not be considered in detail here. Hysteresis effects and the i^2R loss encountered in electrical circuits are both factors that make processes irreversible. Ordinary combustion is also an irreversible process.

It is frequently advantageous to distinguish between internal and external irreversibility. Figure 7.17 shows two identical systems to which heat is transferred. Assuming each system to be a pure substance, the temperature remains constant during the heat-transfer process. In one system the heat is transferred from a reservoir at a temperature $T + dT$, and in the other the reservoir is at a much higher temperature, $T + \Delta T$, than the system. The first is a reversible heat-transfer process, and the second is an irreversible heat-transfer process. However, as far as the system itself is concerned, it passes through exactly the same states in both processes, which we assume are reversible. Thus, we can say for the second system that the process is internally reversible but externally irreversible because the irreversibility occurs outside the system.

We should also note the general interrelation of reversibility, equilibrium, and time. In a reversible process, the deviation from equilibrium is infinitesimal, and therefore it oc-

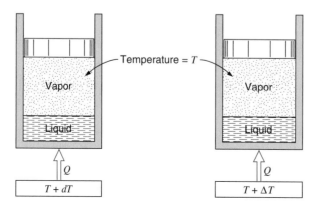

FIGURE 7.17
Illustration of the difference between an internally and externally reversible process.

curs at an infinitesimal rate. Since it is desirable that actual processes proceed at a finite rate, the deviation from equilibrium must be finite, and therefore the actual process is irreversible in some degree. The greater the deviation from equilibrium, the greater the irreversibility, and the more rapidly the process will occur. It should also be noted that the quasi-equilibrium process, which was described in Chapter 2, is a reversible process, and hereafter the term reversible process will be used.

7.5 THE CARNOT CYCLE

Having defined the reversible process and considered some factors that make processes irreversible, let us again pose the question raised in Section 7.3. If the efficiency of all heat engines is less than 100%, what is the most efficient cycle we can have? Let us answer this question for a heat engine that receives heat from a high-temperature reservoir and rejects heat to a low-temperature reservoir. Since we are dealing with reservoirs, we recognize that both the high temperature and the low temperature of the reservoirs are constant and remain constant regardless of the amount of heat transferred.

Let us assume that this heat engine, which operates between the given high-temperature and low-temperature reservoirs, does so in a cycle in which every process is reversible. If every process is reversible, the cycle is also reversible; and if the cycle is reversed, the heat engine becomes a refrigerator. In the next section we will show that this is the most efficient cycle that can operate between two constant-temperature reservoirs. It is called the Carnot cycle and is named after a French engineer, Nicolas Leonard Sadi Carnot (1796–1832), who expressed the foundations of the second law of thermodynamics in 1824.

We now turn our attention to the Carnot cycle. Figure 7.18 shows a power plant that is similar in many respects to a simple steam power plant and, we assume, operates on the Carnot cycle. Consider the working fluid to be a pure substance, such as steam. Heat is transferred from the high-temperature reservoir to the water (steam) in the boiler. For this

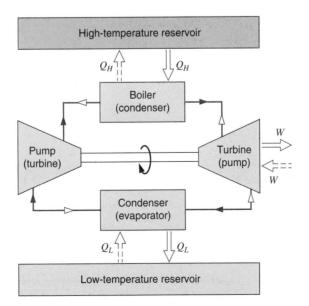

FIGURE 7.18

Example of a heat engine that operates on a Carnot cycle.

process to be a reversible heat transfer, the temperature of the water (steam) must be only infinitesimally lower than the temperature of the reservoir. This result also implies, since the temperature of the reservoir remains constant, that the temperature of the water must remain constant. Therefore, the first process in the Carnot cycle is a reversible isothermal process in which heat is transferred from the high-temperature reservoir to the working fluid. A change of phase from liquid to vapor at constant pressure is of course an isothermal process for a pure substance.

The next process occurs in the turbine without heat transfer and is therefore adiabatic. Since all processes in the Carnot cycle are reversible, this must be a reversible adiabatic process, during which the temperature of the working fluid decreases from the temperature of the high-temperature reservoir to the temperature of the low-temperature reservoir.

In the next process heat is rejected from the working fluid to the low-temperature reservoir. This must be a reversible isothermal process in which the temperature of the working fluid is infinitesimally higher than that of the low-temperature reservoir. During this isothermal process some of the steam is condensed.

The final process, which completes the cycle, is a reversible adiabatic process in which the temperature of the working fluid increases from the low temperature to the high temperature. If this were to be done with water (steam) as the working fluid, a mixture of liquid and vapor would have to be taken from the condenser and compressed. (This would be very inconvenient in practice, and therefore in all power plants the working fluid is completely condensed in the condenser. The pump handles only the liquid phase.)

Sine the Carnot heat engine cycle is reversible, every process could be reversed, in which case it would become a refrigerator. The refrigerator is shown by the dotted lines and parentheses in Fig. 7.18. The temperature of the working fluid in the evaporator would be infinitesimally lower than the temperature of the low-temperature reservoir, and in the condenser it is infinitesimally higher than that of the high-temperature reservoir.

It should be emphasized that the Carnot cycle can, in principle, be executed in many different ways. Many different working substances can be used, such as a gas or a thermoelectric device such as described in Chapter 1. There are also various possible arrangements of machinery. For example, a Carnot cycle can be devised that takes place entirely within a cylinder, using a gas as a working substance, as shown in Fig. 7.19.

The important point to be made here is that the Carnot cycle, regardless of what the working substance may be, always has the same four basic processes. These processes are:

1. A reversible isothermal process in which heat is transferred to or from the high-temperature reservoir.

2. A reversible adiabatic process in which the temperature of the working fluid decreases from the high temperature to the low temperature.

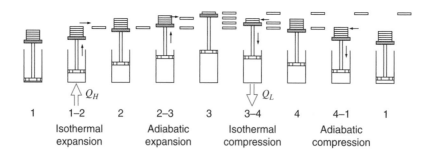

FIGURE 7.19
Example of a gaseous system operating on a Carnot cycle.

3. A reversible isothermal process in which heat is transferred to or from the low-temperature reservoir.

4. A reversible adiabatic process in which the temperature of the working fluid increases from the low temperature to the high temperature.

7.6 TWO PROPOSITIONS REGARDING THE EFFICIENCY OF A CARNOT CYCLE

There are two important propositions regarding the efficiency of a Carnot cycle.

First Proposition

It is impossible to construct an engine that operates between two given reservoirs and is more efficient than a reversible engine operating between the same two reservoirs.

The proof of this statement is accomplished through a "thought experiment." An initial assumption is made, and it is then shown that this assumption leads to impossible conclusions. The only possible conclusion is that the initial assumption was incorrect.

Let us assume that there is an irreversible engine operating between two given reservoirs that has a greater efficiency than a reversible engine operating between the same two reservoirs. Let the heat transfer to the irreversible engine be Q_H, the heat rejected be Q'_L, and the work be W_{IE} (which equals $Q_H - Q'_L$) as shown in Fig. 7.20. Let the reversible engine operate as a refrigerator (this is possible since it is reversible). Finally, let the heat transfer with the low-temperature reservoir be Q_L, the heat transfer with the high-temperature reservoir be Q_H, and the work required be W_{RE} (which equals $Q_H - Q_L$).

Since the initial assumption was that the irreversible engine is more efficient, it follows (because Q_H is the same for both engines) that $Q'_L < Q_L$ and $W_{IE} > W_{RE}$. Now the irreversible engine can drive the reversible engine and still deliver the net work W_{net}, which equals $W_{IE} - W_{RE} = Q_L - Q'_L$. If we consider the two engines and the high-temperature reservoir as a system, as indicated in Fig. 7.20, we have a system that operates in a cycle, exchanges heat with a single reservoir, and does a certain amount of work. However, this

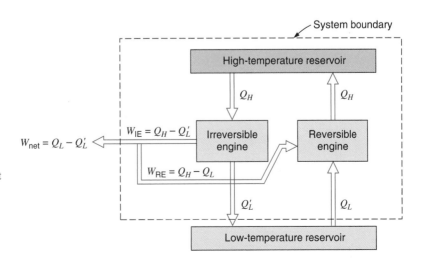

FIGURE 7.20
Demonstration of the fact that the Carnot cycle is the most efficient cycle operating between two fixed-temperature reservoirs.

would constitute a violation of the second law, and we conclude that our initial assumption (that the irreversible engine is more efficient than a reversible engine) is incorrect. Therefore, we cannot have an irreversible engine that is more efficient than a reversible engine operating between the same two reservoirs.

Second Proposition

All engines that operate on the Carnot cycle between two given constant-temperature reservoirs have the same efficiency. The proof of this proposition is similar to the proof just outlined, which assumes that there is one Carnot cycle that is more efficient than another Carnot cycle operating between the same temperature reservoirs. Let the Carnot cycle with the higher efficiency replace the irreversible cycle of the previous argument, and let the Carnot cycle with the lower efficiency operate as the refrigerator. The proof proceeds with the same line of reasoning as in the first proposition. The details are left as an exercise for the student.

7.7 THE THERMODYNAMIC TEMPERATURE SCALE

In discussing the matter of temperature in Chapter 2, we pointed out that the zeroth law of thermodynamics provides a basis for temperature measurement, but that a temperature scale must be defined in terms of a particular thermometer substance and device. A temperature scale that is independent of any particular substance, which might be called an absolute temperature scale, would be most desirable. In the last paragraph we noted that the efficiency of a Carnot cycle is independent of the working substance and depends only on the temperature. This fact provides the basis for such an absolute temperature scale, which we call the thermodynamic scale.

The concept of this temperature scale may be developed with the help of Fig. 7.21, which shows three reservoirs and three engines that operate on the Carnot cycle. T_1 is the highest temperature, T_3 is the lowest temperature, and T_2 is an intermediate temperature, and the engines operate between the various reservoirs as indicated. Q_1 is the same

FIGURE 7.21
Arrangement of heat engines to demonstrate the thermodynamic temperature scale.

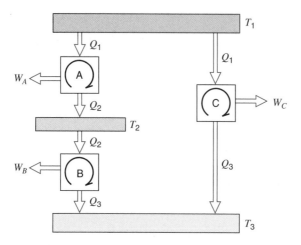

for both A and C and, since we are dealing with reversible cycles, Q_3 is the same for B and C.

Since the efficiency of a Carnot cycle is a function only of the temperature, we can write

$$\eta_{\text{thermal}} = 1 - \frac{Q_L}{Q_H} = 1 - \psi(T_L, T_H) \tag{7.3}$$

where ψ designates a functional relation.

Let us apply this functional relation to the three Carnot cycles of Fig. 7.21:

$$\frac{Q_1}{Q_2} = \psi(T_1, T_2)$$

$$\frac{Q_2}{Q_3} = \psi(T_2, T_3)$$

$$\frac{Q_1}{Q_3} = \psi(T_1, T_3)$$

Since

$$\frac{Q_1}{Q_3} = \frac{Q_1 Q_2}{Q_2 Q_3}$$

it follows that

$$\psi(T_1, T_3) = \psi(T_1, T_2) \times \psi(T_2, T_3) \tag{7.4}$$

Note that the left side is a function of T_1 and T_3 (and not of T_2), and therefore the right side of this equation must also be a function of T_1 and T_3 (and not of T_2). From this fact we can conclude that the form of the function ψ must be such that

$$\psi(T_1, T_2) = \frac{f(T_1)}{f(T_2)}$$

$$\psi(T_2, T_3) = \frac{f(T_2)}{f(T_3)}$$

for in this way $f(T_2)$ will cancel from the product of $\psi(T_1, T_2) \times \psi(T_2, T_3)$. Therefore, we conclude that

$$\frac{Q_1}{Q_3} = \psi(T_1, T_3) = \frac{f(T_1)}{f(T_3)} \tag{7.5}$$

In general terms,

$$\frac{Q_H}{Q_L} = \frac{f(T_H)}{f(T_L)} \tag{7.6}$$

Now there are several functional relations that will satisfy this equation. For the thermodynamic scale of temperature, which was originally proposed by Lord Kelvin, the selected relation is

$$\frac{Q_H}{Q_L} = \frac{(T_H)}{(T_L)} \tag{7.7}$$

With absolute temperatures so defined, the efficiency of a Carnot cycle may be expressed in terms of the absolute temperatures.

$$\eta_{\text{thermal}} = 1 - \frac{Q_L}{Q_H} = 1 - \frac{T_L}{T_H} \tag{7.8}$$

This means that if the thermal efficiency of a Carnot cycle operating between two given constant-temperature reservoirs is known, the ratio of the two absolute temperatures is also known.

It should be noted that Eq. 7.7 gives us a ratio of absolute temperatures, but it does not give us information about the magnitude of the degree. Let us first consider a qualitative approach to this matter and then a more rigorous statement.

Suppose we had a heat engine operating on the Carnot cycle that received heat at the temperature of the steam point and rejected heat at the temperature of the ice point. (Because a Carnot cycle involves only reversible processes, it is impossible to construct such a heat engine and perform the proposed experiment. However, we can follow the reasoning as a "thought experiment" and gain additional understanding of the thermodynamic temperature scale.) If the efficiency of such an engine could be measured, we would find it to be 26.80%. Therefore, from Eq. 7.8,

$$\eta_{th} = 1 - \frac{T_L}{T_H} = 1 - \frac{T_{\text{ice point}}}{T_{\text{steam point}}} = 0.2680$$

$$\frac{T_{\text{ice point}}}{T_{\text{steam point}}} = 0.7320$$

This gives us one equation concerning the two unknowns T_H and T_L. The second equation comes from an arbitrary decision regarding the magnitude of the degree on the thermodynamic temperature scale. If we wish to have the magnitude of the degree on the absolute scale correspond to the magnitude of the degree on the Celsius scale, we can write

$$T_{\text{steam point}} - T_{\text{ice point}} = 100$$

Solving these two equations simultaneously, we find

$$T_{\text{steam point}} = 373.15 \text{ K}, \qquad T_{\text{ice point}} = 273.15 \text{ K}$$

It follows that

$$T(°C) + 273.15 = T(K)$$

The absolute scale related to the Fahrenheit scale is the Rankine scale, designated by R. On both these scales there are 180 degrees between the ice point and the steam point. Therefore, for a Carnot cycle heat engine operating between the steam point and the ice point, we would have the two relations

$$T_{\text{steam point}} - T_{\text{ice point}} = 180$$

$$\frac{T_{\text{ice point}}}{T_{\text{steam point}}} = 0.7320$$

Solving these two equations simultaneously, we find

$$T_{\text{steam point}} = 671.67 \text{ R}, \qquad T_{\text{ice point}} = 491.67 \text{ R}$$

It follows that temperatures on the Fahrenheit and Rankine scales are related as follows:

$$T(F) + 459.67 = T(R)$$

As already noted, the measurement of efficiencies of Carnot cycles is, however, not a practical way to approach the problem of temperature measurement on the thermodynamic scale of temperature. The actual approach is based on the ideal-gas thermometer and an assigned value for the triple point of water. At the tenth Conference on Weights and Measures, which was held in 1954, the temperature of the triple point of water was assigned the value 273.16 K. [The triple point of water is approximately 0.01°C above the ice point. The ice point is defined as the temperature of a mixture of ice and water at a pressure of 1 atm (101.3 kPa) of air that is saturated with water vapor.] The ideal-gas thermometer is discussed in the following section.

7.8 THE IDEAL-GAS TEMPERATURE SCALE

In this section we reconsider in greater detail the ideal-gas temperature scale introduced in Section 3.6. This scale is based on the observation that as the pressure of a real gas approaches zero, its equation of state approaches that of an ideal gas:

$$Pv = RT$$

It will be shown that the ideal-gas temperature scale satisfies the definition of thermodynamic temperature given in the preceding section by Eq. 7.7, but first let us consider how an ideal gas might be used to measure temperature in a constant-volume gas thermometer, shown schematically in Fig. 7.22.

Let the gas bulb be placed in the location where the temperature is to be measured, and let the mercury column be adjusted so that the level of mercury stands at the reference mark A. Thus, the volume of the gas remains constant. Assume that the gas in the capillary tube is at the same temperature as the gas in the bulb. Then the pressure of the gas, which is indicated by the height L of the mercury column, is a measure of the temperature.

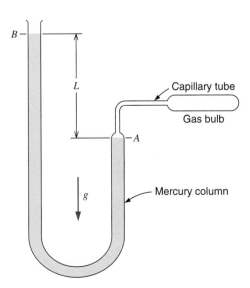

FIGURE 7.22

Schematic diagram of a constant-volume gas thermometer.

Let the pressure that is associated with the temperature of the triple point of water (273.16 K) also be measured, and let us designate this pressure $P_{t.p.}$. Then, from the definition of an ideal gas, any other temperature T could be determined from a pressure measurement P by the relation

$$T = 273.16 \left(\frac{P}{P_{t.p.}} \right)$$

From a practical point of view, we have the problem that no gas behaves exactly like an ideal gas. However, we do know that as the pressure approaches zero, the behavior of all gases approaches that of an ideal gas. Suppose then that a series of measurements is made with varying amounts of gas in the gas bulb. This means that the pressure measured at the triple point, and also the pressure at any other temperature, will vary. If the indicated temperature T_i (obtained by assuming that the gas is ideal) is plotted against the pressure of gas with the bulb at the triple point of water, a curve like the one shown in Fig. 7.23 is obtained. When this curve is extrapolated to zero pressure, the correct ideal-gas temperature is obtained. Different curves might result from different gases, but they would all indicate the same temperature at zero pressure.

We have outlined only the general features and principles for measuring temperature on the ideal-gas scale of temperatures. Precision work in this field is difficult and laborious, and there are only a few laboratories in the world where such work is carried on. The International Temperature Scale, which was mentioned in Chapter 2, closely approximates the thermodynamic temperature scale and is much easier to work with in actual temperature measurement.

We now demonstrate that the ideal-gas temperature scale discussed earlier is, in fact, identical to the thermodynamic temperature scale, which was defined in the discussion of the Carnot cycle and the second law. Our objective can be achieved by using an ideal gas as the working fluid for a Carnot-cycle heat engine and analyzing the four processes that make up the cycle. The four state points, 1, 2, 3, and 4, and the four processes are as shown in Fig. 7.24. For convenience, let us consider a unit mass of gas inside the cylinder. Now for each of the four processes, the reversible work done at the moving boundary is given by Eq. 4.2:

$$\delta w = P \, dv$$

Similarly, for each process the gas behavior is, from the ideal-gas relation, Eq. 3.5,

$$Pv = RT$$

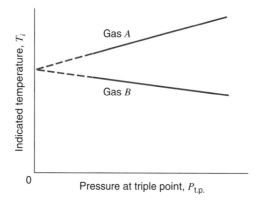

FIGURE 7.23 Sketch showing how the ideal-gas temperature is determined.

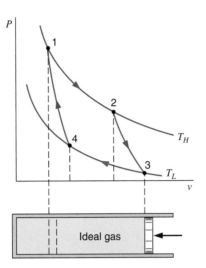

FIGURE 7.24 The ideal-gas Carnot cycle.

and the internal energy change, from Eq. 5.20, is

$$du = C_{v0}\, dT$$

Assuming no changes in kinetic or potential energies, the first law is, from Eq. 5.7 at unit mass,

$$\delta q = du + \delta w$$

Substituting the three previous expressions into this equation, we have for each of the four processes

$$\delta q = C_{v0}\, dT + \frac{RT}{v}\, dv \tag{7.9}$$

The shape of the two isothermal processes shown in Fig. 7.24 is known, since Pv is constant in each case. The process 1–2 is an expansion at T_H, such that v_2 is larger than v_1. Similarly, the process 3–4 is a compression at a lower temperature, T_L, and v_4 is smaller than v_3. The adiabatic process 2–3 is an expansion from T_H to T_L, with an increase in specific volume, while the adiabatic process 4–1 is a compression from T_L to T_H, with a decrease in specific volume. The area below each process line represents the work for that process, as given by Eq. 4.2.

We now proceed to integrate Eq. 7.9 for each of the four processes that make up the Carnot cycle. For the isothermal heat addition process 1–2, we have

$$q_H = {}_1q_2 = 0 + RT_H\, \ln\frac{v_2}{v_1} \tag{7.10}$$

For the adiabatic expansion process 2–3,

$$0 = \int_{T_H}^{T_L} \frac{C_{v0}}{T}\, dT + R\ln\frac{v_3}{v_2} \tag{7.11}$$

For the isothermal heat rejection process 3–4,

$$q_L = -{}_3q_4 = -0 - RT_L\ln\frac{v_4}{v_3}$$

$$= +RT_L\ln\frac{v_3}{v_4} \tag{7.12}$$

and for the adiabatic compression process 4–1,

$$0 = \int_{T_L}^{T_H} \frac{C_{v0}}{T} \, dT + R \ln \frac{v_1}{v_4} \tag{7.13}$$

From Eqs. 7.11 and 7.13, we get

$$\int_{T_L}^{T_H} \frac{C_{v0}}{T} \, dT = R \ln \frac{v_3}{v_2} = -R \ln \frac{v_1}{v_4}$$

Therefore,

$$\frac{v_3}{v_2} = \frac{v_4}{v_1}, \quad \text{or} \quad \frac{v_3}{v_4} = \frac{v_2}{v_1} \tag{7.14}$$

Thus, from Eqs. 7.10 and 7.12 and substituting Eq. 7.14, we find that

$$\frac{q_H}{q_L} = \frac{RT_H \ln \frac{v_2}{v_1}}{RT_L \ln \frac{v_3}{v_4}} = \frac{T_H}{T_L}$$

which is Eq. 7.7, the definition of the thermodynamic temperature scale in connection with the second law.

7.9 IDEAL VERSUS REAL MACHINES

Following the definition of the thermodynamic temperature scale by Eq. 7.7, it was noted that the thermal efficiency of a Carnot cycle heat engine is given by Eq. 7.8. It also follows that a Carnot cycle operating as a refrigerator or heat pump will have a coefficient of performance expressed as

$$\beta = \frac{Q_L}{Q_H - Q_L} \bigg|_{\text{Carnot}} = \frac{T_L}{T_H - T_L} \tag{7.15}$$

$$\beta' = \frac{Q_H}{Q_H - Q_L} \bigg|_{\text{Carnot}} = \frac{T_H}{T_H - T_L} \tag{7.16}$$

For all three "efficiencies" in Eqs. 7.8, 7.15, and 7.16, the first equality sign is the definition with the use of the energy equation and thus is always true. The second equality sign is valid only if the cycle is reversible, that is, a Carnot cycle. Any real heat engine, refrigerator, or heat pump will be less efficient, such that

$$\eta_{\text{real thermal}} = 1 - \frac{Q_L}{Q_H} \leq 1 - \frac{T_L}{T_H}$$

$$\beta_{\text{real}} = \frac{Q_L}{Q_H - Q_L} \leq \frac{T_L}{T_H - T_L}$$

$$\beta'_{\text{real}} = \frac{Q_H}{Q_H - Q_L} \leq \frac{T_H}{T_H - T_L}$$

A final point needs to be made about the significance of absolute-zero temperature in connection with the second law and the thermodynamic temperature scale. Consider a Carnot-cycle heat engine that receives a given amount of heat from a given high-temperature reservoir. As the temperature at which heat is rejected from the cycle is lowered, the net work output increases and the amount of heat rejected decreases. In the limit, the heat rejected is zero, and the temperature of the reservoir corresponding to this limit is absolute zero.

Similarly, for a Carnot-cycle refrigerator, the amount of work required to produce a given amount of refrigeration increases as the temperature of the refrigerated space decreases. Absolute zero represents the limiting temperature that can be achieved, and the amount of work required to produce a finite amount of refrigeration approaches infinity as the temperature at which refrigeration is provided approaches zero.

EXAMPLE 7.3   Let us consider the heat engine, shown schematically in Fig. 7.25, that receives a heat-transfer rate of 1 MW at a high temperature of 550°C and rejects energy to the ambient surroundings at 300 K. Work is produced at a rate of 450 kW. We would like to know how much energy is discarded to the ambient surroundings and the engine efficiency and compare both of these to a Carnot heat engine operating between the same two reservoirs.

Solution

If we take the heat engine as a control volume, the energy equation gives

$$\dot{Q}_L = \dot{Q}_H - \dot{W} = 1000 - 450 = 550 \text{ kW}$$

and from the definition of the efficiency

$$\eta_{\text{thermal}} = \dot{W}/\dot{Q}_H = 450/1000 = 0.45$$

For the Carnot heat engine, the efficiency is given by the temperature of the reservoirs

$$\eta_{\text{Carnot}} = 1 - \frac{T_L}{T_H} = 1 - \frac{300}{550 + 273} = 0.635$$

The rates of work and heat rejection become

$$\dot{W} = \eta_{\text{Carnot}}\dot{Q}_H = 0.635 \times 1000 = 635 \text{ kW}$$
$$\dot{Q}_L = \dot{Q}_H - \dot{W} = 1000 - 635 = 365 \text{ kW}$$

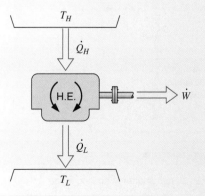

FIGURE 7.25 A heat engine operating between two constant temperature energy reservoirs for Example 7.1.

The actual heat engine thus has a lower efficiency than the Carnot (ideal) heat engine, with a value of 45% typical for a modern steam power plant. This also implies that the actual engine rejects a larger amount of energy to the ambient surroundings (55%) compared with the Carnot heat engine (36%).

EXAMPLE 7.4   As one mode of operation of an air conditioner is the cooling of a room on a hot day, it works as a refrigerator, shown in Fig. 7.26. A total of 4 kW should be removed from a room at 24°C to the outside atmosphere at 35°C. We would like to estimate the magnitude of the required work. To do this we will not analyze the processes inside the refrigerator, which is deferred to Chapter 11, but we can give a lower limit for the rate of work assuming it is a Carnot-cycle refrigerator.

Solution

The coefficient of performance (COP) is

$$\beta = \frac{\dot{Q}_L}{\dot{W}} = \frac{\dot{Q}_L}{\dot{Q}_H - \dot{Q}_L} = \frac{T_L}{T_H - T_L} = \frac{273 + 24}{35 - 24} = 27$$

so the rate of work or power input will be

$$\dot{W} = \dot{Q}_L/\beta = 4/27 = 0.15 \text{ kW}$$

Since the power was estimated assuming a Carnot refrigerator, it is the smallest amount possible. Recall also the expressions for heat-transfer rates in Chapter 4. If the refrigerator should push 4.15 kW out to the atmosphere at 35°C, the high-temperature side of it should be at a higher temperature, maybe 45°C, to have a reasonably small-sized heat

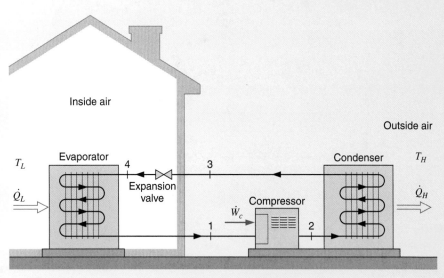

FIGURE 7.26 An air conditioner in cooling mode where T_L is the room.

An air conditioner in cooling mode

exchanger. As it cools the room, a flow of air of less than, say, 18°C would be needed. Redoing the COP with a high of 45°C and a low of 18°C gives 10.8, which is more realistic. A real refrigerator would operate with a COP of the order of 5 or less.

In the previous discussion and examples we considered the constant-temperature energy reservoirs and used those temperatures to calculate the Carnot-cycle efficiency. However, if we recall the expressions for the rate of heat transfer by conduction, convection, or radiation in Chapter 4, they can all be shown as

$$\dot{Q} = C \,\Delta T \tag{7.17}$$

The constant C depends on the mode of heat transfer as

$$\text{Conduction:} \quad C = \frac{kA}{\Delta x} \quad \text{Convection:} \quad C = hA$$

$$\text{Radiation:} \quad C = \epsilon \sigma A (T_s^2 + T_\infty^2)(T_s + T_\infty)$$

For more complex situations with combined layers and modes, we also recover the form in Eq. 7.17, but with a value of C that depends on the geometry, materials, and modes of heat transfer. To have a heat transfer, we therefore must have a temperature difference so that the working substance inside a cycle cannot attain the reservoir temperature unless the area is infinitely large.

SUMMARY

The classical presentation of the second law of thermodynamics starts with the concept of heat engines and refrigerators. A heat engine produces work from a heat transfer obtained from a thermal reservoir, and its operation is limited by the Kelvin–Planck statement. Refrigerators are functionally the same as heat pumps, and they drive energy by heat transfer from a colder environment to a hotter environment, something that will not happen by itself. The Clausius statement says in effect that the refrigerator or heat pump does need work input to accomplish the task. To approach the limit of these cyclic devices, the idea of reversible processes is discussed and further explained by the opposite, namely, irreversible processes and impossible machines. A perpetual motion machine of the first kind violates the first law (energy equation), and a perpetual machine of the second kind violates the second law of thermodynamics.

The limitations for the performance of heat engines (thermal efficiency) and heat pumps or refrigerators (coefficient of performance or COP) are expressed by the corresponding Carnot-cycle device. Two propositions about the Carnot cycle device are another way of expressing the second law of thermodynamics instead of the statements of Kelvin–Planck or Clausius. These propositions lead to the establishment of the thermodynamic absolute temperature, done by Lord Kelvin, and the Carnot-cycle efficiency. We show this temperature to be the same as the ideal-gas temperature introduced in Chapter 3.

You should have learned a number of skills and acquired abilities from studying this chapter that will allow you to

- Understand the concepts of heat engines, heat pumps, and refrigerators.
- Have an idea about reversible processes.
- Know a number of irreversible processes and recognize them.

- Know what a Carnot-cycle is.
- Understand the definition of thermal efficiency of a heat engine.
- Understand the definition of coefficient of performance of a heat pump.
- Know the difference between the absolute and relative temperature.
- Know the limits of thermal efficiency as dictated by the thermal reservoirs and the Carnot-cycle device.
- Have an idea about the thermal efficiency of real heat engines.
- Know the limits of coefficient of performance as dictated by the thermal reservoirs and the Carnot-cycle device.
- Have an idea about the coefficient of performance of real refrigerators.

KEY CONCEPTS AND FORMULAS

(All W, Q can also be rates $\dot{W}, \dot{Q}$)

Heat engine	$W_{HE} = Q_H - Q_L;$ $\eta_{HE} = \dfrac{W_{HE}}{Q_H} = 1 - \dfrac{Q_L}{Q_H}$
Heat pump	$W_{HP} = Q_H - Q_L;$ $\beta_{HP} = \dfrac{Q_H}{W_{HP}} = \dfrac{Q_H}{Q_H - Q_L}$
Refrigerator	$W_{REF} = Q_H - Q_L;$ $\beta_{REF} = \dfrac{Q_L}{W_{REF}} = \dfrac{Q_L}{Q_H - Q_L}$

Factors that make processes irreversible

Friction, unrestrained expansion ($W = 0$), Q over ΔT, mixing, current through a resistor, combustion, or valve flow (throttle).

Carnot cycle

1–2 Isothermal heat addition Q_H in at T_H
2–3 Adiabatic expansion process T does down
3–4 Isothermal heat rejection Q_L out at T_L
4–1 Adiabatic compression process T goes up

Proposition I	$\eta_{any} \leq \eta_{reversible}$	Same T_H, T_L
Proposition II	$\eta_{Carnot\ 1} = \eta_{Carnot\ 2}$	Same T_H, T_L

Absolute temperature $\dfrac{T_L}{T_H} = \dfrac{Q_L}{Q_H}$

Real heat engine $\eta_{HE} = \dfrac{W_{HE}}{Q_H} \leq \eta_{Carnot\ HE} = 1 - \dfrac{T_L}{T_H}$

Real heat pump $\beta_{HP} = \dfrac{Q_H}{W_{HP}} \leq \beta_{Carnot\ HP} = \dfrac{T_H}{T_H - T_L}$

Real refrigerator $\beta_{REF} = \dfrac{Q_L}{W_{REF}} \leq \beta_{Carnot\ REF} = \dfrac{T_L}{T_H - T_L}$

Heat-transfer rates $\dot{Q} = C\,\Delta T$

CONCEPT-STUDY GUIDE PROBLEMS

7.1 Electrical appliances (TV, stereo) use electric power as input. What happens to the power? Are those heat engines? What does the second law say about those devices?

7.2 A gasoline engine produces 20 hp using 35 kW of heat transfer from burning fuel. What is its thermal

efficiency, and how much power is rejected to the ambient?

7.3 A refrigerator removes 1.5 kJ from the cold space using 1 kJ work input. How much energy goes into the kitchen, and what is its coefficient of performance?

7.4 Assume we have a refrigerator operating at steady state using 500 W of electric power with a COP of 2.5. What is the net effect on the kitchen air?

7.5 A window air-conditioner unit is placed on a laboratory bench and tested in cooling mode using 750 W of electric power with a COP of 1.75. What is the cooling power capacity, and what is the net effect on the laboratory?

7.6 Geothermal underground hot water or steam can be used to generate electric power. Does that violate the second law?

7.7 A car engine takes atmospheric air in at 20°C, no fuel, and exhausts the air at −20°C producing work in the process. What do the first and second laws say about that?

7.8 A windmill produces power on a shaft taking kinetic energy out of the wind. Is it a heat engine? Is it a perpetual machine? Explain.

7.9 Ice cubes in a glass of liquid water will eventually melt and all the water will approach room temperature. Is this a reversible process? Why?

7.10 A room is heated with a 1500 W electric heater. How much power can be saved if a heat pump with a COP of 2.0 is used instead?

7.11 If the efficiency of a power plant goes up as the low temperature drops, why do they not just reject energy at say −40°C?

7.12 If the efficiency of a power plant goes up as the low temperature drops, why not let the heat rejection go to a refrigerator at, say, −10°C instead of ambient 20°C?

7.13 A coal-fired power plant operates with a high T of 600°C, whereas a jet engine has about 1400 K. Does that mean we should replace all power plants with jet engines?

7.14 A heat transfer requires a temperature difference (see Chapter 4) to push the Q. What implications does that have for a real heat engine? a refrigerator?

7.15 A large stationary diesel engine produces 15 MW with a thermal efficiency of 40%. The exhaust gas, which we assume is air, flows out at 800 K, and the intake is 290 K. How large a mass flow rate is that if that accounts for half the $\dot{Q}_L$? Can the exhaust flow energy be used?

7.16 Hot combustion gas (air) at 1500 K is used as the heat source in a heat engine where the gas is cooled to 750 K and the ambient is at 300 K. This is not a constant T source. How does that affect the efficiency?

7.17 A remote location without electricity operates a refrigerator with a bottle of propane feeding a burner to create hot gases. Sketch the setup in terms of cyclic devices and give a relation for the ratio of $\dot{Q}_L$ in the refrigerator to $\dot{Q}_{fuel}$ in the burner in terms of the various reservoir temperatures.

HOMEWORK PROBLEMS

Heat Engines and Refrigerators

7.18 Calculate the thermal efficiency of the steam power plant given in Example 6.9.

7.19 Calculate the coefficient of performance of the R-134a refrigerator given in Example 6.10.

7.20 Calculate the thermal efficiency of the steam power plant cycle described in Problem 6.99.

7.21 Calculate the coefficient of performance of the R-12 heat pump cycle described in Problem 6.106.

7.22 A farmer runs a heat pump with a 2 kW motor. It should keep a chicken hatchery at 30°C, which loses energy at a rate of 10 kW to the colder ambient T_{amb}. What is the minimum coefficient of performance that will be acceptable for the heat pump?

7.23 A power plant generates 150 MW of electrical power. It uses a supply of 1000 MW from a geothermal source and rejects energy to the atmosphere. Find the power to the air and how much air should be flowed to the cooling tower (kg/s) if its temperature cannot be increased more than 10°C.

7.24 A car engine delivers 25 hp to the driveshaft with a thermal efficiency of 30%. The fuel has a heating value of 40 000 kJ/kg. Find the rate of fuel consumption and the combined power rejected through the radiator and exhaust.

7.25 For each of the cases below, determine if the heat engine satisfies the first law (energy equation) and if it violates the second law.

    a. $\dot{Q}_H = 6\,\text{kW}$ $\dot{Q}_L = 4\,\text{kW}$ $\dot{W} = 2\,\text{kW}$

    b. $\dot{Q}_H = 6\,\text{kW}$ $\dot{Q}_L = 0\,\text{kW}$ $\dot{W} = 6\,\text{kW}$

c. $\dot{Q}_H = 6$ kW $\dot{Q}_L = 2$ kW $\dot{W} = 5$ kW
d. $\dot{Q}_H = 6$ kW $\dot{Q}_L = 6$ kW $\dot{W} = 0$ kW

7.26 In a steam power plant 1 MW is added in the boiler, 0.58 MW is taken out in the condenser, and the pump work is 0.02 MW. Find the plant thermal efficiency. If everything could be reversed, find the coefficient of performance as a refrigerator?

7.27 Electric solar cells can produce electricity with 15% efficiency. Compare that to a heat engine driving an electric generator of efficiency 80%. What should the heat engine efficiency be to have the same overall efficiency as the solar cells?

7.28 For each of the cases in Problem 7.25 determine if a heat pump satisfies the first law (energy equation) and if it violates the second law.

7.29 An air conditioner discards 5.1 kW to the ambient surroundings with a power input of 1.5 kW. Find the rate of cooling and the coefficient of performance.

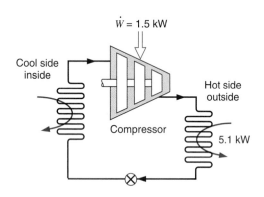

FIGURE P7.29

7.30 Calculate the amount of work input a refrigerator needs to make ice cubes out of a tray of 0.25 kg

liquid water at 10°C. Assume the refrigerator has $\beta = 3.5$ and a motor-compressor of 750 W. How much time does it take if this is the only cooling load?

7.31 A house needs to be heated by a heat pump, with $\beta' = 2.2$, and maintained at 20°C at all times. It is estimated that it loses 0.8 kW per degree the ambient temperature is lower than the 20°C. Assume an outside temperature of -10°C and find the needed power to drive the heat pump.

7.32 Refrigerant R-12 at 95°C with $x = 0.1$ flowing at 2 kg/s is brought to saturated vapor in a constant-pressure heat exchanger. The energy is supplied by a heat pump with a coefficient of performance of $\beta' = 2.5$. Find the required power to drive the heat pump.

Second Law and Processes

7.33 Prove that a cyclic device that violates the Kelvin–Planck statement of the second law also violates the Clausius statement of the second law.

7.34 Discuss the factors that would make the power plant cycle described in Problem 6.99 an irreversible cycle.

7.35 Assume a cyclic machine that exchanges 6 kW with a 250°C reservoir and has
a. $\dot{Q}_L = 0$ kW $\dot{W} = 6$ kW
b. $\dot{Q}_L = 6$ kw $\dot{W} = 0$ kW
and $\dot{Q}_L$ is exchanged with a 30°C ambient surroundings. What can you say about the processes in the two cases a and b if the machine is a heat engine? Repeat the question for the case of a heat pump.

7.36 Discuss the factors that would make the heat pump cycle described in Problem 6.106 an irreversible cycle.

7.37 The water in a shallow pond heats up during the day and cools down during the night. Heat transfer by radiation, conduction, and convection with the ambient surroundings thus cycles the water temperature. Is such a cyclic process reversible or irreversible?

7.38 Consider a heat engine and heat pump connected as shown in Fig. P7.38. Assume $T_{H1} = T_{H2} > T_{amb}$ and determine for each of the three cases if the setup satisfies the first law and/or violates the second law.

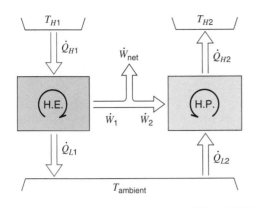

	$\dot{Q}_{H1}$	$\dot{Q}_{L1}$	$\dot{W}_1$	$\dot{Q}_{H2}$	$\dot{Q}_{L2}$	$\dot{W}_2$
a	6	4	2	3	2	1
b	6	4	2	5	4	1
c	3	2	1	4	3	1

FIGURE P7.38

7.39 Consider the four cases of a heat engine in Problem 7.25 and determine if any of those are perpetual-motion machines of the first or second kind.

Carnot Cycles and Absolute Temperature

7.40 Calculate the thermal efficiency of a Carnot-cycle heat engine operating between reservoirs at 300°C and 45°C. Compare the result to that of Problem 7.18.

7.41 At a few places where the air is very cold in the winter, for example, −30°C, it is possible to find a temperature of 13°C down below ground. What efficiency will a heat engine have operating between these two thermal reservoirs?

7.42 Calculate the thermal efficiency of a Carnot-cycle heat pump operating between reservoirs at 0°C and 45°C. Compare the result to that of Problem 7.21.

7.43 Find the power output and the low T heat rejection rate for a Carnot-cycle heat engine that receives 6 kW at 250°C and rejects heat at 30°C as in Problem 7.35.

7.44 A car engine burns 5 kg of fuel (equivalent to adding Q_H) at 1500 K and rejects energy to the radiator and exhaust at an average temperature of 750 K. Assume the fuel has a heating value of 40 000 kJ/kg and find the maximum amount of work the engine can provide.

7.45 Differences in surface water and deep water temperature can be utilized for power generation. It is proposed to construct a cyclic heat engine that will operate near Hawaii, where the ocean temperature is 20°C near the surface and 5°C at some depth. What is the possible thermal efficiency of such a heat engine?

7.46 Find the maximum coefficient of performance for the refrigerator in your kitchen, assuming it runs in a Carnot cycle.

7.47 An air conditioner provides 1 kg/s of air at 15°C cooled from outside atmospheric air at 35°C. Estimate the amount of power needed to operate the air conditioner. Clearly state all assumptions made.

7.48 We propose to heat a house in the winter with a heat pump. The house is to be maintained at 20°C at all times. When the ambient temperature outside drops to −10°C, the rate at which heat is lost from the house is estimated to be 25 kW. What is the minimum electrical power required to drive the heat pump?

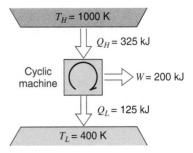

FIGURE P7.48

7.49 A salesperson selling refrigerators and deep freezers will guarantee a minimum coefficient of performance of 4.5 year round. How would the performance of these machines compare? Would it be steady throughout the year?

7.50 A cyclic machine, shown in Fig. P7.50, receives 325 kJ from a 1000 K energy reservoir. It rejects

FIGURE P7.50

125 kJ to a 400 K energy reservoir, and the cycle produces 200 kJ of work as output. Is this cycle reversible, irreversible, or impossible?

7.51 An inventor has developed a refrigeration unit that maintains the cold space at −10°C, while operating in a 25°C room. A coefficient of performance of 8.5 is claimed. How do you evaluate this?

7.52 A household freezer operates in a room at 20°C. Heat must be transferred from the cold space at a rate of 2 kW to maintain its temperature at −30°C. What is the theoretically smallest (power) motor required to operate this freezer?

7.53 In a cryogenic experiment you need to keep a container at −125°C although it gains 100 W due to heat transfer. What is the smallest motor you would need for a heat pump absorbing heat from the container and rejecting heat to the room at 20°C?

7.54 A temperature of about 0.01 K can be achieved by magnetic cooling. In this process a strong magnetic field is imposed on a paramagnetic salt, maintained at 1 K by transfer of energy to liquid helium boiling at low pressure. The salt is then thermally isolated from the helium, the magnetic field is removed, and the salt temperature drops. Assume that 1 mJ is removed at an average temperature of 0.1 K to the helium by a Carnot-cycle heat pump. Find the work input to the heat pump and the coefficient of performance with an ambient at 300 K.

7.55 The lowest temperature that has been achieved is about 1×10^{-6} K. To achieve this an additional stage of cooling is required beyond that described in the previous problem, namely, nuclear cooling. This process is similar to magnetic cooling, but it involves the magnetic moment associated with the nucleus rather than that associated with certain ions in the paramagnetic salt. Suppose that 10 μJ is to be removed from a specimen at an average temperature of 10^{-5} K (10 microjoules is about the potential energy loss of a pin dropping 3 mm). Find the work input to a Carnot heat pump and its coefficient of performance to do this assuming the ambient is at 300 K.

7.56 A certain solar-energy collector produces a maximum temperature of 100°C. The energy is used in a cyclic heat engine that operates in a 10°C environment. What is the maximum thermal efficiency? What is it if the collector is redesigned to focus the incoming light to produce a maximum temperature of 300°C?

7.57 Helium has the lowest normal boiling point of any of the elements at 4.2 K. At this temperature the enthalpy of evaporation is 83.3 kJ/kmol. A Carnot refrigeration cycle is analyzed for the production of 1 kmol of liquid helium at 4.2 K from saturated vapor at the same temperature. What is the work input to the refrigerator and the coefficient of performance for the cycle with an ambient temperature at 300 K?

7.58 Calculate the amount of work input a refrigerator needs to make ice cubes out of a tray of 0.25 kg liquid water at 10°C. Assume the refrigerator works in a Carnot cycle between −8°C and 35°C with a motor-compressor of 750 W. How much time does it take if this is the only cooling load?

7.59 A steel bottle of $V = 0.1$ m³ contains R-134a at 20°C and 200 kPa. It is placed in a deep freezer where it is cooled to −20°C. The deep freezer sits in a room with ambient temperature of 20°C and has an inside temperature of −20°C. Find the amount of energy the freezer must remove from the R-134a and the extra amount of work input to the freezer to do the process.

7.60 Liquid sodium leaves a nuclear reactor at 800°C and is used as the energy source in a steam power plant. The condenser cooling water comes from a cooling tower at 15°C. Determine the maximum thermal efficiency of the power plant. Is it misleading to use the temperatures given to calculate this value?

7.61 A thermal storage device is made with a rock (granite) bed of 2 m³ that is heated to 400 K using solar energy. A heat engine receives a Q_H from the bed and rejects heat to the ambient surroundings at 290 K. The rock bed therefore cools down, and as it reaches 290 K the process stops. Find the energy the rock bed can give out. What is the heat engine efficiency at the beginning of the process, and what is it at the end of the process?

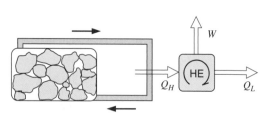

FIGURE P7.61

7.62 A heat engine has a solar collector receiving 0.2 kW/m^2, inside of which a transfer media is heated to 450 K. The collected energy powers a heat engine that rejects heat at 40°C. If the heat engine should deliver 2.5 kW, what is the minimum size (area) solar collector?

7.63 Sixty kilograms per hour of water runs through a heat exchanger, entering as saturated liquid at 200 kPa and leaving as saturated vapor. The heat is supplied by a Carnot heat pump operating from a low-temperature reservoir at 16°C. Find the rate of work into the heat pump.

7.64 A heat pump is driven by the work output of a heat engine as shown in Figure P7.64. If we assume ideal devices, find the ratio of the total power $\dot{Q}_{L1} + \dot{Q}_{H2}$ that heats the house to the power from the hot energy source $\dot{Q}_{H1}$ in terms of the temperatures.

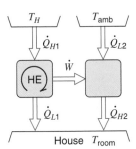

FIGURE P7.64

7.65 It is proposed to build a 1000 MW electric power plant with steam as the working fluid. The condensers are to be cooled with river water (see Fig. P7.65). The maximum steam temperature is 550°C,

and the pressure in the condensers will be 10 kPa. Estimate the temperature rise of the river downstream from the power plant.

7.66 Two different fuels can be used in a heat engine operating between the fuel burning temperature and a low temperature of 350 K. Fuel A burns at 2200 K delivering 30 000 kJ/kg and costs $1.50/kg. Fuel B burns at 1200 K, delivering 40 000 kJ/kg and costs $1.30/kg. Which fuel would you buy and why?

Finite ΔT Heat Transfer

7.67 A refrigerator keeping 5°C inside is located in a 30°C room. It must have a high temperature ΔT above room temperature and a low temperature ΔT below the refrigerated space in the cycle to actually transfer the heat. For a ΔT of 0, 5, and 10°C, respectively, calculate the COP assuming a Carnot cycle.

7.68 A refrigerator uses a power input of 2.5 kW to cool a 5°C space with the high temperature in the cycle as 50°C. The Q_H is pushed to the ambient air at 35°C in a heat exchanger where the transfer coefficient is 50 W/m^2 K. Find the required minimum heat transfer area.

7.69 A house is heated by a heat pump driven by an electric motor using the outside as the low-temperature reservoir. The house loses energy in direct proportion to the temperature difference as $\dot{Q}_{loss} = K(T_H - T_L)$. Determine the minimum electric power required to drive the heat pump as a function of the two temperatures.

FIGURE P7.69

7.70 A farmer runs a heat pump with a motor of 2 kW. It should keep a chicken hatchery at 30°C, which loses energy at a rate of 0.5 kW per degree difference to the colder ambient T_{amb}. The heat pump has a coefficient of performance that is 50% of a Carnot heat pump. What is the minimum ambient temperature for which the heat pump is sufficient?

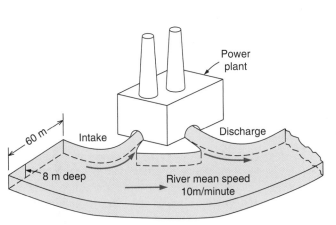

FIGURE P7.65

7.71 Consider a Carnot-cycle heat engine operating in outer space. Heat can be rejected from this engine only by thermal radiation, which is proportional to the radiator area and the fourth power of absolute temperature, $\dot{Q}_{rad} = KAT^4$. Show that for a given engine work output and given T_H, the radiator area will be minimum when the ratio $T_L/T_H = \frac{3}{4}$.

7.72 A house is heated by an electric heat pump using the outside as the low-temperature reservoir. For several different winter outdoor temperatures, estimate the percent savings in electricity if the house is kept at 20°C instead of 24°C. Assume that the house is losing energy to the outside as in Eq. 7.17.

7.73 A house is cooled by an electric heat pump using the outside as the high-temperature reservoir. For several different summer outdoor temperatures, estimate the percent savings in electricity if the house is kept at 25°C instead of 20°C. Assume that the house is gaining energy from the outside in direct proportion to the temperature difference, as in Eq. 7.17.

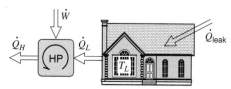

FIGURE P7.73

7.74 A heat pump has a COP of $\beta' = 0.5\,\beta'_{CARNOT}$ and maintains a house at $T_H = 20$°C, while it leaks energy out as $\dot{Q} = 0.6(T_H - T_L)$[kW]. For a maximum of 1.0 kW power input, find the minimum outside temperature, T_L, for which the heat pump is a sufficient heat source.

7.75 An air conditioner cools a house at $T_L = 20$°C with a maximum of 1.2 kW power input. The house gains energy as $\dot{Q} = 0.6(T_H - T_L)$[kW] and the refrigeration COP is $\beta = 0.6\,\beta_{CARNOT}$. Find the maximum outside temperature, T_H, for which the air conditioner unit provides sufficient cooling.

7.76 A Carnot heat engine, shown in Fig. P7.76 receives energy from a reservoir at T_{res} through a heat exchanger where the heat transferred is proportional to the temperature difference as $\dot{Q}_H = K(T_{res} - T_H)$. It rejects heat at a given low temperature T_L. To design the heat engine for maximum work output, show that the high temperature, T_H, in the cycle should be selected as $T_H = (T_L T_{res})^{1/2}$.

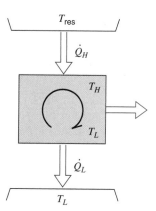

FIGURE P7.76

Ideal-Gas Carnot Cycles

7.77 Hydrogen gas is used in a Carnot cycle having an efficiency of 60% with a low temperature of 300 K. During the heat rejection the pressure changes from 90 kPa to 120 kPa. Find the high- and low-temperature heat transfer and the net cycle work per unit mass of hydrogen.

7.78 An ideal-gas Carnot cycle with air in a piston cylinder has a high temperature of 1200 K and a heat rejection at 400 K. During the heat addition, the volume triples. Find the two specific heat transfers (q) in the cycle and the overall cycle efficiency.

7.79 Air in a piston/cylinder setup goes through a Carnot cycle with the $P-v$ diagram shown in Fig. 7.24. The high and low temperatures are 600 K and 300 K, respectively. The heat added at the high temperature is 250 kJ/kg, and the lowest pressure in the cycle is 75 kPa. Find the specific volume and pressure after heat rejection and the net work per unit mass.

Review Problems

7.80 A car engine operates with a thermal efficiency of 35%. Assume the air conditioner has a coefficient of performance of $\beta = 3$ working as a refrigerator cooling the inside using engine shaft work to drive it. How much extra fuel energy should be spent to remove 1 kJ from the inside?

7.81 An air conditioner with a power input of 1.2 kW is working as a refrigerator ($\beta = 3$) or as a heat pump ($\beta' = 4$). It maintains an office at 20°C year round, which exchanges 0.5 kW per degree temperature difference with the atmosphere. Find the maximum and the minimum outside temperature for which this unit is sufficient.

7.82 A rigid insulated container has two rooms separated by a membrane. Room A contains 1 kg of air at 200°C, and room B has 1.5 kg of air at 20°C; both rooms are at 100 kPa. Consider two different cases:

1. Heat transfer between A and B creates a final uniform T.

2. The membrane breaks, and the air comes to a uniform state.

For both cases find the final temperature. Are the two processes reversible and different? Explain.

7.83 At a certain location, geothermal energy in underground water is available and used as an energy source for a power plant. Consider a supply of saturated liquid water at 150°C. What is the maximum possible thermal efficiency of a cyclic heat engine using this source as energy with the ambient surroundings at 20°C? Would it be better to locate a source of saturated vapor at 150°C than to use the saturated liquid?

7.84 We wish to produce refrigeration at −30°C. A reservoir, shown in Fig. P7.84, is available at 200°C, and the ambient temperature is 30°C. Thus, work can be done by a cyclic heat engine operating between the 200°C reservoir and the ambient surroundings. This work is used to drive the refrigerator. Determine the ratio of the heat transferred from the 200°C reservoir to the heat transferred from the −30°C reservoir, assuming all processes are reversible.

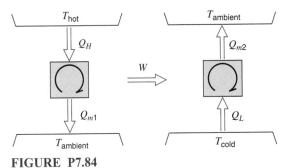

FIGURE P7.84

7.85 A 4 L jug of milk at 25°C is placed in your refrigerator where it is cooled down to 5°C. The high temperature in the Carnot refrigeration cycle is 45°C, and the properties of milk are the same as for liquid water. Find the amount of energy that must be removed from the milk and the additional work needed to drive the refrigerator.

7.86 A combination of a heat engine driving a heat pump (see Fig. P7.86) takes waste energy at 50°C as a source Q_{w1}, to the heat engine rejecting heat at 30°C. The remainder, Q_{w2}, goes into the heat pump that delivers a Q_H at 150°C. If the total waste energy is 5 MW, find the rate of energy delivered at the high temperature.

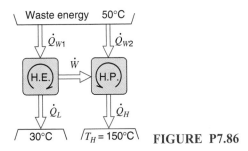

FIGURE P7.86

7.87 Air in a rigid 1 m³ box is at 300 K and 200 kPa. It is heated to 600 K by heat transfer from a reversible heat pump that receives energy from the ambient surroundings at 300 K besides the work input. Use constant specific heat at 300 K. Since the coefficient of performance changes write $\delta Q = m_{air}C_v\,dT$ and find δW. Integrate δW with temperature to find the required heat pump work.

7.88 Consider the rock bed thermal storage in Problem 7.88. Use the specific heat so you can write δQ_H in terms of dT_{rock} and find the expression for δW out of the heat engine. Integrate this expression over temperature and find the total heat engine work output.

7.89 A heat pump heats a house in the winter and then reverses to cool it in the summer. The interior temperature should be 20°C in the winter and 25°C in the summer. Heat transfer through the walls and ceilings is estimated to be 2400 kJ per hour per degree temperature difference between the inside and outside.

a. If the outside winter temperature is 0°C, what is the minimum power required to drive the heat pump?

b. For the same power as in part (a), what is the maximum outside summer temperature for which the house can be maintained at 25°C?

7.90 A furnace, shown in Fig. P7.90, can deliver heat, Q_{H1}, at T_{H1}, and it is proposed to use this to drive a heat engine with a rejection at T_{atm} instead of direct room heating. The heat engine drives a heat pump

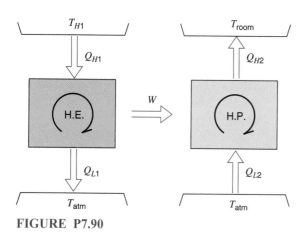

FIGURE P7.90

that delivers Q_{H2} at T_{room} using the atmosphere as the cold reservoir. Find the ratio Q_{H2}/Q_{H1} as a function of the temperatures. Is this a better setup than direct room heating from the furnace?

7.91 A 10 m³ tank of air at 500 kPa and 600 K acts as the high-temperature reservoir for a Carnot heat engine that rejects heat at 300 K. A temperature difference of 25°C between the air tank and the Carnot-cycle high temperature is needed to transfer the heat. The heat engine runs until the air temperature has dropped to 400 K and then stops. Assume constant specific heat capacities for air and determine how much work is given out by the heat engine.

ENGLISH UNIT PROBLEMS

Concept Problems

7.92E A gasoline engine produces 20 hp using 35 Btu/s of heat transfer from burning fuel. What is its thermal efficiency, and how much power is rejected to the ambient?

7.93E A refrigerator removes 1.5 Btu from the cold space using 1 Btu work input. How much energy goes into the kitchen, and what is its coefficient of performance?

7.94E A window air-conditioner unit is placed on a laboratory bench and tested in cooling mode using 0.75 Btu/s of electric power with a COP of 1.75. What is the cooling power capacity, and what is the net effect on the laboratory?

7.95E A car engine takes atmospheric air in at 70 F, no fuel, and exhausts the air at 0 F producing work in the process. What do the first and second laws say about that?

7.96E A large stationary diesel engine produces 20 000 hp with a thermal efficiency of 40%. The exhaust gas, which we assume is air, flows out at 1400 R, and the intake is 520 R. How large a mass flow rate is that if that accounts for half the Q_L? Can the exhaust flow energy be used?

English Unit Problems

7.97E Calculate the thermal efficiency of the steam power plant described in Problem 6.167.

7.98E A farmer runs a heat pump with a 2 kW motor. It should keep a chicken hatchery at 90 F which

loses energy at a rate of 10 Btu/s to the colder T_{amb}. What is the minimum coefficient of performance that will be acceptable for the heat pump?

7.99E Calculate the amount of work input a refrigerator needs to make ice cubes out of a tray of 0.5 lbm liquid water at 50 F. Assume the refrigerator has $\beta = 3.5$ and a motor-compressor of 750 W. How much time does it take if this is the only cooling load?

7.100E In a steam power plant 1000 Btu/s is added at 1200 F in the boiler, 580 Btu/s is taken out at 100 F in the condenser, and the pump work is 20 Btu/s. Find the plant thermal efficiency. Assuming the same pump work and heat transfer to the boiler as given, how much turbine power could be produced if the plant were running in a Carnot cycle?

7.101E Calculate the thermal efficiency of a Carnot-cycle heat engine operating between reservoirs at 920 F and 110 F. Compare the result with that of Problem 7.97E.

7.102E A car engine burns 10 lbm of fuel (equivalent to addition of Q_H) at 2600 R and rejects energy to the radiator and the exhaust at an average temperature of 1300 R. If the fuel provides 17 200 Btu/lbm, what is the maximum amount of work the engine can provide?

7.103E An air-conditioner provides 1 lbm/s of air at 60 F cooled from outside atmospheric air at 95 F. Estimate the amount of power needed to operate the air-conditioner. Clearly state all assumptions made.

7.104E We propose to heat a house in the winter with a heat pump. The house is to be maintained at 68 F at all times. When the ambient temperature outside drops to 15 F, the rate at which heat is lost from the house is estimated to be 80 000 Btu/h. What is the minimum electrical power required to drive the heat pump?

7.105E An inventor has developed a refrigeration unit that maintains the cold space at 14 F, while operating in a 77 F room. A coefficient of performance of 8.5 is claimed. How do you evaluate this?

7.106E Liquid sodium leaves a nuclear reactor at 1500 F and is used as the energy source in a steam power plant. The condenser cooling water comes from a cooling tower at 60 F. Determine the maximum thermal efficiency of the power plant. Is it misleading to use the temperatures given to calculate this value?

7.107E A house is heated by an electric heat pump using the outside as the low-temperature reservoir. For several different winter outdoor temperatures, estimate the percent savings in electricity if the house is kept at 68 F instead of 75 F. Assume that the house is losing energy to the outside directly proportional to the temperature difference as $Q_{loss} = K(T_H - T_L)$.

7.108E Refrigerant-22 at 180 F, $x = 0.1$ flowing at 4 lbm/s is brought to saturated vapor in a constant-pressure heat exchanger. The energy is supplied by a heat pump with a low temperature of 50 F. Find the required power input to the heat pump.

7.109E A heat engine has a solar collector receiving 600 Btu/h per square foot inside which a transfer media is heated to 800 R. The collected energy powers a heat engine that rejects heat at 100 F. If the heat engine should deliver 8500 Btu/h, what is the minimum size (area) solar collector?

7.110E Six-hundred pound-mass per hour of water runs through a heat exchanger, entering as saturated liquid at 30 lbf/in.² and leaving as saturated vapor. The heat is supplied by a Carnot heat pump operating from a low-temperature reser-

voir at 60 F. Find the rate of work into the heat pump.

7.111E A car engine operates with a thermal efficiency of 35%. Assume the air-conditioner has a coefficient of performance that is one-third the theoretical maximum and it is mechanically pulled by the engine. How much fuel energy should you spend extra to remove 1 Btu at 60 F when the ambient is at 95 F?

7.112E A heat pump cools a house at 70 F with a maximum of 4000 Btu/h power input. The house gains 2000 Btu/h per degree temperature difference to the ambient, and the refrigerator coefficient of performance is 60% of the theoretical maximum. Find the maximum outside temperature for which the heat pump provides sufficient cooling.

7.113E A house is cooled by an electric heat pump using the outside as the high-temperature reservoir. For several different summer outdoor temperatures estimate the percent savings in electricity if the house is kept at 77 F instead of 68 F. Assume that the house is gaining energy from the outside directly proportional to the temperature difference.

7.114E A thermal storage is made with a rock (granite) bed of 70 ft³, which is heated to 720 R using solar energy. A heat engine receives a Q_H from the bed and rejects heat to the ambient at 520 R. The rock bed therefore cools down, and as it reaches 520 R the process stops. Find the energy the rock bed can give out. What is the heat engine efficiency at the beginning of the process, and what is it at the end of the process?

7.115E We wish to produce refrigeration at −20 F. A reservoir is available at 400 F and the ambient temperature is 80 F, as shown in Fig. P7.84. Thus, work can be done by a cyclic heat engine operating between the 400 F reservoir and the ambient. This work is used to drive the refrigerator. Determine the ratio of the heat transferred from the 400 F reservoir to the heat transferred from the −20 F reservoir, assuming all processes are reversible.

7.116E Air in a rigid 40 ft³ box is at 540 R, 30 lbf/in.². It is heated to 1100 R by heat transfer from a reversible heat pump that receives energy from the ambient at 540 R besides the work input. Use constant specific heat at 540 R.

Since the coefficient of performance changes, write $\delta Q = m_{air} C_v \, dT$ and find δW. Integrate δW with temperature to find the required heat pump work.

7.117E A 350 ft^3 tank of air at 80 lbf/in.^{2}, 1080 R acts as the high-temperature reservoir for a Carnot heat engine that rejects heat at 540 R. A temperature difference of 45 F between the air tank and the Carnot cycle high temperature is needed to transfer the heat. The heat engine runs until the air temperature has dropped to 700 R and then stops. Assume constant specific heat capacities for air, and find how much work is given out by the heat engine.

7.118E Air in a piston/cylinder goes through a Carnot cycle with the P–v diagram shown in Fig. 7.24. The high and low temperatures are 1200 R and 600 R, respectively. The heat added at the high temperature is 100 Btu/lbm, and the lowest pressure in the cycle is 10 lbf/in.^{2}. Find the specific volume and pressure at all four states in the cycle assuming constant specific heats at 80 F.

ENTROPY 8

Up to this point in our consideration of the second law of thermodynamics, we have dealt only with thermodynamic cycles. Although this is a very important and useful approach, we are often concerned with processes rather than cycles. Thus, we might be interested in the second-law analysis of processes we encounter daily, such as the combustion process in an automobile engine, the cooling of a cup of coffee, or the chemical processes that take place in our bodies. It would also be beneficial to be able to deal with the second law quantitatively as well as qualitatively.

In our consideration of the first law, we initially stated the law in terms of a cycle, but we then defined a property, the internal energy, that enabled us to use the first law quantitatively for processes. Similarly, we have stated the second law for a cycle, and we now find that the second law leads to a property, entropy, that enables us to treat the second law quantitatively for processes. Energy and entropy are both abstract concepts that help to describe certain observations. As we noted in Chapter 2, thermodynamics can be described as the science of energy and entropy. The significance of this statement will become increasingly evident.

8.1 THE INEQUALITY OF CLAUSIUS

The fist step in our consideration of the property we call entropy is to establish the inequality of Clausius, which is

$$\oint \frac{\delta Q}{T} \leq 0$$

The inequality of Clausius is a corollary or a consequence of the second law of thermodynamics. It will be demonstrated to be valid for all possible cycles, including both reversible and irreversible heat engines and refrigerators. Since any reversible cycle can be represented by a series of Carnot cycles, in this analysis we need consider only a Carnot cycle that leads to the inequality of Clausius.

Consider first a reversible (Carnot) heat engine cycle operating between reservoirs at temperatures T_H and T_L, as shown in Fig. 8.1. For this cycle, the cyclic integral of the heat transfer, $\oint \delta Q$, is greater than zero.

$$\oint \delta Q = Q_H - Q_L > 0$$

Since T_H and T_L are constant, from the definition of the absolute temperature scale and from the fact this is a reversible cycle, it follows that

$$\oint \frac{\delta Q}{T} = \frac{Q_H}{T_H} - \frac{Q_L}{T_L} = 0$$

251

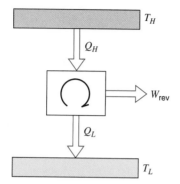

FIGURE 8.1
Reversible heat engine cycle for demonstration of the inequality of Clausius.

If $\oint \delta Q$, the cyclic integral of δQ, approaches zero (by making T_H approach T_L) and the cycle remains reversible, the cyclic integral of $\delta Q/T$ remains zero. Thus, we conclude that for all reversible heat engine cycles

$$\oint \delta Q \geq 0$$

and

$$\oint \frac{\delta Q}{T} = 0$$

Now consider an irreversible cyclic heat engine operating between the same T_H and T_L as the reversible engine of Fig. 8.1 and receiving the same quantity of heat Q_H. Comparing the irreversible cycle with the reversible one, we conclude from the second law that

$$W_{irr} < W_{rev}$$

Since $Q_H - Q_L = W$ for both the reversible and irreversible cycles, we conclude that

$$Q_H - Q_{L\ irr} < Q_H - Q_{L\ rev}$$

and therefore

$$Q_{L\ irr} > Q_{L\ rev}$$

Consequently, for the irreversible cyclic engine,

$$\oint \delta Q = Q_H - Q_{L\ irr} > 0$$
$$\oint \frac{\delta Q}{T} = \frac{Q_H}{T_H} - \frac{Q_{L\ irr}}{T_L} < 0$$

Suppose that we cause the engine to become more and more irreversible, but keep Q_H, T_H, and T_L fixed. The cyclic integral of δQ then approaches zero, and that for $\delta Q/T$ becomes a progressively larger negative value. In the limit, as the work output goes to zero,

$$\oint \delta Q = 0$$
$$\oint \frac{\delta Q}{T} < 0$$

Thus, we conclude that for all irreversible heat engine cycles

$$\oint \delta Q \geq 0$$

$$\oint \frac{\delta Q}{T} < 0$$

To complete the demonstration of the inequality of Clausius, we must perform similar analyses for both reversible and irreversible refrigeration cycles. For the reversible refrigeration cycle shown in Fig. 8.2,

$$\oint \delta Q = -Q_H + Q_L < 0$$

and

$$\oint \frac{\delta Q}{T} = -\frac{Q_H}{T_H} + \frac{Q_L}{T_L} = 0$$

As the cyclic integral of δQ approaches zero reversibly (T_H approaches T_L), the cyclic integral of $\delta Q/T$ remains at zero. In the limit,

$$\oint \delta Q = 0$$

$$\oint \frac{\delta Q}{T} = 0$$

Thus, for all reversible refrigeration cycles,

$$\oint \delta Q \leq 0$$

$$\oint \frac{\delta Q}{T} = 0$$

Finally, let an irreversible cyclic refrigerator operate between temperatures T_H and T_L and receive the same amount of heat Q_L as the reversible refrigerator of Fig. 8.2. From the second law, we conclude that the work input required will be greater for the irreversible refrigerator, or

$$W_{\text{irr}} > W_{\text{rev}}$$

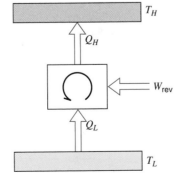

FIGURE 8.2
Reversible refrigeration
cycle for demonstration of
the inequality of Clausius.

Since $Q_H - Q_L = W$ for each cycle, it follows that

$$Q_{H\,irr} - Q_L > Q_{H\,rev} - Q_L$$

and therefore,

$$Q_{H\,irr} > Q_{H\,rev}$$

That is, the heat rejected by the irreversible refrigerator to the high-temperature reservoir is greater than the heat rejected by the reversible refrigerator. Therefore, for the irreversible refrigerator,

$$\oint \delta Q = -Q_{H\,irr} + Q_L < 0$$

$$\oint \frac{\delta Q}{T} = -\frac{Q_{H\,irr}}{T_H} + \frac{Q_L}{T_L} < 0$$

As we make this machine progressively more irreversible, but keep Q_L, T_H, and T_L constant, the cyclic integrals of δQ and $\delta Q/T$ both become larger in the negative direction. Consequently, a limiting case as the cyclic integral of δQ approaches zero does not exist for the irreversible refrigerator.

Thus, for all irreversible refrigeration cycles,

$$\oint \delta Q < 0$$

$$\oint \frac{\delta Q}{T} < 0$$

Summarizing, we note that, in regard to the sign of $\oint \delta Q$, we have considered all possible reversible cycles (i.e., $\oint \delta Q \gtreqless 0$), and for each of these reversible cycles

$$\oint \frac{\delta Q}{T} = 0$$

We have also considered all possible irreversible cycles for the sign of $\oint \delta Q$ (that is, $\oint \delta Q \gtreqless 0$), and for all these irreversible cycles

$$\oint \frac{\delta Q}{T} < 0$$

Thus, for all cycles we can write

$$\oint \frac{\delta Q}{T} \leq 0 \qquad (8.1)$$

where the equality holds for reversible cycles and the inequality for irreversible cycles. This relation, Eq. 8.1, is known as the inequality of Clausius.

The significance of the inequality of Clausius may be illustrated by considering the simple steam power plant cycle shown in Fig. 8.3. This cycle is slightly different from the usual cycle for steam power plants in that the pump handles a mixture of liquid and vapor in such proportions that saturated liquid leaves the pump and enters the boiler. Suppose that someone reports that the pressure and quality at various points in the cycle are as given in Fig. 8.3. Does this cycle satisfy the inequality of Clausius?

Heat is transferred in two places, the boiler and the condenser. Therefore,

$$\oint \frac{\delta Q}{T} = \int \left(\frac{\delta Q}{T}\right)_{boiler} + \int \left(\frac{\delta Q}{T}\right)_{condenser}$$

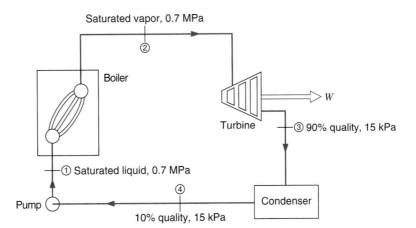

FIGURE 8.3 A simple steam power plant that demonstrates the inequality of Clausius.

Since the temperature remains constant in both the boiler and condenser, this may be integrated as follows:

$$\oint \frac{\delta Q}{T} = \frac{1}{T_1} \int_1^2 \delta Q + \frac{1}{T_3} \int_3^4 \delta Q = \frac{{}_1 Q_2}{T_1} + \frac{{}_3 Q_4}{T_3}$$

Let us consider a 1 kg mass as the working fluid. We have then

$${}_1 q_2 = h_2 - h_1 = 2066.3 \text{ kJ/kg}, \qquad T_1 = 164.97°\text{C}$$

$${}_3 q_4 = h_4 - h_3 = 463.4 - 2361.8 = -1898.4 \text{ kJ/kg}, \qquad T_3 = 53.97°\text{C}$$

Therefore,

$$\oint \frac{\delta Q}{T} = \frac{2066.3}{164.97 + 273.15} - \frac{1898.4}{53.97 + 273.15} = -1.087 \text{ kJ/kg-K}$$

Thus, this cycle satisfies the inequality of Clausius, which is equivalent to saying that it does not violate the second law of thermodynamics.

8.2 ENTROPY—A PROPERTY OF A SYSTEM

By applying Eq. 8.1 and Fig. 8.4, we can demonstrate that the second law of thermodynamics leads to a property of a system that we call entropy. Let a system (control mass) undergo a reversible process from state 1 to state 2 along a path A, and let the cycle be completed along path B, which is also reversible.

Because this is a reversible cycle, we can write

$$\oint \frac{\delta Q}{T} = 0 = \int_1^2 \left(\frac{\delta Q}{T} \right)_A + \int_2^1 \left(\frac{\delta Q}{T} \right)_B$$

Now consider another reversible cycle, which proceeds first along path C and is then completed along path B. For this cycle we can write

$$\oint \frac{\delta Q}{T} = 0 = \int_1^2 \left(\frac{\delta Q}{T} \right)_C + \int_2^1 \left(\frac{\delta Q}{T} \right)_B$$

Subtracting the second equation from the first, we have

$$\int_1^2 \left(\frac{\delta Q}{T} \right)_A = \int_1^2 \left(\frac{\delta Q}{T} \right)_C$$

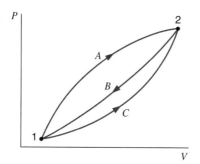

Since the $\oint \delta Q/T$ is the same for all reversible paths between states 1 and 2, we conclude that this quantity is independent of the path and it is a function of the end states only; it is therefore a property. This property is called entropy and is designated S. It follows that entropy may be defined as a property of a substance in accordance with the relation

$$dS \equiv \left(\frac{\delta Q}{T}\right)_{\text{rev}} \tag{8.2}$$

Entropy is an extensive property, and the entropy per unit mass is designated s. It is important to note that entropy is defined here in terms of a reversible process.

The change in the entropy of a system as it undergoes a change of state may be found by integrating Eq. 8.2. Thus,

$$S_2 - S_1 = \int_1^2 \left(\frac{\delta Q}{T}\right)_{\text{rev}} \tag{8.3}$$

To perform this integration, we must know the relation between T and Q, and illustrations will be given subsequently. The important point is that since entropy is a property, the change in the entropy of a substance in going from one state to another is the same for all processes, both reversible and irreversible, between these two states. Equation 8.3 enables us to find the change in entropy only along a reversible path. However, once the change has been evaluated, this value is the magnitude of the entropy change for all processes between these two states.

Equation 8.3 enables us to calculate changes of entropy, but it tells us nothing about absolute values of entropy. From the third law of thermodynamics, which is based on observations of low-temperature chemical reactions, it is concluded that the entropy of all pure substances (in the appropriate structural form) can be assigned the absolute value of zero at the absolute zero of temperature. It also follows from the subject of statistical thermodynamics that all pure substances in the (hypothetical) ideal-gas state at absolute zero temperature have zero entropy.

However, when there is no change of composition, as would occur in a chemical reaction, for example, it is quite adequate to give values of entropy relative to some arbitrarily selected reference state, such as was done earlier when tabulating values of internal energy and enthalpy. In each case, whatever reference value is chosen, it will cancel out when the change of property is calculated between any two states. This is the procedure followed with the thermodynamic tables to be discussed in the following section.

A word should be added here regarding the role of T as an integrating factor. We noted in Chapter 4 that Q is a path function, and therefore δQ in an inexact differential. However, since $(\delta Q/T)_{\text{rev}}$ is a thermodynamic property, it is an exact differential. From a mathematical perspective, we note that an inexact differential may be converted to an exact differential by the introduction of an integrating factor. Therefore, $1/T$ serves as the integrating factor in converting the inexact differential δQ to the exact differential $\delta Q/T$ for a reversible process.

8.3 THE ENTROPY OF A PURE SUBSTANCE

Entropy is an extensive property of a system. Values of specific entropy (entropy per unit mass) are given in tables of thermodynamic properties in the same manner as specific volume and specific enthalpy. The units of specific entropy in the steam tables, refrigerant tables, and ammonia tables are kJ/kg K, and the values are given relative to an arbitrary reference state. In the steam tables the entropy of saturated liquid at 0.01°C is given the value of zero. For many refrigerants, the entropy of saturated liquid at −40°C is assigned the value of zero.

In general, we use the term entropy to refer to both total entropy and entropy per unit mass, since the context or appropriate symbol will clearly indicate the precise meaning of the term.

In the saturation region the entropy may be calculated using the quality. The relations are similar to those for specific volume, internal energy and enthalpy.

$$s = (1 - x)s_f + xs_g$$

$$s = s_f + xs_{fg} \tag{8.4}$$

The entropy of a compressed liquid is tabulated in the same manner as the other properties. These properties are primarily a function of the temperature and are not greatly different from those for saturated liquid at the same temperature. Table 4 of the steam tables, which is summarized in Table B.1.4, give the entropy of compressed liquid water in the same manner as for other properties.

The thermodynamic properties of a substance are often shown on a temperature–entropy diagram and on an enthalpy–entropy diagram, which is also called a Mollier diagram, after Richard Mollier (1863–1935) of Germany. Figures 8.5 and 8.6 show the

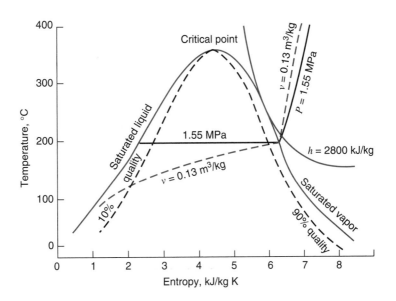

FIGURE 8.5
Temperature–entropy diagram for steam.

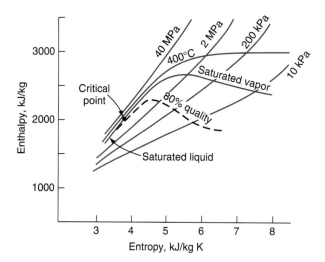

FIGURE 8.6
Enthalpy–entropy
diagram for steam.

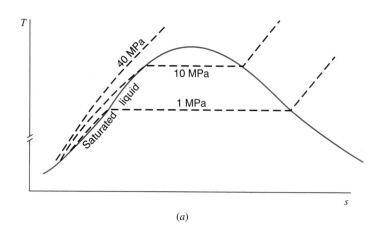

(a)

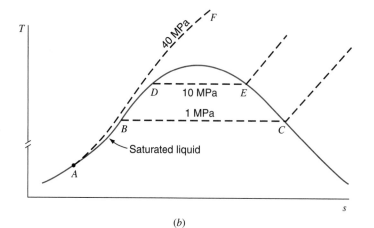

FIGURE 8.7
Temperature–entropy
diagram to show
properties of a
compressed liquid, water.

(b)

essential elements of temperature–entropy and enthalpy–entropy diagrams for steam. The general features of such diagrams are the same for all pure substances. A more complete temperature–entropy diagram for steam is shown in Fig. E.1 in Appendix E.

These diagrams are valuable both because they present thermodynamic data and because they enable us to visualize the changes of state that occur in various processes. As our study progresses, the student should acquire facility in visualizing thermodynamic processes on these diagrams. The temperature–entropy diagram is particularly useful for this purpose.

For most substances the difference in the entropy of a compressed liquid and a saturated liquid at the same temperature is so small that a process in which liquid is heated at constant pressure nearly coincides with the saturated-liquid line until the saturation temperature is reached (Fig. 8.7). Thus, if water at 10 MPa is heated from 0°C to the saturation temperature, it would be shown by line *ABD*, which coincides with the saturated-liquid line.

8.4 ENTROPY CHANGE IN REVERSIBLE PROCESSES

Having established that entropy is a thermodynamic property of a system, we now consider its significance in various processes. In this section we will limit ourselves to systems that undergo reversible processes and consider the Carnot cycle, reversible heat-transfer processes, and reversible adiabatic processes.

Let the working fluid of a heat engine operating on the Carnot cycle make up the system. The first process is the isothermal transfer of heat to the working fluid from the high-temperature reservoir. For this process we can write

$$S_2 - S_1 = \int_1^2 \left(\frac{\delta Q}{T}\right)_{rev}$$

Since this is a reversible process in which the temperature of the working fluid remains constant, the equation can be integrated to give

$$S_2 - S_1 = \frac{1}{T_H}\int_1^2 \delta Q = \frac{_1Q_2}{T_H}$$

This process is shown in Fig. 8.8*a*, and the area under line 1–2, area 1–2–*b*–*a*–1, represents the heat transferred to the working fluid during the process.

FIGURE 8.8 The Carnot cycle on the temperature–entropy diagram.

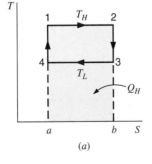

(a)

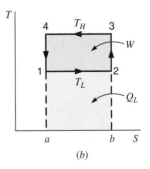

(b)

The second process of a Carnot cycle is a reversible adiabatic one. From the definition of entropy,

$$dS = \left(\frac{\delta Q}{T}\right)_{rev}$$

it is evident that the entropy remains constant in a reversible adiabatic process. A constant-entropy process is called an isentropic process. Line 2–3 represents this process, and this process is concluded at state 3 when the temperature of the working fluid reaches T_L.

The third process is the reversible isothermal process in which heat is transferred from the working fluid to the low-temperature reservoir. For this process we can write

$$S_4 - S_3 = \int_3^4 \left(\frac{\delta Q}{T}\right)_{rev} = \frac{{}_3Q_4}{T_L}$$

Because during this process the heat transfer is negative (in regard to the working fluid), the entropy of the working fluid decreases. Moreover, because the final process 4–1, which completes the cycle, is a reversible adiabatic process (and therefore isentropic), it is evident that the entropy decrease in process 3–4 must exactly equal the entropy increase in process 1–2. The area under line 3–4, area 3–4–a–b–3, represents the heat transferred from the working fluid to the low-temperature reservoir.

Since the net work of the cycle is equal to the net heat transfer, then area 1–2–3–4–1 must represent the net work of the cycle. The efficiency of the cycle may also be expressed in terms of areas:

$$\eta_{th} = \frac{W_{net}}{Q_H} = \frac{\text{area } 1–2–3–4–1}{\text{area } 1–2–b–a–1}$$

Some statements made earlier about efficiencies may now be understood graphically. For example, increasing T_H while T_L remains constant increases the efficiency. Decreasing T_L as T_H remains constant increases the efficiency. It is also evident that the efficiency approaches 100% as the absolute temperature at which heat is rejected approaches zero.

If the cycle is reversed, we have a refrigerator or heat pump. The Carnot cycle for a refrigerator is shown in Fig. 8.8b. Notice that the entropy of the working fluid increases at T_L, since heat is transferred to the working fluid at T_L. The entropy decreases at T_H because of heat transfer from the working fluid.

Let us next consider reversible heat-transfer processes. Actually, we are concerned here with processes that are internally reversible, that is, processes that have no irreversibilities within the boundary of the system. For such processes the heat transfer to or from a system can be shown as an area on a temperature–entropy diagram. For example, consider the change of state from saturated liquid to saturated vapor at constant pressure. This process would correspond to the process 1–2 on the T–s diagram of Fig. 8.9 (note that absolute temperature is required here), and the area 1–2–b–a–1 represents the heat transfer. Since this is a constant-pressure process, the heat transfer per unit mass is equal to h_{fg}. Thus,

$$s_2 - s_1 = s_{fg} = \frac{1}{m} \int_1^2 \left(\frac{\delta Q}{T}\right)_{rev} = \frac{1}{mT} \int_1^2 \delta Q = \frac{{}_1q_2}{T} = \frac{h_{fg}}{T}$$

This relation gives a clue about how s_{fg} is calculated for tabulation in tables of thermodynamic properties. For example, consider steam at 10 MPa. From the steam tables we have

$$h_{fg} = 1317.1 \text{ kJ/kg}$$

$$T = 311.06 + 273.15 = 584.21 \text{ K}$$

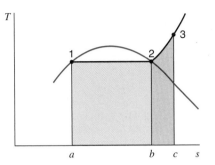

FIGURE 8.9 A temperature–entropy diagram to show areas that represent heat transfer for an internally reversible process.

Therefore,

$$s_{fg} = \frac{h_{fg}}{T} = \frac{1317.1}{584.21} = 2.2544 \text{ kJ/kg K}$$

This is the value listed for s_{fg} in the steam tables.

If heat is transferred to the saturated vapor at constant pressure, the steam is super-heated along line 2–3. For this process we can write

$$_2q_3 = \frac{1}{m} \int_2^3 \delta Q = \int_2^3 T \, ds$$

Since T is not constant, this equation cannot be integrated unless we know a relation between temperature and entropy. However, we do realize that the area under line 2–3, area 2–3–c–b–2, represents $\int_2^3 T \, ds$ and therefore represents the heat transferred during this reversible process.

The important conclusion to draw here is that for processes that are internally reversible, the area underneath the process line on a temperature–entropy diagram represents the quantity of heat transferred. This is not true for irreversible processes, as will be demonstrated later.

EXAMPLE 8.1 Consider a Carnot-cycle heat pump with R-134a as the working fluid. Heat is absorbed into the R-134a at 0°C, during which process it changes from a two-phase state to saturated vapor. The heat is rejected from the R-134a at 60°C so that it ends up as saturated liquid. Find the pressure after compression, before the heat rejection process, and determine the coefficient of performance for the cycle.

Solution

From the definition of the Carnot cycle we have two constant-temperature (isothermal) processes that involve heat transfer and two adiabatic processes in which the temperature changes. The variation in s follows from Eq. 8.2

$$ds = \delta q / T$$

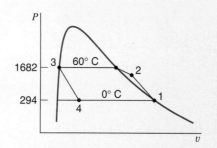
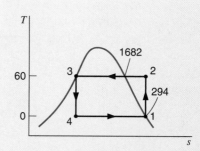

FIGURE 8.10
Diagram for Example 8.1.

and the Carnot cycle is shown in Fig. 8.8 and for this case in Fig. 8.10. We therefore have

State 4 Table B.5.1: $s_4 = s_3 = s_{f@60deg} = 1.2857$ kJ/kg K

State 1 Table B.5.1: $s_1 = s_2 = s_{g@0deg} = 1.7262$ kJ/kg K

State 2 Table B.5.2: $60°C, s_2 = s_1 = s_{g@0deg}$

Interpolate between 1400 kPa and 1600 kPa in Table B.5.2:

$$P_2 = 1400 + (1600 - 1400)\frac{1.7262 - 1.736}{1.7135 - 1.736} = 1487.1 \text{ kPa}$$

From the fact that it is a Carnot cycle the COP becomes, from Eq. 7.13,

$$\beta' = \frac{q_H}{w_{IN}} = \frac{T_H}{T_H - T_L} = \frac{333.15}{60} = 5.55$$

Remark: Notice how much the pressure varies during the heat rejection process. Because this process is very difficult to accomplish in a real device, no heat pump or refrigerator is designed to attempt to approach a Carnot cycle.

EXAMPLE 8.2   A cylinder/piston setup contains 1 L of saturate liquid refrigerant R-12 at 20°C. The piston now slowly expands, maintaining constant temperature to a final pressure of 400 kPa in a reversible process. Calculate the required work and heat transfer to accomplish this process.

Solution

C.V. The refrigerant R-12, which is a control mass.

Continuity Eq.: $m_2 = m_1 = m$;

Energy Eq. 5.11: $m(u_2 - u_1) = {}_1Q_2 - {}_1W_2$

Entropy Eq. 8.3: $m(s_2 - s_1) \geq \int \delta Q/T$

Process: $T =$ constant, reversible so equal sign applies in entropy equation.

State 1 (T, P) Table B.3.1: $u_1 = 54.45$ kJ/kg, $s_1 = 0.2078$ kJ/kg K

$$m = V/v_1 = 0.001/0.000752 = 1.33 \text{ kg}$$

State 2 (T, P) Table B.3.2: $u_2 = 180.57$ kJ/kg, $s_2 = 0.7204$ kJ/kg K

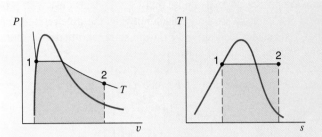

FIGURE 8.11
Diagram for Example 8.2.

As T is constant we have $\int \delta Q/T = {}_1Q_2/T$, so from the entropy equation

$${}_1Q_2 = mT(s_2 - s_1) = 1.33 \times 293.15 \times (0.7204 - 0.2078) = 200 \text{ kJ}$$

The work is then, from the energy equation,

$${}_1W_2 = m(u_1 - u_2) + {}_1Q_2 = 1.33 \times (54.45 - 180.57) + 200 = 32.3 \text{ kJ}$$

Note from Fig. 8.11 that it would be difficult to calculate the work as the area in the P–v diagram due to the shape of the process curve. The heat transfer is the area in the T–s diagram.

8.5 THE THERMODYNAMIC PROPERTY RELATION

At this point we derive two important thermodynamic relations for a simple compressible substance. These relations are

$$T \, dS = dU + P \, dV$$

$$T \, dS = dH - V \, dP$$

The first of these relations can be derived by considering a simple compressible substance in the absence of motion or gravitational effects. The first law for a change of state under these conditions can be written

$$\delta Q = dU + \delta W$$

The equations we are deriving here deal first with the changes of state in which the state of the substance can be identified at all times. Thus, we must consider a quasi-equilibrium process or, to use the term introduced in the last chapter, a reversible process. For a reversible process of a simple compressible substance, we can write

$$\delta Q = T \, dS \qquad \text{and} \qquad \delta W = P \, dV$$

Substituting these relations into the first-law equation, we have

$$T \, dS = dU + P \, dV \qquad (8.5)$$

which is the first equation we set out to derive. Note that this equation was derived by assuming a reversible process. This equation can therefore be integrated for any reversible process, for during such a process the state of the substance can be identified at any point during the process. We also note that Eq. 8.5 deals only with properties. Suppose we have an irreversible process taking place between the given initial and final states. The properties of a substance depend only on the state, and therefore the changes in the properties during a given change of state are the same for an irreversible process as for a reversible process. Therefore, Eq. 8.5 is often applied to an irreversible process between two given states, but the integration of Eq. 8.5 is performed along a reversible path between the same two states.

Since enthalpy is defined as

$$H = U + PV$$

it follows that

$$dH = dU + P\,dV + V\,dP$$

Substituting this relation into Eq. 8.5, we have

$$T\,dS = dH - V\,dP \tag{8.6}$$

which is the second relation that we set out to derive. These two expressions, Eqs. 8.5 and 8.6, are two forms of the thermodynamic property relation and are frequently called Gibbs equations.

These equations can also be written for a unit mass,

$$T\,ds = du + P\,dv \tag{8.7}$$

$$T\,ds = dh - v\,dP \tag{8.8}$$

The Gibbs equations will be used extensively in certain subsequent sections of this book.

If we consider substances of fixed composition other than a simple compressible substance, we can write "$T\,dS$" equations other than those just given for a simple compressible substance. In Eq. 4.15 we noted that for a reversible process we can write the following expression for work:

$$\delta W = P\,dV - \mathscr{T}\,dL - \mathscr{S}\,dA - \mathscr{E}\,dZ + \cdots$$

It follows that a more general expression for the thermodynamic property relation would be

$$T\,dS = dU + P\,dV - \mathscr{T}dL - \mathscr{S}\,dA - \mathscr{E}\,dZ + \cdots \tag{8.9}$$

8.6 ENTROPY CHANGE OF A CONTROL MASS DURING AN IRREVERSIBLE PROCESS

Consider a control mass that undergoes the cycles shown in Fig. 8.12. The cycle made up of the reversible processes A and B is a reversible cycle. Therefore, we can write

$$\oint \frac{\delta Q}{T} = \int_1^2 \left(\frac{\delta Q}{T}\right)_A + \int_2^1 \left(\frac{\delta Q}{T}\right)_B = 0$$

The cycle made of the irreversible process C and the reversible process B is an irreversible cycle. Therefore, for this cycle the inequality of Clausius may be applied, giving the result

$$\oint \frac{\delta Q}{T} = \int_1^2 \left(\frac{\delta Q}{T}\right)_C + \int_2^1 \left(\frac{\delta Q}{T}\right)_B < 0$$

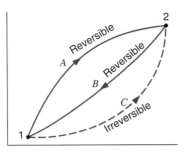

FIGURE 8.12
Entropy change of a control mass during an irreversible process.

Subtracting the second equation from the first and rearranging, we have

$$\int_1^2 \left(\frac{\delta Q}{T}\right)_A > \int_1^2 \left(\frac{\delta Q}{T}\right)_C$$

Since path A is reversible, and since entropy is a property,

$$\int_1^2 \left(\frac{\delta Q}{T}\right)_A = \int_1^2 dS_A = \int_1^2 dS_C$$

Therefore,

$$\int_1^2 dS_C > \int_1^2 \left(\frac{\delta Q}{T}\right)_C$$

As path C was arbitrary, the general result is

$$dS \geq \frac{\delta Q}{T}$$

$$S_2 - S_1 \geq \int_1^2 \frac{\delta Q}{T} \qquad (8.10)$$

In these equations the equality holds for a reversible process and the inequality for an irreversible process.

This is one of the most important equations of thermodynamics. It is used to develop a number of concepts and definitions. In essence, this equation states the influence of irreversibility on the entropy of a control mass. Thus, if an amount of heat δQ is transferred to a control mass at temperature T in a reversible process, the change of entropy is given by the relation

$$dS = \left(\frac{\delta Q}{T}\right)_{rev}$$

If any irreversible effects occur while the amount of heat δQ is transferred to the control mass at temperature T, however, the change of entropy will be greater than for the reversible process. We would then write

$$dS > \left(\frac{\delta Q}{T}\right)_{irr}$$

Equation 8.10 holds when $\delta Q = 0$, when $\delta Q < 0$, and when $\delta Q > 0$. If δQ is negative, the entropy will tend to decrease as a result of the heat transfer. However, the influence of irreversibilities is still to increase the entropy of the mass, and from the absolute numerical perspective we can still write for δQ:

$$dS \geq \frac{\delta Q}{T}$$

8.7 ENTROPY GENERATION

The conclusion from the previous considerations is that the entropy change for an irreversible process is larger than the change in a reversible process for the same δQ and T. This can be written out in a common form as an equality

$$dS = \frac{\delta Q}{T} + \delta S_{\text{gen}} \tag{8.11}$$

provided the last term is positive,

$$\delta S_{\text{gen}} \geq 0 \tag{8.12}$$

The amount of entropy, δS_{gen}, is the entropy generation in the process due to irreversibilities occurring inside the system, a control mass for now but later extended to the more general control volume. This internal generation can be caused by the processes mentioned in Section 7.4, such as friction, unrestrained expansions, and the internal transfer of energy (redistribution) over a finite temperature difference. In addition to this internal entropy generation, external irreversibilities are possible by heat transfer over finite temperature differences as the δQ is transferred from a reservoir or by the mechanical transfer of work.

Equation 8.12 is then valid with the equal sign for a reversible process and the greater than sign for an irreversible process. Since the entropy generation is always positive and the smallest in a reversible process, namely zero, we may deduce some limits for the heat transfer and work terms.

Consider a reversible process, for which the entropy generation is zero, and the heat transfer and work terms therefore are

$$\delta Q = T \, dS \qquad \text{and} \qquad \delta W = P \, dV$$

For an irreversible process with a nonzero entropy generation, the heat transfer from Eq. 8.11 becomes

$$\delta Q_{\text{irr}} = T \, dS - T \, \delta S_{\text{gen}}$$

and thus is smaller than that for the reversible case for the same change of state, dS. we also note that for the irreversible process, the work is no longer equal to $P \, dV$ but is smaller. Furthermore, since the first law is

$$\delta Q_{\text{irr}} = dU + \delta W_{\text{irr}}$$

and the property relation is valid,

$$T \, dS = dU + P \, dV$$

it is found that

$$\delta W_{\text{irr}} = P \, dV - T \, \delta S_{\text{gen}} \tag{8.13}$$

showing that the work is reduced by an amount proportional to the entropy generation. For this reason the term $T\,\delta S_{\text{gen}}$ is often called "lost work," although it is not a real work or energy quantity lost but rather a lost opportunity to extract work.

Equation 8.11 can be integrated between initial and final states to

$$S_2 - S_1 = \int_1^2 dS = \int_1^2 \frac{\delta Q}{T} + {}_1S_{2\,\text{gen}} \tag{8.14}$$

Thus, we have an expression for the change of entropy for an irreversible process as an equality, whereas in the last section we had an inequality. In the limit of a reversible process, with a zero-entropy generation, the change in S expressed in Eq. 8.14 becomes identical to Eq. 8.10 as the equal sign applies and the work term becomes $\int P\,dV$. Equation 8.14 is now the entropy balance equation for a control mass in the same form as the energy equation in Eq. 5.5, and it could include several subsystems. The equation can also be written in the general form

$$\Delta\ \text{Entropy} = +\ \text{in} - \text{out} + \text{gen}$$

expressing that we can generate but not destroy entropy. This is in contrast to energy which we can neither generate nor destroy.

Some important conclusions can be drawn from Eqs. 8.11 to 8.14. First, there are two ways in which the entropy of a system can be increased—by transferring heat to it and by having an irreversible process. Since the entropy generation cannot be less than zero, there is only one way in which the entropy of a system can be decreased, and that is to transfer heat from the system. These changes are illustrated in a T–s diagram in Fig. 8.13 showing the halfplane into which the state moves due to a heat transfer or an entropy generation.

Second, as we have already noted for an adiabatic process, $\delta Q = 0$, and therefore the increase in entropy is always associated with the irreversibilities.

Finally, the presence of irreversibilities will cause the work to be smaller than the reversible work. This means less work out in an expansion process and more work into the control mass ($\delta W < 0$) in a compression process.

One other point concerning the representation of irreversible processes on P–V and T–S diagrams should be made. The work for an irreversible process is not equal to $\int P\,dV$, and the heat transfer is not equal to $\int T\,dS$. Therefore, the area underneath the path does not represent work and heat on the P–V and T–S diagrams, respectively. In fact, in many situations we are not certain of the exact state through which a system passes when it undergoes an irreversible process. For this reason it is advantageous to show irreversible processes as dashed lines and reversible processes as solid lines. Thus, the area underneath the dashed line will never represent work or heat. Figure 8.14a shows an irreversible process, and, because the heat transfer and work for this process are zero, the area underneath the dashed line has no significance. Figure 8.14b shows the reversible process,

FIGURE 8.13 Change of entropy due to heat transfer and entropy generation.

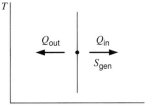

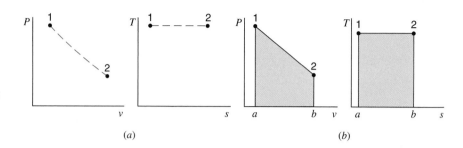

FIGURE 8.14
Reversible and irreversible processes on pressure–volume and temperature–entropy diagrams.

and area 1–2–b–a–1 represents the work on the P–V diagram and the heat transfer on the T–S diagram.

8.8 PRINCIPLE OF THE INCREASE OF ENTROPY

In the previous section, we considered irreversible processes in which the irreversibilities occurred inside the system or control mass. We also found that the entropy change of a control mass could be either positive or negative, since entropy can be increased by internal entropy generation and either increased or decreased by heat transfer, depending on the direction of that transfer. In this section, we examine the effect of heat transfer on the change of state in the surroundings, as well as on the control mass itself.

Consider the process shown in Fig. 8.15 in which a quantity of heat δQ is transferred from the surroundings at temperature T_0 to the control mass at temperature T. Let the work done during this process be δW. For this process we can apply Eq. 8.10 to the control mass and write

$$dS_{\text{c.m.}} \geq \frac{\delta Q}{T}$$

For the surroundings at T_0, δQ is negative, and we assume a reversible heat extraction so

$$dS_{\text{surr}} = \frac{-\delta Q}{T_0}$$

The total net change of entropy is therefore

$$dS_{\text{net}} = dS_{\text{c.m.}} + dS_{\text{surr}} \geq \frac{\delta Q}{T} - \frac{\delta Q}{T_0}$$

$$\geq \delta Q \left(\frac{1}{T} - \frac{1}{T_0} \right) \tag{8.15}$$

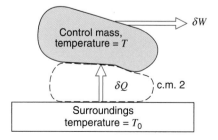

FIGURE 8.15
Entropy change for the control mass plus surroundings.

Since $T_0 > T$, the quantity $[(1/T) - (1/T_0)]$ is positive, and we conclude that

$$dS_{net} = dS_{c.m.} + dS_{surr} \geq 0$$

If $T > T_0$, the heat transfer is from the control mass to the surroundings, and both δQ and the quantity $[(1/T) - (1/T_0)]$ are negative, thus yielding the same result.

It should be noted that the right-hand side of Eq. 8.15 represents an external entropy generation due to heat transfer through a finite temperature difference. To amplify this point, take as a control mass the system that connects the surroundings at T_0 with the previous control mass at T, which typically is the walls. This mass does not experience any change of state, yet heat transfer causes fluxes of entropy to flow in and out in an irreversible process. For this mass, Eq. 8.11 gives

$$dS_{c.m.2} = 0 = \frac{\delta Q}{T_0} - \frac{\delta Q}{T} + \delta S_{gen,2}$$

and we see that the difference in the two $\delta Q/T$ terms (fluxes of S) is the entropy generated in this control mass.

$$\delta S_{gen,2} = \delta Q \left(\frac{1}{T} - \frac{1}{T_0} \right)$$

This is also precisely the location in space where the heat transfer takes place over the finite temperature difference $T_0 - T$. This term is always positive (or zero for an adiabatic process), but as the temperature difference is made to approach zero, this term also approaches zero.

There could also be additional entropy-generation terms in the surroundings of the types discussed in the previous section, if those factors were also present in the surroundings, and those will be positive, as well. Thus, we conclude that the net entropy change is the sum of a number of terms, each of which is positive, due to a specific cause of irreversible entropy generation, such that the net entropy change could also be termed the total entropy generation:

$$dS_{net} = dS_{c.m.} + dS_{surr} = \sum \delta S_{gen} \geq 0 \qquad (8.16)$$

where the equality holds for reversible processes and the inequality for irreversible processes. This is a very important equation, not only for thermodynamics but also for philosophical thought. This equation is referred to as the principle of the increase of entropy. The great significance is that the only processes that can take place are those in which the net change in entropy of the control mass plus its surroundings increases (or in the limit, remains constant). The reverse process, in which both the control mass and surroundings are returned to their original state, can never be made to occur. In other words, Eq. 8.16 dictates the single direction in which any process can proceed. Thus, the principle of the increase of entropy can be considered a quantitative general statement of the second law from the macroscopic point of view and applies to the combustion of fuel in our automobile engines, the cooling of our coffee, and the processes that take place in our body.

Sometimes this principle of the increase of entropy is stated in terms of an isolated system, one in which there is no interaction between the system and its surroundings. Then there is no change in the surroundings, and we then conclude that

$$dS_{isolated\ system} = \delta S_{gen,\ system} \geq 0 \qquad (8.17)$$

That is, in an isolated system, the only processes that can occur are those that have an associated increase in entropy.

The development of Eq. 8.16 as the principle of the increase of entropy was made for an infinitesimal change of state. When we wish to test a claimed process to see whether it satisfies the second law of thermodynamics, it will necessarily be for a finite change of state. Consider a control mass undergoing a process from initial state 1 to final state 2, with an associated heat transfer $_1Q_2$, which may be known or calculated from the first law. The heat transfer is to or from a reservoir at temperature T_0. For this process,

$$\Delta S_{c.m.} = S_2 - S_1, \qquad \Delta S_{surr} = -\frac{_1Q_2}{T_0}$$

$$\Delta S_{net} = \Delta S_{c.m.} + \Delta S_{surr} \qquad (8.18)$$

and the net entropy change as calculated from Eq. 8.18 must be greater than zero (irreversible process), or in the limit be equal to zero (completely reversible process, internally and externally). This type of calculation is illustrated in the following example.

EXAMPLE 8.3 Suppose that 1 kg of saturated water vapor at 100°C is condensed to a saturated liquid at 100°C in a constant-pressure process by heat transfer to the surrounding air, which is at 25°C. What is the net increase in entropy of the water plus surroundings?

Solution

For the control mass (water), from the steam tables, we obtain

$$\Delta S_{c.m.} = -ms_{fg} = -1 \times 6.0480 = -6.0480 \text{ kJ/K}$$

Concerning the surroundings, we have

$$Q_{\text{to surroundings}} = mh_{fg} = 1 \times 2257.0 = 2257 \text{ kJ}$$

$$\Delta S_{surr} = \frac{Q}{T_0} = \frac{2257}{298.15} = 7.5700 \text{ kJ/K}$$

$$\Delta S_{net} = \Delta S_{c.m.} + \Delta S_{surr} = -6.0480 + 7.5700 = 1.5220 \text{ kJ/K}$$

This increase in entropy is in accordance with the principle of the increase of entropy and tells us, as does our experience, that this process can take place.

It is interesting to note how this heat transfer from the water to the surroundings might have taken place reversibly. Suppose that an engine operating on the Carnot cycle received heat from the water and rejected heat to the surroundings, as shown in Fig. 8.16. The decrease in the entropy of the water is equal to the increase in the entropy of the surroundings.

$$\Delta S_{c.m.} = -6.0480 \text{ kJ/K}$$

$$\Delta S_{surr} = 6.0480 \text{ kJ/K}$$

$$Q_{\text{to surroundings}} = T_0 \Delta S = 298.15(6.0480) = 1803.2 \text{ kJ}$$

$$W = Q_H - Q_L = 2257 - 1803.2 = 453.8 \text{ kJ}$$

Since this is a reversible cycle, the engine could be reversed and operated as a heat pump. For this cycle the work input to the heat pump would be 453.8 kJ.

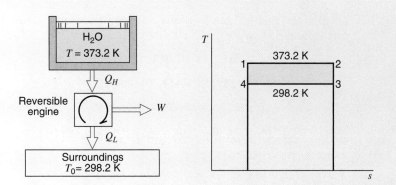

FIGURE 8.16
Reversible heat transfer with the surroundings.

EXAMPLE 8.3E

Suppose that 1 lbm of saturated water vapor at 212 F is condensed to a saturated liquid at 212 F in a constant-pressure process by heat transfer to the surrounding air, which is at 80 F. What is the net increase in entropy of the water plus surroundings?

Solution

For the control mass (water), from the steam tables,

$$\Delta S_{system} = -s_{fg} = -1.4446 \text{ Btu/lbm R}$$

Considering the surroundings, we have

$$Q_{\text{to surroundings}} = h_{fg} = 970.3 \text{ Btu/lbm}$$

$$\Delta S_{surr} = \frac{Q}{T_0} = \frac{970.3}{540} = 1.7980 \text{ Btu/lbm R}$$

$$\Delta S_{system} + \Delta S_{surr} = -1.4446 + 1.7980 = 0.3534 \text{ Btu/lbm R}$$

This increase in entropy is in accordance with the principle of the increase of entropy and tells us, as does our experience, that this process can take place.

It is interesting to note how this heat transfer from the water to the surroundings might have taken place reversibly. Suppose that an engine operating on the Carnot cycle received heat from the water and rejected heat to the surroundings, as shown in Fig. 8.16E.

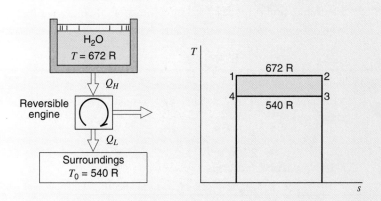

FIGURE 8.16E
Reversible heat transfer with the surroundings.

The decrease in the entropy of the water is equal to the increase in the entropy of the surroundings.

$$\Delta S_{c.m} = -1.4446 \text{ Btu/lbm R}$$

$$\Delta S_{surr} = 1.4446 \text{ Btu/lbm R}$$

$$Q_{\text{to surroundings}} = T_0\, \Delta S = 540(1.4446) = 780.1 \text{ Btu/lbm}$$

$$W = Q_H - Q_L = 970.3 - 780.1 = 190.2 \text{ Btu/lbm}$$

Since this is a reversible cycle, the engine could be reversed and operated as a heat pump. For this cycle the work input to the heat pump would be 190.2 Btu/lbm.

8.9 ENTROPY CHANGE OF A SOLID OR LIQUID

In Section 5.6 we considered the calculation of the internal energy and enthalpy changes with temperature for solids and liquids and found that, in general, it is possible to express both in terms of the specific heat, in the simple manner of Eq. 5.17, and in most instances in the integrated form of Eq. 5.18. We can now use this result and the thermodynamic property relation, Eq. 8.7, to calculate the entropy change for a solid or liquid. Note that for such a phase the specific volume term in Eq. 8.7 is very small, so that substituting Eq. 5.17 yields

$$ds \simeq \frac{du}{T} \simeq \frac{C}{T}\, dT \tag{8.19}$$

Now, as was mentioned in Section 5.6, for many processes involving a solid or liquid, we may assume that the specific heat remains constant, in which case Eq. 8.19 can be integrated. The result is

$$s_2 - s_1 \simeq C \ln \frac{T_2}{T_1} \tag{8.20}$$

If the specific heat is not constant, then commonly C is known as a function of T, in which case Eq. 8.19 can also be integrated to find the entropy change. Equation 8.20 illustrates what happens in a reversible ($ds_{\text{gen}} = 0$) adiabatic ($dq = 0$) process, which therefore is isentropic. In this process, the approximation of constant v leads to constant temperature, which explains why pumping liquid does not change the temperature.

EXAMPLE 8.4   One kilogram of liquid water is heated from 20°C to 90°C. Calculate the entropy change, assuming constant specific heat, and compare the result with that found when using the steam tables.

Control mass:	Water.
Initial and final states:	Known.
Model:	Constant specific heat, value at room temperature.

Solution

For constant specific heat, from Eq. 8.20,

$$s_2 - s_1 = 4.184 \ln \left(\frac{363.2}{293.2} \right) = 0.8958 \text{ kJ/kg K}$$

Comparing this result with that obtained by using the steam tables, we have

$$s_2 - s_1 = s_{f\ 90°C} - s_{f\ 20°C} = 1.1925 - 0.2966$$
$$= 0.8959 \text{ kJ/kg K}$$

8.10 ENTROPY CHANGE OF AN IDEAL GAS

Two very useful equations for computing the entropy change of an ideal gas can be developed from Eq. 8.7 by substituting Eqs. 5.20 and 5.24:

$$T\, ds = du + P\, dv$$

For an ideal gas

$$du = C_{v0}\, dT \qquad \text{and} \qquad \frac{P}{T} = \frac{R}{v}$$

Therefore,

$$ds = C_{v0} \frac{dT}{T} + \frac{R\, dv}{v} \tag{8.21}$$

$$s_2 - s_1 = \int_1^2 C_{v0} \frac{dT}{T} + R \ln \frac{v_2}{v_1} \tag{8.22}$$

Similarly,

$$T\, ds = dh - v\, dP$$

For an ideal gas

$$dh = C_{p0}\, dT \qquad \text{and} \qquad \frac{v}{T} = \frac{R}{P}$$

Therefore,

$$ds = C_{p0} \frac{dT}{T} - R \frac{dP}{P} \tag{8.23}$$

$$s_2 - s_1 = \int_1^2 C_{p0} \frac{dT}{T} - R \ln \frac{P_2}{P_1} \tag{8.24}$$

To integrate Eqs. 8.22 and 8.24, we must know the temperature dependence of the specific heats. However, if we recall that their difference is always constant as expressed by Eq. 5.27, we realize that we need to examine the temperature dependence of only one of the specific heats.

As in Section 5.7, let us consider the specific heat C_{p0}. Again, there are three possibilities to examine, the simplest of which is the assumption of constant specific heat. In this instance it is possible to integrate Eq. 8.24 directly to

$$s_2 - s_1 = C_{p0} \ln \frac{T_2}{T_1} - R \ln \frac{P_2}{P_1} \qquad (8.25)$$

Similarly, integrating Eq. 8.22 for constant specific heat, we have

$$s_2 - s_1 = C_{v0} \ln \frac{T_2}{T_1} + R \ln \frac{v_2}{v_1} \qquad (8.26)$$

The second possibility for the specific heat is to use an analytical equation for C_{p0} as a function of temperature, for example, one of those listed in Table A.6. The third possibility is to integrate the results of the calculations of statistical thermodynamics from reference temperature T_0 to any other temperature T and define the standard entropy

$$s_T^0 = \int_{T_0}^{T} \frac{C_{p0}}{T} dT \qquad (8.27)$$

This function can then be tabulated in the single-entry (temperature) ideal-gas table, as for air in Table A.7 or for other gases in Table A.8. The entropy change between any two states 1 and 2 is then given by

$$s_2 - s_1 = (s_{T2}^0 - s_{T1}^0) - R \ln \frac{P_2}{P_1} \qquad (8.28)$$

As with the energy functions discussed in Section 5.7, the ideal-gas tables, Tables A.7 and A.8, would give the most accurate results, and the equations listed in Table A.6 would give a close empirical approximation. Constant specific heat would be less accurate, except for monatomic gases and for other gases below room temperature. Again, it should be remembered that all these results are part of the ideal-gas model, which may or may not be appropriate in any particular problem.

EXAMPLE 8.5 Consider Example 5.6, in which oxygen is heated from 300 to 1500 K. Assume that during this process the pressure dropped from 200 to 150 kPa. Calculate the change in entropy per kilogram.

Solution

The most accurate answer for the entropy change, assuming ideal-gas behavior, would be from the ideal-gas tables, Table A.8. This result is, using Eq. 8.28,

$$s_2 - s_1 = (8.0649 - 6.4168) - 0.2598 \ln \left(\frac{150}{200} \right)$$

$$= 1.7228 \text{ kJ/kg K}$$

The empirical equation from Table A.6 should give a good approximation to this result. Integrating Eq. 8.24, we have

$$s_2 - s_1 = \int_{T_1}^{T_2} C_{p0} \frac{dT}{T} - R \ln \frac{P_2}{P_1}$$

$$s_2 - s_1 = \left[0.88 \ln \theta - 0.0001\theta + \frac{0.54}{2} \theta^2 - \frac{0.33}{3} \theta^3 \right]_{\theta_1 = 0.3}^{\theta_2 = 1.5}$$

$$- 0.2598 \ln \left(\frac{150}{200} \right)$$

$$= 1.7058 \text{ kJ/kg K}$$

which is within 1.0% of the previous value. For constant specific heat, using the value at 300 K from Table A.5, we have

$$s_2 - s_1 = 0.922 \ln \left(\frac{1500}{300} \right) - 0.2598 \ln \left(\frac{150}{200} \right)$$

$$= 1.5586 \text{ kJ/kg K}$$

which is too low by 9.5%. If, however, we assume that the specific heat is constant at its value at 900 K, the average temperature, as in Example 5.6, then

$$s_2 - s_1 = 1.0767 \ln \left(\frac{1500}{300} \right) + 0.0747 = 1.8076 \text{ kJ/kg K}$$

which is high by 4.9%.

EXAMPLE 8.6   Calculate the change in entropy per kilogram as air is heated from 300 to 600 K while pressure drops from 400 to 300 kPa. Assume:

1. Constant specific heat.
2. Variable specific heat.

Solution

1. From Table A.5 for air at 300 K,

$$C_{p0} = 1.004 \text{ kJ/kg K}$$

Therefore, using Eq. 8.25, we have

$$s_2 - s_1 = 1.004 \ln \left(\frac{600}{300} \right) - 0.287 \ln \left(\frac{300}{400} \right) = 0.7785 \text{ kJ/kg K}$$

2. From Table A.7,

$$s_{T1}^0 = 6.8693 \text{ kJ/kg K}, \qquad s_{T2}^0 = 7.5764 \text{ kJ/kg K}$$

Using Eq. 8.28 gives

$$s_2 - s_1 = 7.5764 - 6.8693 - 0.287 \ln\left(\frac{300}{400}\right) = 0.7897 \text{ kJ/kg K}$$

EXAMPLE 8.6E Calculate the change in entropy per pound as air is heated from 540 R to 1200 R while pressure drops from 50 lbf/in.^{2} to 40 lbf/in.^{2}. Assume:

1. Constant specific heat.
2. Variable specific heat.

Solution

1. From Table F.4 for air at 80 F,

$$C_{p0} = 0.24 \text{ Btu/lbm R}$$

Therefore, using Eq. 8.25, we have

$$s_2 - s_1 = 0.24 \ln\left(\frac{1200}{540}\right) - \frac{53.34}{778} \ln\left(\frac{40}{50}\right) = 0.2068 \text{ Btu/lbm R}$$

2. From Table F.5

$$s_{T1}^0 = 0.6008 \text{ Btu/lbm R} \qquad s_{T2}^0 = 0.7963 \text{ Btu/lbm R}$$

Using Eq. 8.28 gives

$$s_2 - s_1 = 0.7963 - 0.6008 - \frac{53.34}{778} \ln\frac{40}{50} = 0.2108 \text{ Btu/lbm R}$$

Let us now consider the case of an ideal gas undergoing an isentropic process, a situation that is analyzed frequently. We conclude that Eq. 8.24 with the left side of the equation equal to zero then expresses the relation between the pressure and temperature at the initial and final states, with the specific relation depending on the nature of the specific heat as a function of T. As was discussed following Eq. 8.24, there are three possibilities to examine. Of these, the most accurate is the third, that is, the ideal-gas Tables A.7 or A.8 and Eq. 8.28, with the integrated temperature function s_T^0 defined by Eq. 8.27. The following Example illustrates the procedure.

EXAMPLE 8.7 One kilogram of air is contained in a cylinder fitted with a piston at a pressure of 400 kPa and a temperature of 600 K. The air is expanded to 150 kPa in a reversible, adiabatic process. Calculate the work done by the air.

> *Control mass:* Air.
> *Initial state:* P_1, T_1; state 1 fixed.
> *Final state:* P_2.
> *Process:* Reversible and adiabatic.
> *Model:* Ideal gas and air tables, Table A.7.

Analysis

From the first law we have

$$0 = u_2 - u_1 + w$$

The second law gives us

$$s_2 = s_1$$

Solution

From Table A.7,

$$u_1 = 435.10 \text{ kJ/kg}, \qquad s^0_{T1} = 7.5764 \text{ kJ/kg K}$$

From Eq. 8.28,

$$s_2 - s_1 = 0 = (s^0_{T2} - s^0_{T1}) - R \ln \frac{P_2}{P_1}$$

$$= (s^0_{T2} - 7.5764) - 0.287 \ln \left(\frac{150}{400} \right)$$

$$s^0_{T2} = 7.2949 \text{ kJ/kg K}$$

From Table A.7,

$$T_2 = 457 \text{ K}, \qquad u_2 = 328.14 \text{ kJ/kg}$$

Therefore,

$$w = 435.10 - 328.14 = 106.96 \text{ kJ/kg}$$

The first of the three possibilities, constant specific heat, is also worth analyzing as a special case. In this instance, the result is Eq. 8.25 with the left side equal to zero, or

$$s_2 - s_1 = 0 = C_{p0} \ln \frac{T_2}{T_1} - R \ln \frac{P_2}{P_1}$$

This expression can also be written as

$$\ln\left(\frac{T_2}{T_1}\right) = \frac{R}{C_{p0}} \ln\left(\frac{P_2}{P_1}\right)$$

or

$$\frac{T_2}{T_1} = \left(\frac{P_2}{P_1}\right)^{R/C_{p0}} \tag{8.29}$$

However,

$$\frac{R}{C_{p0}} = \frac{C_{p0} - C_{v0}}{C_{p0}} = \frac{k-1}{k} \tag{8.30}$$

where k, the ratio of the specific heats, is defined as

$$k = \frac{C_{p0}}{C_{v0}} \tag{8.31}$$

Equation (8.29) is now conveniently written as

$$\frac{T_2}{T_1} = \left(\frac{P_2}{P_1}\right)^{(k-1)/k} \tag{8.32}$$

From this expression and the ideal-gas equation of state, it also follows that

$$\frac{T_2}{T_1} = \left(\frac{v_1}{v_2}\right)^{k-1} \tag{8.33}$$

and

$$\frac{P_2}{P_1} = \left(\frac{v_1}{v_2}\right)^{k} \tag{8.34}$$

From this last expression, we note that for this process

$$Pv^k = \text{constant} \tag{8.35}$$

This is a special case of a polytropic process in which the polytropic exponent n is equal to the specific heat ratio k.

8.11 THE REVERSIBLE POLYTROPIC PROCESS FOR AN IDEAL GAS

When a gas undergoes a reversible process in which there is heat transfer, the process frequently takes place in such a manner that a plot of log P versus log V is a straight line, as shown in Fig. 8.17. For such a process PV^n is a constant.

A process having this relation between pressure and volume is called a polytropic process. An example is the expansion of the combustion gases in the cylinder of a water-cooled reciprocating engine. If the pressure and volume are measured during the expansion stroke of polytropic process, as might be done with an engine indicator, and the

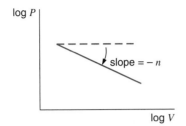

FIGURE 8.17

Example of a polytropic process.

logarithms of the pressure and volume are plotted, the result would be similar to the straight line in Fig. 8.17. From this figure it follows that

$$\frac{d \ln P}{d \ln V} = -n$$

$$d \ln P + n d \ln V = 0$$

If n is a constant (which implies a straight line on the log P versus log V plot), this equation can be integrated to give the following relation:

$$PV^n = \text{constant} = P_1 V_1^n = P_2 V_2^n \tag{8.36}$$

From this equation the following relations can be written for a polytropic process:

$$\frac{P_2}{P_1} = \left(\frac{V_1}{V_2}\right)^n$$

$$\frac{T_2}{T_1} = \left(\frac{P_2}{P_1}\right)^{(n-1)/n} = \left(\frac{V_1}{V_2}\right)^{n-1} \tag{8.37}$$

For a control mass consisting of an ideal gas, the work done at the moving boundary during a reversible polytropic process can be derived from the relations

$$_1W_2 = \int_1^2 P\,dV \qquad \text{and} \qquad PV^n = \text{constant}$$

$$_1W_2 = \int_1^2 P\,dV = \text{constant} \int_1^2 \frac{dV}{V^n}$$

$$= \frac{P_2 V_2 - P_1 V_1}{1 - n} = \frac{mR(T_2 - T_1)}{1 - n} \tag{8.38}$$

for any value of n except $n = 1$.

The polytropic processes for various values of n are shown in Fig. 8.18 on $P\text{-}v$ and $T\text{-}s$ diagrams. The values of n for some familiar processes are

Isobaric process:	$n = 0$,	$P = \text{constant}$
Isothermal process:	$n = 1$,	$T = \text{constant}$
Isentropic process:	$n = k$,	$s = \text{constant}$
Isochoric process:	$n = \infty$,	$v = \text{constant}$

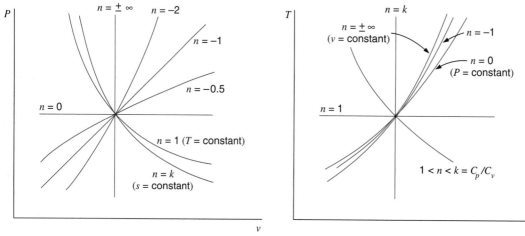

FIGURE 8.18 Polytropic process on P–v and T–s diagrams.

EXAMPLE 8.8 In a reversible process, nitrogen is compressed in a cylinder from 100 kPa and 20°C to 500 kPa. During this compression process, the relation between pressure and volume is $PV^{1.3}$ = constant. Calculate the work and heat transfer per kilogram, and show this process on P–V and T–S diagrams.

> *Control mass:* Nitrogen.
> *Initial state:* P_1, T_1; state 1 known.
> *Final state:* P_2.
> *Process:* Reversible, polytropic with exponent $n < k$.
> *Diagram:* Fig. 8.19
> *Model:* Ideal gas, constant specific heat—value at 300 K.

Analysis

We need to find the boundary movement work. From Eq. 8.38, we have

$$_1W_2 = \int_1^2 P\,dV = \frac{P_2V_2 - P_1V_1}{1 - n} = \frac{mR(T_2 - T_1)}{1 - n}$$

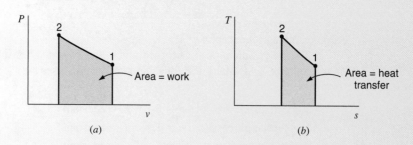

FIGURE 8.19
Diagram for Example 8.8.

The first law is

$$_1q_2 = u_2 - u_1 + {}_1w_2 = C_{v0}(T_2 - T_1) + {}_1w_2$$

Solution

From Eq. 8.37

$$\frac{T_2}{T_1} = \left(\frac{P_2}{P_1}\right)^{(n-1)/n} = \left(\frac{500}{100}\right)^{(1.3-1)/1.3} = 1.4498$$

$$T_2 = 293.2 \times 1.4498 = 425 \text{ K}$$

Then

$$_1w_2 = \frac{R(T_2 - T_1)}{1 - n} = \frac{0.2968(425 - 293.2)}{(1 - 1.3)} = -130.4 \text{ kJ/kg}$$

and from the first law,

$$_1q_2 = C_{v0}(T_2 - T_1) + {}_1w_2$$

$$= 0.745(425 - 293.2) - 130.4 = -32.2 \text{ kJ/kg}$$

The reversible isothermal process for an ideal gas is of particular interest. In this process

$$PV = \text{constant} = P_1V_1 = P_2V_2 \tag{8.39}$$

The work done at the boundary of a simple compressible mass during a reversible isothermal process can be found by integrating the equation

$$_1W_2 = \int_1^2 P\,dV$$

The integration is

$$_1W_2 = \int_1^2 P\,dV = \text{constant} \int_1^2 \frac{dV}{V} = P_1V_1 \ln\frac{V_2}{V_1} = P_1V_1 \ln\frac{P_1}{P_2} \tag{8.40}$$

or

$$_1W_2 = mRT \ln\frac{V_2}{V_1} = mRT \ln\frac{P_1}{P_2} \tag{8.41}$$

Because there is no change in internal energy or enthalpy in an isothermal process, the heat transfer is equal to the work (neglecting changes in kinetic and potential energy). Therefore, we could have derived Eq. 8.40 by calculating the heat transfer.

For example, using Eq. 8.7, we have

$$\int_1^2 T\,ds = {}_1q_2 = \int_1^2 du + \int_1^2 P\,dv$$

But $du = 0$ and $Pv = \text{constant} = P_1 v_1 = P_2 v_2$, such that

$$_1q_2 = \int_1^2 P\, dv = P_1 v_1 \ln \frac{v_2}{v_1}$$

which yields the same result as Eq. 8.40.

8.12 ENTROPY AS A RATE EQUATION

The second law of thermodynamics was used to write the balance of entropy in Eq. 8.11 for a variation and in Eq. 8.14 for a finite change. In some cases the equation is needed in a rate form so that a given process can be tracked in time. The rate form is also the basis for the development of the entropy balance equation in the general control volume analysis for an unsteady situation.

Take the incremental change in S from Eq. 8.11 and divide by δt. We get

$$\frac{dS}{\delta t} = \frac{1}{T}\frac{\delta Q}{\delta t} + \frac{\delta S_{\text{gen}}}{\delta t} \tag{8.42}$$

For a given control volume we may have more than one source of heat transfer, each at a certain surface temperature (semidistributed situation). Since we did not have to consider the temperature at which the heat transfer crossed the control surface for the energy equation, all the terms were added into a net heat transfer in a rate form in Eq. 5.31. Using this and a dot to indicate a rate, the final form for the entropy equation in the limit is

$$\frac{dS_{\text{c.m.}}}{dt} = \sum \frac{1}{T}\dot{Q} + \dot{S}_{\text{gen}} \tag{8.43}$$

expressing the rate of entropy change as due to the flux of entropy into the control mass from heat transfer and an increase due to irreversible processes inside the control mass. If only reversible processes take place inside the control volume, the rate of change of entropy is determined by the rate of heat transfer divided by the temperature terms alone.

EXAMPLE 8.9   Consider an electric space heater that converts 1 kW of electric power into a heat flux of 1 kW delivered at 600 K from the hot wire surface. Let us look at the process of the energy conversion from electricity to heat transfer and find the rate of total entropy generation.

> *Control mass:*   The electric heater wire.
> *State:*   Constant wire temperature 600 K.

Analysis

The first and the second laws of thermodynamics in rate form become

$$\frac{dE_{\text{c.m.}}}{dt} = \frac{dU_{\text{c.m.}}}{dt} = 0 = \dot{W}_{\text{el.in}} - \dot{Q}_{\text{out}}$$

$$\frac{dS_{\text{c.m.}}}{dt} = 0 = -\dot{Q}_{\text{out}}/T_{\text{surface}} + \dot{S}_{\text{gen}}$$

Notice that we neglected kinetic and potential energy changes in going from a rate of E to a rate of U. Then the left-hand side is zero since it is steady state and the right-hand side of the energy equation is electric work in minus the heat transfer out. For the entropy equation the left-hand side is zero because of steady state and the right-hand side has a flux of entropy out due to heat transfer, and entropy is generated in the wire.

Solution

We now get the entropy generation as

$$\dot{S}_{gen} = \dot{Q}_{out}/T = 1/600 = 0.001\,67 \text{ kW/K}$$

SUMMARY

The inequality of Clausius and the property entropy (s) are modern statements of the second law. The final statement of the second law is the entropy balance equation that includes generation of entropy. All the results that were derived from the classical formulation of the second law in Chapter 7 can be re-derived with the entropy balance equation applied to the cyclic devices. For all reversible processes, entropy generation is zero and all real (irreversible) processes have a positive entropy generation. How large the entropy generation is depends on the actual process.

Thermodynamic property relations for s are derived from consideration of a reversible process and leads to Gibbs relations. Changes in the property s are covered through general tables, approximations for liquids and solids, as well as ideal gases. Changes of entropy in various processes are examined in general together with special cases of polytropic processes. Just as a reversible specific boundary work is the area below the process curve in a P–v diagram, the reversible heat transfer is the area below the process curve in a T–s diagram.

You should have learned a number of skills and acquired abilities from studying this chapter that will allow you to

- Know that Clausius inequality is an alternative statement of the second law.
- Know the relation between the entropy and the reversible heat transfer.
- Locate states in the tables involving entropy.
- Understand how a Carnot cycle looks in a T–s diagram.
- Know how different simple process curves look in a T–s diagram.
- Understand how to apply the entropy balance equation for a control mass.
- Recognize processes that generate entropy and where the entropy is made.
- Evaluate changes in s for liquids, solids, and ideal gases.
- Know the various property relations for a polytropic process in an ideal gas.
- Know the application of the unsteady entropy equation and what a flux of s is.

KEY CONCEPTS AND FORMULAS

Clausius inequality $\quad \oint \dfrac{dQ}{T} \leq 0$

Entropy $\quad ds = \dfrac{dq}{T} + ds_{gen}; \qquad ds_{gen} \geq 0$

Rate equation for entropy	$\dot{S}_{\text{c.m.}} = \sum \dfrac{\dot{Q}_{\text{c.m.}}}{T} + \dot{S}_{\text{gen}}$
Entropy equation	$m(s_2 - s_1) = \displaystyle\int_1^2 \dfrac{\delta Q}{T} + {}_1S_{2\,\text{gen}}; \qquad {}_1S_{2\,\text{gen}} \geq 0$
Total entropy change	$\Delta S_{\text{net}} = \Delta S_{\text{cm}} + \Delta S_{\text{surr}} = \Delta S_{\text{gen}} \geq 0$
Lost work	$W_{\text{lost}} = \int T\, dS_{\text{gen}}$
Actual boundary work	${}_1W_2 = \int P\, dV - W_{\text{lost}}$
Gibbs relations	$T\, ds = du + P\, dv$
	$T\, ds = dh - v\, dP$

Solids, Liquids

$$v = \text{constant}, \qquad dv = 0$$

Change in s

$$s_2 - s_1 = \int \frac{du}{T} = \int C \frac{dT}{T} \approx C \ln \frac{T_2}{T_1}$$

Ideal Gas

Standard entropy

$$s_T^0 = \int_{T_0}^T \frac{C_{p0}}{T}\, dT \qquad \text{(Function of } T\text{)}$$

Change in s

$$s_2 - s_1 = s_{T2}^0 - s_{T1}^0 - R \ln \frac{P_2}{P_1} \quad \text{(Using Table A.7 or A.8)}$$

$$s_2 - s_1 = C_{p0} \ln \frac{T_2}{T_1} - R \ln \frac{P_2}{P_1} \quad \text{(For constant } C_p, C_v\text{)}$$

$$s_2 - s_1 = C_v = \ln \frac{T_2}{T_1} + R \ln \frac{v_2}{v_1} \quad \text{(For constant } C_p, C_v\text{)}$$

Ratio of specific heats

$$k = C_{p0}/C_{v0}$$

Polytropic processes

$$Pv^n = \text{constant}; \qquad PV^n = \text{constant}$$

$$\frac{P_2}{P_1} = \left(\frac{V_1}{V_2}\right)^n = \left(\frac{v_1}{v_2}\right)^n = \left(\frac{T_2}{T_1}\right)^{\frac{n}{n-1}}$$

$$\frac{T_2}{T_1} = \left(\frac{v_1}{v_2}\right)^{n-1} = \left(\frac{P_2}{P_1}\right)^{\frac{n-1}{n}}$$

$$\frac{v_2}{v_1} = \left(\frac{P_1}{P_2}\right)^{\frac{1}{n}} = \left(\frac{T_1}{T_2}\right)^{\frac{1}{n-1}}$$

Specific work

$$_1w_2 = \frac{1}{1-n}(P_2 v_2 - P_1 v_1) = \frac{R}{1-n}(T_2 - T_1) \qquad n \neq 1$$

$$_1w_2 = P_1 v_1 \ln \frac{v_2}{v_1} = RT_1 \ln \frac{v_2}{v_1} = RT_1 \ln \frac{P_1}{P_2} \qquad n = 1$$

The work is moving boundary work $w = \displaystyle\int P\, dv$

Identifiable processes	$n = 0$;	P = constant;	Isobaric
	$n = 1$;	T = constant;	Isothermal
	$n = k$;	s = constant;	Isentropic
	$n = \infty$;	v = constant;	Isochoric or isometric

CONCEPT-STUDY GUIDE PROBLEMS

8.1 Does Clausius say anything about the sign for $\oint Q$?

8.2 When a substance has completed a cycle, v, u, h, and s are unchanged. Did anything happen? Explain.

8.3 Assume a heat engine with a given Q_H. Can you say anything about Q_L if the engine is reversible? if it is irreversible?

8.4 How can you change s of a substance going through a reversible process?

8.5 Does the statement of Clausius require a constant T for the heat transfer as in a Carnot cycle?

8.6 A reversible process adds heat to a substance. If T is varying, does that influence the change in s?

8.7 Water at 100 kPa, 150°C, receives 75 kJ/kg in a reversible process by heat transfer. Which process changes s the most: constant T, constant v, or constant P?

8.8 A substance has heat transfer out. Can you say anything about changes in s if the process is reversible? if it is irreversible?

8.9 A substance is compressed adiabatically so P and T go up. Does that change s?

8.10 Saturated water vapor at 200 kPa is compressed to 600 kPa in a reversible adiabatic process. Find the new in v and T.

8.11 A computer chip dissipates 2 kJ of electric work over time and rejects that as heat transfer from its 50°C surface to 25°C air. How much entropy is generated in the chip? How much, if any, is generated outside the chip?

8.12 A car uses an average power of 25 hp for a one-hour round trip. With a thermal efficiency of 35% how much fuel energy was used? What happened to all the energy? What change in entropy took place if we assume ambient at 20°C?

8.13 A liquid is compressed in a reversible adiabatic process. What is the change in T?

8.14 Two 5-kg blocks of steel, one at 250°C, the other at 25°C, come in thermal contact. Find the final temperature and the total entropy generation in the process.

8.15 One kg of air at 300 K is mixed with one kg of air at 400 K in a process at a constant 100 kPa and $Q = 0$. Find the final T and the entropy generation in the process.

8.16 One kg of air at 100 kPa is mixed with one kg of air at 200 kPa, both at 300 K, in a rigid insulated tank. Find the final state (P, T) and the entropy generation in the process.

8.17 An ideal gas goes through a constant T reversible heat addition process. How do the properties (v, u, h, s, P) change (up, down, or constant)?

8.18 Carbon dioxide is compressed to a smaller volume in a polytropic process with $n = 1.2$. How do the properties (u, h, s, P, T) change (up, down, or constant)?

8.19 Hot combustion air at 1500 K expands in a polytropic process to a volume six times as large with $n = 1.5$. Find the specific boundary work and the specific heat transfer.

8.20 A window receives 200 W of heat transfer at the inside surface of 20°C and transmits the 200 W from its outside surface at 2°C continuing to ambient air at −5°C. Find the flux of entropy at all three surfaces and the window's rate of entropy generation.

HOMEWORK PROBLEMS

Inequality of Clausius

8.21 Consider the steam power plant in Example 6.9 and assume an average T in the line between 1 and 2. Show that this cycle satisfies the inequality of Clausius.

8.22 Assume the heat engine in Problem 7.25 has a high temperature of 1200 K and a low temperature of 400 K. What does the inequality of Clausius say about each of the four cases?

8.23 Let the steam power plant in Problem 7.26 have 700°C in the boiler and 40°C during the heat rejection in the condenser. Does that satisfy the inequality of Clausius? Repeat the question for the cycle operated in reverse as a refrigerator.

8.24 A heat engine receives 6 kW from a 250°C source and rejects heat at 30°C. Examine each of three cases with respect to the inequality of Clausius.
   a. $\dot{W} = 6$ kW   b. $\dot{W} = 0$ kW   c. Carnot cycle

8.25 Examine the heat engine given in Problem 7.50 to see if it satisfies the inequality of Clausius.

Entropy of a Pure Substance

8.26 Find the entropy for the following water states and indicate each state on a T–s diagram relative to the two-phase region.
   a. 250°C, $v = 0.02$ m³/kg
   b. 250°C, 2000 kPa
   c. −2°C, 100 kPa
   d. 20°C, 100 kPa
   e. 20°C, 10 000 kPa

8.27 Find the missing properties and give the phase of the substance.
   a. H₂O $s = 7.70$ kJ/kg K, $h = ? T = ? x = ?$
             $P = 25$ kPa
   b. H₂O $u = 3400$ kJ/kg, $T = ? x = ? s = ?$
             $P = 10$ MPa
   c. R-12 $T = 0°C$, $s = ? x = ?$
             $P = 200$ kPa
   d. R-134a $T = −10°C$, $v = ? s = ?$
             $x = 0.45$
   e. NH₃ $T = 20°C$, $u = ? x = ?$
             $s = 5.50$ kJ/kg K

8.28 Saturated liquid water at 20°C is compressed to a higher pressure with constant temperature. Find the changes in u and s when the final pressure is
   a. 500 kPa   b. 2000 kPa   c. 20 000 kPa

8.29 Saturated vapor water at 150°C is expanded to a lower pressure with constant temperature. Find the changes in u and s when the final pressure is
   a. 100 kPa   b. 50 kPa   c. 10 kPa

8.30 Determine the missing property among P, T, s, and x for the following states:
   a. Ammonia   25°C, $v = 0.10$ m³/kg
   b. Ammonia   1000 kPa, $s = 5.2$ kJ/kg K

   c. R-134a   5°C, $s = 1.7$ kJ/kg K
   d. R-134a   50°C, $s = 1.9$ kJ/kg K
   e. R-22   100 kPa, $v = 0.3$ m³/kg

Reversible Processes

8.31 Consider a Carnot-cycle heat engine with water as the working fluid. The heat transfer to the water occurs at 300°C, during which process the water changes from saturated liquid to saturated vapor. The heat is rejected from the water at 40°C. Show the cycle on a T–s diagram and find the quality of the water at the beginning and the end of the heat rejection process. Determine the net work output per kg water and the cycle thermal efficiency.

8.32 In a Carnot engine with ammonia as the working fluid the high temperature is $T_H = 60°C$, and as Q_H is received the ammonia changes from saturated liquid to saturated vapor. The ammonia pressure at the low temperature is $P_{low} = 190$ kPa. Find T_L, the cycle thermal efficiency, the heat added per kg, and the entropy, s, at the beginning of the heat rejection process.

8.33 Water is used as the working fluid in a Carnot-cycle heat engine, where it changes from saturated liquid to saturated vapor at 200°C as heat is added. Heat is rejected in a constant-pressure process (also constant T) at 20 kPa. The heat engine powers a Carnot-cycle refrigerator that operates between −15°C and +20°C, shown in Fig. P8.33. Find the heat added to the water per kg water. How much heat should be added to the water in the heat engine so the refrigerator can remove 1 kJ from the cold space?

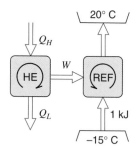

FIGURE P8.33

8.34 Consider a Carnot-cycle heat pump with R-22 as the working fluid. Heat is rejected from the R-22 at

40°C, during which process the R-22 changes from saturated vapor to saturated liquid. The heat is transferred to the R-22 at 0°C.

a. Show the cycle on a *T–s* diagram.

b. Find the quality of the R-22 at the beginning and end of the isothermal heat addition process at 0°C.

c. Determine the coefficient of performance for the cycle.

8.35 Do Problem 8.34 using refrigerant R-134a instead of R-22.

8.36 Water at 200 kPa with $x = 1.0$ is compressed in a piston cylinder to 1 MPa and 250°C in a reversible process. Find the sign for the work and the sign for the heat transfer.

8.37 Water at 200 kPa with $x = 1.0$ is compressed in a piston cylinder at 1 MPa and 350°C in a reversible process. Find the sign for the work and the sign for the heat transfer.

8.38 Ammonia at 1 MPa and 50°C is expanded in a piston cylinder to 500 kPa and 20°C in a reversible process. Find the sign for both the work and the heat transfer.

8.39 One kilogram of ammonia in a piston cylinder at 50°C and 1000 kPa is expanded in a reversible isothermal process to 100 kPa. Find the work and heat transfer for this process.

8.40 One kilogram of ammonia in a piston cylinder at 50°C and 1000 kPa is expanded in a reversible isobaric process to 140°C, shown in Fig. 8.40. Find the work and heat transfer for this process.

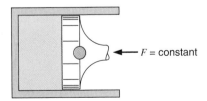

FIGURE P8.40

8.41 One kilogram of ammonia in a piston cylinder at 50°C and 1000 kPa is expanded in a reversible adiabatic process to 100 kPa, shown in Fig. 8.41. Find the work and heat transfer for this process.

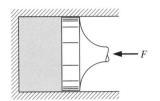

FIGURE P8.41

8.42 A cylinder fitted with a piston contains ammonia at 50°C and 20% quality with a volume of 1 L. The ammonia expands slowly, and during this process heat is transferred to maintain a constant temperature. The process continues until all the liquid is gone. Determine the work and heat transfer for this process.

8.43 An insulated cylinder fitted with a piston contains 0.1 kg of water at 100°C with 90% quality. The piston is moved, compressing the water until it reaches a pressure of 1.2 MPa. How much work is required in the process?

8.44 Compression and heat transfer brings R-134a in a piston cylinder from 500 kPa and 50°C to saturated vapor in an isothermal process. Find the specific heat transfer and the specific work.

8.45 One kilogram of water at 300°C expands against a piston in a cylinder until it reaches ambient pressure, 100 kPa, at which point the water has a quality of 90.2%. It may be assumed that the expansion is reversible and adiabatic. What was the initial pressure in the cylinder and how much work is done by the water?

8.46 Water in a piston/cylinder device at 400°C and 2000 kPa is expanded in a reversible adiabatic process. The specific work is measured to be 415.72 kJ/kg out. Find the final *P* and *T* and show the *P–v* and the *T–s* diagrams for the process.

8.47 A piston cylinder with 2 kg ammonia at 50°C and 100 kPa is compressed to 1000 kPa. The process happens so slowly that the temperature is constant. Find the heat transfer and the work for the process assuming it to be reversible.

8.48 A piston cylinder with R-134a at −20°C and 100 kPa is compressed to 500 kPa in a reversible adiabatic process. Find the final temperature and the specific work.

8.49 A closed tank, with $V = 10$ L, containing 5 kg of water initially at 25°C is heated to 175°C in a reversible process. Find the heat transfer to the water and its change in entropy.

8.50 A cylinder containing R-134a at 10°C and 150 kPa has an initial volume of 20 L. A piston compresses the R-134a in a reversible, isothermal process until it reaches the saturated vapor state. Calculate the required work and heat transfer to accomplish this process.

8.51 A heavily insulated cylinder fitted with a frictionless piston, as shown in Fig. P8.51, contains ammonia at 5°C and 92.9% quality, at which point the volume is 200 L. The external force on the piston is now increased slowly, compressing the ammonia until its temperature reaches 50°C. How much work is done on the ammonia during this process?

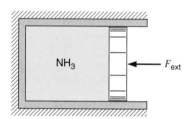

FIGURE P8.51

8.52 A piston/cylinder device with 2 kg water at 1000 kPa and 250°C is cooled with a constant loading on the piston. This isobaric process ends when the water has reached a state of saturated liquid. Find the work and heat transfer and sketch the process in both a P–v and a T–s diagram.

8.53 Water at 1000 kPa and 250°C is brought to saturated vapor in a piston/cylinder assembly with an isothermal process. Find the specific work and heat transfer. Estimate the specific work from the area in the P–v diagram and compare it to the correct value.

8.54 Water at 1000 kPa and 250°C is brought to saturated vapor in a rigid container, shown in Fig. P8.54. Find the final T and the specific heat transfer in this isometric process.

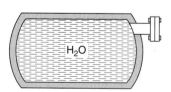

FIGURE P8.54

8.55 Estimate the specific heat transfer from the area in the T–s diagram and compare it to the correct value for the states and process in Problem 8.54.

8.56 Water at 1000 kPa and 250°C is brought to saturated vapor in a piston/cylinder setup with an isobaric process. Find the specific work and heat transfer. Estimate the specific heat transfer from the area in the T–s diagram and compare it to the correct value.

8.57 A heavily insulated cylinder/piston contains ammonia at 1200 kPa, 60°C. The piston is moved, expanding the ammonia in a reversible process until the temperature is −20°C. During the process 600 kJ of work is given out by the ammonia. What was the initial volume of the cylinder?

8.58 Water at 1000 kPa and 250°C is brought to saturated vapor in a piston/cylinder device with an adiabatic process. Find the final T and the specific work. Estimate the specific work from the area in the P–v diagram and compare it to the correct value.

8.59 A rigid, insulated vessel contains superheated vapor steam at 3 MPa, 400°C. A valve on the vessel is opened, allowing steam to escape as shown in Fig. P8.59. The overall process is irreversible, but the steam remaining inside the vessel goes through a reversible adiabatic expansion. Determine the fraction of steam that has escaped when the final state inside is saturated vapor.

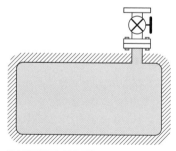

FIGURE P8.59

8.60 A piston/cylinder setup contains 2 kg of water at 200°C and 10 MPa. The piston is slowly moved to expand the water in an isothermal process to a pressure of 200 kPa. Heat transfer takes place with an ambient surrounding at 200°C, and the whole process may be assumed reversible. Sketch the process in a *P–V* diagram and calculate both the heat transfer and the total work.

Entropy Generation

8.61 One kg of water at 500°C and 1 kg of saturated water vapor, both at 200 kPa, are mixed in a constant-pressure and adiabatic process. Find the final temperature and the entropy generation for the process.

8.62 The unrestrained expansion of the reactor water in Problem 5.48 has a final state in the two-phase region. Find the entropy generated in the process.

8.63 A mass- and atmosphere-loaded piston/cylinder device contains 2 kg of water at 5 MPa and 100°C. Heat is added from a reservoir at 700°C to the water until it reaches 700°C. Find the work, heat transfer, and total entropy production for the system and surroundings.

8.64 Ammonia is contained in a rigid sealed tank of unknown quality at 0°C. When heated in boiling water to 100°C, its pressure reaches 1200 kPa. Find the initial quality, the heat transfer to the ammonia, and the total entropy generation.

8.65 An insulated cylinder/piston arrangement contains R-134a at 1 MPa and 50°C, with a volume of 100 L. The R-134a expands, moving the piston until the pressure in the cylinder has dropped to 100 kPa. It is claimed that the R-134a does 190 kJ of work against the piston during the process. Is that possible?

8.66 A piece of hot metal should be cooled rapidly (quenched) to 25°C, which requires removal of 1000 kJ from the metal. There are three possible ways to remove this energy: (1) Submerge the metal into a bath of liquid water and ice, thus melting the ice. (2) Let saturated liquid R-22 at −20°C absorb the energy so that it becomes saturated vapor. (3) Absorb the energy by vaporizing liquid nitrogen at 101.3 kPa pressure.

   a. Calculate the change of entropy of the cooling media for each of the three cases.

   b. Discuss the significance of the results.

8.67 A piston/cylinder setup has 2.5 kg of ammonia at 50 kPa and −20°C. Now it is heated to 50°C at constant pressure through the bottom of the cylinder from external hot gas at 200°C. Find the heat transfer to the ammonia and the total entropy generation.

8.68 A cylinder fitted with a movable piston contains water at 3 MPa with 50% quality, at which point the volume is 20 L. The water now expands to 1.2 MPa as a result of receiving 600 kJ of heat from a large source at 300°C. It is claimed that the water does 124 kJ of work during this process. Is this possible?

8.69 A piston/cylinder device loaded so it gives constant pressure has 0.75 kg of saturated vapor water at 200 kPa. It is now cooled so that the volume becomes half the initial volume by heat transfer to the ambient surroundings at 20°C. Find the work, heat transfer, and total entropy generation.

8.70 A piston/cylinder setup contains 1 kg of water at 150 kPa and 20°C. The piston is loaded so that pressure is linear in volume. Heat is added from a 600°C source until the water is at 1 MPa and 500°C. Find the heat transfer and total change in entropy.

8.71 A piston/cylinder has ammonia at 2000 kPa, 80°C with a volume of 0.1 m³. The piston is loaded with a linear spring, and the outside ambient is at 20°C, shown in Fig. P8.71. The ammonia now cools down to 20°C at which point it has a quality of 10%. Find the work, heat transfer, and total entropy generation in the process.

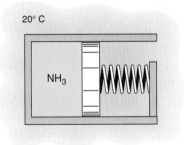

FIGURE P8.71

8.72 A cylinder/piston assembly contains water at 200 kPa and 200°C with a volume of 20 L. The piston is moved slowly, compressing the water to a pressure of 800 kPa. The loading on the piston is such that the product *PV* is a constant. Assuming that the room temperature is 20°C, show that this process does not violate the second law.

8.73 One kilogram of ammonia (NH₃) is contained in a spring-loaded piston/cylinder, Fig. P8.73, as

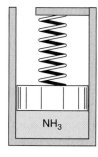

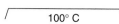

NH

100° C   **FIGURE P8.73**

saturated liquid at $-20°C$. Heat is added from a reservoir at $100°C$ until a final condition of 800 kPa and $70°C$ is reached. Find the work, heat transfer, and entropy generation, assuming the process is internally reversible.

8.74 A piston/cylinder device keeping a constant pressure of 500 kPa has 1 kg of water at $20°C$ and 1 kg of water at $100°C$ separated by a membrane, shown in Fig. P8.74. The membrane is broken and the water comes to a uniform state with no external heat transfer. Find the final temperature and the entropy generation.

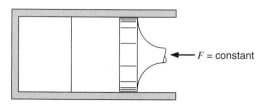

FIGURE P8.74

Entropy of a Liquid or Solid

8.75 A piston/cylinder setup has constant pressure of 2000 kPa with water at $20°C$. It is now heated up to $100°C$. Find the heat transfer and the entropy change using the steam tables. Repeat the calculation using constant heat capacity and incompressibility.

8.76 A large slab of concrete, 5 m $\times$ 8 m $\times$ 0.3 m, is used as a thermal storage mass in a solar-heated house. If the slab cools overnight from $23°C$ to $18°C$ in an $18°C$ house, what is the net entropy change associated with this process?

8.77 A 4-L jug of milk at $25°C$ is placed in your refrigerator where it is cooled down to the refrigerator's

inside constant temperature of $5°C$. Assume the milk to have the properties of liquid water and find the entropy generated in the cooling process.

8.78 A foundry form box with 25 kg of $200°C$ hot sand is dumped into a bucket with 50 L of water at $15°C$. Assuming no heat transfer with the surroundings and no boiling away of liquid water, calculate the net entropy change for the process.

8.79 A 5-kg steel container is cured at $500°C$. An amount of liquid water at $15°C$, 100 kPa, is added to the container so that a final uniform temperature of the steel and the water becomes $75°C$. Neglect any water that might evaporate during the process and any air in the container. How much water should be added, and how much entropy was generated?

8.80 A pan in an autoshop contains 5 L of engine oil at $20°C$ and 100 kPa. Now 2 L of hot $100°C$ oil is mixed into the pan. Neglect any work term and find the final temperature and the entropy generation.

8.81 Find the total work the heat engine can give out as it receives energy from the rock bed as described in Problem 7.61 (see Fig. P8.81). *Hint*: Write the entropy balance equation for the control volume that is the combination of the rockbed and the heat engine.

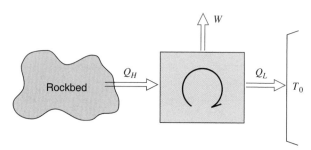

FIGURE P8.81

8.82 Two kg of liquid lead initially at $500°C$ are poured into a form. It then cools at constant pressure down to room temperature of $20°C$ as heat is transferred to the room. The melting point of lead is $327°C$, and the enthalpy change between the phases, h_{if}, is 24.6 kJ/kg. The specific heats are in Tables A.3 and A.4. Calculate the net entropy change for this process.

8.83 A 12-kg steel container has 0.2 kg of superheated water vapor at 1000 kPa and $200°C$. The total mass is now cooled to the ambient temperature of $30°C$.

How much heat transfer occurs, and what is the total entropy generation?

8.84 A 5-kg aluminum radiator holds 2 kg of liquid R-134a at $-10°C$. The setup is brought indoors and heated with 220 kJ from a heat source at $100°C$. Find the total entropy generation for the process assuming the R-134a remains a liquid.

8.85 A piston/cylinder of total 1 kg steel contains 0.5 kg ammonia at 1600 kPa with both masses at $120°C$. Some stops are placed so that a minimum volume is 0.02 m^3, shown in Fig. P8.85. Now the whole system is cooled down to $30°C$ by heat transfer to the ambient at $20°C$, and during the process the steel keeps the same temperature as the ammonia. Find the work, heat transfer, and total entropy generation in the process.

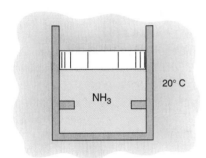

FIGURE P8.85

8.86 A hollow steel sphere with a 0.5-m inside diameter and a 2-mm thick wall contains water at 2 MPa, $250°C$. The system (steel plus water) cools to the ambient temperature, $30°C$. Calculate the net entropy change of the system and surroundings for this process.

Entropy of Ideal Gases

8.87 A mass of 1 kg of air contained in a cylinder at 1.5 MPa and 1000 K expands in a reversible isothermal process to a volume 10 times larger. Calculate the heat transfer during the process and the change of entropy of the air.

8.88 A piston/cylinder setup containing air at 100 kPa and 400 K is compressed to a final pressure of 1000 kPa. Consider two different processes: (1) a reversible adiabatic process and (2) a reversible isothermal process. Show both processes in a P–v diagram and a T–s diagram. Find the final temperature and the specific work for both processes.

8.89 Consider a Carnot-cycle heat pump having 1 kg of nitrogen gas in a cylinder/piston arrangement. This heat pump operates between reservoirs at 300 K and 400 K. At the beginning of the low-temperature heat addition, the pressure is 1 MPa. During this process the volume triples. Analyze each of the four processes in the cycle and determine

a. the pressure, volume, and temperature at each point.

b. the work and heat transfer for each process.

8.90 Consider a small air pistol with a cylinder volume of 1 cm^3 at 250 kPa and $27°C$. The bullet acts as a piston initially held by a trigger, shown in Fig. P8.90. The bullet is released so that the air expands in an adiabatic process. If the pressure should be 100 kPa as the bullet leaves the cylinder, find the final volume and the work done by the air.

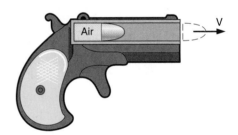

FIGURE P8.90

8.91 Oxygen gas in a piston/cylinder assembly at 300 K and 100 kPa with volume 0.1 m^3 is compressed in a reversible adiabatic process to a final temperature of 700 K. Find the final pressure and volume using Table A.5.

8.92 Oxygen gas in a piston/cylinder device at 300 K and 100 kPa with volume 0.1 m^3 is compressed in a reversible adiabatic process to a final temperature of 700 K. Find the final pressure and volume using Table A.8.

8.93 A hand-held pump for a bicycle has a volume of 25 cm^3 when fully extended. You now press the plunger (piston) in while holding your thumb over the exit hole so that an air pressure of 300 kPa is obtained. The outside atmosphere is at P_0 and T_0.

FIGURE P8.93

Consider two cases: (1) It is done quickly (~1 s) and (2) it is done very slowly (~1 h).

a. State assumptions about the process for each case.

b. Find the final volume and temperature for both cases.

8.94 An insulated cylinder/piston setup contains carbon dioxide gas at 120 kPa and 400 K. The gas is compressed to 2.5 MPa in a reversible adiabatic process. Calculate the final temperature and the work per unit mass, assuming

a. Variable specific heat (Table A.8).

b. Constant specific heat (value from Table A.5).

c. Constant specific heat (value at an intermediate temperature from Table A.6).

8.95 A piston/cylinder assembly shown in Fig. P8.95, contains air at 1380 K and 15 MPa, with $V_1 = 10$ cm^3 and $A_{cyl} = 5$ cm^2. The piston is released, and just before the piston exits the end of the cylinder, the pressure inside is 200 kPa. If the cylinder is insulated, what is its length? How much work is done by the air inside?

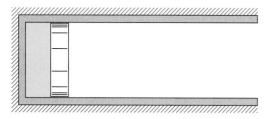

FIGURE P8.95

8.96 Two rigid tanks shown in Fig. P8.96 each contain 10 kg of N_2 gas at 1000 K and 500 kPa. They are now thermally connected to a reversible heat pump, which heats one and cools the other with no heat transfer to the surroundings. When one tank is heated to 1500 K the process stops. Find the final (P, T) in both tanks and the work input to the heat pump, assuming constant heat capacities.

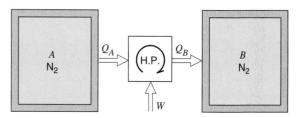

FIGURE P8.96

8.97 A spring-loaded piston/cylinder setup contains 1.5 kg of air at 27°C and 160 kPa. It is now heated in a process where pressure is linear in volume, $P = A + BV$, to twice the initial volume where it reaches 900 K. Find the work, heat transfer, and total entropy generation assuming a source at 900 K.

8.98 A rigid storage tank of 1.5 m^3 contains 1 kg of argon at 30°C. Heat is then transferred to the argon from a furnace operating at 1300°C until the specific entropy of the argon has increased by 0.343 kJ/kg K. Find the total heat transfer and the entropy generated in the process.

8.99 A rigid tank contains 2 kg of air at 200 kPa and ambient temperature, 20°C. An electric current now passes through a resistor inside the tank. After a total of 100 kJ of electrical work has crossed the boundary, the air temperature inside is 80°C. Is this possible?

8.100 Argon in a light bulb is at 90 kPa and heated from 20°C to 60°C with electrical power. Do not consider any radiation, or the glass mass. Find the total entropy generation per unit mass of argon.

8.101 We wish to obtain a supply of cold helium gas by applying the following technique. Helium contained in a cylinder at ambient conditions, 100 kPa and 20°C, is compressed in a reversible isothermal process to 600 kPa, after which the gas is expanded back to 100 kPa in a reversible adiabatic process.

a. Show the process on a T–s diagram.

b. Calculate the final temperature and the net work per kilogram of helium.

8.102 A 1 m^3 insulated, rigid tank contains air at 800 kPa and 25°C. A valve on the tank is opened, and the pressure inside quickly drops to 150 kPa, at which point the valve is closed. Assuming that the air remaining inside has undergone a reversible adiabatic expansion, calculate the mass withdrawn during the process.

8.103 Nitrogen at 200°C and 300 kPa is in a piston/cylinder device of volume 5 L, with the piston locked with a pin. The forces on the piston require a pressure inside of 200 kPa to balance it without the pin. The pin is removed and the piston quickly comes to its equilibrium position without any heat transfer. Find the final P, T, and V and the entropy generation due to this partly unrestrained expansion.

8.104 A rigid container with a volume of 200 L is divided into two equal volumes by a partition, shown in Fig. P8.104. Both sides contain nitrogen; one side is at 2 MPa and 200°C, while the other is at 200 kPa and 100°C. The partition ruptures, and the nitrogen comes to a uniform state at 70°C. Assume the temperature of the surroundings to be 20°C. Determine the work done and the net entropy change for the process.

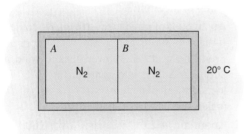

FIGURE P8.104

8.105 Nitrogen at 600 kPa and 127°C is in a 0.5 m³ insulated tank connected to a pipe with a valve to a second insulated initially empty tank with a volume of 0.5 m³, shown in Fig. P8.105. The valve is

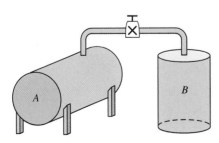

FIGURE P8.105

opened, and the nitrogen fills both tanks at a uniform state. Find the final pressure and temperature and the entropy generation this process causes. Why is the process irreversible?

Polytropic Processes

8.106 Neon at 400 kPa and 20°C is brought to 100°C in a polytropic process with $n = 1.4$. Give the sign for the heat transfer and work terms and explain.

8.107 A mass of 1 kg of air contained in a cylinder at 1.5 MPa and 1000 K expands in a reversible adiabatic process to 100 kPa. Calculate the final temperature and the work done during the process, using

a. Constant specific heat (value from Table A.5).

b. The ideal gas tables (Table A.7).

8.108 An ideal gas having a constant specific heat undergoes a reversible polytropic expansion with exponent $n = 1.4$. If the gas is carbon dioxide, will the heat transfer for this process be positive, negative, or zero?

8.109 A cylinder/piston setup contains 1 kg of methane gas at 100 kPa and 20°C. The gas is compressed reversibly to a pressure of 800 kPa. Calculate the work required if the process is

a. Adiabatic

b. Isothermal

c. Polytropic, with exponent $n = 1.15$

8.110 Helium in a piston/cylinder assembly at 20°C and 100 kPa is brought to 400 K in a reversible polytropic process with exponent $n = 1.25$. You may assume helium is an ideal gas with constant specific heat. Find the final pressure and both the specific heat transfer and specific work.

8.111 The power stroke in an internal combustion engine can be approximated with a polytropic expansion. Consider air in a cylinder volume of 0.2 L at 7 MPa and 1800 K, shown in Fig. P8.111.

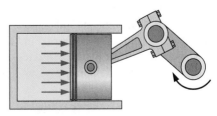

FIGURE P8.111

It now expands in a reversible polytropic process with exponent $n = 1.5$, through a volume ratio of $8:1$. Show this process on $P-v$ and $T-s$ diagrams, and calculate the work and heat transfer for the process.

8.112 A piston/cylinder contains air at 300 K, 100 kPa. It is now compressed in a reversible adiabatic process to a volume seven times as small. Use constant heat capacity and find the final pressure and temperature, the specific work, and specific heat transfer for the process.

8.113 A cylinder/piston device contains carbon dioxide at 1 MPa and 300°C with a volume of 200 L. The total external force acting on the piston is proportional to V^3. This system is allowed to cool to room temperature, 20°C. What is the total entropy generation for the process?

8.114 A device brings 2 kg of ammonia from 150 kPa and $-20°C$ to 400 kPa and 80°C in a polytropic process. Find the polytropic exponent, n, the work, and the heat transfer. Find the total entropy generated assuming a source at 100°C.

8.115 A cylinder/piston setup contains 100 L of air at 110 kPa and 25°C. The air is compressed in a reversible polytropic process to a final state of 800 kPa and 200°C. Assume the heat transfer is with the ambient surroundings at 25°C and determine the polytropic exponent n and the final volume of the air. Find the work done by the air, heat transfer, and total entropy generation for the process.

8.116 A mass of 2 kg of ethane gas at 500 kPa and 100°C undergoes a reversible polytropic expansion with exponent $n = 1.3$ to a final temperature of the ambient surroundings, 20°C. Calculate the total entropy generation for the process if the heat is exchanged with the ambient surroundings.

8.117 A piston/cylinder contains air at 300 K, 100 kPa. A reversible polytropic process with $n = 1.3$ brings the air to 500 K. Any heat transfer if it comes in is from a 325°C reservoir, and if it goes out it is to the ambient at 300 K. Sketch the process in a $P-v$ and a $T-s$ diagram. Find the specific work and specific heat transfer in the process. Find the specific entropy generation (external to the air) in the process.

8.118 A cylinder/piston device contains saturated vapor R-22 at 10°C; the volume is 10 L. The R-22 is compressed to 2 MPa at 60°C in a reversible (internally) polytropic process. If all the heat transfer during the process is with the ambient surroundings at 10°C, calculate the net entropy change.

8.119 A cylinder/piston setup contains air at ambient conditions, 100 kPa and 20°C, with a volume of 0.3 m³. The air is compressed to 800 kPa in a reversible polytropic process with exponent $n = 1.2$, after which it is expanded back to 100 kPa in a reversible adiabatic process.

a. Show the two processes in $P-v$ and $T-s$ diagrams.

b. Determine the final temperature and the net work.

Rates or Fluxes of Entropy

8.120 A reversible heat pump uses 1 kW of power input to heat a 25°C room, drawing energy from the outside at 15°C. Assuming every process is reversible, what are the total rates of entropy into the heat pump from the outside and from the heat pump to the room?

8.121 An amount of power, say 1000 kW, comes from a furnace at 800°C going into water vapor at 400°C. From the water the power goes to solid metal at 200°C and then into some air at 70°C. For each location calculate the flux of s as $(\dot{Q}/T)$. What makes the flux larger and larger?

8.122 Room air at 23°C is heated by a 2000 W space heater with a surface filament temperature of 700 K, shown in Fig. P8.122. The room at steady state loses heat to the outside, which is at 7°C. Find the rate(s) of entropy generation and specify where it is made.

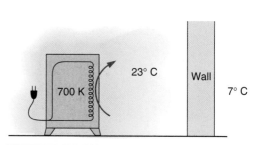

FIGURE P8.122

8.123 A small halogen light bulb receives an electrical power of 50 W. The small filament is at 1000 K and gives out 20% of the power as light and the rest as heat transfer to the gas, which is at 500 K; the glass is at 400 K. All the power is absorbed by the room walls at 25°C. Find the rate of generation of entropy in the filament, in the entire bulb including glass, and in the entire room including the bulb.

8.124 A farmer runs a heat pump using 2 kW of power input. It keeps a chicken hatchery at a constant 30°C, while the room loses 10 kW to the colder outside ambient at 10°C. What is the rate of entropy generated in the heat pump? What is the rate of entropy generated in the heat loss process?

8.125 The automatic transmission in a car receives 25 kW shaft work and gives out 24 kW to the drive shaft. The balance is dissipated in the hydraulic fluid and metal casing, all at 45°C, which in turn transmits it to the outer atmosphere at 20°C. What is the rate of entropy generation inside the transmission unit? What is it outside the unit?

Review Problems

8.126 An insulated cylinder/piston arrangement has an initial volume of 0.15 m³ and contains steam at 400 kPa and 200°C. The steam is expanded adiabatically, and the work output is measured very carefully to be 30 kJ. It is claimed that the final state of the water is in the two-phase (liquid and vapor) region. What is your evaluation of the claim?

8.127 A closed tank, V = 10 L, containing 5 kg of water initially at 25°C, is heated to 175°C by a heat pump that is receiving heat from the surroundings at 25°C. Assume that this process is reversible. Find the heat transfer to the water and the work input to the heat pump.

8.128 Two tanks contain steam, and they are both connected to a piston cylinder, as shown in Fig. P8.128. Initially, the piston is at the bottom, and the mass of the piston is such that a pressure of 1.4 MPa below it will be able to lift it. Steam in A has a mass of 4 kg at 7 MPa and 700°C, and B has 2 kg at 3 MPa and 350°C. The two valves are opened, and the water comes to a uniform state. Find the final temperature and the total entropy generation, assuming no heat transfer.

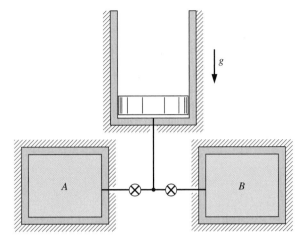

FIGURE P8.128

8.129 A piston/cylinder with constant loading of piston contains 1 L water at 400 kPa, quality 15%. It has some stops mounted so the maximum possible volume is 11 L. A reversible heat pump extracting heat from the ambient at 300 K, 100 kPa, heats the water to 300°C. Find the total work and heat transfer for the water and the work input to the heat pump.

8.130 Water in a piston/cylinder shown in Fig. P8.130 is at 1 MPa, 500°C. There are two stops: a lower one at which V_{min} = 1 m³ and an upper one at V_{max} = 3 m³. The piston is loaded with a mass and outside atmosphere such that it floats when the pressure is 500 kPa. This setup is now cooled to 100°C by rejecting heat to the surroundings at 20°C. Find the total entropy generated in the process.

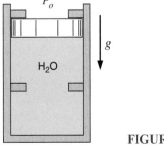

FIGURE P8.130

8.131 A cylinder fitted with a frictionless piston contains water, as shown in Fig. P8.131. A constant

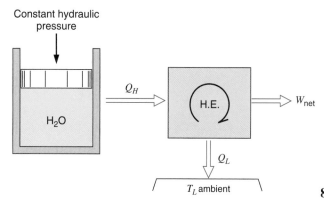

FIGURE P8.131

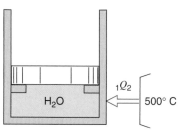

FIGURE P8.134

hydraulic pressure on the back face of the piston maintains a cylinder pressure of 10 MPa. Initially, the water is at 700°C, and the volume is 100 L. The water is now cooled and condensed to saturated liquid. The heat released during this process is the Q supply to a cyclic heat engine that in turn rejects heat to the ambient at 30°C. If the overall process is reversible, what is the net work output of the heat engine?

8.132 A cylinder/piston contains 3 kg of water at 500 kPa, 600°C. The piston has a cross-sectional area of 0.1 m² and is restrained by a linear spring with spring constant 10 kN/m. The setup is allowed to cool down to room temperature due to heat transfer to the room at 20°C. Calculate the total (water and surroundings) change in entropy for the process.

8.133 An insulated cylinder fitted with a frictionless piston contains saturated vapor R-12 at ambient temperature, 20°C. The initial volume is 10 L. The R-12 is now expanded to a temperature of −30°C. The insulation is then removed from the cylinder, allowing it to warm at constant pressure to ambient temperature. Calculate the net work and the net entropy change for the overall process.

8.134 A piston/cylinder assembly contains 2 kg of liquid water at 20°C, 100 kPa, and it is now heated to 300°C by a source at 500°C. A pressure of 1000 kPa will lift the piston off the lower stops, shown in Fig. P8.134. Find the final volume, work, heat transfer, and total entropy generation.

8.135 An uninsulated cylinder fitted with a piston contains air at 500 kPa, 200°C, at which point the volume is 10 L. The external force on the piston is now varied in such a manner that the air expands to 150 kPa, 25 L volume. It is claimed that in this process the air produces 70% of the work that would have resulted from a reversible, adiabatic expansion from the same initial pressure and temperature to the same final pressure. Room temperature is 20°C

a. What is the amount of work claimed?

b. Is this claim possible?

8.136 A cylinder fitted with a piston contains 0.5 kg of R-134a at 60°C, with a quality of 50 percent. The R-134a now expands in an internally reversible polytropic process to ambient temperature 20°C, at which point the quality is 100%. Any heat transfer is with a constant-temperature source, which is at 60°C. Find the polytropic exponent n and show that this process satisfies the second law of thermodynamics.

8.137 A cylinder with a linear spring-loaded piston contains carbon dioxide gas at 2 MPa with a volume of 50 L. The device is of aluminum and has a mass of 4 kg. Everything (Al and gas) is initially at 200°C. By heat transfer the whole system cools to the ambient temperature of 25°C, at which point the gas pressure is 1.5 MPa. Find the total entropy generation for the process.

8.138 A vertical cylinder/piston contains R-22 at −20°C, 70% quality, and the volume is 50 L, shown in Fig. P8.138. This cylinder is brought into a 20°C room, and an electric current of 10 A is passed through a resistor inside the cylinder. The voltage drop across the resistor is 12 V. It is claimed that after 30 min the temperature inside the cylinder is 40°C. Is this possible?

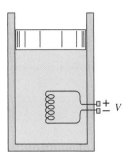

FIGURE P8.138

8.139 A gas in a rigid vessel is at ambient temperature and at a pressure, P_1, slightly higher than ambient pressure, P_0. A valve on the vessel is opened, so gas escapes and the pressure drops quickly to ambient pressure. The valve is closed, and after a long time the remaining gas returns to ambient temperature at which point the pressure is P_2. Develop an expression that allows a determination of the ratio of specific heats, k, in terms of the pressures.

ENGLISH UNIT PROBLEMS

Concept Problems

8.140E Water at 20 psia, 240 F, receives 40 Btu/lbm in a reversible process by heat transfer. Which process changes s the most: constant T, constant v, or constant P?

8.141E Saturated water vapor at 20 psia is compressed to 60 psia in a reversible adiabatic process. Find the change in v and T.

8.142E A computer chip dissipates 2 Btu of electric work over time and rejects that as heat transfer from its 125 F surface to 70 F air. How much entropy is generated in the chip? How much, if any, is generated outside the chip?

8.143E Two 10-lbm blocks of steel, one at 400 F and the other at 70 F, come in thermal contact. Find the final temperature and the total entropy generation in the process.

8.144E One kg of air at 540 R is mixed with one kg air at 720 R in a process at a constant 15 psia and $Q = 0$. Find the final T and the entropy generation in the process.

8.145E One lbm of air at 15 psia is mixed with one lbm air at 30 psia, both at 540 R, in a rigid insulated tank. Find the final state (P, T) and the entropy generation in the process.

8.146E A window receives 600 Btu/h of heat transfer at the inside surface of 70 F and transmits the 600 Btu/h from its outside surface at 36 F, continuing to ambient air at 23 F. Find the flux of entropy at all three surfaces and the window's rate of entropy generation.

English Unit Problems

8.147E Consider the steam power plant in Problem 7.100E and show that this cycle satisfies the inequality of Clausius.

8.148E Find the missing properties and give the phase of the substance.

a. H_2O $s = 1.75$ Btu/lbm R, $h = ?$ $T = ?$
         $P = 4$ lbf/in.^{2} $x = ?$

b. H_2O $u = 1350$ Btu/lbm, $T = ?$ $x = ?$
         $P = 1500$ lbf/in.^{2} $s = ?$

c. R-22 $T = 30$ F, $s = ?$ $x = ?$
        $P = 60$ lbf/in.^{2}

d. R-134a $T = 10$ F, $v = ?$ $s = ?$
         $x = 0.45$

e. NH_3 $T = 60$ F, $u = ?$ $x = ?$
         $s = 1.35$ Btu/lbm R

8.149E In a Carnot engine with water as the working fluid, the high temperature is 450 F, and as Q_H is received, the water changes from saturated liquid to saturated vapor. The water pressure at the low temperature is 14.7 lbf/in.^{2}. Find T_L, cycle thermal efficiency, heat added per pound-mass, and entropy, s, at the beginning of the heat rejection process.

8.150E Consider a Carnot-cycle heat pump with R-22 as the working fluid. Heat is rejected from the R-22 at 100 F, during which process the R-22 changes from saturated vapor to saturated liquid. The heat is transferred to the R-22 at 30 F.

a. Show the cycle on a T–s diagram.

b. Find the quality of the R-22 at the beginning and end of the isothermal heat addition process at 30 F.

c. Determine the coefficient of performance for the cycle.

8.151E Do Problem 8.150 using refrigerant R-134a instead of R-22.

8.152E Water at 30 lbf/in.^{2}, $x = 1.0$ is compressed in a piston/cylinder to 140 lbf/in.^{2}, 600 F in a reversible process. Find the sign for the work and the sign for the heat transfer.

8.153E Two pound-mass of ammonia in a piston/cylinder at 120 F, 150 lbf/in.^{2} is expanded in a reversible adiabatic process to 15 lbf/in.^{2}. Find the work and heat transfer for this process.

8.154E A cylinder fitted with a piston contains ammonia at 120 F, 20% quality with a volume of 60 in.^{3}. The ammonia expands slowly, and during this process heat is transferred to maintain a constant temperature. The process continues until all the liquid is gone. Determine the work and heat transfer for this process.

8.155E One pound-mass of water at 600 F expands against a piston in a cylinder until it reaches ambient pressure, 14.7 lbf/in.^{2}, at which point the water has a quality of 90%. It may be assumed that the expansion is reversible and adiabatic.
a. What was the initial pressure in the cylinder?
b. How much work is done by the water?

8.156E A closed tank, $V = 0.35$ ft^3, containing 10 lbm of water initially at 77 F is heated to 350 F by a heat pump that is receiving heat from the surroundings at 77 F. Assume that this process is reversible. Find the heat transfer to the water and the work input to the heat pump.

8.157E A cylinder containing R-134a at 50 F, 20 lbf/in.^{2} has an initial volume of 1 ft^3. A piston compresses the R-134a in a reversible, isothermal process until it reaches the saturated vapor state. Calculate the required work and heat transfer to accomplish this process.

8.158E A rigid, insulated vessel contains superheated vapor steam at 450 lbf/in.^{2}, 700 F. A valve on the vessel is opened, allowing steam to escape. It may be assumed that the steam remaining inside the vessel goes through a reversible adiabatic expansion. Determine the fraction of steam that has escaped, when the final state inside is saturated vapor.

8.159E An insulated cylinder/piston contains R-134a at 150 lbf/in.^{2}, 120 F, with a volume of 3.5 ft^3. The R-134a expands, moving the piston until the pressure in the cylinder has dropped to 15 lbf/in.^{2}. It is claimed that the R-134a does 180 Btu of work against the piston during the process. Is that possible?

8.160E A mass- and atmosphere-loaded piston/cylinder contains 4 lbm of water at 500 lbf/in.^{2}, 200 F. Heat is added from a reservoir at 1200 F to the water until it reaches 1200 F. Find the work, heat transfer, and total entropy production for the system and surroundings.

8.161E A 1-gallon jug of milk at 75 F is placed in your refrigerator where it is cooled down to the refrigerator's inside temperature of 40 F. Assume the milk has the properties of liquid water and find the entropy generated in the cooling process.

8.162E A cylinder/piston contains water at 30 lbf/in.^{2}, 400 F with a volume of 1 ft^3. The piston is moved slowly, compressing the water to a pressure of 120 lbf/in.^{2}. The loading on the piston is such that the product PV is a constant. Assuming that the room temperature is 70 F, show that this process does not violate the second law.

8.163E One pound-mass of ammonia (NH$_3$) is contained in a linear spring-loaded piston/cylinder as saturated liquid at 0 F. Heat is added from a reservoir at 225 F until a final condition of 125 lbf/in.^{2}, 160 F, is reached. Find the work, heat transfer, and entropy generation, assuming the process is internally reversible.

8.164E A foundry form box with 50 lbm of 400 F hot sand is dumped into a bucket with 2 ft^3 water at 60 F. Assuming no heat transfer with the surroundings and no boiling away of liquid water, calculate the net entropy change for the process.

8.165E Four pounds of liquid lead at 900 F are poured into a form. It then cools at constant pressure down to room temperature at 68 F as heat is transferred to the room. The melting point of lead is 620 F, and the enthalpy change between the phases h_{if} is 10.6 Btu/lbm. The specific heats are in Tables F.2 and F.3. Calculate the net entropy change for this process.

8.166E A hollow steel sphere with a 2-ft inside diameter and a 0.1-in. thick wall contains water at 300 lbf/in.^{2}, 500 F. The system (steel plus water) cools to the ambient temperature, 90 F. Calculate the net entropy change of the system and surroundings for this process.

8.167E Oxygen gas in a piston/cylinder at 500 R and 1 atm with a volume of 1 ft^3 is compressed in a reversible adiabatic process to a final temperature of 1000 R. Find the final pressure and volume using constant heat capacity from Table F.4.

8.168E Oxygen gas in a piston/cylinder at 500 R and 1 atm with a volume of 1 ft^3 is compressed in a reversible adiabatic process to a final temperature of 1000 R. Find the final pressure and volume using Table F.6.

8.169E A handheld pump for a bicycle has a volume of 2 in.^{3} when fully extended. You now press the plunger (piston) in while holding your thumb over the exit hole so an air pressure of 45 lbf/in.^{2} is obtained. The outside atmosphere is at P_0, T_0. Consider two cases: (1) it is done quickly (~1 s), and (2) it is done very slowly (~1 h).

   a. State assumptions about the process for each case.

   b. Find the final volume and temperature for both cases.

8.170E A piston/cylinder contains air at 2500 R, 2200 lbf/in.^{2},with $V_1 = 1$ in.^{3}, $A_{cyl} = 1$ in.^{2}, as shown in Fig. P8.95. The piston is released, and just before the piston exits the end of the cylinder the pressure inside is 30 lbf/in.^{2}. If the cylinder is insulated, what is its length? How much work is done by the air inside?

8.171E A 25-ft^3 insulated, rigid tank contains air at 110 lbf/in.^{2}, 75 F. A valve on the tank is opened, and the pressure inside quickly drops to 15 lbf/in.^{2}, at which point the valve is closed. Assuming that the air remaining inside has undergone a reversible adiabatic expansion, calculate the mass withdrawn during the process.

8.172E A rigid container with volume 7 ft^3 is divided into two equal volumes by a partition. Both sides contain nitrogen, one side is at 300 lbf/in.^{2}, 400 F, and the other at 30 lbf/in.^{2}, 200 F. The partition ruptures, and the nitrogen comes to a uni-

form state at 160 F. Assuming the temperature of the surroundings is 68 F, determine the work done and the net entropy change for the process.

8.173E Nitrogen at 90 lbf/in.^{2}, 260 F, is in a 20 ft^3 insulated tank connected to a pipe with a valve to a second insulated, initially empty tank of volume 20 ft^3. The valve is opened, and the nitrogen fills both tanks. Find the final pressure and temperature and the entropy generation this process causes. Why is the process irreversible?

8.174E Helium in a piston/cylinder at 70 F, 15 lbf/in.^{2}, is brought to 720 R in a reversible polytropic process with exponent $n = 1.25$. You may assume helium is an ideal gas with constant specific heat. Find the final pressure and both the specific heat transfer and specific work.

8.175E A cylinder/piston contains air at ambient conditions, 14.7 lbf/in.^{2} and 70 F, with a volume of 10 ft^3. The air is compressed to 100 lbf/in.^{2} in a reversible polytropic process with exponent, $n = 1.2$, after which it is expanded back to 14.7 lbf/in.^{2} in a reversible adiabatic process.

   a. Show the two processes in P–v and T–s diagrams.

   b. Determine the final temperature and net work.

   c. What is the potential refrigeration capacity (in British thermal units) of the air at the final state?

8.176E A cylinder/piston contains carbon dioxide at 150 lbf/in.^{2}, 600 F, with a volume of 7 ft^3. The total external force acting on the piston is proportional to V^3. This system is allowed to cool to room temperature, 70 F. What is the total entropy generation for the process?

8.177E A cylinder/piston contains 4 ft^3 of air at 16 lbf/in.^{2}, 77 F. The air is compressed in a reversible polytropic process to a final state of 120 lbf/in.^{2}, 400 F. Assume the heat transfer is with the ambient at 77 F and determine the polytropic exponent n and the final volume of the air. Find the work done by the air, heat transfer, and total entropy generation for the process.

8.178E A reversible heat pump uses 1 kW of power input to heat a 78 F room, drawing energy from the outside at 60 F. Assume every process is reversible, what are the total rates of entropy into the heat pump from the outside and from the heat pump to the room?

8.179E A farmer runs a heat pump using 2.5 hp of power input. It keeps a chicken hatchery at a constant 86 F, while the room loses 20 Btu/s to the colder outside ambient at 50 F. What is the rate of entropy generated in the heat pump? What is the rate of entropy generated in the heat loss process?

8.180E A cylinder/piston contains 5 lbm of water at 80 lbf/in.^{2}, 1000 F. The piston has a cross-sectional area of 1 ft^2 and is restrained by a linear spring with spring constant 60 lbf/in. The setup is allowed to cool down to room temperature due to heat transfer to the room at 70 F. Calculate the total (water and surroundings) change in entropy for the process.

8.181E Water in a piston/cylinder is at 150 lbf/in.^{2}, 900 F, as shown in Fig. P8.130. There are two stops: a lower one at which $V_{min} = 35$ ft^3 and an upper one at $V_{max} = 105$ ft^3. The piston is loaded with a mass and outside atmosphere such that it floats when the pressure is 75 lbf/in.^{2}. This setup is now cooled to 210 F by rejecting heat to the surroundings at 70 F. Find the total entropy generated in the process.

8.182E A cylinder with a linear spring-loaded piston contains carbon dioxide gas at 300 lbf/in.^{2} with a volume of 2 ft^3. The device is of aluminum and has a mass of 8 lbm. Everything (Al and gas) is initially at 400 F. By heat transfer the whole system cools to the ambient temperature of 77 F, at which point the gas pressure is 220 lbf/in.^{2}. Find the total entropy generation for the process.

COMPUTER, DESIGN, AND OPEN-ENDED PROBLEMS

8.183 Write a computer program to solve Problem 8.78 using constant specific heat for both the sand and the liquid water. Let the amount and the initial temperatures be input variables.

8.184 Write a program to solve Problem 8.81 with the thermal storage rock bed in Problem 7.68. Let the size and temperatures be input variables so that the heat engine work output can be studied as a function of the system parameters.

8.185 Write a program to solve the following problem. One of the gases listed in Table A.6 undergoes a reversible adiabatic process in a cylinder from P_1, T_1 to P_2. We wish to calculate the final temperature and the work for the process by three methods:

a. Integrating the specific heat equation.

b. Assuming constant specific heat at temperature, T_1.

c. Assuming constant specific heat at the average temperature (by iteration).

8.186 Write a program to solve Problem 8.87. Let the initial state and the expansion ratio be input variables.

8.187 Write a program to solve a problem similar to Problem 8.107, but instead of the ideal gas tables use the formula for the specific heat as a function of temperature in Table A.6.

8.188 Write a program to study a general polytropic process in an ideal gas with constant specific heat. Take Problem 8.106 as an example.

8.189 Write a program to solve the general case of Problem 8.111, in which the initial state and the expansion ratio are input variables.

8.190 A piston/cylinder maintaining constant pressure contains 0.5 kg of water at room temperature 20°C and 100 kPa. An electric heater of 500 W heats the water up to 500°C. Assume no heat losses to the ambient and plot the temperature and total accumulated entropy production as a function of time. Investigate the first part of the process, namely, bringing the water to the boiling point, by measuring it in your kitchen and knowing the rate of power added.

8.191 Air in a piston/cylinder is used as a small air-spring that should support a steady load of 200 N. Assume that the load can vary with ±10% over a period of 1 s and that the displacement should be limited to ±0.01 m. For some choice of sizes show the spring displacement, x, as a function of load and compare that to an elastic linear coil spring designed for the same conditions.

8.192 Consider a piston/cylinder arrangement with ammonia at −10°C, 50 kPa that is compressed to

200 kPa. Examine the effect of heat transfer to/from the ambient at 15°C on the process and the required work. Some limiting processes are a reversible adiabatic compression giving an exit temperature of about 90°C and as mentioned in the text an isothermal compression. Evaluate the work and heat transfer for both cases and for cases in between assuming a polytropic process. Which processes are actually possible and how would they proceed?

9 SECOND-LAW ANALYSIS FOR A CONTROL VOLUME

In the preceding two chapters we discussed the second law of thermodynamics and the thermodynamic property entropy. As was done with the first-law analysis, we now consider the more general application of these concepts, the control volume analysis, and a number of cases of special interest. We will also discuss usual definitions of thermodynamic efficiencies.

9.1 THE SECOND LAW OF THERMODYNAMICS FOR A CONTROL VOLUME

The second law of thermodynamics can be applied to a control volume by a procedure similar to that used in Section 6.1, where the first law was developed for a control volume. We start with the second law expressed as a change of the entropy for a control mass in a rate form from Eq. 8.43,

$$\frac{dS_{\text{c.m.}}}{dt} = \sum \frac{\dot{Q}}{T} + \dot{S}_{\text{gen}} \qquad (9.1)$$

to which we now will add the contributions from the mass flow rates in and out of the control volume, A simple example of such a situation is illustrated in Fig. 9.1. The flow of mass does carry an amount of entropy, s, per unit mass flowing, but it does not give rise to any other contributions. As a process may take place in the flow, entropy can be generated, but this is attributed to the space it belongs to (i.e., either inside or outside of the control volume).

The balance of entropy as an equation then states that the rate of change in total entropy inside the control volume is equal to the net sum of fluxes across the control surface plus the generation rate. That is,

$$\text{rate of change} = + \text{in} - \text{out} + \text{generation}$$

or

$$\frac{dS_{\text{c.v.}}}{dt} = \sum \dot{m}_i s_i - \sum \dot{m}_e s_e + \sum \frac{\dot{Q}_{\text{c.v.}}}{T} + \dot{S}_{\text{gen}} \qquad (9.2)$$

These fluxes are mass flow rates carrying a level of entropy and the rate of heat transfer that takes place at a certain temperature (the temperature right at the control surface). The

302

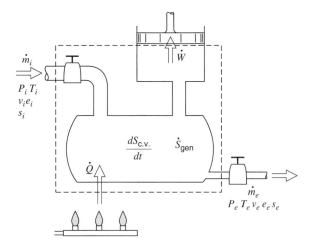

FIGURE 9.1 The
entropy balance for a
control volume on a rate
form.

accumulation and generation terms cover the total control volume and are expressed in
the lumped (integral form) so that

$$S_{c.v.} = \int \rho s \, dV = m_{c.v.}s = m_A s_A + m_B s_B + m_C s_C + \cdots$$

$$\dot{S}_{gen} = \int \rho \dot{s}_{gen} \, dV = \dot{S}_{gen.A} + \dot{S}_{gen.B} + \dot{S}_{gen.C} + \cdots \qquad (9.3)$$

If the control volume has several different accumulation units with different fluid states
and processes occurring in them, we may have to sum the various contributions over the
different domains. If the heat transfer is distributed over the control surface, then an integral has to be done over the total surface area using the local temperature and rate of heat
transfer per unit area, $\dot{Q}/A$, as

$$\sum \frac{\dot{Q}_{c.v.}}{T} = \int \frac{d\dot{Q}}{T} = \int_{surface} \frac{(\dot{Q}/A)}{T} \, dA \qquad (9.4)$$

These distributed cases typically require a much more detailed analysis, which is beyond
the scope of the current presentation of the second law.

The generation term(s) in Eq. 9.2 from a summation of individual positive internal-
irreversibility entropy-generation terms in Eq. 9.3 is necessarily positive (or zero), such
that an inequality is often written as

$$\frac{dS_{c.v.}}{dt} \geq \sum \dot{m}_i s_i - \sum \dot{m}_e s_e + \sum \frac{\dot{Q}_{c.v.}}{T} \qquad (9.5)$$

Now the equality applies to internally reversible processes and the inequality to internally
irreversible processes. The form of the second law in Eq. 9.2 or 9.5 is general, such that
any particular case results in a form that is a subset (simplification) of this form. Examples of various classes of problems are illustrated in the following sections.

If there is no mass flow into or out of the control volume, it simplifies to a control
mass and the equation for the total entropy reverts back to Eq. 8.43. Since that version of
the second law has been covered in Chapter 8 here we will consider the remaining cases
that were done for the first law of thermodynamics in Chapter 6.

9.2 The Steady-State Process and the Transient Process

We now consider in turn the application of the second-law control volume equation, Eq. 9.2 or 9.5, to the two control volume model processes developed in Chapter 6.

Steady-State Process

For the steady-state process, which has been defined in Section 6.3, we conclude that there is no change with time of the entropy per unit mass at any point within the control volume, and therefore the first term of Eq. 9.2 equals zero. That is,

$$\frac{dS_{c.v.}}{dt} = 0 \tag{9.6}$$

so that, for the steady-state process,

$$\sum \dot{m}_e s_e - \sum \dot{m}_i s_i = \sum_{c.v.} \frac{\dot{Q}_{c.v.}}{T} + \dot{S}_{gen} \tag{9.7}$$

in which the various mass flows, heat transfer and entropy generation, rates, and states are all constant with time.

If in a steady-state process there is only one area over which mass enters the control volume at a uniform rate and only one area over which mass leaves the control volume at a uniform rate, we can write

$$\dot{m}(s_e - s_i) = \sum_{c.v.} \frac{\dot{Q}_{c.v.}}{T} + \dot{S}_{gen} \tag{9.8}$$

and dividing the mass flow rate out gives

$$s_e = s_i + \sum \frac{q}{T} + s_{gen}$$

Since s_{gen} is always greater than or equal to zero, for an adiabatic process it follows that

$$s_e = s_i + s_{gen} \geq s_i \tag{9.9}$$

where the equality holds for a reversible adiabatic process.

EXAMPLE 9.1   Steam enters a steam turbine at a pressure of 1 MPa, a temperature of 300°C, and a velocity of 50 m/s. The steam leaves the turbine at a pressure of 150 kPa and a velocity of 200 m/s. Determine the work per kilogram of steam flowing through the turbine, assuming the process to be reversible and adiabatic.

Control volume:	Turbine.
Sketch:	Fig. 9.2.
Inlet state:	Fixed (Fig. 9.2).
Exit state:	P_e, $\mathbf{V}_e$ known.
Process:	Steady state, reversible and adiabatic.
Model:	Steam tables.

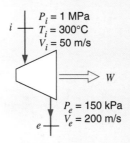
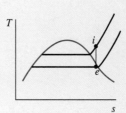

FIGURE 9.2 Sketch for Example 9.1.

Analysis

The continuity equation gives us

$$\dot{m}_e = \dot{m}_i = \dot{m}$$

From the first law we have

$$h_i + \frac{\mathbf{V}_i^2}{2} = h_e + \frac{\mathbf{V}_e^2}{2} + w$$

and the second law is

$$s_e = s_i$$

Solution

From the steam tables, we get

$$h_i = 3051.2 \text{ kJ/kg}, \qquad s_i = 7.1228 \text{ kJ/kg K}$$

The two properties known in the final state are pressure and entropy:

$$P_e = 0.15 \text{ MPa}, \qquad s_e = s_i = 7.1228 \text{ kJ/kg K}$$

The quality and enthalpy of the steam leaving the turbine can be determined as follows:

$$s_e = 7.1228 = s_f + x_e s_{fg} = 1.4335 + x_e 5.7897$$

$$x_e = 0.9827$$

$$h_e = h_f + x_e h_{fg} = 467.1 + 0.9827(2226.5)$$

$$= 2655.0 \text{ kJ/kg}$$

Therefore, the work per kilogram of steam for this isentropic process may be found using the equation for the first law:

$$w = 3051.2 + \frac{50 \times 50}{2 \times 1000} - 2655.0 - \frac{200 \times 200}{2 \times 1000} = 377.5 \text{ kJ/kg}$$

EXAMPLE 9.2 Consider the reversible adiabatic flow of steam through a nozzle. Steam enters the nozzle at 1 MPa and 300°C, with a velocity of 30 m/s. The pressure of the steam at the nozzle

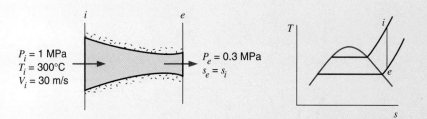

FIGURE 9.3 Sketch for Example 9.2.

exit is 0.3 MPa. Determine the exit velocity of the steam from the nozzle, assuming a reversible, adiabatic, steady-state process.

Control volume:	Nozzle.
Sketch:	Fig. 9.3.
Inlet state:	Fixed (Fig. 9.3).
Exit State:	P_e known.
Process:	Steady state, reversible, and adiabatic.
Model:	Steam tables.

Analysis

Because this is a steady-state process in which the work, heat transfer, and changes in potential energy are zero, we can write

Continuity equation: $\dot{m}_e = \dot{m}_i = \dot{m}$

First law: $h_i + \dfrac{\mathbf{V}_i^2}{2} = h_e + \dfrac{\mathbf{V}_e^2}{2}$

Second law: $s_e = s_i$

Solution

From the steam tables, we have

$$h_i = 3051.2 \text{ kJ/kg}, \qquad s_i = 7.1228 \text{ kJ/kg K}$$

The two properties that we know in the final state are entropy and pressure:

$$s_e = s_i = 7.1228 \text{ kJ/kg K}, \qquad P_e = 0.3 \text{ MPa}$$

Therefore,

$$T_e = 159.1°C, \qquad h_e = 2780.2 \text{ kJ/kg}$$

Substituting into the equation for the first law, we have

$$\frac{\mathbf{V}_e^2}{2} = h_i - h_e + \frac{\mathbf{V}_i^2}{2}$$

$$= 3051.2 - 2780.2 + \frac{30 \times 30}{2 \times 1000} = 271.5 \text{ kJ/kg}$$

$$\mathbf{V}_e = \sqrt{2000 \times 271.5} = 737 \text{ m/s}$$

EXAMPLE 9.2E Consider the reversible adiabatic flow of steam through a nozzle. Steam enters the nozzle at 100 lbf/in.^{2}, 500 F, with a velocity of 100 ft/s. The pressure of the steam at the nozzle exit is 40 lbf/in.^{2}. Determine the exit velocity of the steam from the nozzle, assuming a reversible adiabatic, steady-state process.

Control volume: Nozzle.
Sketch: Fig. 9.3E.
Inlet state: Fixed (Fig. 9.3E).
Exit state: P_e known.
Process: Steady state, reversible, and adiabatic.
Model: Steam tables.

Analysis

Because this is a steady-state process in which the work, the heat transfer, and changes in potential energy are zero, we can write

Continuity equation: $\dot{m}_e = \dot{m}_i = \dot{m}$

First law: $h_i + \dfrac{V_i^2}{2} = h_e + \dfrac{V_e^2}{2}$

Second law: $s_e = s_i$

Solution

From the steam tables, we have

$$h_i = 1279.1 \text{ Btu/lbm} \qquad s_i = 1.7085 \text{ Btu/lbm R}$$

The two properties that we know in the final state are entropy and pressure.

$$s_e = s_i = 1.7085 \text{ Btu/lbm R}, P_e = 40 \text{ lbf/in.}^2$$

Therefore,

$$T_e = 314.2 \text{ F} \qquad h_e = 1193.9 \text{ Btu/lbm}$$

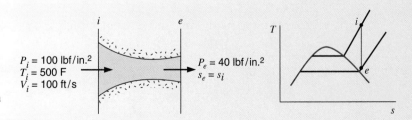

FIGURE 9.3E Sketch for Example 9.2E.

Substituting into the equation for the first law, we have

$$\frac{\mathbf{V}_e^2}{2} = h_i - h_e + \frac{\mathbf{V}_i^2}{2}$$

$$= 1279.1 - 1193.9 + \frac{100 \times 100}{2 \times 32.17 \times 778} = 85.4 \text{ Btu/lbm}$$

$$\mathbf{V}_e = \sqrt{2 \times 32.17 \times 778 \times 85.4} = 2070 \text{ ft/s}$$

EXAMPLE 9.3

An inventor reports having a refrigeration compressor that receives saturated R-134a vapor at $-20°C$ and delivers the vapor at 1 MPa and 40°C. The compression process is adiabatic. Does the process described violate the second law?

Control volume:	Compressor.
Inlet state:	Fixed (saturated vapor at T_i).
Exit state:	Fixed (P_e, T_e known).
Process:	Steady state, adiabatic.
Model:	R-134a tables.

Analysis

Because this is a steady-state adiabatic process, we can write the second law as

$$s_e \geq s_i$$

Solution

From the R-134a tables, we read

$$s_e = 1.7148 \text{ kJ/kg K}, \qquad s_i = 1.7395 \text{ kJ/kg K}$$

Therefore, $s_e < s_i$, whereas for this process the second law requires that $s_e \geq s_i$. The process described involves a violation of the second law and thus would be impossible.

EXAMPLE 9.4

An air compressor in a gas station, see Fig. 9.4, takes in a flow of ambient air at 100 kPa, 290 K, and compresses it to 1000 kPa in a reversible adiabatic process. We want to know the specific work required and the exit air temperature.

Solution

C.V. air compressor, steady state, single flow through it, and we assume adiabatic $\dot{Q} = 0$.

Continuity Eq. 6.11:	$\dot{m}_i = \dot{m}_e = \dot{m}$,
Energy Eq. 6.12:	$\dot{m}h_i = \dot{m}h_e + \dot{W}_C$,
Entropy Eq. 9.8:	$\dot{m}s_i + \dot{S}_{gen} = \dot{m}s_e$
Process:	Reversible $\dot{S}_{gen} = 0$

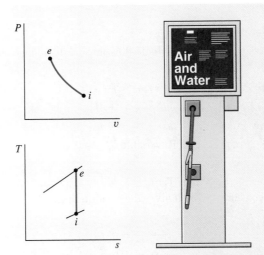

FIGURE 9.4 Diagram for Example 9.4.

Use constant specific heat from Table A.5, $C_{P0} = 1.004$ kJ/kg K, $k = 1.4$. Entropy equation gives constant s, which gives the relation in Eq. 8.32:

$$s_i = s_e \Rightarrow T_e = T_i \left(\frac{P_e}{P_i}\right)^{\frac{k-1}{k}}$$

$$T_e = 290 \left(\frac{1000}{100}\right)^{0.2857} = 559.9 \text{ K}$$

The energy equation per unit mass gives the work term

$$w_c = h_i - h_e = C_{P0}(T_i - T_e) = 1.004(290 - 559.9) = -271 \text{ kJ/kg}$$

EXAMPLE 9.4E An air compressor in a gas station, see Fig. 9.4E, takes in a flow of ambient air at 14.7 lbf/in.^{2}, 520 R, and compresses it to 147 lbf/in.^{2} in a reversible adiabatic process. We want to know the specific work required and the exit air temperature.

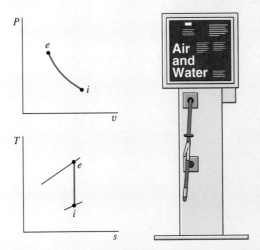

FIGURE 9.4E Diagram for Example 9.4E.

Solution

C.V. air compressor, steady state, single flow through it, and we assume adiabatic $\dot{Q} = 0$.

$$
\begin{aligned}
\text{Continuity Eq. 6.11:} \quad & \dot{m}_i = \dot{m}_e = \dot{m}, \\
\text{Energy Eq. 6.12:} \quad & \dot{m}h_i = \dot{m}h_e + \dot{W}_C, \\
\text{Entropy Eq. 9.8:} \quad & \dot{m}s_i + \dot{S}_{gen} = \dot{m}s_e \\
\text{Process:} \quad & \text{Reversible } \dot{S}_{gen} = 0
\end{aligned}
$$

Use constant specific heat from Table F.4, $C_{P0} = 0.24$ Btu/lbm R, $k = 1.4$. The entropy equation gives constant s, which gives the relation in Eq. 8.32:

$$
s_i = s_e \Rightarrow T_e = T_i \left(\frac{P_e}{P_i} \right)^{\frac{k-1}{k}}
$$

$$
T_e = 520 \left(\frac{147}{14.7} \right)^{0.2857} = 1003.9 \text{ R}
$$

The energy equation per unit mass gives the work term

$$
w_c = h_i - h_e = C_{P0}(T_i - T_e) = 0.24(520 - 1003.9) = -116.1 \text{ Btu/lbm}
$$

EXAMPLE 9.5

A de-superheater works by injecting liquid water into a flow of superheated steam. With 2 kg/s at 300 kPa, 200°C, steam flowing in, what mass flow rate of liquid water at 20°C should be added to generate saturated vapor at 300 kPa? We also want to know the rate of entropy generation in the process.

Solution

C.V. De-superheater, see Fig. 9.5, no external heat transfer, and no work.

$$
\begin{aligned}
\text{Continuity Eq. 6.9:} \quad & \dot{m}_1 + \dot{m}_2 = \dot{m}_3; \\
\text{Energy Eq. 6.10:} \quad & \dot{m}_1 h_1 + \dot{m}_2 h_2 = \dot{m}_3 h_3 = (\dot{m}_1 + \dot{m}_2)h_3 \\
\text{Entropy Eq. 9.7:} \quad & \dot{m}_1 s_1 + \dot{m}_2 s_2 = \dot{S}_{gen} = \dot{m}_3 s_3 \\
\text{Process:} \quad & P = \text{constant}, \; \dot{W} = 0, \text{ and } \dot{Q} = 0
\end{aligned}
$$

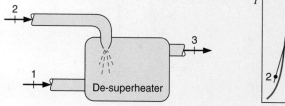 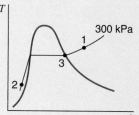

FIGURE 9.5 Sketch and diagram for Example 9.5.

All the states are specified (approximate state 2 with saturated liquid 20°C)

B.1.3: $h_1 = 2865.54 \frac{kJ}{kg}$, $s_1 = 7.3115 \frac{kJ}{kg\ K}$; $h_3 = 2725.3 \frac{kJ}{kg}$, $s_3 = 6.9918 \frac{kJ}{kg\ K}$

B.1.2: $h_2 = 83.94 \frac{kJ}{kg}$, $s_2 = 0.2966 \frac{kJ}{kg\ K}$

Now we can solve for the flow rate $\dot{m}_2$ from the energy equation, having eliminated $\dot{m}_3$ by the continuity equation

$$\dot{m}_2 = \dot{m}_1 \frac{h_1 - h_3}{h_3 - h_2} = 2\frac{2865.54 - 2725.3}{2725.3 - 83.94} = 0.1062 \text{ kg/s}$$
$$\dot{m}_3 = \dot{m}_1 + \dot{m}_2 = 2.1062 \text{ kg/s}$$

Generation is from the entropy equation

$$\dot{S}_{gen} = \dot{m}_3 s_3 - \dot{m}_1 s_1 - \dot{m}_2 s_2$$
$$\dot{S}_{gen} = 2.1062 \times 6.9918 - 2 \times 7.3115 - 0.1062 \times 0.2966 = 0.072 \text{ kW/K}$$

Transient Process

For the transient process, which was described in Section 6.5, the second law for a control volume, Eq. 9.2, can be written in the following form:

$$\frac{d}{dt}(ms)_{c.v.} = \sum \dot{m}_i s_i - \sum \dot{m}_e s_e + \sum \frac{\dot{Q}_{c.v.}}{T} + \dot{S}_{gen} \qquad (9.10)$$

If this is integrated over the time interval t, we have

$$\int_0^t \frac{d}{dt}(ms)_{c.v.}\, dt = (m_2 s_2 - m_1 s_1)_{c.v.}$$

$$\int_0^t \left(\sum \dot{m}_i s_i\right) dt = \sum m_i s_i, \qquad \int_0^t \left(\sum \dot{m}_e s_e\right) dt = \sum m_e s_e, \qquad \int_0^t \dot{S}_{gen}\, dt = {}_1 S_{2gen}$$

Therefore, for this period of time t, we can write the second law for the transient process as

$$(m_2 s_2 - m_1 s_1)_{c.v.} = \sum m_i s_i - \sum m_e s_e + \int_0^t \sum_{c.v.} \frac{\dot{Q}_{c.v.}}{T}\, dt + {}_1 S_{2gen} \qquad (9.11)$$

Since in this process the temperature is uniform throughout the control volume at any instant of time, the integral on the right reduces to

$$\int_0^t \sum_{c.v.} \frac{\dot{Q}_{c.v.}}{T} \, dt = \int_0^t \frac{1}{T} \sum_{c.v.} \dot{Q}_{c.v.} \, dt = \int_0^t \frac{\dot{Q}_{c.v.}}{T} \, dt$$

and therefore the second law for the transient process can be written

$$(m_2 s_2 - m_1 s_1)_{c.v.} = \sum m_i s_i - \sum m_e s_e + \int_0^t \frac{\dot{Q}_{c.v.}}{T} \, dt + {}_1 S_{2\text{gen}} \qquad (9.12)$$

EXAMPLE 9.6 Assume an air tank has 40 L of 100 kPa air at ambient temperature 17°C. The adiabatic and reversible compressor is started so that it charges the tank up to a pressure of 1000 kPa and then it shuts off. We want to know how hot the air in the tank gets and the total amount of work required to fill the tank.

Solution

C.V. compressor and air tank in Fig. 9.6.

Continuity Eq. 6.15: $m_2 - m_1 = m_{\text{in}}$
Energy Eq. 6.16: $m_2 u_2 - m_1 u_1 = {}_1 Q_2 - {}_1 W_2 + m_{\text{in}} h_{\text{in}}$
Entropy Eq. 9.12: $m_2 s_2 - m_1 s_1 = \int dQ/T + {}_1 S_{2\text{gen}} + m_{\text{in}} s_{\text{in}}$
Process: Adiabatic ${}_1 Q_2 = 0$,   Process ideal ${}_1 S_{2\text{gen}} = 0$, $s_1 = s_{\text{in}}$
$\Rightarrow m_2 s_2 = m_1 s_1 + m_{\text{in}} s_{\text{in}} = (m_1 + m_{\text{in}}) s_1 = m_2 s_1 \Rightarrow s_2 = s_1$
Constant $s \Rightarrow$ Eq. 8.28 $s_{T2}^0 = s_{T1}^0 + R \ln(P_2/P_i)$
$s_{T2}^0 = 6.83521 + 0.287 \ln (10) = 7.49605$ kJ/kg K
Interpolate in Table A.7 $\Rightarrow T_2 = 555.7$ K, $u_2 = 401.49$ kJ/kg
$m_1 = P_1 V_1/RT_1 = 100 \times 0.04/(0.287 \times 290) = 0.04806$ kg
$m_2 = P_2 V_2/RT_2 = 1000 \times 0.04/(0.287 \times 555.7) = 0.2508$ kg
$\Rightarrow m_{\text{in}} = 0.2027$ kg
${}_1 W_2 = m_{\text{in}} h_{\text{in}} + m_1 u_1 - m_2 u_2$
$= m_{\text{in}}(290.43) + m_1(207.19) - m_2(401.49) = -31.9$ kJ

Remark: The high final temperature makes the assumption of zero heat transfer poor. The charging process does not happen rapidly so there will be a heat transfer loss. We need to know this to make a better approximation about the real process.

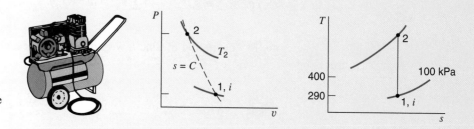

FIGURE 9.6 Sketch and diagram for Example 9.6.

9.3 THE REVERSIBLE STEADY-STATE PROCESS

An expression can be derived for the work in a reversible, adiabatic, steady-state process that is of great help in understanding its significant variables. We have noted that when a steady-state process involves a single flow of fluid into and out of the control volume, the first law, Eq. 6.13, can be written.

$$q + h_i + \frac{\mathbf{V}_i^2}{2} + gZ_i = h_e + \frac{\mathbf{V}_e^2}{2} + gZ_e + w$$

and the second law, Eq. 9.8, is

$$\dot{m}(s_e - s_i) = \sum_{c.v.} \frac{\dot{Q}_{c.v.}}{T} + \dot{S}_{gen}$$

Let us now consider two types of flow, a reversible adiabatic process and a reversible isothermal process.

If the process is reversible and adiabatic, the second-law equation reduces to

$$s_e = s_i$$

It follows from the property relation

$$T\,ds = dh - v\,dP$$

that

$$h_e - h_i = \int_i^e v\,dP \tag{9.13}$$

Substituting these relations into Eq. 6.13 and noting that $q = 0$, we have for the reversible, adiabatic process

$$w = (h_i - h_e) + \frac{\mathbf{V}_i^2 - \mathbf{V}_e^2}{2} + g(Z_i - Z_e)$$

$$= -\int_i^e v\,dP + \frac{\mathbf{V}_i^2 - \mathbf{V}_e^2}{2} + g(Z_i - Z_e) \tag{9.14}$$

If, instead, the process is reversible and isothermal, the second law reduces to

$$m(s_e - s_i) = \frac{1}{T}\sum_{c.v.} \dot{Q}_{c.v.} = \frac{\dot{Q}_{c.v.}}{T} \tag{9.15}$$

or

$$T(s_e - s_i) = \frac{\dot{Q}_{c.v.}}{\dot{m}} = q$$

and the property relation can be integrated to give

$$T(s_e - s_i) = (h_e - h_i) - \int_I^e v\,dP \tag{9.16}$$

Substituting Eqs. 9.15 and 9.16 into the first law, Eq. 6.13, gives us the same expression as for the reversible adiabatic process, Eq. 9.14. We further note that any other reversible process can be constructed, in the limit, from a series of alternate adiabatic and isothermal processes. Thus, we may conclude that Eq. 9.14 is valid for any reversible, steady-state process without the restriction that it be either adiabatic or isothermal.

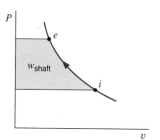

FIGURE 9.7 Shaft work from Eq. 9.18.

This expression has a wide range of application. If we consider a reversible steady-state process in which the work is zero (such as flow through a nozzle) and the fluid is incompressible (v = constant), Eq. 9.14 can be integrated to give

$$v(P_e - P_i) + \frac{\mathbf{V}_e^2 - \mathbf{V}_i^2}{2} + g(Z_e - Z_i) = 0 \tag{9.17}$$

This equation, known as the Bernoulli equation (after Daniel Bernoulli), is very important in fluid mechanics.

Equation 9.14 is also frequently applied to the large class of flow processes involving work (such as turbines and compressors) in which changes in kinetic and potential energies of the working fluid are small. The model process for these machines is then a reversible, steady-state process with no change in kinetic or potential energy (and commonly, though not necessarily, adiabatic as well). For this process Eq. 9.14 reduces to the form

$$w = -\int_i^e v \, dP \tag{9.18}$$

From this result, we conclude that the shaft work associated with this type of process is given by the area shown in the diagram of Fig. 9.7. It is important to note that this result applies to a very special case—the area $\int_i^e v \, dP$ is not the same as the area $\int_i^e P \, dv$—and is applicable only in entirely different circumstances. We also note that the shaft work associated with this type of process is closely related to the specific volume of the fluid during the process. To amplify this point further, consider the simple steam power plant shown in Fig. 9.8. Suppose that this is an ideal power plant with no pressure

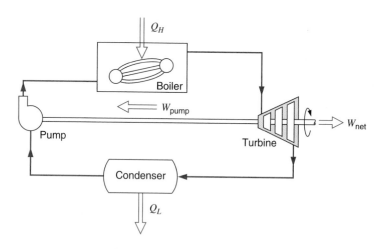

FIGURE 9.8 Simple steam power plant.

drop in the piping, the boiler, or the condenser. Thus, the pressure increase in the pump is equal to the pressure decrease in the turbine. Neglecting kinetic and potential energy changes, the work done in each of these processes is given by Eq. 9.18. Since the pump handles liquid, which has a very small specific volume compared to that of the vapor that flows through the turbine, the power input to the pump is much less than the power output of the turbine. The difference is the net power output of the power plant.

This same line of reasoning can be qualitatively applied to actual devices that involve steady-state processes, even though the processes are not exactly reversible and adiabatic.

EXAMPLE 9.7   Calculate the work per kilogram to pump water isentropically from 100 kPa and 30°C to 5 MPa.

$$
\begin{aligned}
\textit{Control volume:} &\quad \text{Pump.}\\
\textit{Inlet state:} &\quad P_i, T_i \text{ known; state fixed.}\\
\textit{Exit state:} &\quad P_e \text{ known.}\\
\textit{Process:} &\quad \text{Steady-state, isentropic.}\\
\textit{Model:} &\quad \text{Steam tables.}
\end{aligned}
$$

Analysis

Since the process is steady, state, reversible, and adiabatic, and because changes in kinetic and potential energies can be neglected, we have

$$\text{First law: } h_i = h_e + w$$

$$\text{Second law: } s_e - s_i = 0$$

Solution

Since P_e and s_e are known, state e is fixed and therefore h_e is known and w can be found from the first law. However, the process is reversible and steady state, with negligible changes in kinetic and potential energies, so that Eq. 9.18 is also valid. Furthermore, since a liquid is being pumped, the specific volume will change very little during the process.

From the steam tables, $v_i = 0.001\ 004$ m³/kg. Assuming that the specific volume remains constant and using Eq. 9.18, we have

$$-w = \int_1^2 v\, dP = v(P_2 - P_1) = 0.001\ 004(5000 - 100) = 4.92 \text{ kJ/kg}$$

As a final application of Eq. 9.14, we recall the reversible polytropic process for an ideal gas, discussed in Section 8.11 for a control mass process. For the steady-state process with no change in kinetic and potential energies, we have the relations

$$w = -\int_i^e v\, dP \quad \text{and} \quad Pv^n = \text{constant} = C^n$$

$$w = -\int_i^e v\, dP = -C\int_i^e \frac{dP}{P^{1/n}}$$

$$= -\frac{n}{n-1}(P_e v_e - P_i v_i) = -\frac{nR}{n-1}(T_e - T_i) \qquad (9.19)$$

If the process is isothermal, then $n = 1$ and the integral becomes

$$w = -\int_i^e v \, dP = -\text{constant} \int_i^e \frac{dP}{P} = -P_i v_i \ln \frac{P_e}{P_i} \quad (9.20)$$

Note that the P–v and T–s diagrams of Fig. 8.17 are applicable to represent the slope of polytropic processes in this case as well.

These evaluations of the integral

$$\int_i^e v \, dP$$

may also be used in conjunction with Eq. 9.14 for instances in which kinetic and potential energy changes are not negligibly small.

9.4 PRINCIPLE OF THE INCREASE OF ENTROPY

The principle of the increase of entropy for a control mass analysis was discussed in Section 8.8. The same general conclusion is reached for a control volume analysis. To demonstrate this, consider a control volume, Fig. 9.9, that exchanges both mass and heat with its surroundings. At the point in the surroundings where the heat transfer occurs, the temperature is T_0. From Eq. 9.5, the second law for this process is

$$\frac{dS_{c.v.}}{dt} + \sum \dot{m}_e s_e - \sum \dot{m}_i s_i \geq \sum_{c.v.} \frac{\dot{Q}_{c.v.}}{T}$$

We recall that the first term represents the rate of change of entropy within the control volume, and the next terms the net entropy flow out of the control volume resulting from the mass flow. Therefore, for the surroundings, we can write

$$\frac{dS_{surr}}{dt} = \sum \dot{m}_e s_e - \sum \dot{m}_i s_i - \frac{\dot{Q}_{c.v.}}{T_0} \quad (9.21)$$

Adding Eqs. 9.5 and 9.21, we have

$$\frac{dS_{net}}{dt} = \frac{dS_{c.v.}}{dt} + \frac{dS_{surr}}{dt} \geq \sum_{c.v.} \frac{\dot{Q}_{c.v.}}{T} - \frac{\dot{Q}_{c.v.}}{T_0} \quad (9.22)$$

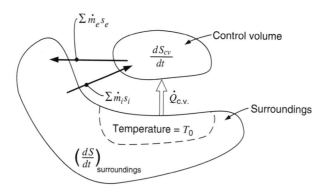

FIGURE 9.9 Entropy change for a control volume plus surroundings.

Because $\dot{Q}_{c.v.} > 0$ when $T_0 > T$ and $\dot{Q}_{c.v.} < 0$ when $T_0 < T$, it follows that

$$\frac{dS_{net}}{dt} = \frac{dS_{c.v.}}{dt} + \frac{dS_{surr}}{dt} = \sum \dot{S}_{gen} \geq 0 \qquad (9.23)$$

which can be termed the general statement of the principle of the increase of entropy.

When we use Eq. 9.23 to check any particular process for a possible violation of the second law, it will be in connection with one of our model processes. For example, in a steady-state process, as we consider the two terms in Eq. 9.23, we realize that, in accordance with Eq. 9.6, the first term is zero. As a result, all the entropy change that is due to irreversibilities in this type of process is observed in the surroundings. This term may then be evaluated using Eq. 9.21. In contrast, for the transient process, there are both control volume and surroundings terms to evaluate. Each term is integrated over the time t of the process, as was done in Section 9.2. Thus, Eq. 9.23 is integrated to

$$\Delta S_{net} = \Delta S_{c.v.} + \Delta S_{surr} \qquad (9.24)$$

in which the control volume term is

$$\Delta S_{c.v.} = (m_2 s_2 - m_1 s_1)_{c.v.} \qquad (9.25)$$

The term for the surroundings is, after applying Eq. 9.21 to the surroundings and integrating,

$$\Delta S_{surr} = \frac{-Q_{c.v.}}{T_0} + \sum m_e s_e - \sum m_i s_i \qquad (9.26)$$

9.5 EFFICIENCY

In Chapter 7 we noted that the second law of thermodynamics led to the concept of thermal efficiency for a heat engine cycle, namely

$$\eta_{th} = \frac{W_{net}}{Q_H}$$

where W_{net} is the net work of the cycle and Q_H is the heat transfer from the high-temperature body.

In this chapter we have extended our consideration of the second law to control volume processes, which leads us now to consider the efficiency of a process. For example, we might be interested in the efficiency of a turbine in a steam power plant or the compressor in a gas turbine engine.

In general, we can say that to determine the efficiency of a machine in which a process takes place, we compare the actual performance of the machine under given conditions to the performance that would have been achieved in an ideal process. It is in the definition of this ideal process that the second law becomes a major consideration. For example, a steam turbine is intended to be an adiabatic machine. The only heat transfer is the unavoidable heat transfer that takes place between the given turbine and the surroundings. We also note that for a given steam turbine operating in a steady-state manner, the state of the steam entering the turbine and the exhaust pressure are fixed. Therefore, the ideal process is a reversible adiabatic process, which is an isentropic process, between the inlet state and the turbine exhaust pressure. In other words, the variables P_i, T_i, and P_e are the design variables—the first two because the working fluid has been prepared in

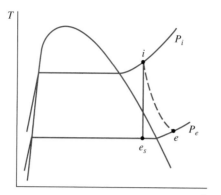
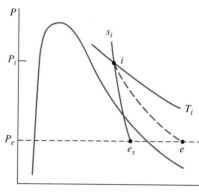

FIGURE 9.10 The process in a reversible adiabatic steam turbine and an actual turbine.

prior processes to be at these conditions at the turbine inlet, while the exit pressure is fixed by the environment into which the turbine exhausts. Thus, the ideal turbine process would go from state i to state e_s, as shown in Fig. 9.10, whereas the real turbine process is irreversible, with the exhaust at a larger entropy at the real exit state e. Figure 9.10 shows typical states for a steam turbine, where state e_s is in the two-phase region, and state e may be as well, or may be in the superheated vapor region, depending on the extent of irreversibility of the real process. Denoting the work done in the real process i to e as w, and that done in the ideal, isentrophic process from the same P_i, T_i to the same P_e as w_s, we define the efficiency of the turbine as

$$\eta_{\text{turbine}} = \frac{w}{w_s} = \frac{h_i - h_e}{h_i - h_{es}} \tag{9.27}$$

The same definition applies to a gas turbine, where all states are in the gaseous phase. Typical turbine efficiencies are 0.70–0.88, with large turbines usually having higher efficiencies than small ones.

EXAMPLE 9.8   A steam turbine receives steam at a pressure of 1 MPa and a temperature of 300°C. The steam leaves the turbine at a pressure of 15 kPa. The work output of the turbine is measured and is found to be 600 kJ/kg of steam flowing through the turbine. Determine the efficiency of the turbine.

Control volume:	Turbine.
Inlet state:	P_i, T_i known; state fixed.
Exit state:	P_e known.
Process:	Steady-state.
Model:	Steam tables.

Analysis

The efficiency of the turbine is given by Eq. 9.27:

$$\eta_{\text{turbine}} = \frac{w_a}{w_s}$$

Thus, to determine the turbine efficiency, we calculate the work that would be done in an isentropic process between the given inlet state and final pressure. For this isentropic process, we have

Continuity equation: $\qquad \dot{m}_i = \dot{m}_e = \dot{m}$

First law: $\qquad\qquad\; h_i = h_{e_s} + w_s$

Second law: $\qquad\quad s_i = s_{e_s}$

Solution

From the steam tables, we get

$$h_i = 3051.2 \text{ kJ/kg}, \qquad s_i = 7.1228 \text{ kJ/kg K}$$

Therefore, at $P_e = 15$ kPa,

$$s_{es} = s_i = 7.1228 = 0.7548 + x_{es}\, 7.2536$$

$$x_{es} = 0.8779$$

$$h_{es} = 225.9 + 0.8779(2373.1) = 2309.3 \text{ kJ/kg}$$

From the first law for the isentropic process,

$$w_s = h_i - h_{es} = 3051.2 - 2309.3 = 741.9 \text{ kJ/kg}$$

But, since

$$w_a = 600 \text{ kJ/kg}$$

we find that

$$\eta_{\text{turbine}} = \frac{w_a}{w_s} = \frac{600}{741.9} = 0.809 = 80.9\%$$

In connection with this example, it should be noted that to find the actual state e of the steam exiting the turbine, we need to analyze the real process taking place. For the real process

$$\dot{m}_i = \dot{m}_e = \dot{m}$$

$$h_i = h_e + w_a$$

$$s_e > s_i$$

Therefore, from the first law for the real process, we have

$$h_e = 3051.2 - 600 = 2451.2 \text{ kJ/kg}$$

$$2451.2 = 225.9 + x_e\, 2373.1$$

$$x_e = 0.9377$$

It is important to keep in mind that the turbine efficiency is defined in terms of an ideal, isentropic process from P_i and T_i to P_e, even when one or more of these variables is unknown. This is illustrated in the following example.

EXAMPLE 9.9   Air enters a gas turbine at 1600 K and exits at 100 kPa and 830 K. The turbine efficiency is estimated to be 85%. What is the turbine inlet pressure?

> *Control volume:*   Turbine.
> *Inlet state:* T_i known.
> *Exit state:* P_e, T_e known; state fixed.
> *Process:*   Steady state.
> *Model:*   Air tables, Table A.7.

Analysis

The efficiency, which is 85%, is given by Eq. 9.27,

$$\eta_{turbine} = \frac{w}{w_s}$$

The first law for the real, irreversible process is

$$h_i = h_e + w$$

For the ideal, isentropic process from P_i, T_i to P_e, the first law is

$$h_i = h_{es} + w_s$$

and the second law is, from Eq. 8.28,

$$s_{es} - s_i = 0 = s_{es}^0 - s_i^0 - R \ln \frac{P_e}{P_i}$$

(Note that this equation is only for the ideal, isentropic process and not for the real process, for which $s_e - s_i > 0$.)

Solution

From the air tables, Table A.7, at 1600 K, we get

$$h_i = 1757.3 \text{ kJ/kg}, \quad s_i^0 = 8.6905 \text{ kJ/kg K}$$

From the air tables at 830 K (the actual turbine exit temperature),

$$h_e = 855.3 \text{ kJ/kg}$$

Therefore, from the first law for the real process.

$$w = 1757.3 - 855.3 = 902.0 \text{ kJ/kg}$$

Using the definition of turbine efficiency,

$$w_s = 902.0/0.85 = 1061.2 \text{ kJ/kg}$$

From the first law for the isentropic process,

$$h_{es} = 1757.3 - 1061.2 = 696.1 \text{ kJ/kg}$$

so that, from the air tables,

$$T_{es} = 683.7 \text{ K}, \qquad s_{es}^0 = 7.7148 \text{ kJ/kg K}$$

and the turbine inlet pressure is determined from

$$0 = 7.7148 - 8.6905 - 0.287 \ln \frac{100}{P_i}$$

or

$$P_i = 2995 \text{ kPa}$$

As was discussed in Section 6.4, unless specifically noted to the contrary, we normally assume compressors or pumps to be adiabatic. In this case the fluid enters the compressor at P_i and T_i, the condition at which it exists, and exits at the desired value of P_e, the reason for building the compressor. Thus, the ideal process between the given inlet state i and the exit pressure would be an isentropic process between state i and state e_s, as shown in Fig. 9.11 with a work input of w_s. The real process, however, is irreversible, and the fluid exits at the real state e with a larger entropy, and a larger amount of work input w is required. The compressor (or pump, in the case of a liquid) efficiency is defined as

$$\eta_{comp} = \frac{w_s}{w} = \frac{h_i - h_{es}}{h_i - h_e} \tag{9.28}$$

Typical compressor efficiencies are in the range of 0.70–0.88 with large compressors usually having higher efficiencies than small ones.

If an effort is made to cool a gas during compression by using a water jacket or fins, the ideal process is considered a reversible isothermal process, the work input for which is w_T, compared to the larger required work w for the real compressor. The efficiency of the cooled compressor is then

$$\eta_{cooled\ comp} = \frac{w_T}{w} \tag{9.29}$$

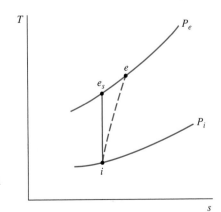 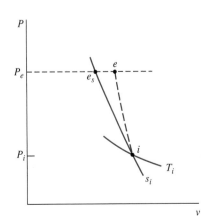

FIGURE 9.11 The compression process in an ideal and actual adiabatic compressor.

EXAMPLE 9.10   Air enters an automotive superhcharger at 100 kPa and 300 K and is compressed to 150 kPa. The efficiency is 70%. What is the required work input per kg of air? What is the exit temperature?

> *Control volume:*   Supercharger (compressor).
> *Inlet state:* P_i, T_i known; state fixed.
> *Exit state:* P_e known.
> *Process:*   Steady-state.
> *Model:*   Ideal gas, 300 K specific heat, Table A.5.

Analysis

The efficiency, which is 70%, is given by Eq. 9.28,

$$\eta_{\text{comp}} = \frac{w_s}{w}$$

The first law for the real, irreversible process is

$$h_i = h_e + w, \qquad w = C_{p0}(T_i - T_e)$$

For the ideal, isentropic process from P_i, T_i to P_e, the first law is

$$h_i = h_{es} + w_s, \qquad w_s = C_{p0}(T_i - T_{es})$$

and the second law is, from Eq. 8.32

$$\frac{T_{es}}{T_i} = \left(\frac{P_e}{P_i}\right)^{(k-1)/k}$$

Solution

Using C_{p0} and k from Table A.5, from the second law, we get

$$T_{es} = 300 \left(\frac{150}{100}\right)^{0.286} = 336.9 \text{ K}$$

From the first law for the isentropic process, we have

$$h_{es} = 1.004(300 - 336.9) = -37.1 \text{ kJ/kg}$$

so that, from the efficiency, the real work input is

$$w = -37.1/0.70 = -53.0 \text{ kJ/kg}$$

and from the first law for the real process, the temperature at the supercharger exit is

$$T_e = 300 - \frac{-53.0}{1.004} = 352.8 \text{ K}$$

Our final example is that of nozzle efficiency. As discussed in Section 6.4, the purpose of a nozzle is to produce a high-velocity fluid stream, or in terms of energy, a large kinetic energy, at the expense of the fluid pressure. The design variables are the same as for a turbine, P_i, T_i, and P_e. A nozzle is usually assumed to be adiabatic, such that the

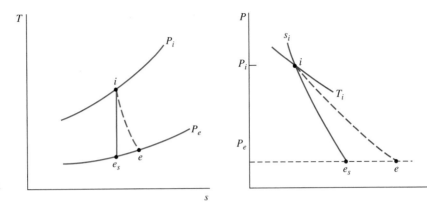

FIGURE 9.12 The ideal and actual process in an adiabatic nozzle.

ideal process is an isentropic process from state i to state e_s, as shown in Fig. 9.12, with the production of velocity $\mathbf{V}_{es}$. The real process is irreversible, with the exit state e having a larger entropy, and a smaller exit velocity $\mathbf{V}_e$. The nozzle efficiency is defined in terms of the corresponding kinetic energies,

$$\eta_{\text{nozz}} = \frac{\mathbf{V}_e^2/2}{\mathbf{V}_{es}^2/2} \tag{9.30}$$

Nozzles are simple devices with no moving parts. As a result, nozzle efficiency may be very high, typically 0.90–0.97.

In summary, to determine the efficiency of a device that carries out a process (rather than a cycle), we compare the actual performance to what would be achieved in a related, but well-defined ideal process.

9.6 SOME GENERAL COMMENTS REGARDING ENTROPY

It is quite possible at this point that a student may have a good grasp of the material that has been covered and yet may have only a vague understanding of the significance of entropy. In fact, the question "What is entropy?" is frequently raised by students with the implication that no one really knows! This section has been included in an attempt to give insight into the qualitative and philosophical aspects of the concept of entropy, and to illustrate the broad application of entropy to many different disciplines.

First, we recall that the concept of energy rises from the first law of thermodynamics and the concept of entropy from the second law of thermodynamics. Actually, it is just as difficult to answer the question "What is energy?" as it is to answer the question "What is entropy?" However, since we regularly use the term energy and are able to relate this term to phenomena that we observe every day, the word energy has a definite meaning to us and thus serves as an effective vehicle for thought and communication. The word entropy could serve in the same capacity. If, when we observed a highly irreversible process (such as cooling coffee by placing an ice cube in it), we said, "That surely increases the entropy," we would soon be as familiar with the word entropy as we are with the word energy. In many cases when we speak about a higher efficiency we are actually speaking about accomplishing a given objective with a smaller total increase in entropy.

A second point to be made regarding entropy is that in statistical thermodynamics, the property entropy is defined in terms of probability. Although this topic will not be examined in detail in this text, a few brief remarks regarding entropy and probability may prove helpful. From this point of view, the net increase in entropy that occurs during an irreversible process can be associated with a change of state from a less probable state to a more probable state. For instance, to use a previous example, one is more likely to find gas on both sides of ruptured membrane in Fig. 7.15 than to find a gas on one side and a vacuum on the other. Thus, when the membrane ruptures, the direction of the process is from a less probable state to a more probable state and associated with this process is an increase in entropy. Similarly, the more probable state is that a cup of coffee will be at the same temperature as its surroundings than at a higher (or lower) temperature. Therefore, as the coffee cools as the result of a transferring of heat to the surroundings, there is a change from a less probable to a more probable state, and associated with this is an increase in entropy.

SUMMARY

The second law of thermodynamics is extended to a general control volume with mass flow rates in or out of steady and transient processes. The vast majority of common devices and complete systems can be treated as nearly steady-state operation even if they have slower transients as in a car engine or jet engine. Simplification of the entropy equation arises when applied to steady-state and single-flow devices like a turbine, nozzle, compressor, or pump. The second law and Gibbs property relation are used to develop a general expression for reversible shaft work in a single flow that is useful in understanding the importance of the specific volume (or density) that influences the magnitude of the work. For a flow with no shaft work, consideration of the reversible process also leads to the derivation of the energy equation for an incompressible fluid as the Bernoulli equation. This covers the flows of liquids such as water or hydraulic fluid as well as airflow at low speeds, which can be considered incompressible for velocities less than a third of the speed of sound.

Many actual devices operate with some irreversibility in the processes that occur, so we also have entropy generation in the flow processes and the total entropy is always increasing. The characterization of performance of actual devices can be done with a comparison to a corresponding ideal device, giving efficiency as the ratio of two energy terms (work or kinetic energy).

You should have learned a number of skills and acquired abilities from studying this chapter that will allow you to

- Apply the second law to more general control volumes.
- Analyze steady-state, single-flow devices such as turbines, nozzles, compressors, and pumps, both reversible and irreversible.
- Know how to extend the second law to transient processes.
- Analyze complete systems as a whole or divide them into individual devices.
- Apply the second law to multiple-flow devices such as heat exchangers, mixing chambers, and turbines with several inlets and outlets.
- Recognize when you have an incompressible flow where you can apply the Bernoulli equation or the expression for reversible shaft work.
- Know when you can apply the Bernoulli equation and when you cannot.
- Know how to evaluate the shaft work for a polytropic process.

- Know how to apply the analysis to an actual device using an efficiency and identify the closest ideal approximation to the actual device.
- Know the difference between a cycle efficiency and a device efficiency.
- Have a sense of entropy as a measure of disorder or chaos.

KEY CONCEPTS AND FORMULAS

Rate equation for entropy   rate of change = + in − out + generation

$$\dot{S}_{c.v.} = \sum \dot{m}_i s_i - \sum \dot{m}_e s_e + \sum \frac{\dot{Q}_{c.v.}}{T} + \dot{S}_{gen}$$

Steady state single flow
$$s_e = s_i + \int_i^e \frac{\delta q}{T} + s_{gen}$$

Reversible shaft work
$$w = -\int_i^e v\, dP + \frac{1}{2}\mathbf{V}_i^2 - \frac{1}{2}\mathbf{V}_e^2 + gZ_i - gZ_e$$

Reversible heat transfer
$$q = \int_i^e T\, ds = h_e - h_i - \int_i^e v\, dP \quad \text{(from Gibbs relation)}$$

Bernoulli equation
$$v(P_i - P_e) + \frac{1}{2}\mathbf{V}_i^2 - \frac{1}{2}\mathbf{V}_e^2 + gZ_i - gZ_e = 0 \quad (v = \text{constant})$$

Polytropic process work
$$w = -\frac{n}{n-1}(P_e v_e - P_i v_i) = -\frac{nR}{n-1}(T_e - T_i) \quad n \neq 1$$
$$w = -P_i v_i \ln\frac{P_e}{P_i} = -RT_i \ln\frac{P_e}{P_i} = RT_i \ln\frac{v_e}{v_i} \quad n = 1$$

The work is shaft work $w = -\int_i^e v\, dP$ and for ideal gas

Isentropic efficiencies
$$\eta_{turbine} = w_{Tac}/w_{Ts} \quad \text{(Turbine work is out)}$$
$$\eta_{compressor} = w_{Cs}/w_{Cac} \quad \text{(Compressor work is in)}$$
$$\eta_{pump} = w_{Ps}/w_{Pac} \quad \text{(Pump work is in)}$$
$$\eta_{nozzle} = \Delta\frac{1}{2}\mathbf{V}_{ac}^2/\Delta\frac{1}{2}\mathbf{V}_s^2 \quad \text{(Kinetic energy is out)}$$

CONCEPT-STUDY GUIDE PROBLEMS

9.1 In a steady state single flow s is either constant or it increases. Is that true?

9.2 Which process will make the previous statement true?

9.3 A reversible adiabatic flow of liquid water in a pump has increasing P. How about T?

9.4 A reversible adiabatic flow of air in a compressor has increasing P. How about T?

9.5 An irreversible adiabatic flow of liquid water in a pump has higher P. How about T?

9.6 A compressor receives R-134a at $-10°C$, 200 kPa, with an exit of 1200 kPa, 50°C. What can you say about the process?

9.7 An air compressor has a significant heat transfer out. See Example 9.4 for how high T becomes if there is no heat transfer. Is that good, or should the compressor be insulated?

9.8 A large condenser in a steam power plant dumps 15 MW at 45°C with an ambient at 25°C. What is the entropy generation rate?

9.9 Air at 1000 kPa, 300 K, is throttled to 500 kPa. What is the specific entropy generation?

9.10 Friction in a pipe flow causes a slight pressure decrease and a slight temperature increase. How does that affect entropy?

9.11 A flow of water at some velocity out of a nozzle is used to wash a car. The water then falls to the ground. What happens to the water state in terms of **V**, T, and s?

9.12 The shaft work in a pump to increase the pressure is small compared to the shaft work in an air compressor for the same pressure increase. Why?

9.13 If the pressure in a flow is constant, can you have shaft work?

9.14 A pump has a 2 kW motor. How much liquid water at 15°C can I pump to 250 kPa from 100 kPa?

9.15 Liquid water is sprayed into the hot gases before they enter the turbine section of a large gasturbine power plant. It is claimed that the larger mass flow rate produces more work. Is that the reason?

9.16 A polytropic flow process with $n = 0$ might be which device?

9.17 A steam turbine inlet is at 1200 kPa, 500°C. The exit is at 200 kPa. What is the lowest possible exit temperature? Which efficiency does that correspond to?

9.18 A steam turbine inlet is at 1200 kPa, 500°C. The exit is at 200 kPa. What is the highest possible exit temperature? Which efficiency does that correspond to?

9.19 A steam turbine inlet is at 1200 kPa, 500°C. The exit is at 200 kPa, 275°C. What is the isentropic efficiency?

9.20 The exit velocity of a nozzle is 500 m/s. If $\eta_{nozzle} = 0.88$, what is the ideal exit velocity?

HOMEWORK PROBLEMS

Steady-State Reversible Processes

Single Flow

9.21 A first stage in a turbine receives steam at 10 MPa and 800°C, with an exit pressure of 800 kPa. Assume the stage is adiabatic and neglect kinetic energies. Find the exit temperature and the specific work.

9.22 Steam enters a turbine at 3 MPa and 450°C, expands in a reversible adiabatic process, and exhausts at 10 kPa. Changes in kinetic and potential energies between the inlet and the exit of the turbine are small. The power output of the turbine is 800 kW. What is the mass flow rate of steam through the turbine?

9.23 A reversible adiabatic compressor receives 0.05 kg/s saturated vapor R-22 at 200 kPa and has an exit pressure of 800 kPa. Neglect kinetic energies and find the exit temperature and the minimum power needed to drive the unit.

9.24 In a heat pump that uses R-134a as the working fluid, the R-134a enters the compressor at 150 kPa and −10°C at a rate of 0.1 kg/s. In the compressor the R-134a is compressed in an adiabatic process to 1 MPa. Calculate the power input required to the compressor, assuming the process to be reversible.

9.25 A boiler section boils 3 kg/s saturated liquid water at 2000 kPa to saturated vapor in a reversible constant-pressure process. Assume you do not know that there is no work. Prove that there is no shaftwork using the first and second laws of thermodynamics.

9.26 Consider the design of a nozzle in which nitrogen gas flowing in a pipe at 500 kPa and 200°C at a velocity of 10 m/s is to be expanded to produce a velocity of 300 m/s. Determine the exit pressure and cross-sectional area of the nozzle if the mass flow rate is 0.15 kg/s and the expansion is reversible and adiabatic.

9.27 Atmospheric air at −45°C and 60 kPa enters the front diffuser of a jet engine, shown in Fig. P9.27, with a velocity of 900 km/h and frontal area of 1 m². After leaving the adiabatic diffuser, the velocity is 20 m/s. Find the diffuser exit temperature and the maximum pressure possible.

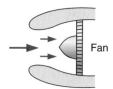

Fan

FIGURE P9.27

9.28 A compressor receives air at 290 K and 100 kPa and a shaft work of 5.5 kW from a gasoline engine. It should deliver a mass flow rate of 0.01 kg/s air to a pipeline. Find the maximum possible exit pressure of the compressor.

9.29 A compressor is surrounded by cold R-134a so it works as an isothermal compressor. The inlet state is 0°C, 100 kPa, and the exit state is saturated vapor. Find the specific heat transfer and specific work.

9.30 A diffuser is a steady-state device in which a fluid flowing at high velocity is decelerated such that the pressure increases in the process. Air at 120 kPa and 30°C enters a diffuser with a velocity of 200 m/s and exits with a velocity of 20 m/s. Assuming the process is reversible and adiabatic, what are the exit pressure and temperature of the air?

9.31 The exit nozzle in a jet engine receives air at 1200 K and 150 kPa with negligible kinetic energy. The exit pressure is 80 kPa, and the process is reversible and adiabatic. Use constant heat capacity at 300 K to find the exit velocity.

9.32 Do the previous problem using the air tables in Table A.7.

9.33 An expander receives 0.5 kg/s air at 2000 kPa, 300 K with an exit state of 400 kPa, 300 K. Assume the process is reversible and isothermal. Find the rates of heat transfer and work neglecting kinetic and potential energy changes.

9.34 Air enters a turbine at 800 kPa and 1200 K and expands in a reversible adiabatic process to 100 kPa. Calculate the exit temperature and the work output per kilogram of air, using
  a. The ideal gas tables (Table A.7).
  b. Constant specific heat (value at 300 K from Table A.5).

9.35 A flow of 2 kg/s saturated vapor R-22 at 500 kPa is heated at constant pressure to 60°C. The heat is supplied by a heat pump that receives heat from the ambient at 300 K and work input shown in Fig. P9.35. Assume everything is reversible and find the rate of work input.

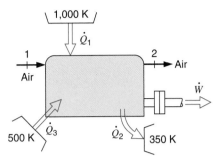

FIGURE P9.36

at 600 K and 100 kPa. Heat transfer of 800 kW is added from a 1000 K reservoir, 100 kW is rejected at 350 K, and some heat transfer takes place at 500 K. Find the heat transferred at 500 K and the rate of work produced.

Multiple Devices and Cycles

9.37 Air at 100 kPa and 17°C is compressed to 400 kPa, after which it is expanded through a nozzle back to the atmosphere. The compressor and the nozzle are both reversible and adiabatic, and kinetic energy in and out of the compressor can be neglected. Find the compressor work and its exit temperature, and find the nozzle exit velocity.

9.38 A small turbine delivers 150 kW and is supplied with steam at 700°C and 2 MPa. The exhaust passes through a heat exchanger where the pressure is 10 kPa and exits as saturated liquid. The turbine is reversible and adiabatic. Find the specific turbine work and the heat transfer in the heat exchanger.

9.39 One technique for operating a steam turbine in part-load power output is to throttle the steam to a lower pressure before it enters the turbine, as shown in Fig. P9.39. The steamline conditions are 2 MPa and 400°C, and the turbine exhaust pressure

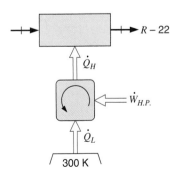

FIGURE P9.35

9.36 A reversible steady-state device receives a flow of 1 kg/s air at 400 K and 450 kPa, and the air leaves

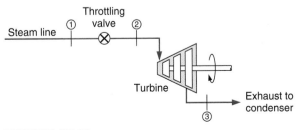

FIGURE P9.39

is fixed at 10 kPa. Assume the expansion inside the turbine to be reversible and adiabatic.

a. Determine the full-load specific work output of the turbine.

b. Find the pressure the steam must be throttled to for 80% of full-load output.

c. Show both processes in a T–s diagram.

9.40 Two flows of air are both at 200 kPa; one has 1 kg/s at 400 K, and the other has 2 kg/s at 290 K. The two lines exchange energy through a number of ideal heat engines, taking energy from the hot line and rejecting it to the colder line. The two flows then leave at the same temperature. Assume the whole setup is reversible and find the exit temperature and the total power out of the heat engines.

9.41 A certain industrial process requires a steady supply of saturated vapor steam at 200 kPa, at a rate of 0.5 kg/s. Also required is a steady supply of compressed air at 500 kPa, at a rate of 0.1 kg/s. Both are to be supplied by the process shown in Fig. P9.41. Steam is expanded in a turbine to supply the power needed to drive the air compressor, and the exhaust steam exits the turbine at the desired state. Air into the compressor is at the ambient conditions, 100 kPa and 20°C. Give the required steam inlet pressure and temperature, assuming that both the turbine and the compressor are reversible and adiabatic.

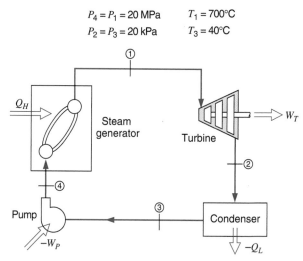

$$P_4 = P_1 = 20 \text{ MPa} \qquad T_1 = 700°C$$
$$P_2 = P_3 = 20 \text{ kPa} \qquad T_3 = 40°C$$

FIGURE P9.42

bine and the pump processes are reversible and adiabatic. Neglecting any changes in kinetic and potential energies, calculate

a. The specific turbine work output and the turbine exit state.

b. The pump work input and enthalpy at the pump exit state.

c. The thermal efficiency of the cycle.

9.43 A turbocharger boosts the inlet air pressure to an automobile engine. It consists of an exhaust gas-driven turbine directly connected to an air compressor, as shown in Fig. P9.43. For a certain engine load, the conditions are given in the figure. Assume that both the turbine and the compressor are reversible and adiabatic, having also the same

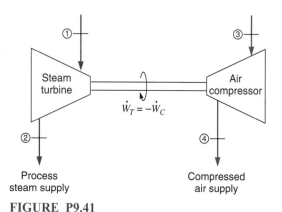

FIGURE P9.41

9.42 Consider a steam turbine power plant operating near critical pressure, as shown in Fig. P9.42. As a first approximation, it may be assumed that the tur-

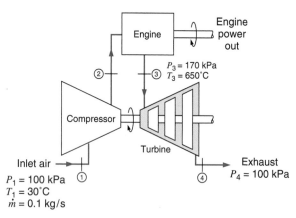

FIGURE P9.43

mass flow rate. Calculate the turbine exit temperature and power output. Find also the compressor exit pressure and temperature.

9.44 A two-stage compressor having an interstage cooler takes in air, 300 K and 100 kPa, and compresses it to 2 MPa, as shown in Fig. P9.44. The cooler then cools the air to 340 K, after which it enters the second stage, which has an exit pressure of 15.74 MPa. Both stages are adiabatic and reversible. Find the specific heat transfer in the intercooler and the total specific work. Compare this to the work required with no intercooler.

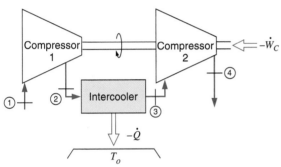

FIGURE P9.44

9.45 A heat-powered portable air compressor consists of three components: (a) an adiabatic compressor; (b) a constant-pressure heater (heat supplied from an outside source); and (c) an adiabatic turbine (see Fig. P9.45). Ambient air enters the compressor at 100 kPa and 300 K and is compressed to 600 kPa. All of the power from the turbine goes into the compressor, and the turbine exhaust is the supply of compressed air. If this pressure is required to be 200 kPa, what must the temperature be at the exit of the heater?

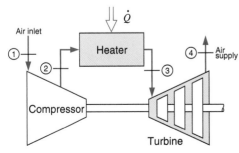

FIGURE P9.45

9.46 A certain industrial process requires a steady 0.5 kg/s supply of compressed air at 500 kPa, at a maximum temperature of 30°C, as shown in Fig. P9.46. This air is to be supplied by installing a compressor and aftercooler. Local ambient conditions are 100 kPa and 20°C. Using a reversible compressor, determine the power required to drive the compressor and the rate of heat rejection in the aftercooler.

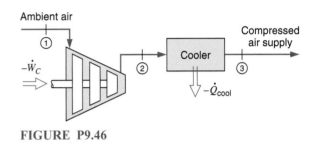

FIGURE P9.46

Steady-State Irreversible Processes

9.47 Analyze the steam turbine described in Problem 6.78. Is it possible?

9.48 Carbon dioxide at 300 K and 200 kPa flows through a steady device where it is heated to 500 K by a 600 K reservoir in a constant-pressure process. Find the specific work, specific heat transfer, and specific entropy generation.

9.49 Consider the steam turbine in Example 6.6. Is this a reversible process?

9.50 The throttle process described in Example 6.5 is an irreversible process. Find the entropy generation per kg of ammonia in the throttling process.

9.51 A geothermal supply of hot water at 500 kPa and 150°C is fed to an insulated flash evaporator at the rate of 1.5 kg/s, shown in Fig. P9.51. A stream of

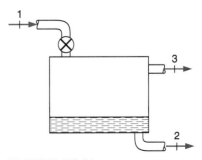

FIGURE P9.51

saturated liquid at 200 kPa is drained from the bottom of the chamber, and a stream of saturated vapor at 200 kPa is drawn from the top and fed to a turbine. Find the rate of entropy generation in the flash evaporator.

9.52 Two flowstreams of water, one of saturated vapor at 0.6 MPa, and the other at 0.6 MPa and 600°C, mix adiabatically in a steady flow to produce a single flow out at 0.6 MPa and 400°C. Find the total entropy generation for this process.

9.53 A condenser in a power plant receives 5 kg/s steam at 15 kPa with a quality of 90% and rejects the heat to cooling water with an average temperature of 17°C. Find the power given to the cooling water in this constant-pressure process, shown in Fig. P9.53, and the total rate of entropy generation when saturated liquid exits the condenser.

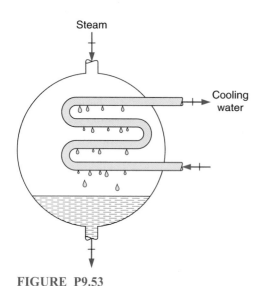

FIGURE P9.53

9.54 A mixing chamber receives 5 kg/min of ammonia as saturated liquid at −20°C from one line and ammonia at 40°C and 250 kPa from another line through a valve. The chamber also receives 325 kJ/min of energy as heat transferred from a 40°C reservoir, shown in Fig. P9.54. This should produce saturated ammonia vapor at −20°C in the exit line. What is the mass flow rate in the second line, and what is the total entropy generation in the process?

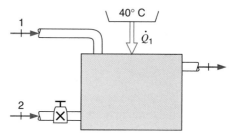

FIGURE P9.54

9.55 A heat exchanger that follows a compressor receives 0.1 kg/s air at 1000 kPa and 500 K and cools it in a constant-pressure process to 320 K. The heat is absorbed by ambient air at 300 K. Find the total rate of entropy generation.

9.56 Air at 327°C and 400 kPa with a volume flow 1 m³/s runs through an adiabatic turbine with exhaust pressure of 100 kPa. Neglect kinetic energies and use constant specific heats. Find the lowest and highest possible exit temperature. For each case find also the rate of work and the rate of entropy generation.

9.57 In a heat-driven refrigerator with ammonia as the working fluid, a turbine with inlet conditions of 2.0 MPa and 70°C is used to drive a compressor with inlet saturated vapor at −20°C. The exhausts, both at 1.2 MPa, are then mixed together, shown in Fig. P9.57. The ratio of the mass flow rate to the turbine to the total exit flow was measured to be 0.62. Can this be true?

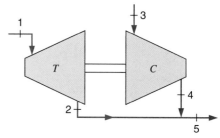

FIGURE P9.57

9.58 Two flows of air are both at 200 kPa; one has 1 kg/s at 400 K, and the other has 2 kg/s at 290 K. The two flows are mixed together in an insulated box to produce a single exit flow at 200 kPa. Find the exit temperature and the total rate of entropy generation.

9.59 One type of feedwater heater for preheating the water before entering a boiler operates on the principle of mixing the water with steam that has been bled from the turbine. For the states as shown in Fig. P9.59, calculate the rate of net entropy increase for the process, assuming the process to be steady flow and adiabatic.

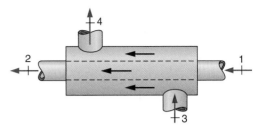

FIGURE P9.62

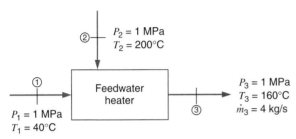

FIGURE P9.59

9.60 A supply of 5 kg/s ammonia at 500 kPa and 20°C is needed. Two sources are available: One is saturated liquid at 20°C, and the other is at 500 kPa and 140°C. Flows from the two sources are fed through valves to an insulated mixing chamber, which then produces the desired output state. Find the two source mass flow rates and the total rate of entropy generation by this setup.

9.61 A counterflowing heat exchanger has one line with 2 kg/s air at 125 kPa and 1000 K entering, and the air is leaving at 100 kPa and 400 K. The other line has 0.5 kg/s water coming in at 200 kPa and 20°C and leaving at 200 kPa. What is the exit temperature of the water and the total rate of entropy generation?

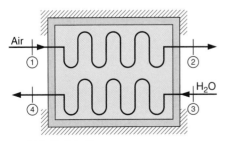

FIGURE P9.61

9.62 A coflowing (same direction) as shown in Fig. P9.62, heat exchanger has one line with 0.25 kg/s

oxygen at 17°C and 200 kPa entering, and the other line has 0.6 kg/s nitrogen at 150 kPa and 500 K entering. The heat exchanger is long enough so that the two flows exit at the same temperature. Use constant heat capacities and find the exit temperature and the total rate of entropy generation.

Transient Processes

9.63 Calculate the specific entropy generated in the filling process given in Example 6.11.

9.64 Calculate the total entropy generated in the filling process given in Example 6.12.

9.65 An initially empty 0.1 m³ canister is filled with R-12 from a line flowing saturated liquid at −5°C. This is done quickly such that the process is adiabatic. Find the final mass, and determine liquid and vapor volumes, if any, in the canister. Is the process reversible?

9.66 A 1 m³ rigid tank contains 100 kg of R-22 at ambient temperature, 15°C. A valve on top of the tank is opened, and saturated vapor is throttled to ambient pressure, 100 kPa, and flows to a collector system, shown in Fig. P9.66. During the process, the temperature inside the tank remains at 15°C. The valve is closed when no more liquid remains inside. Calculate the heat transfer to the tank and the total entropy generation in the process.

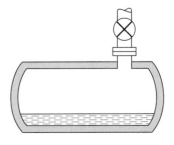

FIGURE P9.66

9.67 Air in a tank is at 300 kPa and 400 K with a volume of 2 m³. A valve on the tank is opened to let some air escape to the ambient surroundings to leave a final pressure inside of 200 kPa. Find the final temperature and mass assuming a reversible adiabatic process for the air remaining inside the tank.

9.68 An empty canister of 0.002 m³ is filled with R-134a from a line flowing saturated liquid R-134a at 0°C. The filling is done quickly so it is adiabatic. Find the final mass in the canister and the total entropy generation.

9.69 An old abandoned salt mine, 100 000 m³ in volume, contains air at 290 K and 100 kPa. The mine is used for energy storage so the local power plant pumps it up to 2.1 MPa using outside air at 290 K and 100 kPa. Assume the pump is ideal and the process is adiabatic. Find the final mass and temperature of the air and the required pump work.

9.70 Air in a tank is at 300 kPa and 400 K with a volume of 2 m³. A valve on the tank is opened to let some air escape to the ambient surroundings to leave a final pressure inside of 200 kPa. At the same time the tank is heated so the air remaining has a constant temperature. What is the mass average value (Table A.7 reference) of the s leaving, assuming this is an internally reversible process?

9.71 An insulated 2 m³ tank is to be charged with R-134a from a line flowing the refrigerant at 3 MPa. The tank is initially evacuated, and the valve is closed when the pressure inside the tank reaches 3 MPa. The line is supplied by an insulated compressor that takes in R-134a at 5°C, with a quality of 96.5%, and compresses it to 3 MPa in a reversible process. Calculate the total work input to the compressor to charge the tank.

9.72 A 0.2 m³ initially empty container is filled with water from a line at 500 kPa and 200°C until there is no more flow. Assume the process is adiabatic and find the final mass, final temperature, and total entropy generation.

9.73 Air from a line at 12 MPa and 15°C flows into a 500 L rigid tank that initially contained air at ambient conditions, 100 kPa and 15°C. The process occurs rapidly and is essentially adiabatic. The valve is closed when the pressure inside reaches some value, P_2. The tank eventually cools to room temperature, at which time the pressure inside is 5 MPa. What is the pressure P_2? What is the net entropy change for the overall process?

9.74 An initially empty canister with a volume of 0.2 m³ is filled with carbon dioxide from a line at 1000 kPa and 500 K. Assume the process is adiabatic and the flow continues until it stops by itself. Use constant heat capacity to solve for the final mass and temperature of the carbon dioxide in the canister and the total entropy generation by the process.

9.75 A cook filled a pressure cooker with 3 kg water at 20°C and a small amount of air and forgot about it. The pressure cooker has a vent valve so if $P > 200$ kPa, steam escapes to maintain the pressure at 200 kPa. How much entropy was generated in the throttling of the steam through the vent to 100 kPa when half the original mass has escaped?

Reversible Shaft Work, Bernoulli Equation

9.76 A large storage tank contains saturated liquid nitrogen at ambient pressure, 100 kPa; it is to be pumped to 500 kPa and fed to a pipeline at the rate of 0.5 kg/s. How much power input is required for the pump, assuming it to be reversible?

9.77 Liquid water at ambient conditions, 100 kPa and 25°C, enters a pump at the rate of 0.5 kg/s. Power input to the pump is 3 kW. Assuming the pump process to be reversible, determine the pump exit pressure and temperature.

9.78 A small dam has a 0.5-m-diameter pipe carrying liquid water at 150 kPa and 20°C with a flow rate of 2000 kg/s. The pipe runs to the bottom of the dam 15 m lower into a turbine with pipe diameter 0.35 m, shown in Fig. P9.78. Assume no friction or heat transfer in the pipe and find the pressure of the turbine inlet. If the turbine exhausts to 100 kPa with negligible kinetic energy, what is the rate of work?

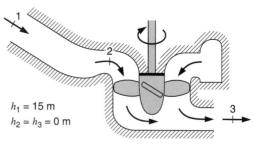

$h_1 = 15$ m
$h_2 = h_3 = 0$ m

FIGURE P9.78

9.79 A firefighter on a ladder 25 m above ground should be able to spray water an additional 10 m up with the hose nozzle of exit diameter 2.5 cm. Assume a water pump on the ground and a reversible flow (hose, nozzle included) and find the minimum required power.

9.80 A small pump is driven by a 2 kW motor with liquid water at 150 kPa and 10°C entering. Find the maximum water flow rate you can get with an exit pressure of 1 MPa and negligible kinetic energies. The exit flow goes through a small hole in a spray nozzle out to the atmosphere at 100 kPa, shown in Fig. P9.80. Find the spray velocity.

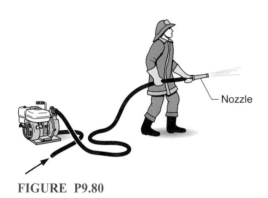

FIGURE P9.80

9.81 A garden water hose has liquid water at 200 kPa and 15°C. How high a velocity can be generated in a small ideal nozzle? If you direct the water spray straight up how high will it go?

9.82 Saturated R-134a at −10°C is pumped/compressed to a pressure of 1.0 MPa at the rate of 0.5 kg/s in a reversible adiabatic process. Calculate the power required and the exit temperature for the two cases of inlet state of the R-134a:

a. Quality of 100%   b. Quality of 0%.

9.83 A small water pump on ground level has an inlet pipe down into a well at a depth H with the water at 100 kPa and 15°C. The pump delivers water at 400 kPa to a building. The absolute pressure of the water must be at least twice the saturation pressure to avoid cavitation. What is the maximum depth this setup will allow?

9.84 A small pump takes in water at 20°C and 100 kPa and pumps it to 2.5 MPa at a flow rate of 100 kg/min. Find the required pump power input.

9.85 A pump/compressor pumps a substance from 100 kPa and 10°C to 1 MPa in a reversible adiabatic process. The exit pipe has a small crack, so that a small amount leaks to the atmosphere at 100 kPa. If the substance is (a) water, (b) R-12, find the temperature after compression and the temperature of the leak flow as it enters the atmosphere, neglecting kinetic energies.

9.86 Atmospheric air at 100 kPa and 17°C blows at 60 km/h toward the side of a building. Assuming the air is nearly incompressible, find the pressure and the temperature at the stagnation (zero-velocity) point on the wall.

9.87 You drive on the highway at 120 km/h on a day with 17°C, 100 kPa atmosphere. When you put your hand out of the window flat against the wind you feel the force from the air stagnating (i.e., it comes to relative zero velocity on your skin). Assume that the air is nearly incompressible and find the air temperature and pressure right on your hand.

9.88 An airflow at 100 kPa, 290 K, and 200 m/s is directed toward a wall. At the wall the flow stagnates (comes to zero velocity) without any heat transfer, as shown in Fig. P9.88. Find the stagnation pressure (a) assuming incompressible flow, (b) assuming an adiabatic compression. *Hint*: T comes from the energy equation.

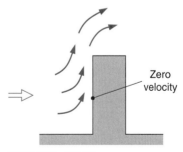

FIGURE P9.88

9.89 Calculate the air temperature and pressure at the stagnation point right in front of a meteorite entering the atmosphere (−50°C, 50 kPa) with a velocity of 2000 m/s. Do this assuming air is incompressible at the given state and repeat for air being a compressible substance going through an adiabatic compression.

9.90 Helium gas enters a steady-flow expander at 800 kPa and 300°C and exits at 120 kPa. The mass flow rate is 0.2 kg/s, and the expansion process can be considered as a reversible polytropic process with exponent $n = 1.3$. Calculate the power output of the expander.

9.91 Air at 100 kPa and 300 K flows through a device at steady state with the exit at 1000 K during which it went through a polytropic process with $n = 1.3$. Find the exit pressure, the specific work, and heat transfer.

9.92 A 4 kg/s flow of ammonia goes through a device in a polytropic process with an inlet state of 150 kPa, −20°C and an exit state of 400 kPa, 80°C. Find the polytropic exponent n, the specific work, and the specific heat transfer.

9.93 Carbon dioxide flows through a device entering at 300 K and 200 kPa and leaving at 500 K. The process is steady-state polytropic with $n = 3.8$, and heat transfer comes from a 600 K source. Find the specific work, specific heat transfer, and specific entropy generation due to this process.

9.94 An expansion in a gas turbine can be approximated with a polytropic process with exponent $n = 1.25$. The inlet air is at 1200 K, 800 kPa, and the exit pressure is 125 kPa with a mass flow rate of 0.75 kg/s. Find the turbine heat transfer and power output.

Device Efficiency

9.95 Find the isentropic efficiency of the R-134a compressor in Example 6.10, assuming the ideal compressor is adiabatic.

9.96 A compressor is used to bring saturated water vapor at 1 MPa up to 17.5 MPa, where the actual exit temperature is 650°C. Find the isentropic compressor efficiency and the entropy generation.

9.97 Liquid water enters a pump at 15°C and 100 kPa and exits at a pressure of 5 MPa. If the isentropic efficiency of the pump is 75%, determine the enthalpy (steam table reference) of the water at the pump exit.

9.98 A centrifugal compressor takes in ambient air at 100 kPa and 15°C and discharges it at 450 kPa. The compressor has an isentropic efficiency of 80%. What is your best estimate for the discharge temperature?

9.99 An emergency drain pump, shown in Fig. P9.99, should be able to pump 0.1 m³/s of liquid water at 15°C, 10 m vertically up delivering it with a velocity of 20 m/s. It is estimated that the pump, pipe, and nozzle have a combined isentropic efficiency expressed for the pump as 60%. How much power is needed to drive the pump?

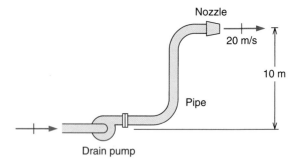

FIGURE P9.99

9.100 A pump receives water at 100 kPa and 15°C and has a power input of 1.5 kW. The pump has an isentropic efficiency of 75%, and it should flow 1.2 kg/s delivered at 30 m/s exit velocity. How high an exit pressure can the pump produce?

9.101 A small air turbine with an isentropic efficiency of 80% should produce 270 kJ/kg of work. The inlet temperature is 1000 K, and the turbine exhausts to the atmosphere. Find the required inlet pressure and the exhaust temperature.

9.102 Repeat Problem 9.42 assuming the turbine and the pump each have an isentropic efficiency of 85%.

9.103 Repeat Problem 9.41 assuming the steam turbine and the air compressor each have an isentropic efficiency of 80%.

9.104 Steam enters a turbine at 300°C, 600 kPa, and exhausts as saturated vapor at 20 kPa. What is the isentropic efficiency?

9.105 A turbine receives air at 1500 K and 1000 kPa and expands it to 100 kPa. The turbine has an isentropic efficiency of 85%. Find the actual turbine exit air temperature and the specific entropy increase in the actual turbine.

9.106 The small turbine in Problem 9.38 was ideal. Assume instead that the isentropic turbine efficiency is 88%. Find the actual specific turbine work and the entropy generated in the turbine.

9.107 Air enters an insulated turbine at 50°C and exits the turbine at −30°C and 100 kPa. The isentropic turbine efficiency is 70%, and the inlet volumetric flow rate is 20 L/s. What is the turbine inlet pressure and the turbine power output?

9.108 Carbon dioxide, CO_2, enters an adiabatic compressor at 100 kPa and 300 K and exits at 1000 kPa and 520 K. Find the compressor efficiency and the entropy generation for the process.

9.109 Air enters an insulated compressor at ambient conditions, 100 kPa and 20°C, at the rate of 0.1 kg/s and exits at 200°C. The isentropic efficiency of the compressor is 70%. Assume the ideal and actual compressor have the same exit pressure. What is the exit pressure? How much power is required to drive the unit?

9.110 Assume an actual compressor has the same exit pressure and specific heat transfer as the ideal isothermal compressor in Problem 9.29 with an isothermal efficiency of 80%. Find the specific work and exit temperature for the actual compressor.

9.111 A water-cooled air compressor takes air in at 20°C and 90 kPa and compresses it to 500 kPa. The isothermal efficiency is 80%, and the actual compressor has the same heat transfer as the ideal one. Find the specific compressor work and the exit temperature.

9.112 A nozzle in a high-pressure liquid water sprayer has an area of 0.5 cm². It receives water at 250 kPa, 20°C, and the exit pressure is 100 kPa. Neglect the inlet kinetic energy and assume a nozzle isentropic efficiency of 85%. Find the ideal nozzle exit velocity and the actual nozzle mass flow rate.

9.113 A nozzle should produce a flow of air with 200 m/s at 20°C and 100 kPa. It is estimated that the nozzle has an isentropic efficiency of 92%. What nozzle inlet pressure and temperature are required assuming the inlet kinetic energy is negligible?

9.114 Redo Problem 9.79 if the water pump has an isentropic efficiency of 85% including hose and nozzle.

9.115 Find the isentropic efficiency of the nozzle in Example 6.4.

9.116 Air flows into an insulated nozzle at 1 MPa and 1200 K with 15 m/s and a mass flow rate of 2 kg/s. It expands to 650 kPa, and the exit tempera-

ture is 1100 K. Find the exit velocity and the nozzle efficiency.

Review Problems

9.117 A coflowing heat exchanger has one line with 2 kg/s saturated water vapor at 100 kPa entering. The other line is 1 kg/s air at 200 kPa, 1200 K. The heat exchanger is very long so the two flows exit at the same temperature. Find the exit temperature by trial and error. Calculate the rate of entropy generation.

9.118 A vortex tube has an air inlet flow at 20°C, 200 kPa, and two exit flows of 100 kPa: one at 0°C and the other at 40°C. The tube, shown in Fig. P9.118, has no external heat transfer and no work, and all the flows are steady and have negligible kinetic energy. Find the fraction of the inlet flow that comes out at 0°C. Is this setup possible?

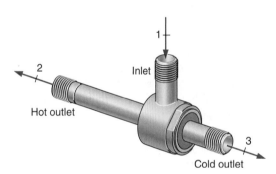

FIGURE P9.118

9.119 An initially empty spring-loaded piston/cylinder requires 100 kPa to float the piston. A compressor with a line and valve now charges the cylinder with water to a final pressure of 1.4 MPa at which point the volume is 0.6 m³, state 2. The inlet condition to the reversible adiabatic compressor is saturated vapor at 100 kPa. After charging, the valve is closed, and the water eventually cools to room temperature, 20°C, state 3. Find the final mass of water, the piston work from 1 to 2, the required compressor work, and the final pressure, P_3.

9.120 In a heat-powered refrigerator, a turbine is used to drive the compressor using the same working fluid. Consider the combination shown in Fig. P9.120 where the turbine produces just enough power to drive the compressor and the two exit flows are mixed together. List any assumptions

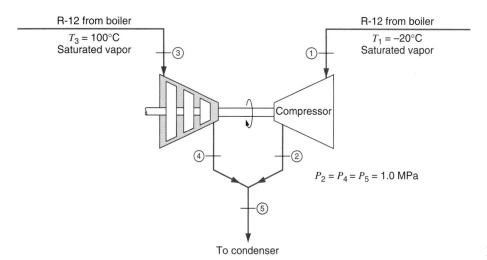

R-12 from boiler
$T_3 = 100°C$
Saturated vapor

R-12 from boiler
$T_1 = -20°C$
Saturated vapor

Compressor

$P_2 = P_4 = P_5 = 1.0$ MPa

To condenser

FIGURE P9.120

made and find the ratio of mass flow rates $\dot{m}_3/\dot{m}_1$ and T_5 (x_5 if in two-phase region) if the turbine and the compressor are reversible and adiabatic.

9.121 A stream of ammonia enters a steady flow device at 100 kPa and 50°C, at the rate of 1 kg/s. Two streams exit the device at equal mass flow rates; one is at 200 kPa and 50°C, and the other is a saturated liquid at 10°C. It is claimed that the device operates in a room at 25°C on an electrical power input of 250 kW. Is this possible?

9.122 A frictionless piston/cylinder is loaded with a linear spring, spring constant 100 kN/m, and the piston cross-sectional area is 0.1 m². The cylinder initial volume of 20 L contains air at 200 kPa and ambient temperature, 10°C. The cylinder has a set of stops that prevents its volume from exceeding 50 L. A valve connects to a line flowing air at 800 kPa, 50°C, as shown in Fig. P9.122. The valve is

now opened, allowing air to flow in until the cylinder pressure reaches 800 kPa, at which point the temperature inside the cylinder is 80°C. The valve is then closed and the process ends.

a. Is the piston at the stops at the final state?

b. Taking the inside of the cylinder as a control volume, calculate the heat transfer during the process.

c. Calculate the net entropy change for this process.

9.123 An insulated piston/cylinder contains R-22 at 20°C, 85% quality, at a cylinder volume of 50 L. A valve at the closed end of the cylinder is connected to a line flowing R-22 at 2 MPa, 60°C. The valve is now opened, allowing R-22 to flow in, and at the same time the external force on the piston is decreased, and the piston moves. When the valve is closed, the cylinder contents are at 800 kPa, 20°C, and a positive work of 50 kJ has been done against the external force. What is the final volume of the cylinder? Does this process violate the second law of thermodynamics?

9.124 Air enters an insulated turbine at 50°C and exits the turbine at −30°C, 100 kPa. The isentropic turbine efficiency is 70%, and the inlet volumetric flow rate is 20 L/s. What is the turbine inlet pressure and the turbine power output?

9.125 A certain industrial process requires a steady 0.5 kg/s supply of compressed air at 500 kPa, at a maximum temperature of 30°C, as shown in Fig. P9.46. This air is to be supplied by installing a

Air line

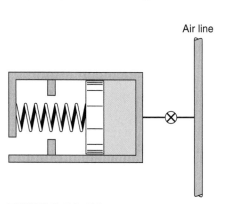

FIGURE P9.122

compressor and aftercooler. Local ambient conditions are 100 kPa, 20°C. Using an isentropic compressor efficiency of 80%, determine the power required to drive the compressor and the rate of heat rejection in the aftercooler.

9.126 Consider the scheme shown in Fig. P9.126 for producing fresh water from salt water. The conditions are as shown in the figure. Assume that the properties of salt water are the same as for pure water, and that the pump is reversible and adiabatic.

    a. Determine the ratio $(\dot{m}_7/\dot{m}_1)$, the fraction of salt water purified.

    b. Determine the input quantities, w_p and q_H.

    c. Make a second law analysis of the overall system.

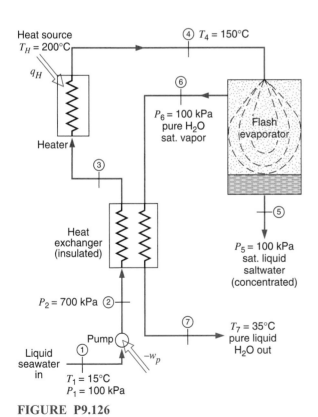

FIGURE P9.126

9.127 Supercharging of an engine is used to increase the inlet air density so that more fuel can be added, the result of which is an increased power output. Assume that ambient air, 100 kPa and 27°C, enters the supercharger at a rate of 250 L/s. The su-

percharger (compressor) has an isentropic efficiency of 75%, and uses 20 kW of power input. Assume that the ideal and actual compressor have the same exit pressure. Find the ideal specific work and verify that the exit pressure is 175 kPa. Find the percent increase in air density entering the engine due to the supercharger and the entropy generation.

9.128 A jet-ejector pump, shown schematically in Fig. P9.128, is a device in which a low-pressure (secondary) fluid is compressed by entrainment in a high-velocity (primary) fluid stream. The compression results from the deceleration in a diffuser. For purposes of analysis, this can be considered as equivalent to the turbine-compressor unit shown in Fig. P9.120 with the states 1, 3, and 5 corresponding to those in Fig. P9.128. Consider a stream jet-pump with state 1 as saturated vapor at 35 kPa; state 3 is 300 kPa, 150°C; and the discharge pressure, P_5, is 100 kPa.

    a. Calculate the ideal mass flow ratio, $\dot{m}_1, \dot{m}_3$.

    b. The efficiency of a jet pump is defined as

$$\eta_{\text{jet pump}} = \frac{(\dot{m}_1/\dot{m}_3)_{\text{actual}}}{(\dot{m}_1/\dot{m}_3)_{\text{ideal}}}$$

for the same inlet conditions and discharge pressure. Determine the discharge temperature of the jet pump if its efficiency is 10%.

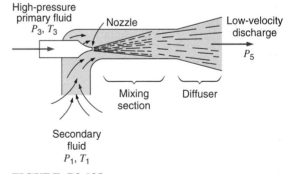

FIGURE P9.128

9.129 A rigid steel bottle, with $V = 0.25$ m³, contains air at 100 kPa and 300 K. The bottle is now charged with air from a line at 260 K and 6 MPa to a bottle pressure of 5 MPa, state 2, and the valve is closed. Assume that the process is adiabatic and that the charge always is uniform. In storage, the bottle slowly returns to room temperature at 300 K, state 3. Find the final mass, the temperature T_2, the final pressure P_3, the heat transfer $_1Q_3$, and the total entropy generation.

9.130 A horizontal, insulated cylinder has a frictionless piston held against stops by an external force of 500 kN, as shown in Fig. P9.130. The piston cross-sectional area is 0.5 m², and the initial volume is 0.25 m³. Argon gas in the cylinder is at 200 kPa and 100°C. A valve is now opened to a line flowing argon at 1.2 MPa and 200°C, and gas flows in until the cylinder pressure just balances the external force, at which point the valve is closed. Use constant head capacity to verify that the final temperature is 645 K and find the total entropy generation.

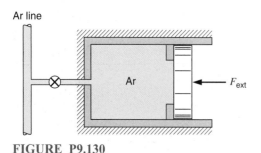

FIGURE P9.130

9.131 A rigid 1.0 m³ tank contains water initially at 120°C, with 50% liquid and 50% vapor, by volume. A pressure-relief valve on the top of the tank is set to 1.0 MPa (the tank pressure cannot exceed 1.0 MPa—water will be discharged instead). Heat is now transferred to the tank from a 200°C heat source until the tank contains saturated vapor at 1.0 MPa. Calculate the heat transfer to the tank and show that this process does not violate the second law.

9.132 A certain industrial process requires a steady 0.5 kg/s of air at 200 m/s, at the condition of 150 kPa, 300 K, as shown in Fig. P9.132. This air is to be the exhaust from a specially designed turbine whose inlet pressure is 400 kPa. The turbine process may be assumed to be reversible and polytropic, with polytropic exponent $n = 1.20$.

a. What is the turbine inlet temperature?

b. What are the power output and heat transfer rate for the turbine?

c. Calculate the rate of net entropy increase, if the heat transfer comes from a source at a temperature 100°C higher than the turbine inlet temperature.

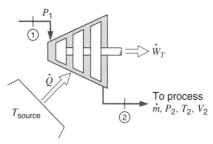

FIGURE P9.132

9.133 Assume both the compressor and the nozzle in Problem 9.37 have an isentropic efficiency of 90%, the rest being unchanged. Find the actual compressor work and its exit temperature and find the actual nozzle exit velocity.

ENGLISH UNIT PROBLEMS

Concept Problems

9.134E A compressor receives R-134a at 20 F, 30 psia, with an exit of 200 psia, $x = 1$. What can you say about the process?

9.135E A large condenser in a steam power plant dumps 15 000 Btu/s at 115 F with an ambient at 77 F. What is the entropy generation rate?

9.136E Air at 150 psia, 540 R, is throttled to 75 psia. What is the specific entropy generation?

9.137E A pump has a 2 kW motor. How much liquid water at 60 F can I pump to 35 psia from 14.7 psia?

9.138E A steam turbine inlet is at 200 psia, 900 F. The exit is at 40 psia. What is the lowest possible exit temperature? Which efficiency does that correspond to?

9.139E A steam turbine inlet is at 200 psia, 900 F. The exit is at 40 psia. What is the highest possible exit temperature? Which efficiency does that correspond to?

9.140E A steam turbine inlet is at 200 psia, 900 F. The exit is at 40 psia, 600 F. What is the isentropic efficiency?

9.141E The exit velocity of a nozzle is 1500 ft/s. If $\eta_{nozzle} = 0.88$, what is the ideal exit velocity?

9.142E Steam enters a turbine at 450 lbf/in.^{2}, 900 F, expands in a reversible adiabatic process, and exhausts at 2 lbf/in.^{2}. Changes in kinetic and potential energies between the inlet and the exit of the turbine are small. The power output of the turbine is 800 Btu/s. What is the mass flow rate of steam through the turbine?

9.143E In a heat pump that uses R-134a as the working fluid, the R-134a enters the compressor at 30 lbf/in.^{2}, 20 F, at a rate of 0.1 lbm/s. In the compressor the R-134a is compressed in an adiabatic process to 150 lbf/in.^{2}. Calculate the power input required to the compressor, assuming the process to be reversible.

9.144E A diffuser is a steady-state, steady-flow device in which a fluid flowing at high velocity is decelerated such that the pressure increases in the process. Air at 18 lbf/in.^{2}, 90 F, enters a diffuser with velocity 600 ft/s and exits with a velocity of 60 ft/s. Assuming the process is reversible and adiabatic, what are the exit pressure and temperature of the air?

9.145E The exit nozzle in a jet engine receives air at 2100 R, 20 psia, with negligible kinetic energy. The exit pressure is 10 psia, and the process is reversible and adiabatic. Use constant heat capacity at 77 F to find the exit velocity.

9.146E Air at 1 atm, 60 F, is compressed to 4 atm, after which it is expanded through a nozzle back to the atmosphere. The compressor and the nozzle are both reversible and adiabatic, and kinetic energy in/out of the compressor can be neglected. Find the compressor work and its exit temperature, and find the nozzle exit velocity.

9.147E An expander receives 1 lbm/s air at 300 psia, 540 R, with an exit state of 60 psia, 540 R. Assume the process is reversible and isothermal. Find the rates of heat transfer and work, neglecting kinetic and potential energy changes.

9.148E A flow of 4 lbm/s saturated vapor R-22 at 100 psia is heated at constant pressure to 140 F. The heat is supplied by a heat pump that receives

heat from the ambient at 540 R and work input as shown in Fig. P9.35. Assume everything is reversible and find the rate of work input.

9.149E One technique for operating a steam turbine in part-load power output is to throttle the steam to a lower pressure before it enters the turbine, as shown in Fig. P9.39. The steamline conditions are 200 lbf/in.^{2}, 600 F, and the turbine exhaust pressure is fixed at 1 lbf/in.^{2}. Assuming the expansion inside the turbine to be reversible and adiabatic,

a. Determine the full-load specific work output of the turbine.

b. Determine the pressure the steam must be throttled to for 80% of full-load output.

c. Show both processes in a T–s diagram.

9.150E Analyze the steam turbine described in Problem 6.160E. Is it possible?

9.151E Two flowstreams of water, one at 100 lbf/in.^{2}, saturated vapor, and the other at 100 lbf/in.^{2}, 1000 F, mix adiabatically in a steady flow process to produce a single flow out at 100 lbf/in.^{2}, 600 F. Find the total entropy generation for this process.

9.152E A mixing chamber receives 10 lbm/min ammonia as saturated liquid at 0 F from one line and ammonia at 100 F, 40 lbf/in.^{2} from another line through a valve. The chamber also receives 340 Btu/min energy as heat transferred from a 100-F reservoir. This should produce saturated ammonia vapor at 0 F in the exit line. What is the mass flow rate at state 2, and what is the total entropy generation in the process?

9.153E A condenser in a power plant receives 10 lbm/s steam at 130 F, quality 90%, and rejects the heat to cooling water with an average temperature of 62 F. Find the power given to the cooling water in this constant pressure process and the total rate of entropy generation when condenser exit is saturated liquid.

9.154E Air at 540 F, 60 lbf/in.^{2}, with a volume flow 40 ft^3/s runs through an adiabatic turbine with exhaust pressure of 15 lbf/in.^{2}. Neglect kinetic energies and use constant specific heats. Find the lowest and highest possible exit temperature. For each case find also the rate of work and the rate of entropy generation.

9.155E A supply of 10 lbm/s ammonia at 80 lbf/in.^{2}, 80 F, is needed. Two sources are available: one is saturated liquid at 80 F, and the other is at 80 lbf/in.^{2}, 260 F. Flows from the two sources are fed through valves to an insulated SSSF mixing chamber, which then produces the desired output state. Find the two source mass flow rates and the total rate of entropy generation by this setup.

9.156E An old abandoned saltmine, 3.5×10^6 ft^3 in volume, contains air at 520 R, 14.7 lbf/in.^{2}. The mine is used for energy storage so the local power plant pumps it up to 310 lbf/in.^{2} using outside air at 520 R, 14.7 lbf/in.^{2}. Assume the pump is ideal and the process is adiabatic. Find the final mass and temperature of the air and the required pump work. Overnight, the air in the mine cools down to 720 R. Find the final pressure and heat transfer.

9.157E Air from a line at 1800 lbf/in.^{2}, 60 F, flows into a 20-ft^3 rigid tank that initially contained air at ambient conditions, 14.7 lbf/in.^{2}, 60 F. The process occurs rapidly and is essentially adiabatic. The valve is closed when the pressure inside reaches some value, P_2. The tank eventually cools to room temperature, at which time the pressure inside is 750 lbf/in.^{2}. What is the pressure P_2? What is the net entropy change for the overall process?

9.158E Liquid water at ambient conditions, 14.7 lbf/in.^{2}, 75 F, enters a pump at the rate of 1 lbm/s. Power input to the pump is 3 Btu/s. Assuming the pump process to be reversible, determine the pump exit pressure and temperature.

9.159E A fireman on a ladder 80 ft above ground should be able to spray water an additional 30 ft up with the hose nozzle of exit diameter 1 in. Assume a water pump on the ground and a reversible flow (hose, nozzle included) and find the minimum required power.

9.160E Saturated R-134a at 10 F is pumped/compressed to a pressure of 150 lbf/in.^{2} at the rate of 1.0 lbm/s in a reversible adiabatic SSSF process. Calculate the power required and the exit temperature for the two cases of inlet state of the R-134a:
a. Quality of 100%
b. Quality of 0%

9.161E A small pump takes in water at 70 F, 14.7 lbf/in.^{2}, and pumps it to 250 lbf/in.^{2} at a flow rate of 200 lbm/min. Find the required pump power input.

9.162E An expansion in a gas turbine can be approximated with a polytropic process with exponent $n = 1.25$. The inlet air is at 2100 R, 120 psia, and the exit pressure is 18 psia with a mass flow rate of 2 lbm/s. Find the turbine heat transfer and power output.

9.163E Helium gas enters a steady-flow expander at 120 lbf/in.^{2}, 500 F, and exits at 18 lbf/in.^{2}. The mass flow rate is 0.4 lbm/s, and the expansion process can be considered as a reversible polytropic process with exponent, $n = 1.3$. Calculate the power output of the expander.

9.164E A compressor is used to bring saturated water vapor at 150 lbf/in.^{2} up to 2500 lbf/in.^{2}, where the actual exit temperature is 1200 F. Find the isentropic compressor efficiency and the entropy generation.

9.165E A small air turbine with an isentropic efficiency of 80% should produce 120 Btu/lbm of work. The inlet temperature is 1800 R, and it exhausts to the atmosphere. Find the required inlet pressure and the exhaust temperature.

9.166E Air enters an insulated compressor at ambient conditions, 14.7 lbf/in.^{2}, 70 F, at the rate of 0.1 lbm/s and exits at 400 F. The isentropic efficiency of the compressor is 70%. What is the exit pressure? How much power is required to drive the compressor?

9.167E A watercooled air compressor takes air in at 70 F, 14 lbf/in.^{2}, and compresses it to 80 lbf/in.^{2}. The isothermal efficiency is 80%, and the actual compressor has the same heat transfer as the ideal one. Find the specific compressor work and the exit temperature.

9.168E A nozzle is required to produce a steady stream of R-134a at 790 ft/s at ambient conditions, 14.7 lbf/in.^{2}, 70 F. The isentropic efficiency may be assumed to be 90%. What pressure and temperature are required in the line upstream of the nozzle?

9.169E Redo Problem 9.159E if the water pump has an isentropic efficiency of 85% (hose, nozzle included).

9.170E Repeat Problem 9.160E for a pump/compressor isentropic efficiency of 70%.

9.171E A rigid 35 ft^3 tank contains water initially at 250 F, with 50% liquid and 50% vapor, by volume. A pressure-relief valve on the top of the tank is set to 140 lbf/in.^{2}. (The tank pressure cannot exceed 140 lbf/in.^{2}—water will be discharged instead.) Heat is now transferred to the tank from a 400 F heat source until the tank contains saturated vapor at 140 lbf/in.^{2}. Calculate the heat transfer to the tank and show that this process does not violate the second law.

9.172E Air at 1 atm, 60 F, is compressed to 4 atm, after which it is expanded through a nozzle back to the atmosphere. The compressor and the nozzle both have efficiency of 90%, and kinetic energy in/out of the compressor can be neglected. Find the actual compressor work and its exit temperature, and find the actual nozzle exit velocity.

COMPUTER, DESIGN, AND OPEN-ENDED PROBLEMS

9.173E Use the menu-driven software to get the properties for the calculation of the isentropic efficiency of the pump in the steam power plant of Problem 6.99.

9.174E Write a program to solve the general case of Problem 9.27, in which the states, velocities, and area are input variables. Use a constant specific heat and find diffuser exit area, temperature, and pressure.

9.175E Write a program to solve Problem 9.118 in which the inlet and exit flow states are input variables. Use a constant specific heat and let the program calculate the split of the mass flow and the overall entropy generation.

9.176E Write a program to solve the general version of Problem 9.84. Initial state, flow rate, and final pressure are input variables. Compute the required pump power from the assumption of constant specific volume equal to the inlet state value.

9.177E Write a program to solve Problem 9.129 with the final bottle pressure as an input variable. Print out the temperature right after charging and the temperature, pressure, and heat transfer after state 3 is reached.

9.178E Consider a small air compressor taking atmospheric air in and compressing it to 1 MPa in a steady flow process. For a maximum flow rate of 0.1 kg/s, discuss the necessary sizes for the piping and the motor to drive the unit.

9.179E Small gasoline engine or electric motor-driven air compressors are used to supply compressed air to power tools, machine shops, and so on. The compressor charges air into a tank that acts as a storage buffer. Find examples of these and discuss their sizes in terms of tank volume, charging pressure, engine, or motor power. Also find the time it will take to charge the system from startup and its continuous supply capacity.

9.180E A coflowing heat exchanger receives air at 800 K, 15 MPa, and water at 15°C, 100 kPa. The two flows exchange energy as they flow alongside each other to the exit, where the air should be cooled to 350 K. Investigate the range of water flows necessary per kilogram per second airflow and the possible water exit temperatures, with the restriction that the minimum temperature difference between the water and air should be 25°C. Include an estimation for the overall entropy generation in the process per kilogram of airflow.

9.181E Consider a geothermal supply of hot water available as saturated liquid at $P_1 = 1.5$ MPa. The liquid is to be flashed (throttled) to some lower pressure, P_2. The saturated liquid and saturated vapor at this pressure are separated, and the vapor is expanded through a reversible adiabatic turbine to the exhaust pressure, $P_3 = 10$ kPa. Study the turbine power output per unit initial mass, m_1 as a function of the pressure, P_2.

9.182E A reversible adiabatic compressor receives air at the state of the surroundings, 20°C, 100 kPa. It should compress the air to a pressure of 1.2 MPa in two stages with a constant pressure intercooler between the two stages. Investigate the work input as a function of the pressure between the two stages assuming the intercooler brings the air down to 50°C.

9.183E (Adv.) Investigate the optimal pressure, P_2, for a constant pressure intercooler between two stages in a compressor. Assume the compression process in each stage follows a polytropic process and that the intercooler brings the substance to the original inlet temperature, T_1. Show that the minimal work for the combined stages arises when

$$P_2 = (P_1 P_3)^{1/2}$$

where P_3 is the final exit pressure.

9.184E (Adv.) Reexamine the previous problem when the intercooler cools the substance to a tempera-ture, $T_2 > T_1$, due to finite heat-transfer rates. What is the effect of having isentropic efficiencies for the compressor stages less than 100% on the total work and selection of P_2?

9.185E Investigate the sizes of turbochargers and superchargers available for automobiles. Look at their boost pressures and check if they also have intercoolers mounted. Analyze an example with respect to the power input and the air it can deliver to the engine and estimate its isentropic efficiency if enough data are found.

IRREVERSIBILITY AND AVAILABILITY 10

We now turn our attention to irreversibility and availability, two additional concepts that have found increasing use in recent years. These concepts are particularly applicable in the analysis of complex thermodynamic systems, for with the aid of a digital computer, irreversibility and availability are very powerful tools in design and optimization studies of such systems.

10.1 AVAILABLE ENERGY, REVERSIBLE WORK, AND IRREVERSIBILITY

In the previous chapter, we introduced the concept of the efficiency of a device, such as a turbine, nozzle, or compressor (perhaps more correctly termed a first-law efficiency, since it is given as the ratio of two energy terms). We proceed now to develop concepts that include more meaningful second-law analysis. Our ultimate goal is to use this analysis to manage our natural resources and environment better.

We first focus our attention on the potential for producing useful work from some source or supply of energy. Consider the simple situation shown in Fig. 10.1a, in which there is an energy source Q in the form of heat transfer from a very large and, therefore, constant-temperature reservoir at temperature T. What is the ultimate potential for producing work?

To answer this question, we imagine that a cyclic heat engine is available, as shown in Fig. 10.1b. To convert the maximum fraction of Q to work requires that the engine be completely reversible, that is, a Carnot cycle, and that the lower-temperature reservoir be at the lowest temperature possible, often, but not necessarily, at the ambient temperature. From the first and second laws for the Carnot cycle and the usual consideration of all the Q's as positive quantities, we find

$$W_{revH.E.} = Q - Q_0$$

$$\frac{Q}{T} = \frac{Q_0}{T_0}$$

so that

$$W_{revH.E.} = Q\left(1 - \frac{T_0}{T}\right) \tag{10.1}$$

343

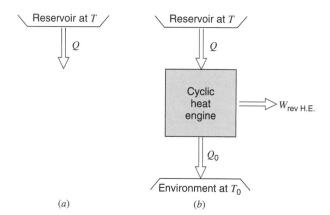

FIGURE 10.1
Constant-temperature
energy source.

(a)   (b)

We might say that the fraction of Q given by the right side of Eq. 10.1 is the available portion of the total energy quantity Q. To carry this thought one step further, consider the situation shown on the T–S diagram in Fig. 10.2. The total shaded area is Q. The portion of Q that is below T_0, the environment temperature, cannot be converted into work by the heat engine and must instead be thrown away. This portion is therefore the unavailable portion of total energy Q, and the portion lying between the two temperatures T and T_0 is the available energy.

Let us next consider the same situation, except that the heat transfer Q is available from a constant-pressure source, for example, a simple heat exchanger as shown in Fig. 10.3a. The Carnot cycle must now be replaced by a sequence of such engines, with the result shown in Fig. 10.3b. The only difference between the first and second examples is that the second includes an integral, which corresponds to ΔS.

$$\Delta S = \int \frac{\delta Q_{\text{rev}}}{T} = \frac{Q_0}{T_0} \tag{10.2}$$

Substituting into the first law, we have

$$W_{\text{revH.E.}} = Q - T_0\,\Delta S \tag{10.3}$$

Note that this ΔS quantity does not include the standard sign convention. It corresponds to the amount of change of entropy shown in Fig. 10.3b. Equation 10.2 specifies the available portion of the quantity Q. The portion unavailable for producing work in this circumstance lies below T_0 in Fig. 10.3b.

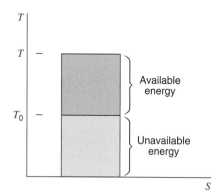

FIGURE 10.2 T–S
diagram for constant-
temperature energy
source.

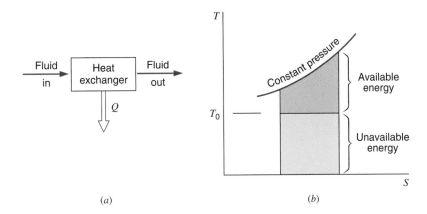

FIGURE 10.3
Changing-temperature
energy source.

(a)

(b)

The Steady-State Process

We now proceed to extend our analysis to real, irreversible processes. In doing so, we will consider first the case of the steady-state control volume process, since the vast majority of applications of this type of analysis refer to components of industrial systems such as power plants or refrigerators. It should be kept in mind that a parallel development for a thermodynamic system analysis or for transient control volume processes will also be made, giving analogous results.

Consider the real steady-state process shown in Fig. 10.4, in which a control volume has a single fluid stream entering at state i, has a single stream exiting at state e, receives an amount of heat q (per unit flow mass) from a reservoir at temperature T_H, and does an amount of work per unit mass w. The first law is, assuming no changes in kinetic or potential energies,

$$q = (h_e - h_i) + w \tag{10.4}$$

This real process is irreversible, such that

$$\frac{1}{\dot{m}}\frac{dS_{net}}{dt} = (s_e - s_i) - \frac{q}{T_H} = s_{gen} > 0 \tag{10.5}$$

We wish to establish a quantitative measure in energy terms of the extent or degree to which any particular real process is irreversible. This can be accomplished by comparison with a control volume undergoing a steady-state process with the same inlet state i, the same exit state e, the same amount of heat transfer with the reservoir, with everything being reversible. That is, the control volume undergoes a steady-state change with inlet h_i,

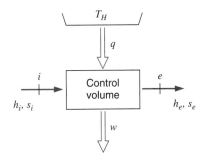

FIGURE 10.4 A real
irreversible process.

s_i and exit h_e, s_e with heat transfer q leaving the reservoir at T_H. If this entire process is reversible, then the net rate of entropy change must equal zero. Noting that all the terms in Eq. 10.5 are the same as for the real irreversible process, we conclude that there must be an additional negative term in the second law for the reversible case. This can only be a heat transfer from the surroundings, divided by the temperature at which it leaves the surroundings. To make this heat transfer as small as possible, it should come from the lowest possible temperature, usually the ambient temperature T_0. This is the situation shown in Fig. 10.5, such that the second law for the reversible process is

$$\frac{1}{\dot{m}} \frac{dS_{net}}{dt} = (s_e - s_i) - \frac{q}{T_H} - \frac{q_0^{rev}}{T_0} = 0 \qquad (10.6)$$

Since any reversible heat transfer must occur over only an infinitesimal temperature difference, we recognize that both heat transfers shown in Fig. 10.5 must be transferred through reversible heat engines or heat pumps. These are located inside the control volume boundary of Fig. 10.5, which includes the original control volume of Fig. 10.4 plus any necessary heat engines and pumps. Only the two fluid flow streams, the net work, and the two heat transfers cross this extended control surface. Equation 10.6 can be rewritten as

$$q_0^{rev} = T_0(s_e - s_i) - q \frac{T_0}{T_H} \qquad (10.7)$$

The first law for the reversible process of Fig. 10.5 is

$$w^{rev} = q + q_0^{rev} - (h_e - h_i) \qquad (10.8)$$

and substituting Eq. 10.7, we get

$$w^{rev} = T_0(s_e - s_i) - (h_e - h_i) + q\left(1 - \frac{T_0}{T_H}\right) \qquad (10.9)$$

This expression establishes the theoretical upper limit for the work per unit mass flow that could be produced by a control volume undergoing a steady-state process from i to e in which heat q is transferred from a reservoir at T_H, with all processes occurring in the environment T_0. The difference between this quantity and the work actually done in the real process of Eqs. 10.4 and 10.5 is a measure of the extent of the irreversibility i (per unit mass flow) of the real process. That is,

$$i = w^{rev} - w \qquad (10.10)$$

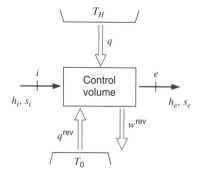

FIGURE 10.5 An ideal reversible process.

The irreversibility i of the real process can also be expressed in another form, by substituting Eqs. 10.4 and 10.9 into 10.10, which results in

$$i = T_0(s_e - s_i) - \frac{T_0}{T_H}q$$

$$= T_0\left[(s_e - s_i) - \frac{q}{T_H}\right]$$

$$= T_0\left[\frac{1}{\dot{m}}\frac{dS_{\text{net/real}}}{dt}\right] = T_0 s_{\text{gen}} \qquad (10.11)$$

And we note that the irreversibility of a real process is another way of expressing the second law for that process, in energy units instead of entropy units and directly proportional to the entropy generation.

EXAMPLE 10.1   A feedwater heater has 5 kg/s water at 5 MPa and 40°C flowing through it, being heated from two sources as shown in Fig. 10.6. One source adds 900 kW from a 100°C reservoir, and the other source transfers heat from a 200°C reservoir such that the water exit condition is 5 MPa, 180°C. Find the reversible work and the irreversibility.

Control volume: Feedwater heater extending out to the two reservoirs.

Inlet state: P_i, T_i known; state fixed.

Exit state: P_e, T_e known; state fixed.

Process: Constant-pressure heat addition with no change in kinetic or potential energy.

Model: Steam tables.

Analysis

This control volume has a single inlet and exit flow with two heat-transfer rates coming from reservoirs different from the ambient surroundings. There is no actual work or actual heat transfer with the surroundings at 25°C. For the actual feedwater heater, the energy equation becomes

$$h_i + q_1 + q_2 = h_e$$

The reversible work for the given change of state is, from Eq. 10.9, with heat transfer q_1 from reservoir T_1 and heat transfer q_2 from reservoir T_2,

$$w^{\text{rev}} = T_0(s_e - s_i) - (h_e - h_i) + q_1\left(1 - \frac{T_0}{T_1}\right) + q_2\left(1 - \frac{T_0}{T_2}\right)$$

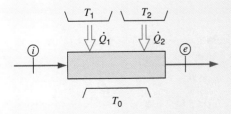

FIGURE 10.6 The feedwater heater for Example 10.1.

From Eq. 10.10, since the actual work is zero, we have

$$i = w^{\text{rev}} - w = w^{\text{rev}}$$

Solution

From the steam tables the inlet and exit state properties are

$$h_i = 171.95 \text{ kJ/kg}, \qquad s_i = 0.5705 \text{ kJ/kg K}$$

$$h_e = 765.24 \text{ kJ/kg}, \qquad s_e = 2.1341 \text{ kJ/kg K}$$

The second heat transfer is found from the energy equation as

$$q_2 = h_e - h_i - q_1 = 765.24 - 171.95 - 900/5 = 413.29 \text{ kJ/kg}$$

The reversible work is

$$w^{\text{rev}} = T_0(s_e - s_i) - (h_e - h_i) + q_1\left(1 - \frac{T_0}{T_1}\right) + q_2\left(1 - \frac{T_0}{T_2}\right)$$

$$= 298.2(2.1341 - 0.5705) - (765.24 - 171.95)$$

$$+ 180\left(1 - \frac{298.2}{373.2}\right) + 413.29\left(1 - \frac{298.2}{473.2}\right)$$

$$= 466.27 - 593.29 + 36.17 + 152.84 = 62.0 \text{ kJ/kg}$$

The irreversibility is

$$i = w^{\text{rev}} = 62.0 \text{ kJ/kg}$$

EXAMPLE 10.2   Consider an air compressor that receives ambient air at 100 kPa and 25°C. It compresses the air to a pressure of 1 MPa, where it exits at a temperature of 540 K. Since the air and compressor housing are hotter than the ambient surroundings, 50 kJ per kilogram air flowing through the compressor are lost. Find the reversible work, and the irreversibility in the process.

> *Control volume:*   The air compressor.
> *Sketch:*   Fig. 10.7.
> *Inlet state:* P_i, T_i known; state fixed.
> *Exit state:* P_e, T_e known; state fixed.
> *Process:*   Nonadiabatic compression with no change in kinetic or potential energy.
> *Model:*   Ideal gas.

Analysis

This steady-state process has a single inlet and exit flow so all quantities are done on a mass basis as specific quantities. From the ideal gas air tables, we obtain

$$h_i = 298.6 \text{ kJ/kg}, \qquad s^0_{T_i} = 6.8631 \text{ kJ/kg K}$$

$$h_e = 544.7 \text{ kJ/kg}, \qquad s^0_{T_e} = 7.4664 \text{ kJ/kg K}$$

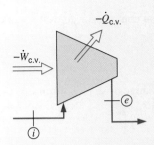

FIGURE 10.7
Illustration for Example 10.2.

so the energy equation for the actual compressor gives the work as

$$q = -50 \text{ kJ/kg}$$

$$w = h_i - h_e + q = 298.6 - 544.7 - 50 = -296.1 \text{ kJ/kg}$$

The reversible work for the given change of state is, from Eq. 10.9, with $T_H = T_0$

$$w^{\text{rev}} = T_0(s_e - s_i) - (h_e - h_i) + q\left(1 - \frac{T_0}{T_H}\right)$$

$$= 298.2(7.4664 - 6.8631 - 0.287 \ln 10) - (544.7 - 298.6) + 0$$

$$= -17.2 - 246.1 = -263.3 \text{ kJ/kg}$$

From Eq. 10.10, we get

$$i = w^{\text{rev}} - w$$

$$= -263.3 - (-296.1) = 32.8 \text{ kJ/kg}$$

EXAMPLE 10.2E Consider an air compressor that receives ambient air at 14.7 lbf/in.^{2}, 80 F. It compresses the air to a pressure of 150 lbf/in.^{2}, where it exits at a temperature of 960 R. Since the air and compressor housing are hotter than the ambient, it loses 22 Btu/lbm air flowing through the compressor. Find the reversible work and the irreversibility in the process.

Control volume:	The air compressor.
Inlet state:	P_i, T_i known; state fixed.
Exit state:	P_e, T_e known; state fixed.
Process:	Nonadiabatic compression with no change kinetic or potential energy.
Model:	Ideal gas.

Analysis

The steady-state process has a single inlet and exit flow so all quantities are done on a mass basis as specific quantities. From the ideal gas air tables, we obtain

$$h_i = 129.18 \text{ Btu/lbm} \qquad s_{T_i}^0 = 1.6405 \text{ Btu/lbm R}$$

$$h_e = 231.20 \text{ Btu/lbm} \qquad s_{T_e}^0 = 1.7803 \text{ Btu/lbm R}$$

so the energy equations for the actual compressor gives the work as

$$q = -22 \text{ Btu/lbm}$$

$$w = h_i - h_e + q = 129.18 - 231.20 - 22 = -124.02 \text{ Btu/lbm}$$

The reversible work for the given change of state is, from Eq. 10.9, with $T_H = T_0$

$$w^{\text{rev}} = T_0(s_e - s_i) - (h_e - h_i) + q\left(1 - \frac{T_0}{T_H}\right)$$

$$= 539.7(1.7803 - 1.6405 - 0.06855 \ln 10.2) - (231.20 - 129.18) + 0$$

$$= -10.47 - 192.02 = -112.49 \text{ Btu/lbm}$$

From Eq. 10.10, we get

$$i = w^{\text{rev}} - w$$

$$= -112.49 - (-124.02) = 11.53 \text{ Btu/lbm}$$

The expression for reversible work for the steady-state process, Eq. 10.9, was derived without including kinetic and potential energy terms. Whenever necessary, especially in nozzles and diffusers where obtaining kinetic energy change is the reason for building the device, these terms can be included along with the enthalpy terms of the fluid stream in and out of the control volume. Another way of including these terms would be to say that the enthalpies in Eq. 10.9 are the total enthalpies, as used in Eq. 6.8. There are also steady-state processes involving more than one fluid stream entering or exiting the control volume. In such cases, it is necessary to rewrite Eq. 10.9 on a rate basis, including the mass flow rates of the different streams involved in the process.

The Control Mass Process

Consider the real process shown in Fig. 10.8, in which a control mass receives an amount of heat $_1Q_2$ from a reservoir at temperature T_H, undergoes a change of state from 1–2, and does an amount of work $_1W_2$. The first law is, assuming no changes in kinetic or potential energies,

$$_1Q_2 = (U_2 - U_1) + {_1W_2} \tag{10.12}$$

This real process is irreversible, such that

$$\Delta S_{\text{net/real}} = \Delta S_{\text{c.m.}} + \Delta S_{\text{surr}}$$

$$= (S_2 - S_1) - \frac{_1Q_2}{T_H} = {_1S_{2\text{gen}}} > 0 \tag{10.13}$$

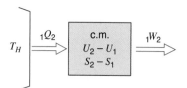

FIGURE 10.8 A real irreversible process.

As with the steady-state process, we wish to compare this irreversible process with an ideal reversible process for the same change of state from U_1, S_1 to U_2, S_2 with heat transfer $_1Q_2$ from reservoir at T_H. This is represented in Fig. 10.9, in which the extended control surface includes the original control mass plus any necessary heat engines and pumps required for reversible heat transfer, analogous to the case for the steady-state process analyzed earlier. The second law for the reversible process is

$$\Delta S_{\text{net/rev}} = (S_2 - S_1) - \frac{_1Q_2}{T_H} - \frac{Q_0^{\text{rev}}}{T_0} = 0 \tag{10.14}$$

which can be rewritten as

$$Q_0^{\text{rev}} = T_0(S_2 - S_1) - {_1Q_2}\frac{T_0}{T_H} \tag{10.15}$$

The first law for the reversible process of Fig. 10.9 is

$$_1W_2^{\text{rev}} = {_1Q_2} + Q_0^{\text{rev}} - (U_2 - U_1) \tag{10.16}$$

and substituting Eq. 10.15,

$$_1W_2^{\text{rev}} = T_0(S_2 - S_1) - (U_2 - U_1) + {_1Q_2}\left(1 - \frac{T_0}{T_H}\right) \tag{10.17}$$

This expression establishes the theoretical upper limit for the work that could be produced by a control mass undergoing the change of state 1–2 in which heat $_1Q_2$ is transferred from a reservoir at T_H, all occurring in the environment T_0. The difference between this quantity and the work actually done in the real process of Eqs. 10.12 and 10.13 is a measure of the extent of the irreversibility I of the real process, or

$$_1I_2 = {_1W_2^{\text{rev}}} - {_1W_2} \tag{10.18}$$

The irreversibility $_1I_2$ of the real process can also be expressed in another form, by substituting Eqs. 10.12 and 10.17 into 10.18_0 which results in

$$
\begin{aligned}
_1I_2 &= T_0(S_2 - S_1) - \frac{T_0}{T_H}{_1Q_2} \\
&= T_0\left[(S_2 - S_1) - \frac{_1Q_2}{T_H}\right] \\
&= T_0[\Delta S_{\text{net/real}}] = T_{01}S_{\text{2gen}} \tag{10.19}
\end{aligned}
$$

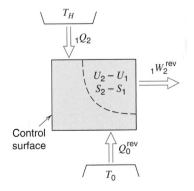

FIGURE 10.9 An ideal reversible process.

EXAMPLE 10.3 An insulated rigid tank is divided into two parts A and B by a diaphragm. Each part has a volume of 1 m^3. Initially, part A contains water at room temperature, 20°C, with a quality of 50%, while part B is evacuated. The diaphragm then ruptures and the water fills the total volume. Determine the reversible work for this change of state and the irreversibility of the process.

$$\begin{array}{rl} Control\ mass: & \text{Water} \\ Initial\ state: & T_1, x_1 \text{ known; state fixed.} \\ Final\ state: & V_2 \text{ known.} \\ Process: & \text{Adiabatic, no change in kinetic or potential energy.} \\ Model: & \text{Steam tables.} \end{array}$$

Analysis

There is a boundary movement for the water, but since it occurs against no resistance, there is no work done. Therefore, the first law reduces to

$$m(u_2 - u_1) = 0$$

From Eq. 10.17 with no change in internal energy and no heat transfer,

$$_1W_2^{\text{rev}} = T_0(S_2 - S_1) = T_0 m(s_2 - s_1)$$

From Eq. 10.18

$$_1I_2 = {_1W_2^{\text{rev}}} - {_1W_2} = {_1W_2^{\text{rev}}}$$

Solution

From the steam tables at state 1,

$$u_1 = 1243.5 \text{ kJ/kg} \qquad v_1 = 28.895 \text{ m}^3/\text{kg} \qquad s_1 = 4.4819 \text{ kJ/kg K}$$

Therefore,

$$v_2 = V_2/m = 2 \times v_1 = 57.79 \qquad u_2 = u_1 = 1243.5$$

These two independent properties, v_2 and u_2, fix state 2. The final temperature T_2 must be found by trial and error in the steam tables.

For $\qquad\qquad T_2 = 5°C \quad$ and $\quad v_2 \Rightarrow x = 0.3928, \qquad u = 948.5 \text{ kJ/kg}$

For $\qquad\qquad T_2 = 10°C \quad$ and $\quad v_2 \Rightarrow x = 0.5433, \qquad u = 1317 \text{ kJ/kg}$

so the final interpolation in u gives a temperature of 9°C. If the software is used, the final state is interpolated to be

$$T_2 = 9.1°C \qquad x_2 = 0.513 \qquad s_2 = 4.644 \text{ kJ/kg K}$$

with the given u and v. Since the actual work is zero we have

$$_1I_2 = {_1W_2^{\text{rev}}} = T_0 m(s_2 - s_1)$$

$$= 293.2(1/28.895)(4.644 - 4.4819) = 1.645 \text{ kJ}$$

For processes in which kinetic and potential energy changes are significant, the development of the expressions for work, reversible work, and irreversibility are all the same, substituting $E = U + KE + PE$ for U in any equation involving energy.

The Transient Process

The transient process has a control volume change from state 1 to state 2 with possible mass flow in at state i and/or flow out at state e. The procedure for developing an expression for reversible work for this process is analogous to the previous examples followed for the steady state and the control mass. In this case, assuming no kinetic or potential energy terms are included, the equation for reversible work will contain control volume entropy and energy terms of the same form as in Eq. 10.17 (but recognizing that the masses at states 1 and 2 are different), and will also contain entropy and enthalpy flow terms the same as in Eq. 10.9 (each one including the appropriate mass flow). The result is

$$W_{c.v.}^{rev} = T_0(m_2s_2 - m_1s_1) - (m_2u_2 - m_1u_1)$$
$$+ T_0(m_es_e - m_is_i) - (m_eh_e - m_ih_i)$$
$$+ Q_{c.v.}\left(1 - \frac{T_0}{T_H}\right) \tag{10.20}$$

This expression can also be grouped as

$$W_{c.v.}^{rev} = m_i(h_i - T_0s_i) - m_e(h_e - T_0s_e)$$
$$+ m_1(u_1 - T_0s_1) - m_2(u_2 - T_0s_2) + Q_{c.v.}\left(1 - \frac{T_0}{T_H}\right) \tag{10.21}$$

As in the previous developments, h_i and h_e can e replaced by h_{TOTi} and h_{TOTe}, and u_1 and u_2 can be replaced by e_1 and e_2, whenever kinetic and potential energies are significant. Also, summations can be added to the flow terms in cases where there is more than one flow stream in or out of the control volume.

The irreversibility for the transient process is found from the general definition.

$$I_{c.v.} = W_{c.v.}^{rev} - W_{c.v.} \tag{10.22}$$

which, by substitution of Eq. 10.20 and the first law, can also be expressed as

$$I_{c.v.} = T_0\left[(m_2s_2 - m_1s_1) + m_es_e - m_is_i - \frac{Q_{c.v.}}{T_H}\right]$$
$$= T_0[\Delta S_{c.v.} + \Delta S_{surr}] = T_0[\Delta S_{net/real}] = T_0S_{gen} \tag{10.23}$$

EXAMPLE 10.4   A 1-m³ rigid tank, Fig. 10.10, contains ammonia at 200 kPa and the ambient temperature 20°C. The tank is connected with a valve to a line flowing saturated liquid ammonia at −10°C. The valve is opened, and the tank is charged quickly until the flow stops and the valve is closed. As the process happens very quickly, there is no heat transfer. Determine the final mass in the tank and the irreversibility in the process.

Control volume: The tank and the valve.
Initial state: T_1, P_1 known; state fixed.

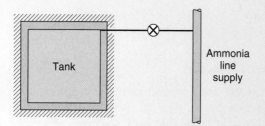

FIGURE 10.10
Ammonia tank and line
for Example 10.4.

Ammonia
line
supply

Tank

Inlet state: T_i, x_i known; state fixed.
Final state: $P_2 = P_{line}$ known.
Process: Adiabatic, no kinetic or potential energy change.
Model: Ammonia tables.

Analysis

Since the line pressure is higher than the initial pressure inside the tank, flow is going
into the tank and the flow stops when the tank pressure has increased to the line pressure. The continuity, energy, and entropy equations are

$$m_2 - m_1 = m_i$$

$$m_2 u_2 - m_1 u_1 = m_i h_i = (m_2 - m_1)h_i$$

$$m_2 s_2 - m_1 s_1 = m_i s_i + {}_1 S_{2gen}$$

where kinetic and potential energies are zero for the initial and final states and neglected
for the inlet flow.

Solution

From the ammonia tables, the initial and line state properties are

$$v_1 = 0.6995 \text{ m}^3/\text{kg} \qquad u_1 = 1369.5 \text{ kJ/kg} \qquad s_1 = 5.927 \text{ kJ/kg K}$$

$$h_i = 134.41 \text{ kJ/kg} \qquad s_i = 0.5408 \text{ kJ/kg K}$$

The initial mass is therefore

$$m_1 = V/v_1 = 1/0.6995 = 1.4296 \text{ kg}$$

It is observed that only the final pressure is known, so one property is needed. The unknowns are the final mass and final internal energy in the energy equation. Since only
one property is unknown, the two quantities are not independent. From the energy equation we have

$$m_2(u_2 - h_i) = m_1(u_1 - h_i)$$

from which it is seen that $u_2 > h_i$ and the state therefore is two-phase or superheated
vapor. Assume that the state is two phase, then

$$m_2 = V/v_2 = 1/(0.001534 + x_2 \times 0.41684)$$

$$u_2 = 133.964 + x_2 \times 1175.257$$

so the energy equation is

$$\frac{133.964 + x_2 \times 1175.257 - 134.41}{0.001534 + x_2 \times 0.041684} = 1.4296(1369.5 - 134.41) = 1765.67 \text{ kJ}$$

This equation is solved for the quality and the rest of the properties to give

$$x_2 = 0.007182 \qquad v_2 = 0.0045276 \text{ m}^3/\text{kg} \qquad s_2 = 0.5762 \text{ kJ/kg}$$

Now the final mass and the irreversibility are found

$$m_2 = V/v_2 = 1/0.0045276 = 220.87 \text{ kg}$$

$$_1S_{2\text{gen}} = m_2 s_2 - m_1 s_1 - m_i s_i = 127.265 - 8.473 - 118.673 = 0.119 \text{ kJ/K}$$

$$I_{\text{c.v.}} = T_0 \,_1S_{2\text{gen}} = 293.15 \times 0.119 = 34.885 \text{ kJ}$$

10.2 AVAILABILITY AND SECOND-LAW EFFICIENCY

What is the maximum reversible work that can be done by a given mass in a given state? In the previous section, we developed expressions for the reversible work for a given change of state for a control mass and control volume undergoing specific types of processes. For any given case, what final state will give the maximum reversible work?

The answer to this question is that, for any type of process, when the mass comes into equilibrium with the environment, no spontaneous change of state will occur and the mass will be incapable of doing any work. Therefore, if a mass in a given state undergoes a completely reversible process until it reaches a state in which it is in equilibrium with the environment, the maximum reversible work will have been done by the mass. In this sense, we refer to the availability at the original state in terms of the potential for achieving the maximum possible work by the mass.

If a control mass is in equilibrium with the surroundings, it must certainly be in pressure and temperature equilibrium with the surroundings, that is, at pressure P_0 and temperature T_0. It must also be in chemical equilibrium with the surroundings, which implies that no further chemical reaction will take place. Equilibrium with the surroundings also requires that the system have zero velocity and minimum potential energy. Similar requirements can be set forth regarding electrical and surface effects if these are relevant to a given problem.

The same general remarks can be made about a quantity of mass that undergoes a steady-state process. With a given state for the mass entering the control volume, the reversible work will be a maximum when this mass leaves the control volume in equilibrium with the surroundings. This means that as the mass leaves the control volume, it must be at the pressure and temperature of the surroundings, be in chemical equilibrium with the surroundings, and have minimum potential energy and zero velocity. (The mass leaving the control volume must of necessity have some velocity, but it can be made to approach zero.)

Let us first consider the availability associated with a steady-state process. For a control volume with a single-flow stream, the reversible work is given by Eq. 10.9. Including kinetic and potential energies, this expression is rewritten in the form,

$$w^{\text{rev}} = (h_{\text{TOT}i} - T_0 s_i) - (h_{\text{TOT}e} - T_0 s_e) + q\left(1 - \frac{T_0}{T_H}\right)$$

From the discussion of the heat engine that led to Eq. 10.1, it is clear that the last term in the expression for the reversible work is the contribution to the net reversible work from the heat transfers. These can be viewed as transfer of availability associated with q, which gives a potential to do work as in a heat engine. Such contributions are separate from the availability in the flow itself. The steady-state flow reversible work will be maximum, relative to the surroundings, when the mass leaving the control volume is in equilibrium with the surroundings. The state in which the fluid is in equilibrium with the surroundings is designated with subscript 0, and the reversible work will be maximum when $h_e = h_0$, $s_e = s_0$, $\mathbf{V}_e = 0$, and $Z_e = Z_0$. This maximum reversible work per unit mass flow without the additional heat transfers is the flow availability or exergy and is assigned the symbol ψ

$$\psi = \left(h - T_0 s + \frac{1}{2}\mathbf{V}^2 + gZ\right) - (h_0 - T_0 s_0 + gZ_0) \qquad (10.24)$$

It is written without subscript for the inlet state to indicate that this is the flow availability associated with a substance in any state as it enters the control volume in a steady-state process. The reversible work is therefore seen to be equal to the decrease in flow availability plus the reversible work that can be extracted from heat engines operating with the heat transfer at T_H and the ambient.

Irreversibility is as usual defined as the difference between the reversible work and the actual work. If we write this as a general expression on a rate basis, including summations to account for the possibility of more than one flow stream and also more than one heat transfer, the result is

$$\dot{I}_{c.v.} = \left(\sum \dot{m}_i \psi_i - \sum \dot{m}_e \psi_e\right) + \sum \left(1 - \frac{T_0}{T_j}\right)\dot{Q}_{c.v.,j} - \dot{W}_{c.v.} \qquad (10.25)$$

In this form, the irreversibility is equal to the decrease in the availability of the mass flows plus the decrease of availability of each heat transfer rate j at reservoir T_j minus the increase in availability of the surroundings that receive the actual work. The rate of irreversibility is thus seen to be the rate of destruction of availability, which is also directly proportional to the net rate of entropy increase, as noted in Eq. 10.11.

For a control mass, a similar consideration of the maximum reversible work will lead to a nonflow availability concept. In this case the volume may change, and some work is exchanged with the ambient, which is not available as useful work. The reversible work for a control mass is given by Eq. 10.17. Including kinetic and potential energies (e instead of u), this expressions is rewritten as

$$_1 w_2^{\text{rev}} = (e_1 - T_0 s_1) - (e_2 - T_0 s_2) + {}_1 q_2 \left(1 - \frac{T_0}{T_H}\right)$$

which is the maximum between the two given states. This work is available if the final state is in equilibrium with the surroundings, for which we must have $e_2 = e_0 = u_0 + gZ_0$, the kinetic energy being zero, and $s_2 = s_0$. The work done against the surroundings, w_{surr}, is

$$w_{\text{surr}} = P_0(v_0 - v_1) = -P_0(v_1 - v_0)$$

such that the maximum available work is

$$w_{\text{avail}}^{\text{max}} = w_{\text{max}}^{\text{rev}} - w_{\text{surr}}$$

$$= (e - T_0 s) - (e_0 - T_0 s_0) + P_0(v - v_0) + {}_1 q_2 \left(1 - \frac{T_0}{T_H}\right) \qquad (10.26)$$

The subscript is again dropped to indicate that this is the maximum available work at a given state having also $_1q_2$ available from a source at T_H. The nonflow availability is defined as the maximum available work from a state without the heat transfers included as

$$\phi = (e - T_0 s) - (e_0 - T_0 s_0) + P_0(v - v_0)$$

$$\phi = (e + P_0 v - T_0 s) - (e_0 + P_0 v_0 - T_0 s_0) \qquad (10.27)$$

Sometimes the definition excludes the kinetic and potential energies, in which case u should be used instead of e.

The irreversibility may again be related to the changes in availability through the difference between the reversible work and the actual work. The reversible work from above is expressed with the availability from which the actual work is subtracted to give

$$_1 I_2 = m(\phi_1 - \phi_2) + \sum \left(1 - \frac{T_0}{T_j}\right) Q_j - (_1 W_2^{\text{ac}} - P_0(V_2 - V_1)) \qquad (10.28)$$

including the possibility of having heat transfer with more than one reservoir. The irreversibility is then equal to the decrease in availability of the control mass plus the decrease in availability of the heat transfers at reservoirs T_j minus the increase in availability of the surroundings that received the actual work. It is noted again that the irreversibility expresses the net destruction of availability of the control mass and surroundings, which is proportional to the net entropy increase, as given in Eq. 10.19.

The less the irreversibility associated with a given change of state, the greater the amount of work that will be done (or the smaller the amount of work that will be required). This relation is significant in at least two regards. The first is that availability is one of our natural resources. This availability is found in such forms as oil reserves, coal reserves, and uranium reserves. Suppose we wish to accomplish a given objective that requires a certain amount of work. If this work is produced reversibly while drawing on one of the availability reserves, the decrease in availability is exactly equal to the reversible work. However, since there are irreversibilities in producing this required amount of work, the actual work will be less than the reversible work, and the decrease in availability will be greater (by the amount of the irreversibility) than if this work had been produced reversibly. Thus, the more irreversibilities we have in all our processes, the greater will be the decrease in our availability reserves.[1] The conservation and effective use of these availability reserves is an important responsibility for all of us.

The second reason that it is desirable to accomplish a given objective with the smallest irreversibility is an economic one. Work costs money, and in many cases a given objective can be accomplished at less cost when the irreversibility is less. It should be noted, however, that many factors enter into the total cost of accomplishing a given objective, and an optimization process that considers many factors is often necessary to arrive at the most economical design. For example, in a heat-transfer process, the smaller the

---

[1]In many popular talks, reference is made to our energy reserves. From a thermodynamic point of view, availability reserves would be a much more acceptable term. There is much energy in the atmosphere and the ocean, but relatively little availability.

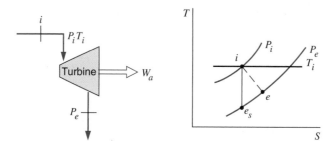

FIGURE 10.11
Irreversible turbine.

temperature difference across which the heat is transferred, the less the irreversibility. However, for a given rate of heat transfer, a smaller temperature difference will require a larger (and therefore more expensive) heat exchanger. These various factors must all be considered in developing the optimum and most economical design.

In many engineering decisions, other factors, such as the impact on the environment (for example, air pollution and water pollution) and the impact on society must be considered in developing the optimum design.

Along with the increased use of availability analysis in recent years, a term called the second-law efficiency has come into more common use. This term refers to comparison of the desired output of a process with the cost, or input, in terms of the thermodynamics availability. Thus, the isentropic turbine efficiency defined by Eq. 9.27 as the actual work output divided by the work for a hypothetical isentropic expansion from the same inlet state to the same exit pressure might well be called a first-law efficiency, in that it is a comparison of two energy quantities. The second-law efficiency, as just described, would be the actual work output of the turbine divided by the decrease in availability from the same inlet state to the same exit state. For the turbine shown in Fig. 10.11, the second-law efficiency is

$$\eta_{\text{2nd law}} = \frac{w_a}{\psi_i - \psi_e} \tag{10.29}$$

In this sense, this concept provides a rating or measure of the real process in terms of the actual change of state and is simply another convenient way of utilizing the concept of thermodynamic availability. In a similar manner, the second-law efficiency of a pump or compressor is the ratio of the increase in availability to the work input to the device.

EXAMPLE 10.5   An insulated steam turbine, Fig. 10.12, receives 30 kg of steam per second at 3 MPa, 350°C. At the point in the turbine where the pressure is 0.5 MPa, steam is bled off for processing equipment at the rate of 5 kg/s. The temperature of this steam is 200°C. The balance of the steam leaves the turbine at 15 kPa, 90% quality. Determine the availability per kilogram of the steam entering and at both points at which steam leaves the turbine, the isentropic efficiency and the second-law efficiency for this process.

> *Control volume:*   Turbine.
> *Inlet state:* P_1, T_1 known; state fixed.
> *Exit state:* P_2, T_2 known; P_3, x_3 known; both states fixed.

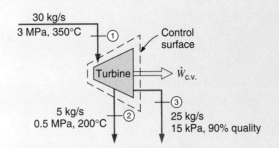

FIGURE 10.12
Sketch for Example 10.5.

Process: Steady state.
Model: Steam tables.

Analysis

The availability at any point for the steam entering or leaving the turbine is given by Eq. 10.24,

$$\psi = (h - h_0) - T_0(s - s_0) + \frac{\mathbf{V}^2}{2} + g(Z - Z_0)$$

Since there are no changes in kinetic and potential energy in this problem, this equation reduces to

$$\psi = (h - h_0) - T_0(s - s_0)$$

For the ideal isentropic turbine,

$$\dot{W}_s = \dot{m}_1 h_1 - \dot{m}_2 h_{2s} - \dot{m}_3 h_{3s}$$

For the actual turbine,

$$\dot{W} = \dot{m}_1 h_1 - \dot{m}_2 h_2 - \dot{m}_3 h_3$$

Solution

At the pressure and temperature of the surroundings, 0.1 MPa, 25°C, the water is a slightly compressed liquid, and the properties of the water are essentially equal to those for saturated liquid at 25°C.

$$h_0 = 104.9 \text{ kJ/kg} \qquad s_0 = 0.3674 \text{ kJ/kg K}$$

From Eq. 10.24

$$\psi_1 = (3115.3 - 104.9) - 298.15(6.7428 - 0.3674) = 1109.6 \text{ kJ/kg}$$

$$\psi_2 = (2855.4 - 104.9) - 298.15(7.0592 - 0.3674) = 755.3 \text{ kJ/kg}$$

$$\psi_3 = (2361.8 - 104.9) - 298.15(7.2831 - 0.3674) = 195.0 \text{ kJ/kg}$$

$$\dot{m}_1 \psi_1 - \dot{m}_2 \psi_2 - \dot{m}_3 \psi_3 = 30(1109.6) - 5(755.3) - 25(195.0) = 24\,637 \text{ kW}$$

For the ideal isentropic turbine,

$$s_{2s} = 6.7428 = 1.8606 + x_{2s} \times 4.9606, \qquad x_{2s} = 0.9842$$

$$h_{2s} = 640.2 + 0.9842 \times 2108.5 = 2715.4$$

$$s_{3s} = 6.7428 = 0.7549 + x_{3s} \times 7.2536, \qquad x_{3s} = 0.8255$$

$$h_{3s} = 225.9 + 0.8255 \times 2373.1 = 2184.9$$

$$\dot{W}_s = 30(3115.3) - 5(2715.4) - 25(2184.9) = 25\,260 \text{ kW}$$

For the actual turbine,

$$\dot{W} = 30(3115.3) - 5(2855.4) - 25(2361.8) = 20\,137 \text{ kW}$$

The isentropic efficiency is

$$\eta_s = \frac{20\,137}{25\,260} = 0.797$$

and the second-law efficiency is

$$\eta_{\text{2nd law}} = \frac{20\,137}{24\,637} = 0.817$$

For a device that does not involve the production or the input of work, the definition of second-law efficiency refers to the accomplishment of the goal of the process relative to the process input, in terms of availability changes or transfers. For example, in a heat exchanger, energy is transferred from a high-temperature fluid stream to a low-temperature fluid stream, as shown in Fig. 10.13, in which case the second-law efficiency is defined as

$$\eta_{\text{2nd law}} = \frac{\dot{m}_1(\psi_2 - \psi_1)}{\dot{m}_3(\psi_3 - \psi_4)} \tag{10.30}$$

The previous expressions for the second-law efficiency can be presented by a single expression. First, notice that the actual work from Eq. 10.25 is

$$\dot{W}_{\text{c.v.}} = \dot{\Phi}_{\text{source}} - \dot{I}_{\text{c.v.}} = \dot{\Phi}_{\text{source}} - T\dot{S}_{\text{gen c.v.}} \tag{10.31}$$

where $\dot{\Phi}_{\text{source}}$ is the total rate of availability supplied from all sources: flows, heat transfers, and work inputs. In other words, the outgoing availability, $\dot{W}_{\text{c.v.}}$, equals the incoming availability less the irreversibility. Then for all cases we may write

$$\eta_{\text{2nd law}} = \frac{\dot{\Phi}_{\text{wanted}}}{\dot{\Phi}_{\text{source}}} = \frac{\dot{\Phi}_{\text{source}} - \dot{I}_{\text{c.v.}}}{\dot{\Phi}_{\text{source}}} \tag{10.32}$$

FIGURE 10.13 A two-fluid heat exchanger.

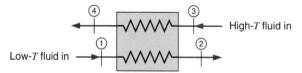

and the wanted quantity is then expressed as availability whether it actually is a work term or a heat transfer. We can verify that this covers the turbine, Eq. 10.29, the pump or compressor, where work input is the source, and the heat exchanger efficiency in Eq. 10.30.

EXAMPLE 10.6

In a boiler, heat is transferred from the products of combustion to the steam. The temperature of the products of combustion decreases from 1100°C to 550°C, while the pressure remains constant at 0.1 MPa. The average constant-pressure specific heat of the products of combustion is 1.09 kJ/kg K. The water enters at 0.8 MPa, 150°C, and leaves at 0.8 MPa, 250°C. Determine the second-law efficiency for this process and the irreversibility per kilogram of water evaporated.

Control volume: Overall heat exchanger.
Sketch: Fig. 10.14.
Inlet states: Both known, given in Fig. 10.14.
Exit states: Both known, given in Fig. 10.14.
Process: Overall, adiabatic.
Diagram: Fig. 10.15.
Model: Products—ideal gas, constant specific heat. Water—steam tables.

Analysis

For the products, the entropy change for this constant-pressure process is

$$(s_e - s_i)_{\text{prod}} = C_{po} \ln \frac{T_e}{T_i}$$

For this control volume we can write the following governing equations:
Continuity equation:

$$(\dot{m}_i)_{\text{H}_2\text{O}} = (\dot{m}_e)_{\text{H}_2\text{O}} \tag{a}$$

$$(\dot{m}_i)_{\text{prod}} = (\dot{m}_e)_{\text{prod}} \tag{b}$$

First law (a steady-state process):

$$(\dot{m}_i h_i)_{\text{H}_2\text{O}} + (\dot{m}_i h_i)_{\text{prod}} = (\dot{m}_e h_e)_{\text{H}_2\text{O}} + (\dot{m}_e h_e)_{\text{prod}} \tag{c}$$

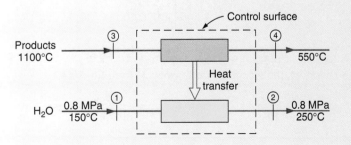

FIGURE 10.14
Sketch for Example 10.6.

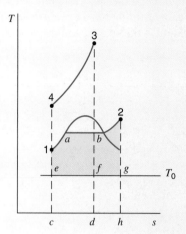

FIGURE 10.15
Temperature-entropy diagram for Example 10.6.

Second law (the process is adiabatic for the control volume shown):

$$(\dot{m}_e s_e)_{H_2O} + (\dot{m}_e s_e)_{prod} \geq (\dot{m}_i s_i)_{H_2O} + (\dot{m}_i s_i)_{prod}$$

Solution

From Eqs. a, b, and c, we can calculate the ratio of the mass flow of products to the mass flow of water.

$$\dot{m}_{prod}(h_i - h_e)_{prod} = \dot{m}_{H_2O}(h_e - h_i)_{H_2O}$$

$$\frac{\dot{m}_{prod}}{\dot{m}_{H_2O}} = \frac{(h_e - h_i)_{H_2O}}{(h_i - h_e)_{prod}} = \frac{2950 - 632.2}{1.09(1100 - 550)} = 3.866$$

The increase in availability of the water is, per kilogram of water,

$$\psi_2 - \psi_1 = (h_2 - h_1) - T_0(s_2 - s_1)$$

$$= (2950 - 632.2) - 298.15(7.0384 - 1.8418)$$

$$= 768.4 \text{ kJ/kg } H_2O$$

The decrease in availability of the products, per kilogram of water, is

$$\frac{\dot{m}_{prod}}{\dot{m}_{H_2O}}(\psi_3 - \psi_4) = \frac{\dot{m}_{prod}}{\dot{m}_{H_2O}}[(h_3 - h_4) - T_0(s_3 - s_4)]$$

$$= 3.866\left[1.09(1100 - 550) - 298.15\left(1.09 \ln \frac{1373.15}{823.15}\right)\right]$$

$$= 1674.7 \text{ kJ/kg } H_2O$$

Therefore, the second-law efficiency is, from Eq. 10.30,

$$\eta_{2nd \, law} = \frac{768.4}{1674.7} = 0.459$$

From Eq. 10.25, the process irreversibility per kilogram of water is

$$\frac{\dot{I}}{\dot{m}_{H_2O}} = \sum_i \frac{\dot{m}_i}{\dot{m}_{H_2O}} \psi_i - \sum_2 \frac{\dot{m}_e}{\dot{m}_{H_2O}} \psi_e$$

$$= (\psi_1 - \psi_2) + \frac{\dot{m}_{prod}}{\dot{m}_{H_2O}}(\psi_3 - \psi_4)$$

$$= (-768.4 + 1674.7) = 906.3 \text{ kJ/kg } H_2O$$

It is also of interest to determine the net change of entropy. The change in the entropy of the water is

$$(s_2 - s_1)_{H_2O} = 7.0384 - 1.8418 = 5.1966 \text{ kJ/kg } H_2O \text{ K}$$

The change in the entropy of the products is

$$\frac{\dot{m}_{prod}}{\dot{m}_{H_2O}}(s_4 - s_3)_{prod} = -3.866\left(1.09 \ln \frac{1373.15}{823.15}\right) = -2.1564 \text{ kJ/kg } H_2O \text{ K}$$

Thus, there is a net increase in entropy during the process. The irreversibility could also have been calculated from Eq. 10.11:

$$\dot{I} = \sum \dot{m}_e T_0 s_e - \sum \dot{m}_i T_0 s_i - \dot{Q}_{c.v.} = T_0 \dot{S}_{gen}$$

For the control volume selected, $\dot{Q}_{c.v.} = 0$, and therefore

$$\frac{\dot{I}}{\dot{m}_{H_2O}} = T_0(s_2 - s_1)_{H_2O} + \frac{\dot{m}_{prod}}{\dot{m}_{H_2O}} T_0(s_4 - s_3)_{prod}$$

$$= 298.15(5.1966) + 298.15(-2.1564)$$

$$= 906.3 \text{ kJ/kg } H_2O$$

These two processes are shown on the T–s diagram of Fig. 10.15. Line 3–4 represents the process for the 3.866 kg of products. Area 3–4–c–d–3 represents the heat transferred from the 3.866 kg of products of combustion, and area 3–4–e–f–3 represents the decrease in availability of these products. Area 1–a–b–2–h–c–1 represents the heat transferred to the water, and this is equal to area 3–4–c–d–3, which represents the heat transferred from the products of combustion. Area 1–a–b–2–g–e–1 represents the increase in availability of the water. The difference between area 3–4–e–f–3 and area 1–a–b–2–g–e–1 represents the net decrease in availability. It is readily shown that this net change is equal to area f–g–h–d–f, or $T_0(\Delta s)_{net}$. Since the actual work is zero, this area also represents the irreversibility, which agrees with our calculation.

10.3 EXERGY BALANCE EQUATION

The previous treatment of availability or exergy in different situations was done separately for the steady-flow, the control mass, and the transient processes. For each case, an actual process was compared to an ideal counterpart, which led to the reversible work and the irreversibility. When the reference was made with respect to the ambient state, we found the flow availability, ψ in Eq. 10.24, and the no-flow availability, ϕ in Eq. 10.27. We want to show that these forms of availability are consistent with one another. The whole concept is unified by a formulation of the exergy for a general control volume from which we will recognize all the previous forms of availability as special cases of the more general form.

In this analysis, we start out with the definition of exergy, $\Phi = m\phi$, as the maximum available work at a given state of a mass from Eq. 10.27, as

$$\Phi = m\phi = m(e - e_0) + P_0 m(v - v_0) - T_0 m(s - s_0) \tag{10.33}$$

Here subscript "0" refers to the ambient state with zero kinetic energy, the dead state, from which we take our reference. Because the properties at the reference state are constants, the rate of change for Φ becomes

$$\frac{d\Phi}{dt} = \frac{dme}{dt} - e_0 \frac{dm}{dt} + P_0 \frac{dV}{dt} - P_0 v_0 \frac{dm}{dt} - T_0 \frac{dms}{dt} + T_0 s_0 \frac{dm}{dt}$$

$$= \frac{dme}{dt} + P_0 \frac{dV}{dt} - T_0 \frac{dms}{dt} - (h_0 - T_0 s_0) \frac{dm}{dt} \tag{10.34}$$

and we used, $h_0 = e_0 + P_0 v_0$, to shorten the expression. Now we substitute the rate of change of mass from the continuity equation, Eq. 6.1,

$$\frac{dm}{dt} = \sum \dot{m}_i - \sum \dot{m}_e$$

the rate of change of total energy from the energy equation, Eq. 6.8,

$$\frac{dE}{dt} = \frac{dme}{dt} = \sum \dot{Q}_{c.v.} - \dot{W}_{c.v.} + \sum \dot{m}_i h_{\text{tot } i} - \sum \dot{m}_e h_{\text{tot } e}$$

and the rate of change of entropy from the entropy equation, Eq. 9.2

$$\frac{dS}{dt} = \frac{dms}{dt} = \sum \dot{m}_i s_i - \sum \dot{m}_e s_e + \sum \frac{\dot{Q}_{c.v.}}{T} + \dot{S}_{\text{gen}}$$

into the rate of exergy equation Eq. 10.34. When that is done, we get

$$\frac{d\Phi}{dt} = \sum \dot{Q}_{c.v.} - \dot{W}_{c.v.} + \sum \dot{m}_i h_{\text{tot } i} - \sum \dot{m}_e h_{\text{tot } e} + P_0 \frac{dV}{dt}$$

$$- T_0 \sum \dot{m}_i s_i + T_0 \sum \dot{m}_e s_e - \sum T_0 \frac{\dot{Q}_{c.v.}}{T} - T_0 \dot{S}_{\text{gen}}$$

$$- (h_0 - T_0 s_0) \left[\sum \dot{m}_i - \sum \dot{m}_e \right] \tag{10.35}$$

Now collect the terms relating to the heat transfer together and those relating to the flow together and group them as

$$\frac{d\Phi}{dt} = \sum \left(1 - \frac{T_0}{T} \right) \dot{Q}_{c.v.} \qquad \text{Transfer by heat at } T$$

$$- \dot{W}_{c.v.} + P_0 \frac{dV}{dt} \qquad \text{Transfer by shaft/boundary work}$$

$$+ \sum \dot{m}_i \psi_i - \sum \dot{m}_e \psi_e \qquad \text{Transfer by flow}$$

$$- T_0 \dot{S}_{\text{gen}} \qquad \text{Exergy destruction} \tag{10.36}$$

Here we recognize the appearance of the specific flow exergies, ψ from Eq. 10.24,

$$\psi_i = h_{\text{tot }i} - h_0 - T_0(s_i - s_0)$$

$$\psi_e = h_{\text{tot }e} - h_0 - T_0(s_e - s_0) \qquad (10.37)$$

as the terms associated with the flow. The rate equation for exergy can verbally be stated as all the other balance equations

Rate of exergy storage = Transfer by heat + Transfer by shaft/boundary work

+ Transfer by flow − Exergy destruction

and we notice that all the transfers take place with some surroundings and thus do not add up to any net change when the total world is considered. Only the exergy destruction due to entropy generation lowers the overall exergy level, and we can thus identify the regions in space where this occurs as the locations that have entropy generation. The exergy destruction is identical to the previously defined term, irreversibility.

Let us briefly consider some of the special cases covered in Section 10.1. First look at the steady-flow control volume with a single flow in and out. For steady state we have no storage of mass, energy, entropy, or exergy, and the volume is constant, so solve for the work term in Eq. 10.36 as

$$\dot{W}_{\text{c.v.}} = \sum \left(1 - \frac{T_0}{T}\right) \dot{Q}_{\text{c.v.}} + \dot{m}_i \psi_i - \dot{m}_e \psi_e - T_0 \dot{S}_{\text{gen}} \qquad (10.38)$$

Now divide by the mass flow rate to give the specific work

$$w = \sum \left(1 - \frac{T_0}{T}\right) q_{\text{c.v.}} + \psi_i - \psi_e - T_0 s_{\text{gen}} \qquad (10.39)$$

The reference state offset in the two specific flow exergies drop out as

$$\psi_i - \psi_e = h_{\text{tot }i} - h_0 - T_0(s_i - s_0) - [h_{\text{tot }e} - h_0 - T_0(s_e - s_0)]$$

$$= h_{\text{tot }i} - h_{\text{tot }e} - T_0(s_i - s_e)$$

and the specific work becomes

$$w = \sum \left(1 - \frac{T_0}{T}\right) q_{\text{c.v.}} + h_{\text{tot }i} - h_{\text{tot }e} - T_0(s_i - s_e) - T_0 s_{\text{gen}}$$

$$= w^{\text{rev}} - T_0 s_{\text{gen}} \qquad (10.40)$$

It reduces to the expression for reversible work in Eq. 10.9 when there is a single heat-transfer term and no entropy generation. We can also see how much lower the actual specific work is due to the entropy generation; namely, it equals the reversible specific work minus the specific exergy destruction.

EXAMPLE 10.7   Let us look at the flows and fluxes of exergy for the feedwater heater in Example 10.1. The feedwater heater has a single flow, two heat transfers, and no work involved. When we do the balance of terms in Eq. 10.39 and evaluate the flow exergies from Eq. 10.37, we need the reference properties (take saturated liquid instead of 100 kPa at 25°C):

Table B.1.1: $h_0 = 104.87$ kJ/kg, $s_0 = 0.3673$ kJ/kg K

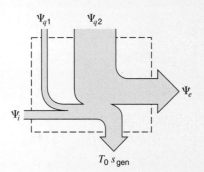

FIGURE 10.16
Fluxes, flows, and destruction of exergy in the feedwater heater.

The flow exergies become

$$\psi_i = h_{\text{tot } i} - h_0 - T_0(s_i - s_0)$$

$$= 171.97 - 104.87 - 298.2 \times (0.5705 - 0.3687) = 6.92 \text{ kJ/kg}$$

$$\psi_e = h_{\text{tot } e} - h_0 - T_0(s_e - s_0)$$

$$= 765.25 - 104.87 - 298.2 \times (2.1341 - 0.3687) = 133.94 \text{ kJ/kg}$$

and the exergy fluxes from each of the heat transfers are

$$\left(1 - \frac{T_0}{T_1}\right) q_1 = \left(1 - \frac{298.2}{373.2}\right) 180 = 36.17 \text{ kJ/kg}$$

$$\left(1 - \frac{T_0}{T_2}\right) q_2 = \left(1 - \frac{298.2}{473.2}\right) 413.28 = 152.84 \text{ kJ/kg}$$

The destruction of exergy is then the balance ($w = 0$) of Eq. 10.39 as

$$T_0 s_{\text{gen}} = \sum \left(1 - \frac{T_0}{T}\right) q_{\text{c.v.}} + \psi_i - \psi_e$$

$$= 36.17 + 152.84 + 6.92 - 133.94 = 62.0 \text{ kJ/kg}$$

We can now express the heater's second-law efficiency as

$$\eta_{\text{2nd law}} = \frac{\dot{\Phi}_{\text{source}} - \dot{I}_{\text{c.v.}}}{\dot{\Phi}_{\text{source}}} = \frac{36.17 + 152.84 - 62.0}{36.17 + 152.84} = 0.67$$

The exergy fluxes are shown in Fig. 10.16, and the second-law efficiency shows that there is a potential for improvement. We should lower the temperature difference between the source and the water flow by adding more energy from the low-temperature source, thus decreasing the irreversibility.

The no flow steady-state control volume of a heat engine, heat pump, electric heating element, or refrigerator has no storage, and the work is given by Eq. 10.38 with $\dot{m} = 0$

$$\dot{W}_{\text{c.v.}} = \sum \left(1 - \frac{T_0}{T}\right) \dot{Q}_{\text{c.v.}} - T_0 \dot{S}_{\text{gen}} \qquad (10.41)$$

The work equals the Carnot heat engine work reduced by the exergy destruction of any irreversible process occurring inside the control volume. For the case of a single heat transfer and a reversible process, the work reduces to that given in Eq. 10.1 except it is expressed as a rate in Eq. 10.41.

EXAMPLE 10.8   Assume a 500 W heating element in a stove with an element surface temperature of 1000 K. On top of the element is a ceramic top with a top surface temperature of 500 K, where both of these are shown in Fig. 10.17. Let us disregard any heat transfer downwards and follow the flux of exergy and find the exergy destruction in the process.

Solution

Take just the heating element as a control volume in steady state with electrical work going in and heat transfer going out.

$$\text{Energy Eq.:} \qquad 0 = \dot{W}_{electrical} - \dot{Q}_{out}$$

$$\text{Entropy Eq.:} \qquad 0 = -\frac{\dot{Q}_{out}}{T_{sur}} + \dot{S}_{gen}$$

$$\text{Exergy Eq.:} \qquad 0 = -\left(1 - \frac{T_0}{T}\right)\dot{Q}_{out} - (-\dot{W}_{electrical}) - T_0\dot{S}_{gen}$$

From the balance equations we get

$$\dot{Q}_{out} = \dot{W}_{electrical} = 500 \text{ W}$$

$$\dot{S}_{gen} = \dot{Q}_{out}/T_{surf} = 500 \text{ W}/1000 \text{ K} = 0.5 \text{ W/K}$$

$$\dot{\Phi}_{destruction} = T_0\dot{S}_{gen} = 298.15 \text{ K} \times 0.5 \text{ W/K} = 149 \text{ W}$$

$$\dot{\Phi}_{transfer\ out} = \left(1 - \frac{T_0}{T}\right)\dot{Q}_{out} = \left(1 - \frac{298.15}{1000}\right)500 = 351 \text{ W}$$

so the heating element receives 500 W of exergy flux, destroys 149 W, and gives out the balance of 351 W with the heat transfer at 1000 K.

FIGURE 10.17 The electric heating element and ceramic top of a stove.

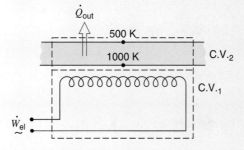

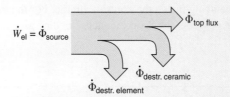

FIGURE 10.18 The fluxes and destruction terms of exergy.

Take a second control volume from the heating element surface to the ceramic stove top. Here heat transfer comes in at 1000 K and leaves at 500 K with no work involved.

$$\text{Energy Eq.:} \qquad 0 = \dot{Q}_{in} - \dot{Q}_{out}$$

$$\text{Entropy Eq.:} \qquad 0 = \frac{\dot{Q}_{in}}{T_{surf}} - \frac{\dot{Q}_{out}}{T_{top}} + \dot{S}_{gen}$$

$$\text{Exergy Eq.:} \qquad 0 = \left(1 - \frac{T_0}{T_{surf}}\right)\dot{Q}_{in} - \left(1 - \frac{T_0}{T_{top}}\right)\dot{Q}_{out} - T_0\dot{S}_{gen}$$

From the energy equation we see that the two heat transfers are equal, and the entropy generation then becomes

$$\dot{S}_{gen} = \frac{\dot{Q}_{out}}{T_{top}} - \frac{\dot{Q}_{in}}{T_{surf}} = 500\left(\frac{1}{500} - \frac{1}{1000}\right)\text{W/K} = 0.5 \text{ W/K}$$

The terms in the exergy equation become

$$0 = \left(1 - \frac{298.15}{1000}\right)500 \text{ W} - \left(1 - \frac{298.15}{500}\right)500 \text{ W} - 298.15 \text{ K} \times 0.5 \text{ W/K}$$

or

$$0 = 351 \text{ W} - 202 \text{ W} - 149 \text{ W}$$

This means that the top layer receives 351 W of exergy from the electric heating element and gives out 202 W from the top surface, having destroyed 149 W of exergy in the process. The flow of exergy and its destruction are illustrated in Fig. 10.18.

A control mass going through a process starting at state 1 and ending at state 2 is described in terms of finite changes by integration of the rate equation over time. With no mass flow rates, time integration of Eq. 10.36 gives

$$\Phi_2 - \Phi_1 = \int_1^2 \sum\left(1 - \frac{T_0}{T}\right)\dot{Q}_{c.v.}\, dt - {}_1W_2 + P_0(V_2 - V_1) - T_{01}S_{2\,gen} \qquad (10.42)$$

The reference state properties (e, v, and s) drop out in the storage term, which becomes

$$\Phi_2 - \Phi_1 = m(e_2 - e_1) + P_0(V_2 - V_1) - T_0 m(s_2 - s_1) \qquad (10.43)$$

With a single heat transfer taking place at a constant T_H, the available work becomes

$${}_1W_2 - P_0(V_2 - V_1) = \left(1 - \frac{T_0}{T_H}\right){}_1Q_2 - (\Phi_2 - \Phi_1) - T_{01}S_{2\,gen} \qquad (10.44)$$

so it is equal to the exergy from the heat transfer lowered by the increase in the stored exergy and the amount of exergy destroyed in the process. This result for a reversible process is identical to the reversible work in Eq. 10.17 minus the work to the ambient.

SUMMARY

Work out of a Carnot-cycle heat engine is the available energy in the heat transfer from the hot source; the heat transfer to the ambient is unavailable. When an actual device is compared to an ideal device with the same flows and states in and out, we get to the concept of reversible work and exergy (availability). The reversible work is the maximum work we can get out of a given set of flows and heat transfers or, alternatively, the minimum work we have to put into the device. The comparison between the actual work and the theoretical maximum work gives a second-law efficiency. When exergy (availability) is used, the second-law efficiency can also be used for devices that do not involve shaft-work such as heat exchangers. In that case, we compare the exergy given out by one flow to the exergy gained by the other flow, giving a ratio of exergies instead of energies used for the first-law efficiency. Any irreversibility (entropy generation) in a process does destroy exergy (availability) and is undesirable. The concept of available work can be used to make a general definition of exergy as being the reversible work minus the work that must go to the ambient. From this definition we can construct the exergy balance equation and apply it to different control volumes. From a design perspective we can then focus on the flows and fluxes of exergy and improve the processes that destroy exergy.

You should have learned a number of skills and acquired abilities from studying this chapter that will allow you to

- Understand the concept of available energy.
- Understand that energy and availability are different concepts.
- Be able to conceptualize the ideal counterpart to an actual system and find the reversible work and heat transfer in the ideal system.
- Understand the difference between a first-law and a second-law efficiency.
- Relate the second-law efficiency to the transfer and destruction of availability.
- Be able to look at flows (fluxes) of exergy.
- Determine irreversibilities as the destruction of exergy.
- Know that destruction of exergy is due to entropy generation.
- Know that transfers of exergy do not change total or net exergy in the world.
- Know that the exergy equation is based on the energy and entropy equations and thus does not add another law.

KEY CONCEPTS AND FORMULAS

Available work from heat

$$W = Q\left(1 - \frac{T_0}{T_H}\right)$$

Reversible flow work with extra q_0^{rev}

from ambient at T_0 and q in at T_H

$$q_0^{\text{rev}} = T_0(s_e - s_i) - q\frac{T_0}{T_H}$$

$$w^{\text{rev}} = h_i - h_e - T_0(s_i - s_e) + q\left(1 - \frac{T_0}{T_H}\right)$$

Flow irreversibility	$i = w^{\text{rev}} - w = q_0^{\text{rev}} = T_0 \dot{S}_{\text{gen}}/\dot{m} = T_0 s_{\text{gen}}$

Reversible work C.M.
$$_1W_2^{\text{rev}} = T_0(S_2 - S_1) - (U_2 - U_1) + {}_1Q_2\left(1 - \frac{T_0}{T_H}\right)$$

Irreversibility C.M.
$$_1I_2 = T_0(S_2 - S_1) - {}_1Q_2\frac{T_0}{T_H} = T_{01}S_{2\,\text{gen}}$$

Second-law efficiency
$$\eta_{\text{2nd law}} = \frac{\dot{\Phi}_{\text{gained}}}{\dot{\Phi}_{\text{supplied}}} = \frac{\dot{\Phi}_{\text{supplied}} - \dot{\Phi}_{\text{destroyed}}}{\dot{\Phi}_{\text{supplied}}}$$

Exergy, flow availability
$$\psi = [h - T_0 s + \tfrac{1}{2}\mathbf{V}^2 + gZ] - [h_0 - T_0 s_0 + gZ_0]$$

Exergy, stored
$$\phi = (e - e_0) + P_0(v - v_0) - T_0(s - s_0); \quad \Phi = m\phi$$

Exergy transfer by heat
$$\phi_{\text{transfer}} = q\left(1 - \frac{T_0}{T_H}\right)$$

Exergy transfer by flow
$$\phi_{\text{transfer}} = h_{\text{tot } i} - h_{\text{tot } e} - T_0(s_i - s_e)$$

Exergy rate Eq.
$$\frac{d\Phi}{dt} = \sum\left(1 - \frac{T_0}{T}\right)\dot{Q}_{\text{c.v.}} - \dot{W}_{\text{c.v.}} + P_0\frac{dV}{dt}$$
$$+ \sum \dot{m}_i \psi_i - \sum \dot{m}_e \psi_e - T_0 \dot{S}_{\text{gen}}$$

Exergy Eq. C.M.
$$\Phi_2 - \Phi_1 = \left(1 - \frac{T_0}{T_H}\right){}_1Q_2 - {}_1W_2$$
$$+ P_0(V_2 - V_1) - {}_1I_2$$

CONCEPT-STUDY GUIDE PROBLEMS

10.1 Can I have any energy transfer as heat transfer that is 100% available?

10.2 Is energy transfer as work 100% available?

10.3 We cannot create or destroy energy, but how about available energy?

10.4 Energy can be stored as internal energy, potential energy, or kinetic energy. Are those energy forms all 100% available?

10.5 Is all the energy in the ocean available?

10.6 Does a reversible process change the availability if there is no work involved?

10.7 Is the reversible work between two states the same as ideal work for the device?

10.8 When is the reversible work the same as the isentropic work?

10.9 If I heat some cold liquid water to T_0, do I increase its availability?

10.10 Are reversible work and availability (exergy) connected?

10.11 Consider availability (exergy) associated with a flow. The total exergy is based on the thermodynamic state and the kinetic and potential energies. Can they all be negative?

10.12 A flow of air at 1000 kPa, 300 K, is throttled to 500 kPa. What is the irreversibility? What is the drop in flow availability?

10.13 A steam turbine inlet is at 1200 kPa, 500°C. The actual exit is at 300 kPa, with an actual work of 407 kJ/kg. What is its second-law efficiency?

10.14 A heat exchanger increases the availability of 3 kg/s water by 1650 kJ/kg using 10 kg/s air coming in at 1400 K and leaving with 600 kJ/kg less availability. What are the irreversibility and the second-law efficiency?

10.15 A heat engine receives 1 kW heat transfer at 1000 K and gives out 600 W as work with the rest as heat transfer to the ambient. What are the fluxes of exergy in and out?

10.16 A heat engine receives 1 kW heat transfer at 1000 K and gives out 600 W as work with the rest as heat transfer to the ambient. Find its first- and second-law efficiencies.

10.17 Is the exergy equation independent of the energy and entropy equations?

10.18 A heat pump has a coefficient of performance of 2 using a power input of 2 kW. Its low temperature is 20°C and the high temperature is 80°C, with an ambient at T_0. Find the fluxes of exergy associated with the energy fluxes in and out.

10.19 Use the exergy balance equation to find the efficiency of a steady-state Carnot heat engine operating between two fixed temperature reservoirs.

10.20 Find the second-law efficiency of the heat pump in Problem 10.18.

HOMEWORK PROBLEMS
Available Energy, Reversible Work

10.21 Find the availability of 100 kW delivered at 500 K when the ambient temperature is 300 K.

10.22 A control mass gives out 10 kJ of energy in the form of

a. Electrical work from a battery.

b. Mechanical work from a spring.

c. Heat transfer at 500°C.

Find the change in availability of the control mass for each of the three cases.

10.23 A heat engine receives 5 kW at 800 K and 10 kW at 1000 K, rejecting energy by heat transfer at 600 K. Assume it is reversible and find the power output. How much power could be produced if it could reject energy at $T_0 = 298$ K?

10.24 The compressor in a refrigerator takes refrigerant R-134a in at 100 kPa, −20°C, and compresses it to 1 MPa, 40°C. With the room at 20°C find the minimum compressor work.

10.25 Find the specific reversible work for a steam turbine with inlet at 4 MPa and 500°C and an actual exit state of 100 kPa, $x = 1.0$ with 25°C ambient surroundings.

10.26 Calculate the reversible work out of the two-stage turbine shown in Problem 6.82, assuming the ambient is at 25°C. Compare this to the actual work that was found to be 18.08 MW.

10.27 A household refrigerator has a freezer at T_F and a cold space at T_C from which energy is removed and rejected to the ambient at T_A as shown in Fig. P10.27. Assuming that the rate of heat transfer from the cold space, $\dot{Q}_c$, is the same as from the freezer, $\dot{Q}_F$, find an expression for the minimum power into the heat pump. Evaluate this power when $T_A = 20°C$, $T_C = 5°C$, $T_F = -10°C$, and $\dot{Q}_F = 3$ kW.

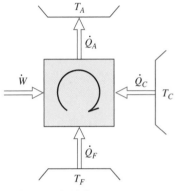

FIGURE P10.27

10.28 Find the specific reversible work for an R-134a compressor with inlet state of −20°C, 100 kPa and an exit state of 600 kPa, 50°C. Use a 25°C ambient temperature.

10.29 An air compressor takes air in at the state of the surroundings, 100 kPa, 300 K. The air exits at 400 kPa, 200°C, at the rate of 2 kg/s. Determine the minimum compressor work input.

10.30 A steam turbine receives steam at 6 MPa, 800°C. It has a heat loss of 49.7 kJ/kg and an isentropic efficiency of 90%. For an exit pressure of 15 kPa and surroundings at 20°C, find the actual work and the reversible work between the inlet and the exit.

10.31 An air compressor receives atmospheric air at $T_0 = 17°C$, 100 kPa, and compresses it up to 1400 kPa. The compressor has an isentropic efficiency of 88%, and it loses energy by heat transfer to the atmosphere as 10% of the isentropic work. Find the actual exit temperature and the reversible work.

10.32 Air flows through a constant pressure heating device, shown in Fig. P10.32. It is heated up in a reversible process with a work input of 200 kJ/kg air flowing. The device exchanges heat with the ambient at 300 K. The air enters at 300 K, 400 kPa. Assuming constant specific heat, develop an expression for the exit temperature and solve for it by iterations.

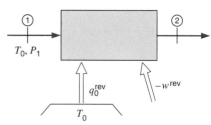

FIGURE P10.32

10.33 A piston/cylinder has forces on the piston so it keeps constant pressure. It contains 2 kg of ammonia at 1 MPa, 40°C, and is now heated to 100°C by a reversible heat engine that receives heat from a 200°C source. Find the work out of the heat engine.

10.34 A rock bed consists of 6000 kg granite and is at 70°C. A small house with lumped mass of 12 000 kg wood and 1000 kg iron is at 15°C. They are now brought to a uniform final temperature with no external heat transfer by connecting the house and rock bed through some heat engines. If the process is reversible, find the final temperature and the work done during the process.

10.35 An airflow of 5 kg/min at 1500 K, 125 kPa, goes through a constant pressure heat exchanger, giving energy to a heat engine shown in Fig. P10.35. The air exits at 500 K, and the ambient is at 298 K, 100 kPa. Find the rate of heat transfer delivered to the engine and the power the engine can produce.

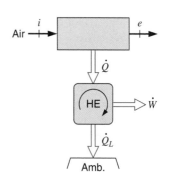

FIGURE P10.35

Irreversibility

10.36 Calculate the irreversibility for the condenser in Problem 9.53 assuming an ambient temperature at 17°C.

10.37 A constant-pressure piston/cylinder contains 2 kg of water at 5 MPa and 100°C. Heat is added from a reservoir at 700°C to the water until it reaches 700°C. We want to find the total irreversibility in the process.

10.38 Calculate the reversible work and irreversibility for the process described in Problem 5.97, assuming that the heat transfer is with the surroundings at 20°C.

10.39 A supply of steam at 100 kPa, 150°C, is needed in a hospital for cleaning purposes at a rate of 15 kg/s. A supply of steam at 150 kPa, 250°C, is available from a boiler, and tap water at 100 kPa, 15°C, is also available. The two sources are then mixed in a mixing chamber to generate the desired state as output. Determine the rate of irreversibility of the mixing process.

10.40 The throttle process in Example 6.5 is an irreversible process. Find the reversible work and irreversibility assuming an ambient temperature at 25°C.

10.41 Two flows of air both at 200 kPa of equal flow rates mix in an insulated mixing chamber. One flow is at 1500 K, and the other is at 300 K. Find the irreversibility in the process per kilogram of air flowing out.

10.42 Fresh water can be produced from saltwater by evaporation and subsequent condensation. An example is shown in Fig. P10.42 where 150-kg/s saltwater, state 1, comes from the condenser in a large power plant. The water is throttled to the saturated pressure in the flash evaporator, and the vapor, state 2, is then condensed by cooling with sea water. As the evaporation takes place below atmospheric pressure, pumps must bring the liquid water flows back up to P_0. Assume that the saltwater has the same properties as pure water, the ambient is at 20°C, and there are no external heat transfers. With the states as shown in the following table find the irreversibility in the throttling valve and in the condenser.

State	1	2	3	4	5	6	7	8
T[°C]	30	25	25	—	23	—	17	20

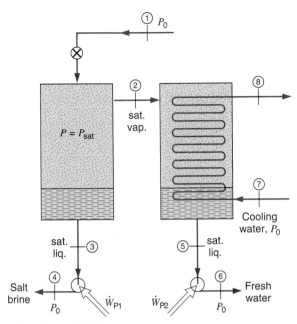

FIGURE P10.42

10.43 Calculate the irreversibility for the process described in Problem 6.132, assuming that heat transfer is with the surroundings at 17°C.

10.44 A 2-kg piece of iron is heated from room temperature 25°C to 400°C by a heat source at 600°C. What is the irreversibility in the process?

10.45 Air enters the turbocharger compressor (see Fig. P10.45) of an automotive engine at 100 kPa, 30°C, and exits at 170 kPa. The air is cooled by 50°C in an intercooler before entering the engine. The isentropic efficiency of the compressor is 75%. Determine the temperature of the air entering the engine and the irreversibility of the compression-cooling process.

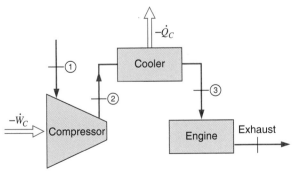

FIGURE P10.45

10.46 A 2 kg/s flow of steam at 1 MPa and 700°C should be brought to 500°C by spraying in liquid water at 1 MPa and 20°C in a steady flow. Find the rate of irreversibility, assuming that the surroundings are at 20°C.

10.47 A car air-conditioning unit has a 0.5-kg aluminum storage cylinder that is sealed with a valve, and it contains 2 L of refrigerant R-134a at 500 kPa and both are at room temperature 20°C. It is now installed in a car sitting outside where the whole system cools down to ambient temperature at −10°C. What is the irreversibility of this process?

10.48 The high-temperature heat source for a cyclic heat engine is a steady-flow heat exchanger where R-134a enters at 80°C, saturated vapor, and exits at 80°C, saturated liquid at a flow rate of 5 kg/s. Heat is rejected from the heat engine to a steady-flow heat exchanger where air enters at 150 kPa and ambient temperature 20°C, and exits at 125 kPa, 70°C. The rate of irreversibility for the overall process is 175 kW. Calculate the mass flow rate of the air and the thermal efficiency of the heat engine.

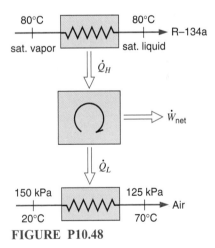

FIGURE P10.48

10.49 A rigid container with volume 200 L is divided into two equal volumes by a partition. Both sides contain nitrogen, one side is at 2 MPa, 300°C, and the other at 1 MPa, 50°C. The partition ruptures, and the nitrogen comes to a uniform state at 100°C. Assuming the surroundings are at 25°C, find the actual heat transfer and the irreversibility in the process.

10.50 A rock bed consists of 6000 kg granite and is at 70°C. A small house with lumped mass of 12 000 kg wood and 1000 kg iron is at 15°C. They are now brought to a uniform final temperature by circulating water between the rock bed and the house. Find the final temperature and the irreversibility in the process, assuming an ambient at 15°C.

Availability (exergy)

10.51 A steady stream of R-22 at ambient temperature, 10°C, and at 750 kPa enters a solar collector. The stream exits at 80°C and 700 kPa. Calculate the change in availability of the R-22 between these two states.

10.52 Consider the springtime melting of ice in the mountains, which gives cold water running in a river at 2°C while the air temperature is 20°C. What is the availability of the water relative to the ambient temperature?

10.53 A geothermal source provides 10 kg/s of hot water at 500 kPa and 150°C flowing into a flash evaporator that separates vapor and liquid at 200 kPa. Find the three fluxes of availability (inlet and two outlets) and the irreversibility rate.

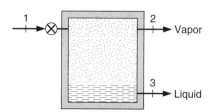

FIGURE P10.53

10.54 Find the availability at all four states in the power plant of Problem 9.42 with an ambient temperature at 298 K.

10.55 Air flows at 1500 K and 100 kPa through a constant-pressure heat exchanger giving energy to a heat engine and comes out at 500 K. At what constant temperature should the same heat transfer be delivered to provide the same availability?

10.56 Calculate the change in availability (kW) of the two flows in Problem 9.61.

10.57 Nitrogen flows in a pipe with a velocity of 300 m/s at 500 kPa and 300°C. What is its availability with respect to ambient surroundings at 100 kPa and 20°C?

10.58 A steady combustion of natural gas yields 0.15 kg/s of products (having approximately the same properties as air) at 1100°C and 100 kPa. The products are passed through a heat exchanger and exit at 550°C. What is the maximum theoretical power output from a cyclic heat engine operating on the heat rejected from the combustion products, assuming that the ambient temperature is 20°C?

10.59 Find the change in availability from inlet to exit of the condenser in Problem 9.42.

10.60 Refrigerant R-12 at 30°C, 0.75 MPa, enters a steady-flow device and exits at 30°C, 100 kPa. Assume the process is isothermal and reversible. Find the change in availability of the refrigerant.

10.61 An air compressor is used to charge an initially empty 200-L tank with air up to 5 MPa. The air inlet to the compressor is at 100 kPa, 17°C, and the compressor isentropic efficiency is 80%. Find the total compressor work and the change in availability of the air.

10.62 Water as saturated liquid at 200 kPa goes through a constant pressure heat exchanger as shown in Fig. P10.62. The heat input is supplied from a reversible heat pump extracting heat from the surroundings at 17°C. The water flow rate is 2 kg/min and the whole process is reversible, that is, there is no overall net entropy change. If the heat pump receives 40 kW of work, find the water exit state by iteration and the increase in availability of the water.

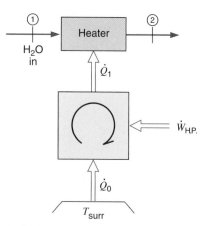

FIGURE P10.62

10.63 An electric stove has one heating element at 300°C getting 500 W of electric power. It transfers 90% of the power to 1 kg water in a kettle initially at ambient 20°C, 100 kPa; the remaining 10% leaks to the room air. The water at a uniform T is brought to the boiling point. At the start of the process, what is the rate of availability transfer by (a) electrical input, (b) from heating element, and (c) into the water at T_{water}?

10.64 Calculate the availability of the water at the initial and final states of Problem 8.70, and the irreversibility of the process.

10.65 A 10-kg iron disk brake on a car is initially at 10°C. Suddenly the brake pad hangs up, increasing the brake temperature by friction to 110°C while the car maintains constant speed. Find the change in availability of the disk and the energy depletion of the car's gas tank due to this process alone. Assume that the engine has a thermal efficiency of 35%.

10.66 A 1-kg block of copper at 350°C is quenched in a 10-kg oil bath initially at ambient temperature of 20°C. Calculate the final uniform temperature (no heat transfer to/from ambient) and the change of availability of the system (copper and oil).

10.67 Calculate the availability of the system (aluminum plus gas) at the initial and final states of Problem 8.137, and also the process irreversibility.

10.68 A wooden bucket (2 kg) with 10 kg hot liquid water, both at 85°C, is lowered 400 m down into a mineshaft. What is the availability of the bucket and water with respect to the surface ambient at 20°C?

Device Second-Law Efficiency

10.69 Air enters a compressor at ambient conditions, 100 kPa and 300 K, and exits at 800 kPa. If the isentropic compressor efficiency is 85%, what is the second-law efficiency of the compressor process?

10.70 A compressor takes in saturated vapor R-134a at −20°C and delivers it at 30°C and 0.4 MPa. Assuming that the compression is adiabatic, find the isentropic efficiency and the second-law efficiency.

10.71 A steam turbine has inlet at 4 MPa and 500°C and actual exit of 100 kPa with $x = 1.0$. Find its first-law (isentropic) and its second-law efficiencies.

10.72 The condenser in a refrigerator receives R-134a at 700 kPa and 50°C, and it exits as saturated liquid at 25°C. The flow rate is 0.1 kg/s, and the condenser has air flowing in at an ambient temperature of 15°C and leaving at 35°C. Find the minimum flow rate of air and the heat exchanger second-law efficiency.

10.73 Steam enters a turbine at 25 MPa, 550°C and exits at 5 MPa, 325°C at a flow rate of 70 kg/s. Determine the total power output of the turbine, its isentropic efficiency, and the second-law efficiency.

10.74 A compressor is used to bring saturated water vapor at 1 MPa up to 17.5 MPa, where the actual exit temperature is 650°C. Find the irreversibility and the second-law efficiency.

10.75 A flow of steam at 10 MPa, 550°C, goes through a two-stage turbine. The pressure between the stages is 2 MPa, and the second stage has an exit at 50 kPa. Assume both stages have an isentropic efficiency of 85%. Find the second-law efficiencies for both stages of the turbine.

10.76 The simple steam power plant shown in Problem 6.99 has a turbine with given inlet and exit states. Find the availability at the turbine exit, state 6. Find the second-law efficiency for the turbine, neglecting kinetic energy at state 5.

10.77 A steam turbine inlet is at 1200 kPa, 500°C. The actual exit is at 200 kPa, 300°C. What are the isentropic efficiency and its second-law efficiency?

10.78 Steam is supplied in a line at 3 MPa, 700°C. A turbine with an isentropic efficiency of 85% is connected to the line by a valve, and it exhausts to the atmosphere at 100 kPa. If the steam is throttled down to 2 MPa before entering the turbine, find the actual turbine specific work. Find the change in availability through the valve and the second-law efficiency of the turbine.

10.79 Air flows into a heat engine at ambient conditions 100 kPa, 300 K, as shown in Fig. P10.79. Energy is supplied as 1200 kJ/kg air from a 1500-K source, and in some part of the process a heat-transfer loss of 300 kJ/kg air occurs at 750 K. The air leaves the engine at 100 kPa, 800 K. Find the first- and second-law efficiencies.

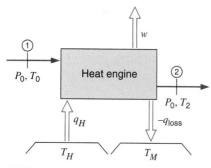

FIGURE P10.79

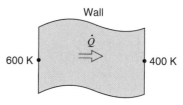

FIGURE P10.86

10.80 Air enters a steady-flow turbine at 1600 K and exhausts to the atmosphere at 1000 K. The second-law efficiency is 85%. What is the turbine inlet pressure?

10.81 Calculate the second-law efficiency of the counterflowing heat exchanger in Problem 9.61 with an ambient temperature at 20°C.

10.82 Calculate the second-law efficiency of the coflowing heat exchanger in Problem 9.62 with an ambient temperature at 17°C.

10.83 A heat exchanger brings 10 kg/s water from 100°C to 500°C at 2000 kPa using air coming in at 1400 K and leaving at 460 K. What is the second-law efficiency?

Exergy Balance Equation

10.84 Find the specific flow exergy in and out of the steam turbine in Example 9.1 assuming an ambient at 293 K. Use the exergy balance equation to find the reversible specific work. Does this calculation of specific work depend on T_0?

10.85 A counterflowing heat exchanger cools air at 600 K, 400 kPa, to 320 K using a supply of water at 20°C, 200 kPa. The water flow rate is 0.1 kg/s, and the airflow rate is 1 kg/s. Assume this can be done in a reversible process by the use of heat engines and neglect kinetic energy changes. Find the water exit temperature and the power out of the heat engine(s).

10.86 Evaluate the steady-state exergy fluxes due to a heat transfer of 250 W through a wall with 600 K on one side and 400 K on the other side, shown in Fig. P10.86. What is the exergy destruction in the wall?

10.87 A heat engine operating with an environment at 298 K produces 5 kW of power output with a first-law efficiency of 50%. It has a second-law efficiency of 80% and $T_L = 310$ K. Find all the energy and exergy transfers in and out.

10.88 Consider the condenser in Problem 9.42. Find the specific energy and exergy that are given out, assuming an ambient at 20°C. Also find the specific exergy destruction in the process.

10.89 The condenser in a power plant cools 10 kg/s water at 10 kPa, quality 90%, so that it comes out as saturated liquid at 100 kPa. The cooling is done by ocean water coming in at ambient 15°C and returned to the ocean at 20°C. Find the transfer out of the water and the transfer into the ocean water of both energy and exergy (4 terms).

10.90 Use the exergy equation to analyze the compressor in Example 6.10 to find its second-law efficiency assuming an ambient at 20°C.

10.91 Consider the car engine in Example 7.1 and assume the fuel energy is delivered at a constant 1500 K. The 70% of the energy that is lost is 40% exhaust flow at 900 K, and the remainder 30% heat transfer to the walls at 450 K goes on to the coolant fluid at 370 K, finally ending up in atmospheric air at ambient 20°C. Find all the energy and exergy flows for this heat engine. Also find the exergy destruction and where that is done.

10.92 Estimate some reasonable temperatures to use and find all the fluxes of exergy in the refrigerator given in Example 7.2.

10.93 Use the exergy equation to evaluate the exergy destruction for Problem 10.44.

10.94 Use the exergy balance equation to solve for the work in Problem 10.33.

Review Problems

10.95 A small air gun has 1 cm³ air at 250 kPa, 27°C. The piston is a bullet of mass 20 g. What is the potential highest velocity with which the bullet can leave?

10.96 Calculate the reversible work and irreversibility for the process described in Problem 5.134 assuming that the heat transfer is with the surroundings at 20°C.

10.97 A piston/cylinder arrangement has a load on the piston so it maintains constant pressure. It contains 1 kg of steam at 500 kPa, 50% quality. Heat from a reservoir at 700°C brings the steam to 600°C. Find the second-law efficiency for this process. Note that no formula is given for this particular case, so determine a reasonable expression for it.

10.98 Consider the high-pressure closed feedwater heater in the nuclear power plant described in Problem 6.102. Determine its second-law efficiency.

10.99 Consider a gasoline engine for a car as a steady device where air and fuel enter at the surrounding conditions 25°C, 100 kPa, and leave the engine exhaust manifold at 1000 K, 100 kPa, as products assumed to be air. The engine cooling system removes 750 kJ/kg air through the engine to the ambient. For the analysis, take the fuel as air where the extra energy of 2200 kJ/kg of air released in the combustion process is added as heat transfer from a 1800-K reservoir. Find the work out of the engine, the irreversibility per kilogram of air, and the first- and second-law efficiencies.

10.100 Consider the nozzle in Problem 9.112. What is the second-law efficiency for the nozzle?

10.101 Air in a piston/cylinder arrangement is at 110 kPa, 25°C, with a volume of 50 L. It goes through a reversible polytropic process to a final state of 700 kPa, 500 K, and exchanges heat with the ambient at 25°C through a reversible device. Find the total work (including the external device) and the heat transfer from the ambient.

10.102 Consider the irreversible process in Problem 8.128. Assume that the process could be done reversibly by adding heat engines/pumps be-

tween tanks A and B and the cylinder. The total system is insulated, so there is no heat transfer to or from the ambient. Find the final state, the work given out to the piston, and the total work to or from the heat engines/pumps.

10.103 Consider the heat engine in Problem 10.79. The exit temperature was given as 800 K, but what are the theoretical limits for this temperature? Find the lowest and the highest, assuming the heat transfers are as given. For each case give the first- and second-law efficiency.

10.104 Air in a piston/cylinder arrangement, shown in Fig. P10.104, is at 200 kPa, 300 K, with a volume of 0.5 m³. If the piston is at the stops, the volume is 1 m³ and a pressure of 400 kPa is required. The air is then heated from the initial state to 1500 K by a 1900 K reservoir. Find the total irreversibility in the process, assuming surroundings are at 20°C.

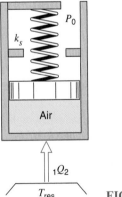

FIGURE P10.104

10.105 A jet of air at 200 m/s flows at 25°C, 100 kPa, towards a wall where the jet flow stagnates and leaves at very low velocity. Consider the process to be adiabatic and reversible. Use the exergy equation and the second law to find the stagnation temperature and pressure.

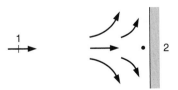

FIGURE P10.105

10.106 Consider the light bulb in Problem 8.123. What are the fluxes of exergy at the various locations mentioned? What is the exergy destruction in the filament, the entire bulb including the glass, and the entire room including the bulb? The light does not affect the gas or the glass in the bulb, but it gets absorbed on the room walls.

ENGLISH UNIT PROBLEMS

Concept Problems

10.107E A flow of air at 150 psia, 540 R, is throttled to 75 psia. What is the irreversibility? What is the drop in flow availability?

10.108E A heat exchanger increases the availability of 6 lbm/s water by 800 Btu/lbm using 20 lbm/s air coming in at 2500 R and leaving with 250 Btu/lbm less availability. What is the irreversibility and the second-law efficiency?

10.109E A heat engine receives 3500 Btu/h heat transfer at 1800 R and gives out 2000 Btu/h as work with the rest as heat transfer to the ambient. What are the fluxes of exergy in and out?

10.110E A heat engine receives 3500 Btu/h heat transfer at 1800 R and gives out 2000 Btu/h as work with the rest as heat transfer to the ambient. Find its first- and second-law efficiencies.

10.111E A heat pump has a coefficient of performance of 2 using a power input of 15000 Btu/h. Its low temperature is T_0, and the high temperature is 180 F, with ambient at T_0. Find the fluxes of exergy associated with the energy fluxes in and out?

10.112E Find the second-law efficiency of the heat pump in Problem 10.111.

English Unit Problems

10.113E A control mass gives out 1000 Btu of energy in the form of

  a. Electrical work from a battery
  b. Mechanical work from a spring
  c. Heat transfer at 700 F

  Find the change in availability of the control mass for each of the three cases.

10.114E The compressor in a refrigerator takes refrigerant R-134a in at 15 lbf/in.^{2}, 0 F, and compresses it to 125 lbf/in.^{2}, 100 F. With the room at 70 F, find the reversible heat transfer and the minimum compressor work.

10.115E A heat engine receives 15 000 Btu/h at 1400 R and 30 000 Btu/h at 1800 R, rejecting energy by heat transfer at 900 R. Assume it is reversible and find the power output. How much power could be produced if it could reject energy at $T_0 = 540$ R?

10.116E Air flows through a constant-pressure heating device as shown in Fig. P10.32. It is heated up in a reversible process with a work input of 85 Btu/lbm air flowing. The device exchanges heat with the ambient at 540 R. The air enters at 540 R, 60 lbf/in.^{2}. Assuming constant specific heat, develop an expression for the exit temperature and solve for it.

10.117E A rock bed consists of 12 000 lbm granite and is at 160 F. A small house with lumped mass of 24 000 lbm wood and 2000 lbm iron is at 60 F. They are now brought to a uniform final temperature with no external heat transfer by connecting the house and rock bed through some heat engines. If the process is reversible, find the final temperature and the work done during the process.

10.118E A constant-pressure piston/cylinder contains 4 lbm of water at 1000 psia and 200 F. Heat is added from a reservoir at 1300 F to the water until it reaches 1300 F. We want to find the total irreversibility in the process.

10.119E A supply of steam at 14.7 lbf/in.^{2}, 320 F, is needed in a hospital for cleaning purposes at a rate of 30 lbm/s. A supply of steam at 20 lbf/in.^{2}, 500 F, is available from a boiler, and tap water at 14.7 lbf/in.^{2}, 60 F, is also available. The two sources are then mixed in a mixing chamber to generate the desired state as output. Determine the rate of irreversibility of the mixing process.

10.120E Fresh water can be produced from saltwater by evaporation and subsequent condensation. An example is shown in Fig. P10.42 where 300-lbm/s saltwater, state 1, comes from the condenser in a large power plant. The water is throttled to the saturated pressure in the flash evaporator, and the vapor, state 2, is then condensed by cooling with sea water. As the evaporation takes place below atmospheric pressure, pumps must bring the liquid water flows back up to P_0. Assume that the saltwater has the same properties as pure water, that the ambient is at 68 F, and that there are no external heat transfers. With the states as shown in the following list find the irreversibility in the throttling valve and in the condenser.

State	1	2	3	4	5	6	7	8
T[F]	86	77	77	—	74	—	63	68

10.121E Calculate the irreversibility for the process described in Problem 6.174, assuming that the heat transfer is with the surroundings at 61 F.

10.122E A 4-lbm piece of iron is heated from room temperature 77 F to 750 F by a heat source at 1100 F. What is the irreversibility in the process?

10.123E Air enters the turbocharger compressor of an automotive engine at 14.7 lbf/in.^{2}, 90 F, and exits at 25 lbf/in.^{2}, as shown in Fig. P10.45. The air is cooled by 90 F in an intercooler before entering the engine. The isentropic efficiency of the compressor is 75%. Determine the temperature of the air entering the engine and the irreversibility of the compression-cooling process.

10.124E A rock bed consists of 12 000 lbm granite and is at 160 F. A small house with lumped mass of 24 000 lbm wood and 2000 lbm iron is at 60 F. They are now brought to a uniform final temperature by circulating water between the rock bed and the house. Find the final temperature and the irreversibility in the process assuming an ambient at 60 F.

10.125E A steady stream of R-22 at ambient temperature, 50 F, and at 110 lbf/in.^{2} enters a solar collector. The stream exits at 180 F, 100 lbf/in.^{2}. Calculate the change in availability of the R-22 between these two states.

10.126E Consider the springtime melting of ice in the mountains, which gives cold water running in a river at 34 F while the air temperature is 68 F. What is the flow availability of the water relative to the temperature of the ambient?

10.127E A geothermal source provides 20 lbm/s of hot water at 80 lbf/in.^{2}, 300 F flowing into a flash evaporator that separates vapor and liquid at 30 lbf/in.^{2}. Find the three fluxes of availability (inlet and two outlets) and the irreversibility rate.

10.128E An air compressor is used to charge an initially empty 7-ft^3 tank with air up to 750 lbf/in.^{2}. The air inlet to the compressor is at 14.7 lbf/in.^{2}, 60 F, and the compressor isentropic efficiency is 80%. Find the total compressor work and the change in energy of the air.

10.129E An electric stove has one heating element at 600 F getting 500 W of electric power. It transfers 90% of the power to 2 lbm water in a kettle initially at ambient 70 F, 1 atm; the rest, 10%, leaks to the room air. The water at a uniform T is brought to the boiling point. At the start of the process, what is the rate of availability transfer by (a) electrical input, (b) from heating element, and (c) into the water at T_{water}?

10.130E A 20-lbm iron disk brake on a car is at 50 F. Suddenly the brake pad hangs up, increasing the brake temperature by friction to 230 F while the car maintains constant speed. Find the change in availability of the disk and the energy depletion of the car's gas tank due to this process alone. Assume that the engine has a thermal efficiency of 35%.

10.131E Calculate the availability of the system (aluminum plus gas) at the initial and final states of Problem 8.183, and also the irreversibility.

10.132E A wood bucket (4 lbm) with 20 lbm hot liquid water, both at 180 F, is lowered 1300 ft down into a mineshaft. What is the availability of the bucket and water with respect to the surface ambient at 70 F?

10.133E A coflowing (same direction) heat exchanger has one line with 0.5 lbm/s oxygen at 68 F and 30 psia entering, and the other line has 1.2 lbm/s nitrogen at 20 psia and 900 R entering. The heat exchanger is long enough so that the two flows exit at the same temperature. Use constant heat capacities and find the exit temperature and the second-law efficiency for the heat exchanger assuming ambient at 68 F.

10.134E A steam turbine has an inlet at 600 psia and 900 F and actual exit of 1 atm with x = 1.0. Find its first-law (isentropic) and second-law efficiencies.

10.135E A compressor is used to bring saturated water vapor at 150 lbf/in.² up to 2500 lbf/in.², where the actual exit temperature is 1200 F. Find the irreversibility and the second-law efficiency.

10.136E The simple steam power plant in Problem 6.166, shown in Fig. P6.99 has a turbine with given inlet and exit states. Find the availability at the turbine exit, state 6. Find the second-law efficiency for the turbine, neglecting kinetic energy at state 5.

10.137E Steam is supplied in a line at 450 lbf/in.², 1200 F. A turbine with an isentropic efficiency of 85% is connected to the line by a valve, and it exhausts to the atmosphere at 14.7 lbf/in.². If the steam is throttled down to 300 lbf/in.² before entering the turbine, find the actual turbine specific work. Find the change in availability through the valve and the second law efficiency of the turbine.

10.138E Air flows into a heat engine at ambient conditions 14.7 lbf/in.², 540 R, as shown in Fig. P.10.79. Energy is supplied as 540 Btu per lbm air from a 2700 R source, and in some part of the process a heat transfer loss of 135 Btu per lbm air happens at 1350 R. The air leaves the engine at 14.7 lbf/in.², 1440 R. Find the first- and second-law efficiencies.

10.139E A heat engine operating with an environment at 540 R produces 17 000 Btu/h of power output with a first-law efficiency of 50%. It has a second-law efficiency of 80% and T_L = 560 R. Find all the energy and exergy transfers in and out.

10.140E The condenser in a power plant cools 20 lbm/s water at 120 F, quality 90%, so that it comes out as saturated liquid at 120 F. The cooling is done by ocean water coming in at 60 F and returned to the ocean at 68 F. Find the transfer out of the water and the transfer into the ocean water of both energy and exergy (4 terms).

10.141E Calculate the reversible work and irreversibility for the process described in Problem 5.168 assuming that the heat transfer is with the surroundings at 68 F.

10.142E A piston/cylinder arrangement has a load on the piston so it maintains constant pressure. It contains 1 lbm of steam at 80 lbf/in.², 50% quality. Heat from a reservoir at 1300 F brings the steam to 1000 F. Find the second-law efficiency for this process. Note that no formula is given for this particular case, so determine a reasonable expression for it.

10.143E Consider a gasoline engine for a car as a steady-flow device where air and fuel enters at the surrounding conditions 77 F, 14.7 lbf/in.², and leaves the engine exhaust manifold at 1800 R, 14.7 lbf/in.² as products assumed to be air. The engine cooling system removes 320 Btu/lbm air through the engine to the ambient. For the analysis, take the fuel as air where the extra energy of 950 Btu/lbm of air released in the combustion process is added as heat transfer from a 3240 R reservoir. Find the work out of the engine, the irreversibility per pound-mass of air, and the first- and second-law efficiencies.

10.144E The exit nozzle in a jet engine receives air at 2100 R, 20 psia, with negligible kinetic energy. The exit pressure is 10 psia, and the actual exit temperature is 1780 R. What is the actual exit velocity and the second-law efficiency?

10.145E Air in a piston/cylinder arrangement, shown in Fig. P.10.104, is at 30 lbf/in.², 540 R, with a volume of 20 ft³. If the piston is at the stops, the volume is 40 ft³ and a pressure of 60 lbf/in.² is required. The air is then heated from the initial state to 2700 R by a 3400 R reservoir. Find the total irreversibility in the process, assuming surroundings are at 70 F.

COMPUTER, DESIGN, AND OPEN-ENDED PROBLEMS

10.146 Use the software to get the properties of water as needed for consideration of the moisture separator in Problem 6.102. Steam comes in at state 3 and leaves as liquid, state 9, with the rest at state 4 going to the low-pressure turbine. Assume no heat transfer to the surroundings at 20°C and find the total entropy generation and irreversibility in the process.

10.147 Use the software to get the properties of water as needed and calculate the second-law efficiency of the low-pressure turbine in Problem 6.102.

10.148 Write a program to solve the general case of Problem 10.44. The initial state is to be the program input variable, and the output should also include the change in availability for both the iron and the source.

10.149 Write a program to solve Problem 10.32. Use constant specific heat and let the work input be a program input variable.

10.150 The maximum power a windmill can possibly extract from the wind is

$$\dot{W} = \frac{16}{27}\rho A \mathbf{V}\frac{1}{2}\mathbf{V}^2 = \frac{16}{27}\dot{m}_{air} \times KE$$

Water flowing through Hoover Dam (see Problem 6.51) produces $\dot{W} = 0.8\dot{m}_{water}\, gh$. Burning 1 kg of coal gives 24 000 kJ delivered at 900 K to a heat engine. Find other examples in the literature and from problems in the previous chapters with steam and gases into turbines. Make a list of the availability (exergy) for a flow of 1 kg/s of substance with the above examples. Use a reasonable choice for the values of the parameters and do the necessary analysis.

10.151 Consider the condenser in the simple steam power plant described in Problem 6.99. The cooling water is lake water at 20°C, and it should not be heated more than 5°C as it goes back to the lake. Assume the heat-transfer rate inside the condenser is 350 W/m²K so $\dot{Q} = 350 \times A\,\Delta T$ in watts. Estimate the flow rate of the cooling water and the needed interface area inside the condenser, A. Find the change in the availability of the cooling water and the steam inside the condenser and compare. Discuss your estimates and the size of the pump for the cooling water.

10.152 Consider the nuclear power plant shown in Problem 6.102. Select one feedwater heater and one pump and make an analysis of their performance. Check the energy balances and do the

second-law analysis. Determine the change of availability in all the flows and discuss measures of performance for both the pump and the feedwater heater.

10.153 Reconsider the use of the geothermal energy as discussed in Problem 6.105. The analysis that was done and the original problem statement specified the turbine exit state as 10 kPa, 90% quality. Reconsider this problem with an adiabatic turbine having an isentropic efficiency of 85% and an exit pressure of 10 kPa. Include a second-law analysis and discuss the changes in availability. Describe another way of using the geothermal energy and make appropriate calculations.

10.154 An air gun should shoot a harpoon of mass 5 kg out so that it has a velocity of 75 m/s as it leaves the gun. The harpoon acts as the piston in a cylinder, and air is trapped below the piston (end of harpoon) that can be initially locked. The air is charged so the initial state is at high pressure and temperature. Determine sizes for the cylinder diameter, cylinder length, air mass, and initial (P,T) of the air. Make reasonable assumptions about the process and include a determination of the state of the air during the process.

10.155 Energy can be stored in many different forms. Thermal energy can be stored as internal energy in a mass like a rock bed, water, or metals. Mechanical energy (potential or kinetic) can be stored in springs, rotating flywheels, elevated masses, and the like. A tank with a compressed gas that can drive a turbine is used. Batteries are used in cars. Make a list with at least five different ways of storing 1000 MJ of energy and size the systems. Note how the energy is taken out and find the availability for each case. Discuss the various alternatives.

10.156 Find from the literature the amount of energy that must be stored in a car to start the engine. Size three different systems to provide that energy and compare those to an ordinary car battery. Discuss the feasibility and cost.

11 POWER AND REFRIGERATION SYSTEMS

Some power plants, such as the simple steam power plant, which we have considered several times, operate in a cycle. That is, the working fluid undergoes a series of processes and finally returns to the initial state. In other power plants, such as the internal-combustion engine and the gas turbine, the working fluid does not go through a thermodynamic cycle, even though the engine itself may operate in a mechanical cycle. In this instance, the working fluid has a different composition or is in a different state at the conclusion of the process than it had or was at the beginning. Such equipment is sometimes said to operate on the open cycle (the word cycle is really a misnomer), whereas the steam power plant operates on a closed cycle. The same distinction between open and closed cycles can be made regarding refrigeration devices. For both the open- and closed-cycle apparatus, however, it is advantageous to analyze the performance of an idealized closed cycle similar to the actual cycle. Such a procedure is particularly advantageous for determining the influence of certain variables on performance. For example, the spark-ignition internal-combustion engine is usually approximated by the Otto cycle. From an analysis of the Otto cycle we conclude that increasing the compression ratio increases the efficiency. This is also true for the actual engine, even though the Otto-cycle efficiencies may deviate significantly from the actual efficiencies.

This chapter is concerned with these idealized cycles for both power and refrigeration apparatus. Both vapors and ideal gases are considered as working fluids. An attempt will be made to point out how the processes in actual apparatus deviate from the ideal. Consideration is also given to certain modifications of the basic cycles that are intended to improve performance. These modifications include the use of devices such as regenerators, multistage compressors and expanders, and intercoolers. Various combinations of these types of systems and also special applications, such as cogeneration of electrical power and energy, combined cycles, topping and bottoming cycles, and binary cycle systems, are also discussed in this chapter and in the problems.

11.1 INTRODUCTION TO POWER SYSTEMS

In introducing the second law of thermodynamics in Chapter 7, we considered cyclic heat engines consisting of four separate processes. We noted that these engines can be operated as steady-state devices involving shaft work, as shown in Fig. 7.16, or instead as

382

cylinder/piston devices involving boundary-movement work, as shown in Fig. 7.17. The former may have a working fluid that changes phase during the processes in the cycle, or may have a single-phase working fluid throughout. The latter type would normally have a gaseous working fluid throughout the cycle.

For a reversible steady-state process involving negligible kinetic and potential energy changes, the shaft work per unit mass is given by Eq. 9.19,

$$w = -\int v \, dP$$

For a reversible process involving a simple compressible substance, the boundary movement work per unit mass is given by Eq. 4.3,

$$w = \int P \, dv$$

The areas represented by these two integrals are shown in Fig. 11.1. It is of interest to note that, in the former case, there is no work involved in a constant-pressure process, while in the latter case, there is no work involved in a constant-volume process.

Let us now consider a power system consisting of four steady-state processes, as in Fig. 7.16. We assume that each process is internally reversible and has negligible changes in kinetic and potential energies, which results in the work for each process being given by Eq. 9.19. For convenience of operation, we will make the two heat-transfer processes (boiler and condenser) constant-pressure processes, such that those are simple heat exchangers involving no work. Let us also assume that the turbine and pump processes are both adiabatic, such that they are therefore isentropic processes. Thus, the four processes comprising the cycle are as shown in Fig. 11.2. Note that if the entire cycle takes place inside the two-phase liquid–vapor dome, the resulting cycle is the Carnot cycle, since the two constant-pressure processes are also isothermal. Otherwise, this cycle is not a Carnot cycle. In either case, we find that the net work output for this power system is given by

$$w_{\text{net}} = -\int_1^2 v \, dP + 0 - \int_3^4 v \, dP + 0 = -\int_1^2 v \, dP + \int_4^3 v \, dP$$

and, since $P_2 = P_3$ and also $P_1 = P_4$, we find that the system produces a net work output because the specific volume is larger during the expansion from 3 to 4 than it is during the compression from 1 to 2. This result is also evident from the areas

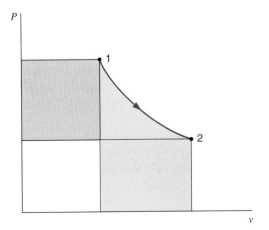

FIGURE 11.1
Comparison of shaft work and boundary-movement work.

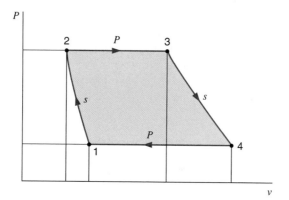

FIGURE 11.2 Four-process power cycle.

$-\int v\, dP$ in Fig. 11.2. We conclude that it would be advantageous to have this difference in specific volume be as large as possible, as, for example, the difference between a vapor and a liquid.

If the four-process cycle shown in Fig. 11.2 were accomplished in a cylinder/piston system involving boundary-movement work, then the net work output for this power system would be given by

$$w_{\text{net}} = \int_1^2 P\, dv + \int_2^3 P\, dv + \int_3^4 P\, dv + \int_4^1 P\, dv$$

and from these four areas on Fig. 11.2, we note that the pressure is higher during any given change in volume in the two expansion processes than in the two compression processes, resulting in a net positive area and a net work output.

For either of the two cases just analyzed, it is noted from Fig. 11.2 that the net work output of the cycle is equal to the area enclosed by the process lines 1–2–3–4–1, and this area is the same for both cases, even though the work terms for the four individual processes are different for the two cases.

In the next several sections, we consider the Rankine cycle, which is the ideal, four-steady-state process cycle shown in Fig. 11.2, utilizing a phase change between vapor and liquid to maximize the difference in specific volume during expansion and compression. This is the idealized model for a steam power plant system.

11.2 THE RANKINE CYCLE

We now consider the idealized four-steady-state-process cycle shown in Fig. 11.2, in which state 1 is saturated liquid and state 3 is either saturated vapor or superheated vapor. This system is termed the Rankine cycle and is the model for the simple steam power plant. It is convenient to show the states and processes on a *T–s* diagram, as given in Fig. 11.3. The four processes are:

1–2: Reversible adiabatic pumping process in the pump

2–3: Constant-pressure transfer of heat in the boiler

3–4: Reversible adiabatic expansion in the turbine (or other prime mover such as a steam engine)

4–1: Constant-pressure transfer of heat in the condenser

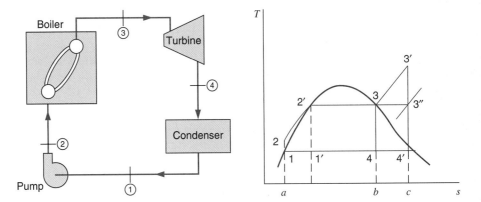

FIGURE 11.3 Simple steam power plant that operates on the Rankine cycle.

As mentioned earlier, the Rankine cycle also includes the possibility of superheating the vapor, as cycle 1–2–3′–4′–1.

If changes of kinetic and potential energy are neglected, heat transfer and work may be represented by various areas on the T–s diagram. The heat transferred to the working fluid is represented by area a–2–2′–3–b–a, and the heat transferred from the working fluid by area a–1–4–b–a. From the first law we conclude that the area representing the work is the difference between these two areas—area 1–2–2′–3–4–1. The thermal efficiency is defined by the relation

$$\eta_{th} = \frac{w_{net}}{q_H} = \frac{\text{area } 1\text{–}2\text{–}2'\text{–}3\text{–}4\text{–}1}{\text{area } a\text{–}2\text{–}2'\text{–}3\text{–}b\text{–}a} \tag{11.1}$$

For analyzing the Rankine cycle, it is helpful to think of efficiency as depending on the average temperature at which heat is supplied and the average temperature at which heat is rejected. Any changes that increase the average temperature at which heat is supplied or decrease the average temperature heat is rejected will increase the Rankine-cycle efficiency.

In analyzing the ideal cycles in this chapter, the changes in kinetic and potential energies from one point in the cycle to another are neglected. In general, this is a reasonable assumption for the actual cycles.

It is readily evident that the Rankine cycle has a lower efficiency than a Carnot cycle with the same maximum and minimum temperatures as a Rankine cycle because the average temperature between 2 and 2′ is less than the temperature during evaporation. We might well ask, why choose the Rankine cycle as the ideal cycle? Why not select the Carnot cycle 1′–2′–3–4–1′? At least two reasons can be given. The first reason concerns the pumping process. State 1′ is a mixture of liquid and vapor. Great difficulties are encountered in building a pump that will handle the mixture of liquid and vapor at 1′ and deliver saturated liquid at 2′. It is much easier to condense the vapor completely and handle only liquid in the pump: The Rankine cycle is based on this fact. The second reason concerns superheating the vapor. In the Rankine cycle the vapor is superheated at constant pressure, process 3–3′. In the Carnot cycle all the heat transfer is at constant temperature, and therefore the vapor is superheated in process 3–3″. Note, however, that during this process the pressure is dropping, which means that the heat must be transferred to the vapor as it undergoes an expansion process in which work is done. This heat transfer is also very difficult to achieve in practice. Thus, the Rankine cycle is the ideal cycle that can be approximated in practice. In the following sections we will consider some variations on the Rankine cycle that enable it to approach more closely the efficiency of the Carnot cycle.

Before we discuss the influence of certain variables on the performance of the Rankine cycle, we will study an example.

EXAMPLE 11.1 Determine the efficiency of a Rankine cycle using steam as the working fluid in which the condenser pressure is 10 kPa. The boiler pressure is 2 MPa. The steam leaves the boiler as saturated vapor.

In solving Rankine-cycle problems, we let w_p denote the work into the pump per kilogram of fluid flowing and q_L denote the heat rejected from the working fluid per kilogram of fluid flowing.

To solve this problem we consider, in succession, a control surface around the pump, the boiler, the turbine, and the condenser. For each the thermodynamic model is the steam tables, and the process is steady state with negligible changes in kinetic and potential energies. First consider the pump:

Control volume: Pump.
Inlet state: P_1 known, saturated liquid; state fixed.
Exit state: P_2 known.

Analysis

From the first law, we have

$$w_p = h_2 - h_1$$

The second law gives

$$s_2 = s_1$$

and so

$$h_2 - h_1 = \int_1^2 v\, dP$$

Solution

Assuming the liquid to be incompressible, we have

$$w_p = v(P_2 - P_1) = (0.001\ 01)(2000 - 10) = 2.0\ \text{kJ/kg}$$
$$h_2 = h_1 + w_p = 191.8 + 2.0 = 193.8\ \text{kJ/kg}$$

Now consider the boiler:

Control volume: Boiler.
Inlet state: P_2, h_2 known; state fixed.
Exit state: P_3 known, saturated vapor; state fixed.

Analysis

From the first law, we write

$$q_H = h_3 - h_2$$

Solution

Substituting, we obtain

$$q_H = h_3 - h_2 = 2799.5 - 193.8 = 2605.7 \text{ kJ/kg}$$

Turning to the turbine next, we have:

Control volume: Turbine.
Inlet state: State 3 known (above).
Exit state: P_4 known.

Analysis

The first law gives

$$w_t = h_3 - h_4$$

The second law gives

$$s_3 = s_4$$

Solution

We can determine the quality at state 4 as follows:

$$s_3 = s_4 = 6.3409 = 0.6493 + x_4 7.5009, \qquad x_4 = 0.7588$$

$$h_4 = 191.8 + 0.7588(2392.8) = 2007.5 \text{ kJ/kg}$$

$$w_t = 2799.5 - 2007.5 = 792.0 \text{ kJ/kg}$$

Finally, we consider the condenser.

Control volume: Condenser.
Inlet state: State 4 known (as given).
Exit state: State 1 known (as given).

Analysis

The first law gives

$$q_L = h_4 - h_1$$

Solution

Substituting, we obtain

$$q_L = h_4 - h_1 = 2007.5 - 191.8 = 1815.7 \text{ kJ/kg}$$

We can now calculate the thermal efficiency:

$$\eta_{th} = \frac{w_{net}}{q_H} = \frac{q_H - q_L}{q_H} = \frac{w_t - w_p}{q_H} = \frac{792.0 - 2.0}{2605.7} = 30.3\%$$

We could also write an expression for thermal efficiency in terms of properties at various points in the cycle:

$$\eta_{th} = \frac{(h_3 - h_2) - (h_4 - h_1)}{h_3 - h_2} = \frac{(h_3 - h_4) - (h_2 - h_1)}{h_3 - h_2}$$

$$= \frac{2605.7 - 1815.7}{2605.7} = \frac{792.0 - 2.0}{2605.7} = 30.3\%$$

11.3 EFFECT OF PRESSURE AND TEMPERATURE ON THE RANKINE CYCLE

Let us first consider the effect of exhaust pressure and temperature on the Rankine cycle. This effect is shown on the T–s diagram of Fig. 11.4. Let the exhaust pressure drop from P_4 to P_4', with the corresponding decrease in temperature at which heat is rejected. The net work is increased by area 1–4–4'–1'–2'–2–1 (shown by the cross-hatching). The heat transferred to the steam is increased by area a'–2'–2–a–a'. Since these two areas are approximately equal, the net result is an increase in cycle efficiency. This is also evident from the fact that the average temperature at which heat is rejected is decreased. Note, however, that lowering the back pressure causes the moisture content of the steam leaving the turbine to increase. This is a significant factor because if the moisture in the low-pressure stages of the turbine exceeds about 10%, not only is there a decrease in turbine efficiency, but erosion of the turbine blades may also be a very serious problem.

Next, consider the effect of superheating the steam in the boiler, as shown in Fig. 11.5. We see that the work is increased by area 3–3'–4'–4–3, and the heat transferred in the boiler is increased by area 3–3'–b'–b–3. Since the ratio of these two areas is greater than the ratio of net work to heat supplied for the rest of the cycle, it is evident that for

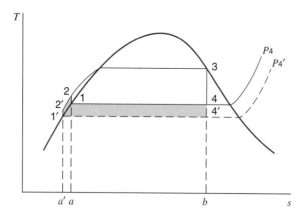

FIGURE 11.4 Effect of exhaust pressure on Rankine-cycle efficiency.

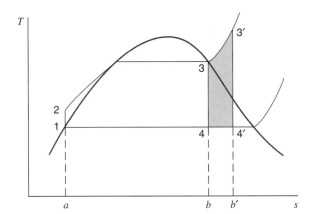

FIGURE 11.5 Effect
of superheating on
Rankine-cycle efficiency.

given pressures, superheating the steam increases the Rankine-cycle efficiency. This increase in efficiency would also follow from the fact that the average temperature at which heat is transferred to the steam is increased. Note also that when the steam is superheated, the quality of the steam leaving the turbine increases.

Finally the influence of the maximum pressure of the steam must be considered, and this is shown in Fig. 11.6. In this analysis the maximum temperature of the steam, as well as the exhaust pressure, is held constant. The heat rejected decreases by area b'–$4'$–4–b–b'. The net work increases by the amount of the single cross-hatching and decreases by the amount of the double cross-hatching. Therefore, the net work tends to remain the same, but the heat rejected decreases, and hence the Rankine-cycle efficiency increases with an increase in maximum pressure. Note that in this instance too the average temperature at which heat is supplied increases with an increase in pressure. The quality of the steam leaving the turbine decreases as the maximum pressure increases.

To summarize this section, we can say that the efficiency of the Rankine cycle can be increased by lowering the exhaust pressure, by increasing the pressure during heat addition, and by superheating the steam. The quality of the steam leaving the turbine is increased by superheating the steam and decreased by lowering the exhaust pressure and by increasing the pressure during heat addition.

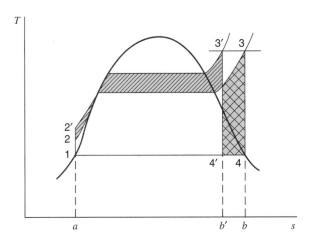

FIGURE 11.6 Effect
of boiler pressure on
Rankine-cycle efficiency.

EXAMPLE 11.2 In a Rankine cycle, steam leaves the boiler and enters the turbine at 4 MPa and 400°C. The condenser pressure is 10 kPa. Determine the cycle efficiency.

To determine the cycle efficiency, we must calculate the turbine work, the pump work, and the heat transfer to the steam in the boiler. We do this by considering a control surface around each of these components in turn. In each case the thermodynamic model is the steam tables, and the process is steady state with negligible changes in kinetic and potential energies.

Control volume: Pump.
Inlet state: P_1 known, saturated liquid; state fixed.
Exit state: P_2 known.

Analysis

From the first law, we have

$$w_p = h_2 - h_1$$

The second law gives

$$s_2 = s_1$$

Since $s_2 = s_1$,

$$h_2 - h_1 = \int_1^2 v\, dP = v(P_2 - P_1)$$

Solution

Substituting, we obtain

$$w_p = v(P_2 - P_1) = (0.001\,01)(4000 - 10) = 4.0 \text{ kJ/kg}$$
$$h_1 = 191.8 \text{ kJ/kg}$$
$$h_2 = 191.8 + 4.0 = 195.8 \text{ kJ/kg}$$

For the turbine we have:

Control volume: Turbine.
Inlet state: P_3, T_3 known; state fixed.
Exit state: P_4 known.

Analysis

The first law is

$$w_t = h_3 - h_4$$

and the second law is

$$s_4 = s_3$$

Solution

Upon substitution we get

$$h_3 = 3213.6 \text{ kJ/kg}, \qquad s_3 = 6.7690 \text{ kJ/kg K}$$

$$s_3 = s_4 = 6.7690 = 0.6493 + x_4 7.5009, \qquad x_4 = 0.8159$$

$$h_4 = 191.8 + 0.8159(2392.8) = 2144.1 \text{ kJ/kg}$$

$$w_t = h_3 - h_4 = 3213.6 - 2144.1 = 1069.5 \text{ kJ/kg}$$

$$w_{\text{net}} = w_t - w_p = 1069.5 - 4.0 = 1065.5 \text{ kJ/kg}$$

Finally, for the boiler we have:

Control volume: Boiler.
Inlet state: P_2, h_2 known; state fixed.
Exit state: State 3 fixed (as given).

Analysis

The first law is

$$q_H = h_3 - h_2$$

Solution

Substituting gives

$$q_H = h_3 - h_2 = 3213.6 - 195.8 = 3017.8 \text{ kJ/kg}$$

$$\eta_{\text{th}} = \frac{w_{\text{net}}}{q_H} = \frac{1065.5}{3017.8} = 35.3\%$$

The net work could also be determined by calculating the heat rejected in the condenser, q_L, and noting, from the first law, that the net work for the cycle is equal to the net heat transfer. Considering a control surface around the condenser, we have

$$q_L = h_4 - h_1 = 2144.1 - 191.8 = 1952.3 \text{ kJ/kg}$$

Therefore,

$$w_{\text{net}} = q_H - q_L = 3017.8 - 1952.3 = 1065.5 \text{ kJ/kg}$$

EXAMPLE 11.2E In a Rankine cycle, steam leaves the boiler and enters the turbine at 600 lbf/in.^{2} and 800 F. The condenser pressure is 1 lbf/in.^{2} Determine the cycle efficiency.

To determine the cycle efficiency, we must calculate the turbine work, the pump work, and the heat transfer to the steam in the boiler. We do this by considering a control surface around each of these components in turn. In each case the thermodynamic model is the steam tables, and the process is steady state with negligible changes in kinetic and potential energies.

Control volume: Pump.
Inlet state: P_1 known, saturated liquid; state fixed.
Exit state: P_2 known.

Analysis

From the first law, we have

$$w_p = h_2 - h_1$$

The second law gives

$$s_2 = s_1$$

Since $s_2 = s_1$,

$$h_2 - h_1 = \int_1^2 v \, dP = v(P_2 - P_1)$$

Solution

Substituting, we obtain

$$w_p = v(P_2 - P_1) = 0.01614(600 - 1) \times \tfrac{144}{778} = 1.8 \text{ Btu/lbm}$$

$$h_1 = 69.70$$

$$h_2 = 69.7 + 1.8 = 71.5 \text{ Btu/lbm}$$

For the turbine we have

> *Control volume:* Turbine.
> *Inlet state:* P_3, T_3 known; state fixed.
> *Exit state:* P_4 known.

Analysis

The first law is

$$w_t = h_3 - h_4$$

and the second law is

$$s_4 = s_3$$

Solution

Upon substitution we get

$$h_3 = 1407.6 \qquad s_3 = 1.6343$$

$$s_3 = s_4 = 1.6343 = 1.9779 - (1 - x)_4 1.8453$$

$$(1 - x)_4 = 0.1861$$

$$h_4 = 1105.8 - 0.1861(1036.0) = 913.0$$

$$w_t = h_3 - h_4 = 1407.6 - 913.0 = 494.6 \text{ Btu/lbm}$$

$$w_{\text{net}} = w_t - w_p = 494.6 - 1.8 = 492.8 \text{ Btu/lbm}$$

Finally, for the boiler we have

Control volume: Boiler.
Inlet state: P_2, h_2 known; state fixed.
Exit state: State 3 fixed (above).

Analysis

The first law is

$$q_H = h_3 - h_2$$

Solution

Substituting gives

$$q_H = h_3 - h_2 = 1407.6 - 71.5 = 1336.1 \text{ Btu/lbm}$$

$$\eta_{th} = \frac{w_{net}}{q_H} = \frac{492.8}{1336.1} = 36.9\%$$

The net work could also be determined by calculating the heat rejected in the condenser, q_L, and noting, from the first law, that the net work for the cycle is equal to the net heat transfer. Considering a control surface around the condenser, we have

$$q_L = h_4 - h_1 = 913.0 - 69.7 = 843.3 \text{ Btu/lbm}$$

Therefore,

$$w_{net} = q_H - q_L = 1336.1 - 843.3 = 492.8 \text{ Btu/lbm}$$

11.4 THE REHEAT CYCLE

In the last section we noted that the efficiency of the Rankine cycle could be increased by increasing the pressure during the addition of heat. However, the increase in pressure also increases the moisture content of the steam in the low-pressure end of the turbine. The reheat cycle has been developed to take advantage of the increased efficiency with higher pressures, and yet avoid excessive moisture in the low-pressure stages of the turbine. This cycle is shown schematically and on a T–s diagram in Fig. 11.7. The unique feature of this cycle is that the steam is expanded to some intermediate pressure in the turbine and is then reheated in the boiler, after which it expands in the turbine to the exhaust pressure. It is evident from the T–s diagram that there is very little gain in efficiency from reheating the steam, because the average temperature at which heat is supplied is not greatly changed. The chief advantage is in decreasing to a safe value the moisture content in the low-pressure stages of the turbine. If metals could be found that would enable us to superheat the steam to $3'$, the simple Rankine cycle would be more efficient than the reheat cycle, and there would be no need for the reheat cycle.

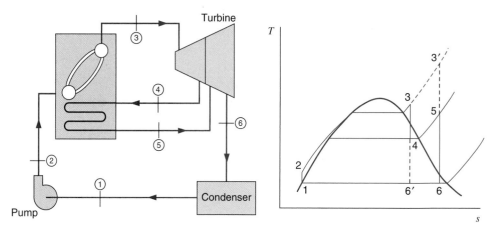

FIGURE 11.7 The ideal reheat cycle.

EXAMPLE 11.3   Consider a reheat cycle utilizing steam. Steam leaves the boiler and enters the turbine at 4 MPa, 400°C. After expansion in the turbine to 400 kPa, the steam is reheated to 400°C and then expanded in the low-pressure turbine to 10 kPa. Determine the cycle efficiency.

For each control volume analyzed, the thermodynamic model is the steam tables, the process is steady state, and changes in kinetic and potential energies are negligible.

For the high pressure turbine,

> *Control volume:*   High-pressure turbine.
> *Inlet state:* P_3, T_3 known; state fixed.
> *Exit state:* P_4 known.

Analysis

The first law is

$$w_{h-p} = h_3 - h_4$$

The second law is

$$s_3 = s_4$$

Solution

Substituting,

$$h_3 = 3213.6, \quad s_3 = 6.7690$$

$$s_4 = s_3 = 6.7690 = 1.7766 + x_4 5.1193, \quad x_4 = 0.9752$$

$$h_4 = 604.7 + 0.9752(2133.8) = 2685.6$$

For the low-pressure turbine,

> *Control volume:*   Low-pressure turbine.
> *Inlet state:* P_5, T_5 known; state fixed.
> *Exit state:* P_6 known.

Analysis

The first law is

$$w_{l-p} = h_5 - h_6$$

For the second law,

$$s_5 = s_6$$

Solution

Upon substituting,

$$h_5 = 3273.4 \qquad s_5 = 7.8985$$

$$s_6 = s_5 = 7.8985 = 0.6493 + x_6 7.5009, \qquad x_6 = 0.9664$$

$$h_6 = 191.8 + 0.9664(2392.8) = 2504.3$$

For the overall turbine, the total work output w_t is the sum of w_{h-p} and w_{l-p}, so that

$$w_t = (h_3 - h_4) + (h_5 - h_6)$$

$$= (3213.6 - 2685.6) + (3273.4 - 2504.3)$$

$$= 1297.1 \text{ kJ/kg}$$

For the pump,

Control volume: Pump.
Inlet state: P_1 known, saturated liquid; state fixed.
Exit state: P_2 known.

Analysis

From the first law, we have

$$w_p = h_2 - h_1$$

and the second law,

$$s_2 = s_1$$

Since $s_2 = s_1$,

$$h_2 - h_1 = \int_1^2 v \, dP = v(P_2 - P_1)$$

Solution

Substituting,

$$w_p = v(P_2 - P_1) = (0.001\,01)(4000 - 10) = 4.0 \text{ kJ/kg}$$

$$h_2 = 191.8 + 4.0 = 195.8$$

Finally, for the boiler

>*Control volume:* Boiler.
>*Inlet states:* States 2 and 4 both known (above).
>*Exit states:* States 3 and 5 both known (as given).

Analysis

The first law gives

$$q_H = (h_3 - h_2) + (h_5 - h_4)$$

Solution

Substituting,

$$q_H = (h_3 - h_2) + (h_5 - h_4)$$
$$= (3213.6 - 195.8) + (3273.4 - 2685.6) = 3605.6 \text{ kJ/kg}$$

Therefore,

$$w_{\text{net}} = w_t - w_p = 1297.1 - 4.0 = 1293.1 \text{ kJ/kg}$$

$$\eta_{\text{th}} = \frac{w_{\text{net}}}{q_H} = \frac{1293.1}{3605.6} = 35.9\%$$

By comparing this example with Example 11.2, we find that through reheating the gain in efficiency is relatively small, but the moisture content of the vapor leaving the turbine is decreased from 18.4 to 3.4%.

11.5 THE REGENERATIVE CYCLE

Another important variation from the Rankine cycle is the regenerative cycle, which uses feedwater heaters. The basic concepts of this cycle can be demonstrated by considering the Rankine cycle without superheat as shown in Fig. 11.8. During the process between states 2 and 2′, the working fluid is heated while in the liquid phase, and the average temperature of the working fluid is much lower than during the vaporization process 2′–3. The process between states 2 and 2′ causes the average temperature at which heat is supplied in the Rankine cycle to be lower than in the Carnot cycle 1′–2′–3–4–1′. Consequently, the efficiency of the Rankine cycle is lower than that of the corresponding Carnot cycle. In the regenerative cycle the working fluid enters the boiler at some state between 2 and 2′; consequently, the average temperature at which heat is supplied is higher.

Consider first an idealized regenerative cycle, as shown in Fig. 11.9. The unique feature of this cycle compared to the Rankine cycle is that after leaving the pump, the liquid circulates around the turbine casing, counterflow to the direction of vapor flow in the turbine. Thus, it is possible to transfer to the liquid flowing around the turbine the heat from the vapor as it flows through the turbine. Let us assume for the moment that this is a reversible heat transfer; that is, at each point the temperature of the vapor is only infinitesimally higher than the temperature of the liquid. In this instance line 4–5 on the T–s dia-

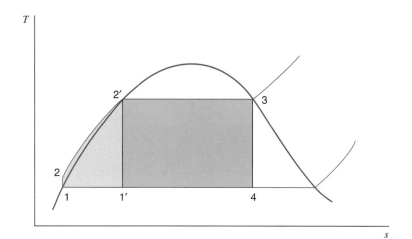

FIGURE 11.8
Temperature-entropy diagram showing the relationships between Carnot-cycle efficiency and Rankine-cycle efficiency.

gram of Fig. 11.9, which represents the states of the vapor flowing through the turbine, is exactly parallel to line 1–2–3, which represents the pumping process (1–2) and the states of the liquid flowing around the turbine. Consequently, areas 2–3–*b*–*a*–2 and 5–4–*d*–*c*–5 are not only equal but congruous, and these areas, respectively, represent the heat transferred to the liquid and from the vapor. Heat is also transferred to the working fluid at constant temperature in process 3–4, and area 3–4–*d*–*b*–3 represents this heat transfer. Heat is transferred from the working fluid in process 5–1, and area 1–5–*c*–*a*–1 represents this heat transfer. This area is exactly equal to area 1′–5′–*d*–*b*–1′, which is the heat rejected in the related Carnot cycle 1′–3–4–5′–1′. Thus, the efficiency of this idealized regenerative cycle is exactly equal to the efficiency of the Carnot cycle with the same heat supply and heat rejection temperatures.

Quite obviously, this idealized regenerative cycle is impractical. First, it would be impossible to effect the necessary heat transfer from the vapor in the turbine to the liquid feedwater. Furthermore, the moisture content of the vapor leaving the turbine increases considerably as a result of the heat transfer. The disadvantage of this has been noted previously. The practical regenerative cycle extracts some of the vapor after it has partially expanded in the turbine and uses feedwater heaters, as shown in Fig. 11.10.

Steam enters the turbine at state 5. After expansion to state 6, some of the steam is extracted and enters the feedwater heater. The steam that is not extracted is expanded in

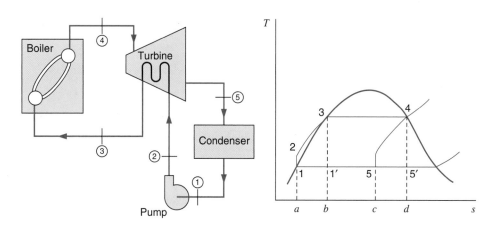

FIGURE 11.9 The ideal regenerative cycle.

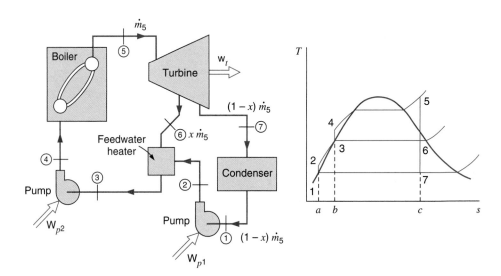

FIGURE 11.10
Regenerative cycle with open feedwater heater.

the turbine to state 7 and is then condensed in the condenser. This condensate is pumped into the feedwater heater where it mixes with the steam extracted from the turbine. The proportion of steam extracted is just sufficient to cause the liquid leaving the feedwater heater to be saturated at state 3. Note that the liquid has not been pumped to the boiler pressure, but only to the intermediate pressure corresponding to state 6. Another pump is required to pump the liquid leaving the feedwater heater to boiler pressure. The significant point is that the average temperature at which heat is supplied has been increased.

Consider a control volume around the open feedwater heater in Fig. 11.10. The conservation of mass requires:

$$\dot{m}_2 + \dot{m}_6 = \dot{m}_3$$

satisfied with the extraction fraction as

$$x = \dot{m}_6/\dot{m}_5 \qquad (11.2)$$

so

$$\dot{m}_7 = (1 - x)\dot{m}_5 = \dot{m}_1 = \dot{m}_2$$

The energy equation with no external heat transfer and no work becomes

$$\dot{m}_2 h_2 + \dot{m}_6 h_6 = \dot{m}_3 h_3 \qquad (11.3)$$

into which we substitute the mass flow rates ($\dot{m}_3 = \dot{m}_5$) as

$$(1 - x)\dot{m}_5 h_2 + x\dot{m}_5 h_6 = \dot{m}_5 h_3 \qquad (11.4)$$

We take state 3 as the limit of saturated liquid (we do not want to heat further as it would move into the two-phase region and damage the pump $P2$) and then solve for x

$$x = \frac{h_3 - h_2}{h_6 - h_2} \qquad (11.5)$$

This establishes the maximum extraction fraction we should take out at this extraction pressure.

This cycle is somewhat difficult to show on a T–s diagram because the masses of steam flowing through the various components vary. The T–s diagram of Fig. 11.10 simply shows the state of the fluid at the various points.

Area 4–5–c–b–4 in Fig. 11.10 represents the heat transferred per kilogram of working fluid. Process 7–1 is the heat rejection process, but since not all the steam passes through the condenser, area 1–7–c–a–1 represents the heat transfer per kilogram flowing through the condenser, which does not represent the heat transfer per kilogram of working fluid entering the turbine. Between states 6 and 7 only part of the steam is flowing through the turbine. The example that follows illustrates the calculations for the regenerative cycle.

EXAMPLE 11.4 Consider a regenerative cycle using steam as the working fluid. Steam leaves the boiler and enters the turbine at 4 MPa, 400°C. After expansion to 400 kPa, some of the steam is extracted from the turbine for the purpose of heating the feedwater in an open feedwater heater. The pressure in the feedwater heater is 400 kPa, and the water leaving it is saturated liquid at 400 kPa. The steam not extracted expands to 10 kPa. Determine the cycle efficiency.

The line diagram and T–s diagram for this cycle are shown in Fig. 11.10

As in previous examples, the model for each control volume is the steam tables, the process is steady state, and kinetic and potential energy changes are negligible.

From Examples 11.2 and 11.3 we have the following properties:

$$h_5 = 3213.6 \qquad h_6 = 2685.6$$
$$h_7 = 2144.1 \qquad h_1 = 191.8$$

For the low-pressure pump,

> *Control volume:* Low-pressure pump.
> *Inlet state:* P_1 known, saturated liquid; state fixed.
> *Exit state:* P_2 known.

Analysis

The first law is

$$w_{p1} = h_2 - h_1$$

For the second law,

$$s_2 = s_1$$

Therefore,

$$h_2 - h_1 = \int_1^2 v\, dP = v(P_2 - P_1)$$

Solution

Substituting,

$$w_{p1} = v(P_2 - P_1) = (0.001\ 01)(400 - 10) = 0.4 \text{ kJ/kg}$$
$$h_2 = h_1 + w_p = 191.8 + 0.4 = 192.2$$

For the turbine,

>*Control volume:* Turbine.
>*Inlet state:* P_5, T_5 known; state fixed.
>*Exit state:* P_6 known; P_7 known.

Analysis

From the first law,

$$w_t = (h_5 - h_6) + (1 - m_1)(h_6 - h_7)$$

and the second law,

$$s_5 = s_6 = s_7$$

Solution

From the second law, the values for h_6 and h_7 given previously were calculated in Examples 11.2 and 11.3.

For the feedwater heater,

>*Control volume:* Feedwater heater.
>*Inlet states:* States 2 and 6 both known (as given).
>*Exit state:* P_3 known, saturated liquid; state fixed.

Analysis

The first law gives

$$m_1(h_6) + (1 - m_1)h_2 = h_3$$

Solution

After substitution,

$$m_1(2685.6) + (1 - m_1)(192.2) = 604.7$$

$$m_1 = 0.1654$$

We can now calculate the turbine work.

$$w_t = (h_5 - h_6) + (1 - m_1)(h_6 - h_7)$$

$$= (3213.6 - 2685.6) + (1 - 0.1654)(2685.6 - 2144.1)$$

$$= 979.9 \text{ kJ/kg}$$

For the high-pressure pump,

>*Control volume:* High-pressure pump.
>*Inlet state:* State 3 known (as given).
>*Exit state:* P_4 known.

Analysis

For the first law,

$$w_{p2} = h_4 - h_3$$

The second law is

$$s_4 = s_3$$

Solution

Substituting,

$$w_{p2} = v(P_4 - P_3) = (0.001\ 084)(4000 - 400) = 3.9 \text{ kJ/kg}$$

$$h_4 = h_3 + w_{p2} = 604.7 + 3.9 = 608.6$$

Therefore,

$$w_{\text{net}} = w_t - (1 - m_1)w_{p1} - w_{p2}$$

$$= 979.9 - (1 - 0.1654)(0.4) - 3.9 = 975.7 \text{ kJ/kg}$$

Finally, for the boiler,

> *Control volume:*   Boiler.
> *Inlet state:* P_4, h_4 known (as given); state fixed.
> *Exit state:*   State 5 known (as given).

Analysis

The first law gives

$$q_H = h_5 - h_4$$

Solution

Substituting,

$$q_H = h_5 - h_4 = 3213.6 - 608.6 = 2605.0 \text{ kJ/kg}$$

$$\eta_{\text{th}} = \frac{w_{\text{net}}}{q_H} = \frac{975.7}{2605.0} = 37.5\%$$

Note the increase in efficiency over the efficiency of the Rankine cycle of Example 11.2.

Up to this point, the discussion and examples have tacitly assumed that the extraction steam and feedwater are mixed in the feedwater heater. Another much-used type of feedwater heater, known as a closed heater, is one in which the steam and feedwater do not mix. Rather, heat is transferred from the extracted steam as it condenses on the outside of tubes while the feedwater flows through the tubes. In a closed heater, a schematic sketch of which is shown in Fig. 11.11, the steam and feedwater may be at considerably different pressures. The condensate may be pumped into the feedwater line, or it may be removed through a trap to a lower-pressure heater or to the condenser. (A trap is a device that permits liquid but not vapor to flow to a region of lower pressure.)

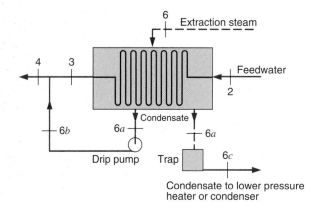

FIGURE 11.11
Schematic arrangement for a closed feedwater heater.

Let us analyze the closed feedwater heater in Fig. 11.11 when a trap with a drain to the condenser is used. We assume we can heat the feedwater up to the temperature of the condensing extraction flow, that is $T_3 = T_4 = T_{6a}$, as there is no drip pump. Conservation of mass for the feedwater heater is

$$\dot{m}_4 = \dot{m}_3 = \dot{m}_2 = \dot{m}_5; \qquad \dot{m}_6 = x\dot{m}_5 = \dot{m}_{6a} = \dot{m}_{6c}$$

Notice that the extraction flow is added to the condenser, so the flowrate at 2 is the same as at state 5. The energy equation is

$$\dot{m}_5 h_2 + x\dot{m}_5 h_6 = \dot{m}_5 h_3 + x\dot{m}_5 h_{6a} \tag{11.6}$$

which we can solve for x as

$$x = \frac{h_3 - h_2}{h_6 - h_{6a}} \tag{11.7}$$

Open feedwater heaters have the advantage of being less expensive and having better heat-transfer characteristics than closed feedwater heaters. They have the disadvantage of requiring a pump to handle the feedwater between each heater.

In many power plants a number of stages of extraction are used, though only rarely more than five. The number is, of course, determined by economics. It is evident that using a very large number of extraction stages and feedwater heaters allows the cycle efficiency to approach that of the idealized regenerative cycle of Fig. 11.9, where the feedwater enters the boiler as saturated liquid at the maximum pressure. In practice, however, this could not be economically justified because the savings effected by the increase in efficiency would be more than offset by the cost of additional equipment (feedwater heaters, piping, and so forth).

A typical arrangement of the main components in an actual power plant is shown in Fig. 11.12. Note that one open feedwater heater is a deaerating feedwater heater; this heater has the dual purpose of heating and removing the air from the feedwater. Unless the air is removed, excessive corrosion occurs in the boiler. Note also that the condensate from the high-pressure heater drains (through a trap) to the intermediate heater, and the intermediate heater drains to the deaerating feedwater heater. The low-pressure heater drains to the condenser.

Many actual power plants combine one reheat stage with a number of extraction stages. The principles already considered are readily applied to such a cycle.

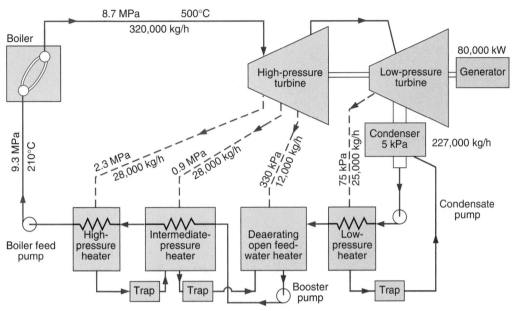

FIGURE 11.12 Arrangement of heaters in an actual power plant utilizing regenerative feedwater heaters.

11.6 DEVIATION OF ACTUAL CYCLES FROM IDEAL CYCLES

Before we leave the matter of vapor power cycles, a few comments are in order regarding the ways in which an actual cycle deviates from an ideal cycle. The most important of these losses are due to the turbine, the pump(s), the pipes, and the condenser. These losses are discussed next.

Turbine Losses

Turbine losses, as described in Section 9.5, represent by far the largest discrepancy between the performance of a real cycle and a corresponding ideal Rankine-cycle power plant. The large positive turbine work is the principal number in the numerator of the cycle thermal efficiency and is directly reduced by the factor of the isentropic turbine efficiency. Turbine losses are primarily those associated with the flow of the working fluid through the turbine blades and passages, with heat transfer to the surroundings also being a loss but of secondary importance. The turbine process might be represented as shown in Fig. 11.13 where state 4_s is the state after an ideal isentropic turbine expansion and state 4 is the actual state leaving the turbine following an irreversible process. The turbine governing procedures may also cause a loss in the turbine, particularly if a throttling process is used to govern the turbine operation.

Pump Losses

The losses in the pump are similar to those of the turbine and are due primarily to the irreversibilities with the fluid flow. Pump efficiency was discussed in Section 9.5, and the ideal exit state 2_s and real exit state 2 are shown in Fig. 11.13. Pump losses are much smaller than those of the turbine, since the associated work is very much smaller.

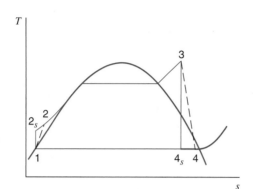

FIGURE 11.13
Temperature–entropy diagram showing effect of turbine and pump inefficiencies on cycle performance.

Piping Losses

Pressure drops caused by frictional effects and heat transfer to the surroundings are the most important piping losses. Consider, for example, the pipe connecting the turbine to the boiler. If only frictional effects occur, states *a* and *b* in Fig. 11.14 would represent the states of the steam leaving the boiler and entering the turbine, respectively. Note that the frictional effects cause an increase in entropy. Heat transferred to the surroundings at constant pressure can be represented by process *bc*. This effect decreases entropy. Both the pressure drop and heat transfer decrease the availability of the steam entering the turbine. The irreversibility of this process can be calculated by the methods outlined in Chapter 10.

A similar loss is the pressure drop in the boiler. Because of this pressure drop, the water entering the boiler must be pumped to a higher pressure than the desired steam pressure leaving the boiler, and this requires additional pump work.

Condenser Losses

The losses in the condenser are relatively small. One of these minor losses is the cooling below the saturation temperature of the liquid leaving the condenser. This represents a loss because additional heat transfer is necessary to bring the water to its saturation temperature.

The influence of these losses on the cycle is illustrated in the following example, which should be compared to Example 11.2.

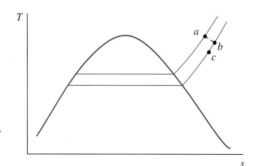

FIGURE 11.14
Temperature–entropy diagram showing effect of losses between boiler and turbine.

EXAMPLE 11.5 A steam power plant operates on a cycle with pressures and temperatures as designated in Fig. 11.15. The efficiency of the turbine is 86%, and the efficiency of the pump is 80%. Determine the thermal efficiency of this cycle.

As in previous examples, for each control volume the model used is the steam tables, and each process is steady state with no changes in kinetic or potential energy. This cycle is shown on the T–s diagram of Fig. 11.16.

> *Control volume:* Turbine.
> *Inlet state:* P_5, T_5 known; state fixed.
> *Exit state:* P_6 known.

Analysis

From the first law, we have

$$w_t = h_5 - h_6$$

The second law is

$$s_{6s} = s_5$$

The efficiency is

$$\eta_t = \frac{w_t}{h_5 - h_{6s}} = \frac{h_5 - h_6}{h_5 - h_{6s}}$$

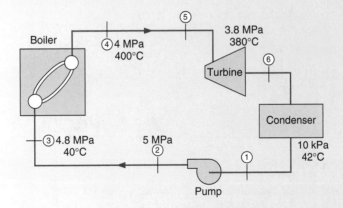

FIGURE 11.15
Schematic diagram for Example 11.5.

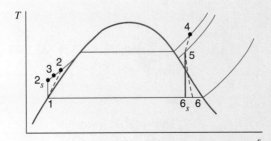

FIGURE 11.16
Temperature–entropy diagram for Example 11.5.

Solution

From the steam tables, we get

$$h_5 = 3169.1 \text{ kJ/kg}, \qquad s_5 = 6.7235$$

$$s_{6_s} = s_5 = 6.7235 = 0.6493 + x_{6_s} 7.5009, \qquad x_{6_s} = 0.8098$$

$$h_{6_s} = 191.8 + 0.8098(2392.8) = 2129.5 \text{ kJ/kg}$$

$$w_t = \eta_t(h_5 - h_{6_s}) = 0.86(3169.1 - 2129.5) = 894.1 \text{ kJ/kg}$$

For the pump, we have:

> *Control volume:* Pump.
> *Inlet state:* P_1, T_1 known; state fixed.
> *Exit state:* P_2 known.

Analysis

The first law gives

$$w_p = h_2 - h_1$$

The second law gives

$$s_{2_s} = s_1$$

The pump efficiency is

$$\eta_p = \frac{h_{2_s} - h_1}{w_p} = \frac{h_{2_s} - h_1}{h_2 - h_1}$$

Since $s_{2_s} = s_1$,

$$h_{2_s} - h_1 = v(P_2 - P_1)$$

Therefore,

$$w_p = \frac{h_{2_s} - h_1}{\eta_p} = \frac{v(P_2 - P_1)}{\eta_p}$$

Solution

Substituting, we obtain

$$w_p = \frac{v(P_2 - P_1)}{\eta_p} = \frac{(0.001\ 009)(5000 - 10)}{0.80} = 6.3 \text{ kJ/kg}$$

Therefore,

$$w_{\text{net}} = w_t - w_p = 894.1 - 6.3 = 887.8 \text{ kJ/kg}$$

Finally, for the boiler:

> *Control volume:* Boiler.
> *Inlet state:* P_3, T_3 known; state fixed.
> *Exit state:* P_4, T_4 known, state fixed.

Analysis

The first law is

$$q_H = h_4 - h_3$$

Solution

Substitution gives

$$q_H = h_4 - h_3 = 3213.6 - 171.8 = 3041.8 \text{ kJ/kg}$$

$$\eta_{th} = \frac{887.8}{3041.8} = 29.2\%$$

This result compares to the Rankine efficiency of 35.3% for the similar cycle of Example 11.2.

EXAMPLE 11.5E   A steam power plant operates on a cycle with pressure and temperatures as designated in Fig. 11.15E. The efficiency of the turbine is 86%, and the efficiency of the pump is 80%. Determine the thermal efficiency of this cycle.

As in previous examples, for each control volume the model used is the steam tables, and each process is steady state with no changes in kinetic or potential energy. This cycle is shown on the T–s diagram of Fig. 11.16.

> *Control volume:* Turbine.
> *Inlet state:* P_5, T_5 known; state fixed.
> *Exit state:* P_6 known.

Analysis

From the first law, we have

$$w_t = h_5 - h_6$$

The second law is

$$s_{6s} = s_5$$

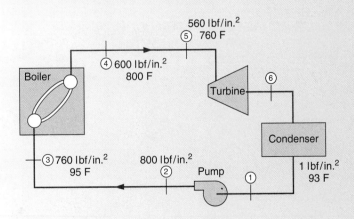

FIGURE 11.15E
Schematic diagram for
Example 11.5E.

The efficiency is

$$\eta_t = \frac{w_t}{h_5 - h_{6s}} = \frac{h_5 - h_6}{h_5 - h_{6s}}$$

Solution

From the steam tables, we get

$$h_5 = 1386.8 \qquad s_5 = 1.6248$$

$$s_{6s} = s_5 = 1.6248 = 1.9779 - (1 - x)_{6s}1.8453$$

$$(1 - x)_{6s} = \frac{0.3531}{1.8453} = 0.1912$$

$$h_{6s} = 1105.8 - 0.1912(1036.0) = 907.6$$

$$w_t = \eta_t(h_5 - h_{6s}) = 0.86(1386.8 - 907.6)$$

$$= 0.86(479.2) = 412.1 \text{ Btu/lbm}$$

For the pump, we have:

> *Control volume:* Pump.
> *Inlet state:* P_1, T_1 known; state fixed.
> *Exit state:* P_2 known.

Analysis

The first law gives

$$w_p = h_2 - h_1$$

The second law gives

$$s_{2s} = s_1$$

The pump efficiency is

$$\eta_p = \frac{h_{2s} - h_1}{w_p} = \frac{h_{2s} - h_1}{h_2 - h_1}$$

Since $s_{2s} = s_1$,

$$h_{2s} - h_1 = v(P_2 - P_1)$$

Therefore,

$$w_p = \frac{h_{2s} - h_1}{\eta_p} = \frac{v(P_2 - P_1)}{\eta_p}$$

Solution

Substituting, we obtain

$$w_p = \frac{v(P_2 - P_1)}{\eta_p} = \frac{0.016\ 15(800 - 1)144}{0.8 \times 778} = 3.0 \text{ Btu/lbm}$$

Therefore,

$$w_{\text{net}} = w_t - w_p = 412.1 - 3.0 = 409.1 \text{ Btu/lbm}$$

Finally, for the boiler:

Control volume: Boiler.
Inlet state: P_3, T_3 known; state fixed.
Exit state: P_4, T_4 known; state fixed.

Analysis

The first law is

$$q_H = h_4 - h_3$$

Solution

Substitution gives

$$q_H = h_4 - h_3 = 1407.6 - 65.1 = 1342.5 \text{ Btu/lbm}$$

$$\eta_{\text{th}} = \frac{409.1}{1342.5} = 30.4\%$$

This result compares to the Rankine efficiency of 36.9% for the similar cycle of Example 11.2E.

11.7 COGENERATION

There are many occasions in industrial settings where the need arises for a specific source or supply of energy within the environment in which a steam power plant is being used to generate electricity. In such cases, it is appropriate to consider supplying this source of energy in the form of steam that has already been expanded through the high-pressure section of the turbine in the power plant cycle, thereby eliminating the construction and use of a second boiler or other energy source. Such an arrangement is shown in Fig. 11.17, in which the turbine is tapped at some intermediate pressure to furnish the necessary amount of process steam required for the particular energy need—perhaps to operate a special process in the plant, or in many cases simply for the purpose of space heating the facilities. This type of application is termed cogeneration. If the system is designed as a package with both the electrical and the process steam requirements in mind, it is possible to achieve a substantial savings in capital cost of equipment and in the operating cost, through careful consideration of all the requirements and optimization of the various parameters involved. Specific examples of cogeneration systems are considered in the problems at the end of the chapter.

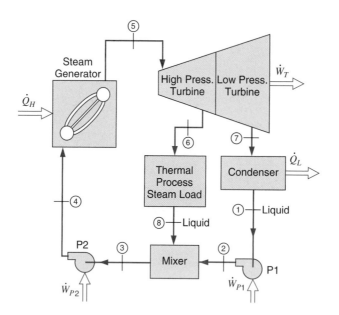

FIGURE 11.17
Example of a
cogeneration system.

11.8 AIR-STANDARD POWER CYCLES

In Section 11.1, we considered idealized four-process cycles, including both steady-state-process and cylinder/piston boundary-movement cycles. The question of phase-change cycles and single-phase cycles was also mentioned. We then proceeded to examine the Rankine power plant cycle in detail, the idealized model of a phase-change power cycle. However, many work-producing devices (engines) utilize a working fluid that is always a gas. The spark-ignition automotive engine is a familiar example, as are the diesel engine and the conventional gas turbine. In all these engines there is a change in the composition of the working fluid, because during combustion it changes from air and fuel to combustion products. For this reason these engines are called internal-combustion engines. In contrast, the steam power plant may be called an external-combustion engine because heat is transferred from the products of combustion to the working fluid. External-combustion engines using a gaseous working fluid (usually air) have been built. To date they have had only limited application, but use of the gas-turbine cycle in conjunction with a nuclear reactor has been investigated extensively. Other external-combustion engines are currently receiving serious attention in an effort to combat air pollution.

Because the working fluid does not go through a complete thermodynamic cycle in the engine (even though the engine operates in a mechanical cycle), the internal-combustion engine operates on the so-called open cycle. However, for analyzing internal-combustion engines, it is advantageous to devise closed cycles that closely approximate the open cycles. One such approach is the air-standard cycle, which is based on the following assumptions:

1. A fixed mass of air is the working fluid throughout the entire cycle, and the air is always an ideal gas. Thus, there is no inlet process or exhaust process.

2. The combustion process is replaced by a process transferring heat from an external source.

3. The cycle is completed by heat transfer to the surroundings (in contrast to the exhaust and intake process of an actual engine).

4. All processes are internally reversible.

5. An additional assumption is often made that air has a constant specific heat, recognizing that this is not the most accurate model.

The principal value of the air-standard cycle is to enable us to examine qualitatively the influence of a number of variables on performance. The quantitative results obtained from the air-standard cycle, such as efficiency and mean effective pressure, will differ from those of the actual engine. Our emphasis, therefore, in our consideration of the air-standard cycle will be primarily on the qualitative aspects.

The term mean effective pressure (mep), which is used in conjunction with reciprocating engines, is defined as the pressure that, if it acted on the piston during the entire power stroke, would do an amount of work equal to that actually done on the piston. The work for one cycle is found by multiplying this mean effective pressure by the area of the piston (minus the area of the rod on the crank end of a double-acting engine) and by the stroke.

11.9 THE BRAYTON CYCLE

In discussing idealized four-steady-state-process power cycles in Section 11.1, a cycle involving two constant-pressure and two isentropic processes was examined, and the results are shown in Fig. 11.2. This cycle used with a condensing working fluid is the Rankine cycle, but when used with a single-phase, gaseous working fluid it is termed the Brayton cycle. The air-standard Brayton cycle is the ideal cycle for the simple gas turbine. The simple open-cycle gas turbine utilizing an internal-combustion process and the simple closed-cycle gas turbine, which utilizes heat-transfer processes, are both shown schematically in Fig. 11.18. The air-standard Brayton cycle is shown on the $P–v$ and $T–s$ diagrams of Fig. 11.19.

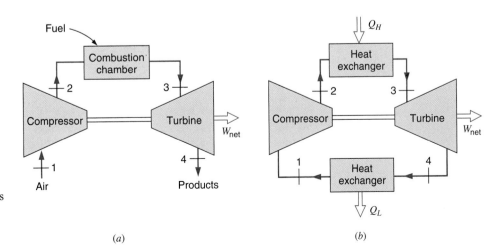

FIGURE 11.18 A gas turbine operating on the Brayton cycle. (*a*) Open cycle. (*b*) Closed cycle.

(*a*)

(*b*)

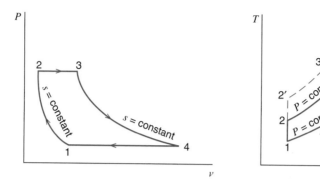

FIGURE 11.19 The air-standard Brayton cycle.

The efficiency of the air-standard Brayton cycle is found as follows:

$$\eta_{th} = 1 - \frac{Q_L}{Q_H} = 1 - \frac{C_p(T_4 - T_1)}{C_p(T_3 - T_2)} = 1 - \frac{T_1(T_4/T_1 - 1)}{T_2(T_3/T_2 - 1)}$$

We note, however, that

$$\frac{P_3}{P_4} = \frac{P_2}{P_1}$$

$$\frac{P_2}{P_1} = \left(\frac{T_2}{T_1}\right)^{k/(k-1)} = \frac{P_3}{P_4} = \left(\frac{T_3}{T_4}\right)^{k/(k-1)}$$

$$\frac{T_3}{T_4} = \frac{T_2}{T_1} \therefore \frac{T_3}{T_2} = \frac{T_4}{T_1} \quad \text{and} \quad \frac{T_3}{T_2} - 1 = \frac{T_4}{T_2} - 1$$

$$\eta_{th} = 1 - \frac{T_1}{T_2} = 1 - \frac{1}{(P_2/P_1)^{(k-1)/k}} \qquad (11.8)$$

The efficiency of the air-standard Brayton cycle is therefore a function of the isentropic pressure ratio. The fact that efficiency increases with pressure ratio is evident from the T–s diagram of Fig. 11.19 because increasing the pressure ratio changes the cycle from 1–2–3–4–1 to 1–2'–3'–4–1. The latter cycle has a greater heat supply and the same heat rejected as the original cycle; therefore, it has a greater efficiency. Note that the latter cycle has a higher maximum temperature, $T_{3'}$, than the original cycle, T_3. In the actual gas turbine, the maximum temperature of the gas entering the turbine is fixed by material considerations. Therefore, if we fix the temperature T_3 and increase the pressure ratio, the resulting cycle is 1–2'–3"–4"–1. This cycle would have a higher efficiency than the original cycle, but the work per kilogram of working fluid is thereby changed.

With the advent of nuclear reactors, the closed-cycle gas turbine has become more important. Heat is transferred, either directly or via a second fluid, from the fuel in the nuclear reactor to the working fluid in the gas turbine. Heat is rejected from the working fluid to the surroundings.

The actual gas-turbine engine differs from the ideal cycle primarily because of irreversibilities in the compressor and turbine, and because of pressure drop in the flow passages and combustion chamber (or in the heat exchanger of a closed-cycle turbine). Thus, the state points in a simple open-cycle gas turbine might be as shown in Fig. 11.20.

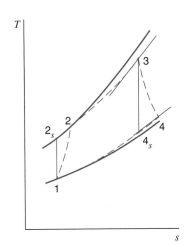

FIGURE 11.20 Effect of inefficiencies on the gas-turbine cycle.

The efficiencies of the compressor and turbine are defined in relation to isentropic processes. With the states designated as in Fig. 11.20, the definitions of compressor and turbine efficiencies are

$$\eta_{comp} = \frac{h_{2s} - h_1}{h_2 - h_1} \qquad (11.9)$$

$$\eta_{turb} = \frac{h_3 - h_4}{h_3 - h_{4s}} \qquad (11.10)$$

Another important feature of the Brayton cycle is the large amount of compressor work (also called back work) compared to turbine work. Thus, the compressor might require from 40 to 80% of the output of the turbine. This is particularly important when the actual cycle is considered because the effect of the losses is to require a larger amount of compression work from a smaller amount of turbine work. Thus, the overall efficiency drops very rapidly with a decrease in the efficiencies of the compressor and turbine. In fact, if these efficiencies drop below about 60%, all the work of the turbine will be required to drive the compressor, and the overall efficiency will be zero. This is in sharp contrast to the Rankine cycle, where only 1 or 2% of the turbine work is required to drive the pump. This demonstrates the inherent advantage of the cycle utilizing a condensing working fluid, such that a much larger difference in specific volume between the expansion and compression processes is utilized effectively.

EXAMPLE 11.6 In an air-standard Brayton cycle the air enters the compressor at 0.1 MPa and 15°C. The pressure leaving the compressor is 1.0 MPa, and the maximum temperature in the cycle is 1100°C. Determine

1. The pressure and temperature at each point in the cycle.
2. The compressor work, turbine work, and cycle efficiency.

For each of the control volumes analyzed, the model is ideal gas with constant specific heat, at 300 K, and each process is steady state with no kinetic or potential energy changes. The diagram for this example is Fig. 11.19.

We consider the compressor, the turbine, and the high-temperature and low-temperature heat exchangers in turn.

> *Control volume:* Compressor.
> *Inlet state:* P_1, T_1 known; state fixed.
> *Exit state:* P_2 known.

Analysis

The first law gives

$$w_c = h_2 - h_1$$

(Note that the compressor work w_c is here defined as work input to the compressor.) The second law is

$$s_2 = s_1$$

so that

$$\frac{T_2}{T_1} = \left(\frac{P_2}{P_1}\right)^{(k-1)/k}$$

Solution

Solving for T_2, we get

$$\left(\frac{P_2}{P_1}\right)^{(k-1)/k} = 10^{0.286} = 1.932, \qquad T_2 = 556.8 \text{ K}$$

Therefore,

$$w_c = h_2 - h_1 = C_p(T_2 - T_1)$$
$$= 1.004(556.8 - 288.2) = 269.5 \text{ kJ/kg}$$

Consider the turbine next.

> *Control volume:* Turbine.
> *Inlet state:* $P_3(= P_2)$ known, T_3 known, state fixed.
> *Exit state:* $P_4(= P_1)$ known.

Analysis

The first law gives

$$w_t = h_3 - h_4$$

The second law is

$$s_3 = s_4$$

so that

$$\frac{T_3}{T_4} = \left(\frac{P_3}{P_4}\right)^{(k-1)/k}$$

Solution

Solving for T_4, we get

$$\left(\frac{P_3}{P_4}\right)^{(k-1)/k} = 10^{0.286} = 1.932, \qquad T_4 = 710.8 \text{ K}$$

Therefore,

$$w_t = h_3 - h_4 = C_p(T_3 - T_4)$$
$$= 1.004(1373.2 - 710.8) = 664.7 \text{ kJ/kg}$$
$$w_{net} = w_t - w_c = 664.7 - 269.5 = 395.2 \text{ kJ/kg}$$

Now we turn to the heat exchangers.

Control volume:	High-temperature heat exchanger.
Inlet state:	State 2 fixed (as given).
Exit state:	State 3 fixed (as given).

Analysis

The first law is

$$q_H = h_3 - h_2 = C_p(T_3 - T_2)$$

Solution

Substitution gives

$$q_H = h_3 - h_2 = C_p(T_3 - T_2) = 1.004(1373.2 - 556.8) = 819.3 \text{ kJ/kg}$$

Control volume:	Low-temperature heat exchanger.
Inlet state:	State 4 fixed (above).
Exit state:	State 1 fixed (above).

Analysis

The first law is

$$q_L = h_4 - h_1 = C_p(T_4 - T_1)$$

Solution

Upon substitution we have

$$q_L = h_4 - h_1 = C_p(T_4 - T_1) = 1.004(710.8 - 288.2) = 424.1 \text{ kJ/kg}$$

Therefore,

$$\eta_{th} = \frac{W_{net}}{q_H} = \frac{395.2}{819.3} = 48.2\%$$

This may be checked by using Eq. 11.8.

$$\eta_{th} = 1 - \frac{1}{(P_2/P_1)^{(k-1)/k}} = 1 - \frac{1}{10^{0.286}} = 48.2\%$$

EXAMPLE 11.7   Consider a gas turbine with air entering the compressor under the same conditions as in Example 11.6 and leaving at a pressure of 1.0 MPa. The maximum temperature is 1100°C. Assume a compressor efficiency of 80%, a turbine efficiency of 85%, and a pressure drop between the compressor and turbine of 15 kPa. Determine the compressor work, turbine work, and cycle efficiency.

As in the previous example, for each control volume the model is ideal gas with constant specific heat, at 300 K, and each process is steady state with no kinetic or potential energy changes. In this example the diagram is Fig. 11.20.

We consider the compressor, the turbine and the high-temperature heat exchanger in turn.

> *Control volume:*   Compressor.
> *Inlet state:* P_1, T_1 known; state fixed.
> *Exit state:* P_2 known.

Analysis

The first law for the real process is

$$w_c = h_2 - h_1$$

The second law for the ideal process is

$$s_{2_s} = s_1$$

so that

$$\frac{T_{2_s}}{T_1} = \left(\frac{P_2}{P_1}\right)^{(k-1)/k}$$

In addition,

$$\eta_c = \frac{h_{2_s} - h_1}{h_2 - h_1} = \frac{T_{2_s} - T_1}{T_2 - T_1}$$

Solution

Solving for T_{2_s}, we get

$$\left(\frac{P_2}{P_1}\right)^{(k-1)/k} = \frac{T_{2_s}}{T_1} = 10^{0.286} = 1.932, \qquad T_{2_s} = 556.8 \text{ K}$$

The efficiency is

$$\eta_c = \frac{h_{2_s} - h_1}{h_2 - h_1} = \frac{T_{2_s} - T_1}{T_2 - T_1} = \frac{556.8 - 288.2}{T_2 - T_1} = 0.80$$

Therefore,

$$T_2 - T_1 = \frac{556.8 - 288.2}{0.80} = 335.8, \qquad T_2 = 624.0 \text{ K}$$

$$w_c = h_2 - h_1 = C_p(T_2 - T_1)$$

$$= 1.004(624.0 - 288.2) = 337.0 \text{ kJ/kg}$$

For the turbine, we have:

Control volume: Turbine.
Inlet state: $P_3(P_2 - \text{drop})$ known, T_3 known; state fixed.
Exit state: P_4 known.

Analysis

The first law for the real process is

$$w_t = h_3 - h_4$$

The second law for the ideal process is

$$s_{4_s} = s_3$$

so that

$$\frac{T_3}{T_{4_s}} = \left(\frac{P_3}{P_4}\right)^{(k-1)/k}$$

In addition,

$$\eta_t = \frac{h_3 - h_4}{h_3 - h_{4_s}} = \frac{T_3 - T_4}{T_3 - T_{4_s}}$$

Solution

Substituting numerical values, we obtain

$$P_3 = P_2 - \text{pressure drop} = 1.0 - 0.015 = 0.985 \text{ MPa}$$

$$\left(\frac{P_3}{P_4}\right)^{(k-1)/k} = \frac{T_3}{T_{4_s}} = 9.85^{0.286} = 1.9236, \qquad T_{4_s} = 713.9 \text{ K}$$

$$\eta_t = \frac{h_3 - h_4}{h_3 - h_{4_s}} = \frac{T_3 - T_4}{T_3 - T_{4_s}} = 0.85$$

$$T_3 - T_4 = 0.85(1373.2 - 713.9) = 560.4 \text{ K}$$

$$T_4 = 812.8 \text{ K}$$
$$w_t = h_3 - h_4 = C_p(T_3 - T_4)$$
$$= 1.004(1373.2 - 812.8) = 562.4 \text{ kJ/kg}$$
$$w_{\text{net}} = w_t - w_c = 562.4 - 337.0 = 225.4 \text{ kJ/kg}$$

Finally, for the heat exchanger:

Control volume: High-temperature heat exchanger.
Inlet state: State 2 fixed (as given).
Exit state: State 3 fixed (as given).

Analysis

The first law is

$$q_H = h_3 - h_2$$

Solution

Substituting, we have

$$q_H = h_3 - h_2 = C_p(T_3 - T_2)$$
$$= 1.004(1373.2 - 624.0) = 751.8 \text{ kJ/kg}$$

so that

$$\eta_{th} = \frac{w_{net}}{q_H} = \frac{225.4}{751.8} = 30.0\%$$

The following comparisons can be made between Examples 11.6 and 11.7:

	w_c	w_t	w_{net}	q_H	η_{th}
Example 11.6 (Ideal)	269.5	664.7	395.2	819.3	48.2
Example 11.7 (Actual)	337.0	562.4	225.4	751.8	30.0

As stated previously, the irreversibilities decrease the turbine work and increase the compressor work. Since the net work is the difference between these two, it decreases very rapidly as compressor and turbine efficiencies decrease. The development of compressors and turbines of high efficiency is therefore an important aspect of the development of gas turbines.

Note that in the ideal cycle (Example 11.6) about 41% of the turbine work is required to drive the compressor and 59% is delivered as net work. In the actual turbine (Example 11.7) 60% of the turbine work is required to drive the compressor and 40% is delivered as net work. Thus, if the net power of this unit is to be 10 000 kW, a 25 000 kW turbine and a 15 000 kW compressor are required. This result demonstrates that a gas turbine has a high back-work ratio.

11.10 THE SIMPLE GAS-TURBINE CYCLE WITH A REGENERATOR

The efficiency of the gas-turbine cycle may be improved by introducing a regenerator. The simple open-cycle gas-turbine cycle with a regenerator is shown in Fig. 11.21, and the corresponding ideal air-standard cycle with a regenerator is shown on the P–v and T–s diagrams. In cycle 1–2–x–3–4–y–1, the temperature of the exhaust gas leaving the turbine in state 4 is higher than the temperature of the gas leaving the compressor. Therefore, heat can be transferred from the exhaust gases to the high-pressure gases leaving the compressor. If this is done in a counterflow heat exchanger (a regenerator), the temperature of the high-pressure gas leaving the regenerator, T_x, may, in the ideal case, have a temperature equal to T_4, the temperature of the gas leaving the turbine. Heat transfer from the external source is necessary only to increase the temperature from T_x to T_3. Area x–3–d–b–x represents the heat transferred, and area y–1–a–c–y represents the heat rejected.

The influence of pressure ratio on the simple gas-turbine cycle with a regenerator is shown by considering cycle 1–2′–3′–4–1. In this cycle the temperature of the exhaust gas leaving the turbine is just equal to the temperature of the gas leaving the compressor; therefore, utilizing a regenerator is not possible. This can be shown more exactly be determining the efficiency of the ideal gas-turbine cycle with a regenerator.

The efficiency of this cycle with regeneration is found as follows, where the states are as given in Fig. 11.21.

$$\eta_{th} = \frac{w_{net}}{q_H} = \frac{w_t - w_c}{q_H}$$

$$q_H = C_p(T_3 - T_x)$$

$$w_t = C_p(T_3 - T_4)$$

But for an ideal regenerator, $T_4 = T_x$, and therefore $q_H = w_t$. Consequently,

$$\eta_{th} = 1 - \frac{w_c}{w_t} = 1 - \frac{C_p(T_2 - T_1)}{C_p(T_3 - T_4)}$$

$$= 1 - \frac{T_1(T_2/T_1 - 1)}{T_3(1 - T_4/T_3)} = 1 - \frac{T_1}{T_3} \frac{[(P_2/P_1)^{(k-1)/k} - 1]}{[1 - (P_1/P_2)^{(k-1)/k}]}$$

$$\eta_{th} = 1 - \frac{T_1}{T_3}\left(\frac{P_2}{P_1}\right)^{(k-1)/k}$$

Thus, for the ideal cycle with regeneration the thermal efficiency depends not only on the pressure ratio but also on the ratio of the minimum to maximum temperature. We note that, in contrast to the Brayton cycle, the efficiency decreases with an increase in pressure ratio.

The effectiveness or efficiency of a regenerator is given by the regenerator efficiency, which can best be defined by reference to Fig. 11.22. State x represents the high-pressure gas leaving the regenerator. In the ideal regenerator there would be only an infinitesimal temperature difference between the two streams, and the high-pressure gas would leave the regenerator at temperature T'_x, and $T'_x = T_4$. In an actual regenerator,

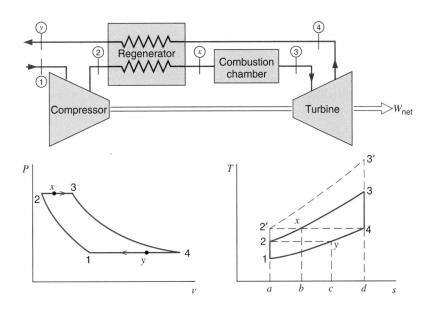

FIGURE 11.21 The ideal regenerative cycle.

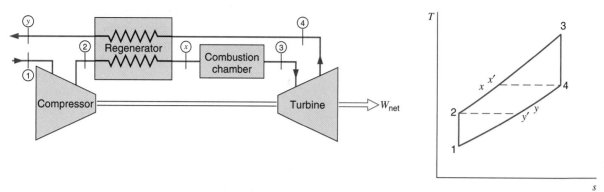

FIGURE 11.22 Temperature–entropy diagram to illustrate the definition of regenerator efficiency.

which must operate with a finite temperature difference T_x, the actual temperature leaving the regenerator is therefore less than T'_x. The regenerator efficiency is defined by

$$\eta_{\text{reg}} = \frac{h_x - h_2}{h'_x - h_2}$$

(11.11)

If the specific heat is assumed to be constant, the regenerator efficiency is also given by the relation

$$\eta_{\text{reg}} = \frac{T_x - T_2}{T'_x - T_2}$$

A higher efficiency can be achieved by using a regenerator with a greater heat-transfer area. However, this also increases the pressure drop, which represents a loss, and both the pressure drop and the regenerator efficiency must be considered in determining which regenerator gives maximum thermal efficiency for the cycle. From an economic point of view, the cost of the regenerator must be weighed against the savings that can be effected by its use.

EXAMPLE 11.8   If an ideal regenerator is incorporated into the cycle of Example 11.6, determine the thermal efficiency of the cycle.

The diagram for this example is Fig. 11.22. Values are from Example 11.6. Therefore, for the analysis of the high-temperature heat exchanger (combustion chamber), from the first law, we have

$$q_H = h_3 - h_x$$

so that the solution is

$$T_x = T_4 = 710.8 \text{ K}$$

$$q_H = h_3 - h_x = C_p(T_3 - T_x) = 1.004(1373.2 - 710.8) = 664.7 \text{ kJ/kg}$$

$$w_{\text{net}} = 395.2 \text{ kJ/kg (from Example 11.6)}$$

$$\eta_{\text{th}} = \frac{395.2}{664.7} = 59.5\%$$

11.11 GAS-TURBINE POWER CYCLE CONFIGURATIONS

The Brayton cycle, being the idealized model for the gas-turbine power plant, has a reversible, adiabatic compressor and a reversible, adiabatic turbine. In the following example, we consider the effect of replacing these components with reversible, isothermal processes.

EXAMPLE 11.9 An air-standard power cycle has the same states as given in Example 11.6. In this cycle, however, the compressor and turbine are both reversible, isothermal processes. Calculate the compressor work and the turbine work, and compare with the results of Example 11.6.

Control volumes: Compressor, turbine.

Analysis

For each reversible, isothermal process, from Eq. 9.18,

$$w = -\int_i^e v\, dP = -P_i v_i \ln \frac{P_e}{P_i} = -RT_i \ln \frac{P_e}{P_i}$$

Solution

For the compressor,

$$w = -0.287 \times 288.2 \times \ln 10 = -190.5 \text{ kJ/kg}$$

compared with -269.5 kJ/kg in the adiabatic compressor.
For the turbine,

$$w = -0.287 \times 1373.2 \times \ln 0.1 = +907.5 \text{ kJ/kg}$$

compared with $+664.7$ kJ/kg in the adiabatic turbine.

It is found that the isothermal process would be preferable to the adiabatic process in both the compressor and turbine. The resulting cycle, called the Ericsson cycle, consists of two reversible, constant-pressure processes and two reversible, constant-temperature processes. The reason that the actual gas turbine does not attempt to emulate this cycle rather than the Brayton cycle is that the compressor and turbine processes are both high-flow rate processes involving work-related devices in which it is not practical to attempt to transfer large quantities of heat. As a consequence, the processes tend to be essentially adiabatic, so that this becomes the process in the model cycle.

There is a modification of the Brayton/gas turbine cycle that tends to change its performance in the direction of the Ericsson cycle. This modification is to use multiple stages of compression with intercooling, and also multiple stages of expansion with reheat. Such a cycle with two stages of compression and expansion, and also incorporating a regenerator, is shown in Fig. 11.23. The air-standard cycle is given on the corresponding *T–s* diagram. It may be shown that for this cycle the maximum efficiency is obtained if equal pressure ratios are maintained across the two compressors and the two turbines. In this

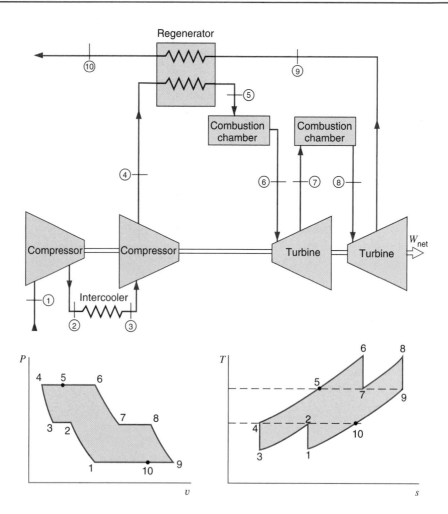

FIGURE 11.23 The ideal gas-turbine cycle utilizing intercooling, reheat, and a regenerator.

ideal cycle, it is assumed that the temperature of the air leaving the intercooler, T_3, is equal to the temperature of the air entering the first stage of compression, T_1, and that the temperature after reheating, T_8, is equal to the temperature entering the first turbine, T_6. Furthermore, in the ideal cycle it is assumed that the temperature of the high-pressure air leaving the regenerator, T_5, is equal to the temperature of the low-pressure air leaving the turbine, T_9.

If a large number of stages of compression and expansion are used, it is evident that the Ericsson cycle is approached. This is shown in Fig. 11.24. In practice, the economical limit to the number of stages is usually two or three. The turbine and compressor losses and pressure drops that have already been discussed would be involved in any actual unit employing this cycle.

The turbines and the compressors using this cycle can be utilized in a variety of ways. Two possible arrangements for closed cycles are shown in Fig. 11.25. One advantage frequently sought in a given arrangement is ease of control of the unit under various loads. Detailed discussion of this point, however, is beyond the scope of this text.

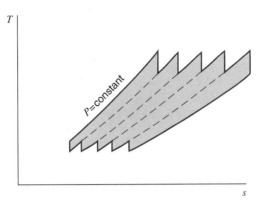

FIGURE 11.24
Temperature–entropy diagram that shows how the gas-turbine cycle with many stages approaches the Ericsson cycle.

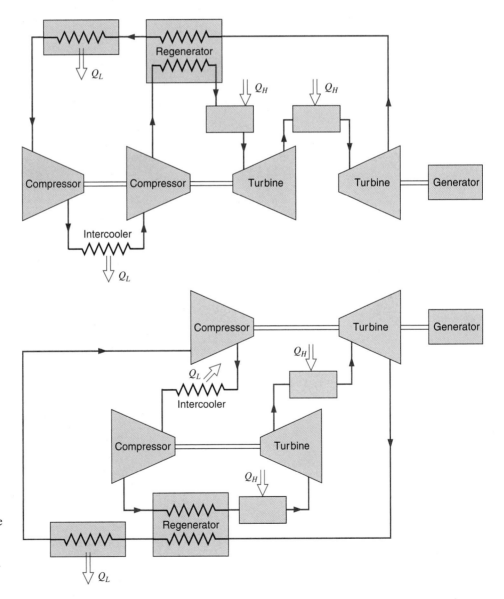

FIGURE 11.25 Some arrangements of components that may be utilized in stationary gas-turbine power plants.

11.12 THE AIR-STANDARD CYCLE FOR JET PROPULSION

The next air-standard power cycle we consider is utilized in jet propulsion. In this cycle the work done by the turbine is just sufficient to drive the compressor. The gases are expanded in the turbine to a pressure for which the turbine work is just equal to the compressor work. The exhaust pressure of the turbine will then be greater than that of the surroundings, and the gas can be expanded in a nozzle to the pressure of the surroundings. Since the gases leave at a high velocity, the change in momentum that the gases undergo gives a thrust to the aircraft in which the engine is installed. A jet engine was shown in Fig. 1.11, and the air-standard cycle for this situation is shown in Fig. 11.26. The principles governing this cycle follow from the analysis of the Brayton cycle plus that for a reversible, adiabatic nozzle.

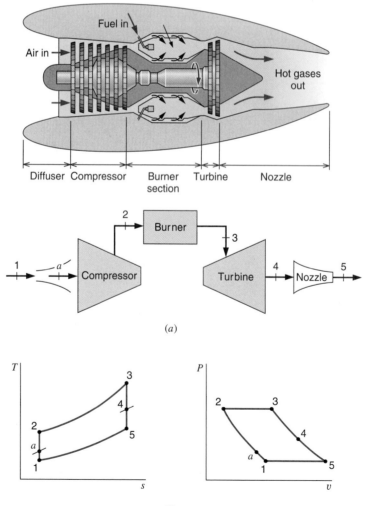

FIGURE 11.26 The ideal gas-turbine cycle for a jet engine.

EXAMPLE 11.10 Consider an ideal jet propulsion cycle in which air enters the compressor at 0.1 MPa and 15°C. The pressure leaving the compressor is 1.0 MPa, and the maximum temperature is 1100°C. The air expands in the turbine to a pressure at which the turbine work is just equal to the compressor work. On leaving the turbine, the air expands in a nozzle to 0.1 MPa. The process is reversible and adiabatic. Determine the velocity of the air leaving the nozzle.

The model used is ideal gas with constant specific heat, at 300 K, and each process is steady state with no potential energy change. The only kinetic energy change occurs in the nozzle. The diagram is shown in Fig. 11.26.

The compressor analysis is the same as in Example 11.6. From the results of that solution, we have

$$P_1 = 0.1 \text{ MPa}, \qquad T_1 = 288.2 \text{ K}$$

$$P_2 = 1.0 \text{ MPa}, \qquad T_2 = 556.8 \text{ K}$$

$$w_c = 269.5 \text{ kJ/kg}$$

The turbine analysis is also the same as in Example 11.6. Here, however,

$$P_3 = 1.0 \text{ MPa}, \qquad T_3 = 1373.2 \text{ K}$$

$$w_c = w_t = C_p(T_3 - T_4) = 269.5 \text{ kJ/kg}$$

$$T_3 - T_4 = \frac{269.5}{1.004} = 268.6, \qquad T_4 = 1104.6 \text{ K}$$

so that

$$\frac{T_3}{T_4} = \left(\frac{P_3}{P_4}\right)^{(k-1)/k} = \frac{1373.2}{1104.6} = 1.2432$$

$$\frac{P_3}{P_4} = 2.142, \qquad P_4 = 0.4668 \text{ MPa}$$

Control volume: Nozzle.
Inlet state: State 4 fixed (above).
Exit state: P_5 known.

Analysis

The first law gives

$$h_4 = h_5 + \frac{\mathbf{V}_5^2}{2}$$

The second law is

$$s_4 = s_5$$

Solution

Since P_5 is 0.1 MPa, from the second law we find that $T_5 = 710.8$ K. Then

$$\mathbf{V}_5^2 = 2C_{p0}(T_4 - T_5)$$

$$\mathbf{V}_5^2 = 2 \times 1000 \times 1.004(1104.6 - 710.8)$$

$$\mathbf{V}_5 = 889 \text{ m/s}$$

11.13 RECIPROCATING ENGINE POWER CYCLES

In Section 11.1, we discussed power cycles incorporating either steady-state processes or cylinder/piston boundary-work processes. In that section, it was noted that for the steady-state process, there is no work in a constant-pressure process. Each of the steady-state power cycles presented in subsequent sections of this chapter incorporated two constant-pressure heat-transfer processes. It was further noted in Section 11.1 that in a boundary-movement work process, there is no work in a constant-volume process. In the next three sections, we will present ideal air-standard power cycles for cylinder/piston boundary-movement work processes, each example of which includes either one or two constant-volume heat-transfer processes.

Before we describe the reciprocating engine cycles, we want to present a few common definitions and terms. Car engines typically have 4, 6, or 8 cylinders, each with a diameter called bore B. The piston is connected to a crankshaft, as shown in Figure 11.27, and as it rotates changing the crank angle, θ, the piston moves up or down with a stroke.

$$S = 2R_{crank} \tag{11.12}$$

This gives a displacement for all cylinders as

$$V_{displ} = N_{cyl}(V_{max} - V_{min}) = N_{cyl}A_{cyl}S \tag{11.13}$$

which is the main characterization of the engine size. The ratio of the largest to the smallest volume is the compression ratio

$$r_v = CR = V_{max}/V_{min} \tag{11.14}$$

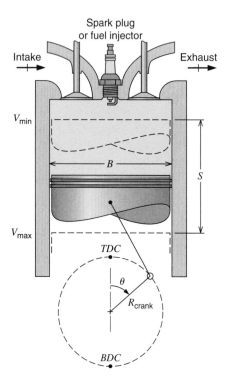

FIGURE 11.27 The piston-cylinder configuration for an internal combustion engine.

and both of these characteristics are fixed with the engine geometry. The net specific work in a complete cycle is used to define a mean effective pressure

$$w_{net} = \oint P \, dv \equiv P_{meff}(v_{max} - v_{min}) \qquad (11.15)$$

or net work per cylinder per cycle

$$W_{net} = mw_{net} = P_{meff}(V_{max} - V_{min}) \qquad (11.16)$$

We now use this to find the rate of work (power) for the whole engine as

$$\dot{W} = N_{cyl} \, mw_{net} \frac{RPM}{60} = P_{meff} V_{displ} \frac{RPM}{60} \qquad (11.17)$$

where RPM is revolutions per minute. This result should be corrected with a factor $\frac{1}{2}$ for a four-stroke engine, where 2 revolutions are needed for a complete cycle to also accomplish the intake and exhaust strokes.

11.14 THE OTTO CYCLE

The air-standard Otto cycle is an ideal cycle that approximates a spark-ignition internal-combustion engine. This cycle is shown on the P–v and T–s diagrams of Fig. 11.28. Process 1–2 is an isentropic compression of the air as the piston moves from crank-end dead center to head-end dead center. Heat is then added at constant volume while the piston is momentarily at rest at head-end dead center. (This process corresponds to the ignition of the fuel–air mixture by the spark and the subsequent burning in the actual engine.) Process 3–4 is an isentropic expansion, and process 4–1 is the rejection of heat from the air while the piston is at crank-end dead center.

The thermal efficiency of this cycle is found as follows, assuming constant specific heat of air:

$$\eta_{th} = \frac{Q_H - Q_L}{Q_H} = 1 - \frac{Q_L}{Q_H} = 1 - \frac{mC_v(T_4 - T_1)}{mC_v(T_3 - T_2)}$$

$$= 1 - \frac{T_1(T_4/T_1 - 1)}{T_1(T_3/T_2 - 1)}$$

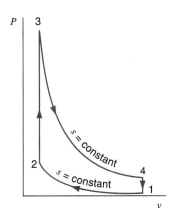

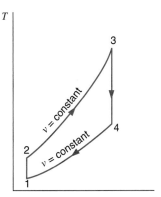

FIGURE 11.28 The air-standard Otto cycle.

We note further that

$$\frac{T_2}{T_1} = \left(\frac{V_1}{V_2}\right)^{k-1} = \left(\frac{V_1}{V_3}\right)^{k-1} = \frac{T_3}{T_4}$$

Therefore,

$$\frac{T_3}{T_2} = \frac{T_4}{T_1}$$

and

$$\eta_{\text{th}} = 1 - \frac{T_1}{T_2} = 1 - (r_v)^{1-k} = 1 - \frac{1}{r_v^{k-1}} \qquad (11.18)$$

where

$$r_v = \text{compression ratio} = \frac{V_1}{V_2} = \frac{V_4}{V_3}$$

It is important to note that the efficiency of the air-standard Otto cycle is a function only of the compression ratio and that the efficiency is increased by increasing the compression ratio. Figure 11.29 shows a plot of the air-standard cycle thermal efficiency versus compression ratio. It is also true of an actual spark-ignition engine that the efficiency can be increased by increasing the compression ratio. The trend toward higher compression ratios is prompted by the effort to obtain higher thermal efficiency. In the actual engine there is an increased tendency for the fuel to detonate as the compression ratio is increased. After detonation the fuel burns rapidly, and strong pressure waves present in the engine cylinder give rise to the so-called spark knock. Therefore, the maximum compression ratio that can be used is fixed by the fact that detonation must be avoided. Advances in compression ratios over the years in actual engines were originally made possible by developing fuels with better antiknock characteristics, primarily through the addition of tetraethyl lead. More recently, however, nonleaded gasolines with good antiknock characteristics have been developed in an effort to reduce atmospheric contamination.

Some of the most important ways in which the actual open-cycle spark-ignition engine deviates from the air-standard cycle are as follows:

1. The specific heats of the actual gases increase with an increase in temperature.
2. The combustion process replaces the heat-transfer process at high temperature, and combustion may be incomplete.

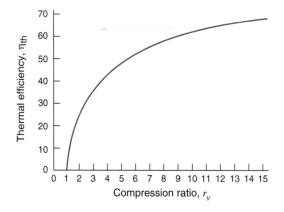

FIGURE 11.29
Thermal efficiency of the Otto cycle as a function of compression ratio.

3. Each mechanical cycle of the engine involves an inlet and an exhaust process and, because of the pressure drop through the valves, a certain amount of work is required to charge the cylinder with air and exhaust the products of combustion.

4. There will be considerable heat transfer between the gases in the cylinder and the cylinder walls.

5. There will be irreversibilities associated with pressure and temperature gradients.

EXAMPLE 11.11   The compression ratio in an air-standard Otto cycle is 10. At the beginning of the compression stoke the pressure is 0.1 MPa and the temperature is 15°C. The heat transfer to the air per cycle is 1800 kJ/kg air. Determine

1. The pressure and temperature at the end of each process of the cycle.
2. The thermal efficiency.
3. The mean effective pressure.

Control mass:	Air inside cylinder.
Diagram:	Fig. 11.29.
State information:	$P_1 = 0.1$ MPa, $T_1 = 288.2$ K.
Process information:	Four processes known (Fig. 11.29). Also, $r_v = 10$ and $q_H = 1800$ kJ/kg.
Model:	Ideal gas, constant specific heat, value at 300 K.

Analysis

The second law for compression process 1–2 is

$$s_2 = s_1$$

so that

$$\frac{T_2}{T_1} = \left(\frac{V_1}{V_2}\right)^{k-1}$$

$$\frac{P_2}{P_1} = \left(\frac{V_1}{V_2}\right)^{k}$$

The first law for heat addition process 2–3 is

$$q_H = {}_2q_3 = u_3 - u_2 = C_v(T_3 - T_2)$$

The second law for expansion process 3–4 is

$$s_4 = s_3$$

so that

$$\frac{T_3}{T_4} = \left(\frac{V_4}{V_3}\right)^{k-1}$$

$$\frac{P_3}{P_4} = \left(\frac{V_4}{V_3}\right)^{k}$$

In addition,

$$\eta_{th} = 1 - \frac{1}{r_v^{k-1}}, \qquad \text{mep} = \frac{w_{net}}{v_1 - v_2}$$

Solution

Substitution yields the following:

$$v_1 = \frac{0.287 \times 288.2}{100} = 0.827 \text{ m}^3/\text{kg}$$

$$\frac{T_2}{T_1} = \left(\frac{V_1}{V_2}\right)^{k-1} = 10^{0.4} = 2.5119, \qquad T_2 = 723.9 \text{ K}$$

$$\frac{P_2}{P_1} = \left(\frac{V_1}{V_2}\right)^{k} = 10^{1.4} = 25.12, \qquad P_2 = 2.512 \text{ MPa}$$

$$v_2 = \frac{0.827}{10} = 0.0827 \text{ m}^3/\text{kg}$$

$$_2q_3 = C_v(T_3 - T_2) = 1800 \text{ kJ/kg}$$

$$T_3 - T_2 = \frac{1800}{0.717} = 2510 \text{ K}, \qquad T_3 = 3234 \text{ K}$$

$$\frac{T_3}{T_2} = \frac{P_3}{P_2} = \frac{3234}{723.9} = 4.467, \qquad P_3 = 11.222 \text{ MPa}$$

$$\frac{T_3}{T_4} = \left(\frac{V_4}{V_3}\right)^{k-1} = 10^{0.4} = 2.5119, \qquad T_4 = 1287.5 \text{ K}$$

$$\frac{P_3}{P_4} = \left(\frac{V_4}{V_3}\right)^{k} = 10^{1.4} = 25.12, \qquad P_4 = 0.4467 \text{ MPa}$$

$$\eta_{th} = 1 - \frac{1}{r_v^{k-1}} = 1 - \frac{1}{10^{0.4}} = 0.602 = 60.2\%$$

This can be checked by finding the heat rejected:

$$_4q_1 = C_v(T_1 - T_4) = 0.717(288.2 - 1287.5) = -716.5 \text{ kJ/kg}$$

$$\eta_{th} = 1 - \frac{716.5}{1800} = 0.602 = 60.2\%$$

$$w_{net} = 1800 - 716.5 = 1083.5 \text{ kJ/kg} = (v_1 - v_2)\text{mep}$$

$$\text{mep} = \frac{1083.5}{(0.827 - 0.0827)} = 1456 \text{ kPa}$$

This is a high value for mean effective pressure, largely because the two constant-volume heat-transfer processes keep the total volume change to a minimum (compared with a Brayton cycle, for example). Thus, the Otto cycle is a good model to emulate in the cylinder-piston internal-combustion engine. At the other extreme, a low mean effective pressure means a large piston displacement for a given power output, which in turn means high frictional losses in an actual engine.

11.15 THE DIESEL CYCLE

The air-standard diesel cycle is shown in Fig. 11.30. This is the ideal cycle for the diesel engine, which is also called the compression-ignition engine.

In this cycle the heat is transferred to the working fluid at constant pressure. This process corresponds to the injection and burning of the fuel in the actual engine. Since the gas is expanding during the heat addition in the air-standard cycle, the heat transfer must be just sufficient to maintain constant pressure. When state 3 is reached, the heat addition ceases and the gas undergoes an isentropic expansion, process 3–4, until the piston reaches crank-end dead center. As in the air-standard Otto cycle, a constant-volume rejection of heat at crank-end dead center replaces the exhaust and intake processes of the actual engine.

The efficiency of the diesel cycle is given by the relation

$$\eta_{th} = 1 - \frac{Q_L}{Q_H} = 1 - \frac{C_v(T_4 - T_1)}{C_p(T_3 - T_2)} = 1 - \frac{T_1(T_4/T_1 - 1)}{kT_2(T_3/T_2 - 1)} \qquad (11.19)$$

The isentropic compression ratio is greater than the isentropic expansion ratio in the diesel cycle. In addition, for a given state before compression and a given compression ratio (that is, given states 1 and 2), the cycle efficiency decreases as the maximum temperature increases. This is evident from the T–s diagram because the constant-pressure and constant-volume lines converge, and increasing the temperature from 3 to 3$'$ requires a large addition of heat (area 3–3$'$–c–b–3) and results in a relatively small increase in work (area 3–3$'$–4$'$–4–3).

A number of comparisons may be made between the Otto cycle and the diesel cycle, but here we will note only two. Consider Otto cycle 1–2–3$''$–4–1 and diesel cycle 1–2–3–4–1, which have the same state at the beginning of the compression stroke and the same piston displacement and compression ratio. From the T–s diagram we see that the Otto cycle has the higher efficiency. In practice, however, the diesel engine can operate on a higher compression ratio than the spark-ignition engine. The reason is that in the spark-ignition engine an air–fuel mixture is compressed, and detonation (spark knock) becomes a serious problem if too high a compression ratio is used. This problem does not exist in the diesel engine because only air is compressed during the compression stroke.

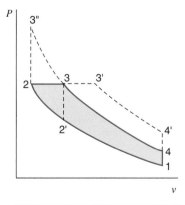

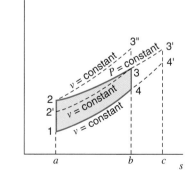

FIGURE 11.30 The air-standard diesel cycle.

Therefore, we might compare an Otto cycle with a diesel cycle and in each case select a compression ratio that might be achieved in practice. Such a comparison can be made by considering Otto cycle 1–2′–3–4–1 and diesel cycle 1–2–3–4–1. The maximum pressure and temperature are the same for both cycles, which means that the Otto cycle has a lower compression ratio than the diesel cycle. It is evident from the *T–s* diagram that in this case the diesel cycle has the higher efficiency. Thus, the conclusions drawn from a comparison of these two cycles must always be related to the basis on which the comparison has been made.

The actual compression-ignition open cycle differs from the air-standard diesel cycle in much the same way that the spark-ignition open cycle differs from the air-standard Otto cycle.

EXAMPLE 11.12 An air-standard diesel cycle has a compression ratio of 20, and the heat transferred to the working fluid per cycle is 1800 kJ/kg. At the beginning of the compression process, the pressure is 0.1 MPa and the temperature is 15°C. Determine

1. The pressure and temperature at each point in the cycle.
2. The thermal efficiency.
3. The mean effective pressure.

Control mass:	Air inside cylinder.
Diagram:	Fig. 11.30.
State information:	$P_1 = 0.1$ MPa, $T_1 = 288.2$ K.
Process information:	Four processes known (Fig. 11.30). Also, $r_v = 20$ and $q_H = 1800$ kJ/kg.
Model:	Ideal gas, constant specific heat, value at 300 K.

Analysis

The second law for compression process 1–2 is

$$s_2 = s_1$$

so that

$$\frac{T_2}{T_1} = \left(\frac{V_1}{V_2}\right)^{k-1}$$

$$\frac{P_2}{P_1} = \left(\frac{V_1}{V_2}\right)^{k}$$

The first law for heat addition process 2–3 is

$$q_H = {}_2q_3 = C_p(T_3 - T_2)$$

and the second law for expansion process 3–4 is

$$s_4 = s_3$$

so that

$$\frac{T_3}{T_4} = \left(\frac{V_4}{V_3}\right)^{k-1}$$

In addition,

$$\eta_{\text{th}} = \frac{w_{\text{net}}}{q_H}, \quad \text{mep} = \frac{w_{\text{net}}}{v_1 - v_2}$$

Solution

Substitution gives

$$v_1 = \frac{0.287 \times 288.2}{100} = 0.827 \text{ m}^3/\text{kg}$$

$$v_2 = \frac{v_2}{20} = \frac{0.827}{20} = 0.04135 \text{ m}^3/\text{kg}$$

$$\frac{T_2}{T_1} = \left(\frac{V_1}{V_2}\right)^{k-1} = 20^{0.4} = 3.3145, \qquad T_2 = 955.2 \text{ K}$$

$$\frac{P_2}{P_1} = \left(\frac{V_1}{V_2}\right)^{k} = 20^{1.4} = 66.29, \qquad P_2 = 6.629 \text{ MPa}$$

$$q_H = {}_2q_3 = C_p(T_3 - T_2) = 1800 \text{ kJ/kg}$$

$$T_3 - T_2 = \frac{1800}{1.004} = 1793 \text{ K}, \qquad T_3 = 2748 \text{ K}$$

$$\frac{V_3}{V_2} = \frac{T_3}{T_2} = \frac{2748}{955.2} = 2.8769, \qquad v_3 = 0.118\,96 \text{ m}^3/\text{kg}$$

$$\frac{T_3}{T_4} = \left(\frac{V_4}{V_3}\right)^{k-1} = \left(\frac{0.827}{0.118\,96}\right)^{0.4} = 2.1719, \qquad T_4 = 1265 \text{ K}$$

$$q_L = {}_4q_1 = C_v(T_1 - T_4) = 0.717(288.2 - 1265) = -700.4 \text{ kJ/kg}$$

$$w_{\text{net}} = 1800 - 700.4 = 1099.6 \text{ kJ/kg}$$

$$\eta_{\text{th}} = \frac{w_{\text{net}}}{q_H} = \frac{1099.6}{1800} = 61.1\%$$

$$\text{mep} = \frac{w_{\text{net}}}{v_1 - v_2} = \frac{1099.6}{0.827 - 0.04135} = 1400 \text{ kPa}$$

11.16 THE STIRLING CYCLE

The final air-standard power cycle to be discussed is the Stirling cycle, which is shown on the *P–v* and *T–s* diagrams of Fig. 11.31. Heat is transferred to the working fluid during the constant-volume process 2–3 and also during the isothermal expansion process 3–4. Heat is rejected during the constant-volume process 4–1 and also during the isothermal

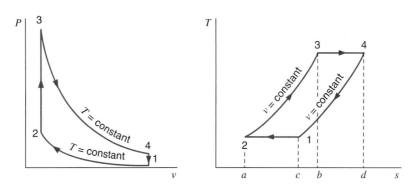

FIGURE 11.31 The air-standard Stirling cycle.

compression process 1–2. Thus, this cycle is the same as the Otto cycle, with the adiabatic processes of that cycle replaced with isothermal processes. Since the Stirling cycle includes two constant-volume heat-transfer processes, keeping the total volume change during the cycle to a minimum, it is a good candidate for a cylinder/piston boundary work application; it should have a high mean effective pressure.

Stirling-cycle engines have been developed in recent years as external-combustion engines with regeneration. The significance of regeneration is noted from the ideal case shown in Fig. 11.31. Note that the heat transfer to the gas between states 2 and 3, area 2–3–b–a–2, is exactly equal to the heat transfer from the gas between states 4 and 1, area 1–4–d–c–1. Thus, in the ideal cycle, all external heat supplied Q_H takes place in the isothermal expansion process 3–4, and all external heat rejection Q_L takes place in the isothermal compression process 1–2. Since all heat is supplied and rejected isothermally, the efficiency of this cycle equals the efficiency of a Carnot cycle operating between the same temperatures. The same conclusions would be drawn in the case of an Ericsson cycle, which was discussed briefly in Section 11.11, if that cycle were to include a regenerator as well.

11.17 INTRODUCTION TO REFRIGERATION SYSTEMS

In Section 11.1, we discussed cyclic heat engines consisting of four separate processes, either steady-state or cylinder–piston boundary-movement work devices. We further allowed for a working fluid that changes phase or for one that remains in a single phase throughout the cycle. We then considered a power system comprised of four reversible steady-state processes, two of which were constant-pressure heat-transfer processes, for simplicity of equipment requirements, since these two processes involve no work. It was further assumed that the other two work-involved processes were adiabatic and therefore isentropic. The resulting power cycle appeared as in Fig. 11.2.

We now consider the basic ideal refrigeration system cycle in exactly the same terms as those described earlier, except that each process is the reverse of that in the power cycle. The result is the ideal cycle shown in Fig. 11.32. Note that if the entire cycle takes place inside the two-phase liquid–vapor dome, the resulting cycle is, as with the power cycle, the Carnot cycle, since the two constant-pressure processes are also isothermal. Otherwise, this cycle is not a Carnot cycle. It is also noted, as before, that the net work input to the cycle is equal to the area enclosed by the process lines 1–2–3–4–1, independently of whether the individual processes are steady state or cylinder/piston boundary movement.

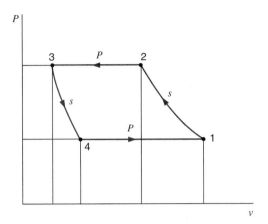

FIGURE 11.32 Four-process refrigeration cycle.

In the next section, we make one modification to this idealized basic refrigeration system cycle in presenting and applying the model of refrigeration and heat pump systems.

11.18 THE VAPOR-COMPRESSION REFRIGERATION CYCLE

In this section, we consider the ideal refrigeration cycle for a working substance that changes phase during the cycle, in a manner equivalent to that done with the Rankine power cycle in Section 11.2. In doing so, we note that state 3 in Fig. 11.32 is saturated liquid at the condenser temperature and state 1 is saturated vapor at the evaporator temperature. This means that the isentropic expansion process from 3–4 will be in the two-phase region, and the substance there will be mostly liquid. As a consequence, there will be very little work output from this process, such that it is not worth the cost of including this piece of equipment in the system. We therefore replace the turbine with a throttling device, usually a valve or a length of small-diameter tubing, by which the working fluid is throttled from the high-pressure to the low-pressure side. The resulting cycle become the ideal model for a vapor-compression refrigeration system, which is shown in Fig. 11.33. Saturated vapor at low pressure enters the compressor and undergoes a reversible adiabatic compression, process 1–2. Heat is then rejected at constant pressure in process 2–3, and the working fluid exits the condenser as saturated liquid. An adiabatic throttling process, 3–4, follows, and the working fluid is then evaporated at constant pressure, process 4–1, to complete the cycle.

The similarity of this cycle to the reverse of the Rankine cycle has already been noted. We also note the difference between this cycle and the ideal Carnot cycle, in which the working fluid always remains inside the two-phase region, $1'$–$2'$–3–$4'$–$1'$. It is much more expedient to have a compressor handle only vapor than a mixture of liquid and vapor, as would be required in process $1'$–$2'$ of the Carnot cycle. It is virtually impossible to compress, at a reasonable rate, a mixture such as that represented by state $1'$ and still maintain equilibrium between liquid and vapor. The other difference, that of replacing the turbine by the throttling process, has already been discussed.

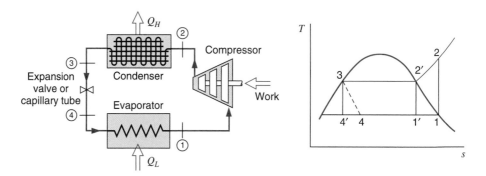

FIGURE 11.33 The ideal vapor-compression refrigeration cycle.

The system described in Fig. 11.33 can be used for either of two purposes. The first use is as a refrigeration system, in which case it is desired to maintain a space at a low temperature T_1 relative to the ambient temperature T_3. (In a real system, it would be necessary to allow a finite temperature difference in both the evaporator and condenser to provide a finite rate of heat transfer in each.) Thus, the reason for building the system in this case is the quantity q_L. The measure of performance of a refrigeration system is given in terms of the coefficient of performance, β, which was defined in Chapter 7 as

$$\beta = \frac{q_L}{w_c} \tag{11.20}$$

The second use of this system described in Fig. 11.33 is as a heat pump system, in which case it is desired to maintain a space at a temperature T_3 above that of the ambient (or other source) T_1. In this case, the reason for building the system is the quantity q_H, and the coefficient of performance for the heat pump, β', is now

$$\beta' = \frac{q_H}{w_c} \tag{11.21}$$

Refrigeration systems and heat pump systems are, of course, different in terms of design variables, but the analysis of the two is the same. When we discuss refrigerators in this and the following two sections, it should be kept in mind that the same comments generally apply to heat pump systems as well.

EXAMPLE 11.13   Consider an ideal refrigeration cycle that uses R-134a as the working fluid. The temperature of the refrigerant in the evaporator is $-20°C$, and in the condenser it is $40°C$. The refrigerant is circulated at the rate of 0.03 kg/s. Determine the coefficient of performance and the capacity of the plant in rate of refrigeration.

The diagram for this example is as shown in Fig. 11.33. For each control volume analyzed, the thermodynamic model is as exhibited in the R-134a tables. Each process is steady state, with no changes in kinetic or potential energy.

> *Control volume:*   Compressor.
> *Inlet state:* T_1 known, saturated vapor; state fixed.
> *Exit state:* P_2 known (saturation pressure at T_3).

Analysis

The first and second laws are

$$w_c = h_2 - h_1$$
$$s_2 = s_1$$

Solution

At $T_3 = 40°C$,

$$P_g = P_2 = 1017 \text{ kPa}$$

From the R-134a tables, we get

$$h_1 = 386.1 \text{ kJ/kg}, \qquad s_1 = 1.7395 \text{ kJ/kg K}$$

Therefore,

$$s_2 = s_1 = 1.7395 \text{ kJ/kg K}$$

so that

$$T_2 = 47.7°C \qquad \text{and} \qquad h_2 = 428.4 \text{ kJ/kg}$$
$$w_c = h_2 - h_1 = 428.4 - 386.1 = 42.3 \text{ kJ/kg}$$

Control volume: Expansion valve.
Inlet state: T_3 known, saturated liquid; state fixed.
Exit state: T_4 known.

Analysis

The first law is

$$h_3 = h_4$$

Solution

Numerically, we have

$$h_4 = h_3 = 256.5 \text{ kJ/kg}$$

Control volume: Evaporator.
Inlet state: State 4 known (as given).
Exit state: State 1 known (as given).

Analysis

The first law is

$$q_L = h_1 - h_4$$

Solution

Substituting, we have

$$q_L = h_1 - h_4 = 386.1 - 256.5 = 129.6 \text{ kJ/kg}$$

Therefore,

$$\beta = \frac{q_L}{w_c} = \frac{129.6}{42.3} = 3.064$$

Refrigeration capacity $= 129.6 \times 0.03 = 3.89$ kW

11.19 WORKING FLUIDS FOR VAPOR-COMPRESSION REFRIGERATION SYSTEMS

A much larger number of different working fluids (refrigerants) are utilized in vapor-compression refrigeration systems than in vapor power cycles. Ammonia and sulfur dioxide were important in the early days of vapor-compression refrigeration, but both are highly toxic and therefore dangerous substances. For many years now, the principal refrigerants have been the halogenated hydrocarbons, which are marketed under the trade names of Freon and Genatron. For example, dichlorodifluoromethane (CCl_2F_2) is known as Freon-12 and Genatron-12, and therefore as refrigerant-12 or R-12. This group of substances, known commonly as chlorofluorocarbons or CFCs, are chemically very stable at ambient temperature, especially those lacking any hydrogen atoms. This characteristic is necessary for a refrigerant working fluid. This same characteristic, however, has devastating consequences if the gas, having leaked from an appliance into the atmosphere, spends many years slowly diffusing upward into the stratosphere. There it is broken down, releasing chlorine, which destroys the protective ozone layer of the stratosphere. It is therefore of overwhelming importance to us all to eliminate completely the widely used but life-threatening CFCs, particularly R-11 and R-12, and to develop suitable and acceptable replacements. The CFCs containing hydrogen (often termed HCFCs), such as R-22, have shorter atmospheric lifetimes and therefore are not as likely to reach the stratosphere before being broken up and rendered harmless. The most desirable fluids, called HFCs, contain no chlorine atoms at all.

There are two important considerations when selecting refrigerant working fluids: the temperature at which refrigeration is needed and the type of equipment to be used.

As the refrigerant undergoes a change of phase during the heat-transfer process, the pressure of the refrigerant will be the saturation pressure during the heat supply and heat rejection processes. Low pressures mean large specific volumes and correspondingly large equipment. High pressures mean smaller equipment, but it must be designed to withstand higher pressure. In particular, the pressures should be well below the critical pressure. For extremely low temperature applications a binary fluid system may be used by cascading two separate systems.

The type of compressor used has a particular bearing on the refrigerant. Reciprocating compressors are best adapted to low specific volumes, which means higher pressures, whereas centrifugal compressors are most suitable for low pressures and high specific volumes.

It is also important that the refrigerants used in domestic appliances be nontoxic. Other beneficial characteristics, in addition to being environmentally acceptable, are miscibility with compressor oil, dielectric strength, stability, and low cost. Refrigerants, however, have an unfortunate tendency to cause corrosion. For given temperatures during evaporation and condensation, not all refrigerants have the same coefficient of performance for the ideal cycle. It is, of course, desirable to use the refrigerant with the highest coefficient of performance, other factors permitting.

11.20 DEVIATION OF THE ACTUAL VAPOR-COMPRESSION REFRIGERATION CYCLE FROM THE IDEAL CYCLE

The actual refrigeration cycle deviates from the ideal cycle primarily because of pressure drops associated with fluid flow and heat transfer to or from the surroundings. The actual cycle might approach the one shown in Fig. 11.34.

The vapor entering the compressor will probably be superheated. During the compression process, there are irreversibilities and heat transfer either to or from the surroundings, depending on the temperature of the refrigerant and the surroundings. Therefore, the entropy might increase or decrease during this process, for the irreversibility and the heat transferred to the refrigerant cause an increase in entropy, and the heat transferred from the refrigerant causes a decrease in entropy. These possibilities are represented by the two dashed lines 1–2 and 1–2′. The pressure of the liquid leaving the condenser will be less than the pressure of the vapor entering, and the temperature of the refrigerant in the condenser will be somewhat higher than that of the surroundings to which heat is being transferred. Usually, the temperature of the liquid leaving the condenser is lower than the saturation temperature. It might drop somewhat more in the piping between the condenser and expansion valve. This represents a gain, however, because as a result of this heat transfer the refrigerant enters the evaporator with a lower enthalpy, which permits more heat to be transferred to the refrigerant in the evaporator.

There is some drop in pressure as the refrigerant flows through the evaporator. It may be slightly superheated as it leaves the evaporator, and through heat transferred from the surroundings its temperature will increase in the piping between the evaporator and the compressor. This heat transfer represents a loss because it increases the work of the compressor, since the fluid entering it has an increased specific volume.

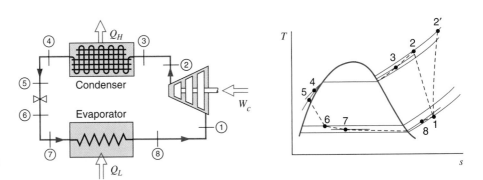

FIGURE 11.34 The actual vapor-compression refrigeration cycle.

EXAMPLE 11.14 A refrigeration cycle utilizes R-12 as the working fluid. The following are the properties at various points of the cycle designated in Fig. 11.34.

$$P_1 = 125 \text{ kPa}, \qquad T_1 = -10°C$$

$$P_2 = 1.2 \text{ MPa}, \qquad T_2 = 100°C$$

$$P_3 = 1.19 \text{ MPa}, \qquad T_3 = 80°C$$

$$P_4 = 1.16 \text{ MPa}, \qquad T_4 = 45°C$$

$$P_5 = 1.15 \text{ MPa}, \qquad T_5 = 40°C$$

$$P_6 = P_7 = 140 \text{ kPa}, \qquad x_6 = x_7$$

$$P_8 = 130 \text{ kPa}, \qquad T_8 = -20°C$$

The heat transfer from R-12 during the compression process is 4 kJ/kg. Determine the coefficient of performance of this cycle.

For each control volume, the R-12 tables are the model. Each process is steady state with no changes in kinetic or potential energy.

As before, we break the process down into stages, treating the compressor, the throttling value and line, and the evaporator in turn.

> *Control volume:* Compressor.
> *Inlet state:* P_1, T_1 known; state fixed.
> *Exit state:* P_2, T_2 known; state fixed.

Analysis

From the first law, we have

$$q + h_1 = h_2 + w$$

$$w_c = -w = h_2 - h_1 - q$$

Solution

From the R-12 tables, we read

$$h_1 = 185.16 \text{ kJ/kg}, \qquad h_2 = 245.52 \text{ kJ/kg}$$

Therefore,

$$w_c = 245.52 - 185.16 - (-4) = 64.36 \text{ kJ/kg}$$

> *Control volume:* Throttling valve plus line.
> *Inlet state:* P_5, T_5 known; state fixed.
> *Exit state:* $P_7 = P_6$ known, $x_7 = x_6$.

Analysis

The first law is

$$h_5 = h_6$$

Since $x_7 = x_6$, it follows that $h_7 = h_6$.

Solution

Numerically, we obtain

$$h_5 = h_6 = h_7 = 74.53$$

Control volume:	Evaporator.
Inlet state:	P_7, h_7 known (above).
Exit state:	P_8, T_8 known; state fixed.

Analysis

The first law is

$$q_L = h_8 - h_7$$

Solution

Substitution gives

$$q_L = h_8 - h_7 = 179.12 - 74.53 = 104.59 \text{ kJ/kg}$$

Therefore,

$$\beta = \frac{q_L}{w_c} = \frac{104.59}{64.36} = 1.625$$

11.21 THE AMMONIA ABSORPTION REFRIGERATION CYCLE

The ammonia absorption refrigeration cycle differs from the vapor-compression cycle in the manner in which compression is achieved. In the absorption cycle the low-pressure ammonia vapor is absorbed in water, and the liquid solution is pumped to a high pressure by a liquid pump. Figure 11.35 shows a schematic arrangement of the essential elements of such a system.

The low-pressure ammonia vapor leaving the evaporator enters the absorber where it is absorbed in the weak ammonia solution. This process takes place at a temperature slightly higher than that of the surroundings. Heat must be transferred to the surroundings during this process. The strong ammonia solution is then pumped through a heat exchanger to the generator where a higher pressure and temperature are maintained. Under these conditions, ammonia vapor is driven from the solution as heat is transferred from a high-temperature source. The ammonia vapor goes to the condenser where it is condensed, as in a vapor-compression system, and then to the expansion valve and evaporator. The weak ammonia solution is returned to the absorber through the heat exchanger.

The distinctive feature of the absorption system is that very little work input is required because the pumping process involves a liquid. This follows from the fact

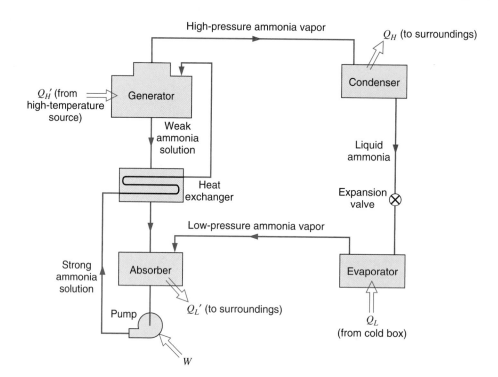

FIGURE 11.35 The ammonia-absorption refrigeration cycle.

that for a reversible steady-state process with negligible changes in kinetic and potential energy, the work is equal to $-\int v\,dP$ and the specific volume of the liquid is much less than the specific volume of the vapor. However, a relatively high-temperature source of heat must be available (100° to 200°C). There is more equipment in an absorption system than in a vapor-compression system, and it can usually be economically justified only when a suitable source of heat is available that would otherwise be wasted. In recent years, the absorption cycle has been given increased attention in connection with alternative energy sources, for example, solar energy or supplies of geothermal energy.

This cycle brings out the important principle that since the shaft work in a reversible steady-state process with negligible changes in kinetic and potential energy is $-\int v\,dP$, a compression process should take place with the smallest possible specific volume.

11.22 THE AIR-STANDARD REFRIGERATION CYCLE

If we consider the original ideal four-process refrigeration cycle of Fig. 11.32 with a noncondensing (gaseous) working fluid, then the work output during the isentropic expansion process is not negligibly small, as was the case with a condensing working fluid. Therefore, we retain the turbine in the four-steady-state process ideal air-standard refrigeration cycle shown in Fig. 11.36. This cycle is seen to be the reverse Brayton cycle, and it is

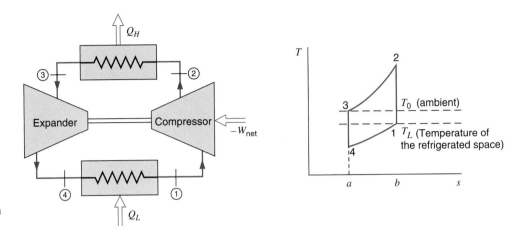

FIGURE 11.36 The air-standard refrigeration cycle.

used in practice in the liquefaction of air and other gases and also in certain special situations that require refrigeration, such as aircraft cooling systems. After compression from states 1 to 2, the air is cooled as heat is transferred to the surroundings at temperature T_0. The air is then expanded in process 3–4 to the pressure entering the compressor, and the temperature drops to T_4 in the expander. Heat may then be transferred to the air until temperature T_L is reached. The work for this cycle is represented by area 1–2–3–4–1, and the refrigeration effect is represented by area 4–1–b–a–4. The coefficient of performance is the ratio of these two areas.

In practice, this cycle has been used to cool aircraft in an open cycle, A simplified form is shown in Fig. 11.37. Upon leaving the expander, the cool air is blown directly into the cabin, thus providing the cooling effect where needed.

When counterflow heat exchangers are incorporated, very low temperatures can be obtained. This is essentially the cycle used in low-pressure air liquefaction plants and in other liquefaction devices such as the Collins helium liquefier. The ideal cycle is as shown in Fig. 11.38. Because the expander operates at very low temperature, the designer is faced with unique problems in providing lubrication and choosing materials.

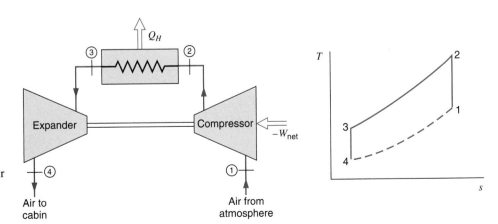

FIGURE 11.37 An air refrigeration cycle that might be utilized for aircraft cooling.

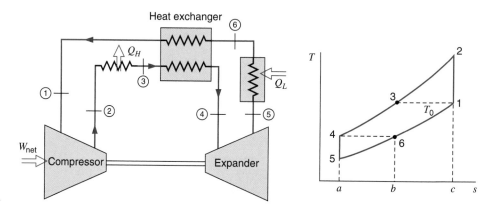

FIGURE 11.38 The air-refrigeration cycle utilizing a heat exchanger.

EXAMPLE 11.15

Consider the simple air-standard refrigeration cycle of Fig. 11.36. Air enters the compressor at 0.1 MPa and −20°C and leaves at 0.5 MPa. Air enters the expander at 15°C. Determine

1. The coefficient of performance for this cycle.
2. The rate at which air must enter the compressor to provide 1 kW of refrigeration.

For each control volume in this example, the model is ideal gas with constant specific heat, at 300 K, and each process is steady state with no kinetic or potential energy changes. The diagram for this example is Fig. 11.36.

Control volume:	Compressor.
Inlet state:	P_1, T_1 known; state fixed.
Exit state:	P_2 known.

Analysis

The first law is

$$w_c = h_2 - h_1$$

(Here w_c designates work into the compressor.)
The second law gives

$$s_1 = s_2$$

so that

$$\frac{T_2}{T_1} = \left(\frac{P_2}{P_1}\right)^{(k-1)/k}$$

Solution

Substituting, we obtain

$$\frac{T_2}{T_1} = \left(\frac{P_2}{P_1}\right)^{(k-1)/k} = 5^{0.286} = 1.5845, \qquad T_2 = 401.2 \text{ K}$$

$$w_c = h_2 - h_1 = C_p(T_2 - T_1)$$

$$= 1.004(401.2 - 253.2) = 148.5 \text{ kJ/kg}$$

Control volume: Expander.
Inlet state: $P_3 (= P_2)$ known, T_3 known; state fixed.
Exit state: $P_4 (= P_1)$ known.

Analysis

From the first law, we have

$$w_t = h_3 - h_4$$

The second law gives

$$s_3 = s_4$$

so that

$$\frac{T_3}{T_4} = \left(\frac{P_3}{P_4}\right)^{(k-1)/k}$$

Solution

Therefore,

$$\frac{T_3}{T_4} = \left(\frac{P_3}{P_4}\right)^{(k-1)/k} = 5^{0.286} = 1.5845, \qquad T_4 = 181.9 \text{ K}$$

$$w_t = h_3 - h_4 = 1.004(288.2 - 181.9) = 106.7 \text{ kJ/kg}$$

Control volume: High-temperature heat exchanger.
Inlet state: State 2 known (as given).
Exit state: State 3 known (as given).

Analysis

The first law is

$$q_H = h_2 - h_3 \text{ (heat rejected)}$$

Solution

Substitution gives

$$q_H = h_2 - h_3 = C_p(T_2 - T_3) = 1.004(401.2 - 288.2) = 113.4 \text{ kJ/kg}$$

Control volume: Low-temperature heat exchanger.
Inlet state: State 4 known (as given).
Exit state: State 1 known (as given).

Analysis

The first law is

$$q_L = h_1 - h_4$$

Solution

Substituting, we obtain

$$q_L = h_1 - h_4 = C_p(T_1 - T_4) = 1.004(253.2 - 181.9) = 71.6 \text{ kJ/kg}$$

Therefore,

$$w_{\text{net}} = w_c - w_t = 148.5 - 106.7 = 41.8 \text{ kJ/kg}$$

$$\beta = \frac{q_L}{w_{\text{net}}} = \frac{71.6}{41.8} = 1.713$$

To provide 1 kW of refrigeration capacity, we have

$$\dot{m} = \frac{\dot{Q}_L}{q_L} = \frac{1}{71.6} = 0.014 \text{ kg/s}$$

11.23 COMBINED-CYCLE POWER AND REFRIGERATION SYSTEMS

In many situations it is desirable to combine two cycles in series, either power systems or refrigeration systems, to take advantage of a very wide temperature range or to utilize what would otherwise be waste heat to improve efficiency. One combined power cycle, shown in Fig. 11.39 as a simple steam cycle with a mercury-topping cycle, is often referred to as a binary cycle. The advantage of this combined system is that mercury has a very low vapor pressure relative to that for water. Therefore, it is possible for an isothermal boiling process in the mercury to take place at a high temperature, much higher than the critical temperature of water, but still at a moderate pressure. The mercury condenser then provides an isothermal heat source as input to the steam boiler, such that the two cycles can be closely matched by proper selection of the cycle variables, with the resulting combined cycle then

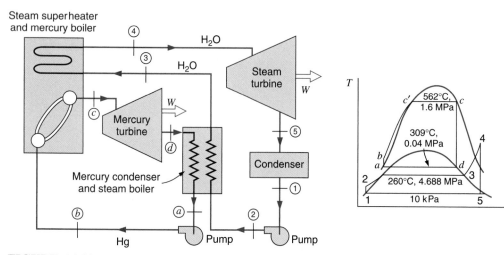

FIGURE 11.39 Mercury–water binary power system.

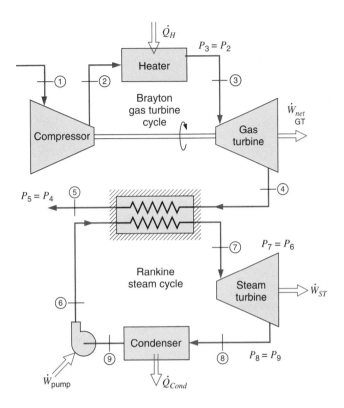

FIGURE 11.40
Combined
Brayton/Rankine cycle
power system.

having a high thermal efficiency. Saturation pressures and temperatures for a typical mercury–water binary cycle are shown in the T–s diagram of Fig. 11.39.

A different type of combined cycle that has seen considerable attention is to use the "waste heat" exhaust from a Brayton cycle gas-turbine engine (or another combustion engine such as a diesel engine) as the heat source for a steam or other vapor power cycle, in which case the vapor cycle acts as a bottoming cycle for the gas engine, in order to improve the overall thermal efficiency of the combined power system. Such a system, utilizing a gas turbine and a steam Rankine cycle, is shown in Fig. 11.40. In such a combination, there is a natural mismatch using the cooling of a noncondensing gas as the energy source to effect an isothermal boiling process plus superheating the vapor, and careful design is required to avoid a pinch point, a condition at which the gas has cooled to the vapor boiling temperature without having provided sufficient energy to complete the boiling process.

One way to take advantage of the cooling exhaust gas in the Brayton-cycle portion of the combined system is to utilize a mixture as the working fluid in the Rankine cycle. An example of this type of application is the Kalina cycle, which uses ammonia–water mixtures as the working fluid in the Rankine-type cycle. Such a cycle can be made very efficient, inasmuch as the temperature differences between the two fluid streams can be controlled through careful design of the combined system.

Combined cycles are used in refrigeration systems in cases in which there is a very large temperature difference between the ambient surroundings and the refrigerated space. Such a refrigeration system is often called a cascade system, an example of which is shown in Fig. 11.41. In this case, the refrigerant R-22 is used in the refrigeration system rejecting heat to the ambient surroundings, while its evaporator picks up the heat rejected in the low-temperature system condenser, the low temperature working fluid in this case being R-23,

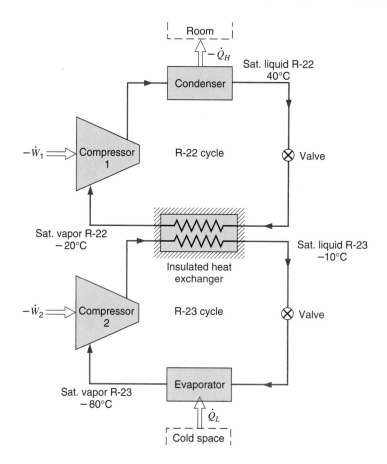

FIGURE 11.41
Combined-cycle cascade
refrigeration system.

whose thermodynamic properties are suited to work as a refrigerant in this low-temperature range. As with the other combined-cycle systems, the working fluids and design variables must be considered very carefully to optimize the performance of each unit.

We have described only a few combined-cycle systems here, as examples of the types of applications that can be dealt with and the resulting improvement in overall performance that can occur. Obviously, many other combinations of power and refrigeration systems are possible. Some of these are discussed in the problems at the end of the chapter.

SUMMARY

A number of standard power-producing cycles and refrigeration cycles are presented. First we cover a number of stationary and mobile power-producing heat engines. The Rankine cycle and its variations represent a steam power plant, which produces most of the world production of electricity. The heat input can come from combustion of fossil fuels, a nuclear reactor, solar radiation, or any other heat source that can generate a temperature high enough to boil water at a high pressure. In low- or very high-temperature applications, substances other than water can be used. Modifications to the basic cycle such as reheat, closed, and open feedwater heaters are covered together with applications where the electricity is cogenerated with a base demand for process steam.

A Brayton cycle is a gas turbine producing electricity and, with a modification, a jet engine producing thrust. This is a high-power, low-mass, and low-volume device and is used where space and weight are at a premium cost. A high back-work ratio makes this cycle sensitive to the compressor efficiency. A number of variations and configurations for the Brayton cycle with regenerators and intercoolers are shown.

Piston/cylinder devices are shown for the Otto and Diesel cycles modeling the gasoline and diesel engines, which can be two- or four-stroke engines. Cold air properties are used to show the influence of compression ratio on the thermal efficiency, and the mean effective pressure is used to relate the engine size to total power output. We give a short mention of the Stirling cycle as an example of an external combustion engine.

Standard refrigeration systems are covered by the vapor-compression refrigeration cycle. This applies to household refrigerators, air conditioners, and heat pumps as well as commercial units to lower temperature ranges. As a special cycle, we mention the ammonia absorption cycle and cover the air-standard refrigeration cycle in detail.

The chapter is completed with a short description of combined cycle applications. This covers stacked or cascade systems for large temperature spans and combinations of different kinds of cycles where one can be added as a topping cycle or a bottoming cycle. Often a Rankine cycle uses exhaust energy from a Brayton cycle in larger stationary applications.

You should have learned a number of skills and acquired abilities from studying this chapter that will allow you to

- Apply the general laws to control volumes with several devices forming a complete system.
- Have a knowledge of how many common power-producing devices work.
- Have a knowledge of how simple refrigerators and heat pumps work.
- Know that most cycle devices do not operate in Carnot cycles.
- Know that real devices have lower efficiencies/COP than ideal cycles.
- Have a sense of the most influential parameters for each type of cycle.
- Have an idea about the importance of the component efficiency for the overall cycle efficiency or COP.
- Know that most real cycles have modifications to the basic cycle setup.
- Know that many of these devices affect our environment.
- Know the principle of combining different cycles.

KEY CONCEPTS AND FORMULAS

Rankine Cycle

Open feedwater heater	Feedwater mixed with extraction steam, exit as saturated liquid
Closed feedwater heater	Feedwater heated by extraction steam, no mixing
Deaerating FWH	Open feedwater heater operating at P_{atm} to vent gas out
Cogeneration	Turbine power is cogenerated with a desired steam supply

Brayton Cycle

Compression ratio	Pressure ratio $r_p = P_{high}/P_{low}$
Regenerator	Dual fluid heat exchanger, uses exhaust flow energy
Intercooler	Cooler between compressor stages, reduces work input
Jet engine	No shaftwork out, kinetic energy generated in exit nozzle
Thrust	$F = \dot{m}(\mathbf{V}_e - \mathbf{V}_i)$ momentum equation
Propulsive power	$\dot{W} = F\mathbf{V}_{aircraft} = \dot{m}(\mathbf{V}_e - \mathbf{V}_i)\mathbf{V}_{aircraft}$

Piston Cylinder Power Cycles

Compression ratio

Volume ratio $r_v = CR = V_{max}/V_{min}$

Displacement (1 cyl.)

$\Delta V = V_{max} - V_{min} = m(v_{max} - v_{min}) = SA_{cyl}$

Stroke

$S = 2 R_{crank}$, piston travel in compression or expansion.

Mean effective pressure

$P_{meff} = w_{net}/(v_{max} - v_{min}) = W_{net}/(V_{max} - V_{min})$

Power by 1 cylinder

$\dot{W} = mw_{net} \dfrac{RPM}{60}$ (times $\frac{1}{2}$ for four-stroke cycle)

Refrigeration Cycle

Coefficient of performance

$COP = \beta_{REF} = \dfrac{\dot{Q}_L}{\dot{W}_C} = \dfrac{q_L}{w_c}$

Combined Cycles

Topping, bottoming cycle

The high- and low-temperature cycles

Cascade system

Stacked refrigeration cycles

CONCEPT-STUDY GUIDE PROBLEMS

11.1 Is a steam power plant running in a Carnot cycle? Name the four processes.

11.2 Consider a Rankine cycle without superheat. How many single properties are needed to determine the cycle? Repeat the answer for a cycle with superheat.

11.3 Which component determines the high pressure in a Rankine cycle? What determines the low pressure?

11.4 Mention two benefits of a reheat cycle.

11.5 What is the difference between an open and a closed feedwater heater?

11.6 Can the energy removed in a power plant condenser be useful?

11.7 In a cogenerating power plant, what is cogenerated?

11.8 Why is the back-work ratio much higher in the Brayton cycle than in the Rankine cycle?

11.9 The Brayton cycle has the same four processes as the Rankine cycle, but the T–s and P–v diagrams look very different; why is that?

11.10 Is it always possible to add a regenerator to the Brayton cycle? What happens when the pressure ratio is increased?

11.11 Why would you use an intercooler between compressor stages?

11.12 The jet engine does not produce shaftwork; how is power produced?

11.13 How is the compression in the Otto cycle different from that in the Brayton cycle?

11.14 Does the inlet state (P_1, T_1) have any influence on the Otto-cycle efficiency? How about the power produced by a real car engine?

11.15 How many parameters do you need to know to completely describe the Otto cycle? How about the Diesel cycle?

11.16 The exhaust and inlet flow processes are not included in the Otto or Diesel cycles. How do these necessary processes affect the cycle performance?

11.17 A refrigerator in my 20°C kitchen uses R-12, and I want to make ice cubes at −5°C. What is the minimum high P and the maximum low P it can use?

11.18 How many parameters are needed to completely determine a standard vapor-compression refrigeration cycle?

11.19 Why would one consider a combined cycle system for a power plant? for a heat pump or refrigerator?

11.20 Since any heat transfer is driven by a temperature difference, how does that affect all the real cycles relative to the ideal cycles?

HOMEWORK PROBLEMS

Rankine Cycles, Power Plants

Simple Cycles

11.21 A steam power plant, as shown in Fig. 11.3, operating in a Rankine cycle has saturated vapor at 3 MPa leaving the boiler. The turbine exhausts to the condenser operating at 10 kPa. Find the specific work and heat transfer in each of the ideal components and the cycle efficiency.

11.22 Consider a solar-energy-powered ideal Rankine cycle that uses water as the working fluid. Saturated vapor leaves the solar collector at 175°C, and the condenser pressure is 10 kPa. Determine the thermal efficiency of this cycle.

11.23 A utility runs a Rankine cycle with a water boiler at 3MPa, and the cycle has the highest and lowest temperatures of 450°C and 45°C, respectively. Find the plant efficiency and the efficiency of a Carnot cycle with the same temperatures.

11.24 A Rankine cycle uses ammonia as the working substance and powered by solar energy. It heats the ammonia to 140°C at 5000 kPa in the boiler/superheater. The condenser is water cooled, and the exit is kept at 25°C. Find (T, P, and x if applicable) for all four states in the cycle.

11.25 A steam power plant operating in an ideal Rankine cycle has a high pressure of 5 MPa and a low pressure of 15 kPa. The turbine exhaust state should have a quality of at least 95%, and the turbine power generated should be 7.5 MW. Find the necessary boiler exit temperature and the total mass flow rate.

11.26 A supply of geothermal hot water is to be used as the energy source in an ideal Rankine cycle, with R-134a as the cycle working fluid. Saturated vapor R-134a leaves the boiler at a temperature of 85°C, and the condenser temperature is 40°C. Calculate the thermal efficiency of this cycle.

11.27 Do Problem 11.26 with R-22 as the working fluid.

11.28 Do Problem 11.26 with ammonia as the working fluid.

11.29 Consider the boiler in Problem 11.26 where the geothermal hot water brings the R-134a to saturated vapor. Assume a counterflowing heat exchanger arrangement. The geothermal water temperature should be equal to or greater than the R-134a temperature at any location inside the heat exchanger. The point with the smallest temperature difference between the source and the working fluid is called the pinch point shown in Fig. P11.29. If 2 kg/s of geothermal water is available at 95°C, what is the maximum power output of this cycle for R-134a as the working fluid? (*Hint*: split the heat exchanger C.V. into two so that the pinch point with $\Delta T = 0$, $T = 85°C$ appears.)

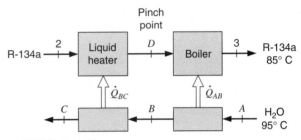

FIGURE P11.29

11.30 Do the previous problem with R-22 as the working fluid.

11.31 Consider the ammonia Rankine-cycle power plant shown in Fig. P11.31, a plant that was designed to operate in a location where the ocean water temperature is 25°C near the surface and 5°C at some greater depth. The mass flow rate of the working fluid is 1000 kg/s.

a. Determine the turbine power output and the pump power input for the cycle.

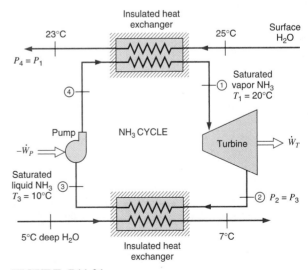

FIGURE P11.31

b. Determine the mass flow rate of water through each heat exchanger.

c. What is the thermal efficiency of this power plant?

11.32 A smaller power plant produces 25 kg/s steam at 3 MPa, 600°C, in the boiler. It cools the condenser with ocean water coming in at 12°C and returned at 15°C so the condenser exit is at 45°C. Find the net power output and the required mass flow rate of ocean water.

11.33 The power plant in Problem 11.21 is modified to have a superheater section following the boiler so that the steam leaves the superheater at 3 MPa and 400°C. Find the specific work and heat transfer in each of the ideal components and the cycle efficiency.

11.34 A steam power plant has a steam generator exit at 4 MPa and 500°C and a condenser exit temperature of 45°C. Assume all components are ideal and find the cycle efficiency and the specific work and heat transfer in the components.

11.35 Consider an ideal Rankine cycle using water with a high-pressure side of the cycle at a supercritical pressure. Such a cycle has a potential advantage of minimizing local temperature differences between the fluids in the steam generator, such as the instance in which the high-temperature energy source is the hot exhaust gas from a gas-turbine engine. Calculate the thermal efficiency of the cycle if the state entering the turbine is 30 MPa, 550°C, and the condenser pressure is 5 kPa. What is the steam quality at the turbine exit?

Reheat Cycles

11.36 A smaller power plant produces steam at 3 MPa, 600°C, in the boiler. It keeps the condenser at 45°C by transfer of 10 MW out as heat transfer. The first turbine section expands to 500 kPa, and then flow is reheated followed by the expansion in the low-pressure turbine. Find the reheat temperature so that the turbine output is saturated vapor. For this reheat, find the total turbine power output and the boiler heat transfer.

11.37 Consider an ideal steam reheat cycle as shown in Fig. 11.7, where steam enters the high-pressure turbine at 3 MPa and 400°C and then expands to 0.8 MPa. It is then reheated at constant pressure 0.8 MPa to 400°C and expands to 10 kPa in the low-pressure turbine. Calculate the thermal efficiency and the moisture content of the steam leaving the low-pressure turbine.

11.38 A smaller power plant produces 25 kg/s steam at 3 MPa, 600°C, in the boiler. It cools the condenser with ocean water so that the condenser exit is at 45°C. There is a reheat done at 500 kPa up to 400°C, and then expansion takes place in the low-pressure turbine. Find the net power output and the total heat transfer in the boiler.

11.39 The reheat pressure affects the operating variables and thus turbine performance. Repeat Problem 11.36 twice, using 0.6 and 1.0 MPa for the reheat pressure.

11.40 The effect of a number of reheat stages on the ideal steam reheat cycle is to be studied. Repeat Problem 11.36 using two reheat stages, one stage at 1.2 MPa and the second at 0.2 MPa, instead of the single reheat stage at 0.8 MPa.

Open Feedwater Heaters

11.41 An open feedwater heater in a regenerative steam power cycle receives 20 kg/s of water at 100°C and 2 MPa. The extraction steam from the turbine enters the heater at 2 MPa and 275°C, and all the feedwater leaves as saturated liquid. What is the required mass flow rate of the extraction steam?

11.42 A power plant with one open feedwater heater has a condenser temperature of 45°C, a maximum pressure of 5 MPa, and boiler exit temperature of 900°C. Extraction steam at 1 MPa to the feedwater heater is mixed with the feedwater line so that the exit is saturated liquid into the second pump. Find the fraction of extraction steam flow and the two specific pump work inputs.

11.43 A Rankine cycle operating with ammonia is heated by some low-temperature source so that the highest T is 120°C at a pressure of 5000 kPa. Its low pressure is 1003 kPa, and it operates with one open feedwater heater at 2033 kPa. The total flow rate is 5 kg/s. Find the extraction flow rate to the feedwater heater assuming its outlet state is saturated liquid at 2033 kPa. Find the total power to the two pumps.

11.44 A steam power plant operates with a boiler output of 20 kg/s steam at 2 MPa and 600°C. The condenser operates at 50°C, dumping energy to a river that has an average temperature of 20°C.

There is one open feedwater heater with extraction from the turbine at 600 kPa, and its exit is saturated liquid. Find the mass flow rate of the extraction flow. If the river water should not be heated more than 5°C, how much water should be pumped from the river to the heat exchanger (condenser)?

11.45 Consider an ideal steam regenerative cycle in which steam enters the turbine at 3 MPa and 400°C and exhausts to the condenser at 10 kPa. Steam is extracted from the turbine at 0.8 MPa for an open feedwater heater. The feedwater leaves the heater as saturated liquid. The appropriate pumps are used for the water leaving the condenser and the feedwater heater. Calculate the thermal efficiency of the cycle and the net work per kilogram of steam.

11.46 In one type of nuclear power plant, heat is transferred in the nuclear reactor to liquid sodium. The liquid sodium is then pumped through a heat exchanger where heat is transferred to boiling water. Saturated vapor steam at 5 MPa exits this heat exchanger and is then superheated to 600°C in an external gas-fired superheater. The steam enters the turbine, which has one (open-type) feedwater extraction at 0.4 MPa. The condenser pressure is 7.5 kPa. Determine the heat transfer in the reactor and in the superheater to produce a net power output of 1 MW.

11.47 A steam power plant has high and low pressures of 20 MPa and 10 kPa, and one open feedwater heater operating at 1 MPa with the exit as saturated liquid. The maximum temperature is 800°C, and the turbine has a total power output of 5 MW. Find the fraction of the flow for extraction to the feedwater and the total condenser heat transfer rate.

Closed Feedwater Heaters

11.48 A closed feedwater heater in a regenerative steam power cycle, as shown in Fig. 11.11, heats 20 kg/s of water from 100°C and 20 MPa to 250°C and 20 MPa. The extraction steam from the turbine enters the heater at 4 MPa and 275°C and leaves as saturated liquid. What is the required mass flow rate of the extraction steam?

11.49 A power plant with one closed feedwater heater has a condenser temperature of 45°C, a maximum

pressure of 5 MPa, and boiler exit temperature of 900°C. Extraction steam at 1 MPa to the feedwater heater condenses and is pumped up to the 5 MPa feedwater line where all the water goes to the boiler at 200°C. Find the fraction of extraction steam flow and the two specific pump work inputs.

11.50 Repeat Problem 11.45, but assume a closed instead of an open feedwater heater. A single pump is used to pump the water leaving the condenser up to the boiler pressure of 3 MPa. Condensate from the feedwater heater is drained through a trap to the condenser.

11.51 Do Problem 11.47 with a closed feedwater heater instead of an open heater and a drip pump to add the extraction flow to the feedwater line at 20 MPa. Assume the temperature is 175°C after the drip pump flow is added to the line. One main pump brings the water to 20 MPa from the condenser.

11.52 Assume the power plant in Problem 11.43 has one closed feedwater heater (FWH) instead of the open FWH. The extraction flow out of the FWH is saturated liquid at 2033 kPa being dumped into the condenser and the feedwater is heated to 50°C. Find the extraction flow rate and the total turbine power output.

Nonideal Cycles

11.53 Steam enters the turbine of a power plant at 5 MPa and 400°C and exhausts to the condenser at 10 kPa. The turbine produces a power output of 20 000 kW with an isentropic efficiency of 85%. What is the mass flow rate of steam around the cycle and the rate of heat rejection in the condenser? Find the thermal efficiency of the power plant. How does this compare with a Carnot cycle?

11.54 A steam power plant has a high pressure of 5 MPa and maintains 50°C in the condenser. The boiler exit temperature is 600°C. All the components are ideal except the turbine, which has an actual exit state of saturated vapor at 50°C. Find the cycle efficiency with the actual turbine and the turbine isentropic efficiency.

11.55 A steam power cycle has a high pressure of 3 MPa and a condenser exit temperature of 45°C. The turbine efficiency is 85%, and other cycle components are ideal. If the boiler superheats to 800°C, find the cycle thermal efficiency.

11.56 A steam power plant operates with a high pressure of 5 MPa and has a boiler exit temperature of 600°C receiving heat from a 700°C source. The ambient air at 20°C provides cooling for the condenser so it can maintain 45°C inside. All the components are ideal except for the turbine, which has an exit state with a quality of 97%. Find the work and heat transfer in all components per kg of water and the turbine isentropic efficiency. Find the rate of entropy generation per kg of water in the boiler/heat source setup.

11.57 For the steam power plant described in Problem 11.21, assume the isentropic efficiencies of the turbine and pump are 85% and 80%, respectively. Find the component specific work and heat transfers and the cycle efficiency.

11.58 A small steam power plant has a boiler exit of 3 MPa and 400° C, while it maintains 50 kPa in the condenser. All the components are ideal except the turbine, which has an isentropic efficiency of 80%, and it should deliver a shaft power of 9.0 MW to an electric generator. Find the specific turbine work, the needed flow rate of steam, and the cycle efficiency.

11.59 Repeat Problem 11.47 assuming the turbine has an isentropic efficiency of 85%.

11.60 Steam leaves a power plant steam generator at 3.5 MPa, 400°C, and enters the turbine at 3.4 MPa, 375°C. The isentropic turbine efficiency is 88%, and the turbine exhaust pressure is 10 kPa. Condensate leaves the condenser and enters the pump at 35°C, 10 kPa. The isentropic pump efficiency is 80%, and the discharge pressure is 3.7 MPa. The feedwater enters the steam generator at 3.6 MPa, 30°C. Calculate the thermal efficiency of the cycle and the entropy generation for the process in the line between the steam generator exit and the turbine inlet, assuming an ambient temperature of 25°C.

11.61 In a particular reheat-cycle power plant, steam enters the high-pressure turbine at 5 MPa, 450°C, and expands to 0.5 MPa, after which it is reheated to 450°C. The steam is then expanded through the low-pressure turbine to 7.5 kPa. Liquid water leaves the condenser at 30°C, is pumped to 5 MPa, and then returned to the steam generator. Each turbine is adiabatic with an isentropic efficiency of 87% and a pump efficiency of 82%. If

the total power output of the turbines is 10 MW, determine the mass flow rate of steam, the pump power input, and the thermal efficiency of the power plant.

11.62 A supercritical steam power plant has a high pressure of 30 MPa and an exit condenser temperature of 50°C. The maximum temperature in the boiler is 1000°C, and the turbine exhaust is saturated vapor. There is one open feedwater heater receiving extraction from the turbine at 1 MPa, and its exit is saturated liquid flowing to pump 2. The isentropic efficiency for the first section and the overall turbine are both 88.5%. Find the ratio of the extraction mass flow to total flow into turbine. What is the boiler inlet temperature with and without the feedwater heater?

Cogeneration

11.63 A cogenerating steam power plant, as in Fig. 11.17, operates with a boiler output of 25 kg/s steam at 7 MPa and 500°C. The condenser operates at 7.5 kPa, and the process heat is extracted at 5 kg/s from the turbine at 500 kPa, state 6, and after use is returned as saturated liquid at 100 kPa, state 8. Assume all components are ideal and find the temperature after pump 1, the total turbine output, and the total process heat transfer.

11.64 A 10-kg/s steady supply of saturated-vapor steam at 500 kPa is required for drying a wood pulp slurry in a paper mill (see Fig. P11.64). It is decided to supply this steam by cogeneration; that is, the steam supply will be the exhaust from a steam turbine. Water at 20°C and 100 kPa is pumped to a pressure of 5 MPa and then fed to a steam generator with an exit at 400°C. What is the additional heat-transfer rate to the steam generator beyond what would have been required to produce only the desired steam supply? What is the difference in net power?

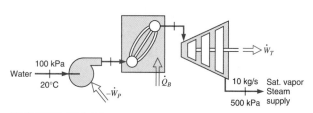

FIGURE P11.64

11.65 In a cogenerating steam power plant the turbine receives steam from a high-pressure steam drum and a low-pressure steam drum, as shown in Fig. P11.65. The condenser is made as two closed heat exchangers used to heat water running in a separate loop for district heating. The high-temperature heater adds 30 MW, and the low-temperature heater adds 31 MW to the district heating water flow. Find the power cogenerated by the turbine and the temperature in the return line to the deaerator.

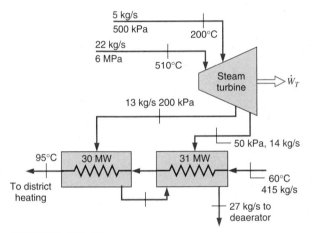

FIGURE P11.65

11.66 A boiler delivers steam at 10 MPa, 550°C to a two-stage turbine as shown in Fig. 11.17. After the first stage, 25% of the steam is extracted at 1.4 MPa for a process application and returned at 1 MPa, 90°C, to the feedwater line. The remainder of the steam continues through the low-pressure turbine stage, which exhausts to the condenser at 10 kPa. One pump brings the feedwater to 1 MPa, and a second pump brings it to 10 MPa. Assume the first and second stages in the steam turbine have isentropic efficiencies of 85% and 80%, respectively, and that both pumps are ideal. If the process application requires 5 MW of power, how much power can then be cogenerated by the turbine?

11.67 A smaller power plant produces 25 kg/s steam at 3 MPa, 600°C, in the boiler. It cools the condenser to an exit of 45°C and the cycle is shown in Fig. P11.67. There is an extraction done at 500 kPa to an open feedwater hater, and in addition a steam supply of 5 kg/s is taken out and not re-

turned. The missing 5 kg/s water is added to the feedwater heater from a 20°C, 500 kPa source. Find the needed extraction flow rate to cover both the feedwater heater and the steam supply. Find the total turbine power output.

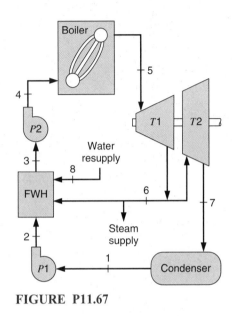

FIGURE P11.67

Brayton Cycles, Gas Turbines

11.68 Consider an ideal air-standard Brayton cycle in which the air into the compressor is at 100 kPa and 20°C, and the pressure ratio across the compressor is 12:1. The maximum temperature in the cycle is 1100°C, and the airflow rate is 10 kg/s. Assume constant specific heat for the air (from Table A.5). Determine the compressor work, the turbine work, and the thermal efficiency of the cycle.

11.69 Repeat Problem 11.68, but assume variable specific heat for the air (Table A.7).

11.70 A Brayton-cycle inlet is at 300 K and 100 kPa, and the combustion adds 670 kJ/kg. The maximum temperature is 1200 K due to material considerations. What is the maximum allowed compression ratio? For this ratio, calculate the net work and cycle efficiency assuming variable specific heat for the air (Table A.7).

11.71 A large stationary Brayton-cycle gas turbine power plant delivers a power output of 100 MW to an electric generator. The minimum temperature in the cycle is 300 K, and the maximum temperature is 1600 K. The minimum pressure in the cycle is 100 kPa, and the compressor pressure ratio is 14 to 1. Calculate the power output of the turbine. What fraction of the turbine output is required to drive the compressor? What is the thermal efficiency of the cycle?

11.72 A Brayton cycle produces 14 MW with an inlet state of 17°C, 100 kPa, and a compression ratio of 16:1. The heat added in the combustion is 960 kJ/kg. What is the highest temperature and the mass flow rate of air, assuming cold air properties?

11.73 Do the previous problem with properties from Table A.7.1 instead of cold air properties.

Regenerators, Intercoolers, and Nonideal Cycles

11.74 An ideal regenerator is incorporated into the ideal air-standard Brayton cycle of Problem 11.68. Find the thermal efficiency of the cycle with this modification.

11.75 The gas-turbine cycle shown in Fig. P11.75 is used as an automotive engine. In the first turbine, the gas expands to pressure P_5, just low enough for this turbine to drive the compressor. The gas is then expanded through the second turbine connected to the drive wheels. The data for the engine are shown in the figure and assume that all processes are ideal. Determine the intermediate pressure P_5, the net specific work output of the engine, and the mass flow rate through the engine. Find also the air temperature entering the burner T_3, and the thermal efficiency of the engine.

11.76 Repeat Problem 11.71, but include a regenerator with 75% efficiency in the cycle.

11.77 A two-stage air compressor has an intercooler between the two stages, as shown in Fig. P11.77. The inlet state is 100 kPa, 290 K, and the final exit pressure is 1.6 MPa. Assume that the constant-pressure intercooler cools the air to the inlet temperature, $T_3 = T_1$. It can be shown that the optimal pressure is $P_2 = (P_1P_4)^{1/2}$, for minimum total compressor work. Find the specific compressor works and the intercooler heat transfer for the optimal P_2.

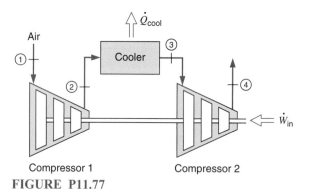

FIGURE P11.77

11.78 A two-stage compressor in a gas turbine brings atmospheric air at 100 kPa and 17°C to 500 kPa, and then cools it in an intercooler to 27°C at constant P. The second stage brings the air to 1000 kPa. Assume that both stages are adiabatic and reversible. Find the combined specific work to the compressor stages. Compare that to the specific work for the case of no intercooler (i.e., one compressor from 100 to 1000 kPa).

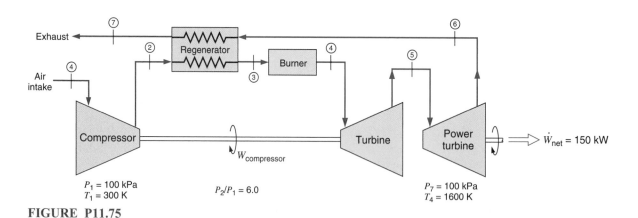

FIGURE P11.75

11.79 A gas turbine with air as the working fluid has two ideal turbine sections, as shown in Fig. P11.79, the first of which drives the ideal compressor, with the second producing the power output. The compressor input is at 290 K, 100 kPa, and the exit is at 450 kPa. A fraction of flow, x, bypasses the burner, and the rest $(1 - x)$ goes through the burner where 1200 kJ/kg is added by combustion. The two flows then mix before entering the first turbine and continue through the second turbine, with exhaust at 100 kPa. If the mixing should result in a temperature of 1000 K into the first turbine, find the fraction x. Find the required pressure and temperature into the second turbine and its specific power output.

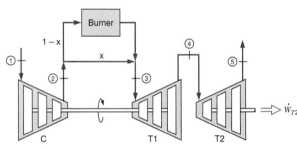

FIGURE P11.79

11.80 Repeat Problem 11.71, but assume that the compressor has an isentropic efficiency of 85% and the turbine an isentropic efficiency of 88%.

11.81 Repeat Problem 11.77 when the intercooler brings the air to $T_3 = 320$ K. The corrected formula for the optimal pressure is $P_2 = [P_1 P_4 (T_3/T_1)^{n/(n-1)}]^{1/2}$. See Problem 9.184, where n is the exponent in the assumed polytropic process.

11.82 Consider an ideal gas-turbine cycle with a pressure ratio across the compressor of 12 to 1. The compressor inlet is at 300 K and 100 kPa, and the cycle has a maximum temperature of 1600 K. An ideal regenerator is also incorporated in the cycle. Find the thermal efficiency of the cycle using cold air (298 K) properties. If the compression ration is raised, $T_4 - T_2$ goes down. At what compression ratio is $T_2 = T_4$ so the regenerator cannot be used?

11.83 A gas turbine has two stages of compression, with an intercooler between the stages (see Fig. P11.77). Air enters the first stage at 100 kPa and

300 K. The pressure ratio across each compressor stage is 5 to 1, and each stage has an isentropic efficiency of 82%. Air exits the intercooler at 330 K. Calculate the exit temperature from each compressor stage and the total specific work required.

11.84 Repeat the questions in Problem 11.75 when we assume that friction causes pressure drops in the burner and on both sides of the regenerator. In each case, the pressure drop is estimated to be 2% of the inlet pressure to that component of the system, so $P_3 = 588$ kPa, $P_4 = 0.98 \, P_3$, and $P_6 = 102$ kPa.

Ericsson Cycles

11.85 Consider an ideal air-standard Ericsson cycle that has an ideal regenerator, as shown in Fig. P11.85. The high pressure is 1 MPa, and the cycle efficiency is 70%. Heat is rejected in the cycle at a temperature of 300 K, and the cycle pressure at the beginning of the isothermal compression process is 100 kPa. Determine the high temperature, the compressor work, and the turbine work per kilogram of air.

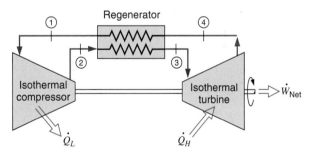

FIGURE P11.85

11.86 An air-standard Ericsson cycle has an ideal regenerator. Heat is supplied at 1000°C, and heat is rejected at 20°C. Pressure at the beginning of the isothermal compression process is 70 kPa. The heat added is 600 kJ/kg. Find the compressor work, the turbine work, and the cycle efficiency.

Jet Engine Cycles

11.87 Consider an ideal air-standard cycle for a gas-turbine, jet propulsion unit, such as that shown in

Fig. 11.26. The pressure and temperature entering the compressor are 90 kPa and 290 K. The pressure ratio across the compressor is 14 to 1, and the turbine inlet temperature is 1500 K. When the air leaves the turbine, it enters the nozzle and expands to 90 kPa. Determine the pressure at the nozzle inlet and the velocity of the air leaving the nozzle.

11.88 The turbine section in a jet engine receives gas (assumed to be air) at 1200 K and 800 kPa with an ambient atmosphere at 80 kPa. The turbine is followed by a nozzle open to the atmosphere, and all the turbine work drives a compressor receiving air at 85 kPa and 270 K with the same flow rate. Find the turbine exit pressure so that the nozzle has an exit velocity of 800 m/s. To what pressure can the compressor bring the incoming air?

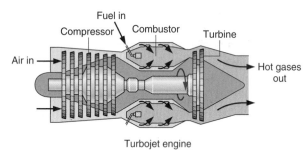

Turbojet engine

FIGURE P11.88

11.89 The turbine in a jet engine receives air at 1250 K and 1.5 MPa. It exhausts to a nozzle at 250 kPa, which in turn exhausts to the atmosphere at 100 kPa. The isentropic efficiency of the turbine is 85%, and the nozzle efficiency is 95%. Find the nozzle inlet temperature and the nozzle exit velocity. Assume negligible kinetic energy out of the turbine.

11.90 Consider an air-standard jet engine cycle operating in a 280-K, 100-kPa environment. The compressor requires a shaft power input of 4000 kW. Air enters the turbine state 3 at 1600 K and 2 MPa, at the rate of 9 kg/s, and the isentropic efficiency of the turbine is 85%. Determine the pressure and temperature entering the nozzle.

11.91 A jet aircraft is flying at an altitude of 4900 m, where the ambient pressure is approximately 55

kPa and the ambient temperature is −18°C. The velocity of the aircraft is 280 m/s, the pressure ratio across the compressor is 14:1, and the cycle maximum temperature is 1450 K. Assume that the inlet flow goes through a diffuser to zero relative velocity at state a, Fig. 11.26. Find the temperature and pressure at state a and the velocity (relative to the aircraft) of the air leaving the engine at 55 kPa.

11.92 An afterburner in a jet engine adds fuel after the turbine, thus raising the pressure and temperature via the energy of combustion. Assume a standard condition of 800 K and 250 kPa after the turbine into the nozzle that exhausts at 95 kPa. Assume the afterburner adds 450 kJ/kg to that state with a rise in pressure for the same specific volume, and neglect any upstream effects on the turbine. Find the nozzle exit velocity before and after the afterburner is turned on.

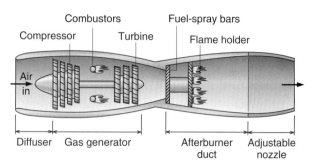

FIGURE P11.92

Otto Cycles

11.93 Air flows into a gasoline engine at 95 kPa and 300 K. The air is then compressed with a volumetric compression ratio of 8:1. The combustion process releases 1300 kJ/kg of energy as the fuel burns. Find the temperature and pressure after combustion using cold air properties.

11.94 A gasoline engine has a volumetric compression ratio of 9. The state before compression is 290 K, 90 kPa, and the peak cycle temperature is 1800 K. Find the pressure after expansion, the cycle net work, and the cycle efficiency using properties from Table A.5.

11.95 To approximate an actual spark-ignition engine, consider an air-standard Otto cycle that has a heat addition of 1800 kJ/kg of air, a compression ratio of 7, and a pressure and temperature at the beginning of the compression process of 90 kPa and 10°C. Assuming constant specific heat, with the value from Table A.5, determine the maximum pressure and temperature of the cycle, the thermal efficiency of the cycle, and the mean effective pressure.

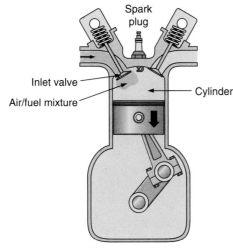

FIGURE P11.95

11.96 A gasoline engine has a volumetric compression ratio of 8 and before compression has air at 280 K and 85 kPa. The combustion generates a peak pressure of 6500 kPa. Find the peak temperature, the energy added by the combustion process, and the exhaust temperature.

11.97 A gasoline engine has a volumetric compression ratio of 10 and before compression has air at 290 K, 85 kPa, in the cylinder. The combustion peak pressure is 6000 kPa. Assume cold air properties. What is the highest temperature in the cycle? Find the temperature at the beginning of the exhaust (heat rejection) and the overall cycle efficiency.

11.98 A four-stroke gasoline engine has a compression ratio of 10:1 with 4 cylinders of total displacement 2.3 L. The inlet state is 280 K, 70 kPa, and the engine is running at 2100 RPM with the fuel adding 1800 kJ/kg in the combustion process. What is the net work in the cycle, and how much power is produced?

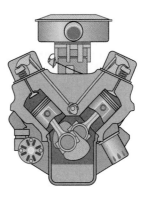

FIGURE P11.98

11.99 A gasoline engine takes air in at 290 K and 90 kPa and then compresses it. The combustion adds 1000 kJ/kg to the air, after which the temperature is 2050 K. Use the cold air properties (i.e., constant heat capacities at 300 K) and find the compression ratio, the compression specific work, and the highest pressure in the cycle.

11.100 Answer the same three questions for the previous problem, but use variable heat capacities (use Table A.7).

11.101 When methanol produced from coal is considered as an alternative fuel to gasoline for automotive engines, it is recognized that the engine can be designed with a higher compression ratio, say 10 instead of 7, but that the energy release with combustion for a stoichiometric mixture with air is slightly smaller, about 1700 kJ/kg. Repeat Problem 11.95 using these values.

11.102 A gasoline engine receives air at 10°C, 100 kPa, having a compression ratio of 9:1 by volume. The heat addition by combustion gives the highest temperature as 2500 K. Use cold air properties to find the highest cycle pressure, the specific energy added by combustion, and the mean effective pressure.

11.103 Repeat Problem 11.95, but assume variable specific heat. The ideal-gas air tables, Table A.7, are recommended for this calculation (and the specific heat from Fig. 5.10 at high temperature).

11.104 It is found experimentally that the power stroke expansion in an internal combustion engine can be approximated, with a polytropic process with a value of the polytropic exponent n somewhat larger than the specific heat ratio k. Repeat Problem 11.95, but assume that the expansion process is reversible and polytropic (instead of the isentropic expansion in the Otto cycle) with n equal to 1.50.

11.105 In the Otto cycle, all the heat transfer q_H occurs at constant volume. It is more realistic to assume that part of q_H occurs after the piston has started its downward motion in the expansion stroke. Therefore, consider a cycle identical to the Otto cycle, except that the first two-thirds of the total q_H occurs at constant volume and the last one-third occurs at constant pressure. Assume that the total q_H is 2100 kJ/kg, that the state at the beginning of the compression process is 90 kPa, 20°C, and that the compression ratio is 9. Calculate the maximum pressure and temperature and the thermal efficiency of this cycle. Compare the results with those of a conventional Otto cycle having the same given variables.

Diesel Cycles

11.106 A diesel engine has a state before compression of 95 kPa, 290 K, a peak pressure of 6000 kPa, and a maximum temperature of 2400 K. Find the volumetric compression ratio and the thermal efficiency.

11.107 A diesel engine has a bore of 0.1 m, a stroke of 0.11 m, and a compression ratio of 19:1 running at 2000 RPM (revolutions per minute). Each cycle takes two revolutions and has a mean effective pressure of 1400 kPa. With a total of 6 cylinders, find the engine power in kW and horsepower, hp.

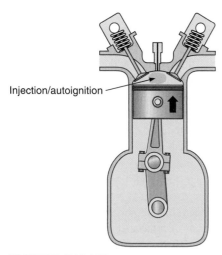

Injection/autoignition

FIGURE P11.107

11.108 A diesel engine has a compression ratio of 20:1 with an inlet of 95 kPa and 290 K, state 1, with volume 0.5 L. The maximum cycle temperature is 1800 K. Find the maximum pressure, the net specific work, and the thermal efficiency.

11.109 At the beginning of compression in a diesel cycle $T = 300$ K and $P = 200$ kPa; after combustion (heat addition) is complete $T = 1500$ K and $P = 7.0$ MPa. Find the compression ratio, the thermal efficiency, and the mean effective pressure.

11.110 Do Problem 11.106 but use the properties from Table A.7 and not the cold air properties.

11.111 A diesel engine has air before compression at 280 K and 85 kPa. The highest temperature is 2200 K, and the highest pressure is 6 MPa. Find the volumetric compression ratio and the mean effective pressure using cold air properties at 300 K.

11.112 Consider an ideal air-standard diesel cycle in which the state before the compression process is 95 kPa, 290 K, and the compression ratio is 20. Find the maximum temperature (by iteration) in the cycle to have a thermal efficiency of 60%.

Stirling and Carnot Cycles

11.113 Consider an ideal Stirling-cycle engine in which the state at the beginning of the isothermal compression process is 100 kPa, 25°C, the compression ratio is 6, and the maximum temperature in the cycle is 1100°C. Calculate the maximum cycle pressure and the thermal efficiency of the cycle with and without regenerators.

11.114 An air-standard Stirling cycle uses helium as the working fluid. The isothermal compression brings helium from 100 kPa, 37°C to 600 kPa. The expansion takes place at 1200 K, and there is no regenerator. Find the work and heat transfer in all of the four processes per kg helium and the thermal cycle efficiency.

11.115 Consider an ideal air-standard Stirling cycle with an ideal regenerator. The minimum pressure and temperature in the cycle are 100 kPa, 25°C, the compression ratio is 10, and the maximum temperature in the cycle is 1000°C. Analyze each of the four processes in this cycle for work and heat transfer, and determine the overall performance of the engine.

11.116 The air-standard Carnot cycle was not shown in the text; show the $T–s$ diagram for this cycle. In an air-standard Carnot cycle, the low temperature is 280 K and the efficiency is 60%. If the pressure before compression and after heat rejection is 100 kPa, find the high temperature and the pressure just before heat addition.

11.117 Air in a piston/cylinder setup goes through a Carnot cycle in which $T_L = 26.8°C$ and the total cycle efficiency is $\eta = 2/3$. Find T_H, the specific work, and the volume ratio in the adiabatic expansion for constant C_p, C_v.

11.118 Do the previous Problem, 11.117, using Table A.7.1.

Refrigeration Cycles

11.119 A refrigerator with R-12 as the working fluid has a minimum temperature of $-10°C$ and a maximum pressure of 1 MPa. Assume an ideal refrigeration cycle as in Fig. 11.33. Find the specific heat transfer from the cold space and that to the hot space, and determine the coefficient of performance.

11.120 Consider an ideal refrigeration cycle that has a condenser temperature of 45°C and an evaporator temperature of $-15°C$. Determine the coefficient of performance of this refrigerator for the working fluids R-12 and R-22.

11.121 The environmentally safe refrigerant R-134a is one of the replacements for R-12 in refrigeration systems. Repeat Problem 11.120 using R-134a and compare the result with that for R-12.

11.122 A refrigerator using R-22 is powered by a small natural gas-fired heat engine with a thermal efficiency of 25%, as shown in Fig. P11.122. The R-22 condenses at 40°C, it evaporates at $-20°C$, and the cycle is standard. Find the two specific heat transfers in the refrigeration cycle. What is the overall coefficient of performance as Q_L/Q_1?

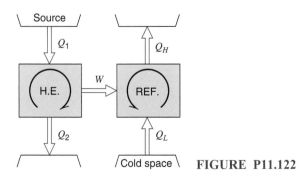

FIGURE P11.122

11.123 A refrigerator in a meat warehouse must keep a low temperature of $-15°C$. It uses R-12 as the refrigerant, which must remove 5 kW from the cold space. Assume that the outside temperature is 20°C. Find the flow rate of the R-12 needed, assuming a standard vapor compression refrigeration cycle with a condenser at 20°C.

11.124 A refrigerator with R-12 as the working fluid has a minimum temperature of $-10°C$ and a maximum pressure of 1 MPa. The actual adiabatic compressor exit temperature is 60°C. Assume no pressure loss in the heat exchangers. Find the specific heat transfer from the cold space and that to the hot space, the coefficient of performance, and the isentropic efficiency of the compressor.

11.125 Consider an ideal heat pump that has a condenser temperature of 50°C and an evaporator temperature of 0°C. Determine the coefficient of performance of this heat pump for the working fluids R-12, R-22, and ammonia.

11.126 The air conditioner in a car uses R-134a, and the compressor power input is 1.5 kW, bringing the R-134a from 201.7 kPa to 1200 kPa by compression. The cold space is a heat exchanger that cools 30°C atmospheric air from the outside down to 10°C and blows it into the car. What is the mass flow rate of the R-134a, and what is the low-temperature heat-transfer rate? What is the mass flow rate of air at 10°C?

11.127 A refrigerator using R-134a is located in a 20°C room. Consider the cycle to be ideal, except that the compressor is neither adiabatic nor reversible. Saturated vapor at $-20°C$ enters the compressor, and the R-134a exits the compressor at 50°C. The condenser temperature is 40°C. The mass flow rate of refrigerant around the cycle is 0.2 kg/s, and the coefficient of performance is measured and found to be 2.3. Find the power input to the compressor and the rate of entropy generation in the compressor process.

11.128 A refrigerator has a steady flow of R-22 as saturated vapor at $-20°C$ into the adiabatic compressor that brings it to 1000 kPa. After the compressor, the temperature is measured to be 60°C. Find the actual compressor work and the actual cycle coefficient of performance.

11.129 A small heat pump unit is used to heat water for a hot-water supply. Assume that the unit uses

R-22 and operates on the ideal refrigeration cycle. The evaporator temperature is 15°C, and the condenser temperature is 60°C. If the amount of hot water needed is 0.1 kg/s, determine the amount of energy saved by using the heat pump instead of directly heating the water from 15 to 60°C.

11.130 The refrigerant R-22 is used as the working fluid in a conventional heat pump cycle. Saturated vapor enters the compressor of this unit at 10°C; its exit temperature from the compressor is measured and found to be 85°C. If the compressor exit is 2 MPa, what is the compressor isentropic efficiency and the cycle COP?

11.131 A refrigerator in a laboratory uses R-22 as the working substance. The high pressure is 1200 kPa, the low pressure is 201 kPa, and the compressor is reversible. It should remove 500 W from a specimen currently at −20°C (not equal to T_L in the cycle) that is inside the refrigerated space. Find the cycle COP and the electrical power required.

11.132 Consider the previous problem and find the two rates of entropy generation in the process and where they occur.

11.133 In an actual refrigeration cycle using R-12 as the working fluid, the refrigerant flow rate is 0.05 kg/s. Vapor enters the compressor at 150 kPa and −10°C and leaves at 1.2 MPa and 75°C. The power input to the nonadiabatic compressor is measured and found to be 2.4 kW. The refrigerant enters the expansion valve at 1.15 MPa and 40°C and leaves the evaporator at 175 kPa and −15°C. Determine the entropy generation in the compression process, the refrigeration capacity, and the coefficient of performance for this cycle.

Ammonia Absorption Cycles

11.134 Consider a small ammonia absorption refrigeration cycle that is powered by solar energy and is to be used as an air conditioner. Saturated vapor ammonia leaves the generator at 50°C, and saturated vapor leaves the evaporator at 10°C. If 7000 kJ of heat is required in the generator (solar collector) per kilogram of ammonia vapor generated, determine the overall performance of this system.

11.135 The performance of an ammonia absorption cycle refrigerator is to be compared with that of a similar vapor-compression system. Consider

an absorption system having an evaporator temperature of −10°C and a condenser temperature of 50°C. The generator temperature in this system is 150°C. In this cycle 0.42 kJ is transferred to the ammonia in the evaporator for each kilojoule transferred from the high-temperature source to the ammonia solution in the generator. To make the comparison, assume that a reservoir is available at 150°C and that heat is transferred from this reservoir to a reversible engine that rejects heat to the surroundings at 25°C. This work is then used to drive an ideal vapor-compression system with ammonia as the refrigerant. Compare the amount of refrigeration that can be achieved per kilojoule from the high temperature source with the 0.42 kJ that can be achieved in the absorption system.

Air-Standard Refrigeration Cycles

11.136 The formula for the coefficient of performance when we use cold air properties is not given in the text. Derive the expression for COP as function of the compression ratio similar to how the Brayton-cycle efficiency was found.

11.137 A heat exchanger is incorporated into an ideal air-standard refrigeration cycle, as shown in Fig. P11.137. It may be assumed that both the compression and the expansion are reversible adiabatic processes in this ideal case. Determine the coefficient of performance for the cycle.

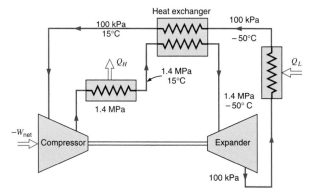

FIGURE P11.137

11.138 Repeat Problem 11.137, but assume that helium is the cycle working fluid instead of air. Discuss the significance of the results.

11.139 Repeat Problem 11.137, but assume an isentropic efficiency of 75% for both the compressor and the expander.

Combined Cycles

11.140 A binary system power plant uses mercury for the high-temperature cycle and water for the low-temperature cycle, as shown in Fig. 11.39. The temperatures and pressures are shown in the corresponding T–s diagram. The maximum temperature in the steam cycle is where the steam leaves the superheater at point 4 where it is 500°C. Determine the ratio of the mass flow rate of mercury to the mass flow rate of water in the heat exchanger that condenses mercury and boils the water and the thermal efficiency of this ideal cycle.

The following saturation properties for mercury are known:

P, MPa	T_g, °C	h_f, kJ/kg	h_g, kJ/kg	s_f, kJ/ kg-K	s_g, kJ/ kg-K
0.04	309	42.21	335.64	0.1034	0.6073
1.60	562	75.37	364.04	0.1498	0.4954

11.141 A Rankine steam power plant should operate with a high pressure of 3 MPa, a low pressure of 10 kPa, and a boiler exit temperature of 500°C. The available high-temperature source is the exhaust of 175 kg/s air at 600°C from a gas turbine. If the boiler operates as a counterflowing heat exchanger where the temperature difference at the pinch point is 20°C, find the maximum water mass flow rate possible and the air exit temperature.

11.142 A simple Rankine cycle with R-22 as the working fluid is to be used as a bottoming cycle for an electrical-generating facility driven by the exhaust gas from a diesel engine as the high-temperature energy source in the R-22 boiler. Diesel inlet conditions are 100 kPa, 20°C, the compression ratio is 20, and the maximum temperature in the cycle is 2800°C. Saturated vapor R-22 leaves the bottoming cycle boiler at 110°C, and the condenser temperature is 30°C. The power output of the diesel engine is 1 MW. Assuming ideal cycles throughout, determine

a. The flow rate required in the diesel engine.

b. The power output of the bottoming cycle, assuming that the diesel exhaust is cooled to 200°C in the R-22 boiler.

11.143 A cascade system is composed of two ideal refrigeration cycles, as shown in Fig. 11.41. The high-temperature cycle uses R-22. Saturated liquid leaves the condenser at 40°C, and saturated vapor leaves the heat exchanger at −20°C. The low-temperature cycle uses a different refrigerant, R-23. Saturated vapor leaves the evaporator at −80°C with $h = 330$ kJ/kg, and saturated liquid leaves the heat exchanger at −10°C with $h = 185$ kJ/kg. R-23 out of the compressor has $h = 405$ kJ/kg. Calculate the ratio of the mass flow rates through the two cycles an the COP of the total system.

11.144 Consider an ideal dual-loop heat-powered refrigeration cycle using R-12 as the working fluid, as shown in Fig. P11.144. Saturated vapor at 105°C leaves the boiler and expands in the turbine to the condenser pressure. Saturated vapor at −15°C leaves the evaporator and is compressed to the condenser pressure. The ratio of the flows through the two loops is such that the turbine produces just enough power to drive the compressor. The two exiting streams mix together and enter the condenser. Saturated liquid

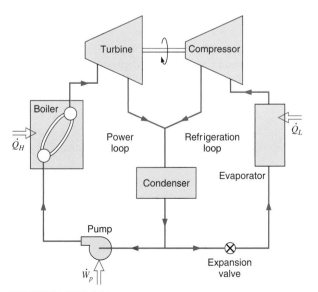

FIGURE P11.144

leaving the condenser at 45°C is then separated into two streams in the necessary proportions. Determine the ratio of mass flow rate through the power loop to that through the refrigeration loop. Find also the performance of the cycle, in terms of the ratio Q_L/Q_H.

11.145 For a cryogenic experiment, heat should be removed from a space at 75 K to a reservoir at 180 K. A heat pump is designed to use nitrogen and methane in a cascade arrangement (see Fig. 11.41), where the high temperature of the nitrogen condensation is at 10 K higher than the low-temperature evaporation of the methane. The two other phase changes take place at the listed reservoir temperatures. Find the saturation temperatures in the heat exchanger between the two cycles that give the best coefficient of performance for the overall system.

Availability or Exergy Concepts

11.146 Find the flows and fluxes of exergy in the condenser of Problem 11.32. Use those to determine the second-law efficiency.

11.147 Find the availability of the water at all four states in the Rankine cycle described in Problem 11.33. Assume that the high-temperature source is 500°C and the low-temperature reservoir is at 25°C. Determine the flow of availability in or out of the reservoirs per kilogram of steam flowing in the cycle. What is the overall cycle second-law efficiency?

11.148 Find the flows of exergy into and out of the feedwater heater in Problem 11.43.

11.149 Find the availability of the water at all the states in the steam power plant described in Problem 11.57. Assume the heat source in the boiler is at 600°C and the low-temperature reservoir is at 25°C. Give the second-law efficiency of all the components.

11.150 Consider the Brayton cycle in Problem 11.72. Find all the flows and fluxes of exergy and find the overall cycle second-law efficiency. Assume the heat transfers are internally reversible processes, and we then neglect any external irreversibility.

11.151 For Problem 11.141, determine the change of availability of the water flow and that of the airflow. Use these to determine a second-law efficiency for the boiler heat exchanger.

Review Problems

11.152 A simple steam power plant is said to have the four states as listed: (1) 20°C, 100 kPa, (2) 25°C, 1 MPa, (3) 1000°C, 1 MPa, (4) 250°C, 100 kPa, with an energy source at 1100°C, and it rejects energy to a 0°C ambient. Is this cycle possible? Are any of the devices impossible?

11.153 Do Problem 11.31 with R-134a as the working fluid in the Rankine cycle.

11.154 An ideal steam power plant is designed to operate on the combined reheat and regenerative cycle and to produce a net power output of 10 MW. Steam enters the high-pressure turbine at 8 MPa, 550°C, and is expanded to 0.6 MPa, at which pressure some of the steam is fed to an open feedwater heater, and the remainder is reheated to 550°C. The reheated steam is then expanded in the low-pressure turbine to 10 kPa. Determine the steam flow rte to the high-pressure turbine and the power required to drive each pump.

11.155 Steam enters the turbine of a power plant at 5 MPa and 400°C and exhausts to the condenser at 10 kPa. The turbine produces a power output of 20 000 kW with an isentropic efficiency of 85%. What is the mass flow rate of steam around the cycle and the rate of heat rejection in the condenser? Find the thermal efficiency of the power plant. How does this compare with a Carnot cycle?

11.156 Consider an ideal steam reheat cycle as shown in Fig. 11.7, where steam enters the high-pressure turbine at 4 MPa and 450°C with a mass flow rate of 20 kg/s. After expansion to 400 kPa, it is reheated to T_5 flowing through the low-pressure turbine out to the condenser operating at 10 kPa. Find T_5 so the turbine exit quality is at least 95%. For this reheat temperature find also the thermal efficiency of the cycle and the net power output.

11.157 In one type of nuclear power plant, heat is transferred in the nuclear reactor to liquid sodium. The liquid sodium is then pumped through a heat exchanger where heat is transferred to boiling water. Saturated vapor steam at 5 MPa exits this heat exchanger and is then superheated to 600°C in an external gas-fired superheater. The steam enters the turbine, which has one (open-type) feedwater extraction at 0.4 MPa. The isentropic turbine efficiency is 87%, and the condenser pressure is 7.5 kPa. Determine the heat transfer in the reactor and in the superheater to produce a net power output of 1 MW.

11.158 An industrial application has the following steam requirement: one 10-kg/s stream at a pressure of 0.5 MPa and one 5-kg/s stream at 1.4 MPa (both saturated or slightly superheated vapor). These are obtained by cogeneration, whereby a high-pressure boiler supplies steam at 10 MPa and 500°C to a turbine. The required amount is withdrawn at 1.4 MPa, and the remainder is expanded in the low-pressure end of the turbine to 0.5 MPa, providing the second required steam flow. Assuming both turbine sections have an isentropic efficiency of 85%.

    a. Determine the power output of the turbine and the heat-transfer rate in the boiler.

    b. Compute the rates needed if the steam were generated is a low-pressure boiler without cogeneration. Assume that for each, 20°C liquid water is pumped to the required pressure and fed to a boiler.

11.159 Repeat Problem 11.75, but assume that the compressor has an efficiency of 82%, that both turbines have efficiencies of 87%, and that the regenerator efficiency is 70%.

11.160 Consider a gas-turbine cycle with two stages of compression and two stages of expansion. The pressure ratio across each compressor stage and each turbine stage is 8 to 1. The pressure at the entrance to the first compressor is 100 kPa, the temperature entering each compressor is 20°C, and the temperature entering each turbine is 1100°C. A regenerator is also incorporated into the cycle and it has an efficiency of 70%. Deter-

mine the compressor work, the turbine work, and the thermal efficiency of the cycle.

11.161 A gas-turbine cycle has two stages of compression, with an intercooler between the stages. Air enters the first stage at 100 kPa, 300 K. The pressure ratio across each compressor stage is 5 to 1, and each stage has an isentropic efficiency of 82%. Air exits the intercooler at 330 K. The maximum cycle temperature is 1500 K, and the cycle has a single-turbine stage with an isentropic efficiency of 86%. The cycle also includes a regenerator with an efficiency of 80%. Calculate the temperature at the exit of each compressor stage, the second-law efficiency of the turbine, and the cycle thermal efficiency.

11.162 A gasoline engine has a volumetric compression ratio of 9. The state before compression is 290 K, 90 kPa, and the peak cycle temperature is 1800 K. Find the pressure after expansion, the cycle net work, and the cycle efficiency using properties from Table A.7.

11.163 The effect of a number of open feedwater heaters on the thermal efficiency of an ideal cycle is to be studied. Steam leaves the steam generator at 20 MPa, 600°C, and the cycle has a condenser pressure of 10 kPa. Determine the thermal efficiency for each of the following cases. **A:** No feedwater heater. **B:** One feedwater heater operating at 1 MPa. **C:** Two feedwater heaters, one operating at 3 MPa and the other at 0.2 MPa.

11.164 The power plant shown in Fig. 11.40 combines a gas-turbine cycle and a steam-turbine cycle. The following data are known for the gas-turbine cycle. Air enters the compressor at 100 kPa,

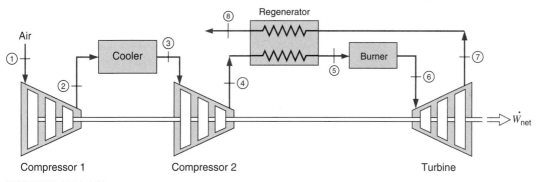

FIGURE P11.161

25°C, the compressor pressure ratio is 14, and the isentropic compressor efficiency is 87%; the heater input rate is 60 MW; the turbine inlet temperature is 1250°C, the exhaust pressure is 100 kPa, and the isentropic turbine efficiency is 87%; the cycle exhaust temperature from the heat exchanger is 200°C. The following data are known for the steam-turbine cycle. The pump inlet state is saturated liquid at 10 kPa, the pump exit pressure is 12.5 MPa, and the isentropic pump efficiency is 85%; turbine inlet temperature is 500°C, and the isentropic turbine efficiency is 87%. Determine

a. The mass flow rate of air in the gas-turbine cycle.

b. The mass flow rate of water in the steam cycle.

c. The overall thermal efficiency of the combined cycle.

11.165 One means of improving the performance of a refrigeration system that operates over a wide temperature range is to use a two-stage compressor. Consider an ideal refrigeration system of this type that uses R-12 as the working fluid, as shown in Fig. P11.165. Saturated liquid leaves the condenser at 40°C and is throttled to −20°C. The liquid and vapor at this temperature are separated, and the liquid is throttled to the evaporator temperature, −70°C. Vapor leaving the evaporator is compressed to the saturation pressure corresponding to −20°C, after which it is mixed with the vapor leaving the flash chamber. It may be assumed that both the flash chamber and the mixing chamber are well insulated to prevent heat transfer from the ambient. Vapor leaving the mixing chamber is compressed in the second stage of the compressor to the saturation pressure corresponding to the condenser temperature, 40°C. Determine the following:

a. The coefficient of performance of the system.

b. The coefficient of performance of a simple ideal refrigeration cycle operating over the same condenser and evaporator ranges as those of the two-stage compressor unit studied in this problem.

11.166 A jet ejector, a device with no moving parts, functions as the equivalent of a coupled turbine-compressor unit (see Problems 9.82 and 9.90). Thus, the turbine-compressor in the dual-loop

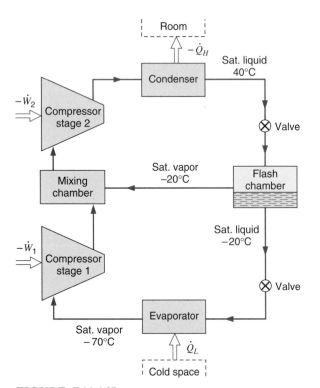

FIGURE P11.165

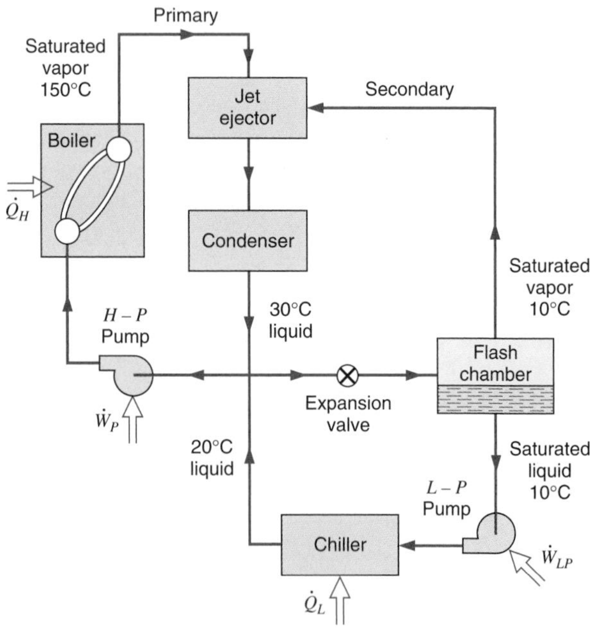

FIGURE P11.166

cycle of Fig. P11.144 could be replaced by a jet ejector. The primary stream of the jet ejector enters from the boiler, the secondary stream enters from the evaporator, and the discharge flows to the condenser. Alternatively, a jet ejector may be used with water as the working fluid. The purpose of the device is to chill water, usually for an air-conditioning system. In this application the

physical setup is as shown in Fig. P11.166. Using the data given on the diagram, evaluate the performance of this cycle in terms of the ratio Q_L/Q_H.

a. Assume an ideal cycle.

b. Assume an ejector efficiency of 20% (see Problem 9.128).

ENGLISH UNIT PROBLEMS

Rankine Cycles

11.167E A steam power plant, as shown in Fig. 11.3, operating in a Rankine cycle has saturated vapor at 600 lbf/in.^{2} leaving the boiler. The turbine exhausts to the condenser operating at 2 lbf/in.^{2}. Find the specific work and heat transfer in each of the ideal components and the cycle efficiency.

11.168E Consider a solar-energy-powered ideal Rankine cycle that uses water as the working fluid. Saturated vapor leaves the solar collector at 350 F, and the condenser pressure is 1 lbf/in.^{2}. Determine the thermal efficiency of this cycle.

11.169E A Rankine cycle uses ammonia as the working substance and powered by solar energy. It heats the ammonia to 320 F at 800 psia in the boiler/superheater. The condenser is water cooled, and the exit is kept at 70 F. Find (T, P, and x if applicable) for all four states in the cycle.

11.170E A supply of geothermal hot water is to be used as the energy source in an ideal Rankine cycle, with R-134a as the cycle working fluid. Saturated vapor R-134a leaves the boiler at a temperature of 180 F, and the condenser temperature is 100 F. Calculate the thermal efficiency of this cycle.

11.171E Do Problem 11.170 with R-22 as the working fluid.

11.172E A smaller power plant produces 50 lbm/s steam at 400 psia, 1100 F, in the boiler. It cools the condenser with ocean water coming in at 55 F and returned at 60 F so that the condenser exit is at 110 F. Find the net power output and the required mass flow rate of ocean water.

11.173E The power plant in Problem 11.167 is modified to have a superheater section following the boiler so the steam leaves the superheater at 600 lbf/in.^{2}, 700 F. Find the specific work and heat transfer in each of the ideal components and the cycle efficiency.

11.174E Consider a simple ideal Rankine cycle using water at a supercritical pressure. Such a cycle has a potential advantage of minimizing local temperature differences between the fluids in the steam generator, such as the instance in which the high-temperature energy source is the hot exhaust gas from a gas-turbine engine. Calculate the thermal efficiency of the cycle if the state entering the turbine is 3500 lbf/in.^{5}, 1100 F, and the condenser pressure is 1 lbf/in.^{2}. What is the steam quality at the turbine exit?

11.175E Consider an ideal steam reheat cycle in which the steam enters the high-pressure turbine at 600 lbf/in.^{2}, 700 F, and then expands to 120 lbf/in.^{2}. It is then reheated to 700 F and expands to 2 lbf/in.^{2} in the low-pressure turbine. Calculate the thermal efficiency of the cycle and the moisture content of the steam leaving the low-pressure turbine.

11.176E Consider an ideal steam regenerative cycle in which steam enters the turbine at 600 lbf/in.^{2}, 700 F, and exhausts to the condenser at 2 lbf/in.^{2}. Steam is extracted from the turbine at 120 lbf/in.^{2} for an open feedwater heater. The feedwater leaves the heater as saturated liquid. The appropriate pumps are used for the water leaving the condenser and the feedwater heater. Calculate the thermal efficiency of the cycle and the net work per pound-mass of steam.

11.177E A closed feedwater heater in a regenerative steam power cycle heats 40 lbm/s of water from 200 F, 2000 lbf/in.^{2} to 450 F, 2000 lbf/in.^{2}. The extraction steam from the turbine enters the heater at 500 lbf/in.^{2}, 550 F and leaves as saturated liquid. What is the required mass flow rate of the extraction steam?

11.178E A steam power cycle has a high pressure of 500 lbf/in.^{2} and a condenser exit temperature of 110 F. The turbine efficiency is 85%, and other cycle components are ideal. If the boiler superheats to 1400 F, find the cycle thermal efficiency.

11.179E The steam power cycle in Problem 11.167 has an isentropic efficiency of the turbine of 85% and that for the pump of 80%. Find the cycle efficiency and the specific work and heat transfer in the components.

11.180E Steam leaves a power plant steam generator at 500 lbf/in.^{2}, 650 F, and enters the turbine at 490 lbf/in.^{2}, 625 F. The isentropic turbine efficiency is 88%, and the turbine exhaust pressure is 1.7 lb/in.^{2}. Condensate leaves the condenser and enters the pump at 110 F, 1.7 lbf/in.^{2}. The isentropic pump efficiency is 80%, and the discharge pressure is 520 lbf/in.^{2}. The feedwater enters the steam generator at 510 lbf/in.^{2}, 100 F. Calculate the thermal efficiency of the cycle and the entropy generation of the flow in the line between the steam generator exit and the turbine inlet, assuming an ambient temperature of 77 F.

11.181E A boiler delivers steam at 1500 lbf/in.^{2}, 1000 F, to a two-stage turbine, as shown in Fig. 11.21. After the first stage, 25% of the steam is extracted at 200 lbf/in.^{2} for a process application and returned at 150 lbf/in.^{2}, 190 F, to the feedwater line. The remainder of the steam continues through the low-pressure turbine stage, which exhausts to the condenser at 2 lbf/in.^{2}. One pump brings the feedwater to 150 lbf/in.^{2} and a second pump brings it to 1500 lbf/in.^{2}. Assume the first and second stages in the steam turbine have isentropic efficiencies of 85% and 80% and that both pumps are ideal. If the process application requires 5000 Btu/s of power, how much power can then be cogenerated by the turbine?

Brayton Cycles

11.182E A large stationary Brayton-cycle gas-turbine power plant delivers a power output of 100 000 hp to an electric generator. The minimum temperature in the cycle is 540 R, and the maximum temperature is 2900 R. The minimum pressure in the cycle is 1 atm, and the compressor pressure ratio is 14 to 1. Calculate the power output of the turbine, the fraction of the turbine output required to drive the compressor, and the thermal efficiency of the cycle.

11.183E A Brayton cycle produces 14 000 Btu/s with an inlet state of 60 F, 14.5 psia, and a compression ratio of 16:1. The heat added in the combustion is 400 Btu/lbm. What is the highest temperature and the mass flow rate of air assuming cold air properties.

11.184E Do the previous problem using properties from Table F.5.

11.185E An ideal regenerator is incorporated into the ideal air-standard Brayton cycle of Problem 11.182. Calculate the cycle thermal efficiency with this modification.

11.186E An air-standard Ericsson cycle has an ideal regenerator, as shown in Fig. P11.85. Heat is supplied at 1800 F, and heat is rejected at 68 F. Pressure at the beginning of the isothermal compression process is 10 lbf/in.^{2}. The heat added is 275 But/lbm. Find the compressor work, the turbine work, and the cycle efficiency.

11.187E The turbine in a jet engine receives air at 2200 R, 220 lbf/in.^{2}. It exhausts to a nozzle at 35 lbf/in.^{2}, which in turn exhausts to the atmosphere at 14.7 lbf/in.^{2}. The isentropic efficiency of the turbine is 85%, and the nozzle efficiency is 95%. Find the nozzle inlet temperature and the nozzle exit velocity. Assume negligible kinetic energy out of the turbine.

Otto, Diesel, Stirling, and Carnot Cycles

11.188E Air flows into a gasoline engine at 14 lbf/in.^{2}, 540 R. The air is then compressed with a volumetric compression ratio of 8:1. In the combustion process 560 Btu/lbm of energy is released as the fuel burns. Find the temperature and pressure after combustion.

11.189E To approximate an actual spark-ignition engine consider an air-standard Otto cycle that has a heat addition of 800 Btu/lbm of air, a compression ratio of 7, and a pressure and temperature at the beginning of the compression process of 13 lbf/in.^{2}, 50 F. Assuming constant specific heat, with the value from Table F.4, determine the maximum pressure and temperature of the cycle, the thermal efficiency of the cycle, and the mean effective pressure.

11.190E A gasoline engine has a volumetric compression ratio of 10 and before compression has air at 520 R, 12.2 psia, in the cylinder. The combustion peak pressure is 900 psia. Assume cold air properties. What is the highest temperature in the cycle? Find the temperature at the beginning of the exhaust (heat rejection) and the overall cycle efficiency.

11.191E A four-stroke gasoline engine has a compression ratio of 10:1 with 4 cylinders of total displacement 75 in.^{3}. The inlet state is 500 R, 10 psia, and the engine is running at 2100 RPM, with the fuel adding 750 Btu/lbm in the combustion process. What is the net work in the cycle, and how much power is produced?

11.192E It is found experimentally that the power stroke expansion in an internal combustion engine can be approximated with a polytropic process with a value of the polytropic exponent n somewhat larger than the specific heat ratio k. Repeat Problem 11.189 but assume the expansion process is reversible and polytropic (instead of the isentropic expansion in the Otto cycle) with n equal to 1.50.

11.193E In the Otto cycle, all the heat transfer q_H occurs at constant volume. It is more realistic to assume that part of q_H occurs after the piston has started its downwards motion in the expansion stroke. Therefore, consider a cycle identical to the Otto cycle, except that the first two-thirds of the total q_H occurs at constant volume and the last one-third occurs at constant pressure. Assume the total q_H is 700 Btu/lbm, that the state at the beginning of the compression process is 13 lbf/in.^{2}, 68 F, and that the compression ratio is 9. Calculate the maximum pressure and temperature and the thermal effi-

ciency of this cycle. Compare the results with those of a conventional Otto cycle having the same given variables.

11.194E A diesel engine has a bore of 4 in., a stroke of 4.3 in., and a compression ratio of 19:1 running at 2000 RPM (revolutions per minute). Each cycle takes two revolutions and has a mean effective pressure of 200 lbf/in.^{2}. With a total of six cylinders find the engine power in Btu/s and horsepower, hp.

11.195E At the beginning of compression in a diesel cycle $T = 540$ R, $P = 30$ lbf/in.^{2}, and the state after combustion (heat addition) is 2600 R and 1000 lbf/in.^{2}. Find the compression ratio, the thermal efficiency, and the mean effective pressure.

11.196E Consider an ideal air-standard diesel cycle where the state before the compression process is 14 lbf/in.^{2}, 63 F, and the compression ratio is 20. Find the maximum temperature (by iteration) in the cycle to have a thermal efficiency of 60%.

11.197E Consider an ideal Stirling-cycle engine in which the pressure and temperature at the beginning of the isothermal compression process are 14.7 lbf/in.^{2}, 80 F, the compression ratio is 6, and the maximum temperature in the cycle is 2000 F. Calculate the maximum pressure in the cycle and the thermal efficiency of the cycle with and without regenerators.

11.198E An ideal air-standard Stirling cycle uses helium as working fluid. The isothermal compression brings the helium from 15 lbf/in.^{2}, 70 F to 90 lbf/in.^{2}. The expansion takes place at 2100 R, and there is no regenerator. Find the work and heat transfer in all four processes per lbm helium and the cycle efficiency.

11.199E The air-standard Carnot cycle was not shown in the text; show the T–s diagram for this cycle. In an air-standard Carnot cycle, the low temperature is 500 R, and the efficiency is 60%. If the pressure before compression and after heat rejection is 14.7 lbf/in.^{2}, find the high temperature and the pressure just before heat addition.

11.200E Air in a piston/cylinder goes through a Carnot cycle in which $T_L = 80.3$ F and the total cycle efficiency is $\eta = 2/3$. Find T_H, the specific work and volume ratio in the adiabatic expansion for constant C_p, C_v.

11.201E Do the previous Problem, 11.200E, using Table F.5.

Refrigeration Cycles

11.202E A car air-conditioner (refrigerator) in 70 F ambient uses R-134a, and I want to have cold air at 20 F produced. What is the minimum high P and the maximum low P it can use?

11.203E Consider an ideal refrigeration cycle that has a condenser temperature of 110 F and an evaporator temperature of 5 F. Determine the coefficient of performance of this refrigerator for the working fluids R-12 and R-22.

11.204E The environmentally safe refrigerant R-134a is one of the replacements for R-12 in refrigeration systems. Repeat Problem 11.203 using R-134a and compare the result with that for R-12.

11.205E Consider an ideal heat pump that has a condenser temperature of 120 F and an evaporator temperature of 30 F. Determine the coefficient of performance of this heat pump for the working fluids R-12, R-22, and ammonia.

11.206E The refrigerant R-22 is used as the working fluid in a conventional heat pump cycle. Saturated vapor enters the compressor of this unit at 50 F; its exit temperature from the compressor is measured and found to be 185 F. If the compressor exit is 300 psia, what is the isentropic efficiency of the compressor and the coefficient of performance of the heat pump?

11.207E Consider an ideal gas-turbine cycle with two stages of compression and two stages of expansion. The pressure ratio across each compressor stage and each turbine stage is 8 to 1. The pressure at the entrance to the first compressor is 14 lbf/in.², the temperature entering each compressor is 70 F, and the temperature entering each turbine is 2000 F. An ideal regenerator is also incorporated into the cycle. Determine the compressor work, the turbine work, and the thermal efficiency of the cycle.

Availability and Combined Cycles

11.208E Find the flows and fluxes of exergy in the condenser of Problem 11.172E. Use those to determine the second-law efficiency.

11.209E (Adv.) Find the availability of the water at all four states in the Rankine cycle described in Problem 11.173E. Assume the high-temperature source is 900 F and the low-temperature reservoir is at 65 F. Determine the flow of availability in or out of the reservoirs per pound-mass of steam flowing in the cycle. What is the overall cycle second-law efficiency?

11.210E Find the flows of exergy into and out of the feedwater heater in Problem 11.176E.

11.211E Consider the Brayton cycle in Problem 11.183E. Find all the flows and fluxes of exergy and find the overall cycle second-law efficiency. Assume the heat transfers are internally reversible processes and we then neglect any external irreversibility.

11.212E Consider an ideal dual-loop heat-powered refrigeration cycle using R-12 as the working fluid, as shown in Fig. P.11.144. Saturated vapor at 220 F leaves the boiler and expands in the turbine to the condenser pressure. Saturated vapor at 0 F leaves the evaporator and is compressed to the condenser pressure. The ratio of the flows through the two loops is such that the turbine produces just enough power to drive the compressor. The two exiting streams mix together and enter the condenser. Saturated liquid leaving the condenser at 110 F is then separated into two streams in the necessary proportions. Determine the ratio of mass flow rate through the power loop to that through the refrigeration loop. Find also the performance of the cycle, in terms of the ratio Q_L/Q_H.

Review Problems

11.213E Consider an ideal combined reheat and regenerative cycle in which steam enters the high-pressure turbine at 500 lbf/in.², 700 F, and is extracted to an open feedwater heater at 120 lbf/in.² with exit as saturated liquid. The remainder of the steam is reheated to 700 F at this pressure, 120 lbf/in.^{2}, and is fed to the low-pressure turbine. The condenser pressure is 2 lbf/in.^{2}. Calculate the thermal efficiency of the cycle and the net work per pound-mass of steam.

11.214E In one type of nuclear power plant, heat is transferred in the nuclear reactor to liquid sodium. The liquid sodium is then pumped through a heat exchanger where heat is transferred to boiling water. Saturated vapor steam at 700 lbf/in.^{2} exits this heat exchanger and is then superheated to 1100 F in an external gas-fired superheater. The steam enters the turbine, which has one (open-type) feedwater extraction at 60 lbf/in.^{2}. The isentropic turbine efficiency is 87%, and the condenser pressure is 1 lbf/in.^{2}. Determine the heat transfer in the reactor and in the superheater to produce a net power output of 1000 Btu/s.

11.215E Consider an ideal gas-turbine cycle with two stages of compression and two stages of expansion. The pressure ratio across each com-

pressor stage and each turbine stage is 8 to 1. The pressure at the entrance to the first compressor is 14 lbf/in.^{2}, the temperature entering each compressor is 70 F, and the temperature entering each turbine is 2000 F. An ideal regenerator is also incorporated into the cycle. Determine the compressor work, the turbine work, and the thermal efficiency of the cycle.

11.216E Repeat Problem 11.215E, but assume that each compressor stage and each turbine stage has an isentropic efficiency of 85%. Also assume that the regenerator has an efficiency of 70%.

11.217E Consider a small ammonia absorption refrigeration cycle that is powered by solar energy and is to be used as an air conditioner. Saturated vapor ammonia leaves the generator at 120 F, and saturated vapor leaves the evaporator at 50 F. If 3000 Btu of heat is required in the generator (solar collector) per pound-mass of ammonia vapor generated, determine the overall performance of this system.

COMPUTER, DESIGN, AND OPEN-ENDED PROBLEMS

11.218 The effect of turbine exhaust pressure on the performance of the ideal steam Rankine cycle given in Problem 11.33 is to be studied. Calculate the thermal efficiency of the cycle and the moisture content of the steam leaving the turbine for turbine exhaust pressures of 5, 10, 50, and 100 kPa. Plot the thermal efficiency versus turbine exhaust pressure for the specified turbine inlet pressure and temperature.

11.219 The effect of turbine inlet pressure on the performance of the ideal steam Rankine cycle given in Problem 11.33 is to be studied. Calculate the thermal efficiency of the cycle and the moisture content of the steam leaving the turbine for turbine inlet pressures of 1, 3.5, 6, and 10 MPa. Plot the thermal efficiency versus turbine inlet pressure for the specified turbine inlet temperature and exhaust pressure.

11.220 The effect of turbine inlet temperature on the performance of the ideal steam Rankine cycle given in Problem 11.33 is to be studied. Calculate the thermal efficiency of the cycle and the

moisture content of the steam leaving the turbine for turbine inlet temperatures of 400°, 500°, 800°C, and saturated vapor (at 3.0 MPa). Plot the thermal efficiency versus turbine inlet temperature for the specified turbine inlet pressure and exhaust pressure.

11.221 Write a program to solve the following problem. The effects of varying parameters on the performance of an air-standard Brayton cycle are to be determined. Consider a compressor inlet condition of 100 kPa, 20°C, and assume constant specific heat. The thermal efficiency of the cycle and the net specific work output should be determined for the combinations of the following variables.

a. Compressor pressure ratio of 6, 9, 12, and 15.

b. Maximum cycle temperature of 900, 1100, 1300, and 1500°C.

c. Compressor and turbine isentropic efficiencies each 100, 90, 80, and 70%.

11.222 The effect of adding a regenerator to the gas-turbine cycle in the previous two problems is to be studied. Repeat one of these problems by including a regenerator with various values of the regenerator efficiency.

11.223 Write a program to simulate the Otto cycle using nitrogen as the working fluid. Use the variable specific heat as given in Table A.6. The beginning of compression has a state of 100 kPa, 20°C. Determine the net specific work output and the cycle thermal efficiency for various combinations of compression ratio and maximum cycle temperature. Compare the result with those found when constant specific heat is assumed.

11.224 A power plant is built to provide district heating of buildings that requires 90°C liquid water at 150 kPa. The district heating water is returned at 50°C, 100 kPa, in a closed loop in an amount such that 20 MW of power is delivered. This hot water is produced from a steam power cycle with a boiler making steam at 5 MPa, 600°C, delivered to the steam turbine. The steam cycle could have its condenser operate at 90°C providing the power to the district heating. It could also be done with extraction of steam from the turbine. Suggest a system and evaluate its performance in terms of the cogenerated amount of turbine work.

11.225 The effect of evaporator temperature on the coefficient of performance of a heat pump is to be studied. Consider an ideal cycle with R-22 as the working fluid and a condenser temperature of 40°C. Plot a curve for the coefficient of performance versus the evaporator temperature for temperatures from +15 to −25°C.

11.226 A hospital requires 2 kg/s steam at 200°C, 125 kPa, for sterilization purposes, and space heating requires 15 kg/s hot water at 90°C, 100 kPa. Both of these requirements are provided by the hospital's steam power plant. Discuss some arrangement that will accomplish this.

11.227 Investigate the maximum power out of a steam power plant with operating conditions as in Problem 11.33. The energy source is 100 kg/s combustion products (air) at 125 kPa, 1200 K. Make sure the air temperature is higher than the water temperature throughout the boiler.

11.228 In Problem 11.141, a steam cycle was powered by the exhaust from a gas turbine. With a single water flow and airflow heat exchanger, the air is leaving with a relatively high temperature. Analyze how some more of the energy in the air can be used before the air is flowing out to the chimney. Can it be used in a feedwater heater?

GAS MIXTURES 12

Up to this point in our development of thermodynamics, we have considered primarily pure substances. A large number of thermodynamic problems involve mixtures of different pure substances. Sometimes these mixtures are referred to as solutions, particularly in the liquid and solid phases.

In this chapter we shall turn our attention to various thermodynamic considerations of gas mixtures. We begin with a consideration of a rather simple problem: mixtures of ideal gases. This leads to a consideration of a simplified but very useful model of certain mixtures, such as air and water vapor, which may involve a condensed (solid or liquid) phase of one of the components.

12.1 GENERAL CONSIDERATIONS AND MIXTURES OF IDEAL GASES

Let us consider a general mixture of N components, each a pure substance, so the total mass and the total number of moles are

$$m_{tot} = m_1 + m_2 + \cdots + m_N = \sum m_i$$

$$n_{tot} = n_1 + n_2 + \cdots + n_N = \sum n_i$$

The mixture is usually described by a mass fraction (concentration)

$$c_i = \frac{m_i}{m_{tot}} \tag{12.1}$$

or a mole fraction for each component as

$$y_i = \frac{n_i}{n_{tot}} \tag{12.2}$$

which are related through the molecular weight, M_i, as $m_i = n_i M_i$. We may then convert from a mole basis to a mass basis as

$$c_i = \frac{m_i}{m_{tot}} = \frac{n_i M_i}{\sum n_j M_j} = \frac{n_i M_i / n_{tot}}{\sum n_j M_j / n_{tot}} = \frac{y_i M_i}{\sum y_j M_j} \tag{12.3}$$

and from a mass basis to a mole basis as

$$y_i = \frac{n_i}{n_{tot}} = \frac{m_i / M_i}{\sum m_j / M_j} = \frac{m_i / (M_i m_{tot})}{\sum m_j / (M_j m_{tot})} = \frac{c_i / M_i}{\sum c_j / M_j} \tag{12.4}$$

473

The molecular weight for the mixture becomes

$$M_{\text{mix}} = \frac{m_{\text{tot}}}{n_{\text{tot}}} = \frac{\sum n_i M_i}{n_{\text{tot}}} = \sum y_i M_i \qquad (12.5)$$

that is also the denominator in Eq. 12.3.

EXAMPLE 12.1   A mole-basis analysis of a gaseous mixture yields the following results:

CO$_2$ 12.0%
O$_2$ 4.0
N$_2$ 82.0
CO   2.0

Determine the analysis on a mass basis and the molecular weight for the mixture.

Control mass: Gas mixture.
State: Composition known.

TABLE 12.1

Constituent	Percent by Mole	Mole Fraction	Molecular Weight	Mass kg per kmol of Mixture	Analysis on Mass Basis, Percent
CO$_2$	12	0.12	× 44.0	= 5.28	$\frac{5.28}{30.08}$ = 17.55
O$_2$	4	0.04	× 32.0	= 1.28	$\frac{1.28}{30.08}$ = 4.26
N$_2$	82	0.82	× 28.0	= 22.96	$\frac{22.96}{30.08}$ = 76.33
CO	2	0.02	× 28.0	= $\frac{0.56}{30.08}$	$\frac{0.56}{30.08}$ = $\frac{1.86}{100.00}$

TABLE 12.2

Constituent	Mass Fraction	Molecular Weight	kmol per kg of Mixture	Mole Fraction	Mole Percent
CO$_2$	0.1755	÷ 44.0	= 0.003 99	0.120	12.0
O$_2$	0.0426	÷ 32.0	= 0.001 33	0.040	4.0
N$_2$	0.7633	÷ 28.0	= 0.027 26	0.820	82.0
CO	0.0186	÷ 28.0	= 0.000 66	0.020	2.0
			0.033 24	1.000	100.0

Solution

It is convenient to set up and solve this problem as shown in Table 12.1. The mass-basis analysis is found using Eq. 12.3, as shown in the table. It is also noted that during this calculation, the molecular weight of the mixture is found to be 30.08.

If the analysis has been given on a mass basis, and the mole fractions or percentages are desired, the procedure shown in Table 12.2 is followed, using Eq. 12.4.

Consider a mixture of two gases (not necessarily ideal gases) such as shown in Fig. 12.1. What properties can we experimentally measure for such a mixture? Certainly we can measure the pressure, temperature, volume, and mass of the mixture. We can also experimentally measure the composition of the mixture, and thus determine the mole and mass fractions.

Suppose that this mixture undergoes a process or a chemical reaction and we wish to perform a thermodynamic analysis of this process or reaction. What type of thermodynamic data would we use in performing such an analysis? One possibility would be to have tables of thermodynamic properties of mixtures. However, the number of different mixtures that is possible, both as regards the substances involved and the relative amounts of each, is such that we would need a library full of tables of thermodynamic properties to handle all possible situations. It would be much simpler if we could determine the thermodynamic properties of a mixture from the properties of the pure components. This is in essence the approach that is used in dealing with ideal gases and certain other simplified models of mixtures.

One exception to this procedure is the case where a particular mixture is encountered very frequently, the most familiar being air. Tables and charts of the thermodynamic properties of air are available. However, even in this case it is necessary to define the composition of the "air" for which the tables are given, because the composition of the atmosphere varies with altitude, with the number of pollutants, and with other variables at a given location. The composition of air on which air tables are usually based is as follows:

Component	% on Mole Basis
Nitrogen	78.10
Oxygen	20.95
Argon	0.92
CO_2 & trace elements	0.03

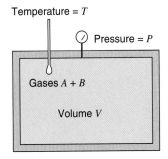

Temperature = T

Pressure = P

Gases A + B

Volume V

FIGURE 12.1 A mixture of two gases.

In this chapter we focus on mixtures of ideal gases. We assume that each component is uninfluenced by the presence of the other components and that each component can be treated as an ideal gas. In the case of a real gaseous mixture at high pressure, this assumption would probably not be accurate because of the nature of the interaction between the molecules of the different components. In this text, we will consider only a single model in analyzing gas mixtures, namely, the Dalton model.

Dalton Model

For the Dalton model of gas mixtures, the properties of each component of the mixture are considered as though each component exists separately and independently at the temperature and volume of the mixture, as shown in Fig. 12.2. We further assume that both the gas mixture and the separated components behave according to the ideal gas model, Eqs. 3.3–3.6. In general, we would prefer to analyze gas mixture behavior on a mass basis. However, in this particular case it is more convenient to use a mole basis, since the gas constant is then the universal gas constant for each component and also for the mixture. Thus, we may write for the mixture (Fig. 12.1)

$$PV = n\overline{R}T$$
$$n = n_A + n_B \tag{12.6}$$

and for the components (Fig. 12.2)

$$P_A V = n_A \overline{R}T$$
$$P_B V = n_B \overline{R}T \tag{12.7}$$

On substituting, we have

$$n = n_A + n_B$$
$$\frac{PV}{\overline{R}T} = \frac{P_A V}{\overline{R}T} + \frac{P_B V}{\overline{R}T} \tag{12.8}$$

or

$$P = P_A + P_B \tag{12.9}$$

where P_A and P_B are referred to as partial pressures. Thus, for a mixture of ideal gases, the pressure is the sum of the partial pressures of the individual components, where, using Eqs. 12.6 and 12.7,

$$P_A = y_A P, \qquad P_B = y_B P \tag{12.10}$$

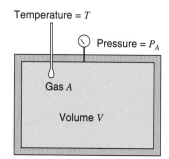
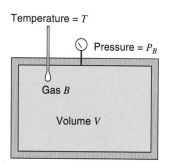

FIGURE 12.2 The Dalton model.

That is, each partial pressure is the product of that component's mole fraction and the mixture pressure.

In determining the internal energy, enthalpy, and entropy of a mixture of ideal gases, the Dalton model proves useful because the assumption is made that each constituent behaves as though it occupies the entire volume by itself. Thus, the internal energy, enthalpy, and entropy can be evaluated as the sum of the respective properties of the constituent gases at the condition at which the component exists in the mixture. Since for ideal gases the internal energy and enthalpy are functions only of temperature, it follows that for a mixture of components A and B, on a mass basis,

$$U = mu = m_A u_A + m_B u_B$$
$$= m(c_A u_A + c_B u_B) \tag{12.11}$$

$$H = mh = m_A h_A + m_B h_B$$
$$= m(c_A h_A + c_B h_B) \tag{12.12}$$

In Eqs. 12.11 and 12.12, the quantities u_A, u_B, h_A, and h_B are the ideal-gas properties of the components at the temperature of the mixture. For a process involving a change of temperature, the changes in these values are evaluated by one of the three models discussed in Section 5.7—involving either the ideal-gas Tables A.7 or the specific heats of the components. In a similar manner to Eqs. 12.11 and 12.12, the mixture energy and enthalpy could be expressed as the sums of the component mole fractions and properties per mole.

The ideal-gas mixture equation of state on a mass basis is

$$PV = mR_{\text{mix}}T \tag{12.13}$$

where

$$R_{\text{mix}} = \frac{1}{m}\left(\frac{PV}{T}\right) = \frac{1}{m}(n\overline{R}) = \overline{R}/M_{\text{mix}} \tag{12.14}$$

Alternatively,

$$R_{\text{mix}} = \frac{1}{m}(n_A \overline{R} + n_B \overline{R})$$
$$= \frac{1}{m}(m_A R_A + m_B R_B)$$
$$= c_A R_A + c_B R_B \tag{12.15}$$

The entropy of an ideal-gas mixture is expressed as

$$S = ms = m_A s_A + m_B s_B$$
$$= m(c_A s_A + c_B s_B) \tag{12.16}$$

It must be emphasized that the component entropies in Eq. 12.16 must each be evaluated at the mixture temperature and the corresponding partial pressure of the component in the mixture, using Eq. 12.10 in terms of the mole fraction.

To evaluate Eq. 12.16 using the ideal-gas entropy expression 8.24, it is necessary to use one of the specific heat models discussed in Section 8.10. The simplest model is constant specific heat, Eq. 8.25, using an arbitrary reference state T_0, P_0, s_0, for each component i in the mixture at T and P,

$$s_i = s_{0i} + C_{p0i}\ln\left(\frac{T}{T_0}\right) - R_i \ln\left(\frac{y_i P}{P_0}\right) \tag{12.17}$$

Consider a process with constant-mixture composition between state 1 and state 2 and let us calculate the entropy change for component i with Eq. 12.17.

$$(s_2 - s_1)_i = s_{0i} - s_{0i} + C_{p0i}\left[\ln\frac{T_2}{T_0} - \ln\frac{T_1}{T_0}\right] - R_i\left[\ln\frac{y_i P_2}{P_0} - \ln\frac{y_i P_1}{P_0}\right]$$

$$= 0 + C_{p0i}\ln\left[\frac{T_2}{T_0} \times \frac{T_0}{T_1}\right] - R_i\ln\left[\frac{y_i P_2}{P_0} \times \frac{P_0}{y_i P_1}\right]$$

$$= C_{p0i}\ln\frac{T_2}{T_1} - R_i\ln\frac{P_2}{P_1}$$

We observe here that this expression is very similar to Eq. 8.25 and that the reference values s_{0i}, T_0, P_0 all cancel out, as does the mole fraction.

An alternative model is to use the s_T^0 function defined in Eq. 8.27, in which case each component entropy in Eq. 12.16 is expressed as

$$s_i = s_{T_i}^0 - R_i\ln\left(\frac{y_i P}{P_0}\right) \tag{12.18}$$

The mixture entropy could also be expressed as the sum of component properties on a mole basis.

EXAMPLE 12.2

Let a mass m_A of ideal gas A at a given pressure and temperature, P and T, be mixed with m_B of ideal gas B at the same P and T, such that the final ideal-gas mixture is also at P and T. Determine the change in entropy for this process.

    Control mass: All gas (A and B).

    Initial states: P, T known for A and B.

    Final state: P, T of mixture known.

Analysis and Solution

The mixture entropy if given by Eq. 12.16. Therefore, the change of entropy can be grouped into changes for A and for B, with each change expressed by Eq. 8.24. Since there is no temperature change for either component, this reduces to

$$\Delta S_{\text{mix}} = m_A\left(0 - R_A\ln\frac{P_A}{P}\right) + m_B\left(0 - R_B\ln\frac{P_B}{P}\right)$$

$$= -m_A R_A\ln y_A - m_B R_B\ln y_B$$

which can also be written in the form

$$\Delta S_{\text{mix}} = -n_A\overline{R}\ln y_A - n_B\overline{R}\ln y_B$$

The result of Example 12.2 can readily be generalized to account for the mixing of any number of components at the same temperature and pressure. The result is

$$\Delta S_{\text{mix}} = -\overline{R}\sum_k n_k\ln y_k \tag{12.19}$$

The interesting thing about this equation is that the increase in entropy depends only on the number of moles of component gases and is independent of the composition of the gas. For example, when 1 mol of oxygen and 1 mol of nitrogen are mixed, the increase in entropy is the same as when 1 mol of hydrogen and 1 mol of nitrogen are mixed. But we also know that if 1 mol of nitrogen is "mixed" with another mole of nitrogen, there is no increase in entropy. The question that arises is, how dissimilar must the gases be in order to have an increase in entropy? The answer lies in our ability to distinguish between the two gases (based on their different molecular weights). The entropy increases whenever we can distinguish between the gases being mixed. When we cannot distinguish between the gases, there is no increase in entropy.

One special case that arises frequently involves an ideal-gas mixture undergoing a process in which there is no change in composition. Let us also assume that the constant specific heat model is reasonable. For this case, from Eq. 12.11 on a unit mass basis, the internal energy change is

$$u_2 - u_1 = c_A C_{v0\,A}(T_2 - T_1) + c_B C_{v0\,B}(T_2 - T_1)$$

$$= C_{v0\,\mathrm{mix}}(T_2 - T_1) \tag{12.20}$$

where

$$C_{v0\,\mathrm{mix}} = c_A C_{v0\,A} + c_B C_{v0B} \tag{12.21}$$

Similarly, from Eq. 12.12, the enthalpy change is

$$h_2 - h_1 = c_A C_{p0\,A}(T_2 - T_1) + c_B C_{p0\,B}(T_2 - T_1)$$

$$= C_{p0\,\mathrm{mix}}(T_2 - T_1) \tag{12.22}$$

where

$$C_{p0\,\mathrm{mix}} = c_A C_{p0\,A} + c_B C_{p0\,B} \tag{12.23}$$

The entropy change for a single component was calculated from Eq. 12.17, so we substitute this result into Eq. 12.16 to evaluate the change as

$$s_2 - s_1 = c_A(s_2 - s_1)_A + c_B(s_2 - s_1)_B$$

$$= c_A C_{p0\,A} \ln \frac{T_2}{T_1} - c_A R_A \ln \frac{P_2}{P_1} + c_B C_{p0\,B} \ln \frac{T_2}{T_1} - c_B R_B \ln \frac{P_2}{P_1}$$

$$= C_{p0\,\mathrm{mix}} \ln \frac{T_2}{T_1} - R_{\mathrm{mix}} \ln \frac{P_2}{P_1} \tag{12.24}$$

The last expression was using Eq. 12.15 for the mixture gas constant and Eq. 12.23 for the mixture heat capacity. We see that Eqs. 12.20, 12.22, and 12.24 are the same as those for the pure substance, Eqs. 5.20, 5.29, and 8.25. So we can treat a mixture similarly to a pure substance once the mixture properties are found from the composition and the component properties in Eqs. 12.15, 12.21, and 12.23.

This also implies that all the polytropic processes in a mixture can be treated similarly to the way it is done for a pure substance (recall Sections 8.10 and 8.11). Specifically the isentropic process where s is constant leads to the power relation between temperature and pressure from Eq. 12.24. This is similar to Eq. 8.29, provided we use the mixture heat capacity and gas constant. The ratio of specific heats becomes

$$k = k_{\mathrm{mix}} = \frac{C_{p\,\mathrm{mix}}}{C_{v\,\mathrm{mix}}} = \frac{C_{p\,\mathrm{mix}}}{C_{p\,\mathrm{mix}} - R_{\mathrm{mix}}}$$

and the relation can then also be written as in Eq. 8.32.

So far we have looked at mixtures of ideal gases as a natural extension to the description of processes involving pure substances. The treatment of mixtures for nonideal (real) gases and multiphase states is important for many technical applications, for instance, in the chemical process industry. It does require a more extensive study of the properties and general equations of state, so we will defer this subject to Chapter 13.

12.2 A SIMPLIFIED MODEL OF A MIXTURE INVOLVING GASES AND A VAPOR

Let us now consider a simplification, which is often a reasonable one, of the problem involving a mixture of ideal gases that is in contact with a solid or liquid phase of one of the components. The most familiar example is a mixture of air and water vapor in contact with liquid water or ice, such as is encountered in air conditioning or in drying. We are all familiar with the condensation of water from the atmosphere when it cools on a summer day.

This problem and a number of similar problems can be analyzed quite simply and with considerable accuracy if the following assumptions are made:

1. The solid or liquid phase contains no dissolved gases.
2. The gaseous phase can be treated as a mixture of ideal gases.
3. When the mixture and the condensed phase are at a given pressure and temperature, the equilibrium between the condensed phase and its vapor is not influenced by the presence of the other component. This means that when equilibrium is achieved, the partial pressure of the vapor will be equal to the saturation pressure corresponding to the temperature of the mixture.

Since this approach is used extensively and with considerable accuracy, let us give some attention to the terms that have been defined and the type of problems for which this approach is valid and relevant. In our discussion we will refer to this as a gas–vapor mixture.

The dew point of a gas–vapor mixture is the temperature at which the vapor condenses or solidifies when it is cooled at constant pressure. This is shown on the T–s diagram for the vapor shown in Fig. 12.3. Suppose that the temperature of the gaseous mixture and the partial pressure of the vapor in the mixture are such that the vapor is ini-

FIGURE 12.3
Temperature–entropy diagram to show definition of the dew point.

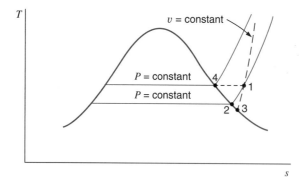

tially superheated at state 1. If the mixture is cooled at constant pressure, the partial pressure of the vapor remains constant until point 2 is reached, and then condensation begins. The temperature at state 2 is the dew-point temperature. Lines 1–3 on the diagram indicate that if the mixture is cooled at constant volume the condensation begins at point 3, which is slightly lower than the dew-point temperature.

If the vapor is at the saturation pressure and temperature, the mixture is referred to as a saturated mixture, and for an air–water vapor mixture, the term saturated air is used.

The relative humidity ϕ is defined as the ratio of the mole fraction of the vapor in the mixture to the mole fraction of vapor in a saturated mixture at the same temperature and total pressure. Since the vapor is considered an ideal gas, the definition reduces to the ratio of the partial pressure of the vapor as it exists in the mixture, P_v, to the saturation pressure of the vapor at the same temperature, P_g:

$$\phi = \frac{P_v}{P_g}$$

In terms of the numbers on the T–s diagram of Fig. 12.3, the relative humidity ϕ would be

$$\phi = \frac{P_1}{P_4}$$

Since we are considering the vapor to be an ideal gas, the relative humidity can also be defined in terms of specific volume or density:

$$\phi = \frac{P_v}{P_g} = \frac{\rho_v}{\rho_g} = \frac{v_g}{v_v} \tag{12.25}$$

The humidity ratio ω of an air–water vapor mixture is defined as the ratio of the mass of water vapor m_v to the mass of dry air m_a. The term dry air is used to emphasize that this refers only to air and not to the water vapor. The term specific humidity is used synonymously with humidity ratio.

$$\omega = \frac{m_v}{m_a} \tag{12.26}$$

This definition is identical for any other gas–vapor mixture, and the subscript a refers to the gas, exclusive of the vapor. Since we consider both the vapor and the mixture to be ideal gases, a very useful expression for the humidity ratio in terms of partial pressures and molecular weights can be developed. Writing

$$m_v = \frac{P_v V}{R_v T} = \frac{P_v V M_v}{\overline{R} T}, \qquad m_a = \frac{P_a V}{R_a T} = \frac{P_a V M_a}{\overline{R} T}$$

we have

$$\omega = \frac{P_v V / R_v T}{P_a V / R_a T} = \frac{R_a P_v}{R_v P_a} = \frac{M_v P_v}{M_a P_a} \tag{12.27}$$

For an air–water vapor mixture, this reduces to

$$\omega = 0.622 \frac{P_v}{P_a} \tag{12.28}$$

The degree of saturation is defined as the ratio of the actual humidity ratio to the humidity ratio of a saturated mixture at the same temperature and total pressure.

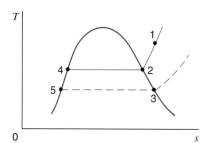

FIGURE 12.4
Temperature–entropy
diagram to show the
cooling of a gas–vapor
mixture at a constant
pressure.

An expression for the relation between the relative humidity ϕ and the humidity ratio ω can be found by solving Eqs. 12.25 and 12.28 for P_v and equating them. The resulting relation for an air–water vapor mixture is

$$\phi = \frac{\omega P_a}{0.622 \, P_g} \tag{12.29}$$

A few words should also be said about the nature of the process that occurs when a gas—vapor mixture is cooled at constant pressure. Suppose that the vapor is initially superheated at state 1 in Fig. 12.4. As the mixture is cooled at constant pressure, the partial pressure of the vapor remains constant until the dew point is reached at point 2, where the vapor in the mixture is saturated. The initial condensate is at state 4 and is in equilibrium with the vapor at state 2. As the temperature is lowered further, more of the vapor condenses, which lowers the partial pressure of the vapor in the mixture. The vapor that remains in the mixture is always saturated, and the liquid or solid is in equilibrium with it. For example, when the temperature is reduced to T_3, the vapor in the mixture is at state 3, and its partial pressure is the saturation pressure corresponding to T_3. The liquid in equilibrium with it is at state 5.

EXAMPLE 12.3

Consider 100 m³ of an air–water vapor mixture at 0.1 MPa, 35°C, and 70% relative humidity. Calculate the humidity ratio, dew point, mass of air, and mass of vapor.

Control mass: Mixture.
State: P, T, ϕ known; state fixed.

Analysis and Solution

From Eq. 12.25 and the steam tables, we have

$$\phi = 0.70 = \frac{P_v}{P_g}$$

$$P_v = 0.70(5.628) = 3.94 \text{ kPa}$$

The dew point is the saturation temperature corresponding to this pressure, which is 28.6°C.
The partial pressure of the air is

$$P_a = P - P_v = 100 - 3.94 = 96.06 \text{ kPa}$$

The humidity ratio can be calculated from Eq. 12.28:

$$\omega = 0.622 \times \frac{P_v}{P_a} = 0.622 \times \frac{3.94}{96.06} = 0.0255$$

The mass of air is

$$m_a = \frac{P_a V}{R_a T} = \frac{96.06 \times 100}{0.287 \times 308.2} = 108.6 \text{ kg}$$

The mass of the vapor can be calculated by using the humidity ratio or by using the ideal-gas equation of state:

$$m_v = \omega m_a = 0.0255(108.6) = 2.77 \text{ kg}$$

$$m_v = \frac{3.94 \times 100}{0.4615 \times 308.2} = 2.77 \text{ kg}$$

EXAMPLE 12.3E   Consider 2000 ft^3 of an air–water vapor mixture at 14.7 lbf/in.^{2}, 90 F, 70% relative humidity. Calculate the humidity ratio, dew point, mass of air, and mass of vapor.

> *Control mass:* Mixture.
> *State:* P, T, ϕ known; state fixed.

Analysis and Solution

From Eq. 12.25 and the steam tables,

$$\phi = 0.70 = \frac{P_v}{P_g}$$

$$P_v = 0.70(0.6988) = 0.4892 \text{ lbf/in}^2$$

The dew point is the saturation temperature corresponding to this pressure, which is 78.9 F.

The partial pressure of the air is

$$P_a = P - P_v = 14.70 - 0.49 = 14.21 \text{ lbf/in}^2$$

The humidity ratio can be calculated from Eq. 12.28.

$$\omega = 0.622 \times \frac{P_v}{P_a} = 0.622 \times \frac{0.4892}{14.21} = 0.02135$$

The mass of air is

$$m_a = \frac{P_a V}{R_a T} = \frac{14.21 \times 144 \times 2000}{53.34 \times 550} = 139.6 \text{ lbm}$$

The mass of the vapor can be calculated by using the humidity ratio or by using the ideal-gas equation of state.

$$m_v = \omega m_a = 0.02135(139.6) = 2.98 \text{ lbm}$$

$$m_v = \frac{0.4892 \times 144 \times 2000}{85.7 \times 550} = 2.98 \text{ lbm}$$

EXAMPLE 12.4 Calculate the amount of water vapor condensed if the mixture of Example 12.3 is cooled to 5°C in a constant-pressure process.

> *Control mass:* Mixture.
> *Initial state:* Known (Example 12.3).
> *Final state:* T known.
> *Process:* Constant pressure.

Analysis

At the final temperature, 5°C, the mixture is saturated, since this is below the dew-point temperature. Therefore,

$$P_{v2} = P_{g2}, \qquad P_{a2} = P - P_{v2}$$

and

$$\omega_2 = 0.622\,\frac{P_{v2}}{P_{a2}}$$

From the conservation of mass, it follows that the amount of water condensed is equal to the difference between the initial and final mass of water vapor, or

$$\text{Mass of vapor condensed} = m_a(\omega_1 - \omega_2)$$

Solution

We have

$$P_{v2} = P_{g2} = 0.8721 \text{ kPa}$$
$$P_{a2} = 100 - 0.8721 = 99.128 \text{ kPa}$$

Therefore,

$$\omega_2 = 0.622 \times \frac{0.8721}{99.128} = 0.0055$$

$$\text{Mass of vapor condensed} = m_a(\omega_1 - \omega_2) = 108.6(0.0255 - 0.0055)$$
$$= 2.172 \text{ kg}$$

EXAMPLE 12.4E Calculate the amount of water vapor condensed if the mixture of Example 12.3E is cooled to 40 F in a constant-pressure process.

> *Control mass:* Mixture.
> *Initial state:* Known (Example 12.3E).
> *Final state:* T known.
> *Process:* Constant pressure.

Analysis

At the final temperature, 40 F, the mixture is saturated, since this is below the dew-point temperature. Therefore,

$$P_{v2} = P_{g2}, \qquad P_{a2} = P - P_{v2}$$

and

$$\omega_2 = 0.622 \frac{P_{v2}}{P_{a2}}$$

From the conservation of mass, it follows that the amount of water condensed is equal to the difference between the initial and final mass of water vapor, or

$$\text{Mass of vapor condensed} = m_a(\omega_1 - \omega_2)$$

Solution

We have

$$P_{v2} = P_{g2} = 0.1217 \text{ lbf/in.}^2$$

$$P_{a2} = 14.7 - 0.12 = 14.58 \text{ lbf/in.}^2$$

Therefore,

$$\omega_2 = 0.622 \times \frac{0.1217}{14.58} = 0.00520$$

$$\text{Mass of vapor condensed} = m_a(\omega_1 - \omega_2) = 139.6(0.02135 - 0.0052)$$

$$= 2.25 \text{ lbm}$$

12.3 THE FIRST LAW APPLIED TO GAS–VAPOR MIXTURES

In applying the first law of thermodynamics to gas–vapor mixtures, it is helpful to realize that because of our assumption that ideal gases are involved, the various components can be treated separately when calculating changes of internal energy and enthalpy. Therefore, in dealing with air–water vapor mixtures, the changes in enthalpy of the water vapor can be found from the steam tables and the ideal-gas relations can be applied to the air. This is illustrated by the examples that follow.

EXAMPLE 12.5   An air-conditioning unit is shown in Fig. 12.5, with pressure, temperature, and relative humidity data. Calculate the heat transfer per kilogram of dry air, assuming that changes in kinetic energy are negligible.

    Control volume: Duct, excluding cooling coils.
        Inlet state: Known (Fig. 12.5).

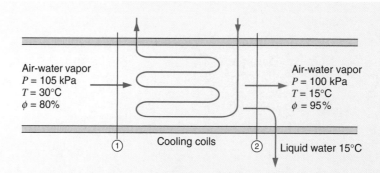

FIGURE 12.5 Sketch for Example 12.5.

Exit state: Known (Fig. 12.5).

Process: Steady state with no kinetic or potential energy changes.

Model: Air—ideal gas, constant specific heat, value at 300 K. Water—steam tables. (Since the water vapor at these low pressures is being considered an ideal gas, the enthalpy of the water vapor is a function of the temperature only. Therefore the enthalpy of slightly superheated water vapor is equal to the enthalpy of saturated vapor at the same temperature.)

Analysis

From the continuity equations for air and water, we have

$$\dot{m}_{a1} = \dot{m}_{a2}$$

$$\dot{m}_{v1} = \dot{m}_{v2} + \dot{m}_{l2}$$

The first law gives

$$\dot{Q}_{c.v.} + \sum \dot{m}_i h_i = \sum \dot{m}_e h_e$$

$$\dot{Q}_{c.v.} + \dot{m}_a h_{a1} + \dot{m}_{v1} h_{v1} = \dot{m}_a h_{a2} + \dot{m}_{v2} h_{v2} + \dot{m}_{l2} h_{l2}$$

If we divide this equation by $\dot{m}_a$, introduce the continuity equation for the water, and note that $\dot{m}_v = \omega \dot{m}_a$, we can write the first law in the form

$$\frac{\dot{Q}_{c.v.}}{\dot{m}_a} + h_{a1} + \omega_1 h_{v1} = h_{a2} + \omega_2 h_{v2} + (\omega_1 - \omega_2) h_{l2}$$

Solution

We have

$$P_{v1} = \phi_1 P_{g1} = 0.80(4.246) = 3.397 \text{ kPa}$$

$$\omega_1 = \frac{R_a}{R_v} \frac{P_{v1}}{P_{a1}} = 0.622 \times \left(\frac{3.397}{105 - 3.4}\right) = 0.0208$$

$$P_{v2} = \phi_2 P_{g2} = 0.95(1.7051) = 1.620 \text{ kPa}$$

$$\omega_2 = \frac{R_a}{R_v} \times \frac{P_{v2}}{P_{a2}} = 0.622 \times \left(\frac{1.62}{100 - 1.62}\right) = 0.0102$$

Substituting, we obtain

$$\dot{Q}_{c.v.}/\dot{m}_a + h_{a1} + \omega_1 h_{v1} = h_{a2} + \omega_2 h_{v2} + (\omega_1 - \omega_2)h_{l2}$$

$$\dot{Q}_{c.v.}/\dot{m}_a = 1.004(15 - 30) + 0.0102(2528.9)$$

$$-0.0208(2556.3) + (0.0208 - 0.0102)(62.99)$$

$$= -41.76 \text{ kJ/kg dry air}$$

EXAMPLE 12.6 A tank has a volume of 0.5 m³ and contains nitrogen and water vapor. The temperature of the mixture is 50°C, and the total pressure is 2 MPa. The partial pressure of the water vapor is 5 kPa. Calculate the heat transfer when the contents of the tank are cooled to 10°C.

Control mass: Nitrogen and water.
Initial state: P_1, T_1 known; state fixed.
Final state: T_2 known.
Process: Constant volume.
Model: Ideal-gas mixture; constant specific heat for nitrogen; steam tables for water.

Analysis

This is a constant-volume process. Since the work is zero, the first law reduces to

$$Q = U_2 - U_1 = m_{N_2}C_{v(N_2)}(T_2 - T_1) + (m_2 u_2)_v + (m_2 u_2)_l - (m_1 u_1)_v$$

This equation assumes that some of the vapor condensed. This assumption must be checked, however, as shown in the solution:

Solution

The mass of nitrogen and water vapor can be calculated using the ideal-gas equation of state:

$$m_{N_2} = \frac{P_{N_2}V}{R_{N_2}T} = \frac{1995 \times 0.5}{0.2968 \times 323.2} = 10.39 \text{ kg}$$

$$m_{v1} = \frac{P_{v1}V}{R_v T} = \frac{5 \times 0.5}{0.4615 \times 323.2} = 0.016\,76 \text{ kg}$$

If condensation takes place, the final state of the vapor will be saturated vapor at 10°C. Therefore,

$$m_{v2} = \frac{P_{v2}V}{R_v T} = \frac{1.2276 \times 0.5}{0.4615 \times 283.2} = 0.004\,70 \text{ kg}$$

Since this amount is less than the original mass of vapor, there must have been condensation.
The mass of liquid that is condensed, m_{l2}, is

$$m_{l2} = m_{v1} - m_{v2} = 0.016\,76 - 0.004\,70 = 0.012\,06 \text{ kg}$$

The internal energy of the water vapor is equal to the internal energy of saturated water vapor at the same temperature. Therefore,

$$u_{v_1} = 2443.5 \text{ kJ/kg}$$

$$u_{v_2} = 2389.2 \text{ kJ/kg}$$

$$u_{l2} = 42.0 \text{ kJ/kg}$$

$$Q_{\text{c.v.}} = 10.39 \times 0.745(10 - 50) + 0.0047(2389.2) \\ + 0.012\,06(42.0) - 0.016\,76(2443.5)$$

$$= -338.8 \text{ kJ}$$

12.4 THE ADIABATIC SATURATION PROCESS

An important process for an air–water vapor mixture is the adiabatic saturation process. In this process, a air–vapor mixture comes in contact with a body of water in a well-insulated duct (Fig. 12.6). If the initial humidity is less than 100%, some of the water will evaporate and the temperature of the air–vapor mixture will decrease. If the mixture leaving the duct is saturated and if the process is adiabatic, the temperature of the mixture on leaving is known as the adiabatic saturation temperature. For this to take place as a steady-state process, makeup water at the adiabatic saturation temperature is added at the same rate at which water is evaporated. The pressure is assumed to be constant.

Considering the adiabatic saturation process to be a steady-state process, and neglecting changes in kinetic and potential energy, the first law reduces to

$$h_{a1} + \omega_1 h_{v1} + (\omega_2 - \omega_1)h_{l2} = h_{a2} + \omega_2 h_{v2}$$

$$\omega_1(h_{v1} - h_{l2}) = C_{pa}(T_2 - T_1) + \omega_2(h_{v2} - h_{l2})$$

$$\omega_1(h_{v1} - h_{l2}) = C_{pa}(T_2 - T_1) + \omega_2 h_{fg2} \tag{12.30}$$

The most significant point to be made about the adiabatic saturation process is that the adiabatic saturation temperature, the temperature of the mixture when it leaves the duct, is a function of the pressure, temperature, and relative humidity of the entering air–vapor mixture and of the exit pressure. Thus, the relative humidity and the humidity ratio of the entering air–vapor mixture can be determined from the measurements of the pressure and temperature of the air–vapor mixture entering and leaving the adiabatic saturator. Since these measurements are relatively easy to make, this is one means of determining the humidity of an air–vapor mixture.

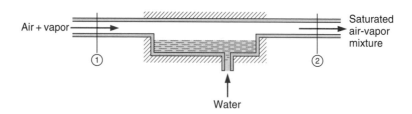

FIGURE 12.6 The adiabatic saturation process.

EXAMPLE 12.7 The pressure of the mixture entering and leaving the adiabatic saturator is 0.1 MPa, the entering temperature is 30°C, and the temperature leaving is 20°C, which is the adiabatic saturation temperature. Calculate the humidity ratio and relative humidity of the air–water vapor mixture entering.

Control volume:	Adiabatic saturator.
Inlet state:	P_1, T_1 known.
Exit state:	P_2, T_2 known; $\phi_2 = 100\%$; state fixed.
Process:	Steady state, adiabatic saturation (Fig. 12.6).
Model:	Ideal-gas mixture; constant specific heat for air; steam tables for water.

Analysis

Use continuity and the first law, Eq. 12.30.

Solution

Since the water vapor leaving is saturated, $P_{v2} = P_{g2}$ and ω_2 can be calculated.

$$\omega_2 = 0.622 \times \left(\frac{2.339}{100 - 2.34}\right) = 0.0149$$

ω_1 can be calculated using Eq. 12.30.

$$\omega_1 = \frac{C_{pa}(T_2 - T_1) + \omega_2 h_{fg2}}{(h_{v1} - h_{l2})}$$

$$\omega_1 = \frac{1.004(20 - 30) + 0.0149 \times 2454.1}{2556.3 - 83.96} = 0.0107$$

$$\omega_1 = 0.0107 = 0.622 \times \left(\frac{P_{v1}}{100 - P_{v1}}\right)$$

$$P_{v1} = 1.691 \text{ kPa}$$

$$\phi_1 = \frac{P_{v1}}{P_{g1}} = \frac{1.691}{4.246} = 0.398$$

EXAMPLE 12.7E The pressure of the mixture entering and leaving the adiabatic saturator is 14.7 lbf/in.², the entering temperature is 84 F, and the temperature leaving is 70 F, which is the adiabatic saturation temperature. Calculate the humidity ratio and relative humidity of the air–water vapor mixture entering.

Control volume:	Adiabatic saturator.
Inlet state::	P_1, T_1 known.
Exit state:	P_2, T_2 known; $\phi_2 = 100\%$; state fixed.
Process:	Steady state, adiabatic saturation (Fig. 12.6).
Model:	Ideal-gas mixture; constant specific heat for air; steam tables for water.

Analysis

Use continuity and the first law, Eq. 12.30.

Solution

Since the water vapor leaving is saturated, $P_{v2} = P_{g2}$ and ω_2 can be calculated.

$$\omega_2 = 0.622 \times \frac{0.3632}{14.7 - 0.36} = 0.01573$$

ω_1 can be calculated using Eq. 12.30.

$$\omega_1 = \frac{C_{pa}(T_2 - T_1) + \omega_2 h_{fg2}}{(h_{v1} - h_{l2})}$$

$$\omega_1 = \frac{0.24(70 - 84) + 0.01573 \times 1054.0}{1098.1 - 38.1} = \frac{-3.36 + 16.60}{1060.0} = 0.0125$$

$$\omega_1 = 0.622 \times \left(\frac{P_{v1}}{14.7 - P_{v1}} \right) = 0.0125$$

$$P_{v1} = 0.289$$

$$\phi_1 = \frac{P_{v1}}{P_{g1}} = \frac{0.289}{0.584} = 0.495$$

12.5 WET-BULB AND DRY-BULB TEMPERATURES

The humidity of air–water vapor mixtures has traditionally been measured with a device called a psychrometer, which uses the flow of air past wet-bulb and dry-bulb thermometers. The bulb of the wet-bulb thermometer is covered with a cotton wick that is saturated with water. The dry-bulb thermometer is used simply to measure the temperature of the air. The airflow can be maintained by a fan, as shown in the continuous-flow psychrometer depicted in Fig. 12.7.

The processes that take place at the wet-bulb thermometer are somewhat complicated. First, if the air–water vapor mixture is not saturated, some of the water in the wick evaporates and diffuses into the surrounding air, which cools the water in the wick. As soon as the temperature of the water drops, however, heat is transferred to the water from both the air and the thermometer, with corresponding cooling. A steady state, determined by heat and mass transfer rates, will be reached, in which the wet-bulb thermometer temperature is lower than the dry-bulb temperature.

It can be argued that this evaporative cooling process is very similar, but not identical, to the adiabatic saturation process described and analyzed in Section 12.4. In fact, the adiabatic saturation temperature is often termed the thermodynamic wet-bulb temperature. It is clear, however, that the wet-bulb temperature as measured by a psychrometer is influenced by heat and mass transfer rates, which depend, for example, on the airflow velocity and not simply on thermodynamic equilibrium properties. It does happen that the

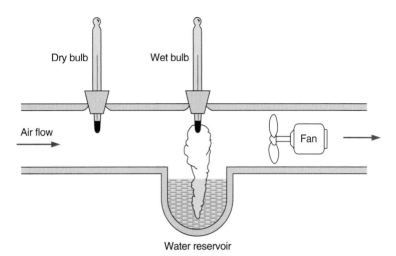

FIGURE 12.7 Steady-flow apparatus for measuring wet- and dry-bulb temperatures.

Dry bulb   Wet bulb

Air flow

Fan

Water reservoir

two temperatures are very close for air–water vapor mixtures at atmospheric temperature and pressure, and they will be assumed to be equivalent in this text.

In recent years, humidity measurements have been made using other phenomena and other devices, primarily electronic devices for convenience and simplicity. For example, some substances tend to change in length, in shape, or in electrical capacitance, or in a number of other ways, when they absorb moisture. They are therefore sensitive to the amount of moisture in the atmosphere. An instrument making use of such a substance can be calibrated to measure the humidity of air–water vapor mixtures. The instrument output can be programmed to furnish any of the desired parameters, such as relative humidity, humidity ratio, or wet-bulb temperature.

12.6 THE PSYCHROMETRIC CHART

Properties of air–water vapor mixtures are given in graphical form on psychrometric charts. These are available in a number of different forms, and only the main features are considered here. It should be recalled that three independent properties—such as pressure, temperature, and mixture composition—will describe the state of this binary mixture.

A simplified version of the chart included in Appendix E, Fig. E.4, is shown in Fig. 12.8. This basic psychrometric chart is a plot of humidity ratio (ordinate) as a function of dry-bulb temperature (abscissa), with relative humidity, wet-bulb temperature, and mixture enthalpy per mass of dry air as parameters. If we fix the total pressure for which the chart is to be constructed (which in our chart is 1 bar, or 100 kPa), lines of constant relative humidity and wet-bulb temperature can be drawn on the chart, because for a given dry-bulb temperature, total pressure, and humidity ratio, the relative humidity and wet-bulb temperature are fixed. The partial pressure of the water vapor is fixed by the humidity ratio and the total pressure, and therefore a second ordinate scale that indicates the partial pressure of the water vapor could be constructed. It would also be possible to include the mixture-specific volume and entropy on the chart.

Most psychrometric charts give the enthalpy of an air–vapor mixture per kilogram of dry air. The values given assume that the enthalpy of the dry air is zero at −20°C, and the enthalpy of the vapor is taken from the steam tables (which are based on the assump-

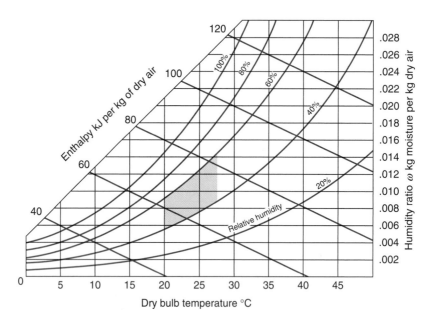

FIGURE 12.8
Psychrometric chart.

tion that the enthalpy of saturated liquid is zero at 0°C). The value used in the psychrometric chart is then

$$\tilde{h} \equiv h_a - h_a(-20°C) + \omega h_v$$

This procedure is satisfactory because we are usually concerned only with differences in enthalpy. That the lines of constant enthalpy are essentially parallel to lines of constant wet-bulb temperature is evident from the fact that the wet-bulb temperature is essentially equal to the adiabatic saturation temperature. Thus, in Fig. 12.6, if we neglect the enthalpy of the liquid entering the adiabatic saturator, the enthalpy of the air–vapor mixture leaving at a given adiabatic saturation temperature fixes the enthalpy of the mixture entering.

The chart plotted in Fig. 12.8 also indicates the human comfort zone, as the range of conditions most agreeable for human well-being. An air conditioner should then be able to maintain an environment within the comfort zone regardless of the outside atmospheric conditions to be considered adequate. Some charts are available that give corrections for

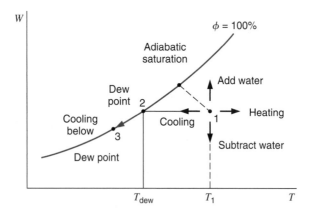

FIGURE 12.9
Processes on a
psychrometric chart.

variation from standard atmospheric pressures. Before using a given chart, one should fully understand the assumptions made in constructing it and should recognize that it is applicable to the particular problem at hand.

The direction in which various processes proceed for an air–water vapor mixture is shown on the psychrometric chart of Fig. 12.9. For example, a constant-pressure cooling process beginning at state 1 proceeds at constant humidity ratio to the dew point at state 2, with continued cooling below that temperature moving along the saturation line (100% relative humidity) to point 3. Other processes could be traced out in a similar manner.

SUMMARY

A mixture of gases is treated from the definition of the mixture composition of the various components based on mass or based on moles. This leads to the mass fractions and mole fractions, both of which can be called concentrations. The mixture has an overall average molecular weight and other mixture properties on a mass or mole basis. Another simple model includes Dalton's model of ideal mixtures of ideal gases, which leads to partial pressures as the contribution from each component to the total pressure given by the mole fraction. As entropy is sensitive to pressure, the mole fraction enters into the entropy generation by mixing. However, for processes other than mixing of different components, we can treat the mixture as we treat a pure substance by using the mixture properties.

Special treatment and nomenclature is used for moist air as a mixture of air and water vapor. The water content is quantified by the relative humidity (how close the water vapor is to a saturated state) or by the humidity ratio (also called absolute humidity). As moist air is cooled down, it eventually reaches the dew point (relative humidity is 100%), where we have saturated moist air. Vaporizing liquid water without external heat transfer gives an adiabatic saturation process also used in a process called evaporative cooling. In an actual apparatus, we can obtain wet-bulb and dry-bulb temperatures, indirectly measuring the humidity of the incoming air. These property relations are shown in a psychrometric chart.

You should have learned a number of skills and acquired abilities from studying this chapter that will allow you to

- Handle the composition of a multicomponent mixture on a mass or mole basis.
- Convert concentrations from a mass to a mole basis and vice versa.
- Compute average properties for the mixture on a mass or mole basis.
- Know partial pressures and how to evaluate them.
- Know how to treat mixture properties such as v, u, h, s, $C_{p\text{mix}}$, R_{mix}.
- Find entropy generation by a mixing process.
- Formulate the general conservation equations for mass, energy, and entropy for the case of a mixture instead of a pure substance.
- Know how to use the simplified formulation of the energy equation using the frozen heat capacities for the mixture.
- Deal with a polytropic process when the substance is a mixture of ideal gases.
- Know the special properties (ϕ, ω) describing humidity in moist air.
- Have a sense of what changes relative humidity and humidity ratio and recognize that you can change one and not the other in a given process.

KEY CONCEPTS AND FORMULAS
COMPOSITION

Mass concentration	$c_i = \dfrac{m_i}{m_{\text{tot}}} = \dfrac{y_i M_j}{\sum y_i M_j}$	
Mole concentration	$y_i = \dfrac{n_i}{n_{\text{tot}}} = \dfrac{c_i/M_j}{\sum c_i/M_j}$	
Molecular weight	$M_{\text{mix}} = \sum y_i M_i$	

Properties

Internal energy	$u_{\text{mix}} = \sum c_i u_i;$	$\bar{u}_{\text{mix}} = \sum y_i \bar{u}_i = u_{\text{mix}} M_{\text{mix}}$
Enthalpy	$h_{\text{mix}} = \sum c_i h_i;$	$\bar{h}_{\text{mix}} = \sum y_i \bar{h}_i = h_{\text{mix}} M_{\text{mix}}$
Gas constant	$R_{\text{mix}} = \bar{R}/M_{\text{mix}} = \sum c_i R_i$	
Heat capacity frozen	$C_{v\,\text{mix}} = \sum c_i C_{v\,i};$	$\bar{C}_{v\,\text{mix}} = \sum y_i \bar{C}_{v\,i}$
	$C_{v\,\text{mix}} = C_{p\,\text{mix}} - R_{\text{mix}};$	$\bar{C}_{v\,\text{mix}} = \bar{C}_{p\,\text{mix}} - \bar{R}$
	$C_{p\,\text{mix}} = \sum c_i C_{p\,i};$	$\bar{C}_{p\,\text{mix}} = \sum y_i \bar{C}_{p\,i}$
Ratio of specific heats	$k_{\text{mix}} = C_{p\,\text{mix}}/C_{v\,\text{mix}}$	
Dalton model	$P_i = y_i P_{\text{tot}}$ &	$V_i = V_{\text{tot}}$
Entropy	$s_{\text{mix}} = \sum c_i s_i;$	$\bar{s}_{\text{mix}} = \sum y_i \bar{s}_i$
Component entropy	$s_i = s_{Ti}^0 - R_i \ln [y_i P/P_0]$	$\bar{s}_i = \bar{s}_{Ti}^0 - \bar{R} \ln [y_i P/P_0]$

Air–Water Mixtures

Relative humidity	$\phi = \dfrac{P_v}{P_g}$
Humidity ratio	$\omega = \dfrac{m_v}{m_a} = 0.622 \dfrac{P_v}{P_a} = 0.622 \dfrac{\phi P_g}{P_{\text{tot}} - \phi P_g}$
Enthalpy per kg dry air	$\tilde{h} = h_a + \omega h_v$

CONCEPT-STUDY GUIDE PROBLEMS

12.1 Are the mass and mole fractions for a mixture ever the same?

12.2 For a mixture how many component concentrations are needed?

12.3 Are any of the properties (P, T, v) for oxygen and nitrogen in air the same?

12.4 If oxygen is 21% by mole of air, what is the oxygen state (P, T, v) in a room at 300 K, 100 kPa, of total volume 60 m³?

12.5 A flow of oxygen and one of nitrogen, both 300 K, are mixed to produce 1 kg/s air at 300 K, 100 kPa. What are the mass and volume flow rates of each line?

12.6 A flow of gas A and a flow of gas B are mixed in a 1:1 mole ratio with same T. What is the entropy generation per kmole flow out?

12.7 A rigid container has 1 kg argon at 300 K and 1 kg argon at 400 K both at 150 kPa. Now they are allowed to mix without any external heat transfer. What is final T, P? Is any s generated?

12.8 A rigid container has 1 kg CO_2 at 300 K and 1 kg argon at 400 K, both at 150 kPa. Now they are allowed to mix without any heat transfer. What is final T, P?

12.9 A flow of 1 kg/s argon at 300 K and another flow of 1 kg/s CO_2 at 1600 K both at 150 kPa are mixed without any heat transfer. What is the exit T, P?

12.10 What is the rate of entropy increase in Problem 12.9?

12.11 If I want to heat a flow of a 4-component mixture from 300 to 310 K at constant P, how many properties and which ones do I need to know to find the heat transfer?

12.12 For a gas mixture in a tank, are the partial pressures important?

12.13 What happens to relative and absolute humidity when moist air is heated?

12.14 I cool moist air; do I reach the dew first in a constant P or constant V process?

12.15 What happens to relative and absolute humidity when moist air is cooled?

12.16 If I have air at 100 kPa and (a) $-10°C$, (b) $45°C$, and (c) $110°C$, what is the maximum absolute humidity I can have?

12.17 Can moist air below the freezing point, say $-5°C$, have a dew point?

12.18 Explain in words what the absolute and relative humidity expresses?

12.19 An adiabatic saturation process changes Φ, ω, and T. In which direction?

12.20 I want to bring air at $35°C$, $\Phi = 40\%$, to a state of $25°C$, $\omega = 0.01$. Do I need to add or subtract water?

HOMEWORK PROBLEMS

Mixture Composition and Properties

12.21 A gas mixture at $120°C$ and 125 kPa is 50% N_2, 30% H_2O, and 20% O_2 on a mole basis. Find the mass fractions, the mixture gas constant, and the volume for 5 kg of mixture.

12.22 A mixture of 60% N_2, 30% argon, and 10% O_2 on a mass basis is in a cylinder at 250 kPa and 310 K with a volume of 0.5 m^3. Find the mole fractions and the mass of argon.

12.23 A mixture of 60% N_2, 30% Ar, and 10% O_2 on a mole basis is in a cylinder at 250 kPa and 310 K with a volume of 0.5 m^3. Find the mass fractions and the mass of argon.

12.24 A new refrigerant R-407 is a mixture of 23% R-32, 25% R-125, and 52% R-134a on a mass basis. Find the mole fractions, the mixture gas constant, and the mixture heat capacities for this new refrigerant.

12.25 A carbureted internal combustion engine is converted to run on methane gas (natural gas). The air–fuel ratio in the cylinder is to be 20 to 1 on a mass basis. How many moles of oxygen per mole of methane are there in the cylinder?

12.26 Weighing of masses gives a mixture at $60°C$, 225 kPa with 0.5 kg O_2, 1.5 kg N_2, and 0.5 kg CH_4. Find the partial pressures of each component, the mixture specific volume (mass basis), mixture molecular weight, and the total volume.

12.27 A 2-kg mixture of 25% N_2, 50% O_2, and 25% CO_2 by mass is at 150 kPa and 300 K. Find the mixture gas constant and the total volume.

12.28 A 100 m^3 storage tank with fuel gases is at $20°C$ and 100 kPa containing a mixture of acetylene C_2H_2, propane C_3H_8, and butane C_4H_{10}. A test shows the partial pressure of the C_2H_2 is 15 kPa and that of C_3H_8 is 65 kPa. How much mass is there of each component?

12.29 A pipe, of cross-sectional area 0.1 m^2, carries a flow of 75% O_2 and 25% N_2 by mole with a velocity of 25 m/s at 200 kPa and 290 K. To install and operate a mass flowmeter, it is necessary to know the mixture density and the gas constant. What are they? What mass flow rate should the meter then show?

12.30 A new refrigerant R-410a is a mixture of R-32 and R-125 in a 1:1 mass ratio. What are the overall molecular weight, the gas constant, and the ratio of specific heats for such a mixture?

Simple Processes

12.31 At a certain point in a coal gasification process, a sample of the gas is taken and stored in a 1 L cylinder. An analysis of the mixture yields the following results:

Component	H_2	CO	CO_2	N_2
Percent by mass	2	45	28	25

Determine the mole fractions and total mass in the cylinder at 100 kPa and 20°C. How much heat must be transferred to heat the sample at constant volume from the initial state to 100°C?

12.32 The mixture in Problem 12.27 is heated to 500 K with constant volume. Find the final pressure and the total heat transfer needed using Table A.5.

12.33 The mixture in problem 12.27 is heated up to 500 K in a constant-pressure process. Find the final volume and the total heat transfer using Table A.5.

12.34 A pipe flows 1.5 kg/s of a mixture with mass fractions of 40% CO_2 and 60% N_2 at 400 kPa and 300 K, shown in Fig. P12.34. Heating tape is wrapped around a section of pipe with insulation added, and 2 kW of electrical power is heating the pipe flow. Find the mixture exit temperature.

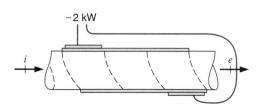

FIGURE P12.34

12.35 A pipe flows 0.05 kmol/s of a mixture with mole fractions of 40% CO_2 and 60% N_2 at 400 kPa, 300 K. Heating tape is wrapped around a section of pipe with insulation added, and 2 kW electrical power is heating the pipe flow. Find the mixture exit temperature.

12.36 A rigid insulated vessel contains 12 kg of oxygen at 200 kPa and 280 K separated by a membrane from 26 kg of carbon dioxide at 400 kPa and 360 K. The membrane is removed, and the mixture comes to a uniform state. Find the final temperature and pressure of the mixture.

12.37 A mixture of 40% water and 60% carbon dioxide by mass is heated from 400 K to 1000 K at a constant pressure of 120 kPa. Find the total change in enthalpy and entropy using Table A.5 values.

12.38 Do Problem 12.37 but with variable heat capacity using values from Table A.8.

12.39 An insulated gas turbine receives a mixture of 10% CO_2, 10% H_2O, and 80% N_2 on a mass basis at 1000 K and 500 kPa. The inlet volume flow rate is 2 m³/s, and the exhaust is at 700 K and 100 kPa. Find the power output in kW using constant specific heat from Table A.5 at 300 K.

12.40 Solve Problem 12.39 using values of enthalpy from Table A.8.

12.41 An insulated gas turbine receives a mixture of 10% CO_2, 10% H_2O, and 80% N_2 on a mole basis at 1000 K, 500 kPa. The inlet volume flow rate is 2 m³/s, and the exhaust is at 700 K, 100 kPa. Find the power output in kW using constant specific heat from A.5 at 300 K.

12.42 Solve Problem 12.41 using values of enthalpy from Table A.9.

12.43 A piston/cylinder device contains 0.1 kg of a mixture of 40% methane and 60% propane by mass at 300 K and 100 kPa. The gas is now slowly compressed in an isothermal (T = constant) process to a final pressure of 250 kPa. Show the process in a P–V diagram and find both the work and heat transfer in the process.

12.44 Consider Problem 12.39 and find the value for the mixture heat capacity, mass basis, and the mixture ratio of specific heats, k_{mix}, both estimated at 850 K from values (differences) of h in Table A.8. With these values make an estimate for the reversible adiabatic exit temperature of the turbine at 100 kPa.

12.45 Consider Problem 12.41 and find the value for the mixture heat capacity, mole basis, and the mixture ratio of specific heats, k_{mix}, both estimated at 850 K from values (differences) of h in Table A.9. With these values make an estimate for the reversible adiabatic exit temperature of the turbine at 100 kPa.

12.46 A mixture of 0.5 kg of nitrogen and 0.5 kg of oxygen is at 100 kPa and 300 K in a piston cylinder keeping constant pressure. Now 800 kJ is added by heating. Find the final temperature and the increase in entropy of the mixture using Table A.5 values.

12.47 Repeat Problem 12.46, but solve using values from Table A.8.

12.48 New refrigerant R-410a is a mixture of R-32 and R-125 in a 1:1 mass ratio. A process brings 0.5

kg R-410a from 270 K to 320 K at a constant pressure of 250 kPa in a piston cylinder. Find the work and heat transfer.

12.49 A piston cylinder contains 0.5 kg of argon and 0.5 kg of hydrogen at 300 K and 100 kPa. The mixture is compressed in an adiabatic process to 400 kPa by an external force on the piston. Find the final temperature, the work, and the heat transfer in the process.

12.50 Natural gas as a mixture of 75% methane and 25% ethane by mass is flowing to a compressor at 17°C and 100 kPa. The reversible adiabatic compressor brings the flow to 250 kPa. Find the exit temperature and the needed work per kg flow.

12.51 A mixture of 2 kg of oxygen and 2 kg of argon is in an insulated piston–cylinder arrangement at 100 kPa and 300 K. The piston now compresses the mixture to half its initial volume. Find the final pressure, final temperature, and the piston work.

12.52 The substance R-410a (see Problem 12.48) is at 100 kPa and 290 K. It is now brought to 250 kPa and 400 K in a reversible polytropic process. Find the change in specific volume, specific enthalpy, and specific entropy for the process.

12.53 Two insulated tanks A and B are connected by a valve, shown in Fig. P12.53. Tank A has a volume of 1 m³ and initially contains argon at 300 kPa and 10°C. Tank B has a volume of 2 m³ and initially contains ethane at 200 kPa and 50°C. The valve is opened and remains open until the resulting gas mixture comes to a uniform state. Determine the final pressure and temperature.

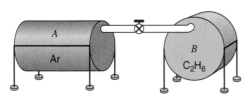

FIGURE P12.53

12.54 A compressor brings R-410a (see Problem 12.48) from −10°C and 125 kPa up to 500 kPa in an adiabatic reversible compression. Assume ideal-gas behavior and find the exit temperature and the specific work.

12.55 A mixture of 50% carbon dioxide and 50% water by mass is brought from 1500 K and 1 MPa to 500 K and 200 kPa in a polytropic process through a steady-state device. Find the necessary heat transfer and work involved using values from Table A.5.

12.56 Solve problem 12.55 using specific heats $C_p = \Delta h/\Delta T$ from Table A.8 at 1000 K.

12.57 A 50/50 (by mass) gas mixture of methane CH_4 and ethylene C_2H_4 is contained in a cylinder piston at the initial state of 480 kPa, 330 K, and 1.05 m³. The piston is now moved, compressing the mixture in a reversible, polytropic process to the final state of 260 K and 0.03 m³. Calculate the final pressure, the polytropic exponent, the work and heat transfer, and entropy change for the mixture.

12.58 The gas mixture from Problem 12.31 is compressed in a reversible adiabatic process from the initial state in the sample cylinder to a volume of 0.2 L. Determine the final temperature of the mixture and the work done during the process.

Entropy Generation

12.59 A flow of 2 kg/s mixture of 50% CO_2 and 50% O_2 by mass is heated in a constant-pressure heat exchanger from 400 K to 1000 K by a radiation source at 1400 K. Find the rate of heat transfer and the entropy generation in the process shown in Fig. P12.59.

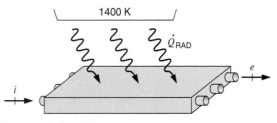

FIGURE P12.59

12.60 Carbon dioxide gas at 320 K is mixed with nitrogen at 280 K in an insulated mixing chamber. Both flows are at 100 kPa, and the mass ratio of carbon dioxide to nitrogen is 2:1. Find the exit temperature and the total entropy generation per kg of the exit mixture.

12.61 Repeat Problem 12.60 with inlet temperatures of 1400 K for the carbon dioxide and 300 K for the nitrogen. First estimate the exit temperature with the specific heats from Table A.5 and use this to start iterations with values from Table A.8.

12.62 Carbon dioxide gas at 320 K is mixed with nitrogen at 280 K in an insulated mixing chamber. Both flows are coming in at 100 kPa and the mole ratio of carbon dioxide to nitrogen is 2:1. Find the exit temperature and the total entropy generation per kmole of the exit mixture.

12.63 Repeat Problem 12.62 with inlet temperature of 1400 K for the carbon dioxide and 300 K for the nitrogen. First estimate the exit temperature with the specific heats from Table A.5 and use this to start iterations with values from A.9.

12.64 The only known sources of helium are the atmosphere (mole fraction approximately 5×10^{-6}) and natural gas. A large unit is being constructed to separate 100 m^3/s of natural gas, assumed to be 0.001 He mole fraction and 0.999 CH_4. The gas enters the unit at 150 kPa, 10°C. Pure helium exits at 100 kPa, 20°C, and pure methane exits at 150 kPa, 30°C. Any heat transfer is with the surroundings at 20°C. Is an electrical power input of 3000 kW sufficient to drive this unit?

12.65 A flow of 1 kg/s carbon dioxide at 1600 K, 100 kPa is mixed with a flow of 2 kg/s water at 800 K, 100 kPa, and after the mixing it goes through a heat exchanger where it is cooled to 500 K by a 400 K ambient. How much heat transfer is taken out in the heat exchanger? What is the entropy generation rate for the whole process?

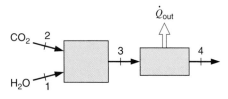

FIGURE P12.65

12.66 A mixture of 60% helium and 40% nitrogen by mass enters a turbine at 1 MPa and 800 K at a rate of 2 kg/s. The adiabatic turbine has an exit pressure of 100 kPa and an isentropic efficiency of 85%. Find the turbine work.

12.67 Repeat Problem 12.50 for an isentropic compressor efficiency of 82%.

12.68 A large air separation plant takes in ambient air (79% N_2, 21% O_2 by mole) at 100 kPa and 20°C at a rate of 25 kg/s. It discharges a stream of pure O_2 gas at 200 kPa and 100°C and a stream of pure N_2 gas at 100 kPa and 20°C. The plant operates on an electrical power input of 2000 kW, shown in Fig. P12.68. Calculate the net rate of entropy change for the process.

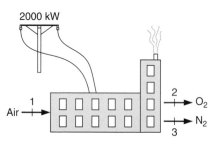

FIGURE P12.68

12.69 A steady flow of 0.3 kg/s of 50% carbon dioxide and 50% water by mass at 1200 K and 200 kPa is used in a heat exchanger where 300 kW is extracted from the flow. Find the flow exit temperature and the rate of change of entropy using Table A.8.

12.70 A steady flow of 0.01 kmol/s of 50% carbon dioxide and 50% water on a mole basis at 1200 K and 200 kPa is used in a heat exchanger where 300 kW is extracted from the flow. Find the flow exit temperature and the rate of change of entropy using Table A.9.

12.71 A flow of 1.8 kg/s steam at 400 kPa, 400°C, is mixed with 3.2 kg/s oxygen at 400 kPa, 400 K, in a steady flow mixing-chamber without any heat transfer. Find the exit temperature and the rate of entropy generation.

12.72 A tank has two sides initially separated by a diaphragm, shown in Fig. P12.72. Side *A* contains 1 kg of water and side *B* contains 1.2 kg of air, both at 20°C and 100 kPa. The diaphragm is now broken, and the whole tank is heated to 600°C by a 700°C reservoir. Find the final total pressure, heat transfer, and total entropy generation.

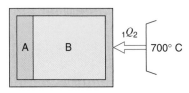

FIGURE P12.72

12.73 Three steady flows are mixed in an adiabatic chamber at 150 kPa. Flow one is 2 kg/s of O_2 at 340 K, flow two is 4 kg/s of N_2 at 280 K, and flow three is 3 kg/s of CO_2 at 310 K. All flows are at 150 kPa, the same as the total exit pressure. Find the exit temperature and the rate of entropy generation in the process.

12.74 Reconsider Problem 12.53, but let the tanks have a small amount of heat transfer so the final mixture is at 400 K. Find the final pressure, the heat transfer, and the entropy change for the process.

Air–Water Vapor Mixtures

12.75 Atmospheric air is at 100 kPa and 25°C with a relative humidity of 75%. Find the absolute humidity and the dew point of the mixture. If the mixture is heated to 30°C, what is the new relative humidity?

12.76 Consider 100 m^3 of atmospheric air, which is an air–water vapor mixture at 100 kPa, 15°C, and 40% relative humidity. Find the mass of water and the humidity ratio. What is the dew point of the mixture?

12.77 The products of combustion are flowing through a heat exchanger with 12% CO_2, 13% H_2O, and 75% N_2 on a volume basis at the rate 0.1 kg/s and 100 kPa. What is the dew-point temperature? If the mixture is cooled 10°C below the dew-point temperature, how long will it take to collect 10 kg of liquid water?

12.78 A 1 kg/s flow of saturated moist air (relative humidity 100%) at 100 kPa and 10°C goes through a heat exchanger and comes out at 25°C. What is the exit relative humidity and how much power is needed?

12.79 A new high-efficiency home heating system includes an air-to-air heat exchanger, which uses energy from outgoing stale air to heat the fresh incoming air. If the outside ambient temperature is −10°C and the relative humidity is 30%, how much water will have to be added to the incoming air, if it flows in at the rate of 1 m^3/s and must eventually be conditioned to 20°C and 40% relative humidity?

12.80 Consider a 1 m^3/s flow of atmospheric air at 100 kPa, 25°C, and 80% relative humidity. Assume this flows into a basement room where it cools to 15°C at 100 kPa. How much liquid water will condense out?

12.81 A 2 kg/s flow of completely dry air at T_1 and 100 kPa is cooled down to 10°C by spraying liquid water at 10°C and 100 kPa into it so it becomes saturated moist air at 10°C. The process is steady state with no external heat transfer or work. Find the exit moist air humidity ratio and the flow rate of liquid water. Find also the dry air inlet temperature T_1.

12.82 A piston cylinder has 100 kg of saturated moist air at 100 kPa and 5°C. If it is heated to 45°C in an isobaric process, find $_1q_2$ and the final relative humidity. If it is compressed from the initial state to 200 kPa in an isothermal process, find the mass of water condensing.

12.83 A saturated air–water vapor mixture at 20°C, 100 kPa, is contained in a 5-m^3 closed tank in equilibrium with 1 kg of liquid water. The tank is heated to 80°C. Is there any liquid water at the final state? Find the heat transfer for the process.

12.84 Ambient moist air enters a steady-flow air-conditioning unit at 102 kPa and 30°C with a 60% relative humidity. The volume flow rate entering the unit is 100 L/s. The moist air leaves the unit at 95 kPa and 15°C with a relative humidity of 100%. Liquid condensate also leaves the unit at 15°C. Determine the rate of heat transfer for this process.

12.85 Consider at 500 L rigid tank containing an air–water vapor mixture at 100 kPa and 35°C with a 70% relative humidity. The system is cooled until the water just begins to condense. Determine the final temperature in the tank and the heat transfer for the process.

12.86 Air in a piston cylinder is at 35°C, 100 kPa, and a relative humidity of 80%. It is now compressed to a pressure of 500 kPa in a constant-temperature process. Find the final relative and specific humidity and the volume ratio V_2/V_1.

12.87 A 300 L rigid vessel initially contains moist air at 150 kPa and 40°C with a relative humidity of 10%. A supply line connected to this vessel by a valve carries steam at 600 kPa and 200°C. The valve is opened, and steam flows into the vessel until the relative humidity of the resultant moist air mixture is 90%. Then the valve is closed. Sufficient heat is transferred from the vessel so that the temperature remains at 40°C during the process. Determine the heat transfer for the process, the mass of steam entering the vessel, and the final pressure inside the vessel.

12.88 A rigid container, 10 m³ in volume, contains moist air at 45°C and 100 kPa with Φ = 40%. The container is now cooled to 5°C. Neglect the volume of any liquid that might be present and find the final mass of water vapor, final total pressure, and the heat transfer.

12.89 A water-filled reactor of 1 m³ is at 20 MPa, 360°C and is located inside an insulated containment room of 100 m³ that contains air at 100 kPa and 25°C. Due to a failure, the reactor ruptures and the water fills the containment room. Find the final quality and pressure by iterations.

Tables and Formulas or Psychrometric Chart

12.90 A flow of moist air at 100 kPa, 40°C, and 40% relative humidity is cooled to 15°C in a constant-pressure device. Find the humidity ratio of the inlet and the exit flow and the heat transfer in the device per kg dry air.

12.91 A flow, 0.2 kg/s dry air, of moist air at 40°C and 50% relative humidity flows from the outside state 1 down into a basement where it cools to 16°C, state 2. Then it flows up to the living room where it is heated to 25°C, state 3. Find the dew point for state 1, any amount of liquid that may appear, the heat transfer that takes place in the basement, and the relative humidity in the living room at state 3.

12.92 Two moist air streams with 85% relative humidity, both flowing at a rate of 0.1 kg/s of dry air, are mixed in a steady-flow setup. One inlet stream is at 32.5°C and the other at 16°C. Find the exit relative humidity.

12.93 The discharge moist air from a clothes dryer is at 35°C, 80% relative humidity. The flow is guided through a pipe up through the roof and a vent to the atmosphere shown in Fig. P12.93. Due to heat transfer in the pipe, the flow is cooled to 24°C by the time it reaches the vent. Find the humidity ratio in the flow out of the clothes dryer and at the vent. Find the heat transfer and any amount of liquid that may be forming per kg dry air for the flow.

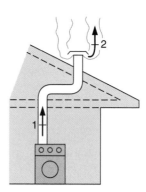

FIGURE P12.93

12.94 A steady supply of 1.0 m³/s air at 25°C, 100 kPa, and 50% relative humidity is needed to heat a building in the winter. The ambient outdoors is at 10°C, 100 kPa, and 50% relative humidity. What are the required liquid water input and heat transfer rates for this purpose?

12.95 A combination air cooler and dehumidification unit receives outside ambient air at 35°C, 100 kPa, and 90% relative humidity. The moist air is first cooled to a low temperature T_2 to condense the proper amount of water; assume all the liquid leaves at T_2. The moist air is then heated and leaves the unit at 20°C, 100 kPa, and 30% relative humidity with a volume flow rate of 0.01 m³/s. Find the temperature T_2, the mass of liquid per kilogram of dry air, and the overall heat transfer rate.

12.96 Use the formulas and the steam tables to find the missing property of: Φ, ω, and T_{dry}, for a total pressure of 100 kPa; repeat the answers using the psychrometric chart.

a. Φ = 50%, ω = 0.010

b. T_{dry} = 25°C, T_{wet} = 21°C

12.97 An insulated tank has an air inlet, ω_1 = 0.0084 and an outlet, T_2 = 22°C, Φ_2 = 90%, both at 100 kPa. A third line sprays 0.25 kg/s of water at

80°C and 100 kPa, as shown in Fig. P12.97. For steady operation, find the outlet specific humidity, the mass flow rate of air needed, and the required air inlet temperature, T_1.

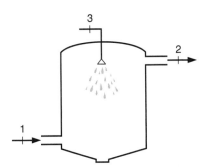

FIGURE P12.97

12.98 A flow of moist air from a domestic furnace, state 1, is at 45°C, 10% relative humidity with a flow rate of 0.05 kg/s dry air. A small electric heater adds steam at 100°C, 100 kPa, generated from tap water at 15°C shown in Fig. P12.98. Up in the living room, the flow comes out at state 4: 30°C, 60% relative humidity. Find the power needed for the electric heater and the heat transfer to the flow from state 1 to state 4.

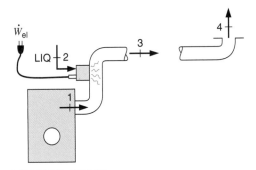

FIGURE P12.98

12.99 A water-cooling tower for a power plant cools 45°C liquid water by evaporation. The tower receives air at 19.5°C, $\Phi = 30\%$, and 100 kPa that is blown through/over the water such that it leaves the tower at 25°C and $\Phi = 70\%$. The remaining liquid water flows back to the condenser at 30°C having given off 1 MW. Find the mass flow rate of air, and determine the amount of water that evaporates.

12.100 A flow of air at 5°C, $\Phi = 90\%$, is brought into a house, where it is conditioned to 25°C, 60% relative humidity. This is done with a combined heater-evaporator where any liquid water is at 10°C. Find any flow of liquid and the necessary heat transfer, both per kilogram dry air flowing. Find the dew point for the final mixture.

12.101 In a car's defrost/defog system atmospheric air at 21°C and 80% relative humidity is taken in and cooled such that liquid water drips out. The now dryer air is heated to 41°C and then blown onto the windshield, where it should have a maximum of 10% relative humidity to remove water from the windshield. Find the dew point of the atmospheric air, specific humidity of air onto the windshield, the lowest temperature, and the specific heat transfer in the cooler.

12.102 Atmospheric air at 35°C with a relative humidity of 10%, is too warm and also too dry. An air conditioner should deliver air at 21°C and 50% relative humidity in the amount of 3600 m³/h. Sketch a setup to accomplish this. Find any amount of liquid (at 20°C) that is needed or discarded and any heat transfer.

12.103 One means of air-conditioning hot summer air is by evaporative cooling, which is a process similar to the adiabatic saturation process. Consider outdoor ambient air at 35°C, 100 kPa, 30% relative humidity. What is the maximum amount of cooling that can be achieved by such a technique? What disadvantage is there to this approach? Solve the problem using a first law analysis and repeat it using the psychrometric chart, Fig. E.4.

12.104 A flow of moist air at 45°C, 10% relative humidity with a flow rate of 0.2 kg/s dry air is mixed with a flow of moist air at 25°C, and absolute humidity of $\omega = 0.018$ with a rate of 0.3 kg/s dry air. The mixing takes place in an air duct at 100 kPa, and there is no significant heat transfer. After the mixing, there is heat transfer to a final temperature of 40°C. Find the temperature and relative humidity after mixing. Find the heat transfer and the final exit relative humidity.

12.105 An indoor pool evaporates 1.512 kg/h of water, which is removed by a dehumidifier to maintain 21°C, $\Phi = 70\%$ in the room. The dehumidifier, shown in Fig. P12.105, is a refrigeration cycle in which air flowing over the evaporator cools such that liquid water drops out, and the air continues flowing over the condenser. For an airflow rate of 0.1 kg/s the unit requires 1.4 kW input to a motor driving a fan and the compressor, and it has a coefficient of performance, $\beta = \dot{Q}_L/\dot{W}_c = 2.0$. Find the state of the air as it returns to the room and the compressor work input.

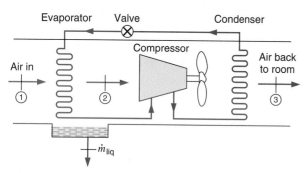

FIGURE P12.105

Psychrometric Chart Only

12.106 Use the psychrometric chart to find the missing property of: Φ, ω, T_{wet}, T_{dry}.
  a. $T_{dry} = 25°C$, $\Phi = 80\%$
  b. $T_{dry} = 15°C$, $\Phi = 100\%$
  c. $T_{dry} = 20°C$, $\omega = 0.008$
  d. $T_{dry} = 25°C$, $T_{wet} = 23°C$

12.107 Use the psychrometric chart to find the missing property of: Φ, ω, T_{wet}, T_{dry}.
  a. $\Phi = 50\%$, $\omega = 0.012$
  b. $T_{wet} = 15°C$, $\Phi = 60\%$
  c. $\omega = 0.008$, $T_{wet} = 17°C$
  d. $T_{dry} = 10°C$, $\omega = 0.006$

12.108 Use the formulas and the steam tables to find the missing property of: Φ, ω, and T_{dry}, total pressure is 100 kPa. Repeat the answers using the psychrometric chart.
  a. $\Phi = 50\%$, $\omega = 0.010$
  b. $T_{wet} = 15°C$, $\Phi = 50\%$
  c. $T_{dry} = 25°C$, $T_{wet} = 21°C$

12.109 For each of the states in Problem 12.107 find the dew-point temperature.

12.110 Compare the weather in two places where it is cloudy and breezy. At beach A the temperature

is 20°C, the pressure is 103.5 kPa, and the relative humidity is 90%; beach B has 25°C, 99 kPa, and 20% relative humidity. Suppose you just took a swim and came out of the water. Where would you feel more comfortable, and why?

12.111 Ambient air at 100 kPa, 30°C, and 40% relative humidity goes through a constant-pressure heat exchanger as a steady flow. In one case it is heated to 45°C, and in another case it is cooled until it reaches saturation. For both cases find the exit relative humidity and the amount of heat transfer per kilogram of dry air.

12.112 A flow of moist air at 21°C with 60% relative humidity should be produced from mixing two different moist airflows. Flow 1 is at 10°C and 80% relative humidity; flow 2 is at 32°C and has $T_{wet} = 27°C$. The mixing chamber can be followed by a heater or a cooler, as shown in Fig. P12.112. No liquid water is added, and $P = 100$ kPa. Find the two controls—one is the ratio of the two mass flow rates $\dot{m}_{a1}/\dot{m}_{a2}$ and the other is the heat transfer in the heater/cooler per kg dry air.

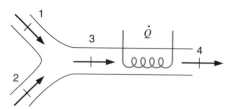

FIGURE P12.112

12.113 In a hot and dry climate, air enters an air-conditioner unit at 100 kPa, 40°C, and 5% relative humidity, at the steady rate of 1.0 m³/s. Liquid water at 20°C is sprayed into the air in the AC unit at the rate of 20 kg/h, and heat is rejected from the unit at the rate 20 kW. The exit pressure is 100 kPa. What are the exit temperature and relative humidity?

12.114 Consider two states of atmospheric air. (1) 35°C, $T_{wet} = 18°C$ and (2) 26.5°C, $\Phi = 60\%$. Suggest a system of devices that will allow air in a steady flow to change from (1) to (2) and from (2) to (1). Heaters, coolers, (de)humidifiers, liquid traps, and the like are available, and any liquid/solid flowing is assumed to be at the lowest temperature seen in the process. Find the specific and relative humidity for state 1, dew point for state 2, and the heat transfer per kilogram dry air in each component in the systems.

12.115 To refresh air in a room, a counterflow heat exchanger, see Fig. P12.115, is mounted in the wall, drawing in outside air at 0.5°C, 80% relative humidity, and pushing out room air, 40°C, 50% relative humidity. Assume an exchange of 3 kg/min dry air in a steady flow, and also assume that the room air exits the heat exchanger to the atmosphere at 23°C. Find the net amount of water removed from the room, any liquid flow in the heat exchanger, and (T, Φ) for the fresh air entering the room.

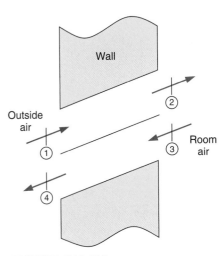

FIGURE P12.115

Availability (Exergy) in Mixtures

12.116 Consider the mixing of a steam flow with an oxygen flow in Problem 12.71. Find the rate of total inflowing availability and the rate of exergy destruction in the process.

12.117 A mixture of 75% carbon dioxide and 25% water by mass is flowing at 1600 K, 100 kPa into a heat exchanger where it is used to deliver energy to a heat engine. The mixture leaves the heat exchanger at 500 K with a mass flow rate of 2 kg/min. Find the rate of energy and the rate of exergy delivered to the heat engine.

12.118 Find the second-law efficiency of the heat exchanger in Problem 12.59.

Review Problems

12.119 A piston/cylinder contains helium at 110 kPa at ambient temperature 20°C, and initial volume of 20 L as shown in Fig. P12.119. The stops are mounted to give a maximum volume of 25 L, and the nitrogen line conditions are 300 kPa, 30°C. The valve is now opened, which allows nitrogen to flow in and mix with the helium. The valve is closed when the pressure inside reaches 200 kPa, at which point the temperature inside is 40°C. Is this process consistent with the second law of thermodynamics?

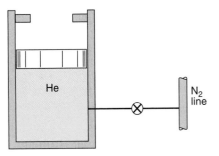

FIGURE P12.119

12.120 A spherical balloon has an initial diameter of 1 m and contains argon gas at 200 kPa, 40°C. The balloon is connected by a valve to a 500-L rigid tank containing carbon dioxide at 100 kPa, 100°C. The valve is opened, and eventually the balloon and tank reach a uniform state in which the pressure is 185 kPa. The balloon pressure is directly proportional to its diameter. Take the balloon and tank as a control volume, and calculate the final temperature and the heat transfer for the process.

12.121 An insulated vertical cylinder is fitted with a frictionless constant loaded piston of cross-sectional area 0.1 m^2 and the initial cylinder height of 1.0 m. The cylinder contains methane gas at 300 K, 150 kPa, and also inside is a 5-L capsule containing neon gas at 300 K, 500 kPa shown in Fig. P12.121. The capsule now breaks, and the two gases mix together in a constant-pressure process. What is the final temperature, final cylinder height, and net entropy change for the process?

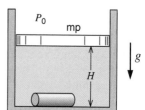

FIGURE P12.121

12.122 An insulated rigid 2 m³ tank A contains CO_2 gas at 200°C, 1 MPa. An uninsulated rigid 1 m³ tank B contains ethane, C_2H_6, gas at 200 kPa, room temperature 20°C. The two are connected by a one-way check valve that will allow gas from A to B, but not from B to A shown in Fig. P12.122. The valve is opened and gas flows from A to B until the pressure in B reaches 500 kPa when the valve is closed. The mixture in B is kept at room temperature due to heat transfer. Find the total number of moles and the ethane mole fraction at the final state in B. Find the final temperature and pressure in tank A and the heat transfer, to/from tank B.

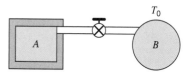

FIGURE P12.122

12.123 A 0.2 m³ insulated, rigid vessel is divided into two equal parts A and B by an insulated partition, as shown in Fig. P.12.123. The partition will support a pressure difference of 400 kPa before breaking. Side A contains methane and side B contains carbon dioxide. Both sides are initially at 1 MPa, 30°C. A valve on side B is opened, and carbon dioxide flows out. The carbon dioxide that remains in B is assumed to undergo a reversible adiabatic expansion while there is flow out. Eventually the partition breaks, and the valve is closed. Calculate the net entropy change for the process that begins when the valve is closed.

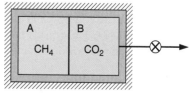

FIGURE P12.123

12.124 An air–water vapor mixture enters a steady-flow heater humidifier unit at state 1: 10°C, 10% relative humidity, at the rate of 1 m³/s. A second air–vapor stream enters the unit at state 2: 20°C, 20% relative humidity, at the rate of 2 m³/s. Liquid water enters at state 3: 10°C, at the rate of 400 kg per hour. A single air–vapor flow exits the unit at state 4: 40°C shown in Fig. P12.124. Calculate the relative humidity of the exit flow and the rate of heat transfer to the unit.

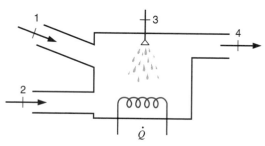

FIGURE P12.124

12.125 You have just washed your hair and now blow dry it in a room with 23°C, $\Phi = 60\%$, (1). The dryer, 500 W, heats the air to 49°C, (2), blows it through your hair where the air becomes saturated (3), and then flows on to hit a window where it cools to 15°C (4). Find the relative humidity at state 2, the heat transfer per kilogram of dry air in the dryer, the airflow rate, and the amount of water condensed on the window, if any.

12.126 Steam power plants often utilize large cooling towers to cool the condenser cooling water so it can be recirculated; see Fig. P12.126. The process is essentially evaporative adiabatic cooling, in which part of the water is lost and must therefore be replenished. Consider the setup shown in Fig. P12.126, in which 1000 kg/s of warm water at 32°C from the condenser enters the top of the cooling tower and the cooled water leaves the bottom at 20°C. The moist ambient air enters the bottom at 100 kPa, dry-bulb temperature of 18°C and a wet-bulb temperature of 10°C. The moist air leaves the tower at 95 kPa, 30°C, and relative humidity of 85%. Determine the required mass flow rate of dry air, and the fraction of the incoming water that evaporates and is lost.

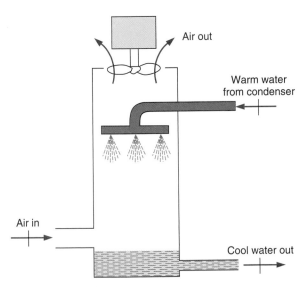

FIGURE P12.126

12.127 Ambient air is at a condition of 100 kPa, 35°C, 50% relative humidity. A steady stream of air at 100 kPa, 23°C, 70% relative humidity is to be produced by first cooling one stream to an appropriate temperature to condense out the proper amount of water and then mix this stream adiabatically with the second one at ambient conditions. What is the ratio of the two flow rates? To what temperature must the first stream be cooled?

12.128 A semipermeable membrane is used for the partial removal of oxygen from air that is blown through a grain elevator storage facility. Ambient air (79% nitrogen, 21% oxygen on a mole basis) is compressed to an appropriate pressure, cooled to ambient temperature 25°C, and then fed through a bundle of hollow polymer fibers that selectively absorb oxygen, so the mixture leaving at 120 kPa, 25°C, contains only 5% oxygen, shown in Fig. P12.128. The absorbed oxygen is bled off through the fiber walls at 40 kPa, 25°C, to a vacuum pump. Assume the process to

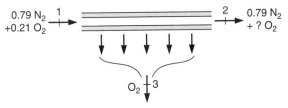

FIGURE P12.128

be reversible and adiabatic and determine the minimum inlet air pressure to the fiber bundle.

12.129 A dehumidifier receives a flow of 0.25 kg/s moist air at 28°C, 80% relative humidity as shown in Figure P12.105. It is cooled down to 20°C as it flows over the evaporator and then heated up again as it flows over the condenser. The standard refrigeration cycle uses R-22 with an evaporator temperature of 5°C and a condensation pressure of 1600 kPa. Find the amount of liquid water removed and the heat transfer in the cooling process. How much compressor work is needed? What is the final air exit temperature and relative humidity?

12.130 A 100-L insulated tank contains N_2 gas at 200 kPa and ambient temperature 25°C. The tank is connected by a valve to a supply line flowing CO_2 at 1.2 MPa, 90°C. A mixture of 50% N_2, 50% CO_2 by mole should be obtained by opening the valve and allowing CO_2 to flow in until an appropriate pressure is reached, when the valve is closed. What is the pressure? The tank eventually cools to ambient temperature. Find the net entropy change for the overall process.

12.131 A cylinder/piston loaded with a linear spring contains saturated moist air at 120 kPa, 0.1 m³ volume and also 0.01 kg of liquid water, all at ambient temperature 20°C. The piston area is 0.2 m², and the spring constant is 20 kN/m. This cylinder is attached by a valve to a line flowing dry air at 800 kPa, 80°C. The valve is opened, and air flows into the cylinder until the pressure reaches 200 kPa, at which point the temperature is 40°C. Determine the relative humidity at the final state, the mass of air entering the cylinder, and the work done during the process.

12.132 Consider the previous problem and additionally determine the heat transfer. Show that the process does not violate the second law.

12.133 The air-conditioning by evaporative cooling in Problem 12.103 is modified by adding a dehumidification process before the water spray cooling process. This dehumidification is achieved as shown in Fig. P12.133 by using a desiccant material, which absorbs water on one side of a rotating drum heat exchanger. The desiccant is regenerated by heating on the other side of the

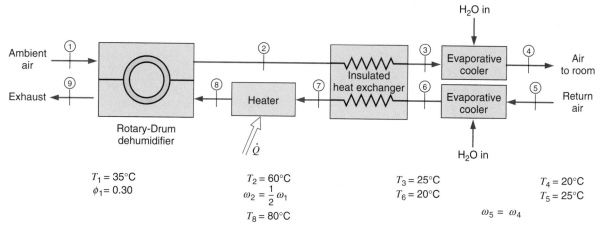

FIGURE P12.133

drum to drive the water out. The pressure is 100 kPa everywhere, and other properties are on the diagram. Calculate the relative humidity of the cool air supplied to the room at state 4, and the heat transfer per unit mass of air that needs to be supplied to the heater unit.

ENGLISH UNIT PROBLEMS

Concept Problems

12.134E If oxygen is 21% by mole of air, what is the oxygen state (P, T, v) in a room at 540 R, 15 psia of total volume 2000 ft³?

12.135E A flow of oxygen and one of nitrogen, both 540 R, are mixed to produce 1 lbm/s air at 540 R, 15 psia. What are the mass and volume flow rates of each line?

12.136E A flow of gas A and a flow of gas B are mixed in a 1:1 mole ratio with same T. What is the entropy generation per kmole flow out?

12.137E A rigid container has 1 kg argon at 300 K and 1 kg argon at 400 K, both at 150 kPa. Now they are allowed to mix without any external heat transfer. What is final T, P? Is any s generated?

12.138E A rigid container has 1 kg CO_2 at 300 K and 1 kg argon at 400 K, both at 150 kPa. Now they are allowed to mix without any heat transfer. What is final T, P?

12.139E A flow of 1 kg/s argon at 300 K and another flow of 1 kg/s CO_2 at 1600 K, both at 150 kPa, are mixed without any heat transfer. What is the exit T, P?

12.140E What is the rate of entropy increase in Problem 12.139?

12.141E If I have air at 14.7 psia and (a) 15 F, (b) 115 F, and (c) 230 F, what is the maximum absolute humidity I can have?

English Unit Problems

12.142E A gas mixture at 250 F, 18 lbf/in.² is 50% N_2, 30% H_2O, and 20% O_2 on a mole basis. Find the mass fractions, the mixture gas constant, and the volume for 10 lbm of mixture.

12.143E Weighing of masses gives a mixture at 80 F, 35 lbf/in² with 1 lbm O_2, 3 lbm N_2, and 1 lbm CH_4. Find the partial pressures of each component, the mixture specific volume (mass basis), the mixture molecular weight, and the total volume.

12.144E A new refrigerant R-410a is a mixture of R-32 and R-125 in a 1:1 mass ratio. What is the overall molecular weight, the gas constant, and the ratio of specific heats for such a mixture?

12.145E A pipe flows 1.5 lbm/s of a mixture with mass fractions of 40% CO_2 and 60% N_2 at 60 lbf/in.², 540 R. Heating tape is wrapped around a section of pipe with insulation added, and 2 Btu/s electrical power is heating the pipe flow. Find the mixture exit temperature.

12.146E An insulated gas turbine receives a mixture of 10% CO_2, 10% H_2O, and 80% N_2 on a mass basis at 1800 R, 75 lbf/in.^{2}. The inlet volume flow rate is 70 ft^3/s, and the exhaust is at 1300 R, 15 lbf/in.^{2}. Find the power output in Btu/s using constant specific heat from F4 at 540 R.

12.147E Solve Problem 12.146 using the values of enthalpy from Table F.6.

12.148E A piston cylinder device contains 0.3 lbm of a mixture of 40% methane and 60% propane by mass at 540 R and 15 psia. The gas is now slowly compressed in an isothermal (T = constant) process to a final pressure of 40 psia. Show the process in a P–V diagram, and find both the work and heat transfer in the process.

12.149E A mixture of 4 lbm oxygen and 4 lbm of argon is in an insulated piston cylinder arrangement at 14.7 lbf/in.^{2}, 540 R. The piston now compresses the mixture to half its initial volume. Find the final pressure, temperature, and the piston work.

12.150E Two insulated tanks A and B are connected by a valve. Tank A has a volume of 30 ft^3 and initially contains argon at 50 lbf/in.^{2}, 50 F. Tank B has a volume of 60 ft^3 and initially contains ethane at 30 lbf/in.^{2}, 120 F. The valve is opened and remains open until the resulting gas mixture comes to a uniform state. Find the final pressure and temperature.

12.151E A mixture of 50% carbon dioxide and 50% water by mass is brought from 2800 R, 150 lbf/in.^{2} to 900 R, 30 lbf/in.^{2} in a polytropic process through a steady-flow device. Find the necessary heat transfer and work involved using values from F.4.

12.152E Carbon dioxide gas at 580 R is mixed with nitrogen at 500 R in an insulated mixing chamber. Both flows are at 14.7 lbf/in.^{2}, and the mole ratio of carbon dioxide to nitrogen is 2:1. Find the exit temperature and the total entropy generation per mole of the exit mixture.

12.153E A mixture of 60% helium and 40% nitrogen by mole enters a turbine at 150 lbf/in.^{2}, 1500 R at a rate of 4 lbm/s. The adiabatic turbine has an exit pressure of 15 lbf/in.^{2} and an insentropic efficiency of 85%. Find the turbine work.

12.154E A large air separation plant, see Fig. P12.68, takes in ambient air (79% N_2, 21% O_2 by volume) at 14.7 lbf/in.^{2}, 70 F, at a rate of 2 lb mol/s. It discharges a stream of pure O_2 gas at 30 lbf/in.^{2}, 200 F, and a stream of pure N_2 gas at 14.7 lbf/in.^{2}, 70 F. The plant operates on an electrical power input of 2000 kW. Calculate the net rate of entropy change for the process.

12.155E A tank has two sides initially separated by a diaphragm. Side A contains 2 lbm of water, and side B contains 2.4 lbm of air—both at 68 F, 14.7 lbf/in.^{2}. The diaphragm is now broken, and the whole tank is heated to 1100 F by a 1300 F reservoir. Find the final total pressure, heat transfer, and total entropy generation.

12.156E Find the entropy generation for the process in Problem 12.150E.

12.157E Consider a volume of 2000 ft^3 that contains an air–water vapor mixture at 14.7 lbf/in.^{2}, 60 F, and 40% relative humidity. Find the mass of water and the humidity ratio. What is the dew point of the mixture?

12.158E A 1 lbm/s flow of saturated moist air (relative humidity 100%) at 14.7 psia and 50 F goes through a heat exchanger and comes out at 80 F. What is the exit relative humidity, and how much power is needed?

12.159E Consider a 10-ft^3 rigid tank containing an air–water vapor mixture at 14.7 lbf/in.^{2}, 90 F, with 70% relative humidity. The system is cooled until the water just begins to condense. Determine the final temperature in the tank and the heat transfer for the process.

12.160E Consider at 35 ft^3/s flow of atmospheric air at 14.7 psia, 80 F, and 80% relative humidity. Assume this flows into a basement room where it cools to 60 F at 14.7 psia. How much liquid will condense out?

12.161E Air in a piston/cylinder is at 95 F, 15 lbf/in.^{2} and relative humidity of 80%. It is now compressed to a pressure of 75 lbf/in.^{2} in a constant-temperature process. Find the final relative and specific humidity and the volume ratio V_2/V_1.

12.162E A 10-ft^3 rigid vessel initially contains moist air at 20 lbf/in.^{2}, 100 F, with a relative humidity of 10%. A supply line connected to this vessel by a valve carries steam at 100 lbf/in.^{2}, 400 F. The valve is opened, and steam flows into the vessel until the relative humidity of the resultant moist air mixture is 90%. Then the valve is closed. Sufficient heat is transferred from the vessel so the temperature remains at 100 F during the process. Determine the heat transfer for the process, the mass of steam entering the vessel, and the final pressure inside the vessel.

12.163E A water-filled reactor of 50 ft³ is at 2000 lbf/in.², 550 F, and located inside an insulated containment room of 5000 ft³ that has air at 1 atm. and 77 F. Due to a failure, the reactor ruptures and the water fills the containment room. Find the final quality and pressure by iterations.

12.164E Two moist air streams with 85% relative humidity, both flowing at a rate of 0.2 lbm/s of dry air are mixed in a steady flow setup. One inlet flowstream is at 90 F, and the other at 61 F. Find the exit relative humidity.

12.165E A flow of moist air from a domestic furnace, state 1 in Fig. P12.98 is at 120 F, 10% relative humidity with a flow rate of 0.1 lbm/s dry air. A small electric heater adds steam at 212 F, 14.7 psia, generated from tap water at 60 F. Up in the living room the flow comes out at state 4: 90 F, 60% relative humidity. Find the power needed for the electric heater and the heat transfer to the flow from state 1 to state 4.

12.166E Atmospheric air at 95 F, relative humidity of 10%, is too warm and also too dry. An air conditioner should deliver air at 70 F, and 50% relative humidity in the amount of 3600 ft³ per hour. Sketch a setup to accomplish this; find any amount of liquid (at 68 F) that is needed or discarded and any heat transfer.

12.167E An indoor pool evaporates 3 lbm/h of water, which is removed by a dehumidifier to maintain 70 F, $\Phi = 70\%$ in the room. The dehumidifier is a refrigeration cycle in which air flowing over the evaporator cools such that liquid water drops out, and the air continues flowing over

the condenser, as shown in Fig. P12.105. For an airflow rate of 0.2 lbm/s, the unit requires 1.2 Btu/s input to a motor driving a fan and the compressor, and it has a coefficient of performance, $\beta = \dot{Q}_L/\dot{W}_c = 2.0$. Find the state of the air after the evaporator, T_2, ω_2, Φ_2, and the heat, rejected. Find the state of the air as it returns to the room and the compressor work input.

12.168E To refresh air in a room, a counterflow heat exchanger is mounted in the wall, as shown in Fig. P12.115. It draws in outside air at 33 F, 80% relative humidity, and draws room air, 104 F, 50% relative humidity, out. Assume an exchange of 6 lbm/min dry air in a steady-flow device, and also that the room air exits the heat exchanger to the atmosphere at 72 F. Find the net amount of water removed from the room, any liquid flow in the heat exchanger, and (T, Φ) for the fresh air entering the room.

12.169E Ambient air is at a condition of 14.7 lbf/in.², 95 F, 50% relative humidity. A steady stream of air at 14.7 lbf/in.², 73 F, 70% relative humidity is to be produced by first cooling one stream to an appropriate temperature to condense out the proper amount of water and then mix this stream adiabatically with the second one at ambient conditions. What is the ratio of the two flow rates? To what temperature must the first stream be cooled?

12.170E A 4-ft³ insulated tank contains nitrogen gas at 30 lbf/in.² and ambient temperature 77 F. The tank is connected by a valve to a supply line flowing carbon dioxide at 180 lbf/in.², 190 F. A mixture of 50 mole percent nitrogen and 50 mole percent carbon dioxide is to be obtained by opening the valve and allowing flow into the tank until an appropriate pressure is reached, when the valve is closed. What is the pressure? The tank eventually cools to ambient temperature. Calculate the net entropy change for the overall process.

COMPUTER, DESIGN, AND OPEN-ENDED PROBLEMS

12.171 Write a program to solve the general case of Problems 12.53/74 in which the two volumes and the initial state properties of the argon and the ethane are input variables. Use constant specific heat from Table A.5.

12.172 Mixing of CO_2 and N_2 in a steady-flow setup was given in Problem 12.62. If the temperatures are very different an assumption of constant specific heat is inappropriate. Study the problem assuming the CO_2 enters at 300 K, 100 kPa, as a function of the N_2 inlet temperature using specific heat from Table A.7 or the formula in A.6. Give the nitrogen inlet temperature for which the constant specific heat assumption starts to be more than 1%, 5%, and 10% wrong for the exit mixture temperature.

12.173 The setup in Problem 12.97 is similar to a process that can be used to produce dry powder from a slurry of water and dry material as coffee or milk. The water flow at state 3 is a mixture of 80% liquid water and 20% dry material on a mass basis with $C_{dry} = 0.4$ kJ/kg K. After the water is evaporated, the dry material falls to the bottom and is removed in an additional line, $\dot{m}_{dry}$ exit at state 4. Assume a reasonable T_4 and that state 1 is heated atmospheric air. Investigate the inlet flow temperature as a function of state 1 humidity ratio.

12.174 A dehumidifier for household applications is similar to the system shown in Fig. P12.105. Study the requirements to the refrigeration cycle as a function of the atmospheric conditions and include a worst case estimation.

12.175 A clothes dryer has a 60°C, $\Phi = 90\%$ airflow out at a rate of 3 kg/min. The atmospheric conditions are 20°C, relative humidity of 50%. How much water is carried away and how much power is needed? To increase the efficiency, a counterflow heat exchanger is installed to preheat the incoming atmospheric air up with the hot exit flow. Estimate suitable exit temperatures from the heat exchanger and investigate the design changes to the clothes dryer. (What happens to the condensed water?) How much energy can be saved this way?

12.176 Addition of steam to combustors in gas turbines and to internal-combustion engines reduces the peak temperatures and lowers emission of NO_x. Consider a modification to a gas turbine, as shown in Fig. P12.176, where the modified cycle is called the Cheng cycle. In this example, it is used for a cogenerating power plant. Assume 12 kg/s air with state 2 at 1.25 MPa, unknown temperature, is mixed with 2.5 kg/s water at 450°C at constant pressure before the inlet to the turbine. The turbine exit temperature is $T_4 = 500$°C, and the pressure is 125 kPa. For a reasonable turbine efficiency, estimate the required

air temperature at state 2. Compare the result to the case where no steam is added to the mixing chamber and only air runs through the turbine.

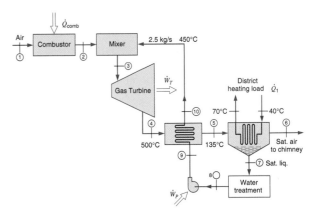

FIGURE P12.176

12.177 Consider the district water heater acting as the condenser for part of the water between states 5 and 6 in Fig. P12.176. If the temperature of the mixture (12 kg/s air, 2.5 kg/s steam) at state 5 is 135°C, make a study of the district heating load, $\dot{Q}_1$, as a function of the exit temperature T_6. Study also the sensitivity of the results with respect to the assumption that state 6 is saturated moist air.

12.178 The cogeneration gas-turbine cycle can be augmented with a heat pump to extract more energy from the turbine exhaust gas, as shown in Fig. P12.178. The heat pump upgrades the energy to

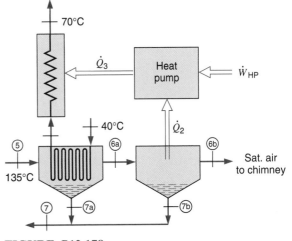

FIGURE P12.178

be delivered at the 70°C line for district heating. In the modified application, the first heat exchanger has exit temperature $T_{6a} = T_{7a} = 45°C$, and the second one has $T_{6b} = T_{7b} = 36°C$. Assume the district heating line has the same exit temperature as before so this arrangement allows for a higher flow rate. Estimate the increase in the district heating load that can be obtained and the necessary work input to the heat pump.

12.179 Several applications of dehumidification do not rely on water condensation by cooling. A desiccant with a greater affinity to water can absorb water directly from the air accompanied by a heat release. The desiccant is then regenerated by heating, driving the water out. Make a list of several such materials as liquids, gels, and solids and show examples of their use.

THERMODYNAMIC RELATIONS 13

We have already defined and used several thermodynamic properties. Among these are pressure, specific volume, density, temperature, mass, internal energy, enthalpy, entropy, constant-pressure and constant-volume specific heats, and the Joule–Thomson coefficient. Two other properties, the Helmholtz function and the Gibbs function, will also be introduced and will be used more extensively in the following chapters. We have also had occasion to use tables of thermodynamic properties for a number of different substances.

One important question is now raised: Which of the thermodynamic properties can be experimentally measured? We can answer this question by considering the measurements we can make in the laboratory. Some of the properties such as internal energy and entropy cannot be measured directly and must be calculated from other experimental data. If we carefully consider all these thermodynamic properties, we conclude that there are only four that can be directly measured: pressure, temperature, volume, and mass.

This leads to a second question: How can values of the thermodynamic properties that cannot be measured be determined from experimental data on those properties that can be measured? In answering this question, we will develop certain general thermodynamic relations. In view of the fact that millions of such equations can be written, our study will be limited to certain basic considerations, with particular reference to the determination of thermodynamic properties from experimental data. We will also consider such related matters as generalized charts and equations of state.

13.1 THE CLAPEYRON EQUATION

In calculating thermodynamic properties such as enthalpy or entropy in terms of other properties that can be measured, the calculations fall into two broad categories: differences in properties between two different phases and changes within a single, homogeneous phase. In this section, we focus our attention on the first of these categories, that of different phases. Let us assume that the two phases are liquid and vapor, but we will see that the results apply to other differences as well.

Consider a Carnot-cycle heat engine operating across a small temperature difference between reservoirs at T and $T - \Delta T$. The corresponding saturation pressures are P and $P - \Delta P$. The Carnot cycle operates with four steady-state devices. In the high temperature heat-transfer process, the working fluid changes from saturated liquid at 1 to saturated vapor at 2, as shown in the two diagrams of Fig. 13.1.

From Fig. 13.1a, for reversible heat transfers,

$$q_H = Ts_{fg}; \qquad q_L = (T - \Delta T)s_{fg}$$

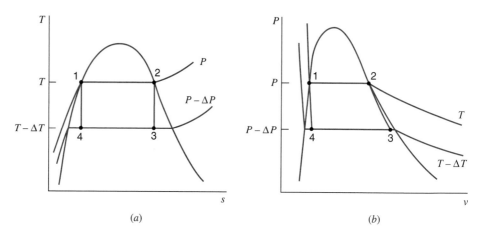

FIGURE 13.1 A Carnot cycle operating across a small temperature difference.

(a) (b)

so that

$$w_{NET} = q_H - q_L = \Delta T s_{fg} \tag{13.1}$$

From Fig. 13.1b, each process is steady-state and reversible, such that the work in each process is given by Eq. 9.19,

$$w = -\int v \, dP$$

Overall, for the four processes in the cycle,

$$w_{NET} = 0 - \int_2^3 v \, dP + 0 - \int_4^1 v \, dP$$

$$\approx -\left(\frac{v_2 + v_3}{2}\right)(P - \Delta P - P) - \left(\frac{v_1 + v_4}{2}\right)(P - P + \Delta P)$$

$$\approx \Delta P\left[\left(\frac{v_2 + v_3}{2}\right) - \left(\frac{v_1 + v_4}{2}\right)\right] \tag{13.2}$$

(The smaller the ΔP, the better the approximation.)

Now, comparing Eqs. 13.1 and 13.2 and rearranging,

$$\frac{\Delta P}{\Delta T} \approx \frac{s_{fg}}{\left(\dfrac{v_2 + v_3}{2}\right) - \left(\dfrac{v_1 + v_4}{2}\right)}$$

In the limit as $\Delta T \to 0$: $v_3 \to v_2 = v_g$, $v_4 \to v_1 = v_f$, which results in

$$\lim_{\Delta T \to 0} \frac{\Delta P}{\Delta T} = \frac{dP_{sat}}{dT} = \frac{s_{fg}}{v_{fg}} \tag{13.3}$$

Since the heat addition process $1 - 2$ is at constant pressure as well as constant temperature,

$$q_H = h_{fg} = T s_{fg}$$

and the general result of Eq. 13.3 is the expression

$$\frac{dP_{sat}}{dT} = \frac{s_{fg}}{v_{fg}} = \frac{h_{fg}}{T v_{fg}} \tag{13.4}$$

which is called the Clapeyron equation. This is a very simple relation and yet an extremely powerful one. We can experimentally determine the left-hand side of Eq. 13.4, which is the slope of the vapor pressure as a function of temperature. We can also measure the specific volumes of saturated vapor and saturated liquid at the given temperature, which means that the enthalpy change and entropy change of vaporization can both be calculated from Eq. 13.4. This establishes the means to cross from one phase to another in first- or second-law calculations, which was the goal of this development.

We could proceed along the same lines for the change of phase solid to liquid or for solid to vapor. In each case, the result is the Clapeyron equation, in which the appropriate saturation pressure, specific volumes, entropy change, and enthalpy change are involved. For solid i to liquid f, the process occurs along the fusion line, and the result is

$$\frac{dP_{\text{fus}}}{dT} = \frac{s_{if}}{v_{if}} = \frac{h_{if}}{Tv_{if}} \tag{13.5}$$

We note that $v_{if} = v_f - v_i$ is typically a very small number, such that the slope of the fusion line is very steep. (In the case of water, v_{if} is a negative number, which is highly unusual, and the slope of the fusion line is not only steep, it is also negative.)

For sublimation, the change from solid i directly to vapor g, the Clapeyron equation has the values

$$\frac{dP_{\text{sub}}}{dT} = \frac{s_{ig}}{v_{ig}} = \frac{h_{ig}}{Tv_{ig}} \tag{13.6}$$

A special case of the Clapeyron equation involving the vapor phase occurs at low temperatures when the saturation pressure becomes very small. The specific volume v_g is then not only much larger than that of the condensed phase, liquid in Eq. 13.4 or solid in Eq. 13.6, but is also closely represented by the ideal-gas equation of state. The Clapeyron equation then reduces to the form

$$\frac{dP_{\text{sat}}}{dT} = \frac{h_{fg}}{Tv_{fg}} = \frac{h_{fg}P_{\text{sat}}}{RT^2} \tag{13.7}$$

At low temperatures (not near the critical temperature), h_{fg} does not change very much with temperature. If it is assumed to be constant, then Eq. 13.7 can be rearranged and integrated over a range of temperatures to calculate a saturation pressure at a temperature at which it is not known. This point is illustrated by the following example.

EXAMPLE 13.1 Determine the sublimation pressure of water vapor at $-60°C$ using data available in the steam tables.

Control mass: Water.

Solution

Appendix Table B.1.5 of the steam tables does not give saturation pressures for temperatures less than $-40°C$. However, we do notice that h_{ig} is relatively constant in this range; therefore, we proceed to Eq. 13.7 and integrate between the limits $-40°C$ and $-60°C$.

$$\int_1^2 \frac{dP}{P} = \int_1^2 \frac{h_{ig}}{R} \frac{dT}{T^2} = \frac{h_{ig}}{R} \int_1^2 \frac{dT}{T^2}$$

$$\ln \frac{P_2}{P_1} = \frac{h_{ig}}{R}\left(\frac{T_2 - T_1}{T_1 T_2}\right)$$

Let

$$P_2 = 0.0129 \text{ kPa} \qquad T_2 = 233.2 \text{ K} \qquad T_1 = 213.2 \text{ K}$$

Then

$$\ln \frac{P_2}{P_1} = \frac{2838.9}{0.461\ 52}\left(\frac{233.2 - 213.2}{233.2 \times 213.2}\right) = 2.4744$$

$$P_1 = 0.001\ 09 \text{ kPa}$$

EXAMPLE 13.1E Determine the sublimation pressure of water vapor at -70 F using data available in the steam tables.

Control mass: Water.

Solution

Appendix Table F.7.4 of the steam tables does not give saturation pressures for temperatures less than -40 F. However, we do notice that h_{ig} is relatively constant in this range; therefore, we proceed to use Eq. 13.7 and integrate between the limits -40 F and -70 F.

$$\int_1^2 \frac{dP}{P} = \int_1^2 \frac{h_{ig}}{R} \frac{dT}{T^2} = \frac{h_{ig}}{R} \int_1^2 \frac{dT}{T^2}$$

$$\ln \frac{P_2}{P_1} = \frac{h_{ig}}{R}\left(\frac{T_2 - T_1}{T_1 T_2}\right)$$

Let

$$P_2 = 0.0019 \text{ lbf/in.}^2 \qquad T_2 = 419.7 \text{ R} \qquad T_1 = 389.7 \text{ R}$$

Then,

$$\ln \frac{P_2}{P_1} = \frac{1218.7 \times 778}{85.76}\left(\frac{419.7 - 389.7}{419.7 \times 389.7}\right) = 2.0279$$

$$P_1 = 0.000\ 25 \text{ lbf/in.}^2$$

13.2 MATHEMATICAL RELATIONS FOR A HOMOGENEOUS PHASE

In the preceding section, we established the means to calculate differences in enthalpy (and therefore internal energy) and entropy between different phases, in terms of properties that are readily measured. In the following sections, we will develop expressions for calculating differences in these properties within a single, homogeneous phase (gas, liquid, or solid), assuming a simple compressible substance. In order to develop such expressions, it is first necessary to present a mathematical relation that will prove useful in this procedure.

Consider a variable (thermodynamic property) that is a continuous function of x and y.

$$z = f(x, y)$$

$$dz = \left(\frac{\partial z}{\partial x}\right)_y dx + \left(\frac{\partial z}{\partial y}\right)_x dy$$

It is convenient to write this function in the form

$$dz = M\, dx + N\, dy \tag{13.8}$$

where

$$M = \left(\frac{\partial z}{\partial x}\right)_y$$

= partial derivative of z with respect to x (the variable y being held constant)

$$N = \left(\frac{\partial z}{\partial y}\right)_x$$

= partial derivative of z with respect to y (the variable x being held constant)

The physical significance of partial derivatives as they relate to the properties of a pure substance can be explained by referring to Fig. 13.2, which shows a P–v–T surface of the superheated vapor region of a pure substance. It shows constant-temperature, constant-pressure, and constant specific volume planes that intersect at point b on the surface. Thus, the partial derivative $(\partial P/\partial v)_T$ is the slope of curve abc at point b. Line de represents the tangent to curve abc at point b. A similar interpretation can be made of the partial derivatives $(\partial P/\partial T)_v$ and $(\partial v/\partial T)_p$.

If we wish to evaluate the partial derivative along a constant-temperature line, the rules for ordinary derivatives can be applied. Thus, we can write for a constant-temperature process:

$$\left(\frac{\partial P}{\partial v}\right)_T = \frac{dP_T}{dv_T}$$

and the integration can be performed as usual. This point will be demonstrated later in a number of examples.

Let us return to the consideration of the relation

$$dz = M\, dx + N\, dy$$

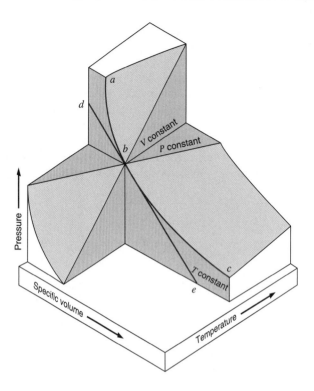

FIGURE 13.2
Schematic representation
of partial derivatives.

If x, y, and z are all point functions (that is, quantities that depend only on the state and are independent of the path), the differentials are exact differentials. If this is the case, the following important relation holds:

$$\left(\frac{\partial M}{\partial y}\right)_x = \left(\frac{\partial N}{\partial x}\right)_y \tag{13.9}$$

The proof of this is

$$\left(\frac{\partial M}{\partial y}\right)_x = \frac{\partial^2 z}{\partial x \partial y}$$

$$\left(\frac{\partial N}{\partial x}\right)_y = \frac{\partial^2 z}{\partial y \partial x}$$

Since the order of differentiation makes no difference when point functions are involved, it follows that

$$\frac{\partial^2 z}{\partial x \partial y} = \frac{\partial^2 z}{\partial y \partial x}$$

$$\left(\frac{\partial M}{\partial y}\right)_x = \left(\frac{\partial N}{\partial x}\right)_y$$

13.3 THE MAXWELL RELATIONS

Consider a simple compressible control mass of fixed chemical composition. The Maxwell relations, which can be written for such a system, are four equations relating the properties P, v, T, and s. These will be found to be useful in the calculation of entropy in terms of the other, measurable properties.

The Maxwell relations are most easily derived by considering the different forms of the thermodynamic property relation, which was the subject of Section 8.5. The two forms of this expression are rewritten here as

$$du = T\,ds - P\,dv \tag{13.10}$$

and

$$dh = T\,ds + v\,dP \tag{13.11}$$

Note that in the mathematical representation of Eq. 13.8, these expressions are of the form

$$u = u(s, v), \qquad h = h(s, P)$$

in both of which entropy is used as one of the two independent properties. This is an undesirable situation in that entropy is one of the properties that cannot be measured. We can, however, eliminate entropy as an independent property by introducing two new properties and thereby two new forms of the thermodynamic property relation. The first of these is the Helmholtz function A,

$$A = U - TS, \qquad a = u - Ts \tag{13.12}$$

Differentiating and substituting Eq. 13.10 results in

$$\begin{aligned} da &= du - T\,ds - s\,dT \\ &= -s\,dT - P\,dv \end{aligned} \tag{13.13}$$

which we note is a form of the property relation utilizing T and v as the independent properties. The second new property is the Gibbs function G,

$$G = H - TS, \qquad g = h - Ts \tag{13.14}$$

Differentiating and substituting Eq. 13.11,

$$\begin{aligned} dg &= dh - T\,ds - s\,dT \\ &= -s\,dT + v\,dP \end{aligned} \tag{13.15}$$

a fourth form of the property relation, this form using T and P as the independent properties.

Since Eqs. 13.10, 13.11, 13.13, and 13.15 are all relations involving only properties, we conclude that these are exact differentials and, therefore, are of the general form of Eq. 13.8,

$$dz = M\,dx + N\,dy$$

in which Eq. 13.9 relates the coefficients M and N,

$$\left(\frac{\partial M}{\partial y}\right)_x = \left(\frac{\partial N}{\partial x}\right)_y$$

It follows from Eq. 13.10 that

$$\left(\frac{\partial T}{\partial v}\right)_s = -\left(\frac{\partial P}{\partial s}\right)_v \tag{13.16}$$

Similarly, from Eqs. 13.11, 13.13, and 13.15 we can write

$$\left(\frac{\partial T}{\partial P}\right)_s = \left(\frac{\partial v}{\partial s}\right)_P \tag{13.17}$$

$$\left(\frac{\partial P}{\partial T}\right)_v = \left(\frac{\partial s}{\partial v}\right)_T \tag{13.18}$$

$$\left(\frac{\partial v}{\partial T}\right)_P = -\left(\frac{\partial s}{\partial P}\right)_T \tag{13.19}$$

These four equations are known as the Maxwell relations for a simple compressible mass, and the great utility of these equations will be demonstrated in later sections of this chapter. As was noted earlier, these relations will enable us to calculate entropy changes in terms of the measurable properties pressure, temperature, and specific volume.

A number of other useful relations can be derived from Eqs. 13.10, 13.11, 13.13, and 13.15. For example, from Eq. 13.10, we can write the relations

$$\left(\frac{\partial u}{\partial s}\right)_v = T, \qquad \left(\frac{\partial u}{\partial v}\right)_s = -P \tag{13.20}$$

Similarly, from the other three equations, we have the following

$$\left(\frac{\partial h}{\partial s}\right)_P = T, \qquad \left(\frac{\partial h}{\partial P}\right)_s = v$$

$$\left(\frac{\partial a}{\partial v}\right)_T = -P, \qquad \left(\frac{\partial a}{\partial T}\right)_v = -s$$

$$\left(\frac{\partial g}{\partial P}\right)_T = v, \qquad \left(\frac{\partial g}{\partial T}\right)_P = -s \tag{13.21}$$

As already noted, the Maxwell relations just presented are written for a simple compressible substance. It is readily evident, however, that similar Maxwell relations can be written for substances involving other effects, such as surface or electrical effects. For example, Eq. 8.9 can be written in the form

$$dU = T\,dS - P\,dV + \mathcal{T}\,dL + \mathcal{S}\,dA + \mathcal{E}\,dZ + \cdots \tag{13.22}$$

Thus, for a substance involving only surface effects, we can write

$$dU = T\,dS + \mathcal{S}\,dA$$

and it follows that for such a substance

$$\left(\frac{\partial T}{\partial A}\right)_S = \left(\frac{\partial \mathcal{S}}{\partial S}\right)_A$$

Other Maxwell relations could also be written for such a substance by writing the property relation in terms of different variables, and this approach could also be extended to systems having multiple effects. This matter also becomes more complex when we consider applying the property relation to a system of variable composition, a topic that will be taken up in Section 13.11.

EXAMPLE 13.2   From an examination of the properties of compressed liquid water, as given in Table B.1.4 of the Appendix, we find that the entropy of compressed liquid is greater than the entropy of saturated liquid for a temperature of 0°C and is less than that of saturated liq-

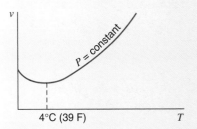

FIGURE 13.3 Sketch for Example 13.2.

uid for all the other temperatures listed. Explain why this follows from other thermodynamic data.

Control mass: Water.

Solution

Suppose we increase the pressure of liquid water that is initially saturated, while keeping the temperature constant. The change of entropy for the water during this process can be found by integrating the following Maxwell relation, Eq. 13.19:

$$\left(\frac{\partial s}{\partial P}\right)_T = -\left(\frac{\partial v}{\partial T}\right)_P$$

Therefore, the sign of the entropy change depends on the sign of the term $(\partial v/\partial T)_P$. The physical significance of this term is that it involves the change in specific volume of water as the temperature changes while the pressure remains constant. As water at moderate pressures and 0°C is heated in a constant-pressure process, the specific volume decreases until the point of maximum density is reached at approximately 4°C, after which it increases. This is shown on a v–T diagram in Fig. 13.3. Thus, the quantity $(\partial v/\partial T)_P$ is the slope of the curve in Fig. 13.3. Since this slope is negative at 0°C, the quantity $(\partial s/\partial P)_T$ is positive at 0°C. At the point of maximum density the slope is zero and, therefore, the constant-pressure line shown in Fig. 8.7 crosses the saturated-liquid line at the point of maximum density.

13.4 THERMODYNAMIC RELATIONS INVOLVING ENTHALPY, INTERNAL ENERGY, AND ENTROPY

Let us first derive two equations, one involving C_p and the other involving C_v.
We have defined C_p as

$$C_p \equiv \left(\frac{\partial h}{\partial T}\right)_P$$

We have also noted that for a pure substance

$$T\, ds = dh - v\, dP$$

Therefore,

$$C_p = \left(\frac{\partial h}{\partial T}\right)_P = T\left(\frac{\partial s}{\partial T}\right)_P \tag{13.23}$$

Similarly, from the definition of C_v,

$$C_v \equiv \left(\frac{\partial u}{\partial T}\right)_v$$

and the relation

$$T\,ds = du + P\,dv$$

it follows that

$$C_v = \left(\frac{\partial u}{\partial T}\right)_v = T\left(\frac{\partial s}{\partial T}\right)_v \qquad (13.24)$$

We will now derive a general relation for the change of enthalpy of a pure substance. We first note that for a pure substance

$$h = h(T, P)$$

Therefore,

$$dh = \left(\frac{\partial h}{\partial T}\right)_P dT + \left(\frac{\partial h}{\partial P}\right)_T dP$$

From the relation

$$T\,ds = dh - v\,dP$$

it follows that

$$\left(\frac{\partial h}{\partial P}\right)_T = v + T\left(\frac{\partial s}{\partial P}\right)_T$$

Substituting the Maxwell relation, Eq. 13.19, we have

$$\left(\frac{\partial h}{\partial P}\right)_T = v - T\left(\frac{\partial v}{\partial T}\right)_P \qquad (13.25)$$

On substituting this equation and Eq. 13.23, we have

$$dh = C_p\,dT + \left[v - T\left(\frac{\partial v}{\partial T}\right)_P\right]dP \qquad (13.26)$$

Along an isobar we have

$$dh_p = C_p\,dT_p$$

and along an isotherm,

$$dh_T = \left[v - T\left(\frac{\partial v}{\partial T}\right)_P\right]dP_T \qquad (13.27)$$

The significance of Eq. 13.26 is that this equation can be integrated to give the change in enthalpy associated with a change of state

$$h_2 - h_1 = \int_1^2 C_p\,dT + \int_1^2\left[v - T\left(\frac{\partial v}{\partial T}\right)_P\right]dP \qquad (13.28)$$

The information needed to integrate the first term is a constant-pressure specific heat along one (and only one) isobar. The integration of the second integral requires that an equation of state giving the relation between P, v, and T be known. Furthermore, it is advantageous to have this equation of state explicit in v, for then the derivative $(\partial v/\partial T)_P$ is readily evaluated.

This matter can be further illustrated by reference to Fig. 13.4. Suppose we wish to know the change of enthalpy between states 1 and 2. We might determine this change along path 1–x–2, which consists of one isotherm, 1–x, and one isobar, x–2. Thus, we could integrate Eq. 13.28:

$$h_2 - h_1 = \int_{T_1}^{T_2} C_p\, dT + \int_{P_1}^{P_2} \left[v - T \left(\frac{\partial v}{\partial T} \right)_P \right] dP$$

Since $T_1 = T_x$ and $P_2 = P_x$, this can be written

$$h_2 - h_1 = \int_{T_x}^{T_2} C_p\, dT + \int_{P_1}^{P_x} \left[v - T \left(\frac{\partial v}{\partial T} \right)_P \right] dP$$

The second term in this equation gives the change in enthalpy along the isotherm 1–x and the first term the change in enthalpy along the isobar x–2. When these are added together, the result is the net change in enthalpy between 1 and 2. Therefore, the constant-pressure specific heat must be known along the isobar passing through 2 and x. The change in enthalpy could also be found by following path 1–y–2, in which case the constant-pressure specific heat must be known along the 1–y isobar. If the constant-pressure specific heat is known at another pressure, say, the isobar passing through m–n, the change in enthalpy can be found by following path 1–m–n–2. This involves calculating the change of enthalpy along two isotherms—1–m and n–2.

Let us now derive a similar relation for the change of internal energy. All the steps in this derivation are given but without detailed comment. Note that the starting point is to write $u = u(T, v)$, whereas in the case of enthalpy the starting point was $h = h(T, P)$.

$$u = f(T, v)$$

$$du = \left(\frac{\partial u}{\partial T} \right)_v dT + \left(\frac{\partial u}{\partial v} \right)_T dv$$

$$T\, ds = du + P\, dv$$

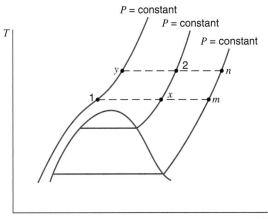

FIGURE 13.4 Sketch showing various paths by which a given change of state can take place.

Therefore,

$$\left(\frac{\partial u}{\partial v}\right)_T = T\left(\frac{\partial s}{\partial v}\right)_T - P \tag{13.29}$$

Substituting the Maxwell relation, Eq. 13.18, we have

$$\left(\frac{\partial u}{\partial v}\right)_T = T\left(\frac{\partial P}{\partial T}\right)_v - P$$

Therefore,

$$du = C_v\,dT + \left[T\left(\frac{\partial P}{\partial T}\right)_v - P\right]dv \tag{13.30}$$

Along an isometric this reduces to

$$du_v = C_v\,dT_v$$

and along an isotherm we have

$$du_T = \left[T\left(\frac{\partial P}{\partial T}\right)_v - P\right]dv_T \tag{13.31}$$

In a manner similar to that outlined earlier for changes in enthalpy, the change of internal energy for a given change of state for a pure substance can be determined from Eq. 13.30 if the constant-volume specific heat is known along one isometric and an equation of state explicit in P [to obtain the derivative $(\partial P/\partial T)_v$] is available in the region involved. A diagram similar to Fig. 13.4 could be drawn, with the isobars replaced with isometrics, and the same general conclusions would be reached.

To summarize, we have derived Eqs. 13.26 and 13.30:

$$dh = C_p\,dT + \left[v - T\left(\frac{\partial v}{\partial T}\right)_P\right]dP$$

$$du = C_v\,dT + \left[T\left(\frac{\partial P}{\partial T}\right)_v - P\right]dv$$

The first of these equations concerns the change of enthalpy, the constant-pressure specific heat, and is particularly suited to an equation of state explicit in v. The second equation concerns the change of internal energy and the constant-volume specific heat, and is particularly suited to an equation of state explicit in P. If the first of these equations is used to determine the change of enthalpy, the internal energy is readily found by noting that

$$u_2 - u_1 = h_2 - h_1 - (P_2 v_2 - P_1 v_1)$$

If the second equation is used to find changes of internal energy, the change of enthalpy is readily found from this same relation. Which of these two equations is used to determine changes in internal energy and enthalpy will depend on the information available for specific heat and an equation of state (or other P–v–T data).

Two parallel expressions can be found for the change of entropy.

$$s = s(T, P)$$

$$ds = \left(\frac{\partial s}{\partial T}\right)_P dT + \left(\frac{\partial s}{\partial P}\right)_T dP$$

Substituting Eqs. 13.19 and 13.23, we have

$$ds = C_p \frac{dT}{T} - \left(\frac{\partial v}{\partial T}\right)_P dP \tag{13.32}$$

$$s_2 - s_1 = \int_1^2 C_p \frac{dT}{T} - \int_1^2 \left(\frac{\partial v}{\partial T}\right)_P dP \tag{13.33}$$

Along an isobar we have

$$(s_2 - s_1)_P = \int_1^2 C_p \frac{dT_P}{T}$$

and along an isotherm

$$(s_2 - s_1)_T = -\int_1^2 \left(\frac{\partial v}{\partial T}\right)_P dP$$

Note from Eq. 13.33 that if a constant-pressure specific heat is known along one isobar and an equation of state explicit in v is available, the change of entropy can be evaluated. This is analogous to the expression for the change of enthalpy given in Eq. 13.26.

$$s = s(T, v)$$

$$ds = \left(\frac{\partial s}{\partial T}\right)_v dT + \left(\frac{\partial s}{\partial v}\right)_T dv$$

Substituting Eqs. 13.18 and 13.24 gives

$$ds = C_v \frac{dT}{T} + \left(\frac{\partial P}{\partial T}\right)_v dv \tag{13.34}$$

$$s_2 - s_1 = \int_1^2 C_v \frac{dT}{T} + \int_1^2 \left(\frac{\partial P}{\partial T}\right)_v dv \tag{13.35}$$

This expression for change of entropy concerns the change of entropy along an isometric where the constant-volume specific heat is known and along an isotherm where an equation of state explicit in P is known. Thus, it is analogous to the expression for change of internal energy given in Eq. 13.30.

EXAMPLE 13.3   Over a certain small range of pressures and temperatures, the equation of state of a certain substance is given with reasonable accuracy by the relation

$$\frac{Pv}{RT} = 1 - C' \frac{P}{T^4}$$

or

$$v = \frac{RT}{P} - \frac{C}{T^3}$$

where C and C' are constants.

Derive an expression for the change of enthalpy and entropy of this substance in an isothermal process.

Control mass: Gas.

Solution

Since the equation of state is explicit in v, Eq. 13.27 is particularly relevant to the change in enthalpy. On integrating this equation, we have

$$(h_2 - h_1)_T = \int_1^2 \left[v - T \left(\frac{\partial v}{\partial T} \right)_P \right] dP_T$$

From the equation of state,

$$\left(\frac{\partial v}{\partial T} \right)_P = \frac{R}{P} + \frac{3C}{T^4}$$

Therefore,

$$(h_2 - h_1)_T = \int_1^2 \left[v - T \left(\frac{R}{P} + \frac{3C}{T^4} \right) \right] dP_T$$

$$= \int_1^2 \left[\frac{RT}{P} - \frac{C}{T^3} - \frac{RT}{P} - \frac{3C}{T^3} \right] dP_T$$

$$(h_2 - h_1)_T = \int_1^2 - \frac{4C}{T^3} \, dP_T = - \frac{4C}{T^3} (P_2 - P_1)_T$$

For the change in entropy we use Eq. 13.33, which is particularly relevant for an equation of state explicit in v.

$$(s_2 - s_1)_T = - \int_1^2 \left(\frac{\partial v}{\partial T} \right)_P dP_T = - \int_1^2 \left(\frac{R}{P} + \frac{3C}{T^4} \right) dP_T$$

$$(s_2 - s_1)_T = -R \ln \left(\frac{P_2}{P_1} \right)_T - \frac{3C}{T^4} (P_2 - P_1)_T$$

13.5 VOLUME EXPANSIVITY AND ISOTHERMAL AND ADIABATIC COMPRESSIBILITY

The student has most likely encountered the coefficient of linear expansion in his or her studies of strength of materials. This coefficient indicates how the length of a solid body is influenced by a change in temperature while the pressure remains constant. In terms of the notation of partial derivatives, the coefficient of linear expansion, δ_T, is defined as

$$\delta_T = \frac{1}{L} \left(\frac{\partial L}{\partial T} \right)_P \qquad (13.36)$$

A similar coefficient can be defined for changes in volume. Such a coefficient is applicable to liquids and gases as well as to solids. This coefficient of volume expansion, α_P, also called the volume expansivity, is an indication of the change in volume as temperature changes while the pressure remains constant. The definition of volume expansivity is

$$\alpha_P \equiv \frac{1}{V} \left(\frac{\partial V}{\partial T} \right)_P = \frac{1}{v} \left(\frac{\partial v}{\partial T} \right)_P = 3\delta_T \qquad (13.37)$$

and it equals three times the coefficient of linear expansion. You differentiate $V = L_x L_y L_z$ with temperature to prove that which is left as a homework exercise. Notice that it is the volume expansivity which enters into the expressions for calculating changes in enthalpy, Eq. 13.26, and entropy, Eq. 13.32.

The isothermal compressibility, β_T, is an indication of the change in volume as pressure changes while the temperature remains constant. The definition of the isothermal compressibility is

$$\beta_T \equiv -\frac{1}{V}\left(\frac{\partial V}{\partial P}\right)_T = -\frac{1}{v}\left(\frac{\partial v}{\partial P}\right)_T \tag{13.38}$$

The reciprocal of the isothermal compressibility is called the isothermal bulk modulus, B_T.

$$B_T \equiv -v\left(\frac{\partial P}{\partial v}\right)_T \tag{13.39}$$

The adiabatic compressibility, β_s, is an indication of the change in volume as pressure changes while the entropy remains constant; it is defined as

$$\beta_s \equiv -\frac{1}{v}\left(\frac{\partial v}{\partial P}\right)_S \tag{13.40}$$

The adiabatic bulk modulus, B_s, is the reciprocal of the adiabatic compressibility.

$$B_s \equiv -v\left(\frac{\partial P}{\partial v}\right)_s \tag{13.41}$$

The velocity of sound, c, in a medium is defined by the relation

$$c^2 = \left(\frac{\partial P}{\partial \rho}\right)_s \tag{13.42}$$

This can also be expressed as

$$c^2 = -v^2\left(\frac{\partial P}{\partial v}\right)_s = vB_s \tag{13.43}$$

in terms of the adiabatic bulk modulus B_s. For a compressible medium such as a gas the speed of sound becomes modest, whereas in an incompressible state such as a liquid or a solid it can be quite large.

The volume expansivity and isothermal and adiabatic compressibility are thermodynamic properties of a substance, and for a simple compressible substance are functions of two independent properties. Values of these properties are found in the standard handbooks of physical properties. The following examples give an indication of the use and significance of the volume expansivity and isothermal compressibility.

EXAMPLE 13.4 The pressure on a block of copper having a mass of 1 kg is increased in a reversible process from 0.1 to 100 MPa while the temperature is held constant at 15°C. Determine the work done on the copper during this process, the change in entropy per kilogram of copper, the heat transfer, and the change of internal energy per kilogram.

Over the range of pressure and temperature in this problem, the following data can be used:

$$\text{Volume expansivity} = \alpha_p = 5.0 \times 10^{-5} \text{K}^{-1}$$

$$\text{Isothermal compressibility} = \beta_T = 8.6 \times 10^{-12} \text{m}^2/\text{N}$$

$$\text{Specific volume} = 0.000\ 114\ \text{m}^3/\text{kg}$$

Analysis

Control mass:	Copper block.
States:	Initial and final states known.
Process:	Constant temperature, reversible.

The work done during the isothermal compression is

$$w = \int P\, dv_T$$

The isothermal compressibility has been defined as

$$\beta_T = -\frac{1}{v}\left(\frac{\partial v}{\partial P}\right)_T$$

$$v\beta_T\, dP_T = -dv_T$$

Therefore, for this isothermal process,

$$w = -\int_1^2 v\beta_T P\, dP_T$$

Since v and β_T remain essentially constant, this is readily integrated:

$$w = -\frac{v\beta_T}{2}(P_2^2 - P_1^2)$$

The change of entropy can be found by considering the Maxwell relation, Eq. 13.19, and the definition of volume expansivity.

$$\left(\frac{\partial s}{\partial P}\right)_T = -\left(\frac{\partial v}{\partial T}\right)_P = -\frac{v}{v}\left(\frac{\partial v}{\partial T}\right)_P = -v\alpha_P$$

$$ds_T = -v\alpha_P\, dP_T$$

This equation can be readily integrated, if we assume that v and α_P remain constant:

$$(s_2 - s_1)_T = -v\alpha_p(P_2 - P_1)_T$$

The heat transfer for this reversible isothermal process is

$$q = T(s_2 - s_1)$$

The change in internal energy follows directly from the first law.

$$(u_2 - u_1) = q - w$$

Solution

$$w = -\frac{v\beta_T}{2}(P_2^2 - P_1^2)$$

$$= -\frac{0.000\ 114 \times 8.6 \times 10^{-12}}{2}(100^2 - 0.1^2) \times 10^{12}$$

$$= -4.9 \text{ J/kg}$$

$$(s_2 - s_1)_T = -v\alpha_p(P_2 - P_1)_T$$

$$= -0.000\ 114 \times 5.0 \times 10^{-5}(100 - 0.1) \times 10^6$$

$$= -0.5694 \text{ J/kg K}$$

$$q = T(s_2 - s_2) = -288.2 \times 0.5694 = -164.1 \text{ J/kg}$$

$$(u_2 - u_1) = q - w = -164.1 - (-4.9) = -159.2 \text{ J/kg}$$

13.6 REAL-GAS BEHAVIOR AND EQUATIONS OF STATE

In Section 3.4, we examined the P–v–T behavior of gases, and we defined the compressibility factor in Eq. 3.6,

$$Z = \frac{Pv}{RT}$$

We then proceeded to develop the generalized compressibility chart, presented in Appendix Fig. D.1 in terms of the reduced pressure and temperature. The generalized chart does not apply specifically to any one substance, but is instead an approximate relation that is reasonably accurate for many substances, especially those that are fairly simple in molecular structure. In this sense, the generalized compressibility chart can be viewed as one aspect of generalized behavior of substances, and also as a graphical form of equation of state representing real behavior of gases and liquids over a broad range of variables.

To gain additional insight into the behavior of gases at low density, let us examine the low-pressure portion of the generalized compressibility chart in greater detail. This behavior is as shown in Fig. 13.5. The isotherms are essentially straight lines in this region, and their slope is of particular importance. Note that the slope increases as T_r increases until a maximum value is reached at a T_r of about 5, and then the slope decreases toward the $Z = 1$ line for higher temperatures. That single temperature, about 2.5 times the critical temperature, for which

$$\lim_{P \to 0}\left(\frac{\partial Z}{\partial P}\right)_T = 0 \tag{13.44}$$

is defined as the Boyle temperature of the substance. This is the only temperature at which a gas behaves exactly as an ideal gas at low, but finite pressures, since all other isotherms

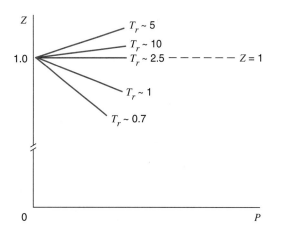

FIGURE 13.5 Low-pressure region of compressibility chart.

go to zero pressure on Fig. 13.5 with a nonzero slope. To amplify this point, let us consider the residual volume α,

$$\alpha = \frac{RT}{P} - v \tag{13.45}$$

Multiplying this equation by P we have

$$\alpha P = RT - Pv$$

Thus, the quantity αP is the difference between $RT-$ and Pv. Now as $P \to 0$, $Pv \to RT$. However, it does not necessarily follow that $\alpha \to 0$ as $P \to 0$. Instead, it is only required that α remain finite. The derivative in Eq. 13.44 can be written as

$$\lim_{P \to 0} \left(\frac{\partial Z}{\partial P} \right)_T = \lim_{P \to 0} \left(\frac{Z - 1}{P - 0} \right)$$

$$= \lim_{P \to 0} \frac{1}{RT} \left(v - \frac{RT}{P} \right)$$

$$= -\frac{1}{RT} \lim_{P \to 0} (\alpha) \tag{13.46}$$

from which we find that α tends to zero as $P \to 0$ only at the Boyle temperature, since that is the only temperature for which the isothermal slope is zero on Fig. 13.5. It is perhaps a somewhat surprising result that in the limit as $P \to 0$, $Pv \to RT$. In general, however, the quantity $(RT/P - v)$ does not go to zero but is instead a small difference between two large values. This does have an effect on certain other properties of the gas.

The compressibility behavior of low-density gases as noted in Fig. 13.5 is the result of intermolecular interactions and can be expressed in the form of equation of state called the virial equation, which is derived from statistical thermodynamics. The result is

$$Z = \frac{Pv}{RT} = 1 + \frac{B(T)}{v} + \frac{C(T)}{v^2} + \frac{D(T)}{v^3} + \cdots \tag{13.47}$$

where $B(T)$, $C(T)$, $D(T)$ are temperature dependent and are called virial coefficients. $B(T)$ is termed the second virial coefficient and is due to binary interactions on the molecular

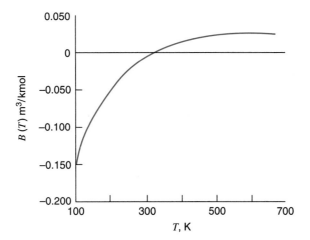

FIGURE 13.6 The second virial coefficient for nitrogen.

level. The general temperature dependence of the second virial coefficient is as shown for nitrogen in Fig. 13.6. If we multiply Eq. 13.47 by RT/P, the result can be rearranged to the form

$$\frac{RT}{P} - v = \alpha = -B(T)\frac{RT}{Pv} - C(T)\frac{RT}{Pv^2}\cdots \qquad (13.48)$$

In the limit, as $P \to 0$,

$$\lim_{P\to 0} \alpha = -B(T) \qquad (13.49)$$

and we conclude from Eqs. 13.44 and 13.46 that the single temperature at which $B(T) = 0$, Fig. 13.6, is the Boyle temperature. The second virial coefficient can be viewed as the first-order correction for nonideality of the gas, and consequently becomes of considerable importance and interest. In fact, the low-density behavior of the isotherms shown in Fig. 13.5 is directly attributable to the second virial coefficient.

Another aspect of generalized behavior of gases is the behavior of isotherms in the vicinity of the critical point. If we plot experimental data on $P–v$ coordinates, it is found that the critical isotherm is unique in that it goes through a horizontal inflection point at the critical point as shown in Fig. 13.7. Mathematically, this means that the first two derivatives are zero at the critical point.

$$\left(\frac{\partial P}{\partial v}\right)_{T_c} = 0 \qquad \text{at C.P.} \qquad (13.50)$$

$$\left(\frac{\partial^2 P}{\partial v^2}\right)_{T_c} = 0 \qquad \text{at C.P.} \qquad (13.51)$$

a feature that is used to constrain many equations of state.

To this point, we have discussed the generalized compressibility chart, a graphical form of equation of state, and the virial equation, a theoretically founded equation of state. We now proceed to discuss other analytical equations of state, which may be either generalized behavior in form, or empirical equations, relying on specific $P–v–T$ data of their constants. The oldest generalized equation, the van der Waals equation, was presented in

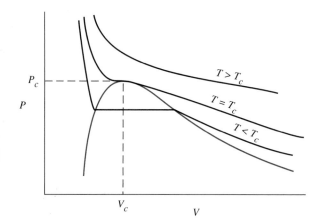

FIGURE 13.7 Plot of isotherms in the region of the critical point on pressure–volume coordinates for a typical pure substance.

1873 as a semitheoretical improvement over the ideal-gas model. The van der Waals equation of state has two constants and is written as

$$P = \frac{RT}{v - b} - \frac{a}{v^2} \qquad (13.52)$$

The constant b is intended to correct for the volume occupied by the molecules, and the term a/v^2 is a correction that accounts for the intermolecular forces of attraction. As might be expected in the case of a generalized equation, the constants a and b are evaluated from the general behavior of gases. In particular, these constants are evaluated by noting that the critical isotherm passes through a point of inflection at the critical point and that the slope is zero at this point. Thus, for the van der Waals equation of state we have

$$\left(\frac{\partial P}{\partial v}\right)_T = -\frac{RT}{(v - b)^2} + \frac{2a}{v^3} \qquad (13.53)$$

$$\left(\frac{\partial^2 P}{\partial v^2}\right)_T = \frac{2RT}{(v - b)^3} - \frac{6a}{v^4} \qquad (13.54)$$

Since both of these derivatives are equal to zero at the critical point we can write

$$-\frac{RT_c}{(v_c - b)^2} + \frac{2a}{v_c^3} = 0$$

$$\frac{2RT_c}{(v_c - b)^3} - \frac{6a}{v_c^4} = 0 \qquad (13.55)$$

$$P_c = \frac{RT_c}{(v_c - b)} - \frac{a}{v_c^2}$$

Solving these three equations, we find

$$v_c = 3b$$

$$a = \frac{27}{64} \frac{R^2 T_c^2}{P_c} \qquad (13.56)$$

$$b = \frac{RT_c}{8P_c}$$

The compressibility factor at the critical point for the van der Waals equation is

$$Z_c = \frac{P_c v_c}{RT_c} = \frac{3}{8}$$

which is considerably higher than the actual value for any substance.

A simple equation of state that is considerably more accurate than the van der Waals equation is that proposed by Redlich and Kwong in 1949.

$$P = \frac{RT}{v - b} - \frac{a}{v(v + b)T^{1/2}} \qquad (13.57)$$

with

$$a = 0.427\,48\,\frac{R^2 T_c^{5/2}}{P_c} \qquad (13.58)$$

$$b = 0.086\,64\,\frac{RT_c}{P_c} \qquad (13.59)$$

The numerical values in the constants have been determined by a procedure similar to that followed in the van der Waals equation. Because of its simplicity, this equation could not be expected to be sufficiently accurate to find use in the calculation of precision tables of thermodynamic properties. It has, however, been used frequently for mixture calculations and phase equilibrium correlations with reasonably good success. A number of modified versions of this equation have also been utilized in recent years.

One of the best known empirical equations of state is the Benedict–Webb–Rubin equation, often termed the BWR equation. The original equation, proposed in 1940, containing eight empirical constants, was given in Chapter 3 as Eq. 3.9. The constants for a number of substances are given in Appendix Table D.2. This equation, and particularly a number of modifications to it, have been widely used over the years.

One particularly interesting modification of the BWR equation of state is the Lee–Kesler equation, which was proposed in 1975. This equation has 12 constants and is written in terms of generalized properties as

$$Z = \frac{P_r v_r'}{T_r} = 1 + \frac{B}{v_r'} + \frac{C}{v_r'^2} + \frac{D}{v_r'^5} + \frac{c_4}{T_r^3 v_r'^2}\left(\beta + \frac{\gamma}{v_r'^2}\right)\exp\left(-\frac{\gamma}{v_r'^2}\right)$$

$$B = b_1 - \frac{b_2}{T_r} - \frac{b_3}{T_r^2} - \frac{b_4}{T_r^3}$$

$$C = c_1 - \frac{c_2}{T_r} + \frac{c_3}{T_r^3} \qquad (13.60)$$

$$D = d_1 + \frac{d_2}{T_r}$$

in which the variable v_r' is not the true reduced specific volume but is instead defined as

$$v_r' = \frac{v}{RT_c/P_c} \qquad (13.61)$$

Empirical constants for simple fluids for this equation are also given in Appendix Table D.3.

13.7 THE GENERALIZED CHART FOR CHANGES OF ENTHALPY AT CONSTANT TEMPERATURE

In Section 13.4, Eq. 13.27 was derived for the change of enthalpy at constant temperature.

$$(h_2 - h_1)_T = \int_1^2 \left[v - T\left(\frac{\partial v}{\partial T}\right)_P \right] dP_T$$

This equation is appropriately used when a volume-explicit equation of state is known. Otherwise, it is more convenient to calculate the isothermal change in internal energy from Eq. 13.31

$$(u_2 - u_1)_T = \int_1^2 \left[T\left(\frac{\partial P}{\partial T}\right)_v - P \right] dv_T$$

and then calculate the change in enthalpy from its definition as

$$(h_2 - h_1) = (u_2 - u_1) + (P_2 v_2 - P_1 v_1)$$
$$= (u_2 - u_1) + RT(Z_2 - Z_1)$$

To determine the change in enthalpy behavior consistent with the generalized chart, Fig. D.1, we follow the second of these approaches, since the Lee–Kesler generalized equation of state, Eq. 13.60, is a pressure-explicit form in terms of specific volume and temperature. Equation 13.60 is expressed in terms of the compressibility factor Z, so we write

$$P = \frac{ZRT}{v}, \qquad \left(\frac{\partial P}{\partial T}\right)_v = \frac{ZR}{v} + \frac{RT}{v}\left(\frac{\partial Z}{\partial T}\right)_v$$

Therefore, substituting into Eq. 13.31, we have

$$du = \frac{RT^2}{v}\left(\frac{\partial Z}{\partial T}\right)_v dv$$

But

$$\frac{dv}{v} = \frac{dv_r'}{v_r'} \qquad \frac{dT}{T} = \frac{dT_r}{T_r}$$

so that, in terms of reduced variables,

$$\frac{1}{RT_c} du = \frac{T_r^2}{v_r'}\left(\frac{\partial Z}{\partial T_r}\right)_{v_r'} dv_r'$$

This expression is now integrated at constant temperature from any given state (P_r, v_r') to the ideal-gas limit $(P_r^* \to 0, v_r'^* \to \infty)$(the superscript * will always denote an ideal-gas state or property), causing an internal energy change or departure from the ideal-gas value at the given state,

$$\frac{u^* - u}{RT_c} = \int_{v_r'}^{\infty} \frac{T_r^2}{v_r'}\left(\frac{\partial Z}{\partial T_r}\right)_{v_r'} dv_r' \qquad (13.62)$$

The integral on the right-hand side of Eq. 13.62 can be evaluated from the Lee–Kesler equation, Eq. 13.60. The corresponding enthalpy departure at the given state (P_r, v_r') is then found from integrating Eq. 13.62 to be

$$\frac{h^* - h}{RT_c} = \frac{u^* - u}{RT_c} + T_r(1 - Z) \tag{13.63}$$

Following the same procedure as for the compressibility factor, we can evaluate Eq. 13.63 with the set of Lee–Kesler simple-fluid constants to give a simple-fluid enthalpy departure. The values for the enthalpy departure are shown graphically in Fig. D.2. Use of the enthalpy departure function is illustrated in the following example.

EXAMPLE 13.5 Nitrogen is throttled from 20 MPa, $-70°C$, to 2 MPa in an adiabatic, steady-state, steady-flow process. Determine the final temperature of the nitrogen.

Control volume:	Throttling valve.
Inlet state:	P_1, T_1 known; state fixed.
Exit state:	P_2 known.
Process:	Steady-state, throttling process.
Diagram:	Figure 13.8.
Model:	Generalized charts, Fig. D.2.

Analysis

First law:

$$h_1 = h_2$$

Solution

Using values from Table A.2, we have

$$P_1 = 20\text{ MPa} \qquad P_{r1} = \frac{20}{3.39} = 5.9$$

$$T_1 = 203.2\text{ K} \qquad T_{r1} = \frac{203.2}{126.2} = 1.61$$

$$P_2 = 2\text{ MPa} \qquad P_{r2} = \frac{2}{3.39} = 0.59$$

From the generalized charts, Fig. D.2, for the change in enthalpy at constant temperature, we have

$$\frac{h_1^* - h_1}{RT_c} = 2.1$$

$$h_1^* - h_1 = 2.1 \times 0.2968 \times 126.2 = 78.7\text{ kJ/kg}$$

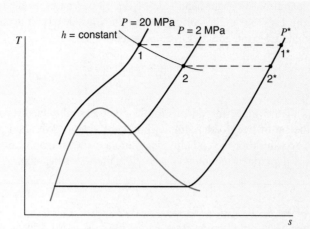

FIGURE 13.8 Sketch for Example 13.5.

It is now necessary to assume a final temperature and to check whether the net change in enthalpy for the process is zero. Let us assume that $T_2 = 146$ K. Then the change in enthalpy between 1* and 2* can be found from the zero-pressure, specific-heat data.

$$h_1^* - h_2^* = C_{p0}(T_1^* - T_2^*) = 1.0416(203.2 - 146) = +59.6 \text{ kJ/kg}$$

(The variation in C_{p0} with temperature can be taken into account when necessary.)
We now find the enthalpy change between 2* and 2.

$$T_{r2} = \frac{146}{126.2} = 1.157 \qquad P_{r2} = 0.59$$

Therefore, from the enthalpy departure chart, Fig. D.2, at this state

$$\frac{h_2^* - h_2}{RT_c} = 0.5$$

$$h_2^* - h_2 = 0.5 \times 0.2968 \times 126.2 = 19.5 \text{ kJ/kg}$$

We now check to see whether the net change in enthalpy for the process is zero.

$$h_1 - h_2 = 0 = -(h_1^* - h_1) + (h_1^* - h_2^*) + (h_2^* - h_2)$$

$$= -78.7 + 59.6 + 19.5 \approx 0$$

It essentially checks. We conclude that the final temperature is approximately 146 K. It is interesting that the thermodynamic tables for nitrogen, Table B.6, give essentially this same value for the final temperature.

13.8 THE GENERALIZED CHART FOR CHANGES OF ENTROPY AT CONSTANT TEMPERATURE

In this section we wish to develop a generalized chart giving entropy departures from ideal–gas values at a given temperature and pressure, in a manner similar to that followed for enthalpy in the previous section. Once again, we have two alternatives. From Eq. 13.32, at constant temperature,

$$ds_T = -\left(\frac{\partial v}{\partial T}\right)_P dP_T$$

which is convenient for use with a volume-explicit equation of state. The Lee–Kesler expression, Eq. 13.60, is, however, a pressure-explicit equation. It is therefore more appropriate to use Eq. 13.34, which is, along an isotherm,

$$ds_T = \left(\frac{\partial P}{\partial T}\right)_v dv_T$$

In the Lee–Kesler form, in terms of reduced properties, this equation becomes

$$\frac{ds}{R} = \left(\frac{\partial P_r}{\partial T_r}\right)_{v_r'} dv_r'$$

When this expression is integrated from a given state (P_r, v_r') to the ideal–gas limit $(P_r^* \to 0, v_r'^* \to \infty)$, there is a problem because ideal–gas entropy is a function of pressure and approaches infinity as the pressure approaches zero. We can eliminate this problem with a two-step procedure. First, the integral is taken only to a certain finite $P_r^*, v_r'^*$, which gives the entropy change

$$\frac{s_{p*}^* - s_p}{R} = \int_{v_r'}^{v_r'^*} \left(\frac{\partial P_r}{\partial T_r}\right)_{v_r'} dv_r' \tag{13.64}$$

This integration by itself is not entirely acceptable, because it contains the entropy at some arbitrary, low-reference pressure. A value for the reference pressure would have to be specified. Let us now repeat the integration over the same change of state, except this time for a hypothetical ideal gas. The entropy change for this integration is

$$\frac{s_{p*}^* - s_p^*}{R} = +\ln\frac{P}{P^*} \tag{13.65}$$

If we now subtract Eq. 13.65 from Eq. 13.64, the result is the difference in entropy of a hypothetical ideal gas at a given state (T_r, P_r) and that of the real substance at the same state, or

$$\frac{s_p^* - s_p}{R} = -\ln\frac{P}{P*} + \int_{v_r'}^{v_r'^* \to \infty} \left(\frac{\partial P_r}{\partial T_r}\right)_{v_r'} dv_r' \tag{13.66}$$

Here the values associated with the arbitrary reference state $P_r^*, v_r'^*$, cancel out of the right-hand side of the equation. (The first term of the integral includes the term $+\ln(P/P^*)$, which cancels the other term. The three different states associated with the development of Eq. 13.66 are shown in Fig. 13.9.

The same procedure that was given in Section 13.7 for enthalpy departure values is followed for generalized entropy departure values. The Lee–Kesler simple-fluid constants are used in evaluating the integral of Eq. 13.66 and yield a simple-fluid entropy departure. The values for the entropy departure are shown graphically in Fig. D.3.

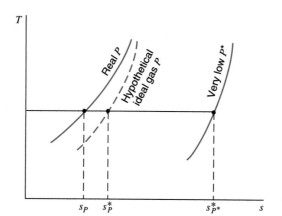

FIGURE 13.9 Real and ideal gas states and entropies.

EXAMPLE 13.6   Nitrogen at 8 MPa, 150 K, is throttled to 0.5 MPa. After the gas passes through a short length of pipe, its temperature is measured and found to be 125 K. Determine the heat transfer and the change of entropy using the generalized charts. Compare these results with those obtained by using the nitrogen tables.

Control volume:	Throttle and pipe.
Inlet state:	P_1, T_1 known; state fixed.
Exit state:	P_2, T_2 known; state fixed.
Process:	Steady state.
Diagram:	Figure 13.10.
Model:	Generalized charts, results to be compared with those obtained with nitrogen tables.

Analysis

There is no work done, and we neglect changes in kinetic and potential energies. Therefore, per kilogram,
First law:

$$q + h_1 = h_2$$

$$q = h_2 - h_1 = -(h_2^* - h_2) + (h_2^* - h_1^*) + (h_1^* - h_1)$$

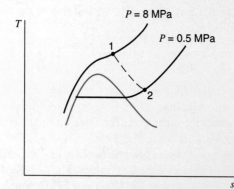

FIGURE 13.10
Sketch for Example 13.6.

Solution

Using values from Table A.2, we have

$$P_{r1} = \frac{8}{3.39} = 2.36 \qquad T_{r1} = \frac{150}{126.2} = 1.189$$

$$P_{r2} = \frac{0.5}{3.39} = 0.147 \qquad T_{r2} = \frac{125}{126.2} = 0.99$$

From Fig. D.2,

$$\frac{h_1^* - h_1}{RT_c} = 2.5$$

$$h_1^* - h_1 = 2.5 \times 0.2968 \times 126.2 = 93.6 \text{ kJ/kg}$$

$$\frac{h_2^* - h_2}{RT_c} = 0.15$$

$$h_2^* - h_2 = 0.15 \times 0.2968 \times 126.2 = 5.6 \text{ kJ/kg}$$

Assuming a constant specific heat for the ideal gas, we have

$$h_2^* - h_1^* = C_{p0}(T_2 - T_1) = 1.0416(125 - 150) = -26.0 \text{ kJ/kg}$$

$$q = -5.6 - 26.0 + 93.6 = 62.0 \text{ kJ/kg}$$

From the nitrogen tables, Table B.6, we can find the change of enthalpy directly.

$$q = h_2 - h_1 = 123.77 - 61.92 = 61.85 \text{ kJ/kg}$$

To calculate the change of entropy using the generalized charts, we proceed as follows:

$$s_2 - s_1 = -(s_{P_2,T_2}^* - s_2) + (s_{P_2,T_2}^* - s_{P_1,T_1}^*) + (s_{P_1,T_1}^* - s_1)$$

From Fig. D.3

$$\frac{s_{P_1,T_1}^* - s_{P_1,T_1}}{R} = 1.6$$

$$s_{P_1,T_1}^* - s_{P_1,T_1} = 1.6 \times 0.2968 = 0.475 \text{ kJ/kg K}$$

$$\frac{s_{P_2,T_2}^* - s_{P_2,T_2}}{R} = 0.1$$

$$s_{P_2,T_2}^* - s_{P_2,T_2} = 0.1 \times 0.2968 = 0.0297 \text{ kJ/kg K}$$

Assuming a constant specific heat for the ideal gas, we have

$$s^*_{P_2, T_2} - s^*_{P_1, T_1} = C_{p0} \ln \frac{T_2}{T_1} - R \ln \frac{P_2}{P_1}$$

$$= 1.0416 \ln \frac{125}{150} - 0.2968 \ln \frac{0.5}{8}$$

$$= 0.6330 \text{ kJ/kg K}$$

$$s_2 - s_1 = -0.0297 + 0.6330 + 0.475$$

$$= 1.078 \text{ kJ/kg K}$$

From the nitrogen tables, Table B.6,

$$s_2 - s_1 = -5.4282 - 4.3522 = 1.0760 \text{ kJ/kg K}$$

13.9 DEVELOPING TABLES OF THERMODYNAMIC PROPERTIES FROM EXPERIMENTAL DATA

For a given pure substance, tables of thermodynamic properties can be developed from experimental data in many ways. This section conveys some general principles and concepts by considering only the liquid and vapor phases.

Let us assume that the following data for a pure substance have been obtained in the laboratory.

1. Vapor-pressure data. That is, saturation pressures and temperatures have been measured over a wide range.
2. Pressure, specific volume, and temperature data in the vapor region. These data are usually obtained by determining the mass of the substance in a closed vessel (which means a fixed specific volume) and then measuring the pressure as the temperature is varied. This is done for a large number of specific volumes.
3. Density of the saturated liquid and the critical pressure and temperature.
4. Zero-pressure specific heat for the vapor. This might be obtained either calorimetrically or from spectroscopic data and statistical thermodynamics (see Appendix C).

From these data a complete set of thermodynamic tables for the saturated liquid, saturated vapor, and superheated vapor can be calculated. The first step is to determine an equation for the vapor-pressure curve that accurately fits the data. It may be necessary to use one equation for one portion of the vapor-pressure curve and a different equation for another portion.

One form of equation that has been used is

$$\ln P_{\text{sat}} = A + \frac{B}{T} + C \ln T + DT$$

Once an equation has been found that accurately represents the data, the saturation pressure for any given temperature can be found by solving this equation. Thus, the saturation pressures in Table B.1.1 of the Steam Tables would be determined for the given temperatures. The second step is to determine an equation of state for the vapor region that accurately represents the P–v–T data. There are many possible forms of the equation of state that may be selected. The important considerations are that the equation of state accurately represents the data, and that it be of such a form that the differentiations required can be performed. That is, though it may be desirable to have an equation of state that is explicit in v, as a function of T and P, in order to use Eq. 13.27, such a representation is inherently not the most accurate. Instead, the most accurate form is one explicit in P, as a function of T and v.

Once an equation of state has been determined, the specific volume of superheated vapor at given pressures and temperatures can be determined by solving the equation and tabulating the results as in the superheat tables for steam, ammonia, and the other substances listed in the appendix. The specific volume of saturated vapor at a given temperature may be found by determining the saturation pressure from the vapor-pressure curve and substituting this saturation pressure and temperature into the equation of state.

The procedure followed in determining enthalpy and entropy is best explained with the aid of Fig. 13.11. Let us assume that the enthalpy and entropy of saturated liquid in state 1 are zero. The enthalpy of saturated vapor in state 2 can be found from the Clapeyron equation.

$$\left(\frac{dP}{dT}\right)_{\text{sat}} = \frac{h_{fg}}{T(v_g - v_f)}$$

The left side of this equation is found by differentiating the vapor-pressure curve. The specific volume of the saturated vapor is found by the procedure outlined in the last paragraph, and it is assumed that the specific volume of the saturated liquid has been measured. Thus, the enthalpy of evaporation, h_{fg}, can be found for this particular temperature, and the enthalpy at state 2 is equal to the enthalpy of evaporation (since the enthalpy in state 1 is assumed to be zero). The entropy at state 2 is readily found, since

$$s_{fg} = \frac{h_{fg}}{T}$$

From state 2 we proceed along this isotherm into the superheated vapor region. The specific volume at 3 is found from the equation of state at this pressure (by iteration, since

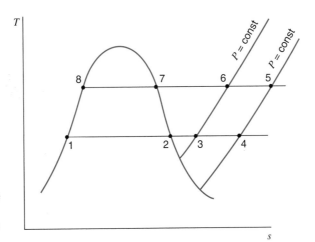

FIGURE 13.11
Sketch showing procedure for developing a table of thermodynamic properties from experimental data.

the equation is explicit in P, not v). The internal energy and entropy are calculated by integrating Eqs. 13.31 and 13.35, and the enthalpy is then calculated from its definition:

$$u_3 - u_2 = \int_2^3 \left[T\left(\frac{\partial P}{\partial T}\right)_v - P \right] dv_T$$

$$s_3 - s_2 = \int_2^3 \left(\frac{\partial P}{\partial T}\right)_v dv$$

$$h_3 - h_2 = u_3 - u_2 + P_3 v_3 - P_2 v_2$$

The properties at point 4 are found in exactly the same manner. Pressure P_4 is sufficiently low that the real superheated vapor behaves essentially as an ideal gas (perhaps 1 kPa). Thus, we use this constant-pressure line to make all temperature changes for our calculations, as, for example, to point 5. Since the specific heat C_{p0} is known as a function of temperature, the enthalpy and entropy at 5 are found by integrating the ideal–gas relations

$$(h_5 - h_4)_P = \int_4^5 C_{p0}\, dT_P$$

$$(s_5 - s_4)_P = \int_4^5 C_{p0}\, \frac{dT_P}{T}$$

The properties at points 6 and 7 are found from those at 5 in the same manner as those at points 3 and 4 were found from 2. (The saturation pressure P_7 is calculated from the vapor-pressure equation.) Finally, the enthalpy and entropy for saturated liquid at point 8 are found from the properties at point 7 by applying the Clapeyron equation.

Thus, values for the pressure, temperature, specific volume, enthalpy, entropy, and internal energy of saturated liquid, saturated vapor, and superheated vapor can be tabulated for the entire region for which experimental data were obtained. The accuracy of such a table depends both on the accuracy of the experimental data and the degree to which the equation for the vapor pressure and the equation of state represent the experimental data.

Finally, it should be noted that most present-day development of thermodynamic property tables follows a somewhat different line, beginning with representation of the Helmholtz function a (Eqs. 13.12 and 13.13) as an empirical function of T and v (or ρ). This representation implicitly includes the ideal-gas specific heat. The function typically includes 40 or 50 terms altogether, and the empirical constants are determined from specific heat terms and P–v–T data. Differentiating this expression with respect to v yields the equation of state explicit in P (see Eq. 13.21), while differentiating with respect to T yields s (also Eq. 13.21). Finally, u can be calculated from Eq. 13.12, and h from its definition. This approach to property calculation requires no mathematical integrations, such as are required in Eqs. 13.31 and 13.35.

13.10 THE PROPERTY RELATION FOR MIXTURES

In Chapter 12 our consideration of mixtures was limited to ideal gases. There was no need at that point for further expansion of the subject. We now continue this subject with a view toward developing the property relations for mixtures. This subject will be particularly relevant to our consideration of chemical equilibrium in Chapter 15.

For a mixture, any extensive property X is a function of the temperature and pressure of the mixture and the number of moles of each component. Thus, for a mixture of two components,

$$X = f(T, P, n_A, n_B)$$

Therefore,

$$dX_{T,P} = \left(\frac{\partial X}{\partial n_A}\right)_{T,P,n_B} dn_A + \left(\frac{\partial X}{\partial n_B}\right)_{T,P,n_A} dn_B \qquad (13.67)$$

Since at constant temperature and pressure an extensive property is directly proportional to the mass, Eq. 13.67 can be integrated to give

$$X_{T,P} = \overline{X}_A n_A + \overline{X}_B n_B \qquad (13.68)$$

where

$$\overline{X}_A = \left(\frac{\partial X}{\partial n_A}\right)_{T,P,n_B}, \qquad \overline{X}_B = \left(\frac{\partial X}{\partial n_B}\right)_{T,P,n_A}$$

Here $\overline{X}$ is defined as the partial molal property for a component in a mixture. It is particularly important to note that the partial molal property is defined under conditions of constant temperature and pressure.

The partial molal property is particularly significant when a mixture undergoes a chemical reaction. Suppose a mixture consists of components A and B, and a chemical reaction takes place so that the number of moles of A is changed by dn_A and the number of moles of B by dn_B. The temperature and the pressure remain constant. What is the change in internal energy of the mixture during this process? From Eq. 13.67 we conclude that

$$dU_{T,P} = \overline{U}_A \, dn_A + \overline{U}_B \, dn_B \qquad (13.69)$$

where $\overline{U}_A$ and $\overline{U}_B$ are the partial molal internal energy of A and B, respectively. Equation 13.69 suggests that the partial molal internal energy of each component can also be defined as the internal energy of the component as it exists in the mixture.

In Section 13.3 we considered a number of property relations for systems of fixed mass such as

$$dU = T \, dS - P \, dV$$

In this equation, temperature is the intensive property or potential function associated with entropy, and pressure is the intensive property associated with volume. Suppose we have a chemical reaction such as described in the last paragraph. How would we modify this property relation for this situation? Intuitively, we might write the equation

$$dU = T \, dS - P \, dV + \mu_A \, dn_A + \mu_B \, dn_B \qquad (13.70)$$

where μ_A is the intensive property or potential function associated with n_A, and similarly μ_B for n_B. This potential function is called the chemical potential.

To derive an expression for this chemical potential, we examine Eq. 13.70 and conclude that it might be reasonable to write an expression for U in the form

$$U = f(S, V, n_A, n_B)$$

Therefore,

$$dU = \left(\frac{\partial U}{\partial S}\right)_{V,n_A,n_B} ds + \left(\frac{\partial U}{\partial V}\right)_{S,n_A,n_B} dV + \left(\frac{\partial U}{\partial n_A}\right)_{S,V,n_B} dn_A + \left(\frac{\partial U}{\partial n_B}\right)_{S,V,n_A} dn_B$$

Since the expressions

$$\left(\frac{\partial U}{\partial S}\right)_{V,n_A,n_B} \quad \text{and} \quad \left(\frac{\partial U}{\partial V}\right)_{S,n_A,n_B}$$

imply constant composition, it follows from Eq. 13.20 that

$$\left(\frac{\partial U}{\partial S}\right)_{V,n_A,n_B} = T \quad \text{and} \quad \left(\frac{\partial U}{\partial V}\right)_{S,n_A,n_B} = -P$$

Thus

$$dU = T\,dS - P\,dV + \left(\frac{\partial U}{\partial n_A}\right)_{S,V,n_B} dn_A + \left(\frac{\partial U}{\partial n_B}\right)_{S,V,n_A} dn_B \qquad (13.71)$$

On comparing this equation with Eq. 13.70, we find that the chemical potential can be defined by the relation

$$\mu_A = \left(\frac{\partial U}{\partial n_A}\right)_{S,V,n_B}, \qquad \mu_B = \left(\frac{\partial U}{\partial n_B}\right)_{S,V,n_A} \qquad (13.72)$$

We can also relate the chemical potential to the partial molal Gibbs function. We proceed as follows.

$$G = U + PV - TS$$

$$dG = dU + P\,dV + V\,dP - T\,dS - S\,dT$$

Substituting Eq. 13.70 into this relation, we have

$$dG = -S\,dT + V\,dP + \mu_A\,dn_A + \mu_B\,dn_B \qquad (13.73)$$

This equation suggests that we write an expression for G in the following form.

$$G = f(T, P, n_A, n_B)$$

Proceeding as we did for a similar expression for internal energy, we have

$$dG = \left(\frac{\partial G}{\partial T}\right)_{P,n_A,n_B} dT + \left(\frac{\partial G}{\partial P}\right)_{T,n_A,n_B} dP + \left(\frac{\partial G}{\partial n_A}\right)_{T,P,n_B} dn_A + \left(\frac{\partial G}{\partial n_B}\right)_{T,P,n_A} dn_B$$

$$= -S\,dT + V\,dP + \left(\frac{\partial G}{\partial n_A}\right)_{T,P,n_B} dn_A + \left(\frac{\partial G}{\partial n_B}\right)_{T,P,n_A} dn_B$$

When this equation is compared with Eq. 13.73, it follows that

$$\mu_A = \left(\frac{\partial G}{\partial n_A}\right)_{T,P,n_B}, \qquad \mu_B = \left(\frac{\partial G}{\partial n_B}\right)_{T,P,n_A}$$

Because partial molal properties are defined at constant temperature and pressure, the quantities $(\partial G/\partial n_A)_{T,P,n_B}$ and $(\partial G/\partial n_B)_{T,P,n_A}$ are the partial molal Gibbs functions for the two components. That is, the chemical potential is equal to the partial molal Gibbs function.

$$\mu_A = \overline{G}_A = \left(\frac{\partial G}{\partial n_A}\right)_{T,P,n_B}, \qquad \mu_B = \overline{G}_B = \left(\frac{\partial G}{\partial n_B}\right)_{T,P,n_A} \tag{13.74}$$

Although μ can also be defined in terms of other properties, such as in Eq. 13.72, this expression is not the partial molal internal energy, since the pressure and temperature are not constant in this partial derivative. The partial molal Gibbs function is an extremely important property in the thermodynamic analysis of chemical reactions, for at constant temperature and pressure (the conditions under which many chemical reactions occur) it is a measure of the chemical potential or the driving force that tends to make a chemical reaction take place.

13.11 PSEUDOPURE SUBSTANCE MODELS FOR REAL-GAS MIXTURES

A basic prerequisite to the treatment of real-gas mixtures in terms of pseudopure substance models is the concept and use of appropriate reference states. As an introduction to this topic, let us consider several preliminary reference state questions for a pure substance undergoing a change of state, for which it is desired to calculate the entropy change. We can express the entropy at the initial state 1 and also at the final state 2 in terms of a reference state 0, in a manner similar to that followed when dealing with the generalized-chart corrections. It follows that

$$s_1 = s_0 + (s^*_{P_0 T_0} - s_0) + (s^*_{P_1 T_1} - s^*_{P_0 T_0}) + (s_1 - s^*_{P_1 T_1}) \tag{13.75}$$

$$s_2 = s_0 + (s^*_{P_0 T_0} - s_0) + (s^*_{P_2 T_2} - s^*_{P_0 T_0}) + (s_2 - s^*_{P_2 T_2}) \tag{13.76}$$

These are entirely general expressions for the entropy at each state in terms of an arbitrary reference state value and a set of consistent calculations from that state to the actual desired state. One simplification of these equations would result from choosing the reference state to be a hypothetical ideal-gas state at P_0 and T_0, thereby making the term

$$(s^*_{P_0 T_0} - s_0) = 0 \tag{13.77}$$

in each equation, which results in

$$s_0 = s^*_0 \tag{13.78}$$

It should be apparent that this choice is a reasonable one, since whatever value is chosen for the correction term, Eq. 13.77, it will cancel out of the two equations when the change $s_2 - s_1$ is calculated, and the simplest value to choose is zero. In a similar manner, the simplest value to choose for the ideal-gas reference value, Eq. 13.78, is zero, and we would commonly do that if there are no restrictions on choice, such as occur in the case of a chemical reaction.

Another point to be noted concerning reference states is related to the choice of P_0 and T_0. For this purpose, let us substitute Eqs. 13.77 and 13.78 into Eqs. 13.75 and 13.76, and also assume constant specific heat, such that those equations can be written in the form

$$s_1 = s_0^* + C_{p0} \ln\left(\frac{T_1}{T_0}\right) - R \ln\left(\frac{P_1}{P_0}\right) + (s_1 - s_{P_1 T_1}^*) \tag{13.79}$$

$$s_2 = s_0^* + C_{p0} \ln\left(\frac{T_2}{T_0}\right) - R \ln\left(\frac{P_2}{P_0}\right) + (s_2 - s_{P_2 T_2}^*) \tag{13.80}$$

Since the choice for P_0 and T_0 is arbitrary if there are no restrictions, such as would be the case with chemical reactions, it should be apparent from examining Eqs. 13.79 and 13.80 that the simplest choice would be for

$$P_0 = P_1 \quad \text{or} \quad P_2 \qquad T_0 = T_1 \quad \text{or} \quad T_2$$

It should be emphasized that inasmuch as the reference state was chosen as a hypothetical ideal gas at P_0, T_0, Eq. 13.77, it is immaterial how the real substance behaves at that pressure and temperature. As a result, there is no need to select a low value for the reference state pressure P_0.

Let us now extend these reference state developments to include real-gas mixtures. Consider the mixing process shown in Fig. 13.12, with the states and amounts of each substance as given on the diagram. Proceeding with entropy expressions as was done earlier, we have

$$\bar{s}_1 = \bar{s}_{A_0}^* + \overline{C}_{p0_A} \ln\left(\frac{T_1}{T_0}\right) - \overline{R} \ln\left(\frac{P_1}{P_0}\right) + (\bar{s}_1 - \bar{s}_{P_1 T_1}^*)_A \tag{13.81}$$

$$\bar{s}_2 = \bar{s}_{B_0}^* + \overline{C}_{p0_B} \ln\left(\frac{T_2}{T_0}\right) - \overline{R} \ln\left(\frac{P_2}{P_0}\right) + (\bar{s}_2 - \bar{s}_{P_2 T_2}^*)_B \tag{13.82}$$

$$\bar{s}_3 = \bar{s}_{mix_0}^* + \overline{C}_{p0_{mix}} \ln\left(\frac{T_3}{T_0}\right) - \overline{R} \ln\left(\frac{P_3}{P_0}\right) + (\bar{s}_3 - \bar{s}_{P_3 T_3}^*)_{mix} \tag{13.83}$$

in which

$$\bar{s}_{mix_0}^* = y_A \bar{s}_{A_0}^* + y_B \bar{s}_{B_0}^* - \overline{R}(y_A \ln y_A + y_B \ln y_B) \tag{13.84}$$

$$\overline{C}_{p0_{mix}} = y_A \overline{C}_{p0_A} + y_B \overline{C}_{p0_B} \tag{13.85}$$

When Eqs. 13.81–13.83 are substituted into the equation for the entropy change,

$$n_3 \bar{s}_3 - n_1 \bar{s}_1 - n_2 \bar{s}_2$$

the arbitrary reference values s_{A0}^*, s_{B0}^*, P_0, and T_0 all cancel out of the result, which is, of course, necessary in view of their arbitrary nature. An ideal-gas entropy of mixing expression, the final term in Eq. 13.84, remains in the result, establishing, in effect, the mixture reference value relative to its components. The remarks made earlier concerning the choices for reference state and the reference state entropies apply in this situation as well.

FIGURE 13.12
Example of mixing process.

To summarize the development to this point, we find that a calculation of real mixture properties, as, for example, using Eq. 13.83, requires the establishment of a hypothetical ideal gas reference state, a consistent ideal-gas calculation to the conditions of the real mixture, and finally a correction that accounts for the real behavior of the mixture at that state. This last term is the only place that the real behavior is introduced, and this is therefore the term that must be calculated by the pseudopure substance model to be used.

In treating a real-gas mixture as a pseudopure substance, we will follow two approaches to represent the P–v–T behavior: use of the generalized charts and use of an analytical equation of state. With the generalized charts, we need to have a model that provides a set of pseudocritical pressure and temperature in terms of the mixture component values. Many such models have been proposed and utilized over the years, but the simplest is that suggested by W. B. Kay in 1936, in which

$$(P_c)_{\text{mix}} = \sum_i y_i P_{ci}, \qquad (T_c)_{\text{mix}} = \sum_i y_i T_{ci} \qquad (13.86)$$

This is the only pseudocritical model that we will consider in this chapter. Other models are somewhat more complicated to evaluate and use, but considerably more accurate.

The other approach to be considered is that of using an analytical equation of state, in which the equation for the mixture must be developed from that for the components. In other words, for an equation in which the constants are known for each of the components, we must develop a set of empirical combining rules that will then give a set of constants for the mixture as though it were a pseudopure substance. This problem has been studied for many equations of state, using experimental data for the real-gas mixtures, and various empirical rules have been proposed. For example, for both the van der Waals equation, Eq. 13.52, and the Redlich–Kwong equation, Eq. 13.57, the two pure substance constants a and b are commonly combined according to the relations

$$a_m = \left(\sum_1 c_i a_i^{1/2} \right)^2 \qquad b_m = \sum_i c_i b_i \qquad (13.87)$$

The following example illustrates the use of these two approaches to treating real-gas mixtures as pseudopure substances.

EXAMPLE 13.7   A mixture of 80% CO_2 and 20% CH_4 (mass basis) is maintained at 310.94 K, 86.19 bar, at which condition the specific volume has been measured as 0.006757 m^3/kg. Calculate the percent deviation if the specific volume had been calculated by (a) Kay's rule and (b) van der Waals' equation of state.

Control mass: Gas mixture.

State: P, v, T known.

Model: (a) Kay's rule. (b) van der Waals' equation.

Solution

Let subscript A denote CO_2 and B denote CH_4; then from Tables A.2 and A.5

$$T_{c_A} = 304.1 \text{ K} \qquad P_{c_A} = 7.38 \text{ MPa} \qquad R_A = 0.1889 \text{ kJ/kg K}$$

$$T_{c_B} = 190.4 \text{ K} \qquad P_{c_B} = 4.60 \text{ MPa} \qquad R_B = 0.5183 \text{ kJ/kg K}$$

The gas constant from Eq. 12.15 becomes

$$R_m = \sum c_i R_i = 0.8 \times 0.1889 + 0.2 \times 0.5183 = 0.2548 \text{ kJ/kg K}$$

and the mole fractions are

$$y_A = (c_A/M_A)/\sum (c_i/M_i) = \frac{0.8/44.01}{(0.8/44.01) + (0.2/16.043)} = 0.5932$$

$$y_B = 1 - y_A = 0.4068$$

a. For Kay's rule, Eq. 13.86,

$$T_{c_m} = \sum_i y_i T_{c_i} = y_A T_{c_A} + y_B T_{c_B}$$

$$= 0.5932(304.1) + 0.4068(190.4)$$

$$= 257.9 \text{ K}$$

$$P_{c_m} = \sum_i y_i P_{c_i} = y_A P_{c_A} + y_B P_{c_B}$$

$$= 0.5932(7.38) + 0.4068(4.60)$$

$$= 6.249 \text{ MPa}$$

Therefore, the pseudoreduced properties of the mixture are

$$T_{r_m} = \frac{T}{T_{c_m}} = \frac{310.94}{257.9} = 1.206$$

$$P_{r_m} = \frac{P}{P_{c_m}} = \frac{8.619}{6.249} = 1.379$$

From the generalized chart, Fig. D.1

$$Z_m = 0.7$$

and

$$v = \frac{Z_m R_m T}{P} = \frac{0.7 \times 0.2548 \times 310.94}{8619} = 0.006435 \text{ m}^3/\text{kg}$$

The percent deviation from the experimental value is

$$\text{Percent deviation} = \left(\frac{0.006757 - 0.006435}{0.006757} \right) \times 100 = 4.8\%$$

The major factor contributing to this 5% error is the use of the linear Kay's rule pseudocritical model, Eq. 13.86. Use of an accurate pseudocritical model and the generalized chart would reduce the error to approximately 1%.

b. For van der Waals' equation, the pure substance constants are

$$a_A = \frac{27 R_A^2 T_{cA}^2}{64 P_{cA}} = 0.18864 \; \frac{\text{kPa m}^6}{\text{kg}^2}$$

$$b_A = \frac{R_A T_{cA}}{8 P_{cA}} = 0.000973 \; \text{m}^3/\text{kg}$$

and

$$a_B = \frac{27 R_B^2 T_{cB}^2}{64 P_{cB}} = 0.8931 \; \frac{\text{kPa m}^6}{\text{kg}^2}$$

$$b_B = \frac{R_B T_{cB}}{8 P_{cB}} = 0.002682 \; \text{m}^3/\text{kg}$$

Therefore, for the mixture, from Eq. 13.87,

$$a_m = (c_A \sqrt{a_A} + c_B \sqrt{a_B})^2$$

$$= (0.8 \sqrt{0.18864} + 0.2 \sqrt{0.8931})^2 = 0.2878 \; \frac{\text{kPa m}^6}{\text{kg}^2}$$

$$b_m = c_A b_A + c_B b_B$$

$$= 0.8 \times 0.000973 + 0.2 \times 0.002682 = 0.001315 \; \text{m}^3/\text{kg}$$

The equation of state for the mixture of this composition is

$$P = \frac{R_m T}{v - b_m} - \frac{a_m}{v^2}$$

$$8619 = \frac{0.2548 \times 310.94}{v - 0.001315} - \frac{0.2878}{v^2}$$

Solving for v by trial and error,

$$v = 0.006326 \; \text{m}^3/\text{kg}$$

$$\text{Percent deviation} = \left(\frac{0.006757 - 0.006326}{0.006757} \right) \times 100 = 6.4\%$$

As a point of interest from the ideal-gas law, $v = 0.00919 \; \text{m}^3/\text{kg}$, which is a deviation of 36% from the measured value. Also, if we use the Redlich–Kwong equation of state, and follow the same procedure as for the van der Waals equation, the calculated specific volume of the mixture is $0.00652 \; \text{m}^3/\text{kg}$, which is in error by 3.5%.

We must be careful not to draw too general a conclusion from the results of this example. We have calculated percent deviation in v at only a single point for only one mixture. We do note, however, that the various methods used give quite different results. From a more general study of these models for a number of mixtures, we find that the results found here are fairly typical, at least qualitatively. Kay's rule is very useful because it is fairly accurate and yet relatively simple. The van der Waals equation is too simplified an expression to accurately represent P–v–T behavior, but it is useful to demonstrate the procedures followed in utilizing more complex analytical equations of state. The Redlich–Kwong equation is considerably better and is still relatively simple to use.

As noted in the example, the more sophisticated generalized behavior models and empirical equations of state will represent mixture P–v–T behavior to within about 1% over a wide range of density, but they are of course more difficult to use than the methods considered in Example 13.7. The generalized models have the advantage of being easier to use, and they are suitable for hand computations. Calculations with the complex empirical equations of state become very involved, but have the advantage of expressing the P–v–T composition relations in analytical form, which is of great value when using a digital computer for such calculations.

SUMMARY

As an introduction to the development of property information that can be obtained experimentally, we derive the Clapeyron equation. This equation relates the slope of the two-phase boundaries in the P–T diagram to the enthalpy and specific volume change going from one phase to the other. If we measure pressure, temperature, and the specific volumes for liquid and vapor in equilibrium, we can calculate the enthalpy of evaporation. Because thermodynamic properties are functions of two variables, a number of relations can be derived from the mixed second derivatives and the Gibbs relations, which are known as Maxwell relations. Many other relations can be derived, and those that are useful let us relate thermodynamic properties to those that can be measured directly like P, v, T, and indirectly like the heat capacities.

Changes of enthalpy, internal energy, and entropy between two states are presented as integrals over properties that can be measured and thus obtained from experimental data. Some of the partial derivatives are expressed as coefficients like expansivity and compressibility, with the process as a qualifier like isothermal or isentropic (adiabatic). These coefficients, as single numbers, are useful when they are nearly constant over some range of interest, which happens for liquids and solids and thus are found in various handbooks. The speed of sound is also a property that can be measured, and it relates to a partial derivative in a nonlinear fashion.

The experimental information about a substance behavior is normally correlated in an equation of state relating P–v–T to represent part of the thermodynamic surface. Starting with the general compressibility and its extension to the virial equation of state, we lead up to other more complex equations of state (EOS). We show the most versatile equations such as the van der Waals EOS, the Redlich–Kwong EOS, and the Lee–Kesler EOS, which is shown as an extension of Benedict–Webb–Rubin (BWR), with others that are presented in Appendix D. The most accurate equations are too complex for hand calculations and are used on computers to generate tables of properties. Therefore, we do not cover those details.

As an application of Lee–Kesler EOS for a simple fluid, we present the development of the generalized charts that can be used for substances for which we do not have a table. The charts express the deviation of the properties from an ideal gas in terms of a compressibility factor (Z) and the enthalpy and entropy departure terms. These charts are in dimensionless properties based on the properties at the critical point.

Properties for mixtures are introduced in general, and the concept of a partial molal property leads to the chemical potential derived from the Gibbs function. Real mixtures are treated on a mole basis, and we realize that a model is required to do so. We present a pseudocritical model of Kay that predicts the critical properties for the mixture and then uses the generalized charts. Other models predict EOS parameters for the mixture and then use the EOS as for a pure substance. Typical examples here are the van der Waals and Redlich–Kwong equations of state.

You should have learned a number of skills and acquired abilities from studying this chapter that will allow you to:

- Apply and understand the assumptions for the Clapeyron equation.
- Use the Clapeyron equation for all three two-phase regions.
- Have a sense of what a partial derivative means.
- Understand why Maxwell relations and other relations are relevant.
- Know that the relations are used to develop expression for changes in h, u, and s.
- Know that coefficients of linear expansion and compressibility are common data useful for describing certain processes.
- Know that speed of sound is also a property.
- Be familiar with various equations of state and their use.
- Know the background for and how to use the generalized charts.
- Know that a model is needed to deal with a mixture.
- Know the pseudocritical model of Kay and the equation of state models for a mixture.

KEY CONCEPTS AND FORMULAS

Clapeyron equation
$$\frac{dP_{sat}}{dT} = \frac{h'' - h'}{T(v'' - v')}; \quad S\text{–}L, S\text{–}V \text{ and } V\text{–}L \text{ regions}$$

Maxwell relations
$$dz = M\,dx + N\,dy \implies \left(\frac{\partial M}{\partial y}\right)_x = \left(\frac{\partial N}{\partial x}\right)_y$$

Change in enthalpy
$$h_2 - h_1 = \int_1^2 C_p\,dT + \int_1^2 \left[v - T\left(\frac{\partial v}{\partial T}\right)_p\right]dP$$

Change in energy
$$u_2 - u_1 = \int_1^2 C_v\,dT + \int_1^2 \left[T\left(\frac{\partial P}{\partial T}\right)_v - P\right]dv$$
$$= h_2 - h_1 - (P_2 v_2 - P_1 v_1)$$

Change in entropy
$$s_2 - s_1 = \int_1^2 \frac{C_p}{T}\,dT - \int_1^2 \left(\frac{\partial v}{\partial T}\right)_p dP$$

Virial equation
$$Z = \frac{Pv}{RT} = 1 + \frac{B(T)}{v} + \frac{C(T)}{v^2} + \frac{D(T)}{v^3} + \cdots \text{ (mass basis)}$$

Van der Waals
$$P = \frac{RT}{v - b} - \frac{a}{v^2} \quad \text{(mass basis)}$$

Redlich Kwong $\quad\quad P = \dfrac{RT}{v - b} - \dfrac{a}{v(v + b)T^{1/2}}$ (mass basis)

Other equations of state $\quad$ See Appendix D.

Generalized charts for h $\quad h_2 - h_1 = (h_2^* - h_1^*)_{ID.G.} - RT_c(\Delta\hat{h}_2 - \Delta\hat{h}_1)$

Enthalpy departure $\quad\quad \Delta\hat{h} = (h^* - h)/RT_c; \quad\quad h^*$ value for ideal gas

Generalized charts for s $\quad s_2 - s_1 = (s_2^* - s_1^*)_{ID.G.} - R(\Delta\hat{s}_2 - \Delta\hat{s}_1)$

Entropy departure $\quad\quad \Delta\hat{s} = (s^* - s)/R; \quad\quad s^*$ value for ideal gas

Pseudocritical pressure $\quad P_{c\;mix} = \sum_i y_i P_{ci}$

Pseudocritical temperature $\quad T_{c\;mix} = \sum_i y_i T_{ci}$

Pseudopure substance $\quad a_m = \left(\sum_i c_i a_i^{1/2}\right)^2; \quad b_m = \sum_i c_i b_i$ (mass basis)

CONCEPT-STUDY GUIDE PROBLEMS

13.1 Mention two uses of the Clapeyron equation.

13.2 The slope dP/dT of the vaporization line is finite as you approach the critical point, yet h_{fg} and v_{fg} both approach zero. How can that be?

13.3 In view of Clapeyron's equation and Fig. 3.7, is there something special about ice I versus the other forms of ice?

13.4 If we take a derivative as $(\partial P/\partial T)_v$ in the two-phase region (see Figs. 3.18 and 3.19), does it matter what v is? How about T?

13.5 Sketch on a P–T diagram how a constant v line behaves in the compressed liquid region, the two-phase L–V region, and the superheated vapor region.

13.6 If I raise the pressure in an isentropic process, does h go up or down? Is that independent upon the phase?

13.7 If I raise the pressure in an isothermal process, does h go up or down for a liquid or solid? What do you need to know if it is a gas phase?

13.8 The equation of state in Example 13.3 was used as explicit in v. Is it explicit in P?

13.9 Over what range of states are the various coefficients in Section 13.5 most useful?

13.10 For a liquid or a solid, is v more sensitive to T or P? How about an ideal gas?

13.11 If I raise the pressure in a solid at constant T, does s go up or down?

13.12 Most equations of state are developed to cover which range of states?

13.13 Is an equation of state valid in the two-phase regions?

13.14 As $P \to 0$, the specific volume $v \to \infty$. For $P \to \infty$, does $v \to 0$?

13.15 Must an equation of state satisfy the two conditions in Eqs. 13.50 and 13.51?

13.16 At which states are the departure terms for h and s small? What is Z there?

13.17 What is the benefit of the generalized charts? Which properties must be known besides the charts themselves?

13.18 What does it imply if the compressibility factor is larger than 1?

13.19 The departure functions for h and s as defined are always positive. What does that imply for the real substance h and s values relative to ideal-gas values?

13.20 What is the benefit of Kay's rule versus a mixture equation of state?

HOMEWORK PROBLEMS

Clapeyron Equation

13.21 A special application requires R-12 at $-140°C$. It is known that the triple-point temperature is $-157°C$. Find the pressure and specific volume of the saturated vapor at the required condition.

13.22 Ice (solid water) at $-3°C$, 100 kPa, is compressed isothermally until it becomes liquid. Find the required pressure.

13.23 An approximation for the saturation pressure can be $\ln P_{sat} = A - B/T$, where A and B are constants. Which phase transition is that suitable for, and what kind of property variations are assumed?

13.24 In a Carnot heat engine, the heat addition changes the working fluid from saturated liquid to saturated vapor at T, P. The heat rejection process occurs at lower temperature and pressure $(T - \Delta T)$, $(P - \Delta P)$. The cycle takes place in a piston cylinder arrangement where the work is boundary work. Apply both the first and second law with simple approximations for the integral equal to work. Then show that the relation between ΔP and ΔT results in the Clapeyron equation in the limit $\Delta T \to dT$.

13.25 Calculate the values h_{fg} and s_{fg} for nitrogen at 70 K and at 110 K from the Clapeyron equation, using the necessary pressure and specific volume values from Table B.6.1.

13.26 Ammonia at $-70°C$ is used in a special application at a quality of 50%. Assume the only table available is B.2 that goes down to $-50°C$. To size a tank to hold 0.5 kg with $x = 0.5$, give your best estimate for the saturated pressure and the tank volume.

13.27 The saturation pressure can be approximated as $\ln P_{sat} = A - B/T$, where A and B are constants. Use the steam tables and determine A and B from properties at 25°C only. Use the equation to predict the saturation pressure at 30°C and compare to table value.

13.28 Using the properties of water at the triple point, develop an equation for the saturation pressure along the fusion line as a function of temperature.

13.29 Helium boils at 4.22 K at atmospheric pressure, 101.3 kPa, with $h_{fg} = 83.3$ kJ/kmol. By pumping

a vacuum over liquid helium, the pressure can be lowered, and it may then boil at a lower temperature. Estimate the necessary pressure to produce a boiling temperature of 1 K and one of 0.5 K.

13.30 A certain refrigerant vapor enters a steady-flow constant-pressure condenser at 150 kPa, 70°C, at a rate of 1.5 kg/s, and it exits as saturated liquid. Calculate the rate of heat transfer from the condenser. It may be assumed that the vapor is an ideal gas and also that at saturation, $v_f \ll v_g$. The following is known:

$$\ln P_g = 8.15 - 1000/T \qquad C_{p0} = 0.7 \text{ kJ/kg K}$$

with pressure in kPa and temperature in K. The molecular weight is 100.

13.31 Using thermodynamic data for water from Tables B.1.1 and B.1.5, estimate the freezing temperature of liquid water at a pressure of 30 MPa.

13.32 Small solid particles formed in combustion should be investigated. We would like to know the sublimation pressure as a function of temperature. The only information available is T, h_{fg} for boiling at 101.3 kPa and T, h_{if} for melting at 101.3 kPa. Develop a procedure that will allow a determination of the sublimation pressure, $P_{sub}(T)$.

13.33 A container has a double wall where the wall cavity is filled with carbon dioxide at room temperature and pressure. When the container is filled with a cryogenic liquid at 100 K, the carbon dioxide will freeze so that the wall cavity has a mixture of solid and vapor carbon dioxide at the sublimation pressure. Assume that we do not have data for CO_2 at 100 K, but it is known that at $-90°C$: $P_{sub} = 38.1$ kPa, $h_{ig} = 574.5$ kJ/kg. Estimate the pressure in the wall cavity at 100 K.

Property Relations, Maxwell, and those for Enthalpy, Internal Energy, and Entropy

13.34 Use Gibbs relation $du = T\, ds - P\, dv$ and one of Maxwell's relations to find an expression for $(\partial u/\partial P)_T$ that only has properties P, v, and T involved. What is the value of that partial derivative if you have an ideal gas?

13.35 Start from Gibbs relation $dh = T\, ds + v\, dP$ and use one of Maxwell's equations to get $(\partial h/\partial v)_T$ in terms of properties P, v, and T. Then use Eq. 13.24 to also find an expression for $(\partial h / \partial T)_v$.

13.36 From Eqs. 13.23 and 13.24 and the knowledge that $C_p > C_v$, what can you conclude about the slopes of constant v and constant P curves in a T–s diagram? Notice that we are looking at functions $T(s)_P$ or $T(s)_v$.

13.37 Derive expressions for $(\partial T / \partial v)_u$ and for $(\partial h / \partial s)_v$ that do not contain the properties h, u, or s. Use Eq. 13.30 with $du = 0$.

13.38 Develop an expression for the variation in temperature with pressure in a constant entropy process, $(\partial T / \partial P)_s$, that only includes the properties P–v–T and the specific heat, C_p. Follow the development of Eq. 13.32.

13.39 Use Eq. 13.34 to get an expression for the derivative $(\partial T / \partial v)_s$. what is the general shape of a constant s process curve in a T–v diagram? For an ideal gas can you say a little more about the shape?

13.40 Evaluate the isothermal changes in the internal energy, the enthalpy, and the entropy for an ideal gas. Confirm the results in Chapters 5 and 8.

Volume Expansivity and Compressibility

13.41 Determine the volume expansivity, α_P, and the isothermal compressibility, β_T, for water at 20°C, 5 MPa and at 300°C, 15 MPa using the steam tables.

13.42 What are the volume expansivity α_p, the isothermal compressibility β_T, and the adiabatic compressibility β_s for an ideal gas?

13.43 Find the speed of sound for air at 20°C, 100 kPa, using the definition in Eq. 13.43 and relations for polytropic processes in ideal gases.

13.44 Assume a substance has uniform properties in all directions with $V = L_x L_y L_z$ and show that volume expansivity $\alpha_p = 3\delta_T$. *Hint*: Differentiate with respect to T and divide by V.

13.45 A cylinder fitted with a piston contains liquid methanol at 20°C, 100 kPa, and volume 10 L. The piston is moved, compressing the methanol to 20 MPa at constant temperature. Calculate the work required for this process. The isothermal compressibility of liquid methanol at 20°C is 1.22×10^{-9} m²/N.

13.46 Use Eq. 13.32 to solve for $(\partial T / \partial P)_s$ in terms of T, v, C_p and α_p. How large a temperature change does 25°C water ($\alpha_p = 2.1 \times 10^{-4}$ K^{-1}) have, when compressed from 100 kPa to 1000 kPa in an isentropic process?

13.47 Sound waves propagate through media as pressure waves that cause the media to go through isentropic compression and expansion processes. The speed of sound c is defined by $c^2 = (\partial P / \partial \rho)_s$ and it can be related to the adiabatic compressibility, which for liquid ethanol at 20°C is 9.4×10^{-10} m²/N. Find the speed of sound at this temperature.

13.48 For commercial copper at 25°C (see Table A.3), the speed of sound is about 4800 m/s. What is the adiabatic compressibility β_s?

13.49 Consider the speed of sound as defined in Eq. 13.43. Calculate the speed of sound for liquid water at 20°C, 2.5 MPa, and for water vapor at 200°C, 300 kPa, using the steam tables.

13.50 Soft rubber is used as a part of a motor mounting. Its adiabatic bulk modulus is $B_s = 2.82 \times 10^6$ kPa, and the volume expansivity is $\alpha_p = 4.86 \times 10^{-4}$ K^{-1}. What is the speed of sound vibrations through the rubber, and what is the relative volume change for a pressure change of 1 MPa?

13.51 Liquid methanol at 25°C has an adiabatic compressibility of 1.05×10^{-9} m²/N. What is the speed of sound? If it is compressed from 100 kPa to 10 MPa in an insulated piston/cylinder, what is the specific work?

13.52 Use Eq. 13.32 to solve for $(\partial T / \partial P)_s$ in terms of T, v, C_p, and α_p. How much higher does the temperature become for the compression of the methanol in Problem 13.51? Use $\alpha_p = 2.4 \times 10^{-4}$ K^{-1} for methanol at 25°C.

Equations of State

13.53 Use the equation of state as shown in Example 13.3 where changes in enthalpy and entropy were found. Find the isothermal change in internal energy in a similar fashion; do not compute it from enthalpy.

13.54 Evaluate changes in an isothermal process for u, h, and s for a gas with an equation of state as $P(v - b) = RT$.

13.55 Two uninsulated tanks of equal volume are connected by a valve. One tank contains a gas at a moderate pressure P_1, and the other tank is evacuated. The valve is opened and remains open for a long time. Is the final pressure P_2 greater than, equal to, or less than $P_1/2$? *Hint:* Recall Fig. 13.5.

13.56 Determine the reduced Boyle temperature as predicted by an equation of state (the experimentally observed value is about 2.5), using the van der Waals equation and the Redlich–Kwong equation. *Note:* It is helpful to use Eqs. 13.45 and 13.46 in addition to Eq. 13.44.

13.57 Develop expressions for isothermal changes in internal energy, enthalpy, and entropy for a gas obeying the van der Waals equation of state.

13.58 Develop expressions for isothermal changes in internal energy, enthalpy, and entropy for a gas obeying the Redlich–Kwong equation of state.

13.59 Consider the following equation of state, expressed in terms of reduced pressure and temperature: $Z = 1 + (P_r/14T_r)[1 - 6T_r^{-2}]$. What does this predict for the reduced Boyle temperature?

13.60 What is the Boyle temperature for the following equation of state: $P = RT/v - b - a/v^2T$ where a and b are constants.

13.61 Show that the van der Waals equation can be written as a cubic equation in the compressibility factor involving the reduced pressure and reduced temperature as

$$Z^3 - \left(\frac{P_r}{8T_r} + 1\right)Z^2 + \left(\frac{27P_r}{64T_r^2}\right)Z - \frac{27P_r^2}{512T_r^3} = 0$$

13.62 Determine the second virial coefficient $B(T)$ using the van der Waals equation of state. Also find its value at the critical temperature where the experimentally observed value is about $-0.34\ RT_c/P_c$.

13.63 Determine the second virial coefficient $B(T)$ using the Redlich–Kwong equation of state. Also find its value at the critical temperature where the experimentally observed value is about $-0.34\ RT_c/P_c$.

13.64 One early attempt to improve on the van der Waals equation of state was an expression of the form

$$P = \frac{RT}{v - b} - \frac{a}{v^2T}$$

Solve for the constants a, b, and v_c using the same procedure as for the van der Waals equation.

13.65 Calculate the difference in internal energy of the ideal-gas value and the real-gas value for carbon dioxide at state 20°C, 1 MPa, as determined using the virial equation of state, including second virial coefficient terms. For carbon dioxide we have: $B = -0.128$ m³/kmol, $T(dB/dT) = 0.266$ m³/kmol, both at 20°C.

13.66 Calculate the difference in entropy of the ideal-gas value and the real-gas value for carbon dioxide at the state 20°C, 1 MPa, as determined using the virial equation of state. Use numerical values given in Problem 13.65.

13.67 A rigid tank contains 1 kg oxygen at 160 K, 4 MPa, Determine the volume of the tank assuming we can use the Redlich–Kwong equation of state for oxygen. Compare the result with the ideal-gas law.

13.68 A flow of oxygen at 230 K, 5 MPa, is throttled to 100 KPa in a steady-flow process. Find the exit temperature and the specific entropy generation using the Redlich–Kwong equation of state and ideal-gas heat capacity. Notice that this becomes iterative due to the nonlinearity coupling h, P, v, and T.

Generalized Charts

13.69 A 200-L rigid tank contains propane at 9 MPa, 280°C. The propane is then allowed to cool to 50°C as heat is transferred with the surroundings. Determine the quality at the final state and the mass of liquid in the tank, using the generalized compressibility chart, Fig. D.1.

13.70 A rigid tank contains 5 kg of ethylene at 3 MPa, 30°C. It is cooled until the ethylene reaches the saturated vapor curve. What is the final temperature?

13.71 Refrigerant-123, dichlorotrifluoroethane, which is currently under development as a potential replacement for environmentally hazardous refrigerants, undergoes an isothermal steady-flow process in which the R-123 enters a heat exchanger as saturated liquid at 40°C and exits at 100 kPa. Calculate the heat transfer per kilogram of R-123, using the generalized charts, Fig. D.2.

13.72 An ordinary lighter is nearly full of liquid propane with a small amount of vapor, the volume is 5 cm³, and temperature is 23°C. The propane is now discharged slowly such that heat transfer keeps the propane and valve flow at 23°C. Find the initial pressure and mass of propane and the total heat transfer to empty the lighter.

13.73 A piston/cylinder contains 5 kg of butane gas at 500 K, 5 MPa. The butane expands in a reversible polytropic process to 3 MPa, 460 K. Determine the polytropic exponent n and the work done during the process.

13.74 Calculate the heat transfer during the process described in Problem 13.73.

13.75 A cylinder contains ethylene, C_2H_4, at 1.536 MPa, −13°C. It is now compressed in a reversible isobaric (constant P) process to saturated liquid. Find the specific work and heat transfer.

13.76 Carbon dioxide collected from a fermentation process at 5°C, 100 kPa, should be brought to 243 K, 4 MPa in a steady-flow process. Find the minimum amount of work required and the heat transfer. What devices are needed to accomplish this change of state?

13.77 Consider the following equation of state, expressed in terms of reduced pressure and temperature: $Z = 1 + (P_r/14T_r)[1 − 6T_r^{-2}]$. What does this predict for the enthalpy departure at $P_r = 0.4$ and $T_r = 0.9$?

13.78 Consider the following equation of state, expressed in terms of reduced pressure and temperature: $Z = 1 + (P_r/14T_r)[1 − 6T_r^{-2}]$. What does this predict for the entropy departure at $P_r = 0.4$ and $T_r = 0.9$?

13.79 A flow of oxygen at 230 K, 5 MPa is throttled to 100 kPa in a steady-flow process. Find the exit temperature and the entropy generation.

13.80 A cylinder contains ethylene, C_2H_4, at 1.536 MPa, −13°C. It is now compressed isothermally in a reversible process to 5.12 MPa. Find the specific work and heat transfer.

13.81 Saturated vapor R-22 at 30°C is throttled to 200 kPa in a steady-flow process. Calculate the exit temperature assuming no changes in the kinetic energy, using the generalized charts, Fig. D.2, and the R-22 tables, Table B.4.

13.82 A 250-L tank contains propane at 30°C, 90% quality. The tank is heated to 300°C. Calculate the heat transfer during the process.

13.83 The new refrigerant fluid R-123 (see Table A.2) is used in a refrigeration system that operates in the

ideal refrigeration cycle, except the compressor is neither reversible nor adiabatic. Saturated vapor at −26.5°C enters the compressor, and superheated vapor exits at 65°C. Heat is rejected from the compressor as 1 kW, and the R-123 flow rate is 0.1 kg/s. Saturated liquid exits the condenser at 37.5°C. Specific heat for R-123 is $C_{p0} = 0.6$ kJ/kg K. Find the coefficient of performance.

13.84 An uninsulated piston/cylinder contains propene, C_3H_6, at ambient temperature, 19°C, with a quality of 50% and a volume of 10 L. The propene now expands slowly until the pressure drops to 460 kPa. Calculate the mass of propene, the work, and heat transfer for this process.

13.85 A geothermal power plant on the Raft River uses isobutane as the working fluid. The fluid enters the reversible adiabatic turbine, as shown in Fig. P13.85, at 160°C, 5.475 MPa, and the condenser exit condition is saturated liquid at 33°C. Isobutane has the properties $T_c = 408.14$ K, $P_c = 3.65$ MPa, $C_{p0} = 1.664$ kJ/kg K, and ratio of specific heats $k = 1.094$ with a molecular weight as 58.124. Find the specific turbine work and the specific pump work.

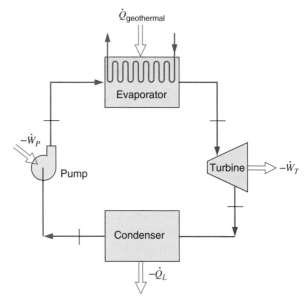

FIGURE P13.85

13.86 A line with a steady supply of octane, C_8H_{18}, is at 400°C, 3 MPa. What is your best estimate for the availability in a steady-flow setup where changes in potential and kinetic energies may be neglected?

13.87 An insulated piston/cylinder contains saturated vapor carbon dioxide at 0°C and a volume of 20 L. The external force on the piston is slowly decreased, allowing the carbon dioxide to expand until the temperature reaches −30°C. Calculate the work done by the CO_2 during this process.

13.88 An evacuated 100-L rigid tank is connected to a line flowing R-142b gas, chlorodifluoroethane, at 2 MPa, 100°C. The valve is opened, allowing the gas to flow into the tank for a period of time, and then it is closed. Eventually, the tank cools to ambient temperature, 20°C, at which point it contains 50% liquid, 50% vapor, by volume. Calculate the quality at the final state and the heat transfer for the process. The ideal-gas specific heat of R-142b is $C_p = 0.787$ kJ/kg K.

13.89 Saturated liquid ethane at 2.44 MPa enters a heat exchanger and is brought to 611 K at constant pressure, after which it enters a reversible adiabatic turbine where it expands to 100 kPa. Find the specific heat transfer in the heat exchanger, the turbine exit temperature, and turbine work.

13.90 A control mass of 10 kg butane gas initially at 80°C, 500 kPa, is compressed in a reversible isothermal process to one-fifth of its initial volume. What is the heat transfer in the process?

13.91 An uninsulated compressor delivers ethylene, C_2H_4, to a pipe, $D = 10$ cm, at 10.24 MPa, 94°C, and velocity 30 m/s. The ethylene enters the compressor at 6.4 MPa, 20.5°C, and the work input required is 300 kJ/kg. Find the mass flow rate, the total heat transfer, and entropy generation, assuming the surroundings are at 25°C.

13.92 A distributor of bottled propane, C_3H_8, needs to bring propane from 350 K, 100 kPa, to saturated liquid at 290 K in a steady-flow process. If this should be accomplished in a reversible setup given the surroundings at 300 K, find the ratio of the volume flow rates $\dot{V}_{in}/\dot{V}_{out}$, the heat specific transfer, and the work involved in the process.

13.93 The environmentally safe refrigerant R-152a is to be evaluated as the working fluid for a heat pump system that will heat a house. It uses an evaporator temperature of −20°C and a condensing temperature of 30°C. Assume all processes are

ideal and R-152a has a heat capacity of $C_p = 0.996$ kJ/kg K. Determine the cycle coefficient of performance.

13.94 Rework the previous problem using an evaporator temperature of 0°C.

Mixtures

13.95 A 2-kg mixture of 50% argon and 50% nitrogen by mole is in a tank at 2 MPa, 180 K. How large is the volume using a model of (a) ideal gas and (b) Kay's rule with generalized compressibility charts.

13.96 A 2-kg mixture of 50% argon and 50% nitrogen by mass is in a tank at 2 MPa, 180 K. How large is the volume using a model of (a) ideal gas and (b) van der Waals equation of state with a, b for a mixture?

13.97 A 2-kg mixture of 50% argon and 50% nitrogen by mass is in a tank at 2 MPa, 180 K. How large is the volume using a model of (a) ideal gas and (b) Redlich–Kwong equation of state with a, b for a mixture.

13.98 Saturated liquid ethane at $T_1 = 14°C$ is throttled into a steady-flow mixing chamber at the rate of 0.25 kmol/s. Argon gas at $T_2 = 25°C$, 800 kPa, enters the chamber at the rate 0.75 kmol/s. Heat is transferred to the chamber from a constant temperature source at 150°C at a rate such that a gas mixture exits the chamber at $T_3 = 120°C$, 800 kPa. Find the rate of heat transfer and the rate of entropy generation.

13.99 A modern jet engine operates so that the fuel is sprayed into air at P, T higher than the fuel critical point. Assume we have a rich mixture of 50% n-octane and 50% air by moles at 500 K and 3.5 MPa near the nozzle exit. Do I need to treat this as a real-gas mixture, or is an ideal-gas assumption reasonable? To answer, find Z and the enthalpy departure for the mixture assuming Kay's rule and the generalized charts.

13.100 A mixture of 60% ethylene and 40% acetylene by moles is at 6 MPa, 300 K. The mixture flows through a preheater where it is heated to 400 K at constant P. Using the Redlich–Kwong equation of state with a, b for a mixture, find the inlet specific volume. Repeat using Kay's rule and the generalized charts.

13.101 For the previous problem, find the specific heat transfer using Kay's rule and the generalized charts.

13.102 One kmol/s of saturated liquid methane, CH_4, at 1 MPa and 2 kmol/s of ethane, C_2H_6, at 250°C, 1 MPa, are fed to a mixing chamber with the resultant mixture exiting at 50°C, 1 MPa. Assume that Kay's rule applies to the mixture and determine the heat transfer in the process.

13.103 A piston/cylinder initially contains propane at $T_1 = -7$°C, quality 50%, and volume 10 L. A valve connecting the cylinder to a line flowing nitrogen gas at $T_i = 20$°C, $P_i = 1$ MPa, is opened and nitrogen flows in. When the valve is closed, the cylinder contains a gas mixture of 50% nitrogen, 50% propane, on a mole basis at $T_2 = 20$°C, $P_2 = 500$ kPa. What is the cylinder volume at the final state, and how much heat transfer took place?

13.104 Consider the following reference state conditions: the entropy of real saturated liquid methane at -100°C is to be taken as 100 kJ/kmol K, and the entropy of hypothetical ideal gas ethane at -100°C is to be taken as 200 kJ/kmol K. Calculate the entropy per kmol of a real-gas mixture of 50% methane, 50% ethane (mole basis) at 20°C, 4 MPa, in terms of the specified reference state values, and assuming Kay's rule for the real mixture behavior.

13.105 A cylinder/piston contains a gas mixture, 50% CO_2 and 50% C_2H_6 (mole basis) at 700 kPa, 35°C, at which point the cylinder volume is 5 L. The mixture is now compressed to 5.5 MPa in a reversible isothermal process. Calculate the heat transfer and work for the process, using the following model for the gas mixture:

a. Ideal-gas mixture.

b. Kay's rule and the generalized charts.

13.106 A cylinder/piston contains a gas mixture, 50% CO_2 and 50% C_2H_6 (mole basis) at 700 kPa, 35°C, at which point the cylinder volume is 5 L. The mixture is now compressed to 5.5 MPa in a reversible isothermal process. Calculate the heat transfer and work for the process, using the following model for the gas mixture:

a. Ideal-gas mixture.

b. The van der Waals equation of state.

Review Problems

13.107 Consider a straight line connecting the point $P = 0$, $Z = 1$ to the critical point $P = P_c$, $Z = Z_c$ on a Z versus P compressibility diagram. This straight line will be tangent to one particular isotherm at low pressure. (The experimentally determined value is about 0.8 T_c.) Determine what value of reduced temperature is predicted by an equation of state, using the van der Waals equation and the Redlich–Kwong equation. See also note for Problem 13.56.

13.108 A 200-L rigid tank contains propane at 400 K, 3.5 MPa. A valve is opened, and propane flows out until half the initial mass has escaped, at which point the valve is closed. During this process the mass remaining inside the tank expands according to the relation $Pv^{1.4} = $ constant. Calculate the heat transfer to the tank during the process.

13.109 A newly developed compound is being considered for use as the working fluid in a small Rankine-cycle power plant driven by a supply of waste heat. Assume the cycle is ideal, with saturated vapor at 200°C entering the turbine and saturated liquid at 20°C exiting the condenser. The only properties known for this compound are molecular weight of 80 kg/kmol, ideal-gas heat capacity $C_{p0} = 0.80$ kJ/kg K and $T_c = 500$ K, $P_c = 5$ MPa. Calculate the work input, per kilogram, to the pump and the cycle thermal efficiency.

13.110 A piston/cylinder contains propane initially at 67°C and 50% quality with a volume of 2 L. The piston cross-sectional area is 0.2 m². The external force on the piston is gradually reduced to a final value of 85 kN during which process the propane expands to ambient temperature, 4°C. Any heat transfer to the propane comes from a constant-temperature reservoir at 67°C, while any heat transfer from the propane goes to the ambient. It is claimed that the propane does 30 kJ of work during the process. Does this violate the second law?

13.111 One kilogram per second water enters a solar collector at 40°C and exits at 190°C, as shown in Fig. P.13.111. The hot water is sprayed into a direct-contact heat exchanger (no mixing of the

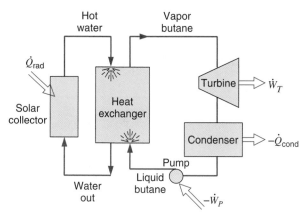

Hot water

Vapor butane

$\dot{Q}_{rad}$

Solar collector

Heat exchanger

Turbine ⇒ $\dot{W}_T$

Condenser ⇒ $-\dot{Q}_{cond}$

Pump

Water out

Liquid butane

$-\dot{W}_P$

FIGURE P13.111

two fluids) used to boil the liquid butane. Pure saturated-vapor butane exits at the top at 80°C and is fed to the turbine. If the butane condenser temperature is 30°C and the turbine and pump isentropic efficiencies are each 80%, determine the net power output of the cycle.

13.112 A piston/cylinder contains ethane gas initially at 500 kPa, 100 L, and at ambient temperature 0°C. The piston is moved, compressing the ethane until it is at 20°C with a quality of 50%. The work required is 25% more than would have been needed for a reversible polytropic process between the same initial and final states. Calculate the heat transfer and the net entropy change for the process.

13.113 An experiment is conducted at −100°C inside a rigid sealed tank containing liquid R-22 with a small amount of vapor at the top. When the experiment is done, the container and the R-22 warm up to room temperature of 20°C. What is the pressure inside the tank during the experiment? If the pressure at room temperature should not exceed 1 MPa, what is the maximum percent of liquid by volume that can be used during the experiment?

13.114 The refrigerant R-152a, difluoroethane, is tested by the following procedure. A 10-L evacuated tank is connected to a line flowing saturated-vapor R-152a at 40°C. The valve is opened, and the fluid flows in rapidly, so the process is essentially adiabatic. The valve is to be closed when the pressure reaches a certain value P_2, and the tank will then be disconnected from the line.

After a period of time, the temperature inside the tank will return to ambient temperature, 25°C, through heat transfer with the surroundings. At this time, the pressure inside the tank must be 500 kPa. What is the pressure P_2 at which the valve should be closed during the filling process? The ideal-gas specific heat of R-152a is $C_p = 0.996$ kJ/kg K.

13.115 Carbon dioxide gas enters a turbine at 5 MPa, 100°C, and exits at 1 MPa. If the isentropic efficiency of the turbine is 75%, determine the exit temperature and the second-law efficiency.

13.116 A 4-m³ uninsulated storage tank, initially evacuated, is connected to a line flowing ethane gas at 10 MPa, 100°C. The valve is opened, and ethane flows into the tank for a period of time, after which the valve is closed. Eventually, the whole system cools to ambient temperature, 0°C, at which time it contains one-fourth liquid and three-fourths vapor, by volume. For the overall process, calculate the heat transfer from the tank and the net change of entropy.

13.117 A 10-m³ storage tank contains methane at low temperature. The pressure inside is 700 kPa, and the tank contains 25% liquid and 75% vapor, on a volume basis. The tank warms very slowly because heat is transferred from the ambient.

a. What is the temperature of the methane when the pressure reaches 10 MPa?

b. Calculate the heat transferred in the process, using the generalized charts.

c. Repeat parts (a) and (b), using the methane tables, Table B.7. Discuss the differences in the results.

13.118 A gas mixture of a known composition is required for the calibration of gas analyzers. It is desired to prepare a gas mixture of 80% ethylene and 20% carbon dioxide (mole basis) at 10 MPa, 25°C in an uninsulated, rigid 50-L tank. The tank is initially to contain CO_2 at 25°C and some pressure P_1. The valve to a line flowing C_2H_4 at 25°C, 10 MPa, is now opened slightly and remains open until the tank reaches 10 MPa, at which point the temperature can be assumed to be 25°C. Assume that the gas mixture so prepared can be represented by Kay's rule and the generalized charts. Given the desired final state, what is the initial pressure of the carbon dioxide, P_1?

13.119 Determine the heat transfer and the net entropy change in the previous problem. Use the initial pressure of the carbon dioxide to be 4.56 MPa before the ethylene is flowing into the tank.

ENGLISH UNIT PROBLEMS

13.120E A special application requires R-22 at -150 F. It is known that the triple-point temperature is less than -150 F. Find the pressure and specific volume of the saturated vapor at the required condition.

13.121E Ice (solid water) at 27 F, 1 atm, is compressed isothermally until it becomes liquid. Find the required pressure.

13.122E The saturation pressure can be approximated as $\ln P_{sat} = A - B/T$, where A and B are constants. Use the steam tables and determine A and B from properties at 70 F only. Use the equation to predict the saturation pressure at 80 F and compare to table value.

13.123E Using thermodynamic data for water from Tables F.7.1 and F.7.4, estimate the freezing temperature of liquid water at a pressure of 5000 lbf/in.^{2}.

13.124E Determine the volume expansivity, α_p, and the isothermal compressibility, β_T, for water at 50 F, 500 lbf/in.^{2} and at 500 F, 1500 lbf/in.^{2} using the steam tables.

13.125E A cylinder fitted with a piston contains liquid methanol at 70 F, 15 lbf/in.^{2} and volume 1 ft^3. The piston is moved, compressing the methanol to 3000 lbf/in.^{2} at constant temperature. Calculate the work required for this process. The isothermal compressibility of liquid methanol at 70 F is 8.3×10^{-6} in.^{2}/lbf.

13.126E Sound waves propagate through media as pressure waves that cause the media to go through isentropic compression and expansion processes. The speed of sound c is defined by $c^2 = (\partial P/\partial \rho)_s$, and it can be related to the adiabatic compressibility, which for liquid ethanol at 70 F is 6.4×10^{-6} in.^{2}/lbf. Find the speed of sound at this temperature.

13.127E Consider the speed of sound as defined in Eq. 13.43. Calculate the speed of sound for liquid water at 50 F, 250 lbf/in.^{2}, and for water vapor at 400 F, 80 lbf/in.^{2}, using the steam tables.

13.128E Liquid methanol at 77 F has an adiabatic compressibility of 7.1×10^{-6} in.^{2}/lbf. What is the speed of sound? If it is compressed from 15 psia to 1500 psia in an insulated piston/cylinder, what is the specific work?

13.129E Calculate the difference in internal energy of the ideal-gas value and the real-gas value for carbon dioxide at the state 70 F, 150 lbf/in.^{2}, as determined using the virial equation of state. At this state $B = -2.036$ ft^3/lb mol, $T(dB/dT) = 4.236$ ft^3/lb mol.

13.130E A 7-ft^3 rigid tank contains propane at 1300 lbf/in.^{2}, 540 F. The propane is then allowed to cool to 120 F as heat is transferred with the surroundings. Determine the quality at the final state and the mass of liquid in the tank, using the generalized compressibility chart.

13.131E A rigid tank contains 5 lbm of ethylene at 450 lbf/in.^{2}, 90 F. It is cooled until the ethylene reaches the saturated vapor curve. What is the final temperature?

13.132E A piston/cylinder contains 10 lbm of butane gas at 900 R, 750 lbf/in.^{2}. The butane expands in a reversible polytropic process to 450 lbf/in^2 and 820 R. Determine the polytropic exponent and the work done during the process.

13.133E Calculate the heat transfer during the process described in Problem 13.132E.

13.134E A cylinder contains ethylene, C_2H_4, at 222.6 lbf/in.^{2}, 8 F. It is now compressed in a reversible isobaric (constant P) process to saturated liquid. Find the specific work and heat transfer.

13.135E Carbon dioxide collected from a fermentation process at 40 F, 15 lbf/in.^{2} should be brought to 438 R, 590 lbf/in.^{2}, in a steady-flow process. Find the minimum amount of work required and the heat transfer. What devices are needed to accomplish this change of state?

13.136E Saturated vapor R-22 at 90 F is throttled to 30 lbf/in.^{2} in a steady-flow process. Calculate the exit temperature assuming no changes in the kinetic energy, using the generalized charts, Fig. D.2, and repeat using the R-22 tables, Table F.9.

13.137E A 10-ft^3 tank contains propane at 90 F, 90% quality. The tank is heated to 600 F. Calculate the heat transfer during the process.

13.138E A cylinder contains ethylene, C_2H_4, at 222.6 lbf/in.^{2}, 8 F. It is now compressed isothermally in a reversible process to 742 lbf/in.^{2}. Find the specific work and heat transfer.

13.139E A geothermal power plant on the Raft River uses isobutane as the working fluid as shown in Fig. P13.85. The fluid enters the reversible adiabatic turbine at 320 F, 805 lbf/in.^{2}, and the condenser exit condition is saturated liquid at 91 F. Isobutane has the properties T_c = 734.65 R, P_c = 537 lbf/in.^{2}, C_{p0} = 0.3974 Btu/lbm R and ratio of specific heats k = 1.094 with a molecular weight as 58.124. Find the specific turbine work and the specific pump work.

13.140E A line with a steady supply of octane, C_8H_{18}, is at 750 F, 440 lbf/in.^{2}. What is your best estimate for the availability in a steady-flow setup where changes in potential and kinetic energies may be neglected?

13.141E A control mass of 10 lbm butane gas initially at 180 F, 75 lbf/in.^{2}, is compressed in a reversible isothermal process to one-fifth of its initial volume. What is the heat transfer in the process?

13.142E A distributor of bottled propane, C_3H_8, needs to bring propane from 630 R, 14.7 lbf/in.^{2} to saturated liquid at 520 R in a steady-flow process. If this should be accomplished in a reversible setup given the surroundings at 540 R, find the ratio of the volume flow rates $\dot{V}_{in}/\dot{V}_{out}$, the heat transfer, and the work involved in the process.

13.143E A 4-lbm mixture of 50% argon and 50% nitrogen by mole is in a tank at 300 psia, 320 R. How large is the volume using a model of (a) ideal gas and (b) Kay's rule with generalized compressibility charts.

13.144E A 7-ft^3 rigid tank contains propane at 730 R, 500 lbf/in.^{2}. A valve is opened, and propane flows out until half the initial mass has escaped, at which point the valve is closed. During this process the mass remaining inside the tank expands according to the relation $Pv^{1.4}$ = constant. Calculate the heat transfer to the tank during the process.

13.145E A newly developed compound is being considered for use as the working fluid in a small Rankine-cycle power plant driven by a supply of waste heat. Assume the cycle is ideal, with saturated vapor at 400 F entering the turbine and saturated liquid at 70 F exiting the condenser. The only properties known for this compound are molecular weight of 80 lbm/lbmol, ideal-gas heat capacity C_{p0} = 0.20 Btu/lbm R and T_c = 900 R, P_c = 750 lbf/in.^{2}. Calculate the work input, per lbm, to the pump and the cycle thermal efficiency.

COMPUTER, DESIGN, AND OPEN-ENDED PROBLEMS

13.146 Write a program to obtain a plot of pressure versus specific volume at various temperatures (all on a generalized reduced basis) as predicted by the van der Waals equation of state. Temperatures less than the critical temperature should be included in the results.

13.147 We wish to determine the isothermal compressibility, β_T, for a range of states of liquid water. Use the menu-driven software or write a program to determine this at a pressure of 1 MPa and at 25 MPa for temperatures of 0°C, 100°C, and 300°C.

13.148 Consider the small Rankine-cycle power plant in Problem 13.109. What single change would

you suggest to make the power plant more realistic?

13.149 Supercritical fluid chromatography is an experimental technique for analyzing compositions of mixtures. It utilizes a carrier fluid, often CO_2, in the dense fluid region just above the critical temperature. Write a program to express the fluid density as a function of reduced temperature and pressure in the region of $1.0 \leq T_r \leq 1.2$ in reduced temperature and $2 \leq P_r \leq 8$ in reduced pressure. The relation should be an expression curve-fitted to values consistent with the generalized compressibility charts.

13.150 It is desired to design a portable breathing system for an average-sized adult. The breather will store liquid oxygen sufficient for a 24-hour supply, and include a heater for delivering oxygen gas at ambient temperature. Determine the size of the system container and the heat exchanger.

13.151 Liquid nitrogen is used in cryogenic experiments and applications where a nonoxidizing gas is desired. Size a tank to hold 500 kg to be placed next to a building and estimate the size of an environmental (to atmospheric air) heat exchanger that can deliver nitrogen gas at a rate of 10 kg/h at roughly ambient temperature.

13.152 List a number of requirements for a substance that should be used as the working fluid in a refrigerator. Discuss the choices and explain the requirements.

13.153 The speed of sound is used in many applications. Make a list of the speed of sound at P_0, T_0 for gases, liquids, and solids. Find at least three different substances for each phase. List a number of applications where knowledge about the speed of sound can be used to estimate other quantities of interest.

13.154 Propane is used as a fuel distributed to the end consumer in a steel bottle. Make a list of design specifications for these bottles and give characteristic sizes and the amount of propane they can hold.

13.155 Carbon dioxide is used in soft drinks and comes in a separate bottle for large-volume users such as restaurants. Find typical sizes of these, the pressure they should withstand, and the amount of carbon dioxide they can hold.

CHEMICAL REACTIONS 14

Many thermodynamic problems involve chemical reactions. Among the most familiar of these is the combustion of hydrocarbon fuels, for this process is utilized in most of our power-generating devices. However, we can all think of a host of other processes involving chemical reactions, including those that occur in the human body.

This chapter considers a first- and second-law analysis of systems undergoing a chemical reaction. In many respects, this chapter is simply an extension of our previous consideration of the first and second laws. However, a number of new terms are introduced, and it will also be necessary to introduce the third law of thermodynamics.

In this chapter the combustion process is considered in detail. There are two reasons for this emphasis. First, the combustion process is of great significance in many problems and devices with which the engineer is concerned. Second, the combustion process provides an excellent vehicle for teaching the basic principles of the thermodynamics of chemical reactions. The student should keep both of these objectives in mind as the study of this chapter progresses.

Chemical equilibrium will be considered in Chapter 15 and, therefore, the matter of dissociation will be deferred until then.

14.1 FUELS

A thermodynamics textbook is not the place for a detailed treatment of fuels. However, some knowledge of them is a prerequisite to a consideration of combustion, and this section is therefore devoted to a brief discussion of some of the hydrocarbon fuels. Most fuels fall into one of three categories—coal, liquid hydrocarbons, or gaseous hydrocarbons.

Coal consists of the remains of vegetation deposits of past geologic ages, after subjection of biochemical actions, high pressure, temperature, and submersion. The characteristics of coal vary considerably with location, and even within a given mine there is some variation in composition.

The analysis of a sample of coal is given on one of two bases: the proximate analysis specifies, on a mass basis, the relative amounts of moisture, volatile matter, fixed carbon, and ash; the ultimate analysis specifies, on a mass basis, the relative amounts of carbon, sulfur, hydrogen, nitrogen, oxygen, and ash. The ultimate analysis may be given on an "as-received" basis or on a dry basis. In the latter case, the ultimate analysis does not include the moisture as determined by the proximate analysis.

A number of other properties of coal are important in evaluating a coal for a given use. Some of these are the fusibility of the ash, the grindability or ease of pulverization, the weathering characteristics, and size.

Table 14.1

Characteristics of Some of the Hydrocarbon Families

Family	Formula	Structure	Saturated
Paraffin	C_nH_{2n+2}	Chain	Yes
Olefin	C_nH_{2n}	Chain	No
Diolefin	C_nH_{2n-2}	Chain	No
Naphthene	C_nH_{2n}	Ring	Yes
Aromatic			
Benzene	C_nH_{2n-6}	Ring	No
Naphthene	C_nH_{2n-12}	Ring	No

Most liquid and gaseous hydrocarbon fuels are a mixture of many different hydrocarbons. For example, gasoline consists primarily of a mixture of about 40 hydrocarbons, with many others present in very small quantities. In discussing hydrocarbon fuels, therefore, brief consideration should be given to the most important families of hydrocarbons, which are summarized in Table 14.1.

Three concepts should be defined. The first pertains to the structure of the molecule. The important types are the ring and chain structures; the difference between the two is illustrated in Fig. 14.1. The same figure illustrates the definition of saturated and unsaturated hydrocarbons. An unsaturated hydrocarbon has two or more adjacent carbon atoms joined by a double or triple bond, whereas in a saturated hydrocarbon all the carbon atoms are joined by a single bond. The third term to be defined is an isomer. Two hydrocarbons with the same number of carbon and hydrogen atoms and different structures are called isomers. Thus, there are several different octanes (C_8H_{18}), each having 8 carbon atoms and 18 hydrogen atoms, but each with a different structure.

The various hydrocarbon families are identified by a common suffix. The compounds comprising the paraffin family all end in "-ane" (as propane and octane). Similarly, the compounds comprising the olefin family end in "-ylene" or "-ene" (as propene and octene), and the diolefin family ends in "-diene" (as butadiene). The naphthene family has the same chemical formula as the olefin family but has a ring rather than chain structure. The hydrocarbons in the naphthene family are named by adding the prefix "cyclo-" (as cyclopentane).

The aromatic family includes the benzene series (C_nH_{2n-6}) and the naphthalene series (C_nH_{2n-12}). The benzene series has a ring structure and is unsaturated.

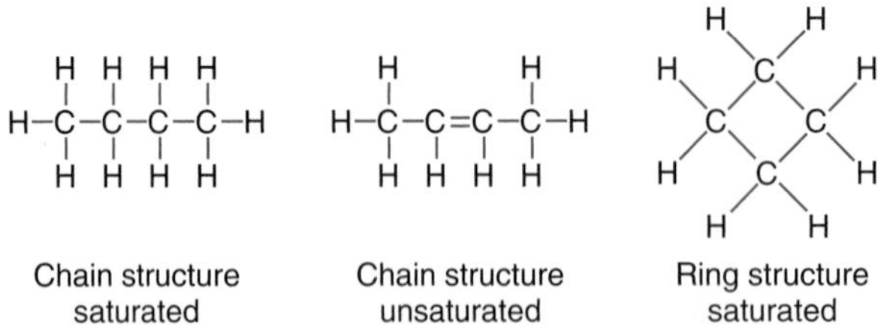

FIGURE 14.1
Molecular structure of
some hydrocarbon fuels.

Chain structure
saturated

Chain structure
unsaturated

Ring structure
saturated

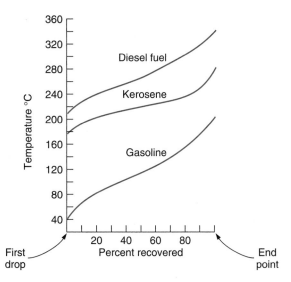

FIGURE 14.2 Typical distillation curves of some hydrocarbon fuels.

Alcohols are sometimes used as a fuel in internal combustion engines. Characteristically, in the alcohol family, one of the hydrogen atoms is replaced by an OH radical. Thus, methyl alcohol, also called methanol, is CH_3OH.

Most liquid hydrocarbon fuels are mixtures of hydrocarbons that are derived from crude oil through distillation and cracking processes. Thus, from a given crude oil, a variety of different fuels can be produced, some of the common ones being gasoline, kerosene, diesel fuel, and fuel oil. Within each of these classifications there is a wide variety of grades, and each is made up of a large number of different hydrocarbons. The important distinction between these fuels is the distillation curve, Fig. 14.2. The distillation curve is obtained by slowly heating a sample of fuel so that it vaporizes. The vapor is then condensed and the amount measured. The more volatile hydrocarbons are vaporized first, and thus the temperature of the nonvaporized fraction increases during the process. The distillation curve, which is a plot of the temperature of the nonvaporized fraction versus the amount of vapor condensed, is an indication of the volatility of the fuel.

For the combustion of liquid fuels, it is convenient to express the composition in terms of a single hydrocarbon, even though it is a mixture of many hydrocarbons. Thus, gasoline is usually considered to be octane, C_8H_{18}, and diesel fuel is considered to be dodecane, $C_{12}H_{26}$. The composition of a hydrocarbon fuel may also be given in terms of percentage of carbon and hydrogen.

The two primary sources of gaseous hydrocarbon fuels are natural gas wells and certain chemical manufacturing processes. Table 14.2 gives the composition of a number of gaseous fuels. The major constituent of natural gas is methane, which distinguishes it from manufactured gas.

At the present time, considerable effort is being devoted to develop more economical processes for producing gaseous and also liquid hydrocarbon fuels from coal, oil shale, and tar sands deposits. Several alternative techniques have been demonstrated to be feasible, and these resources promise to provide an increasing proportion of our fuel supply in future years.

TABLE 14.2
Volumetric Analyses of Some Typical Gaseous Fuels

Constituent	Various Natural Gases				Producer Gas from Bituminous Coal	Carbureted Water Gas	Coke-Oven Gas
	A	B	C	D			
Methane	93.9	60.1	67.4	54.3	3.0	10.2	32.1
Ethane	3.6	14.8	16.8	16.3			
Propane	1.2	13.4	15.8	16.2			
Butanes plus[a]	1.3	4.2		7.4			
Ethene						6.1	3.5
Benzene						2.8	0.5
Hydrogen					14.0	40.5	46.5
Nitrogen		7.5		5.8	50.9	2.9	8.1
Oxygen					0.6	0.5	0.8
Carbon monoxide					27.0	34.0	6.3
Carbon dioxide					4.5	3.0	2.2

[a]This includes butane and all heavier hydrocarbons

14.2 THE COMBUSTION PROCESS

The combustion process consists of the oxidation of constituents in the fuel that are capable of being oxidized and can therefore be represented by a chemical equation. During a combustion process, the mass of each element remains the same. Thus, writing chemical equations and solving problems concerning quantities of the various constituents basically involve the conservation of mass of each element. This chapter presents a brief review of this subject, particularly as it applies to the combustion process.

Consider first the reaction of carbon with oxygen.

$$\text{Reactants} \qquad \text{Products}$$
$$C + O_2 \quad \rightarrow \quad CO_2$$

This equation states that 1 kmol of carbon reacts with 1 kmol of oxygen to form 1 kmol of carbon dioxide. This also means that 12 kg of carbon react with 32 kg of oxygen to form 44 kg of carbon dioxide. All the initial substances that undergo the combustion process are called the reactants, and the substances that result from the combustion process are called the products.

When a hydrocarbon fuel is burned, both the carbon and the hydrogen are oxidized. Consider the combustion of methane as an example.

$$CH_4 + 2O_2 \rightarrow CO_2 + 2H_2O \tag{14.1}$$

Here the products of combustion include both carbon dioxide and water. The water may be in the vapor, liquid, or solid phases, depending on the temperature and pressure of the products of combustion.

In the combustion process, many intermediate products are formed during the chemical reaction. In this book we are concerned with the initial and final products and not with the intermediate products, but this aspect is very important in a detailed consideration of combustion.

In most combustion processes, the oxygen is supplied as air rather than as pure oxygen. The composition of air on a molal basis is approximately 21% oxygen, 78% nitrogen, and 1% argon. We assume that the nitrogen and the argon do not undergo chemical reaction (except for dissociation, which will be considered in Chapter 15). They do leave at the same temperature as the other products, however, and therefore undergo a change of state if the products are at a temperature other than the original air temperature. At the high temperatures achieved in internal-combustion engines, there is actually some reaction between the nitrogen and oxygen, and this gives rise to the air pollution problem associated with the oxides of nitrogen in the engine exhaust.

In combustion calculations concerning air, the argon is usually neglected, and the air is considered to be composed of 21% oxygen and 79% nitrogen by volume. When this assumption is made, the nitrogen is sometimes referred to as atmospheric nitrogen. Atmospheric nitrogen has a molecular weight of 28.16 (which takes the argon into account) as compared to 28.013 for pure nitrogen. This distinction will not be made in this text, and we will consider the 79% nitrogen to be pure nitrogen.

The assumption that air is 21.0% oxygen and 79.0% nitrogen by volume leads to the conclusion that for each mole of oxygen, $79.0/21.0 = 3.76$ moles of nitrogen are involved. Therefore, when the oxygen for the combustion of methane is supplied as air, the reaction can be written

$$CH_4 + 2O_2 + 2(3.76)N_2 \rightarrow CO_2 + 2H_2O + 7.52N_2 \qquad (14.2)$$

The minimum amount of air that supplies sufficient oxygen for the complete combustion of all the carbon, hydrogen, and any other elements in the fuel that may oxidize is called the theoretical air. When complete combustion is achieved with theoretical air, the products contain no oxygen. A general combustion reaction with a hydrocarbon fuel and air is thus written

$$C_xH_y + v_{O_2}(O_2 + 3.76N_2) \rightarrow v_{CO_2}CO_2 + v_{H_2O}H_2O + v_{N_2}N_2 \qquad (14.3)$$

with the coefficients to the substances called stoichiometric coefficients. The balance of atoms yields the theoretical amount of air as

$$\begin{aligned} \text{C:} & \quad v_{CO_2} = x \\ \text{H:} & \quad 2v_{H_2O} = y \\ \text{N}_2\text{:} & \quad v_{N_2} = 3.76 \times v_{O_2} \\ \text{O}_2\text{:} & \quad v_{O_2} = v_{CO_2} + v_{H_2O}/2 = x + y/4 \end{aligned}$$

and the total number of moles of air for 1 mole of fuel becomes

$$n_{air} = v_{O_2} \times 4.76 = 4.76(x + y/4)$$

This amount of air is equal to 100% theoretical air. In practice, complete combustion is not likely to be achieved unless the amount of air supplied is somewhat greater than the theoretical amount. Two important parameters often used to express the ratio of fuel and air are the air–fuel ratio (designed AF) and its reciprocal, the fuel–air ratio (designated FA). These ratios are usually expressed on a mass basis, but a mole basis is used at times.

$$AF_{mass} = \frac{m_{air}}{m_{fuel}} \tag{14.4}$$

$$AF_{mole} = \frac{n_{air}}{n_{fuel}} \tag{14.5}$$

They are related through the molecular weights as

$$AF_{mass} = \frac{m_{air}}{m_{fuel}} = \frac{n_{air} M_{air}}{n_{fuel} M_{fuel}} = AF_{mole} \frac{M_{air}}{M_{fuel}}$$

and a subscript s is used to indicate the ratio for 100% theoretical air, also called a stoichiometric mixture. In an actual combustion process, an amount of air is expressed as a fraction of the theoretical amount, called percent theoretical air. A similar ratio named the equivalence ratio equals the actual fuel–air ratio divided by the theoretical fuel–air ratio as

$$\Phi = FA/FA_s = AF_s/AF \tag{14.6}$$

the reciprocal of percent theoretical air. Since the percent theoretical air and the equivalence ratio are both ratios of the stoichiometric air–fuel ratio and the actual air–fuel ratio, the molecular weights cancel out and they are the same whether a mass basis or a mole basis is used.

Thus, 150% theoretical air means that the air actually supplied is 1.5 times the theoretical air and the equivalence ratio is $\frac{2}{3}$. The complete combustion of methane with 150% theoretical air is written

$$CH_4 + 1.5 \times 2(O_2 + 3.76 N_2) \rightarrow CO_2 + 2 H_2O + O_2 + 11.28 N_2 \tag{14.7}$$

having balanced all the stoichiometric coefficients from conservation of all the atoms.

The amount of air actually supplied may also be expressed in terms of percent excess air. The excess air is the amount of air supplied over and above the theoretical air. Thus, 150% theoretical air is equivalent to 50% excess air. The terms "theoretical air," "excess air," and "equivalence ratio" are all in current usage and give an equivalent information about the reactant mixture of fuel and air.

When the amount of air supplied is less than the theoretical air required, the combustion is incomplete. If there is only a slight deficiency of air, the usual result is that some of the carbon unites with the oxygen to form carbon monoxide (CO) instead of carbon dioxide (CO_2). If the air supplied is considerably less than the theoretical air, there may also be some hydrocarbons in the products of combustion.

Even when some excess air is supplied, small amounts of carbon monoxide may be present, the exact amount depending on a number of factors including the mixing and turbulence during combustion. Thus, the combustion of methane with 110% theoretical air might be as follows:

$$CH_4 + 2(1.1)O_2 + 2(1.1)3.76 N_2 \rightarrow$$
$$+0.95 CO_2 + 0.05 CO + 2 H_2O + 0.225 O_2 + 8.27 N_2 \tag{14.8}$$

The material covered so far in this section is illustrated by the following examples.

EXAMPLE 14.1 Calculate the theoretical air–fuel ratio for the combustion of octane, C_8H_{18}.

Solution

The combustion equation is

$$C_8H_{18} + 12.5\,O_2 + 12.5(3.76)\,N_2 \rightarrow 8\,CO_2 + 9\,H_2O + 47.0\,N_2$$

The air–fuel ratio on a mole basis is

$$AF = \frac{12.5 + 47.0}{1} = 59.5 \text{ kmol air/kmol fuel}$$

The theoretical air–fuel ratio on a mass basis is found by introducing the molecular weight of the air and fuel.

$$AF = \frac{59.5(28.97)}{114.2} = 15.0 \text{ kg air/kg fuel}$$

EXAMPLE 14.2 Determine the molal analysis of the products of combustion when octane, C_8H_{18}, is burned with 200% theoretical air, and determine the dew point of the products if the pressure is 0.1 MPa.

Solution

The equation for the combustion of octane with 200% theoretical air is

$$C_8H_{18} + 12.5(2)\,O_2 + 12.5(2)(3.76)\,N_2 \rightarrow 8\,CO_2 + 9\,H_2O + 12.5\,O_2 + 94.0\,N_2$$

Total kmols of product = 8 + 9 + 12.5 + 94.0 = 123.5
Molal analysis of products:

$$CO_2 = 8/123.5 \quad = \quad 6.47\%$$
$$H_2O = 9/123.5 \quad = \quad 7.29$$
$$O_2 = 12.5/123.5 = \quad 10.12$$
$$N_2 = 94/123.5 \quad = \quad \underline{76.12}$$
$$100.00\%$$

The partial pressure of the water is 100(0.0729) = 7.29 kPa.
The saturation temperature corresponding to this pressure is 39.7°C, which is also the dew-point temperature.
The water condensed from the products of combustion usually contains some dissolved gases and therefore may be quite corrosive. For this reason the products of combustion are often kept above the dew point until discharged to the atmosphere.

EXAMPLE 14.2E Determine the molal analysis of the products of combustion when octane, C_8H_{18}, is burned with 200% theoretical air, and determine the dew point of the products if the pressure is 14.7 lbf/in.[2].

Solution

The equation for the combustion of octane with 200% theoretical air is

$$C_8H_{18} + 12.5(2)O_2 + 12.5(2)(3.76)N_2 \rightarrow 8CO_2 + 9H_2O + 12.5O_2 + 94.0N_2$$

Total moles of product = $8 + 9 + 12.5 + 94.0 = 123.5$
Molal analysis of products:

$$CO_2 = 8/123.5 \quad = \quad 6.47\%$$
$$H_2O = 9/123.5 \quad = \quad 7.29$$
$$O_2 = 12.5/123.5 = \quad 10.12$$
$$N_2 = 94/123.5 \quad = \quad \underline{76.12}$$
$$100.00\%$$

The partial pressure of the H_2O is $14.7(0.0729) = 1.072$ lbf/in.^{2}

The saturation temperature corresponding to this pressure is 104 F, which is also the dew-point temperature.

The water condensed from the products of combustion usually contains some dissolved gases and therefore may be quite corrosive. For this reason the products of combustion are often kept above the dew point until discharged to the atmosphere.

EXAMPLE 14.3 Producer gas from bituminous coal (see Table 14.2) is burned with 20% excess air. Calculate the air–fuel ratio on a volumetric basis and on a mass basis.

Solution

To calculate the theoretical air requirement, let us write the combustion equation for the combustible substances in 1 kmol of fuel.

$$0.14H_2 + 0.070O_2 \rightarrow 0.14H_2O$$
$$0.27CO + 0.135O_2 \rightarrow 0.27CO_2$$
$$0.03CH_4 + \underline{0.06O_2} \rightarrow 0.03CO_2 + 0.06H_2O$$
$$0.265 = \text{kmol oxygen required/kmol fuel}$$
$$\underline{-0.006} = \text{oxygen in fuel/kmol fuel}$$
$$0.259 = \text{kmol oxygen required from air/kmol fuel}$$

Therefore, the complete combustion equation for 1 kmol of fuel is

$$\overbrace{0.14H_2 + 0.27CO + 0.03CH_4 + 0.006O_2 + 0.509N_2 + 0.045CO_2}^{\text{fuel}}$$
$$\overbrace{+0.259O_2 + 0.259(3.76)N_2}^{\text{air}} \rightarrow 0.20H_2O + 0.345CO_2 + 1.482N_2$$
$$\left(\frac{\text{kmol air}}{\text{kmol fuel}}\right)_{\text{theo}} = 0.259 \times \frac{1}{0.21} = 1.233$$

If the air and fuel are at the same pressure and temperature, this also represents the ratio of the volume of air to the volume of fuel.

$$\text{For 20\% excess air, } \frac{\text{kmol air}}{\text{kmol fuel}} = 1.233 \times 1.200 = 1.48$$

The air–fuel ratio on a mass basis is

$$AF = \frac{1.48(28.97)}{0.14(2) + 0.27(28) + 0.03(16) + 0.006(32) + 0.509(28) + 0.045(44)}$$

$$= \frac{1.48(28.97)}{24.74} = 1.73 \text{ kg air/kg fuel}$$

An analysis of the products of combustion affords a very simple method for calculating the actual amount of air supplied in a combustion process. There are various experimental methods by which such an analysis can be made. Some yield results on a "dry" basis, that is, the fractional analysis of all the components, except for water vapor. Other experimental procedures give results that include the water vapor. In this presentation we are not concerned with the experimental devices and procedures, but rather with the use of such information in a thermodynamic analysis of the chemical reaction. The following examples illustrate how an analysis of the products can be used to determine the chemical reaction and the composition of the fuel.

The basic principle in using the analysis of the products of combustion to obtain the actual fuel–air ratio is conservation of the mass of each of the elements. Thus, in changing from reactants to products, we can make a carbon balance, hydrogen balance, oxygen balance, and nitrogen balance (plus any other elements that may be involved). Furthermore, we recognize that there is a definite ratio between the amounts of some of these elements. Thus, the ratio between the nitrogen and oxygen supplied in the air is fixed, as well as the ratio between carbon and hydrogen if the composition of a hydrocarbon fuel is known.

Often the combustion of a fuel uses atmospheric air as the oxidizer, in which case the reactants also hold some water vapor. Assuming we know the humidity ratio for the moist air, ω, then we would like to know the composition of air per mole of oxygen as

$$1 \text{ O}_2 + 3.76 \text{ N}_2 + x \text{ H}_2\text{O}$$

Since the humidity ratio is, $\omega = m_v/m_a$, the number of moles of water is

$$n_v = \frac{m_v}{M_v} = \frac{\omega m_a}{M_v} = \omega n_a \frac{M_a}{M_v}$$

and the number of moles of dry air per mole of oxygen is $(1 + 3.76)/1$, so we get

$$x = \frac{n_v}{n_{\text{oxygen}}} = \omega 4.76 \frac{M_a}{M_v} = 7.655\omega \qquad (14.9)$$

This amount of water is found in the products together with the water produced by the oxidation of the hydrogen in the fuel.

EXAMPLE 14.4   Methane (CH_4) is burned with atmospheric air. The analysis of the products on a dry basis is as follows:

CO_2	10.00%
O_2	2.37
CO	0.53
N_2	87.10
	100.00%

Calculate the air–fuel ratio and the percent theoretical air, and determine the combustion equation.

Solution

The solution consists of writing the combustion equation for 100 kmol of dry products, introducing letter coefficients for the unknown quantities, and then solving for them.

From the analysis of the products, the following equation can be written, keeping in mind that this analysis is on a dry basis.

$$a\,CH_4 + b\,O_2 + c\,N_2 \rightarrow 10.0\,CO_2 + 0.53\,CO + 2.37\,O_2 + d\,H_2O + 87.1\,N_2$$

A balance for each of the elements will enable us to solve for all the unknown coefficients:

Nirogen balance: $c = 87.1$

Since all the nitrogen comes from the air,

$$\frac{c}{b} = 3.76 \qquad b = \frac{87.1}{3.76} = 23.16$$

Carbon balance: $a = 10.00 + 0.53 = 10.53$

Hydrogen balance: $d = 2a = 21.06$

Oxygen balance: All the unknown coefficients have been solved for, and therefore the oxygen balance provides a check on the accuracy. Thus, b can also be determined by an oxygen balance

$$b = 10.00 + \frac{0.53}{2} + 2.37 + \frac{21.06}{2} = 23.16$$

Substituting these values for a, b, c, and d, we have

$$10.53\,CH_4 + 23.16\,O_2 + 87.1\,N_2 \rightarrow$$
$$10.0\,CO_2 + 0.53\,CO + 2.37\,O_2 + 21.06\,H_2O + 87.1\,N_2$$

Dividing through by 10.53 yields the combustion equation per kmol of fuel.

$$CH_4 + 2.2\,O_2 + 8.27\,N_2 \rightarrow 0.95\,CO_2 + 0.05\,CO + 2\,H_2O + 0.225\,O_2 + 8.27\,N_2$$

The air–fuel ratio on a mole basis is

$$2.2 + 8.27 = 10.47 \text{ kmol air/kmol fuel}$$

The air–fuel ratio on a mass basis is found by introducing the molecular weights.

$$AF = \frac{10.47 \times 28.97}{16.0} = 18.97 \text{ kg air/kg fuel}$$

The theoretical air–fuel ratio is found by writing the combustion equation for theoretical air.

$$CH_4 + 2O_2 + 2(3.76)N_2 \rightarrow CO_2 + 2H_2O + 7.52N_2$$

$$AF_{theo} = \frac{(2 + 7.52)28.97}{16.0} = 17.23 \text{ kg air/kg fuel}$$

The percent theoretical air is $\frac{18.97}{17.23} = 110\%$

EXAMPLE 14.5   Coal from Jenkin, Kentucky, has the following ultimate analysis on a dry basis, percent by mass:

Component	Percent by Mass
Sulfur	0.6
Hydrogen	5.7
Carbon	79.2
Oxygen	10.0
Nitrogen	1.5
Ash	3.0

This coal is to be burned with 30% excess air. Calculate the air–fuel ratio on a mass basis.

Solution

One approach to this problem is to write the combustion equation for each of the combustible elements per 100 kg of fuel. The molal composition per 100 kg of fuel is found first.

$$\text{kmol S/100 kg fuel} = \frac{0.6}{32} = 0.02$$

$$\text{kmol } H_2\text{/100 kg fuel} = \frac{5.7}{2} = 2.85$$

$$\text{kmol C/100 kg fuel} = \frac{79.2}{12} = 6.60$$

$$\text{kmol } O_2\text{/100 kg fuel} = \frac{10}{32} = 0.31$$

$$\text{kmol } N_2\text{/100 kg fuel} = \frac{1.5}{28} = 0.05$$

The combustion equations for the combustible elements are now written, which enables us to find the theoretical oxygen required.

$$0.02\,S + 0.02\,O_2 \rightarrow 0.02\,SO_2$$

$$2.85\,H_2 + 1.42\,O_2 \rightarrow 2.85\,H_2O$$

$$6.60\,C + 6.60\,O_2 \rightarrow 6.60\,CO_2$$

$$8.04 \text{ kmol } O_2 \text{ required/100 kg fuel}$$

$$-0.31 \text{ kmol } O_2 \text{ in fuel/100 kg fuel}$$

$$7.73 \text{ kmol } O_2 \text{ from air/100 kg fuel}$$

$$AF_{\text{theo}} = \frac{[7.73 + 7.73(3.76)]28.97}{100} = 10.63 \text{ kg air/kg fuel}$$

For 30% excess air the air–fuel ratio is

$$AF = 1.3 \times 10.63 = 13.82 \text{ kg air/kg fuel}$$

14.3 ENTHALPY OF FORMATION

In the first thirteen chapters of this book, the problems always concerned a fixed chemical composition and never a change of composition through a chemical reaction. Therefore, in dealing with a thermodynamic property, we used tables of thermodynamic properties for the given substance, and in each of these tables the thermodynamic properties were given relative to some arbitrary base. In the steam tables, for example, the internal energy of saturated liquid at 0.01°C is assumed to be zero. This procedure is quite adequate when there is no change in composition because we are concerned with the changes in the properties of a given substance. The properties at the condition of the reference state cancel out in the calculation. When dealing with the matter of reference states in Section 13.11, we noted that for a given substance (perhaps a component of a mixture), we are free to choose a reference state condition, for example, a hypothetical ideal gas, as long as we then carry out a consistent calculation from that state and condition to the real desired state. We also noted that we are free to choose a reference state value, as long as there is no subsequent inconsistency in the calculation of the change in a property because of a chemical reaction with a resulting change in the amount of a given substance. Now that we are to include the possibility of a chemical reaction, it will become necessary to choose these reference state values on a common and consistent basis. We will use as our reference state a temperature of 25°C, a pressure of 0.1 MPa, and a hypothetical ideal-gas condition for those substances that are gases.

Consider the simple steady-state combustion process shown in Fig. 14.3. This idealized reaction involves the combustion of solid carbon with gaseous (ideal gas) oxygen, each of which enters the control volume at the reference state, 25°C and 0.1 MPa. The carbon dioxide (ideal gas) formed by the reaction leaves the chamber at the reference state, 25°C and 0.1 MPa. If the heat transfer could be accurately measured, it would be

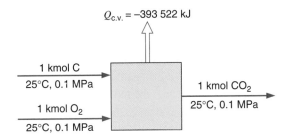

FIGURE 14.3
Example of combustion process.

found to be $-393\,522$ kJ/kmol of carbon dioxide formed. The chemical reaction can be written

$$C + O_2 \rightarrow CO_2$$

Applying the first law to this process we have

$$Q_{c.v.} + H_R = H_P \qquad (14.10)$$

where the subscripts R and P refer to the reactants and products, respectively. W will find it convenient to also write the first law for such a process in the form

$$Q_{c.v.} + \sum_R n_i \bar{h}_i = \sum_P n_e \bar{h}_e \qquad (14.11)$$

where the summations refer, respectively, to all the reactants or all the products.

Thus, a measurement of the heat transfer would give us the difference between the enthalpy of the products and the reactants, where each is at the reference state condition. Suppose, however, that we assign the value of zero to the enthalpy of all the elements at the reference state. In this case, the enthalpy of the reactants is zero, and

$$Q_{c.v.} = H_p = -393\,522 \text{ kJ/kmol}$$

The enthalpy of (hypothetical) ideal-gas carbon dioxide at 25°C, 0.1 MPa pressure (with reference to this arbitrary base in which the enthalpy of the elements is chosen to be zero), is called the enthalpy of formation. We designate this with the symbol $\bar{h}_f$. Thus, for carbon dioxide

$$\bar{h}_f^\circ = -393\,522 \text{ kJ/kmol}$$

The enthalpy of carbon dioxide in any other sate, relative to this base in which the enthalpy of the elements is zero, would be found by adding the change of enthalpy between ideal gas at 25°C, 0.1 MPa, and the given state to the enthalpy of formation. That is, the enthalpy at any temperature and pressure, $h_{T,P}$, is

$$\bar{h}_{T,P} = (\bar{h}_f^\circ)_{298,0.1\text{MPa}} + (\Delta\bar{h})_{298,0.1\text{MPa}\rightarrow T,P} \qquad (14.12)$$

where the term $(\Delta\bar{h})_{298,\,0.1\text{MPa}\rightarrow T,P}$ represents the difference in enthalpy between any given state and the enthalpy of ideal gas at 298.15 K, 0.1 MPa. For convenience we usually drop the subscripts in the examples that follow.

The procedure that we have demonstrated for carbon dioxide can be applied to any compound.

Table A.10 gives values of the enthalpy of formation for a number of substances in the units kJ/kmol (or Btu/lb mol in F.11).

Three further observations should be made in regard to enthalpy of formation.

1. We have demonstrated the concept of enthalpy of formation in terms of the measurement of the heat transfer in an idealized chemical reaction in which a compound is formed from the elements. Actually, the enthalpy of formation is usually found by the application of statistical thermodynamics, using observed spectroscopic data.

2. The justification of this procedure of arbitrarily assigning the value of zero to the enthalpy of the elements at 25°C, 0.1 MPa, rests on the fact that in the absence of nuclear reactions the mass of each element is conserved in a chemical reaction. No conflicts or ambiguities arise with this choice of reference state, and it proves to be very convenient in studying chemical reactions from a thermodynamic point of view.

3. In certain cases an element or compound can exist in more than one state at 25°C, 0.1 MPa. Carbon, for example, can be in the form of graphite or diamond. It is essential that the state to which a given value is related be clearly identified. Thus, in Table A.10, the enthalpy of formation of graphite is given the value of zero, and the enthalpy of each substance that contains carbon is given relative to this base. Another example is that oxygen may exist in the monatomic or diatomic form, and also as ozone, O_3. The value chosen as zero is for the form that is chemically stable at the reference state, which in the case of oxygen is the diatomic form. Then each of the other forms must have an enthalpy of formation consistent with the chemical reaction and heat transfer for the reaction that produces that form of oxygen.

It will be noted from Table A.10 that two values are given for the enthalpy of formation for water; one is for liquid water and the other for gaseous (hypothetical ideal gas) water, both at the reference state of 25°C, 0.1 MPa. It is convenient to use the hypothetical ideal-gas reference in connection with the ideal-gas table property changes given in Table A.9, and to use the real liquid reference in connection with real water property changes as given in the steam tables, Table B.1. The real-liquid reference state properties are obtained from those at the hypothetical ideal-gas reference by following the procedure of calculation described in Section 13.11. The same procedure can be followed for other substances that have a saturation pressure less than 0.1 MPa at the reference temperature of 25°C.

Frequently, students are bothered by the minus sign when the enthalpy of formation is negative. For example, the enthalpy of formation of carbon dioxide is negative. This is quite evident because the heat transfer is negative during the steady-flow chemical reaction, and the enthalpy of the carbon dioxide must be less than the sum of enthalpy of the carbon and oxygen initially, both of which are assigned the value of zero. This is quite analogous to the situation we would have in the steam tables if we let the enthalpy of saturated vapor be zero at 0.1 MPa pressure. In this case the enthalpy of the liquid would be negative, and we would simply use the negative value for the enthalpy of the liquid when solving problems.

14.4 FIRST-LAW ANALYSIS OF REACTING SYSTEMS

The significance of the enthalpy of formation is that it is most convenient in performing a first-law analysis of a reacting system, for the enthalpies of different substances can be added or subtracted, since they are all given relative to the same base.

In such problems, we will write the first law for a steady-state, steady-flow process in the form

$$Q_{\text{c.v.}} + H_R = W_{\text{c.v.}} + H_P$$

or

$$Q_{\text{c.v.}} + \sum_R n_i \overline{h}_i = W_{\text{c.v.}} + \sum_P n_e \overline{h}_e$$

where R and P refer to the reactants and products, respectively. In each problem it is necessary to choose one parameter as the basis of the solution. Usually this is taken as 1 kmol of fuel.

EXAMPLE 14.6   Consider the following reaction, which occurs in a steady-state, steady-flow process.

$$\text{CH}_4 + 2\,\text{O}_2 \rightarrow \text{CO}_2 + 2\,\text{H}_2\text{O}(l)$$

The reactants and products are each at a total pressure of 0.1 MPa and 25°C. Determine the heat transfer per kilomole of fuel entering the combustion chamber.

Control volume:	Combustion chamber.
Inlet state:	P and T known; state fixed.
Exit state:	P and T known, state fixed.
Process:	Steady state.
Model:	Three gases ideal gases; real liquid water.

Analysis

First law:

$$Q_{\text{c.v.}} + \sum_R n_i \overline{h}_i = \sum_P n_e \overline{h}_e$$

Solution

Using values from Table A.10, we have

$$\sum_R n_i \overline{h}_i = (\overline{h}_f^\circ)_{\text{CH}_4} = -74\,873 \text{ kJ}$$

$$\sum_P n_e \overline{h}_e = (\overline{h}_f^\circ)_{\text{CO}_2} + 2(\overline{h}_f^\circ)_{\text{H}_2\text{O}(l)}$$

$$= -393\,522 + 2(-285\,830) = -965\,182 \text{ kJ}$$

$$Q_{\text{c.v.}} = -965\,182 - (-74\,873) = -890\,309 \text{ kJ}$$

In most instances, however, the substances that comprise the reactants and products in a chemical reaction are not at a temperature of 25°C and a pressure of 0.1 MPa (the state at which the enthalpy of formation is given). Therefore, the change of enthalpy between 25°C and 0.1 MPa and the given state must be known. For a solid or liquid, this change of enthalpy can usually be found from a table of thermodynamic properties or from specific heat data. For gases, the change of enthalpy can usually be found by one of the following procedures.

1. Assume ideal-gas behavior between 25°C, 0.1 MPa, and the given state. In this case, the enthalpy is a function of the temperature only and can be found by an equation of $\overline{C}_{p0}$ or from tabulated values of enthalpy as a function of temperature (which assumes ideal-gas behavior). Table A.6 gives an equation for $\overline{C}_{p0}$ for a number of substances and Table A.9 gives values of $\overline{h}° - \overline{h}°_{298}$(that is, the $\Delta\overline{h}$ of Eq. 14.11) in kJ/kmol, ($\overline{h}°_{298}$ refers to 25°C or 298.15 K. For simplicity this is designated $\overline{h}°_{298}$.) The superscript 0 is used to designate that this is the enthalpy at 0.1 MPa pressure, based on ideal-gas behavior, that is, the standard-state enthalpy.

2. If a table of thermodynamic properties is available, $\Delta\overline{h}$ can be found directly from these tables if a real substance behavior reference state is being used, such as that described above for liquid water. If a hypothetical ideal-gas reference state is being used, then it is necessary to account for the real substance correction to properties at that sate to gain entry to the tables.

3. If the deviation from ideal-gas behavior is significant, but no tables of thermodynamic properties are available, the value of $\Delta\overline{h}$ can be found from the generalized tables or charts and the values for $\overline{C}_{p0}$ or $\Delta\overline{h}$ at 0.1 MPa pressure as indicated above.

Thus, in general, for applying the first law to a steady-state process involving a chemical reaction and negligible changes in kinetic and potential energy, we can write

$$Q_{c.v.} + \sum_R n_i(\overline{h}°_f + \Delta\overline{h})_i = W_{c.v.} + \sum_P n_e(\overline{h}°_f + \Delta\overline{h})_e \qquad (14.13)$$

EXAMPLE 14.7 Calculate the enthalpy of water (on a kilomole basis) at 3.5 MPa, 300°C, relative to the 25°C and 0.1 MPa base, using the following procedures.

1. Assume the steam to be an ideal gas with the value of $\overline{C}_{p0}$ given in the Appendix, Table A.6.
2. Assume the steam to be an ideal gas with the value for $\Delta\overline{h}$ as given in the Appendix, Table A.9.
3. The steam tables.
4. The specific heat behavior given in 2 above and the generalized charts.

Solution

For each of these procedures, we can write

$$\overline{h}_{T,P} = (\overline{h}°_f + \Delta\overline{h})$$

The only difference is in the procedure by which we calculate $\Delta\overline{h}$. From Table A.10 we note that

$$(\overline{h}°_f)_{H_2O(g)} = -241\ 826 \text{ kJ/kmol}$$

1. Using the specific heat equation for $H_2O(g)$ from Table A.6,

$$C_{p0} = 1.79 + 0.107\theta + 0.586\theta^2 - 0.20\theta^3, \theta = T/1000$$

The specific heat at the average temperature

$$T_{avg} = \frac{298.15 + 573.15}{2} = 435.65 \text{ K}$$

is

$$C_{p0} = 1.79 + 0.107(0.43565) + 0.586(0.43565)^2 - 0.2(0.43565)^3$$
$$= 1.9313 \frac{\text{kJ}}{\text{kg K}}$$

Therefore,

$$\Delta\bar{h} = MC_{p0} \Delta T$$
$$= 18.015 \times 1.9313(573.15 - 298.15) = 9568 \frac{\text{kJ}}{\text{kmol}}$$
$$\bar{h}_{T,P} = -241\,826 + 9568 = -232\,258 \text{ kJ/kmol}$$

2. Using Table A.9 for $H_2O(g)$,

$$\Delta\bar{h} = 9494 \text{ kJ/kmol}$$

$$\bar{h}_{T,P} = -241\,826 + 9494 = -232\,332 \text{ kJ/kmol}$$

3. Using the steam tables, either the liquid reference or the gaseous reference state may be used.

For the liquid,

$$\Delta\bar{h} = 18.015(2977.5 - 104.9) = 51\,750 \text{ kJ/kmol}$$

$$\bar{h}_{T,P} = -285\,830 + 51\,750 = -234\,080 \text{ kJ/kmol}$$

For the gas,

$$\Delta\bar{h} = 18.015(2977.5 - 2547.2) = 7752 \text{ kJ/kmol}$$

$$\bar{h}_{T,P} = -241\,826 + 7752 = -234\,074 \text{ kJ/kmol}$$

The very small difference results from using the enthalpy of saturated vapor at 25°C (which is almost but not exactly an ideal gas) in calculating the $\Delta\bar{h}$.

4. When using the generalized charts, we use the notation introduced in Chapter 13.

$$\bar{h}_{T,P} = \bar{h}_f^\circ - (\bar{h}_2^* - \bar{h}_2) + (\bar{h}_2^* - \bar{h}_1^*) + (\bar{h}_1^* - \bar{h}_1)$$

where subscript 2 refers to the state at 3.5 MPa, 300°C, and state 1 refers to the state at 0.1 MPa, 25°C.

From part 2, $\bar{h}_2 - \bar{h}_1^* = 9494$ kJ/kmol.

$$\bar{h}_1^* - \bar{h}_1 = 0 \quad \text{(ideal-gas reference)}$$

$$P_{r2} = \frac{3.5}{22.09} = 0.158 \qquad T_{r2} = \frac{573.2}{647.3} = 0.886$$

From the generalized enthalpy chart, Fig. D.2.

$$\frac{\bar{h}_2^* - \bar{h}_2}{\bar{R}T_c} = 0.21, \qquad \bar{h}_2^* - \bar{h}_2 = 0.21 \times 8.3145 \times 647.3 = 1130 \text{ kJ/kmol}$$

$$\bar{h}_{T,P} = -241\,826 - 1130 + 9494 = -233\,462 \text{ kJ/kmol}$$

The particular approach that is used in a given problem will depend on the data available for the given substance.

EXAMPLE 14.8 A small gas-turbine uses $C_8H_{18}(l)$ for fuel and 400% theoretical air. The air and fuel enter at 25°C, and the products of combustion leave at 900 K. The output of the engine and the fuel consumption are measured, and it is found that the specific fuel consumption is 0.25 kg/s of fuel per megawatt output. Determine the heat transfer from the engine per kilomole of fuel. Assume complete combustion.

> *Control volume:*   Gas-turbine engine.
> *Inlet states:* T known for fuel and air.
> *Exit state:* T known for combustion products.
> *Process:*   Steady state.
> *Model:*   All gases ideal gases, Table A.9; liquid octane, Table A.10.

Analysis

The combustion equation is

$$C_8H_{18}(l) + 4(12.5)O_2 + 4(12.5)(3.76)N_2 \rightarrow 8CO_2 + 9H_2O + 37.5O_2 + 188.0N_2$$

First law:

$$Q_{c.v.} + \sum_R n_i(\bar{h}_f^0 + \Delta\bar{h})_i = W_{c.v.} + \sum_P n_e(\bar{h}_f^0 + \Delta\bar{h})_e$$

Solution

Since the air is composed of elements and enters at 25°C, the enthalpy of the reactants is equal to that of the fuel,

$$\sum_R n_i(\bar{h}_f^0 + \Delta\bar{h})_i = (\bar{h}_f^0)_{C_8H_{18}(l)} = -250\,105 \text{ kJ/kmol fuel}$$

Considering the products, we have

$$\begin{aligned}
\sum_P n_e(\bar{h}_f^0 + \Delta\bar{h})_e &= n_{CO_2}(\bar{h}_f^0 + \Delta\bar{h})_{CO_2} + n_{H_2O}(\bar{h}_f^0 + \Delta\bar{h})_{H_2O} \\
&\quad + n_{O_2}(\Delta\bar{h})_{O_2} + n_{N_2}(\Delta\bar{h})_{N_2} \\
&= 8(-393\,522 + 28\,030) + 9(-241\,826 + 21\,892) \\
&\quad + 37.5(19\,249) + 188(18\,222) \\
&= -755\,769 \text{ kJ/kmol fuel}
\end{aligned}$$

$$W_{c.v.} = \frac{1000 \text{ kJ/s}}{0.25 \text{ kg/s}} \times \frac{114.23 \text{ kg}}{\text{kmol}} = 456\,920 \text{ kJ/kmol fuel}$$

Therefore, from the first law,

$$Q_{c.v.} = -755\ 769 + 456\ 920 - (-250\ 105)$$

$$= -48\ 744 \text{ kJ/kmol fuel}$$

EXAMPLE 14.8E	A small gas-turbine uses $C_8H_{18}(l)$ for fuel and 400% theoretical air. The air and fuel enter at 77 F, and the products of combustion leave at 1100 F. The output of the engine and the fuel consumption are measured, and it is found that the specific fuel consumption is one pound of fuel per horsepower-hour. Determine the heat transfer from the engine per pound mole of fuel. Assume complete combustion.

Control volume: Gas-turbine engine.
Inlet states: T known for fuel and air.
Exit state: T known for combustion products.
Process: Steady state.
Model: All gases ideal gases, Table F.6; liquid octane, Table F.11.

Analysis

The combustion equation is

$$C_8H_{18}(l) + 4(12.5)O_2 + 4(12.5)(3.76)N_2 \rightarrow 8CO_2 + 9H_2O + 37.5O_2 + 188.0N_2$$

First law:

$$Q_{c.v.} + \sum_R n_i(\bar{h}_f^0 + \Delta\bar{h})_i = W_{c.v.} + \sum_P n_e(\bar{h}_f^0 + \Delta\bar{h})_e$$

Solution

Since the air is composed of elements and enters at 77 F, the enthalpy of the reactants is equal to that of the fuel.

$$\sum_R n_i[\bar{h}_f^0 + \Delta\bar{h}]_i = (\bar{h}_f^0)_{C_8H_{18}(l)} = -107\ 526 \text{ Btu/lb mol}$$

Considering the products

$$\sum_P n_e(\bar{h}_f^0 + \Delta\bar{h})_e = n_{CO_2}(\bar{h}_f^0 + \Delta\bar{h})_{CO_2} + n_{H_2O}(\bar{h}_f^0 + \Delta\bar{h})_{H_2O} + n_{O_2}(\Delta\bar{h})_{O_2} + n_{N_2}(\Delta\bar{h})_{N_2}$$

$$= 8(-169\ 184 + 11\ 291) + 9(-103\ 966 + 8858)$$

$$+ 37.5(7778) + 188(7374)$$

$$= -441\ 129 \text{ Btu/lb mol fuel}.$$

$$W_{c.v.} = 2544 \times 114.23 = 290\ 601 \text{ Btu/lb mol fuel}$$

Therefore, from the first law,

$$Q_{c.v.} = -441\ 129 + 290\ 601 - (-107\ 526)$$

$$= -43\ 002\ \text{Btu/lb mol fuel}$$

EXAMPLE 14.9   A mixture of 1 kmol of gaseous ethene and 3 kmol of oxygen at 25°C reacts in a constant-volume bomb. Heat is transferred until the products are cooled to 600 K. Determine the amount of heat transfer from the system.

Control mass: Constant-volume bomb.

Initial state: T known.

Final state: T known.

Process: Constant volume.

Model: Ideal-gas mixtures, Tables A.9, A.10.

Analysis

The chemical reaction is

$$C_2H_4 + 3\,O_2 \rightarrow 2\,CO_2 + 2\,H_2O(g)$$

First law:

$$Q + U_R = U_P$$

$$Q + \sum_R n(\overline{h}_f^0 + \Delta\overline{h} - \overline{R}T) = \sum_P n(\overline{h}_f^0 + \Delta\overline{h} - \overline{R}T)$$

Solution

Using values from Tables A.9 and A.10, gives

$$\sum_R n(\overline{h}_f^0 + \Delta\overline{h} - \overline{R}T) = (\overline{h}_f^0 - \overline{R}T)_{C_2H_4} - n_{O_2}(\overline{R}T)_{O_2} = (\overline{h}_f^0)_{C_2H_4} - 4\overline{R}T$$

$$= 52\ 467 - 4 \times 8.3145 \times 298.2 = 42\ 550\ \text{kJ}$$

$$\sum_P n(\overline{h}_f^0 + \Delta\overline{h} - \overline{R}T) = 2[(\overline{h}_f^0)_{CO_2} + \Delta\overline{h}_{CO_2}] + 2[(\overline{h}_f^0)_{H_2O(g)} + \Delta\overline{h}_{H_2O(g)}] - 4\overline{R}T$$

$$= 2(-393\ 522 + 12\ 899) + 2(-241\ 826 + 10\ 463)$$
$$-4 \times 8.3145 \times 600$$

$$= -1\ 243\ 927\ \text{kJ}$$

Therefore,

$$Q = -1\ 243\ 927 - 42\ 550 = -1\ 286\ 477\ \text{kJ}$$

For a real-gas mixture, a pseudocritical method such as Kay's rule, Eq. 13.86 could be used to evaluate the nonideal-gas contribution to enthalpy at the temperature and pressure of the mixture and this value added to the ideal-gas mixture enthalpy at that temperature, as in the procedure developed in Section 13.11.

14.5 ENTHALPY AND INTERNAL ENERGY OF COMBUSTION; HEAT OF REACTION

The enthalpy of combustion, h_{RP}, is defined as the difference between the enthalpy of the products and the enthalpy of the reactants when complete combustion occurs at a given temperature and pressure. That is,

$$\overline{h}_{RP} = H_P - H_R$$

$$\overline{h}_{RP} = \sum_P n_e(\overline{h}_f^0 + \Delta \overline{h})_e - \sum_R n_i(\overline{h}_f^0 + \Delta \overline{h})_i \qquad (14.14)$$

The usual parameter for expressing the enthalpy of combustion is a unit mass of fuel, such as a kilogram (h_{RP}) or a kilomole ($\overline{h}_{RP}$) of fuel.

As the enthalpy of formation is fixed, we can separate the terms as

$$H = H^0 + \Delta H$$

where

$$H_R^0 = \sum_R n_i \overline{h}_{fi}^0; \qquad \Delta H_R = \sum_R n_i \, \Delta \overline{h}_i$$

and

$$H_P^0 = \sum_P n_i \overline{h}_{fi}^0; \qquad \Delta H_P = \sum_P n_i \, \Delta \overline{h}_i$$

Now the difference in enthalpies is written

$$H_P - H_R = H_P^0 - H_R^0 + \Delta H_P - \Delta H_R$$

$$= h_{RP} + \Delta H_P - \Delta H_R \qquad (14.15)$$

explicitly showing the reference enthalpy of combustion, $\overline{h}_{RP0}$, and the two departure terms ΔH_P and ΔH_R. The latter two terms for the products and reactants are nonzero if they exist at a state other than the reference state.

The tabulated values of the enthalpy of combustion of fuels are usually given for a temperature of 25°C and a pressure of 0.1 MPa. The enthalpy of combustion for a number of hydrocarbon fuels at this temperature and pressure, which we designate h_{RP0}, is given in Table 14.3.

The internal energy of combustion is defined in a similar manner.

$$\overline{u}_{RP} = U_P - U_R$$

$$= \sum_P n_e(\overline{h}_f^0 + \Delta \overline{h} - P\overline{v})_e - \sum_R n_i(\overline{h}_f^0 + \Delta \overline{h} - P\overline{v})_i \qquad (14.16)$$

TABLE 14.3

Enthalpy of Combustion of Some Hydrocarbons at 25°C

Hydrocarbon	Formula	UNITS: $\frac{kJ}{kg}$	LIQUID H_2O IN PRODUCTS		GAS H_2O IN PRODUCTS	
			Liq. HC	Gas HC	Liq. HC	Gas HC
Paraffins	C_nH_{2n+2}					
Methane	CH_4			−55 496		−50 010
Ethane	C_2H_6			−51 875		−47 484
Propane	C_3H_8		−49 973	−50 343	−45 982	−46 352
n-Butane	C_4H_{10}		−49 130	−49 500	−45 344	−45 714
n-Pentane	C_5H_{12}		−48 643	−49 011	−44 983	−45 351
n-Hexane	C_6H_{14}		−48 308	−48 676	−44 733	−45 101
n-Heptane	C_7H_{16}		−48 071	−48 436	−44 557	−44 922
n-Octane	C_8H_{18}		−47 893	−48 256	−44 425	−44 788
n-Decane	$C_{10}H_{22}$		−47 641	−48 000	−44 239	−44 598
n-Dodecane	$C_{12}H_{26}$		−47 470	−47 828	−44 109	−44 467
n-Cetane	$C_{16}H_{34}$		−47 300	−47 658	−44 000	−44 358
Olefins	C_nH_{2n}					
Ethene	C_2H_4			−50 296		−47 158
Propene	C_3H_6			−48 917		−45 780
Butene	C_4H_8			−48 453		−45 316
Pentene	C_5H_{10}			−48 134		−44 996
Hexene	C_6H_{12}			−47 937		−44 800
Heptene	C_7H_{14}			−47 800		−44 662
Octene	C_8H_{16}			−47 693		−44 556
Nonene	C_9H_{18}			−47 612		−44 475
Decene	$C_{10}H_{20}$			−47 547		−44 410
Alkylbenzenes	$C_{6+n}H_{6+2n}$					
Benzene	C_6H_6		−41 831	−42 266	−40 141	−40 576
Methylbenzene	C_7H_8		−42 437	−42 847	−40 527	−40 937
Ethylbenzene	C_8H_{10}		−42 997	−43 395	−40 924	−41 322
Propylbenzene	C_9H_{12}		−43 416	−43 800	−41 219	−41 603
Butylbenzene	$C_{10}H_{14}$		−43 748	−44 123	−41 453	−41 828
Other fuels						
Gasoline	C_7H_{17}		−48 201	−48 582	−44 506	−44 886
Diesel T-T	$C_{14.4}H_{24.9}$		−45 700	−46 074	−42 934	−43 308
JP8 jet fuel	$C_{13}H_{23.8}$		−45 707	−46 087	−42 800	−43 180
Methanol	CH_3OH		−22 657	−23 840	−19 910	−21 093
Ethanol	C_2H_5OH		−29 676	−30 596	−26 811	−27 731
Nitromethane	CH_3NO_2		−11 618	−12 247	−10 537	−11 165
Phenol	C_6H_5OH		−32 520	−33 176	−31 117	−31 774
Hydrogen	H_2			−141 781		−119 953

When all the gaseous constituents can be considered as ideal gases, and the volume of the liquid and solid constituents is negligible compared to the value of the gaseous constituents, this relation for $\bar{u}_{RP}$ reduces to

$$\bar{u}_{RP} = \bar{h}_{RP} - \bar{R}T\,(n_{\text{gaseous products}} - n_{\text{gaseous reactants}}) \tag{14.17}$$

Frequently the term "heating value" or "heat of reaction" is used. This represents the heat transferred from the chamber during combustion or reaction at constant temperature. In the case of a constant pressure or steady-flow process, we conclude from the first law of thermodynamics that it is equal to the negative of the enthalpy of combustion. For this reason, this heat transfer is sometimes designated the constant-pressure heating value for combustion processes.

In the case of a constant-volume process, the heat transfer is equal to the negative of the internal energy of combustion. This is sometimes designated the constant-volume heating value in the case of combustion.

When the term heating value is used, the terms "higher" and "lower" heating value are used. The higher heating value is the heat transfer with liquid water in the products, and the lower heating value is the heat transfer with vapor water in the products.

EXAMPLE 14.10   Calculate the enthalpy of combustion of propane at 25°C on both a kilomole and kilogram basis under the following conditions.

1. Liquid propane with liquid water in the products.
2. Liquid propane with gaseous water in the products.
3. Gaseous propane with liquid water in the products.
4. Gaseous propane with gaseous water in the products.

This example is designed to show how the enthalpy of combustion can be determined from enthalpies of formation. The enthalpy of evaporation of propane is 370 kJ/kg.

Analysis and Solution

The basic combustion equation is

$$C_3H_8 + 5\,O_2 \rightarrow 3\,CO_2 + 4\,H_2O$$

From Table A.10 $(\bar{h}_f^0)_{C_3H_8(g)} = -103\,900$ kJ/kmol. Therefore,

$$(\bar{h}_f^0)_{C_3H_8(l)} = -103\,900 - 44.097(370) = -120\,216 \text{ kJ/kmol}$$

1. Liquid propane–liquid water:

$$\bar{h}_{RP_0} = 3(\bar{h}_f^0)_{CO_2} + 4(\bar{h}_f^0)_{H_2O(l)} - (\bar{h}_f^0)_{C_3H_8(l)}$$

$$= 3(-393\,522) + 4(-285\,830) - (-120\,216)$$

$$= -2\,203\,670 \text{ kJ/kmol} = -\frac{2\,203\,670}{44.097} = -49\,973 \text{ kJ/kg}$$

The higher heating value of liquid propane is 49 973 kJ/kg.

2. Liquid propane–gaseous water:

$$\bar{h}_{RP_0} = 3(\bar{h}_f^0)_{CO_2} + 4(\bar{h}_f^0)_{H_2O(g)} - (\bar{h}_f^0)_{C_3H_8(l)}$$

$$= 3(-393\ 522) + 4(-241\ 826) - (-120\ 216)$$

$$= -2\ 027\ 654 \text{ kJ/kmol} = -\frac{2\ 027\ 654}{44.097} = -45\ 982 \text{ kJ/kg}$$

The lower heating value of liquid propane is 45 982 kJ/kg.

3. Gaseous propane–liquid water:

$$\bar{h}_{RP_0} = 3(\bar{h}_f^0)_{CO_2} + 4(\bar{h}_f^0)_{H_2O(l)} - (\bar{h}_f^0)_{C_3H_8(g)}$$

$$= 3(-393\ 522) + 4(-285\ 830) - (-103\ 900)$$

$$= -2\ 219\ 986 \text{ kJ/kmol} = -\frac{2\ 219\ 986}{44.097} = -50\ 343 \text{ kJ/kg}$$

The higher heating value of gaseous propane is 50 343 kJ/kg.

4. Gaseous propane–gaseous water:

$$\bar{h}_{RP_0} = 3(\bar{h}_f^0)_{CO_2} + 4(\bar{h}_f^0)_{H_2O(g)} - (\bar{h}_f^0)_{C_3H_8(g)}$$

$$= 3(-393\ 522) + 4(-241\ 826) - (-103\ 900)$$

$$= -2\ 043\ 970 \text{ kJ/kmol} = -\frac{2\ 043\ 970}{44.097} = -46\ 352 \text{ kJ/kg}$$

The lower heating value of gaseous propane is 46 352 kJ/kg.

Each of the four values calculated in this example corresponds to the appropriate value given in Table 14.3.

EXAMPLE 14.11

Calculate the enthalpy of combustion of gaseous propane at 500 K. (At this temperature all the water formed during combustion will be vapor.) This example will demonstrate how the enthalpy of combustion of propane varies with temperature. The average constant-pressure specific heat of propane between 25°C and 500 K is 2.1 kJ/kg K.

Analysis

The combustion equation is

$$C_3H_8(g) + 5\,O_2 \rightarrow 3\,CO_2 + 4\,H_2O(g)$$

The enthalpy of combustion is, from Eq. 14.13,

$$(\bar{h}_{RP})_T = \sum_P n_e(\bar{h}_f^0 + \Delta\bar{h})_e - \sum_R n_i(\bar{h}_f^0 + \Delta\bar{h})_i$$

Solution

$$\bar{h}_{R_{500}} = [\bar{h}_f^0 + \bar{C}_{p\,av}(\Delta T)]_{C_3H_8(g)} + n_{O_2}(\Delta\bar{h})_{O_2}$$

$$= -103\,900 + 2.1 \times 44.097(500 - 298.2) + 5(6095)$$

$$= -54\,738 \text{ kJ/kmol}$$

$$\bar{h}_{P_{500}} = n_{CO_2}(\bar{h}_f^0 + \Delta\bar{h})_{CO_2} + n_{H_2O}(\bar{h}_f^0 + \Delta\bar{h})_{H_2O}$$

$$= 3(-393\,522 + 8297) + 4(-241\,826 + 6896)$$

$$= -2\,095\,395 \text{ kJ/kmol}$$

$$\bar{h}_{RP_{500}} = -2\,095\,395 - (-54\,738) = -2\,040\,657 \text{ kJ/kmol}$$

$$h_{RP_{500}} = \frac{-2\,040\,657}{44.097} = -46\,277 \text{ kJ/kg}$$

This compares with a value of $-46\,352$ at 25°C.

This problem could also have been solved using the given value of the enthalpy of combustion at 25°C by noting that

$$\bar{h}_{RP_{500}} = (H_P)_{500} - (H_R)_{500}$$

$$= n_{CO_2}(\bar{h}_f^0 + \Delta\bar{h})_{CO_2} + n_{H_2O}(\bar{h}_f^0 + \Delta\bar{h})_{H_2O}$$

$$- [\bar{h}_f^0 + \bar{C}_{p\,av}(\Delta T)]_{C_3H_8(g)} - n_{O_2}(\Delta\bar{h})_{O_2}$$

$$= \bar{h}_{RP_0} + n_{CO_2}(\Delta\bar{h})_{CO_2} + n_{H_2O}(\Delta\bar{h})_{H_2O}$$

$$- \bar{C}_{p\,av}(\Delta T)_{C_3H_8(g)} - n_{O_2}(\Delta\bar{h})_{O_2}$$

$$\bar{h}_{RP_{500}} = -46\,352 \times 44.097 + 3(8297) + 4(6896)$$

$$- 2.1 \times 44.097(500 - 298.2) - 5(6095)$$

$$= -2\,040\,657 \text{ kJ/kmol}$$

$$h_{RP_{500}} = \frac{-2\,040\,657}{44.097} = -46\,277 \text{ kJ/kg}$$

14.6 ADIABATIC FLAME TEMPERATURE

Consider a given combustion process that takes place adiabatically and with no work or changes in kinetic or potential energy involved. For such a process the temperature of the products is referred to as the adiabatic flame temperature. With the assumptions of no work and no changes in kinetic or potential energy, this is the maximum temperature that can be achieved for the given reactants because any heat transfer from the reacting substances and any incomplete combustion would tend to lower the temperature of the products.

For a given fuel and given pressure and temperature of the reactants, the maximum adiabatic flame temperature that can be achieved is with a stoichiometric mixture. The adiabatic flame temperature can be controlled by the amount of excess air that is used. This is important, for example, in gas turbines, where the maximum permissible temperature is determined by metallurgical considerations in the turbine, and close control of the temperature of the products is essential.

Example 14.12 shows how the adiabatic flame temperature may be found. The dissociation that takes place in the combustion products, which has a significant effect on the adiabatic flame temperature, will be considered in the next chapter.

EXAMPLE 14.12

Liquid octane at 25°C is burned with 400% theoretical air at 25°C in a steady-state process. Determine the adiabatic flame temperature.

Control volume:	Combustion chamber.
Inlet states:	T known for fuel and air.
Process:	Steady state.
Model:	Gases ideal gases, Table A.9; liquid octane, Table A.10.

Analysis

The reaction is

$$C_8H_{18}(l) + 4(12.5)O_2 + 4(12.5)(3.76)N_2 \rightarrow$$

$$8CO_2 + 9H_2O(g) + 37.5O_2 + 188.0N_2$$

First law: Since the process is adiabatic,

$$H_R = H_P$$

$$\sum_R n_i(\bar{h}_f^0 + \Delta\bar{h})_i = \sum_P n_e(\bar{h}_f^0 + \Delta\bar{h})_e$$

where $\Delta\bar{h}_e$ refers to each constituent in the products at the adiabatic flame temperature.

Solution

From Tables A.9 and A.10,

$$H_R = \sum_R n_i(\bar{h}_f^0 + \Delta\bar{h})_i = (\bar{h}_f^0)_{C_8H_{18}(l)} = -250\,105 \text{ kJ/kmol fuel}$$

$$H_P = \sum_P n_e(\bar{h}_f^0 + \Delta\bar{h})_e$$

$$= 8(-393\,522 + \Delta\bar{h}_{CO_2}) + 9(-241\,826 + \Delta\bar{h}_{H_2O}) + 37.5\,\Delta\bar{h}_{O_2} + 188.0\,\Delta\bar{h}_{N_2}$$

By trial-and-error solution, a temperature of the products is found that satisfies this equation. Assume that

$$T_P = 900 \text{ K}$$

$$H_P = \sum_P n_e(\bar{h}_f^0 + \Delta\bar{h})_e$$

$$= 8(-393\,522 + 28\,030) + 9(-241\,826 + 21\,892)$$

$$+ 37.5(19\,249) + 188(18\,222)$$

$$= -755\,769 \text{ kJ/kmol fuel}$$

Assume that

$$T_p = 1000 \text{ K}$$

$$H_P = \sum_P n_e(\bar{h}_f^0 + \Delta\bar{h})_e$$

$$= 8(-393\ 522 + 33\ 400) + 9(-241\ 826 + 25\ 956)$$

$$+ 37.5(22\ 710) + 188(21\ 461)$$

$$= 62\ 487 \text{ kJ/kmol fuel}$$

Since $H_P = H_R = -250\ 105$ kJ/kmol, we find by linear interpolation that the adiabatic flame temperature is 961.8 K. Because the ideal-gas enthalpy is not really a linear function of temperature, the true answer will be slightly different from this value.

14.7 THE THIRD LAW OF THERMODYNAMICS AND ABSOLUTE ENTROPY

As we consider a second-law analysis of chemical reactions, we face the same problem we had with the first law: What base should be used for the entropy of the various substances? This problem leads directly to a consideration of the third law of thermodynamics.

The third law of thermodynamics was formulated during the early part of the twentieth century. The initial work was done primarily by W. H. Nernst (1864–1941) and Max Planck (1858–1947). The third law deals with the entropy of substances at the absolute zero of temperature and in essence states that the entropy of a perfect crystal is zero at absolute zero. From a statistical point of view, this means that the crystal structure has the maximum degree of order. Furthermore, because the temperature is absolute zero, the thermal energy is minimum. It also follows that a substance that does not have a perfect crystalline structure at absolute zero, but instead has a degree of randomness, such as a solid solution or a glassy solid, has a finite value of entropy at absolute zero. The experimental evidence on which the third law rests is primarily data on chemical reactions at low temperatures and measurements of heat capacity at temperatures approaching absolute zero. In contrast to the first and second laws, which lead, respectively, to the properties of internal energy and entropy, the third law deals only with the question of entropy at absolute zero. However, the implications of the third law are quite profound, particularly in respect to chemical equilibrium.

The particular relevance of the third law is that it provides an absolute base from which to measure the entropy of each substance. The entropy relative to this base is termed the absolute entropy. The increase in entropy between absolute zero and any given state can be found either from calorimetric data or by procedures based on statistical thermodynamics. The calorimetric method gives precise measurements of specific-heat data over the temperature range, as well as of the energy associated with phase transformations. These measurements are in agreement with the calculations based on statistical thermodynamics and observed molecular data.

Table A.10 gives the absolute entropy at 25°C and 0.1-MPa pressure for a number of substances. Table A.9 gives the absolute entropy for a number of gases at 0.1-MPa pressure and various temperatures. For gases the numbers in all these tables are the hypothetical ideal-gas values. The pressure P^0 of 0.1 MPa is termed the standard-state pressure, and the absolute entropy as given in these tables is designated $\bar{s}^0$. The temperature is designated in kelvins with a subscript such as $\bar{s}^0_{1000}$.

If the value of the absolute entropy is known at the standard-state pressure of 0.1 MPa and a given temperature, it is a straightforward procedure to calculate the entropy change from this state (whether hypothetical ideal gas or real substance) to another desired state following the procedure described in Section 13.11. If the substance is listed in Table A.8, then

$$\bar{s}_{T,P} = \bar{s}^0_T - \bar{R} \ln \frac{P}{P^0} + (\bar{s}_{T,P} - \bar{s}^*_{T,P}) \tag{14.18}$$

In this expression, the first term on the right side is the value from Table A.9, the second is the ideal-gas term to account for a change in pressure from P^0 to P, and the third is the term that corrects for real-substance behavior, as given in the generalized entropy chart in the Appendix. If the real-substance behavior is to be evaluated from an equation of state or thermodynamic table of properties, the term for the change in pressure should be made to a low pressure P^*, at which ideal-gas behavior is a reasonable assumption, but it is also listed in the tables. Then

$$\bar{s}_{T,P} = \bar{s}^0_T - \bar{R} \ln \frac{P^*}{P^0} + (\bar{s}_{T,P} - \bar{s}^*_{T,P*}) \tag{14.19}$$

If the substance is not one of those listed in Table A.9, and the absolute entropy is known only at one temperature T_0, as given in Table A.10 for example, then it will be necessary to calculate $\bar{s}^0_T$ from

$$\bar{s}^0_T = \bar{s}^0_{T_0} + \int_{T_0}^T \frac{\overline{C}_{p0}}{T} dT \tag{14.20}$$

and then proceed with the calculation of Eq. 14.17 or 14.19.

If Eq. 14.18 is being used to calculate the absolute entropy of a substance in a region in which the ideal-gas model is a valid representation of the behavior of that substance, then the last term on the right-side of Eq. 14.18 simply drops out of the calculation.

For calculation of the absolute entropy of a mixture of ideal gases at T, P, the mixture entropy is given in terms of the component partial entropies as

$$\bar{s}^*_{mix} = \sum_i y_i \bar{s}^*_i \tag{14.21}$$

where

$$\bar{s}^*_i = \bar{s}^0_{T_i} - \bar{R} \ln \frac{P}{P^0} - \bar{R} \ln y_i = \bar{s}^0_{T_i} - \bar{R} \ln \frac{y_i P}{P^0} \tag{14.22}$$

For a real-gas mixture, a correction can be added to the ideal-gas entropy calculated from Eqs. 14.21 and 14.22 by using a pseudocritical method such as was discussed in Section 13.11. The corrected expression is

$$\bar{s}_{mix} = \bar{s}^*_{mix} + (\bar{s} - \bar{s}^*)_{T,P} \tag{14.23}$$

in which the second term on the right side is the correction term from the generalized entropy chart.

14.8 SECOND-LAW ANALYSIS OF REACTING SYSTEMS

The concepts of reversible work, irreversibility, and availability (exergy) were introduced in Chapter 10. These concepts included both the first and second laws of thermodynamics. We proceed now to develop this matter further, and we will be particularly concerned with determining the maximum work (availability) that can be done through a combustion process and with examining the irreversibilities associated with such processes.

The reversible work for a steady-state process in which there is no heat transfer with reservoirs other than the surroundings, and also in the absence of changes in kinetic and potential energy is, from Eq. 10.9 on a total mass basis,

$$W^{\text{rev}} = \sum m_i(h_i - T_0 s_i) - \sum m_e(h_e - T_0 s_e)$$

Applying this equation to a steady-state process that involves a chemical reaction, and introducing the symbols from this chapter, we have

$$W^{\text{rev}} = \sum_R n_i(\bar{h}_f^0 + \Delta\bar{h} - T_0\bar{s})_i - \sum_P n_e(\bar{h}_f^0 + \Delta\bar{h} - T_0\bar{s})_e \qquad (14.24)$$

Similarly, the irreversibility for such a process can be written as

$$I = W^{\text{rev}} - W = \sum_P n_e T_0 \bar{s}_e - \sum_R n_i T_0 \bar{s}_i - Q_{\text{c.v.}} \qquad (14.25)$$

The availability, ψ, for a steady-flow process, in the absence of kinetic and potential energy changes, is given by Eq. 10.19 as

$$\psi = (h - T_0 s) - (h_0 - T_0 s_0)$$

We further note that if a steady-state chemical reaction takes place in such a manner that both the reactants and products are in temperature equilibrium with the surroundings, the Gibbs function $(g = h - Ts)$, defined in Eq. 13.14, becomes a significant variable. For such a process, in the absence of changes in kinetic and potential energy, the reversible work is given by the relation

$$W^{\text{rev}} = \sum_R n_i \bar{g}_i - \sum_P n_e \bar{g}_e = -\Delta G \qquad (14.26)$$

in which

$$\Delta G = \Delta H - T\,\Delta S \qquad (14.27)$$

We should keep in mind that Eq. 14.26 is a special case and that the reversible work is given by Eq. 14.24 if the reactants and products are not in temperature equilibrium with the surroundings.

Let us now consider the question of the maximum work that can be done during a chemical reaction. For example, consider 1 kmol of hydrocarbon fuel and the necessary air for complete combustion, each at 0.1-MPa pressure and 25°C, the pressure and temperature of the surroundings. What is the maximum work that can be done as this fuel reacts with the air? From the considerations covered in Chapter 10, we conclude that the maximum work would be done if this chemical reaction took place reversibly and the products were finally in pressure and temperature equilibrium with the surroundings. We conclude that this reversible work could be calculated from the relation in Eq. 14.26,

$$W^{\text{rev}} = \sum_R n_i \bar{g}_i - \sum_p n_e \bar{g}_e = -\Delta G$$

However, since the final state is in equilibrium with the surroundings, we could consider this amount of work to be the availability of the fuel and air.

EXAMPLE 14.13 Ethene (g) at 25°C and 0.1-MPa pressure is burned with 400% theoretical air at 25°C and 0.1-MPa pressure. Assume that this reaction takes place reversibly at 25°C and that the products leave at 25°C and 0.1-MPa pressure. To simplify this problem further, assume that the oxygen and nitrogen are separated before the reaction takes place (each at 0.1 MPa, 25°C), that the constituents in the products are separated, and that each is at 25°C and 0.1 MPa. Thus, the reaction takes place as shown in Fig. 14.4. For purposes of comparison between this and the two subsequent examples, we consider all the water in the products to be a gas (a hypothetical situation in this example and Example 14.15).

Determine the reversible work for this process (that is, the work that would be done if this chemical action took place reversibly and isothermally).

Control volume:	Combustion chamber.
Inlet states:	P, T known for each gas.
Exit states:	P, T known for each gas.
Model:	All ideal gases, Tables A.9 and A.10.
Sketch:	Figure 14.4.

Analysis

The equation for this chemical reaction is

$$\text{C}_2\text{H}_4(g) + 3(4)\,\text{O}_2 + 3(4)(3.76)\,\text{N}_2 \rightarrow 2\,\text{CO}_2 + 2\,\text{H}_2\text{O}(g) + 9\,\text{O}_2 + 45.1\,\text{N}_2$$

The reversible work for this process is equal to the decrease in Gibbs function during this reaction, Eq. 14.26. Since each component is at the standard-state pressure P^0, we write Eqs. 14.26 and 14.27 as

$$W^{\text{rev}} = -\Delta G^0, \qquad \Delta G^0 = \Delta H^0 - T\,\Delta S^0$$

We also note that the 45.1 N_2 cancels out of both sides in these expressions, as does 9 of the 12 O_2.

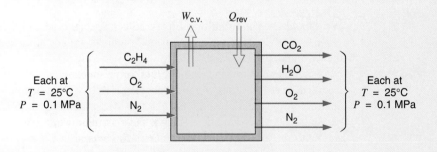

FIGURE 14.4 Sketch for Example 14.13.

Solution

Using values from Tables A.8 and A.9 at 25°C,

$$\Delta H^0 = 2\bar{h}^0_{f\,CO_2} + 2\bar{h}^0_{f\,H_2O(g)} - \bar{h}^0_{f\,C_2H_4} - 3\bar{h}^0_{f\,O_2}$$

$$= 2(-393\ 522) + 2(-241\ 826) - (+52\ 467) - 3(0)$$

$$= -1\ 323\ 163 \text{ kJ/kmol fuel}$$

$$\Delta S = 2\bar{s}^0_{CO_2} + 2\bar{s}^0_{H_2O(g)} - \bar{s}^0_{C_2H_4} - 3\bar{s}^0_{O_2}$$

$$= 2(213.795) + 2(188.843) - (219.330) - 3(205.148)$$

$$= -29.516 \text{ kJ/kmol fuel}$$

$$\Delta G^0 = -1\ 323\ 163 - 298.15(-29.516)$$

$$= -1\ 314\ 363 \text{ kJ/kmol } C_2H_4$$

$$W^{\text{rev}} = -\Delta G^0 = 1\ 314\ 363 \text{ kJ/kmol } C_2H_4$$

$$= \frac{1\ 314\ 363}{28.054} = 46\ 851 \text{ kJ/kg}$$

Therefore, we might say that when 1 kg of ethene is at 25°C, and the standard-state pressure, 0.1 MPa, it has an availability of 46 851 kJ.

Thus, it would seem logical to rate the efficiency of a device designed to do work by utilizing a combustion process, such as an internal-combustion engine or a steam power plant, as the ratio of the actual work to the reversible work, or in Example 14.13, the decrease in Gibbs function for the chemical reaction, instead of comparing the actual work to the heating value, as is commonly done. This is, in fact, the basic principle of the second-law efficiency, which was introduced in connection with availability analysis in Chapter 10. As noted from Example 14.13, the difference between the decrease in Gibbs function and the heating value is small, which is typical for hydrocarbon fuels. The difference in the two types of efficiencies will, therefore, not usually be large. We must always be careful, however, when discussing efficiencies, to note the definition of the efficiency under consideration.

It is of particular interest to study the irreversibility that takes place during a combustion process. The following examples illustrate this matter. We consider the same hydrocarbon fuel that was used in Example 14.13—ethene (g) at 25°C and 0.1 MPa. We determined its availability and found it to be 46 851 kJ/kg. Now let us burn this fuel with 400% theoretical air in a steady-state adiabatic process. We can determine the irreversibility of this process in two ways. The first way is to calculate the increase in entropy for the process. Since the process is adiabatic, the increase in entropy is due entirely to the irreversibilities for the process, and we can find the irreversibility from Eq. 14.25. We can also calculate the availabilities of the products of combustion at the adiabatic flame temperature, and note that they are less than the availability of the fuel and air before the combustion process. The difference is the irreversibility that occurs during the combustion process.

EXAMPLE 14.14 Consider the same combustion process as in Example 14.13, but let it take place adiabatically. Assume that each constituent in the products is at 0.1-MPa pressure and at the adiabatic flame temperature. This combustion process is shown schematically in Fig. 14.5. The temperature of the surroundings is 25°C.

For this combustion process, determine (1) the increase in entropy during combustion and (2) the availability of the products of combustion.

$$\begin{aligned} \textit{Control volume:} &\quad \text{Combustion chamber.}\\ \textit{Inlet states:} &\quad P, T \text{ known for each gas.}\\ \textit{Exit states:} &\quad P \text{ known for each gas.}\\ \textit{Sketch:} &\quad \text{Figure 14.5.}\\ \textit{Model:} &\quad \text{All ideal gases, Table A.9; Table A.10 for ethene.} \end{aligned}$$

Analysis

The combustion equation is

$$C_2H_4(g) + 12\,O_2 + 12(3.76)\,N_2 \rightarrow 2\,CO_2 + 2\,H_2O(g) + 9\,O_2 + 45.1\,N_2$$

The adiabatic flame temperature is determined first.
First law:

$$H_R = H_P$$

$$\sum_R n_i(\bar{h}_f^0)_i = \sum_P n_e(\bar{h}_f^0 + \Delta\bar{h})_e$$

Solution

$$52\,467 = 2(-393\,522 + \Delta\bar{h}_{CO_2}) + 2(-241\,826 + \Delta\bar{h}_{H_2O(g)}) + 9\Delta\bar{h}_{O_2} + 45.1\Delta\bar{h}_{N_2}$$

By a trial-and-error solution we find the adiabatic flame temperature to be 1016 K.

We now proceed to find the change in entropy during this adiabatic combustion process.

$$S_R = \sum_R (n_i\bar{s}_i^0)_{298} = (\bar{s}_{C_2H_4}^0 + 12\bar{s}_{O_2}^0 + 45.1\bar{s}_{N_2}^0)_{298}$$

$$= 219.330 + 12(205.147) + 45.1(191.610)$$

$$= 11\,322.705 \text{ kJ/kmol (fuel) K}$$

$$S_P = \sum_P (n_e\bar{s}_e^0)_{1016} = (2\bar{s}_{CO_2}^0 + 2\bar{s}_{H_2O(g)}^0 + 9\bar{s}_{O_2}^0 + 45.1\bar{s}_{N_2}^0)_{1016}$$

$$= 2(270.194) + 2(233.355) + 9(244.135) + 45.1(228.691)$$

$$= 13\,518.277 \text{ kJ/kmol (fuel) K}$$

$$S_P - S_R = 2195.572 \text{ kJ/kmol (fuel) K}$$

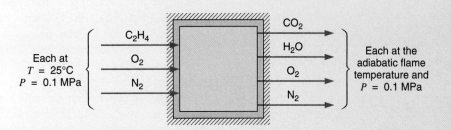

FIGURE 14.5 Sketch for Example 14.14.

Since this is an adiabatic process, the increase in entropy indicates the irreversibility of the adiabatic combustion process. This irreversibility can be found from Eq. 14.25.

$$I = T_0 \left(\sum_P n_e \bar{s}_e - \sum_R n_i \bar{s}_i \right)$$

$$= 298.15 \times 2195.572 = 654\ 610 \text{ kJ/kmol}$$

$$= \frac{654\ 610}{28.054} = 23\ 334 \text{ kJ/kg fuel}$$

Therefore, the availability after the combustion process is

$$\psi_P = 46\ 851 - 23\ 334 = 23\ 517 \text{ kJ/kg}$$

The availability of the products can also be found from the relation

$$\psi_P = \sum_P n_e [(\bar{h}_e - T_0 \bar{s}_e) - (\bar{h}_0 - T_0 \bar{s}_0)]$$

Since in this problem the products are separated, and each is at 0.1-MPa pressure and the adiabatic flame temperature of 1016 K, this equation can be evaluated, yielding

$$\psi_P = \sum_P n_e [(\bar{h}_e^0 - \bar{h}_0^0) - T_0 (\bar{s}_e^0 - \bar{s}_0^0)]$$

$$= 2(34\ 271) + 2(26\ 618) + 9(23\ 268) + 45.1(21\ 985)$$

$$- 298.15[2(270.194 - 213.795) + 2(233.355 - 188.834)$$

$$+ 9(244.135 - 205.147) + 45.1(228.691 - 191.610)]$$

$$= 659\ 746 \text{ kJ/kmol} = 23\ 517 \text{ kJ/kg}$$

In other words, if every process after the adiabatic combustion process is reversible, the maximum amount of work that could be done is 23 517 kJ/kg fuel. This compares to a value of 46 851 kJ/kg for the reversible isothermal reaction. This means that if we had an engine with the indicated adiabatic combustion process, and if all other processes were completely reversible, the efficiency would be about 50%.

What is the effect of having the combustion process of Example 14.14 take place with pure oxygen instead of air? In this case, the first law is the same, except with no nitrogen term in the reactants or products. As a result, the adiabatic flame temperature will be much higher, without having to heat up the 45.1 kmol of nitrogen. By trial and error, this temperature is found to be 2800 K. The entropy of the reactants is now 2681.094 kJ/kmol fuel K, and that of the products at 2800 K is found to be 3760.695 kJ/kmol fuel K. Therefore, the process irreversibility from Eq. 14.25 is, converting to a mass basis, 11474 kJ/kg fuel, such that the availability of the products is 35 377 kJ/kg fuel, significantly higher than for combustion with air in Example 14.14. This is because the products in this case are at a much higher temperature.

In the two prior examples we assumed, for purposes of simplifying the calculation, that the constituents in the reactants and products were separated, and each was at 0.1-MPa pressure. This of course is not a realistic problem. In the following example, Example 14.13 is repeated with the assumption that the reactants and products each consist of a mixture at 0.1-MPa pressure.

EXAMPLE 14.15

Consider the same combustion process as in Example 14.13, but assume that the reactants consist of a mixture at 0.1-MPa pressure and 25°C and that the products also consist of a mixture at 0.1 MPa and 25°C. Thus, the combustion process is as shown in Fig. 14.6.

Determine the work that would be done if this combustion process took place reversibly and in pressure and temperature equilibrium with the surroundings.

Control volume:	Combustion chamber.
Inlet state:	P, T known.
Exit state:	P, T known.
Sketch:	Figure 14.6.
Model:	Reactants—ideal-gas mixture, Table A.9. Products—ideal-gas mixture, Table A.9.

Analysis

The combustion equation, as noted previously, is

$$C_2H_4(g) + 3(4)O_2 + 3(4)(3.76)N_2 \rightarrow 2CO_2 + 2H_2O(g) + 9O_2 + 45.1N_2$$

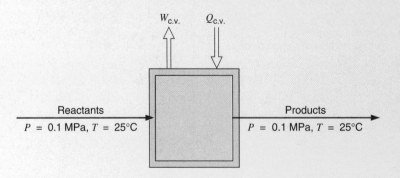

FIGURE 14.6 Sketch for Example 14.15.

In this case we must find the entropy of each substance as it exists in the mixture—that is, at its partial pressure and the given temperature of 25°C. Because the absolute entropies given in Tables A.9 and A.10 are at 0.1-MPa pressure and 25°C, the entropy of each constituent in the mixture can be found, using Eq. 14.20 from the relation

$$\bar{S}^* = \bar{s}^0 - \bar{R} \ln y \frac{P}{P^0}$$

where $\bar{S}^*$ = partial entropy of the constituent in the mixture

$\bar{s}^0$ = absolute entropy at the same temperature and 0.1-MPa pressure

P = pressure of the mixture

P^0 = 0.1-MPa pressure

y = mole fraction of the constituent

Since P^0 and the pressure of the mixture are both 0.1 MPa, the partial entropy of each constituent can be found by the relation

$$\bar{S}^* = \bar{s}^0 - \bar{R} \ln y = \bar{s}^0 + \bar{R} \ln \frac{1}{y}$$

Solution

For the reactants:

	n	$1/y$	$\bar{R} \ln 1/y$	$\bar{s}^0$	$\bar{S}^*$
C_2H_4	1	58.1	33.774	219.330	253.104
O_2	12	4.842	13.114	205.147	218.261
N_2	45.1	1.288	2.104	191.610	193.714
	58.1				

For the products:

	n	$1/y$	$\bar{R} \ln 1/y$	$\bar{s}^0$	$\bar{S}^*$
CO_2	2	29.05	28.011	213.795	241.806
H_2O	2	29.05	28.011	188.834	216.845
O_2	9	6.456	15.506	205.147	220.653
N_2	45.1	1.288	2.104	191.610	193.714
	58.1				

With the assumption of ideal-gas behavior, the enthalpy of each constituent is equal to the enthalpy of formation at 25°C. The values of entropy are as just calculated. Therefore, from Eq. 14.24,

$$
\begin{aligned}
W^{\text{rev}} &= \sum_R n_i(\bar{h}_f^0)_i - \sum_P n_e(\bar{h}_f^0)_e - T_0\left(\sum_R n_i\bar{s}_i - \sum_P n_e\bar{s}_e\right) \\
&= (\bar{h}_f^0)_{\text{C}_2\text{H}_4} - 2(\bar{h}_f^0)_{\text{CO}_2} - 2(\bar{h}_f^0)_{\text{H}_2\text{O}(g)} \\
&\quad -298.15(\bar{S}_{\text{C}_2\text{H}_4}^* + 12\bar{S}_{\text{O}_2}^* + 45.1\bar{S}_{\text{N}_2}^* - 2\bar{S}_{\text{CO}_2}^* - 2\bar{S}_{\text{H}_2\text{O}(g)}^* - 9\bar{S}_{\text{O}_2}^* - 45.1\bar{S}_{\text{N}_2}^*) \\
&= 52\,467 - 2(-393\,522) - 2(-241\,826) \\
&\quad -298.15[253.104 + 12(218.264) + 45.1(193.714) \\
&\quad -2(241.806) - 2(216.845) - 9(220.653) - 45.1(193.714)] \\
&= 1\,332\,378 \text{ kJ/kmol} \\
&= \frac{1\,332\,378}{28.054} = 47\,493 \text{ kJ/kg}
\end{aligned}
$$

Note that this value is essentially the same as the value that was obtained in Example 14.13, when the reactants and products were each separated and at 0.1-MPa pressure.

14.9 FUEL CELLS

The previous examples raise the question of the possibility of a reversible chemical reaction. Some reactions can be made to approach reversibility by having them take place in an electrolytic cell, as described in Chapter 1. When a potential exactly equal to the electromotive force of the cell is applied, no reaction takes place. When the applied potential is increased slightly, the reaction proceeds in one direction, and if the applied potential is decreased slightly, the reaction proceeds in the opposite direction. The work done is the electrical energy supplied or delivered.

Consider a reversible reaction occurring at constant temperature equal to that of its environment. The work output of the fuel cell is

$$
W = -\left(\sum n_e\bar{g}_e - \sum n_i\bar{g}_i\right) = -\Delta G
$$

where ΔG is the change in Gibbs function for the overall chemical reaction. We also realize that the work is given in terms of the charged electrons flowing through an electrical potential $\mathscr{E}$ as

$$
W = \mathscr{E}n_eN_0e
$$

in which n_e is the number of kilomoles of electrons flowing through the external circuit and

$$
\begin{aligned}
N_0e &= 6.022\,136 \times 10^{26} \text{ elec/kmol} \times 1.602\,177 \times 10^{-22} \text{ kJ/elec V} \\
&= 96\,485 \text{ kJ/kmol V}
\end{aligned}
$$

Thus, for a given reaction, the maximum (reversible reaction) electrical potential $\mathscr{E}^0$ of a fuel cell at a given temperature is

$$\mathscr{E}^0 = \frac{-\Delta G}{96\,485 n_e} \qquad (14.28)$$

EXAMPLE 14.16 Calculate the reversible electromotive force (EMF) at 25°C for the hydrogen–oxygen fuel cell described in Section 1.2.

Solution

The anode side reaction was stated to be

$$2\,H_2 \rightarrow 4\,H^+ + 4\,e^-$$

and the cathode side reaction is

$$4\,H^+ + 4\,e^- + O_2 \rightarrow 2\,H_2O$$

Therefore, the overall reaction is, in kilomoles,

$$2\,H_2 + O_2 \rightarrow 2\,H_2O$$

for which 4 kmol of electrons flow through the external circuit. Let us assume that each component is at its standard-state pressure of 0.1 MPa and that the water formed is liquid. Then

$$\Delta H^0 = 2\bar{h}^0_{f_{H_2O(l)}} - 2\bar{h}^0_{f_{H_2}} - \bar{h}^0_{f_{O_2}}$$

$$= 2(-285\,830) - 2(0) - 1(0) = -571\,660 \text{ kJ}$$

$$\Delta S^0 = 2\bar{s}^0_{H_2O_{(l)}} - 2\bar{s}^0_{H_2} - \bar{s}^0_{O_2}$$

$$= 2(69.950) - 2(130.678) - 1(205.148) = -326.604 \text{ kJ/K}$$

$$\Delta G^0 = -571\,660 - 298.15(-326.604) = -474\,283 \text{ kJ}$$

Therefore, from Eq. 14.28,

$$\mathscr{E}^0 = \frac{-(-474\,283)}{96\,485 \times 4} = 1.229 \text{ V}$$

In Example 14.16, we found the shift in the Gibbs function and the reversible EMF at 25°C. In practice, however, many fuel cells operate at an elevated temperature where the water leaves as a gas and not as a liquid; thus, it carries away more energy. The com-

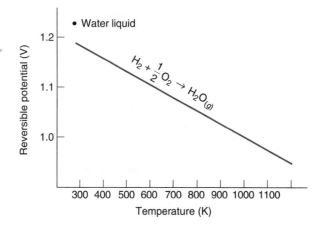

FIGURE 14.7

Hydrogen–oxygen fuel cell ideal EMF as a function of temperature.

putations can be done for a range of temperatures, leading to lower EMF as the temperature increases. This behavior is shown in Fig. 14.7.

A variety of fuel cells are being investigated for use in stationary as well as mobile power plants. The low-temperature fuel cells use hydrogen as the fuel, whereas the higher temperature cells can use methane and carbon monoxide that are then internally reformed into hydrogen and carbon dioxide. The most important fuel cells are listed in Table 14.4 with their main characteristics.

The low-temperature fuel cells are very sensitive to being poisoned by CO gas so that they require an external reformer and purifier to deliver hydrogen gas. The higher temperature fuel cells can reform natural gas, mainly methane, but also ethane and propane as shown in Table 14.2, into hydrogen gas and carbon monoxide inside the cell. The latest research is being done with gasified coal as a fuel and operating the cell at higher pressures like 15 atmospheres. As the fuel cell has exhaust gas with a small amount of fuel in it, additional combustion can occur and then combine the fuel cell with a gas turbine or steam power plant to utilize the exhaust gas energy. These combined cycle power plants strive to have an efficiency of up to 60%.

TABLE 14.4

Fuel Cell Types

FUEL CELL	PEC	PAC	MCC	SOC
	Polymer Electrolyte	Phosphoric Acid	Molten Carbonate	Solid Oxide
T	80°C	200°C	650°C	900°C
Fuel	Hydrogen, H_2	Hydrogen, H_2	CO, Hydrogen	Natural Gas
Carrier	H^+	H^+	CO_3^{--}	O^{--}
Charge, n_e	$2e^-$ per H_2	$2e^-$ per H_2	$2e^-$ per H_2 $2e^-$ per CO	$8e^-$ per CH_4
Catalyst	Pt	Pt	Ni	ZrO_2
Poison	CO	CO		

14.10 EVALUATION OF ACTUAL COMBUSTION PROCESSES

A number of different parameters can be defined for evaluating the performance of an actual combustion process, depending on the nature of the process and the system considered. In the combustion chamber of a gas-turbine, for example, the objective is to raise the temperature of the products to a given temperature (usually the maximum temperature the metals in the turbine can withstand). If we had a combustion process that achieved complete combustion and that was adiabatic, the temperature of the products would be the adiabatic flame temperature. Let us designate the fuel–air ratio needed to reach a given temperature under these conditions as the ideal fuel–air ratio. In the actual combustion chamber, the combustion will be incomplete to some extent, and there will be some heat transfer to the surroundings. Therefore, more fuel will be required to reach the given temperature, and this we designate as the actual fuel–air ratio. The combustion efficiency, η_{comb} is defined here as

$$\eta_{comb} = \frac{FA_{ideal}}{FA_{actual}} \qquad (14.29)$$

On the other hand, in the furnace of a steam generator (boiler), the purpose is to transfer the maximum possible amount of heat to the steam (water). In practice, the efficiency of a steam generator is defined as the ratio of the heat transferred to the steam to the higher heating value of the fuel. For a coal this is the heating value as measured in a bomb calorimeter, which is the constant-volume heating value, and it corresponds to the internal energy of combustion. We observe a minor inconsistency, since the boiler involves a flow process, and the change in enthalpy is the significant factor. In most cases, however, the error thus introduced is less than the experimental error involved in measuring the heating value, and the efficiency of a steam generator is defined by the relation

$$\eta_{steam\ generator} = \frac{\text{heat transferred to steam/kg fuel}}{\text{higher heating value of the fuel}} \qquad (14.30)$$

In an internal-combustion engine the purpose is to do work. The logical way to evaluate the performance of an internal-combustion engine would be to compare the actual work done to the maximum work that would be done by a reversible change of state from the reactants to the products. This, as we noted previously, is called the second-law efficiency.

In practice, however, the efficiency of an internal-combustion engine is defined as the ratio of the actual work to the negative of the enthalpy of combustion of the fuel (that is, the constant-pressure heating value). This ratio is usually called the thermal efficiency, η_{th}:

$$\eta_{th} = \frac{w}{-h_{RP_0}} = \frac{w}{\text{heating value}} \qquad (14.31)$$

The overall efficiency of a gas-turbine or steam power plant is defined in the same way. It should be pointed out that in an internal-combustion engine or fuel-burning steam power plant, the fact that the combustion is itself irreversible is a significant factor in the relatively low thermal efficiency of these devices.

One other factor should be pointed out regarding efficiency. We have noted that the enthalpy of combustion of a hydrocarbon fuel varies considerably with the phase of the water in the products, which leads to the concept of higher and lower heating values. Therefore, when we consider the thermal efficiency of an engine, the heating value used to determine this efficiency must be borne in mind. Two engines made by different manufacturers may have identical performance, but if one manufacturer bases his or her efficiency on the higher heating value and the other on the lower heating value, the latter will be able to claim a higher thermal efficiency. This claim is not significant, of course, as the performance is the same; this would be revealed by consideration of how the efficiency was defined.

The whole matter of the efficiencies of devices that undergo combustion processes is treated in detail in textbooks dealing with particular applications; our discussion is intended only as an introduction to the subject. Two examples are given, however, to illustrate these remarks.

EXAMPLE 14.17

The combustion chamber of a gas-turbine uses a liquid hydrocarbon fuel that has an approximate composition of C_8H_{18}. During testing, the following data are obtained.

$$T_{air} = 400 \text{ K} \qquad T_{products} = 1100 \text{ K}$$

$$\mathbf{V}_{air} = 100 \text{ m/s} \qquad \mathbf{V}_{products} = 150 \text{ m/s}$$

$$T_{fuel} = 50°C \qquad FA_{actual} = 0.0211 \text{ kg fuel/kg air}$$

Calculate the combustion efficiency for this process.

Control volume: Combustion chamber.
Inlet states: T known for air and fuel.
Exit state: T known.
Model: Air and products—ideal gas, Table A.8. Fuel—Table A.9.

Analysis

For the ideal chemical reaction the heat transfer is zero. Therefore, writing the first law for a control volume that includes the combustion chamber, we have

$$H_R + KE_R = H_P + KE_P$$

$$H_R + KE_R = \sum_R n_i \left(\bar{h}_f^0 + \Delta \bar{h} + \frac{M\mathbf{V}^2}{2} \right)_i$$

$$= [\bar{h}_f^0 + \overline{C}_p(50 - 25)]_{C_8H_{18}(l)} + n_{O_2} \left(\Delta \bar{h} + \frac{M\mathbf{V}^2}{2} \right)_{O_2}$$

$$+ 3.76 n_{O_2} \left(\Delta \bar{h} + \frac{M\mathbf{V}^2}{2} \right)_{N_2}$$

$$H_P + KE_P = \sum_P n_e \left(\bar{h}_f^0 + \Delta \bar{h} + \frac{M\mathbf{V}^2}{2} \right)_e$$

$$= 8 \left(\bar{h}_f^0 + \Delta \bar{h} + \frac{M\mathbf{V}^2}{2} \right)_{CO_2} + 9 \left(\bar{h}_f^0 + \Delta \bar{h} + \frac{M\mathbf{V}^2}{2} \right)_{H_2O}$$

$$+ (n_{O_2} - 12.5) \left(\Delta \bar{h} + \frac{M\mathbf{V}^2}{2} \right)_{O_2} + 3.76 n_{O_2} \left(\Delta \bar{h} + \frac{m\mathbf{V}^2}{2} \right)_{N_2}$$

Solution

$$H_R + KE_R = -250\ 105 + 2.23 \times 114.23(50 - 25)$$

$$+ n_{O_2}\left[3034 + \frac{32 \times (100)^2}{2 \times 1000}\right]$$

$$+ 3.76 n_{O_2}\left[2971 + \frac{28.02 \times (100)^2}{2 \times 1000}\right]$$

$$= -243\ 737 + 14\ 892 n_{O_2}$$

$$H_P + KE_P = 8\left[-393\ 522 + 38\ 891 + \frac{44.01 \times (150)^2}{2 \times 1000}\right]$$

$$+ 9\left[-241\ 826 + 30\ 147 + \frac{18.02 \times (150)^2}{2 \times 1000}\right]$$

$$+ (n_{O_2} - 12.5)\left[26\ 218 + \frac{32 \times (150)^2}{2 \times 1000}\right]$$

$$+ 3.76 n_{O_2}\left[24\ 758 + \frac{28.02 \times (150)^2}{2 \times 1000}\right]$$

$$= -5\ 068\ 599 + 120\ 853 n_{O_2}$$

Therefore,

$$-243\ 737 + 14\ 892 n_{O_2} = -5\ 068\ 599 + 120\ 853 n_{O_2}$$

$$n_{O_2} = 45.53 \text{ kmol O}_2/\text{kmol fuel}$$

$$\text{kmol air/kmol fuel } 4.76(45.53) = 216.72$$

$$FA_{\text{ideal}} = \frac{114.23}{216.72 \times 28.97} = 0.0182 \text{ kg fuel/kg air}$$

$$\eta_{\text{comb}} = \frac{0.0182}{0.0211} \times 100 = 86.2 \text{ percent}$$

EXAMPLE 14.18 In a certain steam power plant, 325 000 kg of water per hour enters the boiler at a pressure of 12.5 MPa and a temperature of 200°C. Steam leaves the boiler at 9 MPa, 500°C. The power output of the turbine is 81 000 kW. Coal is used at the rate of 26 700 kg/h and has a higher heating value of 33 250 kJ/kg. Determine the efficiency of the steam generator and the overall thermal efficiency of the plant.

In power plants, the efficiency of the boiler and the overall efficiency of the plant are based on the higher heating value of the fuel.

Solution

The efficiency of the boiler is defined by Eq. 14.30 as

$$\eta_{\text{steam generator}} = \frac{\text{heat transferred to H}_2\text{O/kg fuel}}{\text{higher heating value}}$$

Therefore,

$$\eta_{\text{steam generator}} = \frac{325\ 000(3386.1 - 857.1)}{26\ 700 \times 33\ 250} \times 100 = 92.6\%$$

The thermal efficiency is defined by Eq. 14.31,

$$\eta_{\text{th}} = \frac{w}{\text{heating value}} = \frac{81\ 000 \times 3600}{26\ 700 \times 33\ 250} \times 100 = 32.8\%$$

SUMMARY

An introduction to combustion of hydrocarbon fuels and chemical reactions in general is given. A simple oxidation of a hydrocarbon fuel with pure oxygen or air burns the hydrogen to water and the carbon to carbon dioxide. We apply the continuity equation for each kind of atom to balance the stoichiometric coefficients of the species in the reactants and products. The reactant mixture composition is described by the air–fuel ratio on a mass or mole basis or percent theoretical air or equivalence ratio, according to the practice of the particular area of use. The product of a given fuel for a stoichiometric mixture and complete combustion is unique, whereas actual combustion can lead to incomplete combustion and more complex products described by measurements on a dry or wet basis. As water is part of the products, they have a dew point, so it is possible to see water condensing out from the products as they are cooled.

Because of the chemical changes from the reactants to the products, we need to measure energy from an absolute reference. Chemically pure substances (not compounds like CO) in their ground state (graphite for carbon, not diamond form) are assigned a value of 0 for the formation enthalpy at reference temperature and pressure (25°C, 100 kPa). Stable compounds have a negative formation enthalpy, and unstable compounds have a positive formation enthalpy. The shift in the enthalpy from the reactants to the products is the enthalpy of combustion, which is also the negative of the heating value HV. When a combustion process takes place without any heat transfer, the resulting product temperature is the adiabatic flame temperature. The enthalpy of combustion, the heating value (lower or higher), and the adiabatic flame temperature depend on the mixture (fuel and A/F ratio) and the reactant's supply temperature. When a single unique number for these properties is used, it is understood to be for a stoichiometric mixture at the reference conditions.

Similarly to the enthalpy, an absolute value of entropy is needed for the application of the second law. The absolute entropy is zero for a perfect crystal at 0 K, which is the third law of thermodynamics. The combustion process is an irreversible process, and there is thus a loss of availability (exergy) associated with it. This irreversibility is increased by mixtures different from stoichiometric and by dilution of the oxygen (i.e., nitrogen in air), which lowers the adiabatic flame temperature. From the concept of flow exergy, we apply the second law to find the reversible work given by the change in Gibbs function. A process that has less irreversibility than combustion at high temperature is the chemical conversion in a fuel cell where we approach a chemical equilibrium process (covered in detail in the following chapter). Here the energy release is directly converted into an electrical power output, a system that is currently under intense study and development for future energy conversion systems.

You should have learned a number of skills and acquired abilities from studying this chapter that will allow you to:

- Write the combustion equation for the stoichiometric reaction of any fuel.
- Balance the stoichiometric coefficients for a reaction with a set of products measured on a dry basis.
- Handle the combustion of fuel mixtures as well as moist air oxidicer.
- Apply the energy equation with absolute values of enthalpy or internal energy.
- Use the proper tables for high-temperature products.
- Deal with condensation of water in low-temperature products of combustion.
- Calculate the adiabatic flame temperature for a given set of reactants.
- Know the difference between the enthalpy of formation and the enthalpy of combustion.
- Know the definition of the higher and lower heating values.
- Apply the second law to a combustion problem and find irreversibilities.
- Calculate the change in Gibbs function and the reversible work.
- Know how a fuel cell operates and how to find its electrical potential.
- Know some basic definition of combustion efficiencies.

KEY CONCEPTS AND FORMULAS

Reaction	Fuel + Oxidizer $\Rightarrow$ Products
	hydrocarbon + air $\Rightarrow$ carbon dioxide + water + nitrogen
Stoichiometric ratio	No excess fuel, no excess oxygen
Stoichiometric coefficients	Factors to balance atoms between reactants and products
Stoichiometric reaction	$C_xH_y + v_{O_2}(O_2 + 3.76\,N_2)$
	$\qquad \Rightarrow v_{CO_2}CO_2 + v_{H_2O}H_2O + v_{N_2}N_2$
	$v_{O_2} = x + y/4; \quad v_{CO_2} = x;\ v_{H_2O} = y/2; \quad v_{N_2} = 3.76 v_{O_2}$
Air–fuel ratio	$AF_{mass} = \dfrac{m_{air}}{m_{fuel}} = AF_{mole}\dfrac{M_{air}}{M_{fuel}}$
Equivalence ratio	$\Phi = \dfrac{FA}{FA_s} = \dfrac{AF_s}{AF}$
Enthalpy of formation	$\bar{h}_f^0$, zero for chemically pure substance, ground state
Enthalpy of combustion	$h_{RP} = H_P - H_R$
Heating value HV	$HV = -h_{RP}$
Internal energy of combustion	$u_{RP} = U_P - U_R = h_{RP} - RT(n_P - n_R)$ if ideal gases
Adiabatic flame temperature	$H_P = H_R$ if flow; $U_P = U_R$ if constant volume
Reversible work	$W^{rev} = G_R - G_P = -\Delta G = -(\Delta H - T\,\Delta S)$
	This requires that any Q is transferred at the local T.
Gibbs function	$G = H - TS$
Irreversibility	$i = w^{rev} - w = T_0\dot{S}_{gen}/\dot{m} = T_0 s_{gen}$
	$I = W^{rev} - W = T_0\dot{S}_{gen}/\dot{n} = T_0 s_{gen}$ for 1 kmole fuel

Concept-Study Guide Problems

14.1 How many kmoles of air are needed to burn 1 kmol of carbon?

14.2 If I burn 1 kmol of hydrogen H_2 with 6 kmol air, what is the A/F ratio on a mole basis and what is the percent theoretical air?

14.3 Why would I sometimes need A/F on a mole basis? on a mass basis?

14.4 Why is there no significant difference between the number of moles of reactants versus products in combustion of hydrocarbon fuels with air?

14.5 For the 110% theoretical air in Eq. 14.8, what is the equivalence ratio? Is that mixture rich or lean?

14.6 Why are products measured on a dry basis?

14.7 What is the dew point of hydrogen burned with stoichiometric pure oxygen? air?

14.8 How does the dew point change as the equivalence ratio goes from 0.9 to 1 to 1.1?

14.9 In most cases, combustion products are exhausted above the dew point. Why?

14.10 Why does combustion contribute to global warming?

14.11 What is the enthalpy of formation for oxygen as O_2? if O? for CO_2?

14.12 How is a fuel enthalpy of combustion connected to its enthalpy of formation?

14.13 What is the higher and lower heating value HHV, LHV of n-butane?

14.14 What is the value of h_{fg} for n-octane?

14.15 Why do some fuels not have entries for liquid fuel in Table 14.3?

14.16 Does it make a difference for the enthalpy of combustion whether I burn with pure oxygen or air? What about the adiabatic flame temperature?

14.17 What happens to the adiabatic flame temperature if I burn rich or lean?

14.18 Is the irreversibility in a combustion process significant? Why is that?

14.19 If the A/F ratio is larger than stoichiometric, is it more or less reversible?

14.20 What makes the fuel cell attractive from a power-generating point of view?

Homework Problems

Fuels and the Combustion Process

14.21 Calculate the theoretical air–fuel ratio on a mass and mole basis for the combustion of ethanol, C_2H_5OH.

14.22 A certain fuel oil has the composition $C_{10}H_{22}$. If this fuel is burned with 150% theoretical air, what is the composition of the products of combustion?

14.23 Methane is burned with 200% theoretical air. Find the composition and the dew point of the products.

14.24 In a combustion process with decane, $C_{10}H_{22}$, and air, the dry product mole fractions are 83.61% N_2, 4.91% O_2, 10.56% CO_2, and 0.92% CO. Find the equivalence ratio and the percent theoretical air of the reactants.

14.25 Natural gas B from Table 14.2 is burned with 20% excess air. Determine the composition of the products.

14.26 A Pennsylvania coal contains 74.2% C, 5.1% H, 6.7% O (dry basis, mass percent) plus ash and small percentages of N and S. This coal is fed into a gasifier along with oxygen and steam, as shown in Fig. P14.26. The exiting product gas composition is measured on a mole basis to: 39.9% CO, 30.8% H_2, 11.4% CO_2, 16.4% H_2O plus small percentages of CH_4, N_2, and H_2S. How many kilograms of coal are required to produce 100 kmol of product gas? How much oxygen and steam are required?

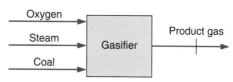

FIGURE P14.26

14.27 Repeat Problem 14.26 for a certain Utah coal that contains, according to the coal analysis, 68.2% C, 4.8% H, and 15.7% O on a mass basis. The exiting product gas contains 30.9% CO, 26.7% H_2, 15.9% CO_2, and 25.7% H_2O on a mole basis.

14.28 For complete stoichiometric combustion of gasoline, C_7H_{17}, determine the fuel molecular weight, the combustion products, and the mass of carbon dioxide produced per kg of fuel burned.

14.29 A sample of pine bark has the following ultimate analysis on a dry basis, percent by mass: 5.6% H, 53.4% C, 0.1% S, 0.1% N, 37.9% O, and 2.9% ash. This bark will be used as a fuel by burning it with 100% theoretical air in a furnace. Determine the air–fuel ratio on a mass basis.

14.30 Liquid propane is burned with dry air. A volumetric analysis of the products of combustion yields the following volume percent composition on a dry basis: 8.6% CO_2, 0.6% CO, 7.2% O_2, and 83.6% N_2. Determine the percent of theoretical air used in this combustion process.

14.31 A fuel, C_xH_y, is burned with dry air, and the product composition is measured on a dry mole basis to be: 9.6% CO_2, 7.3% O_2, and 83.1% N_2. Find the fuel composition (x/y) and the percent theoretical air used.

14.32 For the combustion of methane, 150% theoretical air is used at 25°C, 100 kPa, and relative humidity of 70%. Find the composition and dew point of the products.

14.33 Many coals from the western United States have a high moisture content. Consider the following sample of Wyoming coal, for which the ultimate analysis on an as-received basis is, by mass:

Component	Moisture	H	C	S	N	O	Ash
% mass	28.9	3.5	48.6	0.5	0.7	12.0	5.8

This coal is burned in the steam generator of a large power plant with 150% theoretical air. Determine the air–fuel ratio on a mass basis.

14.34 Pentane is burned with 120% theoretical air in a constant-pressure process at 100 kPa. The products are cooled to ambient temperature, 20°C. How much mass of water is condensed per kilogram of fuel? Repeat the answer, assuming that the air used in the combustion has a relative humidity of 90%.

14.35 The coal gasifier in an integrated gasification combined cycle (IGCC) power plant produces a gas mixture with the following volumetric percent composition:

Product	CH_4	H_2	CO	CO_2	N_2	H_2O	H_2S	NH_3
% vol.	0.3	29.6	41.0	10.0	0.8	17.0	1.1	0.2

This gas is cooled to 40°C, 3 MPa, and the H_2S and NH_3 are removed in water scrubbers. Assuming that the resulting mixture, which is sent to the combustors, is saturated with water, determine the mixture composition and the theoretical air–fuel ratio in the combustors.

14.36 The hot exhaust gas from an internal combustion engine is analyzed and found to have the following percent composition on a volumetric basis at the engine exhaust manifold: 10% CO_2, 2% CO, 13% H_2O, 3% O_2, and 72% N_2. This gas is fed to an exhaust gas reactor and mixed with a certain amount of air to eliminate the carbon monoxide, as shown in Fig. P14.36. It has been determined that a mole fraction of 10% oxygen in the mixture at state 3 will ensure that no CO remains. What must be the ratio of flows entering the reactor?

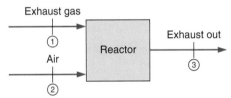

FIGURE P14.36

14.37 Butane is burned with dry air at 40°C, 100 kPa, with $AF = 26$ on a mass basis. For complete combustion find the equivalence ratio, % theoretical air, and the dew point of the products. How much water (kg/kg fuel) is condensed out, if any, when the products are cooled down to ambient temperature 40°C?

14.38 Methanol, CH_3OH, is burned with 200% theoretical air in an engine, and the products are brought to 100 kPa, 30°C. How much water is condensed per kilogram of fuel?

14.39 The output gas mixture of a certain air-blown coal gasifier has the composition of producer gas as listed in Table 14.2. Consider the combustion of this gas with 120% theoretical air at 100 kPa pressure. Determine the dew point of the products and find how many kilograms of water will be condensed per kilogram of fuel if the products are cooled 10°C below the dew-point temperature.

14.40 In an engine, liquid octane and ethanol, mole ratio 9:1, and stoichiometric air are taken in at 298 K, 100 kPa. After complete combustion, the products run out of the exhaust system where they are cooled to 10°C. Find the dew point of the products and the mass of water condensed per kilogram of fuel mixture.

Energy Equation, Enthalpy of Formation

14.41 A rigid vessel initially contains 2 kmol of carbon and 2 kmol of oxygen at 25°C, 200 kPa. Combustion occurs, and the resulting products consist of 1 kmol of carbon dioxide, 1 kmol of carbon monoxide, and excess oxygen at a temperature of 1000 K. Determine the final pressure in the vessel and the heat transfer from the vessel during the process.

14.42 In a test of rocket propellant performance, liquid hydrazine (N_2H_4) at 100 kPa, 25°C, and oxygen gas at 100 kPa, 25°C, are fed to a combustion chamber in the ratio of 0.5 kg O_2/kg N_2H_4. The heat transfer from the chamber to the surroundings is estimated to be 100 kJ/kg N_2H_4. Determine the temperature of the products exiting the chamber. Assume that only H_2O, H_2, and N_2 are present. The enthalpy of formation of liquid hydrazine is +50 417 kJ/kmol.

14.43 The combustion of heptane C_7H_{16} takes place in a steady-flow burner where fuel and air are added as gases at P_0, T_0. The mixture has 125% theoretical air, and the products pass through a heat exchanger where they are cooled to 600 K. Find the heat transfer from the heat exchanger per kmol of heptane burned.

14.44 Butane gas and 200% theoretical air, both at 25°C, enter a steady-flow combustor. The products of combustion exit at 1000 K. Calculate the heat transfer from the combustor per kmol of butane burned.

14.45 One alternative to using petroleum or natural gas as fuels is ethanol (C_2H_5OH), which is commonly produced from grain by fermentation. Consider a combustion process in which liquid ethanol is burned with 120% theoretical air in a steady-flow process. The reactants enter the combustion chamber at 25°C, and the products exit at 60°C, 100 kPa. Calculate the heat transfer per kilomole of ethanol.

14.46 Do the previous problem with the ethanol fuel delivered as a vapor.

14.47 Another alternative to using petroleum or natural gas as fuels is methanol, (CH_3OH), which can be produced from coal. Both methanol and ethanol have been used in automotive engines. Repeat Problem 14.45 using liquid methanol as the fuel instead of ethanol.

14.48 Another alternative fuel to be seriously considered is hydrogen. It can be produced from water by various techniques that are under extensive study. Its biggest problems at the present time are cost, storage, and safety. repeat Problem 14.45 using hydrogen gas as the fuel instead of ethanol.

14.49 In a new high-efficiency furnace, natural gas, assumed to be 90% methane and 10% ethane (by volume) and 110% theoretical air each enter at 25°C, 100 kPa, and the products (assumed to be 100% gaseous) exit the furnace at 40°C, 100 kPa. What is the heat transfer for this process? Compare this to an older furnace where the products exit at 250°C, 100 kPa.

14.50 Repeat the previous problem but take into account the actual phase behavior of the products exiting the furnace.

14.51 Pentene, C_5H_{10}, is burned with pure oxygen in a steady-flow process. The products at one point are brought to 700 K and used in a heat exchanger, where they are cooled to 25°C. Find the specific heat transfer in the heat exchanger.

14.52 Methane, CH_4, is burned in a steady-flow process with two different oxidizers: Case **A**: Pure oxygen, O_2, and case **B**: A mixture of $O_2 + x$Ar. The reactants are supplied at T_0, P_0 and the products for both cases should be at 1800 K. Find the required equivalence ratio in case **A** and the amount of argon, x, for a stoichiometric ratio in case **B**.

14.53 A closed, insulated container is charged with a stoichiometric ratio of oxygen and hydrogen at 25°C and 150 kPa. After combustion, liquid water at 25°C is sprayed in such a way that the final temperature is 1200 K. What is the final pressure?

14.54 Gaseous propane mixes with air, both supplied at 500 K, 0.1 MPa. The mixture goes into a combustion chamber, and products of combustion exit at 1300 K, 0.1 MPa. The products analyzed on a dry basis are 11.42% CO_2, 0.79% CO, 2.68% O_2, and 85.11% N_2 on a volume basis. Find the equivalence ratio and the heat transfer per kmol of fuel.

Enthalpy of Combustion and Heating Value

14.55 Liquid pentane is burned with dry air, and the products are measured on a dry basis as 10.1% CO_2, 0.2% CO, 5.9% O_2 remainder N_2. Find the enthalpy of formation for the fuel and the actual equivalence ratio.

14.56 Phenol has an entry in Table 14.3, but it does not have a corresponding value of the enthalpy of formation in Table A.10. Can you calculate it?

14.57 Do Problem 14.43 using Table 14.3 instead of Table A.10 for the solution.

14.58 Wet biomass waste from a food-processing plant is fed to a catalytic reactor, where in a steady-flow process it is converted into a low-energy fuel gas suitable for firing the processing plant boilers. The fuel gas has a composition of 50% methane, 45% carbon dioxide, and 5% hydrogen on a volumetric basis. Determine the lower heating value of this fuel gas mixture per unit volume.

14.59 Determine the lower heating value of the gas generated from coal, as described in Problem 14.35. Do not include the components removed by the water scrubbers.

14.60 Do Problem 14.45 using Table 14.3 instead of Table A.10 for the solution.

14.61 Propylbenzene, C_9H_{12}, is listed in Table 14.3, but not in Table A.9. No molecular weight is listed in the book. Find the molecular weight, the enthalpy of formation for the liquid fuel, and the enthalpy of evaporation.

14.62 Determine the higher heating value of the sample Wyoming coal as specified in Problem 14.33.

14.63 Do Problem 14.47 using Table 14.3 instead of Table A.10 for the solution.

14.64 A burner receives a mixture of two fuels with mass fraction 40% n-butane and 60% methanol, both vapor. The fuel is burned with stoichiometric air. Find the product composition and the lower heating value of this fuel mixture (kJ/kg fuel mix).

14.65 Consider natural gas A and natural gas D, both of which are listed in Table 14.2. Calculate the enthalpy of combustion of each gas at 25°C, assuming that the products include vapor water. Repeat the answer for liquid water in the products.

14.66 Blast furnace gas in a steel mill is available at 250°C to be burned for the generation of steam. The composition of this gas is, on a volumetric basis,

Component	CH_4	H_2	CO	CO_2	N_2	H_2O
Percent by volume	0.1	2.4	23.3	14.4	56.4	3.4

Find the lower heating value (kJ/m³) of this gas at 250°C and ambient pressure.

14.67 Natural gas, we assume methane, is burned with 200% theoretical air, shown in Fig. P14.67, and the reactants are supplied as gases at the reference temperature and pressure. The products are flowing through a heat exchanger where they give off energy to some water flowing in at 20°C, 500 kPa, and out at 700°C, 500 kPa. The products exit at 400 K to the chimney. How much energy per kmole fuel can the products deliver, and how many kg water per kg fuel can they heat?

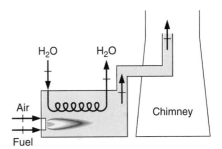

FIGURE P14.67

14.68 Gasoline, C_7H_{17}, is burned in a steady-state burner with stoichiometric air at P_0, T_0, shown in Fig. P14.68. The gasoline is flowing as a liquid at T_0 to a carburetor where it is mixed with air to produce a fuel air–gas mixture at T_0. The carburetor takes some heat transfer from the hot products to do the heating. After the combustion, the products

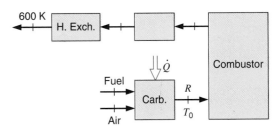

FIGURE P14.68

go through a heat exchanger, which they leave at 600 K. The gasoline consumption is 10 kg per hour. How much power is given out in the heat exchanger, and how much power does the carburetor need?

14.69 In an engine a mixture of liquid octane and ethanol, mole ratio 9:1, and stoichiometric air are taken in at T_0, P_0. In the engine, the enthalpy of combustion is used so that 30% goes out as work, 30% goes out as heat loss, and the rest goes out the exhaust. Find the work and heat transfer per kilogram of fuel mixture and also the exhaust temperature.

14.70 Liquid nitromethane is added to the air in a carburetor to make a stoichiometric mixture where both fuel and air are added at 298 K, 100 kPa. After combustion, a constant-pressure heat exchanger brings the products to 600 K before being exhausted. Assume the nitrogen in the fuel becomes N_2 gas. Find the total heat transfer per kmol fuel in the whole process.

Adiabatic Flame Temperature

14.71 Hydrogen gas is burned with pure oxygen in a steady-flow burner, shown in Fig. P14.71, where both reactants are supplied in a stoichiometric ratio at the reference pressure and temperature. What is the adiabatic flame temperature?

FIGURE P14.71

14.72 In a rocket, hydrogen is burned with air, both reactants supplied as gases at P_0, T_0. The combustion is adiabatic, and the mixture is stoichiometric (100% theoretical air). Find the products' dew point and the adiabatic flame temperature (~2500 K).

14.73 Carbon is burned with air in a furnace with 150% theoretical air, and both reactants are supplied at the reference pressure and temperature. What is the adiabatic flame temperature?

14.74 A stoichiometric mixture of benzene, C_6H_6, and air is mixed from the reactants flowing at 25°C, 100 kPa. Find the adiabatic flame temperature. What is the error if constant-specific heat at T_0 for the products from Table A.5 is used?

14.75 Hydrogen gas is burned with 200% theoretical air in a steady-flow burner where both reactants are supplied at the reference pressure and temperature. What is the adiabatic flame temperature?

14.76 A gas-turbine burns natural gas (assume methane) where the air is supplied to the combustor at 1000 kPa, 500 K, and the fuel is at 298 K, 1000 kPa. What is the equivalence ratio and the percent theoretical air if the adiabatic flame temperature should be limited to 1800 K?

14.77 Liquid *n*-butane at T_0, is sprayed into a gas turbine, as in Fig. P14.77, with primary air flowing at 1.0 MPa, 400 K, in a stoichiometric ratio. After complete combustion, the products are at the adiabatic flame temperature, which is too high, so secondary air at 1.0 MPa, 400 K, is added, with the resulting mixture being at 1400 K. Show that $T_{ad} > 1400$ K and find the ratio of secondary to primary airflow.

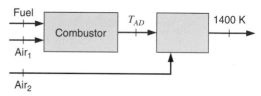

FIGURE P14.77

14.78 Butane gas at 25°C is mixed with 150% theoretical air at 600 K and is burned in an adiabatic steady-flow combustor. What is the temperature of the products exiting the combustor?

14.79 Natural gas, we assume methane, is burned with 200% theoretical air, and the reactants are sup-

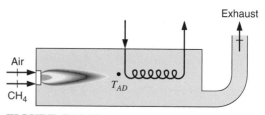

FIGURE P14.79

plied as gases at the reference temperature and pressure. The products are flowing through a heat exchanger and then out the exhaust, as in Fig. P14.79. What is the adiabatic flame temperature right after combustion before the heat exchanger?

14.80 Liquid butane at 25°C is mixed with 150% theoretical air at 600 K and is burned in a steady-flow burner. Use the enthalpy of combustion from Table 14.3 to find the adiabatic flame temperature out of the burner.

14.81 Acetylene gas at 25°C, 100 kPa, is fed to the head of a cutting torch. Calculate the adiabatic flame temperature if the acetylene is burned with

  a. 100% theoretical air at 25°C.

  b. 100% theoretical oxygen at 25°C.

14.82 Ethene, C_2H_4, burns with 150% theoretical air in a steady-flow constant-pressure process with reactants entering at P_0, T_0. Find the adiabatic flame temperature.

14.83 Solid carbon is burned with stoichiometric air in a steady-flow process. The reactants at T_0, P_0 are heated in a preheater to $T_2 = 500$ K, as shown in Fig. P14.83, with the energy given by the product gases before flowing to a second heat exchanger, which they leave at T_0. Find the temperature of the products T_4, and the heat transfer per kmol of fuel (4 to 5) in the second heat exchanger.

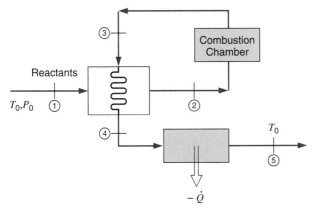

FIGURE P14.83

14.84 Gaseous ethanol, C_2H_5OH, is burned with pure oxygen in a constant-volume combustion bomb. The reactants are charged in a stoichiometric ratio

at the reference condition. Assume no heat transfer and find the final temperature (>5000 K).

14.85 The enthalpy of formation of magnesium oxide, MgO(s), is $-601\ 827$ kJ/kmol at 25°C. The melting point of magnesium oxide is approximately 3000 K, and the increase in enthalpy between 298 and 3000 K is 128 449 kJ/kmol. The enthalpy of sublimation at 3000 K is estimated at 418 000 kJ/kmol, and the specific heat of magnesium oxide vapor above 3000 K is estimated at 37.24 kJ/kmol K.

  a. Determine the enthalpy of combustion per kilogram of magnesium.

  b. Estimate the adiabatic flame temperature when magnesium is burned with theoretical oxygen.

Second Law for the Combustion Process

14.86 Calculate the irreversibility for the process described in Problem 14.41.

14.87 Methane is burned with air, both of which are supplied at the reference conditions. There is enough excess air to give a flame temperature of 1800 K. What are the percent theoretical air and the irreversibility in the process?

14.88 Consider the combustion of hydrogen with pure oxygen in a stoichiometric ratio under steady-flow adiabatic conditions. The reactants enter separately at 298 K, 100 kPa, and the product(s) exit at a pressure of 100 kPa. What is the exit temperature, and what is the irreversibility?

14.89 Pentane gas at 25°C, 150 kPa, enters an insulated steady-flow combustion chamber. Sufficient excess air to hold the combustion products temperature to 1800 K enters separately at 500 K, 150 kPa. Calculate the percent theoretical air required and the irreversibility of the process per kmol of pentane burned.

14.90 Consider the combustion of methanol, CH_3OH, with 25% excess air. The combustion products are passed through a heat exchanger and exit at 200 kPa, 400 K. Calculate the absolute entropy of the products exiting the heat exchanger assuming all the water is vapor.

14.91 Consider the combustion of methanol, CH_3OH, with 25% excess air. The combustion products are passed through a heat exchanger and exit at 200 kPa, 40°C. Calculate the absolute entropy of the products exiting the heat exchanger per kilo-mole of methanol burned, using proper amounts of liquid and vapor water.

14.92 An inventor claims to have built a device that will take 0.001 kg/s of water from the faucet at 10°C, 100 kPa, and produce separate streams of hydrogen and oxygen gas, each at 400 K, 175 kPa. It is stated that this device operates in a 25°C room on 10-kW electrical power input. How do you evaluate this claim?

14.93 Two kilomoles of ammonia are burned in a steady-flow process with x kmol of oxygen. The products, consisting of H_2O, N_2, and the excess O_2, exit at 200°C, 7 MPa.

    a. Calculate x if half the water in the products is condensed.

    b. Calculate the absolute entropy of the products at the exit conditions.

14.94 Graphite, C, at P_0, T_0 is burned with air coming in at P_0, 500 K, in a ratio so the products exit at P_0, 1200 K. Find the equivalence ratio, the percent theoretical air, and the total irreversibility.

14.95 A flow of hydrogen gas is mixed with a flow of oxygen in a stoichiometric ratio, both at 298 K and 50 kPa. The mixture burns without any heat transfer in complete combustion. Find the adiabatic flame temperature and the amount of entropy generated per kmole hydrogen in the process.

14.96 A closed, rigid container is charged with propene, C_3H_6, and 150% theoretical air at 100 kPa, 298 K. The mixture is ignited and burns with complete combustion. Heat is transferred to a reservoir at 500 K so the final temperature of the products is 700 K. Find the final pressure, the heat transfer per kmol fuel, and the total entropy generated per kmol fuel in the process.

Problems Involving Generalized Charts or Real Mixtures

14.97 Repeat Problem 14.42, but assume that satu-rated-liquid oxygen at 90 K is used instead of 25°C oxygen gas in the combustion process. Use the generalized charts to determine the properties of liquid oxygen.

14.98 Hydrogen peroxide, H_2O_2, enters a gas generator at 25°C, 500 kPa, at the rate of 0.1 kg/s and is

decomposed to steam and oxygen exiting at 800 K, 500 kPa. The resulting mixture is expanded through a turbine to atmospheric pressure, 100 kPa, as shown in Fig. P14.98. Determine the power output of the turbine and the heat transfer rate in the gas generator. The enthalpy of forma-tion of liquid H_2O_2 is $-187\,583$ kJ/kmol.

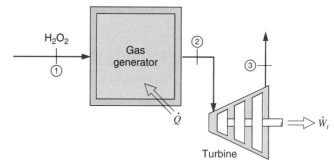

FIGURE P14.98

14.99 Liquid butane at 25°C is mixed with 150% theo-retical air at 600 K and is burned in an adiabatic steady-state combustor. Use the generalized charts for the liquid fuel and find the tempera-ture of the products exiting the combustor.

14.100 Saturated liquid butane enters an insulated con-stant-pressure combustion chamber at 25°C, and x times theoretical oxygen gas enters at the same P and T. The combustion products exit at 3400 K. With complete combustion find x, What is the pressure at the chamber exit? What is the ir-reversibility of the process?

14.101 A gas mixture of 50% ethane and 50% propane by volume enters a combustion chamber at 350 K, 10 MPa. Determine the enthalpy per kilomole of this mixture relative to the thermochemical base of enthalpy using Kay's rule.

14.102 A mixture of 80% ethane and 20% methane on a mole basis is throttled from 10 MPa, 65°C, to 100 kPa and is fed to a combustion chamber where it undergoes complete combustion with air, which enters at 100 kPa, 600 K. The amount of air is such that the products of combustion exit at 100 kPa, 1200 K. Assume that the com-bustion process is adiabatic and that all compo-nents behave as ideal gases except the fuel mixture, which behaves according to the gener-alized charts, with Kay's rule for the pseudocritical constants. Determine the percentage of theoretical air used in the process and the dew-point temperature of the products.

14.103 Liquid hexane enters a combustion chamber at 31°C, 200 kPa, at the rate of 1 kmol/s. 200% theoretical air enters separately at 500 K, 200 kPa, and the combustion products exit at 1000 K, 200 kPa. The specific heat of ideal-gas hexane is $C_{p0} = 143$ kJ/kmol K. Calculate the rate of irreversibility of the process.

Fuel Cells

14.104 In Example 14.16, a basic hydrogen–oxygen fuel cell reaction was analyzed at 25°C, 100 kPa. Repeat this calculation, assuming that the fuel cell operates on air at 25°C, 100 kPa, instead of on pure oxygen at this state.

14.105 Assume that the basic hydrogen–oxygen fuel cell operates at 600 K instead of 298 K as in Example 14.16. Find the change in the Gibbs function and the reversible EMF it can generate.

14.106 Consider a methane–oxygen fuel cell in which the reaction at the anode is

$$CH_4 + 2H_2O \rightarrow CO_2 + 8e^- + 8H^+$$

The electrons produced by the reaction flow through the external load, and the positive ions migrate through the electrolyte to the cathode, where the reaction is

$$8e^- + 8H^+ + 2O_2 \rightarrow 4H_2O$$

Calculate the reversible work and the reversible EMF for the fuel cell operating at 25°C, 100 kPa.

14.107 Redo the previous problem, but assume that the fuel cell operates at 1200 K instead of at room temperature.

Combustion Efficiency

14.108 Consider the steady-state combustion of propane at 25°C with air at 400 K. The products exit the combustion chamber at 1200 K. It may be assumed that the combustion efficiency is 90% and that 95% of the carbon in the propane burns to form carbon dioxide; the remaining 5% forms carbon monoxide. Determine the ideal fuel–air ratio and the heat transfer from the combustion chamber.

14.109 A gasoline engine is converted to run on propane as shown in Fig. P14.109. Assume the propane enters the engine at 25°C, at the rate 40 kg/h. Only 90% theoretical air enters at 25°C, such that 90% of the C burns to form CO_2 and 10% of the C burns to form CO. The combustion products, also including H_2O, H_2, and N_2, exit the exhaust pipe at 1000 K. Heat loss from the engine (primarily to the cooling water) is 120 kW. What is the power output of the engine? What is the thermal efficiency?

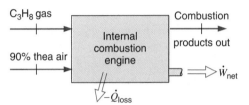

FIGURE P14.109

14.110 A small air-cooled gasoline engine is tested, and the output is found to be 1.0 kW. The temperature of the products is measured as 600 K. The products are analyzed on a dry volumetric basis, with the result: 11.4% CO_2, 2.9% CO, 1.6% O_2, and 84.1% N_2. The fuel may be considered to be liquid octane. The fuel and air enter the engine at 25°C, and the flow rate of fuel to the engine is 1.5×10^{-4} kg/s. Determine the rate of heat transfer from the engine and its thermal efficiency.

14.111 A gasoline engine uses liquid octane and air, both supplied at P_0, T_0, in a stoichiometric ratio. The products (complete combustion) flow out of the exhaust valve at 1100 K. Assume that the heat loss carried away by the cooling water, at 100°C, is equal to the work output. Find the efficiency of the engine expressed as (work/lower heating value) and the second-law efficiency.

Review Problems

14.112 Ethene, C_2H_4, and propane, C_3H_8, in a 1:1 mole ratio as gases are burned with 120% theoretical air in a gas turbine. Fuel is added at 25°C, 1 MPa, and the air comes from the atmosphere, 25°C, 100 kPa, through a compressor to 1 MPa and mixed with the fuel. The turbine work is such that the exit temperature is 800 K with an exit pressure of 100 kPa. Find the mixture temperature before combustion, and also the work, assuming an adiabatic turbine.

14.113 Carbon monoxide, CO, is burned with 150% theoretical air, and both gases are supplied at 150 kPa and 600 K. Find the reference enthalpy of reaction and the adiabatic flame temperature.

14.114 Consider the gas mixture fed to the combustors in the integrated gasification combined cycle power plant, as described in Problem 14.35. If the adiabatic flame temperature should be limited to 1500 K, what percent theoretical air should be used in the combustors?

14.115 A study is to be made using liquid ammonia as the fuel in a gas-turbine engine. Consider the compression and combustion processes of this engine.

  a. Air enters the compressor at 100 kPa, 25°C, and is compressed to 1600 kPa, where the isentropic compressor efficiency is 87%. Determine the exit temperature and the work input per kilomole.

  b. Two kilomoles of liquid ammonia at 25°C and x times theoretical air from the compressor enter the combustion chamber. What is x if the adiabatic flame temperature is to be fixed at 1600 K?

14.116 A rigid container is charged with butene, C_4H_8, and air in a stoichiometric ratio at P_0, T_0. The charge burns in a short time with no heat transfer to state 2. The products then cool with time to 1200 K, state 3. Find the final pressure, P_3, the total heat transfer, $_1Q_3$, and the temperature immediately after combustion, T_2.

14.117 The turbine in Problem 14.112 is adiabatic. Is it reversible, irreversible, or impossible?

14.118 Consider the combustion process described in Problem 14.102.

  a. Calculate the absolute entropy of the fuel mixture before it is throttled into the combustion chamber.

  b. Calculate the irreversibility for the overall process.

14.119 Natural gas (approximate it as methane) at a rate of 0.3 kg/s is burned with 250% theoretical air in a combustor at 1 MPa where the reactants are supplied at T_0. Steam at 1 MPa, 450°C, at a rate of 2.5 kg/s is added to the products before they enter an adiabatic turbine with an exhaust pressure of 150 kPa. Determine the turbine inlet temperature and the turbine work assuming the turbine is reversible.

14.120 Consider one cylinder of a spark-ignition, internal-combustion engine. Before the compression stroke, the cylinder is filled with a mixture of air and methane. Assume that 110% theoretical air has been used and that the state before compression is 100 kPa, 25°C. The compression ratio of the engine is 9 to 1.

  a. Determine the pressure and temperature after compression, assuming a reversible adiabatic process.

  b. Assume that complete combustion takes place while the piston is at top dead center (at minimum volume) in an adiabatic process. Determine the temperature and pressure after combustion, and the increase in entropy during the combustion process.

  c. What is the irreversibility for this process?

14.121 Liquid acetylene, C_2H_2, is stored in a high-pressure storage tank at ambient temperature, 25°C. The liquid is fed to an insulated combustor/steam boiler at the steady rate of 1 kg/s, along with 140% theoretical oxygen, O_2, which enters at

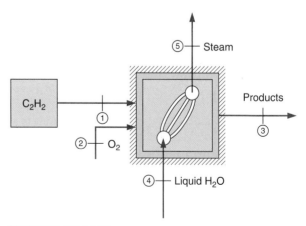

FIGURE P14.121

500 K, as shown in Fig. P14.121. The combustion products exit the unit at 500 kPa, 350 K. Liquid water enters the boiler at 10°C, at the rate of 15 kg/s, and superheated steam exits at 200 kPa.

a. Calculate the absolute entropy, per kmol, of liquid acetylene at the storage tank state.

b. Determine the phase(s) of the combustion products exiting the combustor boiler unit, and the amount of each, if more than one.

c. Determine the temperature of the steam at the boiler exit.

ENGLISH UNIT PROBLEMS

Concept Problems

14.122E What is the enthalpy of formation for oxygen as O_2? if O? for CO_2?

14.123E What is the higher heating value (HHV) of n-butane?

Fuels and the Combustion Process

14.124E Pentane is burned with 120% theoretical air in a constant-pressure process at 14.7 lbf/in.^{2}. The products are cooled to ambient temperature, 70 F. How much mass of water is condensed per pound-mass of fuel? Repeat the problem, assuming that the air used in the combustion has a relative humidity of 90%.

14.125E The output gas mixture of a certain air-blown coal gasifier has the composition of producer gas as listed in Table 14.2. Consider the combustion of this gas with 120% theoretical air at 14.7 lbf/in.^{2} pressure. Find the dew point of the products and the mass of water condensed per pound-mass of fuel if the products are cooled 20 F below the dew-point temperature?

Energy and Enthalpy of Formation

14.126E A rigid vessel initially contains 2-pound moles of carbon and 2-pound moles of oxygen at 77 F, 30 lbf/in.^{2}. Combustion occurs, and the resulting products consist of a 1-pound mole of carbon dioxide, 1-pound mole of carbon monoxide, and excess oxygen at a temperature of 1800 R. Determine the final pressure in the vessel and the heat transfer from the vessel during the process.

14.127E In a test of rocket propellant performance, liquid hydrazine (N_2H_4) at 14.7 lbf/in.^{2}, 77 F, and oxygen gas at 14.7 lbf/in.^{2}, 77 F, are fed to a combustion chamber in the ratio of 0.5 lbm O_2/lbm N_2H_4. The heat transfer from the chamber to the surroundings is estimated to be 45 Btu/lbm N_2H_4. Determine the temperature of the products exiting the chamber. Assume that only H_2O, H_2, and N_2 are present. The enthalpy of formation of liquid hydrazine is +21 647 Btu/lb mole.

14.128E One alternative to using petroleum or natural gas as fuels is ethanol (C_2H_5OH), which is commonly produced from grain by fermentation. Consider a combustion process in which liquid ethanol is burned with 120% theoretical air in a steady-flow process. The reactants enter the combustion chamber at 77 F, and the products exit at 140 F, 14.7 lbf/in.^{2}. Calculate the heat transfer per pound mole of ethanol, using the enthalpy of formation of ethanol gas plus the generalized tables or charts.

14.129E In a new high-efficiency furnace, natural gas, assumed to be 90% methane and 10% ethane (by volume) and 110% theoretical air each enter at 77 F, 14.7 lbf/in.^{2}, and the products (assumed to be 100% gaseous) exit the furnace at 100 F, 14.7 lbf/in.^{2}. What is the heat transfer for this process? Compare this to an older furnace where the products exit at 450 F, 14.7 lbf/in.^{2}.

14.130E Repeat the previous problem, but take into account the actual phase behavior of the products exiting the furnace.

14.131E Pentene, C_5H_{10} is burned with pure oxygen in a steady-state process. The products at one point are brought to 1300 R and used in a heat exchanger, where they are cooled to 77 F. Find the specific heat transfer in the heat exchanger.

14.132E Methane, CH_4, is burned in a steady-state process with two different oxidizers: case **A**—pure oxygen, O_2 and case **B**—a mixture of

$O_2 + x$Ar. The reactants are supplied at T_0, P_0, and the products are at 3200 R in both cases. Find the required equivalence ratio in case **A** and the amount of argon, x, for a stoichiometric ratio in case **B**.

14.133E A closed, insulated container is charged with a stoichiometric ratio of oxygen and hydrogen at 77 F and 20 lbf/in.^{2}. After combustion, liquid water at 77 F is sprayed in such a way that the final temperature is 2100 R. What is the final pressure?

Enthalpy of Combustion and Heating Value

14.134E A burner receives a mixture of two fuels with mass fraction 40% n-butane and 60% methanol, both vapor. The fuel is burned with stoichiometric air. Find the product composition and the lower heating value of this fuel mixture (Btu/lbm fuel mix).

14.135E Blast furnace gas in a steel mill is available at 500 F to be burned for the generation of steam. The composition of this gas is, on a volumetric basis,

Component	CH_4	H_2	CO	CO_2	N_2	H_2O
Percent by volume	0.1	2.4	23.3	14.4	56.4	3.4

Find the lower heating value (Btu/ft^3) of this gas at 500 F and P_0.

Adiabatic Flame Temperature

14.136E Hydrogen gas is burned with pure oxygen in a steady-flow burner where both reactants are supplied in a stoichiometric ratio at the reference pressure and temperature. What is the adiabatic flame temperature?

14.137E Carbon is burned with air in a furnace with 150% theoretical air, and both reactants are supplied at the reference pressure and temperature. What is the adiabatic flame temperature?

14.138E Butane gas at 77 F is mixed with 150% theoretical air at 1000 R and is burned in an adiabatic steady-state combustor. What is the temperature of the products exiting the combustor?

14.139E Liquid *n*-butane at T_0, is sprayed into a gas turbine with primary air flowing at 150 lbf/in.^{2}, 700 R in a stoichiometric ratio. After complete combustion, the products are at the adiabatic

flame temperature, which is too high. Therefore, secondary air at 150 lbf/in.^{2}, 700 R is added, see Fig. P14.77, with the resulting mixture being at 2500 R. Show that $T_{ad} > 2500$ R and find the ratio of secondary to primary airflow.

14.140E Acetylene gas at 77 F, 14.7 lbf/in.^{2} is fed to the head of a cutting torch. Calculate the adiabatic flame temperature if the acetylene is burned with 100% theoretical air at 77 F. Repeat the answer for 100% theoretical oxygen at 77 F.

14.141E Ethene, C_2H_4, burns with 150% theoretical air in a steady-state constant-pressure process, with reactants entering at P_0, T_0. Find the adiabatic flame temperature.

14.142E Solid carbon is burned with stoichiometric air in a steady-state process, as shown in Fig. P14.83. The reactants at T_0, P_0 are heated in a preheater to $T_2 = 900$ R with the energy given by the products before flowing to a second heat exchanger, which they leave at T_0. Find the temperature of the products T_4 and the heat transfer per lb mol of fuel (4 to 5) in the second heat exchanger.

Second Law for the Combustion Process

14.143E Methane is burned with air, both of which are supplied at the reference conditions. There is enough excess air to give a flame temperature of 3200 R. What are the percent theoretical air and the irreversibility in the process?

14.144E Two-pound moles of ammonia are burned in a steady-state process with x lb mol of oxygen. The products, consisting of H_2O, N_2, and the excess O_2, exit at 400 F, 1000 lbf/in.^{2}.

a. Calculate x if half the water in the products is condensed.

b. Calculate the absolute entropy of the products at the exit conditions.

14.145E Graphite, C, at P_0, T_0 is burned with air coming in at P_0, 900 R, in a ratio so the products exit at P_0, 2200 R. Find the equivalence ratio, the percent theoretical air, and the total irreversibility.

Problems Involving Generalized Charts or Real Mixtures

14.146E Repeat Problem 14.127E, but assume that saturated-liquid oxygen at 170 R is used instead of 77 F oxygen gas in the combustion process.

Use the generalized charts to determine the properties of liquid oxygen.

14.147E Hydrogen peroxide, H_2O_2, enters a gas generator at 77 F, 75 lbf/in.^{2}, at the rate of 0.2 lbm/s and is decomposed to steam and oxygen exiting at 1500 R, 75 lbf/in.^{2}. The resulting mixture is expanded through a turbine to atmospheric pressure, 14.7 lbf/in.^{2}, as shown in Fig. P14.98. Determine the power output of the turbine and the heat-transfer rate in the gas generator. The enthalpy of formation of liquid H_2O_2 is −80 541 Btu/lb mol.

Fuel Cells, Efficiency, and Review

14.148E In Example 14.16, a basic hydrogen–oxygen fuel cell reaction was analyzed at 25°C, 100 kPa. Repeat this calculation, assuming that the fuel cell operates on air at 77 F, 14.7 lbf/in.^{2}, instead of on pure oxygen at this state.

14.149E A small, air-cooled gasoline engine is tested, and the output is found to be 2.0 hp. The temperature of the products is measured and found to be 730 F. The products are analyzed on a dry volumetric basis, with the following result:

11.4% CO_2, 2.9% CO, 1.6% O_2, and 84.1% N_2. The fuel may be considered to be liquid octane. The fuel and air enter the engine at 77 F, and the flow rate of fuel to the engine is 1.8 lbm/h. Determine the rate of heat transfer from the engine and its thermal efficiency.

14.150E A gasoline engine uses liquid octane and air, both supplied at P_0, T_0, in a stoichiometric ratio. The products (complete combustion) flow out of the exhaust valve at 2000 R. Assume that the heat loss carried away by the cooling water, at 200 F, is equal to the work output. Find the efficiency of the engine expressed as (work/lower heating value) and the second-law efficiency.

14.151E Ethene, C_2H_4, and propane, C_3H_8, in a 1:1 mole ratio as gases are burned with 120% theoretical air in a gas turbine. Fuel is added at 77 F, 150 lbf/in.^{2}, and the air comes from the atmosphere, 77 F, 15 lbf/in.^{2} through a compressor to 150 lbf/in.^{2} and mixed with the fuel. The turbine work is such that the exit temperature is 1500 R with an exit pressure of 14.7 lbf/in.^{2}. Find the mixture temperature before combustion, and also the work, assuming an adiabatic turbine.

COMPUTER, DESIGN, AND OPEN-ENDED PROBLEMS

14.152 Write a program to solve the general case of Problem 14.31 for any hydrocarbon fuel C_xH_y, where x and y are input parameters. We wish to calculate the percentage of theoretical air for any given percentages of combustion products.

14.153 Write a program to solve the general case of Problem 14.26 for different percentages of the components given in the ultimate analysis of the coal.

14.154 Use the software program for the ideal-gas properties to do Problem 14.83.

14.155 Write a program to study the effect of the percentage of theoretical air on the adiabatic flame temperature for a (variable) hydrocarbon fuel. Assume reactants enter the combustion chamber at 25°C, and complete combustion. Use constant-specific heat of the various products of combustion and let the fuel composition and its enthalpy of formation be program inputs.

14.156 Power plants may use off-peak power to compress air into a large storage facility (see Prob-

lem 9.69). The compressed air is then used as the air supply to a gas-turbine system where it is burned with some fuel, usually natural gas. The system is then used to produce power at peak load times. Investigate such a setup and estimate the power generated with conditions given in Problem 9.69 and combustion with 200–300% theoretical air and exhaust to the atmosphere.

14.157 A car that runs on natural gas has it stored in a heavy tank with a maximum pressure of 3600 psi (25 MPa). Size the tank for a range of 300 miles (500 km) assuming a car engine that has a 30% efficiency requiring about 25 hp (20 kW) to drive the car at 55 mi/h (90 km/h).

14.158 The Cheng cycle, shown in Fig. P12.176, is powered by the combustion of natural gas (essentially methane) being burned with 250–300% theoretical air. In the case with a single water-condensing heat exchanger, where $T_6 = 40°C$ and $\Phi_6 = 100\%$, is any makeup water needed at state 8, or is there a surplus? Does the humidity in the compressed atmospheric air at state 1 make any difference? Study the problem over a range of air–fuel ratios.

14.159 The cogenerating power plant shown in Problem 11.65 burns 170 kg/s air with natural gas, CH_4. The set up is shown in Fig. P14.159 where a fraction of the airflow out of the compressor with pressure ratio 15.8 : 1 is used to preheat the feedwater in the steam cycle. The fuel flow rate is 3.2 kg/s. Make an analysis of the system determining the total heat transfer to the steam cycle from the turbine exhaust gases, the heat transfer in the preheater, and the gas turbine inlet temperature.

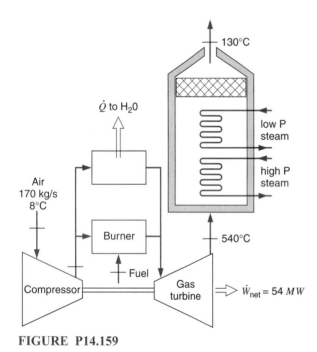

FIGURE P14.159

14.160 Consider the combustor in the Cheng cycle (see Problems 12.176 and 14.119E). Atmospheric air

is compressed to 1.25 MPa, state 1. It is burned with natural gas, CH_4, with the products leaving at state 2. The fuel should add a total of about 15 MW to the cycle, with an airflow of 12 kg/s. For a compressor with an intercooler estimate the temperatures T_1, T_2, and the fuel flow rate.

14.161 Study the coal gasification process that will produce methane, CH_4, or methanol CH_3OH. What is involved in such a process? Compare the heating values of the gas products with those of the original coal. Discuss the merits of this conversion.

14.162 Ethanol, C_2H_5OH, can be produced from corn or biomass. Investigate the process and the chemical reactions that occur. For different raw materials estimate the amount of ethanol that can be obtained per mass of the raw material.

14.163 A Diesel engine is used as a stationary power plant in remote locations such as a ship, oil drilling rig, or farm. Assume diesel fuel is used with 300% theoretical air in a 1000-hp diesel engine. Estimate the amount of fuel used, the efficiency, and the potential use of the exhaust gases for heating of rooms or water. Investigate if other fuels can be used.

14.164 When a power plant burns coal or some blends of oil, the combustion process can generate pollutants as SO_x and NO_x Investigate the use of scrubbers to remove these. Explain the processes that take place and the effect on the power plant operation (energy, exhaust pressures, etc.).

14.165 For a number of fuels listed in Table 14.3 estimate their adiabatic flame temperature when they are burned with 200% theoretical air. Assume a power-generating device like a gasoline or diesel engine or a gas-turbine with reasonable choices for their operating conditions. Find the power that can be generated as a fraction of the enthalpy of combustion. Does a ranking of the fuels follow the magnitude of the enthalpy of combustion?

INTRODUCTION TO PHASE AND CHEMICAL EQUILIBRIUM 15

Up to this point, we have assumed that we are dealing either with systems that are in equilibrium or with those in which the deviation from equilibrium is infinitesimal, as in a quasi-equilibrium or reversible process. For irreversible processes we made no attempt to describe the state of the system during the process but dealt only with the initial and final states of the system, in the case of a control mass, or the inlet and exit states as well in the case of a control volume. For any case, we either considered the system to be in equilibrium throughout or at least made the assumption of local equilibrium.

In this chapter we examine the criteria for equilibrium and from them derive certain relations that will enable us, under certain conditions, to determine the properties of a system when it is in equilibrium. The specific case we will consider is that involving chemical equilibrium in a single phase (homogeneous equilibrium) as well as certain related topics.

15.1 REQUIREMENTS FOR EQUILIBRIUM

As a general requirement for equilibrium, we postulate that a system is in equilibrium when there is no possibility that it can do any work when it is isolated from its surroundings. In applying this criterion, it is helpful to divide the system into two or more subsystems, and consider the possibility of doing work by any conceivable interaction between these two subsystems. For example, in Fig. 15.1 a system has been divided into two systems and an engine, of any conceivable variety, placed between these subsystems. A system may be so defined as to include the immediate surroundings. In this case, we can let the immediate surroundings be a subsystem and thus consider the general case of the equilibrium between a system and its surroundings.

The first requirement for equilibrium is that the two subsystems have the same temperature, for otherwise we could operate a heat engine between the two systems and do work. Thus, we conclude that one requirement for equilibrium is that a system must be at a uniform temperature to be in equilibrium. It is also evident that there must be no unbalanced mechanical forces between the two systems, or else one could operate a turbine or piston engine between the two systems and do work.

We would like to establish general criteria for equilibrium that would apply to all simple compressible substances, including those that undergo chemical reactions. We will find that the Gibbs function is a particularly significant property in defining the criteria for equilibrium.

Let us first consider a qualitative example to illustrate this point. Consider a natural gas well that is 1 km deep, and let us assume that the temperature of the gas is constant

617

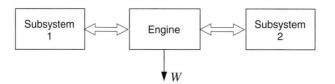

FIGURE 15.1 Two subsystems that communicate through an engine.

throughout the gas well. Suppose we have analyzed the composition of the gas at the top of the well, and we would like to know the composition of the gas at the bottom of the well. Furthermore, let us assume that equilibrium conditions prevail in the well. If this is true we would expect that an engine such as that shown in Fig. 15.2 (which operates on the basis of the pressure and composition change with elevation and does not involve combustion) would not be capable of doing any work.

If we consider a steady-state process for a control volume around this engine, the reversible work for the change of state from i to e is given by Eq. 10.9 on a total mass basis

$$\dot{W}^{\text{rev}} = \dot{m}_i \left(h_i + \frac{\mathbf{V}_i^2}{2} + gZ_i - T_0 s_i \right) - \dot{m}_e \left(h_e + \frac{\mathbf{V}_e^2}{2} + gZ_e - T_0 s_e \right)$$

Furthermore, since $T_i = T_e = T_0 = $ constant, this reduces to the form of the Gibbs function $g = h - Ts$, Eq. 13.14, and the reversible work is

$$\dot{W}^{\text{rev}} = \dot{m}_i \left(g_i + \frac{\mathbf{V}_i^2}{2} + gZ_i \right) - \dot{m}_e \left(g_e + \frac{\mathbf{V}_e^2}{2} + gZ_e \right)$$

However,

$$\dot{W}^{\text{rev}} = 0, \qquad \dot{m}_i = \dot{m}_e \qquad \text{and} \qquad \frac{\mathbf{V}_i^2}{2} = \frac{\mathbf{V}_e^2}{2}$$

Then we can write

$$g_i + gZ_i = g_e + gZ_e$$

and the requirement for equilibrium in the well between two levels that are a distance dZ apart would be

$$dg_T + g \, dZ_T = 0$$

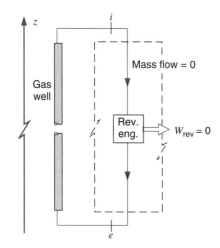

FIGURE 15.2
Illustration showing the relation between reversible work and the criteria for equilibrium.

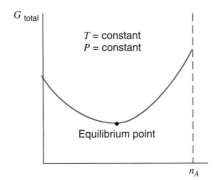

FIGURE 15.3
Illustration of the requirement for chemical equilibrium.

In contrast to a deep gas well, most of the systems that we consider are of such size that ΔZ is negligibly small, and therefore we consider the pressure to be uniform throughout.

This leads to the general statement of equilibrium that applies to simple compressible substances that may undergo a change in chemical composition, namely, that at equilibrium

$$dG_{T,P} = 0 \qquad (15.1)$$

In the case of a chemical reaction, it is helpful to think of the equilibrium state as the state in which the Gibbs function is a minimum. For example, consider a control mass consisting initially of n_A moles of substance A and n_B moles of substance B, which react in accordance with the relation

$$v_A A + v_B B \rightleftharpoons v_C C + v_D D$$

Let the reaction take place at constant pressure and temperature. If we plot G for this control mass as a function of n_A, the number of moles of A present, we would have a curve as shown in Fig. 15.3. At the minimum point on the curve, $dG_{T,P} = 0$, and this will be the equilibrium composition for this system at the given temperature and pressure. The subject of chemical equilibrium will be developed further in Section 15.4.

15.2 EQUILIBRIUM BETWEEN TWO PHASES OF A PURE SUBSTANCE

As another example of this requirement for equilibrium, let us consider the equilibrium between two phases of a pure substance. Consider a control mass consisting of two phases of a pure substance at equilibrium. We know that under these conditions the two phases are at the same pressure and temperature. Consider the change of state associated with a transfer of dn moles from phase 1 to phase 2 while the temperature and pressure remain constant. That is,

$$dn^1 = -dn^2$$

The Gibbs function of this control mass is given by

$$G = f(T, P, n^1, n^2)$$

where n^1 and n^2 designate the number of moles in each phase. Therefore,

$$dG = \left(\frac{\partial G}{\partial T}\right)_{P,n^1,n^2} dT + \left(\frac{\partial G}{\partial P}\right)_{T,n^1,n^2} dP + \left(\frac{\partial G}{\partial n^1}\right)_{T,P,n^2} dn^1 + \left(\frac{\partial G}{\partial n^2}\right)_{T,P,n^1} dn^2$$

By definition,

$$\left(\frac{\partial G}{\partial n^1}\right)_{T,P,n^2} = \bar{g}^1 \qquad \left(\frac{\partial G}{\partial n^2}\right)_{T,P,n^1} = \bar{g}^2$$

Therefore, at constant temperature and pressure,

$$dG = \bar{g}^1\, dn^1 + \bar{g}^2\, dn^2 = dn^1(\bar{g}^1 - \bar{g}^2)$$

Now at equilibrium (Eq. 15.1)

$$dG_{T,P} = 0$$

Therefore, at equilibrium, we have

$$\bar{g}^1 = \bar{g}^2 \tag{15.2}$$

That is, under equilibrium conditions, the Gibbs function of each phase of a pure substance is equal. Let us check this by determining the Gibbs function of saturated liquid (water) and saturated vapor (steam) at 300 kPa. From the steam tables:

For the liquid:

$$g_f = h_f - Ts_f = 561.47 - 406.7 \times 1.6718 = -118.4 \text{ kJ/kg}$$

For the vapor:

$$g_g = h_g - Ts_g = 2725.3 - 406.7 \times 6.9919 = -118.4 \text{ kJ/kg}$$

Equation 15.2 can also be derived by applying the relation

$$T\, ds = dh - v\, dP$$

to the change of phase that takes place at constant pressure and temperature. For this process this relation can be integrated as follows:

$$\int_f^g T\, ds = \int_f^g dh$$

$$T(s_g - s_f) = (h_g - h_f)$$

$$h_f - Ts_f = h_g - Ts_g$$

$$g_f = g_g$$

The Clapeyron equation, which was derived in Section 13.1, can be derived by an alternate method by considering the fact that the Gibbs functions of two phases in equilibrium are equal. In Chapter 13 we considered the relation (Eq. 13.15) for a simple compressible substance:

$$dg = v\, dP - s\, dT$$

Consider a control mass that consists of a saturated liquid and a saturated vapor in equilibrium, and let this system undergo a change of pressure dP. The corresponding change in temperature, as determined from the vapor-pressure curve, is dT. Both phases will undergo the change in Gibbs function, dg, but since the phases always have the same value of the Gibbs function when they are in equilibrium, it follows that

$$dg_f = dg_g$$

But, from Eq. 13.15,

$$dg = v\,dP - s\,dT$$

it follows that

$$dg_f = v_f\,dP - s_f\,dT$$
$$dg_g = v_g\,dP - s_g\,dT$$

Since

$$dg_f = dg_g$$

it follows that

$$v_f dP - s_f\,dT = v_g\,dP - s_g\,dT$$
$$dP(v_g - v_f) = dT(s_g - s_f) \tag{15.3}$$

$$\frac{dP}{dT} = \frac{s_{fg}}{v_{fg}} = \frac{h_{fg}}{Tv_{fg}}$$

In summary, when different phases of a pure substance are in equilibrium, each phase has the same value of the Gibbs function per unit mass. This fact is relevant to different solid phases of a pure substance and is important in metallurgical applications of thermodynamics. Example 15.1 illustrates this principle.

EXAMPLE 15.1 What pressure is required to make diamonds from graphite at a temperature of 25°C? The following data are given for a temperature of 25°C and a pressure of 0.1 MPa.

	Graphite	Diamond
g	0	2867.8 kJ/mol
v	0.000 444 m³/kg	0.000 284 m³/kg
β_T	0.304×10^{-6} 1/MPa	0.016×10^{-6} 1/MPa

Analysis and Solution

The basic principle in the solution is that graphite and diamond can exist in equilibrium when they have the same value of the Gibbs function. At 0.1 MPa pressure the Gibbs function of the diamond is greater than that of the graphite. However, the rate of increase in Gibbs function with pressure is greater for the graphite than for the diamond, therefore, at some pressure they can exist in equilibrium. Our problem is to find this pressure.

We have already considered the relation

$$dg = v\,dP - s\,dT$$

Since we are considering a process that takes place at constant temperature, this reduces to

$$dg_T = v\,dP_T \tag{a}$$

Now at any pressure P and the given temperature, the specific volume can be found from the following relation, which utilizes isothermal compressibility factor.

$$v = v^0 + \int_{P=0.1}^{P} \left(\frac{\partial v}{\partial P}\right)_T dP = v^0 + \int_{P=0.1}^{P} \frac{v}{v}\left(\frac{\partial v}{\partial P}\right)_T dP$$

$$= v^0 - \int_{P=0.1}^{P} v\beta_T dP \qquad\qquad\text{(b)}$$

The superscript ^{0} will be used in this example to indicate the properties at a pressure of 0.1 MPa and a temperature of 25°C.

The specific volume changes only slightly with pressure, so that $v \approx v^0$. Also, we assume that β_T is constant and that we are considering a very high pressure. With these assumptions this equation can be integrated to give

$$v = v^0 - v^0\beta_T P = v^0(1 - \beta_T P) \qquad\qquad\text{(c)}$$

We can now substitute this into Eq. (a) to give the relation

$$dg_T = [v^0(1 - \beta_T P)]\, dP_T$$

$$g - g^0 = v^0(P - P^0) - v^0\beta_T \frac{(P^2 - P^{02})}{2} \qquad\qquad\text{(d)}$$

If we assume that $P^0 \ll P$, this reduces to

$$G - G^0 = V^0\left(P - \frac{\beta_T P^2}{2}\right) \qquad\qquad\text{(E)}$$

For the graphite, $g^0 = 0$ and we can write

$$g_G = v_G^0\left[P - (\beta_T)_G \frac{P^2}{2}\right]$$

For the diamond, g^0 has a definite value and we have

$$g_D = g_D^0 + v_D^0\left[P - (\beta_T)_D \frac{P^2}{2}\right]$$

But at equilibrium the Gibbs function of the graphite and diamond are equal

$$g_G = g_D$$

Therefore,

$$v_G^0\left[P - (\beta_T)_G \frac{P^2}{2}\right] = g_D^0 + v_D^0\left[P - (\beta_T)_D \frac{P^2}{2}\right]$$

$$(v_G^0 - v_D^0)P - [v_G^0(\beta_T)_G - v_D^0(\beta_T)_D]\frac{P^2}{2} = g_D^0$$

$$(0.000\,444 - 0.000\,284)P$$

$$-(0.000\,444 \times 0.304 \times 10^{-6} - 0.000\,284 \times 0.016 \times 10^{-6})P^2/2 = \frac{2867.8}{12.011 \times 1000}$$

Solving this for P we find

$$P = 1493 \text{ MPa}$$

That is, at 1493 MPa, 25°C, graphite and diamond can exist in equilibrium, and the possibility exists for conversion from graphite to diamonds.

15.3 METASTABLE EQUILIBRIUM

Although the limited scope of this book precludes an extensive treatment of metastable equilibrium, a brief introduction to the subject is presented in this section. Let us first consider an example of metastable equilibrium.

Consider a slightly superheated vapor, such as steam, expanding in a convergent-divergent nozzle, as shown in Fig. 15.4. Assuming the process is reversible and adiabatic, the steam will follow path 1-a on the T-s diagram, and at point a we would expect condensation to occur. However, if point a is reached in the divergent section of the nozzle, it is observed that no condensation occurs until point b is reached, and at this point the condensation occurs very abruptly in what is referred to as a condensation shock. Between points a and b the steam exists as a vapor, but the temperature is below the saturation temperature for the given pressure. This is known as a metastable state. The possibility of a metastable state exists with any phase transformation. The dotted lines on the equilibrium diagram shown in Fig. 15.5 represent possible metastable states for solid–liquid–vapor equilibrium.

The nature of a metastable state is often pictured schematically by the kind of diagram shown in Fig. 15.6. The ball is in a stable position (the "metastable state") for small

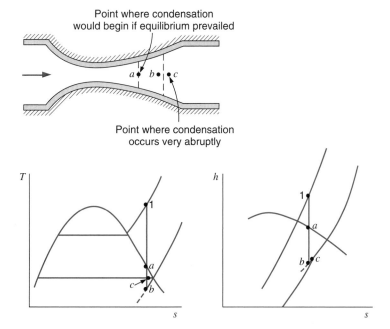

FIGURE 15.4
Illustration of the phenomenon of supersaturation in a nozzle.

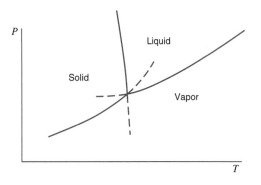

FIGURE 15.5
Metastable states for solid–liquid–vapor equilibrium.

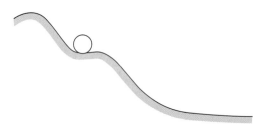

FIGURE 15.6
Schematic diagram illustrating a metastable state.

displacements, but with a large displacement it moves to a new equilibrium position. The steam expanding in the nozzle is in a metastable state between a and b. This means that droplets smaller than a certain critical size will reevaporate, and only when droplets larger than this critical size have formed (this corresponds to moving the ball out of the depression) will the new equilibrium state appear.

15.4 CHEMICAL EQUILIBRIUM

We now turn our attention to chemical equilibrium and consider first a chemical reaction involving only one phase. This is referred to as a homogeneous chemical reaction. It may be helpful to visualize this as a gaseous phase, but the basic considerations apply to any phase.

Consider a vessel, Fig. 15.7, that contains four compounds, A, B, C, and D, which are in chemical equilibrium at a given pressure and temperature. For example, these might consist of CO_2, H_2, CO, and H_2O in equilibrium. Let the number of moles of each component be designated n_A, n_B, n_C and n_D. Furthermore, let the chemical reaction that takes place between these four constituents be

$$\nu_A A + \nu_B B \rightleftharpoons \nu_C C + \nu_D D \qquad (15.4)$$

where the ν's are the stoichiometric coefficients. It should be emphasized that there is a very definite relation between the ν's (the stoichiometric coefficients), whereas the n's (the number of moles present) for any constituent can be varied simply by varying the amount of that component in the reaction vessel.

Let us now consider how the requirement for equilibrium, namely, that $dG_{T,P} = 0$ at equilibrium, applies to a homogeneous chemical reaction. Let us assume that the four

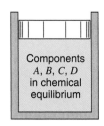

FIGURE 15.7
Schematic diagram for consideration of chemical equilibrium.

Components
A, B, C, D
in chemical
equilibrium

components are in chemical equilibrium and then assume that from this equilibrium state, while the temperature and pressure remain constant, the reaction proceeds an infinitesimal amount toward the right as Eq. 15.4 is written. This results in a decrease in the moles of A and B and an increase in the moles of C and D. Let us designate the degree of reaction by ε, and define the degree of reaction by the relations

$$dn_A = -v_A \, d\varepsilon$$
$$dn_B = -v_B \, d\varepsilon$$
$$dn_C = +v_C \, d\varepsilon$$
$$dn_D = +v_D \, d\varepsilon \qquad (15.5)$$

That is, the change in the number of moles of any component during a chemical reaction is given by the product of the stoichiometric coefficients (the v's) and the degree of reaction.

Let us evaluate the change in the Gibbs function associated with this chemical reaction that proceeds to the right in the amount $d\varepsilon$. In doing so we use, as would be expected, the Gibbs function of each component in the mixture—the partial molal Gibbs function (or its equivalent, the chemical potential):

$$dG_{T,P} = \overline{G}_C \, dn_C + \overline{G}_D \, dn_D + \overline{G}_A \, dn_A + \overline{G}_B \, dn_B$$

Substituting Eq. 15.5, we have

$$dG_{T,P} = (v_C \overline{G}_C + v_D \overline{G}_D - v_A \overline{G}_A - v_B \overline{G}_B) \, d\varepsilon \qquad (15.6)$$

We now need to develop expressions for the partial molal Gibbs functions in terms of properties that we are able to calculate. From the definition of Gibbs function, Eq. 13.14,

$$G = H - TS$$

For a mixture of two components A and B, we differentiate this equation with respect to n_A at constant T, P, and n_B, which results in

$$\left(\frac{\partial G}{\partial n_A}\right)_{T,P,n_B} = \left(\frac{\partial H}{\partial n_A}\right)_{T,P,n_B} - T\left(\frac{\partial S}{\partial n_A}\right)_{T,P,n_B}$$

All three of these quantities satisfy the definition of partial molal properties according to Eq. 13.68, such that

$$\overline{G}_A = \overline{H}_A - T\overline{S}_A \qquad (15.7)$$

For an ideal-gas mixture, enthalpy is not a function of pressure, and

$$\overline{H}_A = \overline{h}_{ATP} = \overline{h}^0_{ATP^0} \qquad (15.8)$$

Entropy is, however, a function of pressure, so that the partial entropy of A can be expressed by Eq. 14.22 in terms of the standard-state value,

$$\overline{S}_A = \overline{s}_{ATP_A = Y_A P}$$
$$= \overline{s}^0_{ATP^0} - \overline{R} \ln\left(\frac{y_A P}{P^0}\right) \qquad (15.9)$$

Now, substituting Eqs. 15.8 and 15.9 into Eq. 15.7,

$$\overline{G}_A = \overline{h}^0_{ATP^0} - T\overline{s}^0_{ATP^0} + \overline{R}T\ln\left(\frac{y_A P}{P^0}\right)$$

$$= \overline{g}^0_{ATP^0} + \overline{R}T\ln\left(\frac{y_A P}{P^0}\right) \tag{15.10}$$

Equation 15.10 is an expression for the partial Gibbs function of a component in a mixture in terms of a specific reference value, the pure-substance standard-state Gibbs function at the same temperature, and a function of the temperature, pressure, and composition of the mixture. This expression can be applied to each of the components in Eq. 15.6, resulting in

$$dG_{TP} = \left\{ v_C\left[\overline{g}^0_C + \overline{R}T\ln\left(\frac{y_C P}{P^0}\right)\right] + v_D\left[\overline{g}^0_D + \overline{R}T\ln\left(\frac{y_D P}{P^0}\right)\right] \right.$$
$$\left. - v_A\left[\overline{g}^0_A + \overline{R}T\ln\left(\frac{y_A P}{P^0}\right)\right] - v_B\left[\overline{g}^0_B + \overline{R}T\ln\left(\frac{y_B P}{P^0}\right)\right] \right\} d\varepsilon \tag{15.11}$$

Let us define ΔG^0 as follows:

$$\Delta G^0 = v_C\overline{g}^0_C + v_D\overline{g}^0_D - v_A\overline{g}^0_A - v_B\overline{g}^0_B \tag{15.12}$$

That is, ΔG^0 is the change in the Gibbs function that would occur if the chemical reaction given by Eq. 15.4 (which involves the stoichiometric amounts of each component) proceeded completely from left to right, with the reactants A and B initially separated and at temperature T and the standard-state pressure and the products C and D finally separated and at temperature T and the standard-state pressure. Note also that ΔG^0 for a given reaction is a function of only the temperature. This will be most important to bear in mind as we proceed with our developments of homogeneous chemical equilibrium. Let us now digress from our development to consider an example involving the calculation of ΔG^0.

EXAMPLE 15.2 Determine the value of ΔG^0 for the reaction $2H_2O \rightleftharpoons 2H_2 + O_2$ at 25°C and at 2000 K, with the water in the gaseous phase.

Solution

At any given temperature, the standard-state Gibbs function change of Eq. 15.12 can be calculated from the relation

$$\Delta G^0 = \Delta H^0 - T\Delta S^0$$

At 25°C,

$$\Delta H^0 = 2\overline{h}^0_{fH_2} + \overline{h}^0_{fO_2} - 2\overline{h}^0_{fH_2O(g)}$$
$$= 2(0) + 1(0) - 2(-241\ 826) = 483\ 652\ \text{kJ}$$
$$\Delta S^0 = 2\overline{s}^0_{H_2} + \overline{s}^0_{O_2} - 2\overline{s}^0_{H_2O(g)}$$
$$= 2(130.678) + 1(205.148) - 2(188.834) = 88.836\ \text{kJ/K}$$

Therefore, at 25°C,

$$\Delta G^0 = 483\ 652 - 298.15(88.836) = 457\ 166\ \text{kJ}$$

At 2000 K,

$$\Delta H^0 = 2(\bar{h}^0_{2000} - \bar{h}^0_{298})_{H_2} + (\bar{h}^0_{2000} - \bar{h}^0_{298})_{O_2} - 2(\bar{h}^0_f + \bar{h}^0_{2000} - \bar{h}^0_{298})_{H_2O}$$

$$= 2(52\,942) + (59\,176) - 2(-241\,826 + 72\,788)$$

$$= 503\,136 \text{ kJ}$$

$$\Delta S^0 = 2(\bar{s}^0_{2000})_{H_2} + (\bar{s}^0_{2000})_{O_2} - 2(\bar{s}^0_{2000})_{H_2O}$$

$$= 2(188.419) + (268.748) - 2(264.769)$$

$$= 116.048 \text{ kJ/K}$$

Therefore,

$$\Delta G^0 = 503\,136 - 2000 \times 116.048 = 271\,040 \text{ kJ}$$

Returning now to our development, substituting Eq. 15.12 into Eq. 15.11 and rearranging we can write

$$dG_{T,P} = \left\{ \Delta G^0 + \bar{R}T \ln \left[\frac{y_C^{\nu_C} y_D^{\nu_D}}{y_A^{\nu_A} y_B^{\nu_B}} \left(\frac{P}{P} \right)^{\nu_C + \nu_D - \nu_A - \nu_B} \right] \right\} d\varepsilon \qquad (15.13)$$

At equilibrium $dG_{T,P} = 0$. Therefore, since $d\varepsilon$ is arbitrary,

$$\ln \left[\frac{y_C^{\nu_C} y_D^{\nu_D}}{y_A^{\nu_A} y_B^{\nu_B}} \left(\frac{P}{P^0} \right)^{\nu_C + \nu_D - \nu_A - \nu_B} \right] = -\frac{\Delta G^0}{\bar{R}T} \qquad (15.14)$$

For convenience, we define the equilibrium constant K as

$$\ln K = -\frac{\Delta G^0}{\bar{R}T} \qquad (15.15)$$

which we note must be a function of temperature only for a given reaction, since ΔG^0 is given by Eq. 15.12 in terms of the properties of the pure substances at a given temperature and the standard-state pressure.

Combining Eqs. 15.14 and 15.15, we have

$$K = \frac{y_C^{\nu_C} y_D^{\nu_D}}{y_A^{\nu_A} y_B^{\nu_B}} \left(\frac{P}{P^0} \right)^{\nu_C + \nu_D - \nu_A - \nu_B} \qquad (15.16)$$

which is the chemical equilibrium equation corresponding to the reaction equation, Eq. 15.4.

From the equilibrium constant definition in Eqs. 15.15 and 15.16 we can draw a few conclusions. If the shift in the Gibbs function is large and positive, $\ln K$ is large and negative leading to a very small value of K. At a given P in Eq. 15.16 this leads to relatively small values of the RHS (component C and D) concentrations relative to the LHS component concentrations; the reaction is shifted to the left. The opposite is the case of a shift in the Gibbs function that is large and negative, giving a large value of K and the reaction is shifted to the right as shown in Fig. 15.8. If the shift in Gibbs function is zero, then $\ln K$ is zero, and K is exactly equal to 1, the reaction is in the middle with all concentrations of the same order of magnitude, unless the stoichiometric coefficients are extreme.

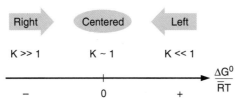

FIGURE 15.8 The shift in the reaction with the change in Gibbs function.

The other trends we can see are the influences of the temperature and pressure. For a higher temperature, but same shift in Gibbs function, the absolute value of $\ln K$ is smaller, which means K is closer to 1 and the reaction is more centered. For low temperatures, the reaction is shifted toward the side with the smallest Gibbs function G^0. The pressure has an influence only if the power in Eq. 15.16 is different from zero. That is so when the number of moles on the RHS ($\nu_C + \nu_D$) is different from the number of moles on the LHS ($\nu_A + \nu_B$). Assuming we have more moles on the RHS, then we see that the power is positive. So if the pressure is larger than the reference pressure, the whole pressure factor is larger than 1, which reduces the RHS concentrations as K is fixed for a given temperature. You can argue all the other combinations, and the result is that a higher pressure pushes the reaction toward the side with fewer moles, and a lower pressure pushes the reaction toward the side with more moles. The reaction tries to counteract the externally imposed pressure variation.

EXAMPLE 15.3

Determine the equilibrium constant K, expressed as in K, for the reaction $2H_2O \rightleftharpoons 2H_2 + O_2$ at 25°C and at 2000 K.

Solution

We have already found, in Example 15.2, $\Delta G°$ for this reaction at these two temperatures. Therefore, at 25°C,

$$(\ln K)_{298} = -\frac{\Delta G^0_{298}}{\overline{R}T} = \frac{-457\ 166}{8.3145 \times 298.15} = -184.42$$

At 2000 K, we have

$$(\ln K)_{2000} = -\frac{\Delta G^0_{2000}}{\overline{R}T} = \frac{-271\ 040}{8.3145 \times 2000} = -16.299$$

Table A.11 gives the values of the equilibrium constant for a number of reactions. Note again that for each reaction the value of the equilibrium constant is determined from the properties of each of the pure constituents at the standard-state pressure and is a function of temperature only.

For other reaction equations, the chemical equilibrium constant can be calculated as in Example 15.3. Sometimes you can write a reaction scheme as a linear combination of the elementary reactions that are already tabulated, as for example in Table A.11. Assume we can write a reaction III as a linear combination of reaction I and reaction II, which means

$$\text{LHS}_{\text{III}} = a\,\text{LHS}_{\text{I}} + b\,\text{LHS}_{\text{II}} \tag{15.17}$$

$$\text{RHS}_{\text{III}} = a\,\text{RHS}_{\text{I}} + b\,\text{RHS}_{\text{II}}$$

From the definition of the shift in the Gibbs function, Eq. 15.12, it follows that

$$\Delta G_{\text{III}}^0 = G_{\text{III RHS}}^0 - G_{\text{III LHS}}^0 = a\,\Delta G_{\text{I}}^0 + b\,\Delta G_{\text{II}}^0$$

Then from the definition of the equilibrium constant in Eq. 15.15 we get

$$\ln K_{\text{III}} = -\frac{\Delta G_{\text{III}}^0}{\overline{R}T} = -a\,\frac{\Delta G_{\text{I}}^0}{\overline{R}T} - b\,\frac{\Delta G_{\text{II}}^0}{\overline{R}T} = a\ln K_{\text{I}} + b\ln K_{\text{II}}$$

or

$$K_{\text{III}} = K_{\text{I}}^a K_{\text{II}}^b \tag{15.18}$$

EXAMPLE 15.4 Show that the equilibrium constant for the reaction called the water-gas reaction

$$\text{III}\ \ H_2 + CO_2 \rightleftharpoons H_2O + CO$$

can be calculated from values listed in Table A.11.

Solution

Using the reaction equations from Table A.11,

$$\text{I}\quad 2CO_2 \rightleftharpoons 2CO + O_2$$

$$\text{II}\quad 2H_2O \rightleftharpoons 2H_2 + O_2$$

It is seen that

$$\text{III} = \tfrac{1}{2}\text{I} - \tfrac{1}{2}\text{II} = \tfrac{1}{2}(\text{I} - \text{II})$$

so that

$$K_{\text{III}} = \left(\frac{K_{\text{I}}}{K_{\text{II}}}\right)^{\frac{1}{2}}$$

where K_{III} is calculated from the Table A.11 values

$$\ln K_{\text{III}} = \tfrac{1}{2}(\ln K_{\text{I}} - \ln K_{\text{II}})$$

We now consider a number of examples that illustrate the procedure for determining the equilibrium composition for a homogeneous reaction and the influence of certain variables on the equilibrium composition.

EXAMPLE 15.5 One kilomole of carbon at 25°C and 0.1 MPa pressure reacts with 1 kmol of oxygen at 25°C and 0.1 MPa pressure to form an equilibrium mixture of CO_2, CO, and O_2 at 3000 K, 0.1 MPa pressure, in a steady-state process. Determine the equilibrium composition and the heat transfer for this process.

> *Control volume:* Combustion chamber.
> *Inlet states:* P, T known for carbon and for oxygen.
> *Exit state:* P, T known.
> *Process:* Steady-state.
> *Sketch:* Figure 15.9.
> *Model:* Table A.10 for carbon; ideal gases, Tables A.9 and A.10.

Analysis and Solution

It is convenient to view the overall process as though it occurs in two separate steps, a combustion process followed by a heating and dissociation of the combustion product carbon dioxide, as indicated in Fig. 15.9. This two-step process is represented as

> Combustion: $C + O_2 \rightarrow CO_2$
>
> Dissociation reaction: $2CO_2 \rightleftharpoons 2CO + O_2$

That is, the energy released by the combustion of C and O_2 heats the CO_2 formed to high temperature, which causes dissociation of part of the CO_2 to CO and O_2. Thus, the overall reaction can be written

$$C + O_2 \rightarrow aCO_2 + bCO + dO_2$$

where the unknown coefficients a, b, and d must be found by solution of the equilibrium equation associated with the dissociation reaction. Once this is accomplished, we can write the first law for a control volume around the combustion chamber to calculate the heat transfer.

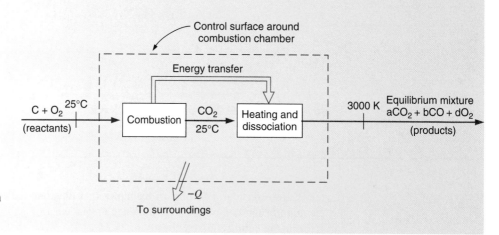

FIGURE 15.9 Sketch for Example 15.5.

From the combustion equation we find that the initial composition for the dissociation reaction is 1 kmol CO_2. Therefore, letting $2z$ be the number of kilomoles of CO_2 dissociated, we find

$$2CO_2 \rightleftharpoons 2CO + O_2$$

Initial:	1	0	0
Change:	$-2z$	$+2z$	$+z$
At equilibrium:	$(1-2z)$	$2z$	z

Therefore, the overall reaction is

$$C + O_2 \rightarrow (1-2z)CO_2 + 2zCO + zO_2$$

and the total number of kilomoles at equilibrium is

$$n = (1-2z) + 2z + z = 1 + z$$

The equilibrium mole fractions are

$$y_{CO_2} = \frac{1-2z}{1+z} \qquad y_{CO} = \frac{2z}{1+z} \qquad y_{O_2} = \frac{z}{1+z}$$

From Table A.11 we find that the value of the equilibrium constant at 3000 K for the dissociation reaction considered here is

$$\ln K = -2.217 \qquad K = 0.1089$$

Substituting these quantities along with $P = 0.1$ MPa into Eq. 15.16, we have the equilibrium equation,

$$K = 0.1089 = \frac{y_{CO}^2 y_{O_2}}{y_{CO_2}^2}\left(\frac{P}{P^0}\right)^{2+1-2} = \frac{\left(\frac{2z}{1+z}\right)^2\left(\frac{z}{1+z}\right)}{\left(\frac{1-2z}{1+z}\right)^2} \quad (1)$$

or, in more convenient form,

$$\frac{K}{P/P^0} = \frac{0.1089}{1} = \left(\frac{2z}{1-2z}\right)^2\left(\frac{z}{1+z}\right)$$

To obtain the physically meaningful root of this mathematical relation, we note that the number of moles of each component must be greater than zero. Thus, the root of interest to us must lie in the range

$$0 \leq z \leq 0.5$$

Solving the equilibrium equation by trial and error, we find

$$z = 0.2189$$

Therefore, the overall process is

$$C + O_2 \rightarrow 0.5622CO_2 + 0.4378CO + 0.2189O_2$$

where the equilibrium mole fractions are

$$y_{CO_2} = \frac{0.5622}{1.2189} = 0.4612$$

$$y_{CO} = \frac{0.4378}{1.2189} = 0.3592$$

$$y_{O_2} = \frac{0.2189}{1.2189} = 0.1796$$

The heat transfer from the combustion chamber to the surroundings can be calculated using the enthalpies of formation and Table A.9. For this process

$$H_R = (\bar{h}_f^0)_C + (\bar{h}_f^0)_{O_2} = 0 + 0 = 0$$

The equilibrium products leave the chamber at 3000 K. Therefore,

$$\begin{aligned}
H_P = &\ n_{CO_2}(\bar{h}_f^0 + \bar{h}_{3000}^0 - \bar{h}_{298}^0)_{CO_2} \\
&+ n_{CO}(\bar{h}_f^0 + \bar{h}_{3000}^0 - \bar{h}_{298}^0)_{CO} \\
&+ n_{O_2}(\bar{h}_f^0 + \bar{h}_{3000}^0 - \bar{h}_{298}^0)_{O_2} \\
= &\ 0.5622(-393\ 522 + 152\ 853) \\
&+ 0.4378(-110\ 527 + 93\ 504) \\
&+ 0.2189(98\ 013) \\
= &\ -121\ 302\ \text{kJ}
\end{aligned}$$

Substituting into the first law gives

$$Q_{c.v.} = H_P - H_g$$

$$= -121\ 302\ \text{kJ/kmol C burned}$$

EXAMPLE 15.6 One kilomole of carbon at 25°C reacts with 2 kmol of oxygen at 25°C to form an equilibrium mixture of CO_2, CO, and O_2 at 3000 K, 0.1 MPa pressure. Determine the equilibrium composition.

Control volume:	Combustion chamber.
Inlet states:	*T* known for carbon and for oxygen.
Exit state:	*P, T* known.
Process:	Steady-state.
Model:	Ideal-gas mixture at equilibrium.

Analysis and Solution

The overall process can be imagined to occur in two steps as in the previous example. The combustion process is

$$C + 2O_2 \rightarrow CO_2 + O_2$$

and the subsequent dissociation reaction is

	$2CO_2 \rightleftharpoons$	$2CO$	$+ O_2$
Initial:	1	0	1
Change:	$-2z$	$+2z$	$+z$
At equilibrium:	$(1 - 2z)$	$2z$	$(1 + z)$

We find that in this case the overall process is

$$C + 2O_2 \rightarrow (1 - 2z)CO_2 + 2zCO + (1 + z)O_2$$

and the total number of kilomoles at equilibrium is

$$n = (1 - 2z) + 2z + (1 + z) = 2 + z$$

The mole fractions are

$$y_{CO_2} = \frac{1 - 2z}{2 + z} \qquad y_{CO} = \frac{2z}{2 + z} \qquad y_{O_2} = \frac{1 + z}{2 + z}$$

The equilibrium constant for the reaction $2CO_2 \rightleftharpoons 2CO + O_2$ at 3000 K was found in Example 15.5 to be 0.1089. Therefore, with these expressions, quantities, and $P = 0.1$ MPa substituted, the equilibrium equation is

$$K = 0.1089 = \frac{y_{CO}^2 y_{O_2}}{y_{CO_2}^2} \left(\frac{P}{P^0}\right)^{2+1-2} = \frac{\left(\dfrac{2z}{2+z}\right)^2 \left(\dfrac{1+z}{2+z}\right)}{\left(\dfrac{1-2z}{2+z}\right)^2} \quad (1)$$

or

$$\frac{K}{P/P^0} = \frac{0.1089}{1} = \left(\frac{2z}{1 - 2z}\right)^2 \left(\frac{1+z}{2+z}\right)$$

We note that in order for the number of kilomoles of each component to be greater than zero,

$$0 \le z \le 0.5$$

Solving the equilibrium equation for z, we find

$$z = 0.1553$$

so that the overall process is

$$C + 2O_2 \rightarrow 0.6894CO_2 + 0.3106CO + 1.1553O_2$$

When we compare this result with that of Example 15.5, we notice that there is more CO_2 and less CO. The presence of additional oxygen shifts the dissociation reaction more to the left side.

The mole fractions of the components in the equilibrium mixture are

$$y_{CO_2} = \frac{0.6894}{2.1553} = 0.320$$

$$y_{CO} = \frac{0.3106}{2.1553} = 0.144$$

$$y_{O_2} = \frac{1.1553}{2.1553} = 0.536$$

The heat transferred from the chamber in this process could be found by the same procedure followed in Example 15.5, considering the overall process.

15.5 SIMULTANEOUS REACTIONS

In developing the equilibrium equation and equilibrium constant expressions of Section 15.4, it was assumed that there was only a single chemical reaction equation relating the substances present in the system. To demonstrate the more general situation in which there is more than one chemical reaction, we will now analyze a case involving two simultaneous reactions by a procedure analogous to that followed in Section 15.4. These results are then readily extended to systems involving several simultaneous reactions.

Consider a mixture of substances A, B, C, D, L, M, and N as indicated in Fig. 15.10. These substances are assumed to exist at a condition of chemical equilibrium at temperature T and pressure P, and are related by the two independent reactions

$$(1)\ \nu_{A1}A + \nu_B B \rightleftharpoons \nu_C C + \nu_D D \tag{15.19}$$

$$(2)\ \nu_{A2}A + \nu_L L \rightleftharpoons \nu_M M + \nu_N N \tag{15.20}$$

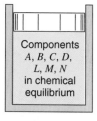

FIGURE 15.10

Sketch demonstrating simultaneous reactions.

We have considered the situation where one of the components (substance A) is involved in each of the reactions in order to demonstrate the effect of this condition on the resulting equations. As in the previous section, the changes in amounts of the components are related by the various stoichiometric coefficients (which are not the same as the number of moles of each substance present in the vessel). We also realize that the coefficients ν_{A1} and ν_{A2} are not necessarily the same. That is, substance A does not in general take part in each of the reactions to the same extent.

Development of the requirement for equilibrium is completely analogous to that of Section 15.4. We consider that each reaction proceeds an infinitesimal amount toward the right side. This results in a decrease in the number of moles of A, B, and L, and an increase in the moles of C, D, M, and N. Letting the degrees of reaction be ε_1 and ε_2 for reactions

1 and 2, respectively, the changes in the number of moles are, for infinitesimal shifts from the equilibrium composition,

$$dn_A = -v_{A_1} d\varepsilon_1 - v_{A_2} d\varepsilon_2$$

$$dn_B = -v_B d\varepsilon_1$$

$$dn_L = -v_L d\varepsilon_2$$

$$dn_C = +v_C d\varepsilon_1$$

$$dn_D = +v_D d\varepsilon_1$$

$$dn_M = +v_M d\varepsilon_2$$

$$dn_N = +v_N d\varepsilon_2 \qquad (15.21)$$

The change in Gibbs function for the mixture in the vessel at constant temperature and pressure is

$$dG_{T,P} = \overline{G}_A \, dn_A + \overline{G}_B \, dn_B + \overline{G}_C \, dn_C + \overline{G}_D \, dn_D + \overline{G}_L \, dn_L + \overline{G}_M \, dn_M + \overline{G}_N \, dn_N$$

Substituting the expressions of Eq. 15.21 and collecting terms,

$$dG_{T,P} = (v_C \overline{G}_C + v_D \overline{G}_D - v_{A_1} \overline{G}_A - v_B \overline{G}_B)\, d\varepsilon_1$$
$$+ (v_M \overline{G}_M + v_N \overline{G}_N - v_{A_2} \overline{G}_A - v_L \overline{G}_L)\, d\varepsilon_2 \qquad (15.22)$$

It is convenient to again express each of the partial molal Gibbs functions in terms of

$$\overline{G}_i = \overline{g}_i^0 + \overline{R}T \ln\left(\frac{y_i P}{P^0}\right)$$

Equation 15.22 written in this form becomes

$$dG_{T,P} = \left\{ \Delta G_1^0 + \overline{R}T \ln\left[\frac{y_C^{v_C} y_D^{v_D}}{y_A^{v_{A_1}} y_B^{v_B}}\left(\frac{P}{P^0}\right)^{v_C+v_D-v_{A1}-v_B}\right] \right\} d\varepsilon_1$$
$$+ \left\{ \Delta G_2^0 + \overline{R}T \ln\left[\frac{y_M^{v_M} y_N^{v_N}}{y_A^{v_{A_2}} y_L^{v_L}}\left(\frac{P}{P^0}\right)^{v_M+v_N-v_{A2}-v_L}\right] \right\} d\varepsilon_2 \qquad (15.23)$$

In this equation the standard-state change in Gibbs function for each reaction is defined as

$$\Delta G_1^0 = v_C \overline{g}_C^0 + v_D \overline{g}_D^0 - v_{A1} \overline{g}_A^0 - v_B \overline{g}_B^0 \qquad (15.24)$$

$$\Delta G_2^0 = v_M \overline{g}_M^0 + v_N \overline{g}_N^0 - v_{A2} \overline{g}_A^0 - v_L \overline{g}_L^0 \qquad (15.25)$$

Equation 15.23 expresses the change in Gibbs function of the system at constant T, P, for infinitesimal degrees of reaction of both reactions 1 and 2, Eqs. 15.19 and 15.20. The requirement for equilibrium is that $dG_{T,P} = 0$. Therefore, since reactions 1 and 2 are independent, $d\varepsilon_1$ and $d\varepsilon_2$ can be independently varied. It follows that at equilibrium each of the bracketed terms of Eq. 15.23 must be zero. Defining equilibrium constants for the two reactions by

$$\ln K_1 = -\frac{\Delta G_1^0}{\overline{R}T} \qquad (15.26)$$

and

$$\ln K_2 = -\frac{\Delta \overline{G}_2^0}{\overline{R}T} \qquad (15.27)$$

we find that, at equilibrium

$$K_1 = \frac{y_C^{\nu_C} y_D^{\nu_D}}{y_{A_1}^{\nu_{A_1}} y_B^{\nu_B}} \left(\frac{P}{P^0}\right)^{\nu_C + \nu_D - \nu_{A_1} - \nu_B} \qquad (15.28)$$

and

$$K_2 = \frac{y_M^{\nu_M} y_N^{\nu_N}}{y_{A_2}^{\nu_{A_2}} y_L^{\nu_L}} \left(\frac{P}{P^0}\right)^{\nu_M + \nu_N - \nu_{A_2} - \nu_L} \qquad (15.29)$$

These expressions for the equilibrium composition of the mixture must be solved simultaneously. The following example demonstrates and clarifies this procedure.

EXAMPLE 15.7   One kilomole of water vapor is heated to 3000 K, 0.1 MPa pressure. Determine the equilibrium composition, assuming that H_2O, H_2, O_2, and OH are present.

Control volume:	Heat exchanger.
Exit state:	P, T known.
Model:	Ideal-gas mixture at equilibrium.

Analysis and Solution

There are two independent reactions relating the four components of the mixture at equilibrium. These can be written as

$$(1)\ 2\,H_2O \rightleftharpoons 2\,H_2 + O_2$$
$$(2)\ 2\,H_2O \rightleftharpoons H_2 + 2\,OH$$

Let $2a$ be the number of kilomoles of water dissociating according to reaction 1 during the heating, and let $2b$ be the number of kilomoles of water dissociating according to reaction 2. Since the initial composition is 1 kmol water, the changes according to the two reactions are

$$(1)\ 2\,H_2O \rightleftharpoons 2\,H_2 + O_2$$
$$\text{Change:}\quad -2a \qquad +2a + a$$

$$(2)\ 2\,H_2O \rightleftharpoons H_2 + 2\,OH$$
$$\text{Change:}\quad -2b \qquad +b + 2b$$

Therefore, the number of kilomoles of each component at equilibrium is its initial number plus the change, so that at equilibrium

$$n_{H_2O} = 1 - 2a - 2b$$

$$n_{H_2} = 2a + b$$

$$n_{O_2} = a$$

$$\underline{n_{OH} = 2b}$$

$$n = 1 + a + b$$

The overall chemical reaction that occurs during the heating process can be written

$$H_2O \rightarrow (1 - 2a - 2b)H_2O + (2a + b)H_2 + aO_2 + 2bOH$$

The right-hand side of this expression is the equilibrium composition of the system. Since the number of kilomoles of each substance must necessarily be greater than zero, we find that the possible values of a and b are restricted to

$$a \geq 0$$

$$b \geq 0$$

$$(a + b) \leq 0.5$$

The two equilibrium equations are, assuming that the mixture behaves as an ideal gas,

$$K_1 = \frac{y_{H_2}^2 \, y_{O_2}}{y_{H_2O}^2} \left(\frac{P}{P^0}\right)^{2+1-2}$$

$$K_2 = \frac{y_{H_2} \, y_{OH}^2}{y_{H_2O}^2} \left(\frac{P}{P^0}\right)^{1+2-2}$$

Since the mole fraction of each component is the ratio of the number of kilomoles of the component to the total number of kilomoles of the mixture, these equations can be written in the form

$$K_1 = \frac{\left(\dfrac{2a + b}{1 + a + b}\right)^2 \left(\dfrac{a}{1 + a + b}\right)}{\left(\dfrac{1 - 2a - 2b}{1 + a + b}\right)^2} \left(\frac{P}{P^0}\right)$$

$$= \left(\frac{2a + b}{1 - 2a - 2b}\right)^2 \left(\frac{a}{1 + a + b}\right) \left(\frac{P}{P^0}\right)$$

and

$$K_2 = \frac{\left(\dfrac{2a + b}{1 + a + b}\right) \left(\dfrac{2b}{1 + a + b}\right)^2}{\left(\dfrac{1 - 2a - 2b}{1 + a + b}\right)^2} \left(\frac{P}{P^0}\right)$$

$$= \left(\frac{2a + b}{1 + a + b}\right) \left(\frac{2b}{1 - 2a - 2b}\right)^2 \left(\frac{P}{P^0}\right)$$

giving two equations in the two unknowns a and b, since $P = 0.1$ MPa and the values of K_1, K_2 are known. From Table A.11 at 3000 K, we find

$$K_1 = 0.002\ 062 \qquad K_2 = 0.002\ 893$$

Therefore, the equations can be solved simultaneously for a and b. The values satisfying the equations are

$$a = 0.0534 \qquad b = 0.0551$$

Substituting these values into the expressions for the number of kilomoles of each component and of the mixture, we find the equilibrium mole fractions to be

$$y_{H_2O} = 0.7063$$

$$y_{H_2} = 0.1461$$

$$y_{O_2} = 0.0482$$

$$y_{OH} = 0.0994$$

The methods used in this section can readily be extended to equilibrium systems having more than two independent reactions. In each case, the number of simultaneous equilibrium equations is equal to the number of independent reactions. The solution of a large set of nonlinear simultaneous equations naturally becomes quite difficult, however, and is not easily accomplished by hand calculations. These problems are normally solved using iterative procedures on a computer.

15.6 IONIZATION

In the final section of this chapter, we consider the equilibrium of systems that are made up of ionized gases, or plasmas, a field that has been studied and applied increasingly in recent years. In previous sections, we discussed chemical equilibrium, with a particular emphasis on molecular dissociation, as for example the reaction

$$N_2 \rightleftharpoons 2N$$

which occurs to an appreciable extent for most molecules only at high temperature, of the order of magnitude 3000 to 10 000 K. At still higher temperatures, such as those found in electric arcs, the gas becomes ionized. That is, some of the atoms lose an electron, according to the reaction

$$N \rightleftharpoons N^+ + e^-$$

where N^+ denotes a singly ionized nitrogen atom, one that has lost one electron and consequently has a positive charge, and e^- represents the free electron. As the temperature rises still higher, many of the ionized atoms lose another electron, according to the reaction

$$N^+ \rightleftharpoons N^{++} + e^-$$

and thus becomes doubly ionized. As the temperature continues to rise, the process continues until a temperature is reached at which all the electrons have been stripped from the nucleus.

Ionization generally is appreciable only at high temperature. However, dissociation and ionization both tend to occur to greater extents at low pressure, and consequently dissociation and ionization may be appreciable in such environments as the upper atmosphere, even at moderate temperature. Other effects such as radiation will also cause ionization, but these effects are not considered here.

The problems of analyzing the composition in a plasma become much more difficult than for an ordinary chemical reaction, for in an electric field the free electrons in the mixture do not exchange energy with the positive ions and neutral atoms at the same rate that they do with the field. Consequently, in a plasma in an electric field, the electron gas is not at exactly the same temperature as the heavy particles. However, for moderate fields, assuming a condition of thermal equilibrium in the plasma is a reasonable approximation, at least for preliminary calculations. Under this condition, we can treat the ionization equilibrium in exactly the same manner as an ordinary chemical equilibrium analysis.

At these extremely high temperatures, we may assume that the plasma behaves as an ideal-gas mixture of neutral atoms, positive ions, and electron gas. Thus, for the ionization of some atomic species A,

$$A \rightleftharpoons A^+ + e^-$$ (15.30)

we may write the ionization equilibrium equation in the form

$$K = \frac{y_{A^+} y_{e^-}}{y_A} \left(\frac{P}{P^0} \right)^{1+1-1}$$ (15.31)

The ionization-equilibrium constant K is defined in the ordinary manner

$$\ln K = -\frac{\Delta G^0}{RT}$$ (15.32)

and is a function of temperature only. The standard-state Gibbs function change for reaction 15.30 is found from

$$\Delta G^0 = \overline{g}_{A^+}^0 + \overline{g}_{e^-}^0 - \overline{g}_A^0$$ (15.33)

The standard-state Gibbs function for each component at the given plasma temperature can be calculated using the procedures of statistical thermodynamics, so that ionization–equilibrium constants can be tabulated as functions of temperature.

Solution of the ionization–equilibrium equation, Eq. 15.31, is then accomplished in the same manner as for an ordinary chemical-reaction equilibrium.

EXAMPLE 15.8   Calculate the equilibrium composition if argon gas is heated in an arc to 10 000 K, 1 kPa, assuming the plasma to consist of Ar, Ar^+, e^-. The ionization–equilibrium constant for the reaction

$$Ar \rightleftharpoons Ar^+ + e^-$$

at this temperature is 0.000 42.

> *Control volume:*   Heating arc.
> *Exit state:* P, T known.
> *Model:*   Ideal-gas mixture at equilibrium.

Analysis and Solution

Consider an initial composition of 1 kmol neutral argon, and let z be the number of kilomoles ionized during the heating process. Therefore,

$$Ar \rightleftharpoons Ar^+ + e^-$$

	Ar	Ar^+	e^-
Initial:	1	0	0
Change:	$-z$	$+z$	$+z$
Equilibrium:	$(1-z)$	z	z

and

$$n = (1-z) + z + z = 1 + z$$

Since the number of kilomoles of each component must be positive, the variable z is restricted to the range

$$0 \leq z \leq 1$$

The equilibrium mole fractions are

$$y_{Ar} = \frac{n_{Ar}}{n} = \frac{1-z}{1+z}$$

$$y_{Ar^+} = \frac{n_{Ar^+}}{n} = \frac{z}{1+z}$$

$$y_{e^-} = \frac{n_{e^-}}{n} = \frac{z}{1+z}$$

The equilibrium equation is

$$K = \frac{y_{Ar^+} y_{e^-}}{y_{Ar}} \left(\frac{P}{P^0}\right)^{1+1-1} = \frac{\left(\frac{z}{1+z}\right)\left(\frac{z}{1+z}\right)}{\left(\frac{1-z}{1+z}\right)} \left(\frac{P}{P^0}\right)$$

so that, at 10 000 K, 1 kPa,

$$0.000\ 42 = \left(\frac{z^2}{1-z^2}\right)(0.01)$$

Solving,

$$z = 0.2008$$

and the composition is found to be

$$y_{Ar} = 0.6656$$

$$y_{Ar^+} = 0.1672$$

$$y_{e^-} = 0.1672$$

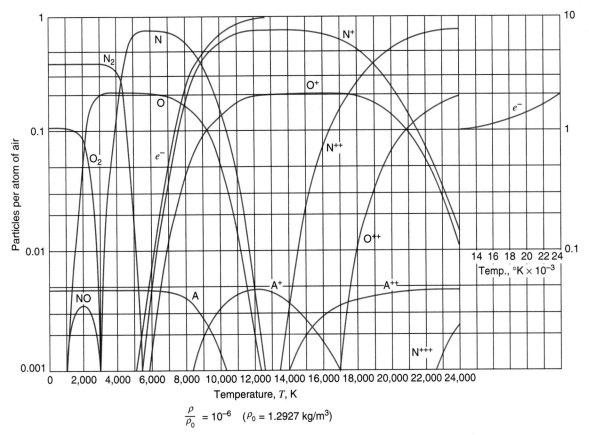

FIGURE 15.11 Equilibrium composition of air [W. E. Moeckel and K. C. Weston, NACA TN 4265 (1958)].

Simultaneous reactions, such as simultaneous molecular dissociation and ionization reactions or multiple ionization reactions, can be analyzed in the same manner as the ordinary simultaneous chemical reactions of Section 15.5. In doing so, we again make the assumption of thermal equilibrium in the plasma, which, as mentioned before, is, in many cases, a reasonable approximation. Figure 15.11 shows the equilibrium composition of air at high temperature and low density, and indicates the overlapping regions of the various dissociation and ionization processes.

SUMMARY

A short introduction is given to equilibrium in general with application to phase equilibrium and chemical equilibrium. From previous analysis with the second law we have found the reversible shaft work as the change in Gibbs function. This is extended to give the equilibrium state as the one with minimum Gibbs function at a given T, P. This also applies to two phases in equilibrium, so each phase has the same Gibbs function.

Chemical equilibrium is formulated for a single-equilibrium reaction assuming the components are all ideal gases. This leads to an equilibrium equation tying together the mole fractions of the components, the pressure, and the reaction constant. The reaction constant is related to the shift in Gibbs function from the reactants (LHS) to the products (RHS) at a temperature T. As temperature or pressure changes, the equilibrium composition will shift according to its sensitivity to T and P. For very large equilibrium constants, the reaction is shifted toward the RHS, and for very small ones it is shifted toward the LHS.

In most real systems of interest, there are multiple reactions coming to equilibrium simultaneously with a fairly large number of species involved. Often species are present in the mixture without participation in the reactions causing a dilution, so all mole fractions are lower than they otherwise would be. As a last example of a reaction, we show an ionization process in which one or more electrons can be separated from an atom.

You should have learned a number of skills and acquired abilities from studying this chapter that will allow you to:

- Apply the principle of minimum Gibbs function to a phase equilibrium.
- Understand that the concept of equilibrium can include other effects such as elevation, surface tension, and electrical potentials, as well as the concept of metastable states.
- Understand that the chemical equilibrium is written for ideal-gas mixtures.
- Understand the meaning of the shift in Gibbs function due to the reaction.
- Know when the absolute pressure has an influence on the composition.
- Know the connection between the reaction scheme and the equilibrium constant.
- Understand that all species are present and influence the mole fractions.
- Know that a dilution with an inert gas has an effect.
- Understand the coupling between the chemical equilibrium and the energy equation.
- Intuitively know that most problems must be solved by iterations.
- Be able to treat a dissociation added on to a combustion process.
- Be able to treat multiple simultaneous reactions.
- Know what an ionization process is and how to treat it.

KEY CONCEPTS AND FORMULAS

Gibbs function	$g = h - Ts$
Equilibrium	Minimum g for given $T, P \Rightarrow dG_{T,P} = 0$
Phase equilibrium	$g_f = g_g$
Equilibrium reaction	$v_A A + v_B B \Leftrightarrow v_C C + v_D D$
Change in Gibbs function	$\Delta G^0 = v_C \bar{g}_C^0 + v_D \bar{g}_D^0 - v_A \bar{g}_A^0 - v_B \bar{g}_B^0$ evaluate at T, P^0
Equilibrium constant	$K = e^{-\Delta G^0/\bar{R}T}$

$$K = \frac{y_C^{v_C} y_D^{v_D}}{y_A^{v_A} y_B^{v_B}} \left(\frac{P}{P^0}\right)^{v_C + v_D - v_A - v_B}$$

Reaction scheme	Reaction scheme III $= a$ I $+ b$ II $\Rightarrow K_{III} = K_I^a K_{II}^b$
Dilution	reaction the same, y's are smaller
Simultaneous reactions	$K_1, K_2, \ldots$ and more y's

CONCEPT-STUDY GUIDE PROBLEMS

15.1 Is the concept of equilibrium limited to thermodynamics?

15.2 How does the Gibbs function vary with quality as you move from liquid to vapor?

15.3 How is a chemical equilibrium process different from a combustion process?

15.4 Must P and T be held fixed to obtain chemical equilibrium?

15.5 The change in Gibbs function for a reaction is a function of which property?

15.6 In a steady-flow burner, T is not controlled; which properties are?

15.7 In a closed, rigid combustion bomb, which properties are held fixed?

15.8 Is the dissociation of water pressure sensitive?

15.9 At 298 K, $K = \exp(-184)$ for the water dissociation, what does that imply?

15.10 For a mixture of O_2 and O the pressure is increased at constant T; what happens to the composition?

15.11 For a mixture of O_2 and O the temperature is increased at constant P; what happens to the composition?

15.12 For a mixture of O_2 and O I add some argon, keeping constant T, P; what happens to the moles of O?

15.13 In a combustion process, is the adiabatic flame temperature affected by reactions?

15.14 When dissociations occur after combustion, does T go up or down?

15.15 In equilibrium, the Gibbs function of the reactants and the products is the same; how about the energy?

15.16 Does a dissociation process require energy, or does it give out energy?

15.17 If I consider the nonfrozen (composition can vary) heat capacity, but still assume all components are ideal gases, does that C become a function of temperature? of pressure?

15.18 What is K for the water-gas reaction in Example 15.4 at 1200 K?

15.19 Which atom in air ionizes first as T increases? What is the explanation?

15.20 At what temperature range does air become a plasma?

HOMEWORK PROBLEMS

Equilibrium and Phase Equilibrium

15.21 Carbon dioxide at 15 MPa is injected into the top of a 5-km-deep well in connection with an enhanced oil-recovery process. The fluid column standing in the well is at a uniform temperature of 40°C. What is the pressure at the bottom of the well, assuming ideal-gas behavior?

15.22 Consider a 2-km-deep gas well containing a gas mixture of methane and ethane at a uniform temperature of 30°C. The pressure at the top of the well is 14 MPa, and the composition on a mole basis is 90% methane, 10% ethane. Each component is in equilibrium (top to bottom) with $dG + g\, dZ = 0$ and assume ideal gas, so for each component Eq. 15.10 applies. Determine the pressure and composition at the bottom of the well.

15.23 A container has liquid water at 20°C, 100 kPa, in equilibrium with a mixture of water vapor and dry air also at 20°C, 100 kPa. How much is the water vapor pressure, and what is the saturated water vapor pressure?

15.24 Using the same assumptions as those in developing Eq. d in Example 15.1, develop an expression for pressure at the bottom of a deep column of liquid in terms of the isothermal compressibility, β_T. For liquid water at 20°C, we know that $\beta_T = 0.0005$ [1/MPa]. Use the result of the first question to estimate the pressure in the Pacific Ocean at the depth of 3 km.

Chemical Equilibrium, Equilibrium Constant

15.25 Calculate the equilibrium constant for the reaction $O_2 \rightleftharpoons 2O$ at temperatures of 298 K and 6000 K. Verify the result with Table A.11.

15.26 For the dissociation of oxygen, $O_2 \Leftrightarrow 2O$, around 2000 K we want a mathematical expression for the equilibrium constant $K(T)$. Assume constant heat capacity, at 2000 K, for O_2 and O from Table A.9 and develop the expression from Eqs. 15.12 and 15.15.

15.27 Calculate the equilibrium constant for the reaction $H_2 \rightleftharpoons 2H$ at a temperature of 2000 K, using properties from Table A.9. Compare the result with the value listed in Table A.11.

15.28 Plot to scale the values of ln K versus $1/T$ for the reaction $2CO_2 \rightleftharpoons 2CO + O_2$. Write an equation for ln K as a function of temperature.

15.29 Calculate the equilibrium constant for the reaction $2CO_2 \rightleftharpoons 2CO + O_2$ at 3000 K using values from Table A.9 and compare the result to Table A.11.

15.30 Consider the dissociation of oxygen, $O_2 \Leftrightarrow 2O$, starting with 1 kmol oxygen at 298 K and heating it at constant pressure 100 kPa. At which temperature will we reach a concentration of monatomic oxygen of 10%?

15.31 Pure oxygen is heated from 25°C to 3200 K in a steady-state process at a constant pressure of 200 kPa. Find the exit composition and the heat transfer.

15.32 Nitrogen gas, N_2, is heated to 4000 K, 10 kPa. What fraction of the N_2 is dissociated to N at this state?

15.33 Hydrogen gas is heated from room temperature to 4000 K, 500 kPa, at which state the diatomic species has partially dissociated to the monatomic form. Determine the equilibrium composition at this state.

15.34 One kilomole Ar and one kilomole O_2 are heated at a constant pressure of 100 kPa to 3200 K, where they come to equilibrium. Find the final mole fractions for Ar, O_2, and O.

15.35 Consider the reaction $2CO_2 \Leftrightarrow 2CO + O_2$ obtained after heating 1 kmol CO_2 to 3000 K. Find the equilibrium constant from the shift in Gibbs function and verify its value with the entry in Table A.11. What is the mole fraction of CO at 3000 K, 100 kPa?

15.36 Air (assumed to be 79% nitrogen and 21% oxygen) is heated in a steady-state process at a constant pressure of 100 kPa, and some NO is formed (disregard dissociations of N_2 and O_2). At what temperature will the mole fraction of NO be 0.001?

15.37 The combustion products from burning pentane, C_5H_{12}, with pure oxygen in a stoichiometric ratio exit at 2400 K, 100 kPa. Consider the dissociation of only CO_2 and find the equilibrium mole fraction of CO.

15.38 Find the equilibrium constant for the reaction $2NO + O_2 \rightleftharpoons 2NO_2$ from the elementary reactions in Table A.11 to answer which of the nitrogen oxides, NO or NO_2, is the more stable at ambient conditions? What about at 2000 K?

15.39 Pure oxygen is heated from 25°C, 100 kPa, to 3200 K in a constant-volume container. Find the final pressure, composition, and the heat transfer.

15.40 A mixture of 1 kmol carbon dioxide, 2 kmol carbon monoxide, and 2 kmol oxygen, at 25°C, 150 kPa, is heated in a constant-pressure steady-state process to 3000 K. Assuming that only these same substances are present in the exiting chemical equilibrium mixture, determine the composition of that mixture.

15.41 Repeat the previous problem for an initial mixture that also includes 2 kmol of nitrogen, which does not dissociate during the process.

15.42 One approach to using hydrocarbon fuels in a fuel cell is to "reform" the hydrocarbon to obtain hydrogen, which is then fed to the fuel cell. As a part of the analysis of such a procedure, consider the reaction $CH_4 + H_2O \rightleftharpoons 3H_2 + CO$. Determine the equilibrium constant for this reaction at a temperature of 800 K.

15.43 Consider combustion of methane with pure oxygen forming carbon dioxide and water as the products. Find the equilibrium constant for the reaction at 1000 K. Use an average heat capacity of $C_p = 52$ kJ/kmol K for the fuel and Table A.9 for the other components.

15.44 Find the equilibrium constant for the reaction: $2NO + O_2 \Leftrightarrow 2NO_2$ from the elementary reaction in Table A.11 to answer these two questions. Which of the nitrogen oxides, NO or NO_2, is the more stable at 25°C, 100 kPa? At what T do we have an equal amount of each?

15.45 The equilibrium reaction with methane as $CH_4 \rightleftharpoons C + 2H_2$ has ln $K = -0.3362$ at 800 K and ln $K = -4.607$ at 600 K. By noting the relation of K to temperature, show how you would interpolate ln K in $(1/T)$ to find K at 700 K and compare that to a linear interpolation.

15.46 Water from the combustion of hydrogen and pure oxygen is at 3800 K and 50 kPa. Assume we only have H_2O, O_2 and H_2 as gases, find the equilibrium composition.

15.47 Complete combustion of hydrogen and pure oxygen in a stoichiometric ratio at P_0, T_0 to form water would result in a computed adiabatic flame temperature of 4990 K for a steady-state setup.

    a. How should the adiabatic flame temperature be found if the equilibrium reaction $2H_2 + O_2 \rightleftharpoons 2H_2O$ is considered? Disregard all other possible reactions (dissociations) and show the final equation(s) to be solved.

    b. Which other reactions should be considered, and which components will be present in the final mixture?

15.48 The van't Hoff equation

$$d \ln K = \frac{\Delta H^0}{RT^2} dT_{p^0}$$

relates the chemical equilibrium constant K to the enthalpy of reaction ΔH^0. From the value of K in Table A.11 for the dissociation of hydrogen at 2000 K and the value of ΔH^0 calculated from Table A.9 at 2000 K, use the van't Hoff equation to predict the equilibrium constant at 2400 K.

15.49 Gasification of char (primarily carbon) with steam following coal pyrolysis yields a gas mixture of 1 kmol CO and 1 kmol H_2. We wish to upgrade the hydrogen content of this syngas fuel mixture, so it is fed to an appropriate catalytic reactor along with 1 kmol of H_2O. Exiting the reactor is a chemical equilibrium gas mixture of CO, H_2, H_2O, and CO_2 at 600 K, 500 kPa. Determine the equilibrium composition. *Note*: See Example 15.4.

15.50 Consider the water gas reaction in Example 15.4. Find the equilibrium constant at 500, 1000, 1200, and 1400 K. What can you infer from the result?

15.51 Catalytic gas generators are frequently used to decompose a liquid, providing a desired gas mixture (spacecraft control systems, fuel cell gas supply, and so forth). Consider feeding pure liquid hydrazine, N_2H_4, to a gas generator, from which exits a gas mixture of N_2, H_2, and NH_3 in chemical equilibrium at 100°C, 350 kPa. Calculate the mole fractions of the species in the equilibrium mixture.

15.52 A piston/cylinder contains 0.1 kmol hydrogen and 0.1 kmol Ar gas at 25°C, 200 kPa. It is heated in a constant-pressure process so the mole fraction of atomic hydrogen is 10%. Find the final temperature and the heat transfer needed.

15.53 A tank contains 0.1 kmol hydrogen and 0.1 kmol of argon gas at 25°C, 200 kPa, and the tank keeps constant volume. To what T should it be heated to have a mole fraction of atomic hydrogen, H, of 10%?

15.54 A gas mixture of 1 kmol carbon monoxide, 1 kmol nitrogen, and 1 kmol oxygen at 25°C, 150 kPa, is heated in a constant-pressure steady-state process. The exit mixture can be assumed to be in chemical equilibrium with CO_2, CO, O_2, and N_2 present. The mole fraction of CO_2 at this point is 0.176. Calculate the heat transfer for the process.

15.55 A rigid container initially contains 2 kmol of carbon monoxide and 2 kmol of oxygen at 25°C, 100 kPa. The content is then heated to 3000 K, at which point an equilibrium mixture of CO_2, CO, and O_2 exists. Disregard other possible species and determine the final pressure, the equilibrium composition, and the heat transfer for the process.

15.56 A coal gasifier produces a mixture of 1CO and $2H_2$ that is fed to a catalytic converter to produce methane. The reaction is $CO + 3H_2 \Leftrightarrow CH_4 + H_2O$. The equilibrium constant at 600 K is $K = 1.83 \times 10^6$. What is the composition of the exit flow assuming a pressure of 600 kPa?

15.57 One approach to using hydrocarbon fuels in a fuel cell is to "reform" the hydrocarbon to obtain hydrogen, which is then fed to the fuel cell. As a part of the analysis of such a procedure, consider the reaction $CH_4 + H_2O \Leftrightarrow CO + 3H_2$. One kilomole each of methane and water are fed to a catalytic reformer. A mixture of CH_4, H_2O, H_2, and CO exits in chemical equilibrium at 800 K, 100 kPa; determine the equilibrium composition of this mixture using an equilibrium constant of $K = 0.0237$.

15.58 Use the information in Problem 15.45 to estimate the enthalpy of reaction, ΔH^0, at 700 K using the van't Hoff equation (see Problem 15.48) with finite differences for the derivatives.

15.59 Acetylene gas at 25°C is burned with 140% theoretical air, which enters the burner at 25°C, 100 kPa, 80% relative humidity. The combustion products form a mixture of CO_2, H_2O, N_2, O_2, and NO in chemical equilibrium at 2200 K, 100 kPa. This mixture is then cooled to 1000 K very rapidly, so that the composition does not change.

Determine the mole fraction of NO in the products and the heat transfer for the overall process.

15.60 A step in the production of a synthetic liquid fuel from organic waste matter is the following conversion process: 1 kmol of ethylene gas (converted from the waste) at 25°C, 5 MPa, and 2 kmol of steam at 300°C, 5 MPa, enter a catalytic reactor. An ideal-gas mixture of ethanol, ethylene, and water in chemical equilibrium leaves the reactor at 700 K, 5 MPa. Determine the composition of the mixture and the heat transfer for the reactor.

15.61 Methane at 25°C, 100 kPa, is burned with 200% theoretical oxygen at 400 K, 100 kPa, in an adiabatic steady-state process, and the products of combustion exit at 100 kPa. Assume that the only significant dissociation reaction in the products is that of carbon dioxide going to carbon monoxide and oxygen. Determine the equilibrium composition of the products and also their temperature at the combustor exit.

15.62 Calculate the irreversibility for the adiabatic combustion process described in the previous problem.

15.63 An important step in the manufacture of chemical fertilizer is the production of ammonia, according to the reaction: $N_2 + 3H_2 \rightleftharpoons 2NH_3$
  a. Calculate the equilibrium constant for this reaction at 150°C
  b. For an initial composition of 25% nitrogen, 75% hydrogen, on a mole basis, calculate the equilibrium composition at 150°C, 5 MPa.

15.64 One kilomole of carbon dioxide, CO_2, and 1 kmol of hydrogen, H_2, at room temperature and 200 kPa is heated to 1200 K, 200 kPa. Use the water gas reaction to determine the mole fraction of CO. Neglect dissociations of H_2 and O_2.

15.65 Consider the production of a synthetic fuel (methanol) from coal. A gas mixture of 50% CO and 50% H_2 leaves a coal gasifier at 500 K, 1 MPa, and enters a catalytic converter. A gas mixture of methanol, CO, and H_2 in chemical equilibrium with the reaction $CO + 2H_2 \rightleftharpoons CH_3OH$ leaves the converter at the same temperature and pressure, where it is known that $\ln K = -5.119$.
  a. Calculate the equilibrium composition of the mixture leaving the converter.
  b. Would it be more desirable to operate the converter at ambient pressure?

15.66 Hydrides are rare earth metals, M, that have the ability to react with hydrogen to form a different substance MH_x with a release of energy. The hydrogen can then be released, the reaction reversed, by heat addition to the MH_x. In this reaction only the hydrogen is a gas, so the formula developed for the chemical equilibrium is inappropriate. Show that the proper expression to be used instead of Eq. 15.14 is

$$\ln (P_{H2}/P_0) = \Delta G^0/RT$$

when the reaction is scaled to 1 kmol of H_2.

Simultaneous Reactions

15.67 Water from the combustion of hydrogen and pure oxygen is at 3800 K and 50 kPa. Assuming we only have H_2O, O_2, OH, and H_2 as gases with the two simple water dissociation reactions active, find the equilibrium composition.

15.68 Ethane is burned with 150% theoretical air in a gas-turbine combustor. The products exiting consist of a mixture of CO_2, H_2O, O_2, N_2, and NO in chemical equilibrium at 1800 K, 1 MPa. Determine the mole fraction of NO in the products. Is it reasonable to ignore CO in the products?

15.69 Butane is burned with 200% theoretical air, and the products of combustion, an equilibrium mixture containing only CO_2, H_2O, O_2, N_2, NO, and NO_2, exits from the combustion chamber at 1400 K, 2 MPa. Determine the equilibrium composition at this state.

15.70 A mixture of 1 kmol water and 1 kmol oxygen at 400 K is heated to 3000 K, 200 kPa, in a steady-state process. Determine the equilibrium composition at the outlet of the heat exchanger, assuming that the mixture consists of H_2O, H_2, O_2, and OH.

15.71 One kilomole of air (assumed to be 78% nitrogen, 21% oxygen, and 1% argon) at room temperature is heated to 4000 K, 200 kPa. Find the equilibrium composition at this state, assuming that only N_2, O_2, NO, O, and Ar are present.

15.72 One kilomole of water vapor at 100 kPa, 400 K, is heated to 3000 K in a constant-pressure flow process. Determine the final composition, assuming that H_2O, H_2, H, O_2, and OH are present at equilibrium.

15.73 Acetylene gas and x times theoretical air ($x > 1$) at room temperature and 500 kPa are burned at constant pressure in an adiabatic flow process. The flame temperature is 2600 K, and the combustion products are assumed to consist of N_2, O_2, CO_2, H_2O, CO, and NO. Determine the value of x.

Ionization

15.74 At 10 000 K the ionization reaction for Ar is: $Ar \Leftrightarrow Ar^+ + e^-$ with equilibrium constant of $K = 4.2 \times 10^{-4}$. What should the pressure be for a mole concentration of argon ions (Ar^+) of 10%?

15.75 Operation of an MHD converter requires an electrically conducting gas. A helium gas "seeded" with 1.0 mole percent cesium, as shown in Fig. P15.75, is used where the cesium is partly ionized ($Cs \rightleftharpoons Cs^+ + e^-$) by heating the mixture to 1800 K, 1 MPa, in a nuclear reactor to provide free electrons. No helium is ionized in this process, so that the mixture entering the converter consists of He, Cs, Cs^+, and e^-. Determine the mole fraction of electrons in the mixture at 1800 K, where $\ln K = 1.402$ for the cesium ionization reaction described.

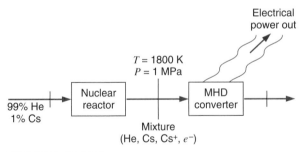

FIGURE P15.75

15.76 One kilomole of argon gas at room temperature is heated to 20 000 K, 100 kPa. Assume that the plasma in this condition consists of an equilibrium mixture of Ar, Ar^+, Ar^{++}, and e^- according to the simultaneous reactions

$$(1)\ Ar \rightleftharpoons Ar^+ + e^- \qquad (2)\ Ar^+ \rightleftharpoons Ar^{++} + e^-$$

The ionization equilibrium constants for these reactions at 20 000 K have been calculated from spectroscopic data as $\ln K_1 = 3.11$ and $\ln K_2 = -4.92$. Determine the equilibrium composition of the plasma.

15.77 At 10 000 K the two ionization reactions for N and Ar as

$$(1)\ Ar \Leftrightarrow Ar^+ + e^- \qquad (2)\ N \Leftrightarrow N^+ + e^-$$

have equilibrium constants of $K_1 = 4.2 \times 10^{-4}$ and $K_2 = 6.3 \times 10^{-4}$, respectively. If we start out with 1 kmol Ar and 0.5 kmol N_2, what is the equilibrium composition at a pressure of 10 kPa?

15.78 Plot to scale the equilibrium composition of nitrogen at 10 kPa over the temperature range 5000 K to 15 000 K, assuming that N_2, N, N^+, and e^- are present. For the ionization reaction $N \rightleftharpoons N^+ + e^-$, the ionization equilibrium constant K has been calculated from spectroscopic data as

T[K]	10 000	12 000	14 000	16 000
$100K$	6.26×10^{-2}	1.51	15.1	92

Review Problems

15.79 Repeat Problem 15.21 using the generalized charts, instead of ideal-gas behavior.

15.80 In a test of a gas-turbine combustor, saturated-liquid methane at 115 K is burned with excess air to hold the adiabatic flame temperature to 1600 K. It is assumed that the products consist of a mixture of CO_2, H_2O, N_2, O_2, and NO in chemical equilibrium. Determine the percent excess air used in the combustion and the percentage of NO in the products.

15.81 A space heating unit in Alaska uses propane combustion as the heat supply. Liquid propane comes from an outside tank at $-44°C$, and the air supply is also taken in from the outside at $-44°C$. The airflow regulator is misadjusted, such that only 90% of the theoretical air enters the combustion chamber, resulting in incomplete combustion. The products exit at 1000 K as a chemical equilibrium gas mixture, including only CO_2, CO, H_2O, H_2, and N_2. Find the composition of the products. *Hint:* Use the water gas reaction in Example 15.4.

15.82 Consider the following coal gasifier proposed for supplying a syngas fuel to a gas-turbine power plant. Fifty kilograms per second of dry coal (represented as 48 kg C plus 2 kg H) enter the gasifier, along with 4.76 kmol/s of air and 2 kmol/s of steam. The output stream from this unit is a gas mixture containing H_2, CO, N_2,

CH$_4$, and CO$_2$ in chemical equilibrium at 900 K, 1 MPa.

a. Set up the reaction and equilibrium equation(s) for this system, and calculate the appropriate equilibrium constant(s).

b. Determine the composition of the gas mixture leaving the gasifier.

15.83 One kilomole of liquid oxygen, O$_2$, at 93 K, and x kmol of gaseous hydrogen, H$_2$, at 25°C, are fed to a combustion chamber, x is greater than 2, such that there is excess hydrogen for the combustion process. There is a heat loss from the chamber of 1000 kJ per kmol of reactants. Products exit the chamber at chemical equilibrium at 3800 K, 400 kPa, and are assumed to include only H$_2$O, H$_2$, and O.

a. Determine the equilibrium composition of the products and also x, the amount of H$_2$ entering the combustion chamber.

b. Should another substance(s) have been included in part (a) as being present in the products? Justify your answer.

15.84 Saturated liquid butane (note: use generalized charts) enters an insulated constant-pressure combustion chamber at 25°C, and x times theoretical oxygen gas enters at the same pressure and temperature. The combustion products exit at 3400 K. Assuming that the products are a chemical equilibrium gas mixture that includes CO, what is x?

15.85 Derive the van't Hoff equation given in Problem 15.48, using Eqs. 15.12 and 15.15. *Note*: The $d(\bar{g}/T)$ at constant P^0 for each component can be expressed using the relations in Eqs. 13.18 and 13.19.

15.86 A coal gasifier produces a mixture of 1CO and 2H$_2$ which is then fed to a catalytic converter to produce methane. A chemical-equilibrium gas mixture containing CH$_4$, CO, H$_2$, and H$_2$O exits the reactor at 600 K, 600 kPa. Determine the mole fraction of methane in the mixture.

15.87 Dry air is heated from 25°C to 4000 K in a 100-kPa constant-pressure process. List the possible reactions that may take place and determine the equilibrium composition. Find the required heat transfer.

15.88 Methane is burned with theoretical oxygen in a steady-state process, and the products exit the combustion chamber at 3200 K, 700 kPa. Calculate the equilibrium composition at this state, assuming that only CO$_2$, CO, H$_2$O, H$_2$, O$_2$, and OH are present.

ENGLISH UNIT PROBLEMS

15.89E Carbon dioxide at 2200 lbf/in.^{2} is injected into the top of a 3-mi-deep well in connection with an enhanced oil recovery process. The fluid column standing in the well is at a uniform temperature of 100 F. What is the pressure at the bottom of the well assuming ideal-gas behavior?

15.90E Calculate the equilibrium constant for the reaction O$_2 \rightleftharpoons$ 2O at temperatures of 537 R and 10 800 R.

15.91E Pure oxygen is heated from 77 F to 5300 F in a steady-state process at a constant pressure of 30 lbf/in.^{2}. Find the exit composition and the heat transfer.

15.92E Air (assumed to be 79% nitrogen and 21% oxygen) is heated in a steady-state process at a constant pressure of 14.7 lbf/in.^{2}, and some NO is formed. At what temperature will the mole fraction of NO be 0.001?

15.93E The combustion products from burning pentane, C$_5$H$_{12}$, with pure oxygen in a stoichiometric ratio exit at 4400 R. Consider the dissociation of only CO$_2$ and find the equilibrium mole fraction of CO.

15.94E Pure oxygen is heated from 77 F, 14.7 lbf/in.^{2}, to 5300 F in a constant-volume container. Find the final pressure, composition, and the heat transfer.

15.95E The equilibrium reaction with methane as CH$_4 \rightleftharpoons$ C + 2H$_2$ has ln $K = -0.3362$ at 1440 R and ln $K = -4.607$ at 1080 R. By noting the relation of K to temperature, show how you would interpolate ln K in (1/T) to find K at 1260 R and compare that to a linear interpolation.

15.96E A gas mixture of 1 pound mol carbon monoxide, 1 pound mol nitrogen, and 1 pound mol oxygen at 77 F, 20 lbf/in.^{2}, is heated in a constant-pressure flow process. The exit mixture can be assumed to be in chemical equilibrium with CO_2, CO, O_2, and N_2 present. The mole fraction of CO_2 at this point is 0.176. Calculate the heat transfer for the process.

15.97E Use the information in Problem 15.95E to estimate the enthalpy of reaction, ΔH^0, at 1260 R using the van't Hoff equation (see Problem 15.48) with finite differences for the derivatives.

15.98E Acetylene gas at 77 F is burned with 140% theoretical air, which enters the burner at 77 F, 14.7 lbf/in.^{2}, 80% relative humidity. The combustion products form a mixture of CO_2, H_2O, N_2, O_2, and NO in chemical equilibrium at 3500 F, 14.7 lbf/in.^{2}. This mixture is then cooled to 1340 F very rapidly, so that the composition does not change. Determine the mole fraction of NO in the products and the heat transfer for the overall process.

15.99E An important step in the manufacture of chemical fertilizer is the production of ammonia, according to the reaction $N_2 + 3H_2 \rightleftharpoons 2NH_3$

a. Calculate the equilibrium constant for this reaction at 300 F.

b. For an initial composition of 25% nitrogen, 75% hydrogen, on a mole basis, calculate the equilibrium composition at 300 F, 750 lbf/in.^{2}.

15.100E Ethane is burned with 150% theoretical air in a gas-turbine combustor. The products exiting consist of a mixture of CO_2, H_2O, O_2, N_2, and NO in chemical equilibrium at 2800 F, 150 lbf/in.^{2}. Determine the mole fraction of NO in the products. Is it reasonable to ignore CO in the products?

15.101E One-pound mole of air (assumed to be 78% nitrogen, 21% oxygen, and 1% argon) at room temperature is heated to 7200 R, 30 lbf/in.^{2}. Find the equilibrium composition at this state, assuming that only N_2, O_2, NO, O, and Ar are present.

15.102E One-pound mole of water vapor at 14.7 lbf/in.^{2}, 720 R, is heated to 5400 R in a constant-pressure flow process. Determine the final composition, assuming that H_2O, H_2, H, O_2, and OH are present at equilibrium.

15.103E Acetylene gas and x times theoretical air ($x > 1$) at room temperature and 75 lbf/in.^{2} are burned at constant pressure in an adiabatic flow process. The flame temperature is 4600 R, and the combustion products are assumed to consist of N_2, O_2, CO_2, H_2O, CO, and NO. Determine the value of x.

15.104E Methane is burned with theoretical oxygen in a steady-state process, and the products exit the combustion chamber at 5300 F, 100 lbf/in.^{2}. Calculate the equilibrium composition at this state, assuming that only CO_2, CO, H_2O, H_2, O_2, and OH are present.

15.105E In a test of a gas-turbine combustor, saturated-liquid methane at 210 R is to be burned with excess air to hold the adiabatic flame temperature to 2880 R. It is assumed that the products consist of a mixture of CO_2, H_2O, N_2, O_2, and NO in chemical equilibrium. Determine the percent excess air used in the combustion, and the percentage of NO in the products.

15.106E Dry air is heated from 77 F to 7200 R in a 14.7 lbf/in.^{2} constant-pressure process. List the possible reactions that may take place and determine the equilibrium composition. Find the required heat transfer.

COMPUTER, DESIGN, AND OPEN-ENDED PROBLEMS

15.107 Write a program to solve the general case of Problem 15.60, in which the relative amount of steam input and the reactor temperature and pressure are program input variables and use constant specific heats.

15.108 Write a program to solve the following problem. One kmol of carbon at 25°C is burned with b kmol of oxygen in a constant-pressure adiabatic process. The products consist of an equilibrium mixture of CO_2, CO, and O_2. We wish to determine the flame temperature for various combinations of b and the pressure P, assuming constant specific heat for the components from Table A.5.

15.109 Study the chemical reactions that take place when CFC-type refrigerants are released into the atmosphere. The chlorine may create compounds as HCl and $ClONO_2$ that react with the ozone O_3.

15.110 Examine the chemical equilibrium that takes place in an engine where CO and various nitrogen oxygen compounds summarized as NO_x may be formed. Study the processes for a range of air–fuel ratios and temperatures for typical fuels. Are there important reactions not listed in the book?

15.111 A number of products may be produced from the conversion of organic waste that can be used as fuel (see Problem 15.60). Study the subject and make a list of the major products that are formed and the conditions at which they are formed in desirable concentrations.

15.112 The hydrides as explained in Problem 15.66 can store large amounts of hydrogen. The penalty for the storage is that energy must be supplied when the hydrogen is released. Investigate the literature for quantitative information about the quantities and energy involved in such a hydrogen storage.

15.113 The hydrides explained in Problem 15.66 can be used in a chemical heat pump. The energy involved in the chemical reaction can be added and removed at different temperatures. For some hydrides these temperatures are low enough to make them feasible for heat pumps for heat upgrade, refrigerators, and air conditioners. Investigate the literature for such applications and give some typical values for these systems.

15.114 Power plants and engines have high peak temperatures in the combustion products where NO is produced. The equilibrium NO level at the high temperature is frozen at that level during the rapid drop in temperature with the expansion. The final exhaust therefore contains NO at a level much higher than the equilibrium value at the exhaust temperature. Study the NO level at equilibrium when natural gas, CH_4, is burned adiabatically with air (at T_0) in various ratios.

15.115 Excess air or steam addition is often used to lower the peak temperature in combustion to limit formation of pollutants like NO. Study the steam addition to the combustion of natural gas as in the Cheng cycle (see Problem 12.176), assuming the steam is added before the combustion. How does this affect the peak temperature and the NO concentration?

CONTENTS OF APPENDIX

APPENDIX A
SI UNITS: SINGLE-STATE PROPERTIES

TABLE A.1
Conversion Factors

Area (A)

$1 \text{ mm}^2 = 1.0 \times 10^{-6} \text{ m}^2$ $1 \text{ ft}^2 = 144 \text{ in.}^2$

$1 \text{ cm}^2 = 1.0 \times 10^{-4} \text{ m}^2 = 0.1550 \text{ in.}^2$ $1 \text{ in.}^2 = 6.4516 \text{ cm}^2 = 6.4516 \times 10^{-4} \text{ m}^2$

$1 \text{ m}^2 = 10.7639 \text{ ft}^2$ $1 \text{ ft}^2 = 0.092\ 903 \text{ m}^2$

Conductivity (k)

$1 \text{ W/m-K} = 1 \text{ J/s-m-K}$

$= 0.577\ 789 \text{ Btu/h-ft-}°\text{R}$ $1 \text{ Btu/h-ft-R} = 1.730\ 735 \text{ W/m-K}$

Density (ρ)

$1 \text{ kg/m}^3 = 0.06242797 \text{ lbm/ft}^3$ $1 \text{ lbm/ft}^3 = 16.018\ 46 \text{ kg/m}^3$

$1 \text{ g/cm}^3 = 1000 \text{ kg/m}^3$

$1 \text{ g/cm}^3 = 1 \text{ kg/L}$

Energy (E, U)

$1 \text{ J} = 1 \text{ N-m} = 1 \text{ kg-m}^2/\text{s}^2$

$1 \text{ J} = 0.737\ 562 \text{ lbf-ft}$ $1 \text{ lbf-ft} = 1.355\ 818 \text{ J}$

$1 \text{ cal (Int.)} = 4.1868 \text{ J}$ $= 1.28507 \times 10^{-3} \text{ Btu}$

$1 \text{ Btu (Int.)} = 1.055\ 056 \text{ kJ}$

$1 \text{ erg} = 1.0 \times 10^{-7} \text{ J}$ $= 778.1693 \text{ lbf-ft}$

$1 \text{ eV} = 1.602\ 177\ 33 \times 10^{-19} \text{ J}$

Force (F)

$1 \text{ N} = 0.224809 \text{ lbf}$ $1 \text{ lbf} = 4.448\ 222 \text{ N}$

$1 \text{ kp} = 9.80665 \text{ N} \ (1 \text{ kgf})$

Gravitation

$g = 9.80665 \text{ m/s}^2$ $g = 32.17405 \text{ ft/s}^2$

Heat capacity (C_p, C_v, C), specific entropy (s)

$1 \text{ kJ/kg-K} = 0.238\ 846 \text{ Btu/lbm-}°\text{R}$ $1 \text{ Btu/lbm-}°\text{R} = 4.1868 \text{ kJ/kg-K}$

Heat flux (per unit area)

$1 \text{ W/m}^2 = 0.316\ 998 \text{ Btu/h-ft}^2$ $1 \text{ Btu/h-ft}^2 = 3.15459 \text{ W/m}^2$

TABLE A.1 (*continued*)
Conversion Factors

Heat-transfer coefficient (h)

1 W/m^2-K = 0.176 11 Btu/h-ft^2-°R   1 Btu/h-ft^2-°R = 5.67826 W/m^2-K

Length (L)

1 mm = 0.001 m = 0.1 cm	1 ft = 12 in.
1 cm = 0.01 m = 10 mm = 0.3970 in.	1 in. = 2.54 cm = 0.0254 m
1 m = 3.28084 ft = 39.370 in.	1 ft = 0.3048 m
1 km = 0.621 371 mi	1 mi = 1.609344 km
1 mi = 1609.3 m (US statute)	1 yd = 0.9144 m

Mass (m)

1 kg = 2.204 623 lbm	1 lbm = 0.453 592 kg
1 tonne = 1000 kg	1 slug = 14.5939 kg
1 grain = 6.47989 × 10^{-5} kg	1 ton = 2000 lbm

Moment (torque, T)

1 N-m = 0.737 562 lbf-ft   1 lbf-ft = 1.355 818 N-m

Momentum (mV)

1 kg-m/s = 7.232 94 lbm-ft/s	1 lbm-ft/s = 0.138 256 kg-m/s
= 0.224809 lbf-s	

Power ($\dot{Q}, \dot{W}$)

1 W	= 1 J/s = 1 N-m/s	1 lbf-ft/s	= 1.355 818 W
	= 0.737 562 lbf-ft/s		= 4.626 24 Btu/h
1 kW	= 3412.14 Btu/h	1 Btu/s	= 1.055 056 kW
1 hp (metric)	= 0.735 499 kW	1 hp (UK)	= 0.7457 kW
			= 550 lbf-ft/s
			= 2544.43 Btu/h

1 ton of		1 ton of	
refrigeration	= 3.516 85 kW	refrigeration	= 12 000 Btu/h

Pressure (P)

1 Pa	= 1 N/m^2 = 1 kg/m-s^2	1 lbf/in.^{2}	= 6.894 757 kPa
1 bar	= 1.0 × 10^5 Pa = 100 kPa		
1 atm	= 101.325 kPa	1 atm	= 14.695 94 lbf/in.^{2}
	= 1.01325 bar		= 29.921 in. Hg [32°F]
	= 760 mm Hg [0°C]		= 33.899 5 ft H$_2$O [4°C]
	= 10.332 56 m H$_2$O [4°C]		
1 torr	= 1 mm Hg [0°C]		
1 mm Hg [0°C] = 0.133 322 kPa		1 in. Hg [0°C] = 0.49115 lbf/in.^{2}	
1 m H$_2$O [4°C] = 9.806 38 kPa		1 in. H$_2$O [4°C] = 0.036126 lbf/in.^{2}	

Specific energy (e, u)

1 kJ/kg = 0.42992 Btu/lbm	1 Btu/lbm = 2.326 kJ/kg
= 334.55 lbf-ft/lbm	1 lbf-ft/lbm = 2.98907 × 10^{-3} kJ/kg
	= 1.28507 × 10^{-3} Btu/lbm

TABLE A.1 (*continued*)
Conversion Factors

Specific kinetic energy ($\frac{1}{2}\mathbf{V}^2$)
$1\ m^2/s^2 = 0.001\ kJ/kg$ $1\ ft^2/s^2 = 3.9941 \times 10^{-5}\ Btu/lbm$
$1\ kJ/kg = 1000\ m^2/s^2$ $1\ Btu/lbm = 25037\ ft^2/s^2$

Specific potential energy (*Zg*)
$1\ m\text{-}g_{std} = 9.80665 \times 10^{-3}\ kJ/kg$ $1\ ft\text{-}g_{std} = 1.0\ lbf\text{-}ft/lbm$
$\quad\quad\quad = 4.21607 \times 10^{-3}\ Btu/lbm$ $\quad\quad = 0.001285\ Btu/lbm$
$\quad\quad = 0.002989\ kJ/kg$

Specific volume (*v*)
$1\ cm^3/g = 0.001\ m^3/kg$
$1\ cm^3/g = 1\ L/kg$
$1\ m^3/kg = 16.018\ 46\ ft^3/lbm$ $1\ ft^3/lbm = 0.062\ 428\ m^3/kg$

Temperature (*T*)
$1\ K = 1°C = 1.8\ R = 1.8\ F$ $1\ R = (5/9)\ K$
$TC = TK - 273.15$ $TF = TR - 459.67$
$\quad = (TF - 32)/1.8$ $\quad = 1.8\ TC + 32$
$TK = TR/1.8$ $TR = 1.8\ TK$

Universal Gas Constant
$\bar{R} = N_0 k = 8.31451\ kJ/kmol\text{-}K$ $\bar{R} = 1.98589\ Btu/lbmol\text{-}R$
$\quad = 1.98589\ kcal/kmol\text{-}K$ $\quad = 1545.36\ lbf\text{-}ft/lbmol\text{-}R$
$\quad = 82.0578\ atm\text{-}L/kmol\text{-}K$ $\quad = 0.73024\ atm\text{-}ft^3/lbmol\text{-}R$
$\quad = 10.7317\ (lbf/in.^2)\text{-}ft^3/lbmol\text{-}R$

Velcoity (V)
$1\ m/s = 3.6\ km/h$ $1\ ft/s = 0.681818\ mi/h$
$\quad = 3.28084\ ft/s$ $\quad = 0.3048\ m/s$
$\quad = 2.23694\ mi/h$ $\quad = 1.09728\ km/h$
$1\ km/h = 0.27778\ m/s$ $1\ mi/h = 1.46667\ ft/s$
$\quad = 0.91134\ ft/s$ $\quad = 0.44704\ m/s$
$\quad = 0.62137\ mi/h$ $\quad = 1.609344\ km/h$

Volume (*V*)
$1\ m^3 = 35.3147\ ft^3$ $1\ ft^3 = 2.831\ 685 \times 10^{-2}\ m^3$
$1\ L = 1\ dm^3 = 0.001\ m^3$ $1\ in.^3 = 1.6387 \times 10^{-5}\ m^3$
$1\ Gal\ (US) = 3.785\ 412\ L$ $1\ Gal\ (UK) = 4.546\ 090\ L$
$\quad = 3.785\ 412 \times 10^{-3}\ m^3$ $1\ Gal\ (US) = 231.00\ in.^3$

Table A.2
Critical Constants

Substance	Formula	Molec. Mass	Temp. (K)	Press. (MPa)	Vol. (m³/kg)
Ammonia	NH_3	17.031	405.5	11.35	0.00426
Argon	Ar	39.948	150.8	4.87	0.00188
Bromine	Br_2	159.808	588	10.30	0.000796
Carbon dioxide	CO_2	44.01	304.1	7.38	0.00212
Carbon monoxide	CO	28.01	132.9	3.50	0.00333
Chlorine	Cl_2	70.906	416.9	7.98	0.00175
Fluorine	F_2	37.997	144.3	5.22	0.00174
Helium	He	4.003	5.19	0.227	0.0143
Hydrogen (normal)	H_2	2.016	33.2	1.30	0.0323
Krypton	Kr	83.80	209.4	5.50	0.00109
Neon	Ne	20.183	44.4	2.76	0.00206
Nitric oxide	NO	30.006	180	6.48	0.00192
Nitrogen	N_2	28.013	126.2	3.39	0.0032
Nitrogen dioxide	NO_2	46.006	431	10.1	0.00365
Nitrous oxide	N_2O	44.013	309.6	7.24	0.00221
Oxygen	O_2	31.999	154.6	5.04	0.00229
Sulfur dioxide	SO_2	64.063	430.8	7.88	0.00191
Water	H_2O	18.015	647.3	22.12	0.00317
Xenon	Xe	131.30	289.7	5.84	0.000902
Acetylene	C_2H_2	26.038	308.3	6.14	0.00433
Benzene	C_6H_6	78.114	562.2	4.89	0.00332
n-Butane	C_4H_{10}	58.124	425.2	3.80	0.00439
Chlorodifluoroethane (142b)	CH_3CClF_2	100.495	410.3	4.25	0.00230
Chlorodifluoromethane (22)	$CHClF_2$	86.469	369.3	4.97	0.00191
Dichlorofluoroethane (141)	CH_3CCl_2F	116.95	481.5	4.54	0.00215
Dichlorotrifluoroethane (123)	$CHCl_2CF_3$	152.93	456.9	3.66	0.00182
Difluoroethane (152a)	CHF_2CH_3	66.05	386.4	4.52	0.00272
Difluoromethane (32)	CF_2H_2	52.024	351.3	5.78	0.00236
Ethane	C_2H_6	30.070	305.4	4.88	0.00493
Ethyl alcohol	C_2H_5OH	46.069	513.9	6.14	0.00363
Ethylene	C_2H_4	28.054	282.4	5.04	0.00465
n-Heptane	C_7H_{16}	100.205	540.3	2.74	0.00431
n-Hexane	C_6H_{14}	86.178	507.5	3.01	0.00429
Methane	CH_4	16.043	190.4	4.60	0.00615
Methyl alcohol	CH_3OH	32.042	512.6	8.09	0.00368
n-Octane	C_8H_{18}	114.232	568.8	2.49	0.00431
Pentafluoroethane (125)	CHF_2CF_3	120.022	339.2	3.62	0.00176
n-Pentane	C_5H_{12}	72.151	469.7	3.37	0.00421
Propane	C_3H_8	44.094	369.8	4.25	0.00454
Propene	C_3H_6	42.081	364.9	4.60	0.00430
Tetrafluoroethane (134a)	CF_3CH_2F	102.03	374.2	4.06	0.00197

TABLE A.3
Properties of Selected Solids at 25°C

Substance	ρ (kg/m^3)	C_p (kJ/kg-K)
Asphalt	2120	0.92
Brick, common	1800	0.84
Carbon, diamond	3250	0.51
Carbon, graphite	2000–2500	0.61
Coal	1200–1500	1.26
Concrete	2200	0.88
Glass, plate	2500	0.80
Glass, wool	200	0.66
Granite	2750	0.89
Ice (0°C)	917	2.04
Paper	700	1.2
Plexiglass	1180	1.44
Polystyrene	920	2.3
Polyvinyl chloride	1380	0.96
Rubber, soft	1100	1.67
Sand, dry	1500	0.8
Salt, rock	2100–2500	0.92
Silicon	2330	0.70
Snow, firm	560	2.1
Wood, hard (oak)	720	1.26
Wood, soft (pine)	510	1.38
Wool	100	1.72
Metals		
Aluminum	2700	0.90
Brass, 60-40	8400	0.38
Copper, commercial	8300	0.42
Gold	19300	0.13
Iron, cast	7272	0.42
Iron, 304 St Steel	7820	0.46
Lead	11340	0.13
Magnesium, 2% Mn	1778	1.00
Nickel, 10% Cr	8666	0.44
Silver, 99.9% Ag	10524	0.24
Sodium	971	1.21
Tin	7304	0.22
Tungsten	19300	0.13
Zinc	7144	0.39

TABLE A.4
Properties of Some Liquids at 25°C*

Substance	ρ (kg/m^3)	C_p (kJ/kg-K)
Ammonia	604	4.84
Benzene	879	1.72
Butane	556	2.47
CCl_4	1584	0.83
CO_2	680	2.9
Ethanol	783	2.46
Gasoline	750	2.08
Glycerine	1260	2.42
Kerosene	815	2.0
Methanol	787	2.55
n-octane	692	2.23
Oil engine	885	1.9
Oil light	910	1.8
Propane	510	2.54
R-12	1310	0.97
R-22	1190	1.26
R-32	961	1.94
R-125	1191	1.41
R-134a	1206	1.43
Water	997	4.18
Liquid metals		
Bismuth, Bi	10040	0.14
Lead, Pb	10660	0.16
Mercury, Hg	13580	0.14
NaK (56/44)	887	1.13
Potassium, K	828	0.81
Sodium, Na	929	1.38
Tin, Sn	6950	0.24
Zinc, Zn	6570	0.50

*Or T_{melt} if higher.

TABLE A.5
Properties of Various Ideal Gases at 25°C, 100 kPa (SI Units)*

Gas	Chemical Formula	Molecular Mass	R (kJ/kg-K)	ρ (kg/m³)	C_{p0} (kJ/kg-K)	C_{v0} (kJ/kg-K)	$k = \dfrac{C_p}{C_v}$
Steam	H_2O	18.015	0.4615	0.0231	1.872	1.410	1.327
Acetylene	C_2H_2	26.038	0.3193	1.05	1.699	1.380	1.231
Air	—	28.97	0.287	1.169	1.004	0.717	1.400
Ammonia	NH_3	17.031	0.4882	0.694	2.130	1.642	1.297
Argon	Ar	39.948	0.2081	1.613	0.520	0.312	1.667
Butane	C_4H_{10}	58.124	0.1430	2.407	1.716	1.573	1.091
Carbon dioxide	CO_2	44.01	0.1889	1.775	0.842	0.653	1.289
Carbon monoxide	CO	28.01	0.2968	1.13	1.041	0.744	1.399
Ethane	C_2H_6	30.07	0.2765	1.222	1.766	1.490	1.186
Ethanol	C_2H_5OH	46.069	0.1805	1.883	1.427	1.246	1.145
Ethylene	C_2H_4	29.054	0.2964	1.138	1.548	1.252	1.237
Helium	He	4.003	2.0771	0.1615	5.193	3.116	1.667
Hydrogen	H_2	2.016	4.1243	0.0813	14.209	10.085	1.409
Methane	CH_4	16.043	0.5183	0.648	2.254	1.736	1.299
Methanol	CH_3OH	32.042	0.2595	1.31	1.405	1.146	1.227
Neon	Ne	20.183	0.4120	0.814	1.03	0.618	1.667
Nitric oxide	NO	30.006	0.2771	1.21	0.993	0.716	1.387
Nitrogen	N_2	28.013	0.2968	1.13	1.042	0.745	1.400
Nitrous oxide	N_2O	44.013	0.1889	1.775	0.879	0.690	1.274
n-octane	C_8H_{18}	114.23	0.07279	0.092	1.711	1.638	1.044
Oxygen	O_2	31.999	0.2598	1.292	0.922	0.662	1.393
Propane	C_3H_8	44.094	0.1886	1.808	1.679	1.490	1.126
R-12	CCl_2F_2	120.914	0.06876	4.98	0.616	0.547	1.126
R-22	$CHClF_2$	86.469	0.09616	3.54	0.658	0.562	1.171
R-32	CF_2H_2	52.024	0.1598	2.125	0.822	0.662	1.242
R-125	CHF_2CF_3	120.022	0.06927	4.918	0.791	0.721	1.097
R-134a	CF_3CH_2F	102.03	0.08149	4.20	0.852	0.771	1.106
Sulfur dioxide	SO_2	64.059	0.1298	2.618	0.624	0.494	1.263
Sulfur trioxide	SO_3	80.053	0.10386	3.272	0.635	0.531	1.196

*Or saturation pressure if it is less than 100 kPa.

TABLE A.6

Constant-Pressure Specific Heats of Various Ideal Gases[†]

	$C_{p0} = C_0 + C_1\theta + C_2\theta^2 + C_3\theta^3$		(kJ/kg K)	$\theta = T(\text{Kelvin})/1000$	
Gas	**Formula**	C_0	C_1	C_2	C_3
Steam	H_2O	1.79	0.107	0.586	−0.20
Acetylene	C_2H_2	1.03	2.91	−1.92	0.54
Air	—	1.05	−0.365	0.85	−0.39
Ammonia	NH_3	1.60	1.4	1.0	−0.7
Argon	Ar	0.52	0	0	0
Butane	C_4H_{10}	0.163	5.70	−1.906	−0.049
Carbon dioxide	CO_2	0.45	1.67	−1.27	0.39
Carbon monoxide	CO	1.10	−0.46	1.0	−0.454
Ethane	C_2H_6	0.18	5.92	−2.31	0.29
Ethanol	C_2H_5OH	0.2	−4.65	−1.82	0.03
Ethylene	C_2H_4	1.36	5.58	−3.0	0.63
Helium	He	5.193	0	0	0
Hydrogen	H_2	13.46	4.6	−6.85	3.79
Methane	CH_4	1.2	3.25	0.75	−0.71
Methanol	CH_3OH	0.66	2.21	0.81	−0.89
Neon	Ne	1.03	0	0	0
Nitric oxide	NO	0.98	−0.031	0.325	−0.14
Nitrogen	N_2	1.11	−0.48	0.96	−0.42
Nitrous oxide	N_2O	0.49	1.65	−1.31	0.42
n-octane	C_8H_{18}	−0.053	6.75	−3.67	0.775
Oxygen	O_2	0.88	−0.0001	0.54	−0.33
Propane	C_3H_8	−0.096	6.95	−3.6	0.73
R-12*	CCl_2F_2	0.26	1.47	−1.25	0.36
R-22*	$CHClF_2$	0.2	1.87	−1.35	0.35
R-32*	CF_2H_2	0.227	2.27	−0.93	0.041
R-125*	CHF_2CF_3	0.305	1.68	−0.284	0
R-134a*	CF_3CH_2F	0.165	2.81	−2.23	1.11
Sulfur dioxide	SO_2	0.37	1.05	−0.77	0.21
Sulfur trioxide	SO_3	0.24	1.7	−1.5	0.46

[†]Approximate forms valid from 250 K to 1200 K.

*Formula limited to maximum 500 K.

TABLE A7.1
Ideal-Gas Properties of Air, Standard Entropy at 0.1-MPa (1-bar) Pressure

T (K)	u (kJ/kg)	h (kJ/kg)	s_T^0 (kJ/kg-K)	T (K)	u (kJ/kg)	h (kJ/kg)	s_T^0 (kJ/kg-K)
200	142.77	200.17	6.46260	1100	845.45	1161.18	8.24449
220	157.07	220.22	6.55812	1150	889.21	1219.30	8.29616
240	171.38	240.27	6.64535	1200	933.37	1277.81	8.34596
260	185.70	260.32	6.72562	1250	977.89	1336.68	8.39402
280	200.02	280.39	6.79998	1300	1022.75	1395.89	8.44046
290	207.19	290.43	6.83521	1350	1067.94	1455.43	8.48539
298.15	213.04	298.62	6.86305	1400	1113.43	1515.27	8.52891
300	214.36	300.47	6.86926	1450	1159.20	1575.40	8.57111
320	228.73	320.58	6.93413	1500	1205.25	1635.80	8.61208
340	243.11	340.70	6.99515	1550	1251.55	1696.45	8.65185
360	257.53	360.86	7.05276	1600	1298.08	1757.33	8.69051
380	271.99	381.06	7.10735	1650	1344.83	1818.44	8.72811
400	286.49	401.30	7.15926	1700	1391.80	1879.76	8.76472
420	301.04	421.59	7.20875	1750	1438.97	1941.28	8.80039
440	315.64	441.93	7.25607	1800	1486.33	2002.99	8.83516
460	330.31	462.34	7.30142	1850	1533.87	2064.88	8.86908
480	345.04	482.81	7.34499	1900	1581.59	2126.95	8.90219
500	359.84	503.36	7.38692	1950	1629.47	2189.19	8.93452
520	374.73	523.98	7.42736	2000	1677.52	2251.58	8.96611
540	389.69	544.69	7.46642	2050	1725.71	2314.13	8.99699
560	404.74	565.47	7.50422	2100	1774.06	2376.82	9.02721
580	419.87	586.35	7.54084	2150	1822.54	2439.66	9.05678
600	435.10	607.32	7.57638	2200	1871.16	2502.63	9.08573
620	450.42	628.38	7.61090	2250	1919.91	2565.73	9.11409
640	465.83	649.53	7.64448	2300	1968.79	2628.96	9.14189
660	481.34	670.78	7.67717	2350	2017.79	2692.31	9.16913
680	496.94	692.12	7.70903	2400	2066.91	2755.78	9.19586
700	512.64	713.56	7.74010	2450	2116.14	2819.37	9.22208
720	528.44	735.10	7.77044	2500	2165.48	2883.06	9.24781
740	544.33	756.73	7.80008	2550	2214.93	2946.86	9.27308
760	560.32	778.46	7.82905	2600	2264.48	3010.76	9.29790
780	576.40	800.28	7.85740	2650	2314.13	3074.77	9.32228
800	592.58	822.20	7.88514	2700	2363.88	3138.87	9.34625
850	633.42	877.40	7.95207	2750	2413.73	3203.06	9.36980
900	674.82	933.15	8.01581	2800	2463.66	3267.35	9.39297
950	716.76	989.44	8.07667	2850	2513.69	3331.73	9.41576
1000	759.19	1046.22	8.13493	2900	2563.80	3396.19	9.43818
1050	802.10	1103.48	8.19081	2950	2613.99	3460.73	9.46025
1100	845.45	1161.18	8.24449	3000	2664.27	3525.36	9.48198

TABLE A7.2
The Isentropic Relative Pressure and Relative Volume Functions

T[K]	P_r	v_r	T[K]	P_r	v_r	T[K]	P_r	v_r
200	0.2703	493.47	700	23.160	20.155	1900	1327.5	0.95445
220	0.3770	389.15	720	25.742	18.652	1950	1485.8	0.87521
240	0.5109	313.27	740	28.542	17.289	2000	1658.6	0.80410
260	0.6757	256.58	760	31.573	16.052	2050	1847.1	0.74012
280	0.8756	213.26	780	34.851	14.925	2100	2052.1	0.68242
290	0.9899	195.36	800	38.388	13.897	2150	2274.8	0.63027
298.15	1.0907	182.29	850	48.468	11.695	2200	2516.2	0.58305
300	1.1146	179.49	900	60.520	9.9169	2250	2777.5	0.54020
320	1.3972	152.73	950	74.815	8.4677	2300	3059.9	0.50124
340	1.7281	131.20	1000	91.651	7.2760	2350	3364.6	0.46576
360	2.1123	113.65	1050	111.35	6.2885	2400	3693.0	0.43338
380	2.5548	99.188	1100	134.25	5.4641	2450	4046.2	0.40378
400	3.0612	87.137	1150	160.73	4.7714	2500	4425.8	0.37669
420	3.6373	77.003	1200	191.17	4.1859	2550	4833.0	0.35185
440	4.2892	68.409	1250	226.02	3.6880	2600	5269.5	0.32903
460	5.0233	61.066	1300	265.72	3.2626	2650	5736.7	0.30805
480	5.8466	54.748	1350	310.74	2.8971	2700	6236.2	0.28872
500	6.7663	49.278	1400	361.62	2.5817	2750	6769.7	0.27089
520	7.7900	44.514	1450	418.89	2.3083	2800	7338.7	0.25443
540	8.9257	40.344	1500	483.16	2.0703	2850	7945.1	0.23921
560	10.182	36.676	1550	554.96	1.8625	2900	8590.7	0.22511
580	11.568	33.436	1600	634.97	1.6804	2950	9277.2	0.21205
600	13.092	30.561	1650	723.86	1.52007	3000	10007.	0.19992
620	14.766	28.001	1700	822.33	1.37858			
640	16.598	25.713	1750	931.14	1.25330			
660	18.600	23.662	1800	1051.05	1.14204			
680	20.784	21.818	1850	1182.9	1.04294			
700	23.160	20.155	1900	1327.5	0.95445			

The relative pressure and relative volume are temperature functions calculated with two scaling constants A_1, A_2.

$$P_r = \exp[s_T^0/R - A_1]; \qquad v_r = A_2 T/P_r$$

such that for an isentropic process ($s_1 = s_2$)

$$\frac{P_2}{P_1} = \frac{P_{r2}}{P_{r1}} = \frac{e^{s_{T2}^0/R}}{e^{s_{T1}^0/R}} \approx \left(\frac{T_2}{T_1}\right)^{C_p/R} \quad \text{and} \quad \frac{v_2}{v_1} = \frac{v_{r2}}{v_{r1}} \approx \left(\frac{T_2}{T_1}\right)^{C_v/R}$$

where the near equalities are for the constant heat capacity approximation.

TABLE A.8
Ideal-Gas Properties of Various Substances, Entropies at 0.1-MPa (1-bar) Pressure, Mass Basis

T (K)	NITROGEN, DIATOMIC (N_2) $R = 0.2968$ kJ/kg-K $M = 28.013$			OXYGEN, DIATOMIC (O_2) $R = 0.2598$ kJ/kg-K $M = 31.999$		
	u (kJ/kg)	h (kJ/kg)	s_T^0 (kJ/kg-K)	u (kJ/kg)	h (kJ/kg)	s_T^0 (kJ/kg-K)
200	148.39	207.75	6.4250	129.84	181.81	6.0466
250	185.50	259.70	6.6568	162.41	227.37	6.2499
300	222.63	311.67	6.8463	195.20	273.15	6.4168
350	259.80	363.68	7.0067	228.37	319.31	6.5590
400	297.09	415.81	7.1459	262.10	366.03	6.6838
450	334.57	468.13	7.2692	296.52	413.45	6.7954
500	372.35	520.75	7.3800	331.72	461.63	6.8969
550	410.52	573.76	7.4811	367.70	510.61	6.9903
600	449.16	627.24	7.5741	404.46	560.36	7.0768
650	488.34	681.26	7.6606	441.97	610.86	7.1577
700	528.09	735.86	7.7415	480.18	662.06	7.2336
750	568.45	791.05	7.8176	519.02	713.90	7.3051
800	609.41	846.85	7.8897	558.46	766.33	7.3728
850	650.98	903.26	7.9581	598.44	819.30	7.4370
900	693.13	960.25	8.0232	638.90	872.75	7.4981
950	735.85	1017.81	8.0855	679.80	926.65	7.5564
1000	779.11	1075.91	8.1451	721.11	980.95	7.6121
1100	867.14	1193.62	8.2572	804.80	1090.62	7.7166
1200	957.00	1313.16	8.3612	889.72	1201.53	7.8131
1300	1048.46	1434.31	8.4582	975.72	1313.51	7.9027
1400	1141.35	1556.87	8.5490	1062.67	1426.44	7.9864
1500	1235.50	1680.70	8.6345	1150.48	1540.23	8.0649
1600	1330.72	1805.60	8.7151	1239.10	1654.83	8.1389
1700	1426.89	1931.45	8.7914	1328.49	1770.21	8.2088
1800	1523.90	2058.15	8.8638	1418.63	1886.33	8.2752
1900	1621.66	2185.58	8.9327	1509.50	2003.19	8.3384
2000	1720.07	2313.68	8.9984	1601.10	2120.77	8.3987
2100	1819.08	2442.36	9.0612	1693.41	2239.07	8.4564
2200	1918.62	2571.58	9.1213	1786.44	2358.08	8.5117
2300	2018.63	2701.28	9.1789	1880.17	2477.79	8.5650
2400	2119.08	2831.41	9.2343	1974.60	2598.20	8.6162
2500	2219.93	2961.93	9.2876	2069.71	2719.30	8.6656
2600	2321.13	3092.81	9.3389	2165.50	2841.07	8.7134
2700	2422.66	3224.03	9.3884	2261.94	2963.49	8.7596
2800	2524.50	3355.54	9.4363	2359.01	3086.55	8.8044
2900	2626.62	3487.34	9.4825	2546.70	3210.22	8.8478
3000	2729.00	3619.41	9.5273	2554.97	3334.48	8.8899

TABLE A.8 (*continued*)
Ideal-Gas Properties of Various Substances, Entropies at 0.1-MPa (1-bar) Pressure, Mass Basis

T (K)	CARBON DIOXIDE (CO₂) $R = 0.1889$ kJ/kg-K $M = 44.010$			WATER (H₂O) $R = 0.4615$ kJ/kg-K $M = 18.015$		
	u (kJ/kg)	h (kJ/kg)	s_T^0 (kJ/kg-K)	u (kJ/kg)	h (kJ/kg)	s_T^0 (kJ/kg-K)
200	97.49	135.28	4.5439	276.38	368.69	9.7412
250	126.21	173.44	4.7139	345.98	461.36	10.1547
300	157.70	214.38	4.8631	415.87	554.32	10.4936
350	191.78	257.90	4.9972	486.37	647.90	10.7821
400	228.19	303.76	5.1196	557.79	742.40	11.0345
450	266.69	351.70	5.2325	630.40	838.09	11.2600
500	307.06	401.52	5.3375	704.36	935.12	11.4644
550	349.12	453.03	5.4356	779.79	1033.63	11.6522
600	392.72	506.07	5.5279	856.75	1133.67	11.8263
650	437.71	560.51	5.6151	935.31	1235.30	11.9890
700	483.97	616.22	5.6976	1015.49	1338.56	12.1421
750	531.40	673.09	5.7761	1097.35	1443.49	12.2868
800	579.89	731.02	5.8508	1180.90	1550.13	12.4244
850	629.35	789.93	5.9223	1266.19	1658.49	12.5558
900	676.69	849.72	5.9906	1353.23	1768.60	12.6817
950	730.85	910.33	6.0561	1442.03	1880.48	12.8026
1000	782.75	971.67	6.1190	1532.61	1994.13	12.9192
1100	888.55	1096.36	6.2379	1719.05	2226.73	13.1408
1200	996.64	1223.34	6.3483	1912.42	2466.25	13.3492
1300	1106.68	1352.28	6.4515	2112.47	2712.46	13.5462
1400	1218.38	1482.87	6.5483	2318.89	2965.03	13.7334
1500	1331.50	1614.88	6.6394	2531.28	3223.57	13.9117
1600	1445.85	1748.12	6.7254	2749.24	3487.69	14.0822
1700	1561.26	1882.43	6.8068	2972.35	3756.95	14.2454
1800	1677.61	2017.67	6.8841	3200.17	4030.92	14.4020
1900	1794.78	2153.73	6.9577	3432.28	4309.18	14.5524
2000	1912.67	2290.51	7.0278	3668.24	4591.30	14.6971
2100	2031.21	2427.95	7.0949	3908.08	4877.29	14.8366
2200	2150.34	2565.97	7.1591	4151.28	5166.64	14.9712
2300	2270.00	2704.52	7.2206	4397.56	5459.08	15.1012
2400	2390.14	2843.55	7.2798	4646.71	5754.37	15.2269
2500	2510.74	2983.04	7.3368	4898.49	6052.31	15.3485
2600	2631.73	3122.93	7.3917	5152.73	6352.70	15.4663
2700	2753.10	3263.19	7.4446	5409.24	6655.36	15.5805
2800	2874.81	3403.79	7.4957	5667.86	6960.13	15.6914
2900	2996.84	3544.71	7.5452	5928.44	7266.87	15.7990
3000	3119.18	3685.95	7.5931	6190.86	7575.44	15.9036

TABLE A.9

Ideal-Gas Properties of Various Substances (SI Units), Entropies at 0.1-MPa (1-bar) Pressure,
Mole Basis

	NITROGEN, DIATOMIC (N_2) $\bar{h}^0_{f,298} = 0$ kJ/kmol $M = 28.013$		NITROGEN, MONATOMIC (N) $\bar{h}^0_{f,298} = 472\ 680$ kJ/kmol $M = 14.007$	
T K	$(\bar{h} - \bar{h}^0_{298})$ kJ/kmol	$\bar{s}^0_T$ kJ/kmol K	$(\bar{h} - \bar{h}^0_{298})$ kJ/kmol	$\bar{s}^0_T$ kJ/kmol
0	−8670	0	−6197	0
100	−5768	159.812	−4119	130.593
200	−2857	179.985	−2040	145.001
298	0	191.609	0	153.300
300	54	191.789	38	153.429
400	2971	200.181	2117	159.409
500	5911	206.740	4196	164.047
600	8894	212.177	6274	167.837
700	11937	216.865	8353	171.041
800	15046	221.016	10431	173.816
900	18223	224.757	12510	176.265
1000	21463	228.171	14589	178.455
1100	24760	231.314	16667	180.436
1200	28109	234.227	18746	182.244
1300	31503	236.943	20825	183.908
1400	34936	239.487	22903	185.448
1500	38405	241.881	24982	186.883
1600	41904	244.139	27060	188.224
1700	45430	246.276	29139	189.484
1800	48979	248.304	31218	190.672
1900	52549	250.234	33296	191.796
2000	56137	252.075	35375	192.863
2200	63362	255.518	39534	194.845
2400	70640	258.684	43695	196.655
2600	77963	261.615	47860	198.322
2800	85323	264.342	52033	199.868
3000	92715	266.892	56218	201.311
3200	100134	269.286	60420	202.667
3400	107577	271.542	64646	203.948
3600	115042	273.675	68902	205.164
3800	122526	275.698	73194	206.325
4000	130027	277.622	77532	207.437
4400	145078	281.209	86367	209.542
4800	160188	284.495	95457	211.519
5200	175352	287.530	104843	213.397
5600	190572	290.349	114550	215.195
6000	205848	292.984	124590	216.926

TABLE A.9 (*continued*)
Ideal-Gas Properties of Various Substances (SI Units), Entropies at 0.1-MPa (1-bar) Pressure, Mole Basis

	OXYGEN, DIATOMIC (O_2) $\bar{h}^0_{f,298} = 0$ kJ/kmol $M = 31.999$		OXYGEN, MONATOMIC (O) $\bar{h}^0_{f,298} = 249\ 170$ kJ/kmol $M = 16.00$	
T K	$(\bar{h} - \bar{h}^0_{298})$ kJ/kmol	$\bar{s}^0_T$ kJ/kmol K	$(\bar{h} - \bar{h}^0_{298})$ kJ/kmol	$\bar{s}^0_T$ kJ/kmol K
0	−8683	0	−6725	0
100	−5777	173.308	−4518	135.947
200	−2868	193.483	−2186	152.153
298	0	205.148	0	161.059
300	54	205.329	41	161.194
400	3027	213.873	2207	167.431
500	6086	220.693	4343	172.198
600	9245	226.450	6462	176.060
700	12499	231.465	8570	179.310
800	15836	235.920	10671	182.116
900	19241	239.931	12767	184.585
1000	22703	243.579	14860	186.790
1100	26212	246.923	16950	188.783
1200	29761	250.011	19039	190.600
1300	33345	252.878	21126	192.270
1400	36958	255.556	23212	193.816
1500	40600	258.068	25296	195.254
1600	44267	260.434	27381	196.599
1700	47959	262.673	29464	197.862
1800	51674	264.797	31547	199.053
1900	55414	266.819	33630	200.179
2000	59176	268.748	35713	201.247
2200	66770	272.366	39878	203.232
2400	74453	275.708	44045	205.045
2600	82225	278.818	48216	206.714
2800	90080	281.729	52391	208.262
3000	98013	284.466	56574	209.705
3200	106022	287.050	60767	211.058
3400	114101	289.499	64971	212.332
3600	122245	291.826	69190	213.538
3800	130447	294.043	73424	214.682
4000	138705	296.161	77675	215.773
4400	155374	300.133	86234	217.812
4800	172240	303.801	94873	219.691
5200	189312	307.217	103592	221.435
5600	206618	310.423	112391	223.066
6000	224210	313.457	121264	224.597

TABLE A.9 (*continued*)
Ideal-Gas Properties of Various Substances (SI Units), Entropies at 0.1-MPa (1-bar) Pressure,
Mole Basis

	CARBON DIOXIDE (CO_2) $\overline{h}^0_{f,298} = -393\ 522$ kJ/kmol $M = 44.01$		CARBON MONOXIDE (CO) $\overline{h}^0_{f,298} = -110\ 527$ kJ/kmol $M = 28.01$	
T K	$(\overline{h} - \overline{h}^0_{298})$ kJ/kmol	$\overline{s}^0_T$ kJ/kmol K	$(\overline{h} - \overline{h}^0_{298})$ kJ/kmol	$\overline{s}^0_T$ kJ/kmol K
0	−9364	0	−8671	0
100	−6457	179.010	−5772	165.852
200	−3413	199.976	−2860	186.024
298	0	213.794	0	197.651
300	69	214.024	54	197.831
400	4003	225.314	2977	206.240
500	8305	234.902	5932	212.833
600	12906	243.284	8942	218.321
700	17754	250.752	12021	223.067
800	22806	257.496	15174	227.277
900	28030	263.646	18397	231.074
1000	33397	269.299	21686	234.538
1100	38885	274.528	25031	237.726
1200	44473	279.390	28427	240.679
1300	50148	283.931	31867	243.431
1400	55895	288.190	35343	246.006
1500	61705	292.199	38852	248.426
1600	67569	295.984	42388	250.707
1700	73480	299.567	45948	252.866
1800	79432	302.969	49529	254.913
1900	85420	306.207	53128	256.860
2000	91439	309.294	56743	258.716
2200	103562	315.070	64012	262.182
2400	115779	320.384	71326	265.361
2600	128074	325.307	78679	268.302
2800	140435	329.887	86070	271.044
3000	152853	334.170	93504	273.607
3200	165321	338.194	100962	276.012
3400	177836	341.988	108440	278.279
3600	190394	345.576	115938	280.422
3800	202990	348.981	123454	282.454
4000	215624	352.221	130989	284.387
4400	240992	358.266	146108	287.989
4800	266488	363.812	161285	291.290
5200	292112	368.939	176510	294.337
5600	317870	373.711	191782	297.167
6000	343782	378.180	207105	299.809

TABLE A.9 (*continued*)
Ideal-Gas Properties of Various Substances (SI Units), Entropies at 0.1-MPa (1-bar) Pressure,
Mole Basis

T K	WATER (H_2O) $\bar{h}^0_{f,298} = -241\ 826$ kJ/kmol $M = 18.015$		HYDROXYL (OH) $\bar{h}^0_{f,298} = 38\ 987$ kJ/kmol $M = 17.007$	
	$(\bar{h} - \bar{h}^0_{298})$ kJ/kmol	$\bar{s}^0_T$ kJ/kmol K	$(\bar{h} - \bar{h}^0_{298})$ kJ/kmol	$\bar{s}^0_T$ kJ/kmol
0	−9904	0	−9172	0
100	−6617	152.386	−6140	149.591
200	−3282	175.488	−2975	171.592
298	0	188.835	0	183.709
300	62	189.043	55	183.894
400	3450	198.787	3034	192.466
500	6922	206.532	5991	199.066
600	10499	213.051	8943	204.448
700	14190	218.739	11902	209.008
800	18002	223.826	14881	212.984
900	21937	228.460	17889	216.526
1000	26000	232.739	20935	219.735
1100	30190	236.732	24024	222.680
1200	34506	240.485	27159	225.408
1300	38941	244.035	30340	227.955
1400	43491	247.406	33567	230.347
1500	48149	250.620	36838	232.604
1600	52907	253.690	40151	234.741
1700	57757	256.631	43502	236.772
1800	62693	259.452	46890	238.707
1900	67706	262.162	50311	240.556
2000	72788	264.769	53763	242.328
2200	83153	269.706	60751	245.659
2400	93741	274.312	67840	248.743
2600	104520	278.625	75018	251.614
2800	115463	282.680	82268	254.301
3000	126548	286.504	89585	256.825
3200	137756	290.120	96960	259.205
3400	149073	293.550	104388	261.456
3600	160484	296.812	111864	263.592
3800	171981	299.919	119382	265.625
4000	183552	302.887	126940	267.563
4400	206892	308.448	142165	271.191
4800	230456	313.573	157522	274.531
5200	254216	318.328	173002	277.629
5600	278161	322.764	188598	280.518
6000	302295	326.926	204309	283.227

TABLE A.9 (*continued*)
Ideal-Gas Properties of Various Substances (SI Units), Entropies at 0.1-MPa (1-bar) Pressure, Mole Basis

	HYDROGEN (H_2) $\bar{h}^0_{f,298} = 0$ kJ/kmol $M = 2.016$		HYDROGEN, MONATOMIC (H) $\bar{h}^0_{f,298} = 217\ 999$ kJ/kmol $M = 1.008$	
T **K**	$(\bar{h} - \bar{h}^0_{298})$ **kJ/kmol**	$\bar{s}^0_T$ **kJ/kmol K**	$(\bar{h} - \bar{h}^0_{298})$ **kJ/kmol**	$\bar{s}^0_T$ **kJ/kmol K**
0	−8467	0	−6197	0
100	−5467	100.727	−4119	92.009
200	−2774	119.410	−2040	106.417
298	0	130.678	0	114.716
300	53	130.856	38	114.845
400	2961	139.219	2117	120.825
500	5883	145.738	4196	125.463
600	8799	151.078	6274	129.253
700	11730	155.609	8353	132.457
800	14681	159.554	10431	135.233
900	17657	163.060	12510	137.681
1000	20663	166.225	14589	139.871
1100	23704	169.121	16667	141.852
1200	26785	171.798	18746	143.661
1300	29907	174.294	20825	145.324
1400	33073	176.637	22903	146.865
1500	36281	178.849	24982	148.299
1600	39533	180.946	27060	149.640
1700	42826	182.941	29139	150.900
1800	46160	184.846	31218	152.089
1900	49532	186.670	33296	153.212
2000	52942	188.419	35375	154.279
2200	59865	191.719	39532	156.260
2400	66915	194.789	43689	158.069
2600	74082	197.659	47847	159.732
2800	81355	200.355	52004	161.273
3000	88725	202.898	56161	162.707
3200	96187	205.306	60318	164.048
3400	103736	207.593	64475	165.308
3600	111367	209.773	68633	166.497
3800	119077	211.856	72790	167.620
4000	126864	213.851	76947	168.687
4400	142658	217.612	85261	170.668
4800	158730	221.109	93576	172.476
5200	175057	224.379	101890	174.140
5600	191607	227.447	110205	175.681
6000	208332	230.322	118519	177.114

TABLE A.9 (*continued*)
Ideal-Gas Properties of Various Substances (SI Units), Entropies at 0.1-MPa (1-bar) Pressure, Mole Basis

| T | NITRIC OXIDE (NO) $\bar{h}_{f,298}^0 = 90\ 291$ kJ/kmol $M = 30.006$ | | NITROGEN DIOXIDE (NO$_2$) $\bar{h}_{f,298}^0 = 33\ 100$ kJ/kmol $M = 46.005$ | |
K	$(\bar{h} - \bar{h}_{298}^0)$ kJ/kmol	$\bar{s}_T^0$ kJ/kmol K	$(\bar{h} - \bar{h}_{298}^0)$ kJ/kmol	$\bar{s}_T^0$ kJ/kmol K
0	−9192	0	−10186	0
100	−6073	177.031	−6861	202.563
200	−2951	198.747	−3495	225.852
298	0	210.759	0	240.034
300	55	210.943	68	240.263
400	3040	219.529	3927	251.342
500	6059	226.263	8099	260.638
600	9144	231.886	12555	268.755
700	12308	236.762	17250	275.988
800	15548	241.088	22138	282.513
900	18858	244.985	27180	288.450
1000	22229	248.536	32344	293.889
1100	25653	251.799	37606	298.904
1200	29120	254.816	42946	303.551
1300	32626	257.621	48351	307.876
1400	36164	260.243	53808	311.920
1500	39729	262.703	59309	315.715
1600	43319	265.019	64846	319.289
1700	46929	267.208	70414	322.664
1800	50557	269.282	76008	325.861
1900	54201	271.252	81624	328.898
2000	57859	273.128	87259	331.788
2200	65212	276.632	98578	337.182
2400	72606	279.849	109948	342.128
2600	80034	282.822	121358	346.695
2800	87491	285.585	132800	350.934
3000	94973	288.165	144267	354.890
3200	102477	290.587	155756	358.597
3400	110000	292.867	167262	362.085
3600	117541	295.022	178783	365.378
3800	125099	297.065	190316	368.495
4000	132671	299.007	201860	371.456
4400	147857	302.626	224973	376.963
4800	163094	305.940	248114	381.997
5200	178377	308.998	271276	386.632
5600	193703	311.838	294455	390.926
6000	209070	314.488	317648	394.926

TABLE A.10

Enthalpy of Formation and Absolute Entropy of Various Substances at 25°C, 100 kPa Pressure

Substance	Formula	M	State	$\bar{h}_f^0$ kJ/kmol	$\bar{s}_f^0$ kJ/kmol K
Water	H_2O	18.015	gas	−241 826	188.834
Water	H_2O	18.015	liq	−285 830	69.950
Hydrogen peroxide	H_2O_2	34.015	gas	−136 106	232.991
Ozone	O_3	47.998	gas	+142 674	238.932
Carbon (graphite)	C	12.011	solid	0	5.740
Carbon monoxide	CO	28.011	gas	−110 527	197.653
Carbon dioxide	CO_2	44.010	gas	−393 522	213.795
Methane	CH_4	16.043	gas	−74 873	186.251
Acetylene	C_2H_2	26.038	gas	+226 731	200.958
Ethene	C_2H_4	28.054	gas	+52 467	219.330
Ethane	C_2H_6	30.070	gas	−84 740	229.597
Propene	C_3H_6	42.081	gas	+20 430	267.066
Propane	C_3H_8	44.094	gas	−103 900	269.917
Butane	C_4H_{10}	58.124	gas	−126 200	306.647
Pentane	C_5H_{12}	72.151	gas	−146 500	348.945
Benzene	C_6H_6	78.114	gas	+82 980	269.562
Hexane	C_6H_{14}	86.178	gas	−167 300	387.979
Heptane	C_7H_{16}	100.205	gas	−187 900	427.805
n-Octane	C_8H_{18}	114.232	gas	−208 600	466.514
n-Octane	C_8H_{18}	114.232	liq	−250 105	360.575
Methanol	CH_3OH	32.042	gas	−201 300	239.709
Methanol	CH_3OH	32.042	liq	−239 220	126.809
Ethanol	C_2H_5OH	46.069	gas	−235 000	282.444
Ethanol	C_2H_5OH	46.069	liq	−277 380	160.554
Ammonia	NH_3	17.031	gas	−45 720	192.572
T-T-Diesel	$C_{14.4}H_{24.9}$	198.06	liq	−174 000	525.90
Sulfur	S	32.06	solid	0	32.056
Sulfur dioxide	SO_2	64.059	gas	−296 842	248.212
Sulfur trioxide	SO_3	80.058	gas	−395 765	256.769
Nitrogen oxide	N_2O	44.013	gas	+82 050	219.957
Nitromethane	CH_3NO_2	61.04	liq	−113 100	171.80

TABLE A.11

Logarithms to the Base e of the Equilibrium Constant K

For the reaction $v_A A + v_B B \rightleftharpoons v_C C + v_D D$, the equilibrium constant K is defined as

$$K = \frac{y_C^{v_C} y_D^{v_D}}{y_A^{v_A} y_B^{v_B}} \left(\frac{P}{P^0}\right)^{v_C + v_D - v_A - v_B}, \quad P^0 = 0.1 \text{ MPa}$$

Temp K	$H_2 \rightleftharpoons 2H$	$O_2 \rightleftharpoons 2O$	$N_2 \rightleftharpoons 2N$	$2H_2O \rightleftharpoons 2H_2 + O_2$	$2H_2O \rightleftharpoons H_2 + 2OH$	$2CO_2 \rightleftharpoons 2CO + O_2$	$N_2 + O_2 \rightleftharpoons 2NO$	$N_2 + 2O_2 \rightleftharpoons 2NO_2$
298	−164.003	−186.963	−367.528	−184.420	−212.075	−207.529	−69.868	−41.355
500	−92.830	−105.623	−213.405	−105.385	−120.331	−115.234	−40.449	−30.725
1000	−39.810	−45.146	−99.146	−46.321	−51.951	−47.052	−18.709	−23.039
1200	−30.878	−35.003	−80.025	−36.363	−40.467	−35.736	−15.082	−21.752
1400	−24.467	−27.741	−66.345	−29.222	−32.244	−27.679	−12.491	−20.826
1600	−19.638	−22.282	−56.069	−23.849	−26.067	−21.656	−10.547	−20.126
1800	−15.868	−18.028	−48.066	−19.658	−21.258	−16.987	−9.035	−19.577
2000	−12.841	−14.619	−41.655	−16.299	−17.406	−13.266	−7.825	−19.136
2200	−10.356	−11.826	−36.404	−13.546	−14.253	−10.232	−6.836	−18.773
2400	−8.280	−9.495	−32.023	−11.249	−11.625	−7.715	−6.012	−18.470
2600	−6.519	−7.520	−28.313	−9.303	−9.402	−5.594	−5.316	−18.214
2800	−5.005	−5.826	−25.129	−7.633	−7.496	−3.781	−4.720	−17.994
3000	−3.690	−4.356	−22.367	−6.184	−5.845	−2.217	−4.205	−17.805
3200	−2.538	−3.069	−19.947	−4.916	−4.401	−0.853	−3.755	−17.640
3400	−1.519	−1.932	−17.810	−3.795	−3.128	0.346	−3.359	−17.496
3600	−0.611	−0.922	−15.909	−2.799	−1.996	1.408	−3.008	−17.369
3800	0.201	−0.017	−14.205	−1.906	−0.984	2.355	−2.694	−17.257
4000	0.934	0.798	−12.671	−1.101	−0.074	3.204	−2.413	−17.157
4500	2.483	2.520	−9.423	0.602	1.847	4.985	−1.824	−16.953
5000	3.724	3.898	−6.816	1.972	3.383	6.397	−1.358	−16.797
5500	4.739	5.027	−4.672	3.098	4.639	7.542	−0.980	−16.678
6000	5.587	5.969	−2.876	4.040	5.684	8.488	−0.671	−16.588

Source: Consistent with thermodynamic data in *JANAF Thermochemical Tables*, third edition, Thermal Group, Dow Chemical U.S.A., Midland, MI, 1985.

APPENDIX B
SI UNITS: THERMODYNAMIC TABLES

TABLE B.1
Thermodynamic Properties of Water

TABLE B.1.1
Saturated Water

Temp. (°C)	Press. (kPa)	SPECIFIC VOLUME, m³/kg			INTERNAL ENERGY, kJ/kg		
		Sat. Liquid v_f	Evap. v_{fg}	Sat. Vapor v_g	Sat. Liquid u_f	Evap. u_{fg}	Sat. Vapor u_g
0.01	0.6113	0.001000	206.131	206.132	0	2375.33	2375.33
5	0.8721	0.001000	147.117	147.118	20.97	2361.27	2382.24
10	1.2276	0.001000	106.376	106.377	41.99	2347.16	2389.15
15	1.705	0.001001	77.924	77.925	62.98	2333.06	2396.04
20	2.339	0.001002	57.7887	57.7897	83.94	2318.98	2402.91
25	3.169	0.001003	43.3583	43.3593	104.86	2304.90	2409.76
30	4.246	0.001004	32.8922	32.8932	125.77	2290.81	2416.58
35	5.628	0.001006	25.2148	25.2158	146.65	2276.71	2423.36
40	7.384	0.001008	19.5219	19.5229	167.53	2262.57	2430.11
45	9.593	0.001010	15.2571	15.2581	188.41	2248.40	2436.81
50	12.350	0.001012	12.0308	12.0318	209.30	2234.17	2443.47
55	15.758	0.001015	9.56734	9.56835	230.19	2219.89	2450.08
60	19.941	0.001017	7.66969	7.67071	251.09	2205.54	2456.63
65	25.03	0.001020	6.19554	6.19656	272.00	2191.12	2463.12
70	31.19	0.001023	5.04114	5.04217	292.93	2176.62	2469.55
75	38.58	0.001026	4.13021	4.13123	313.87	2162.03	2475.91
80	47.39	0.001029	3.40612	3.40715	334.84	2147.36	2482.19
85	57.83	0.001032	2.82654	2.82757	355.82	2132.58	2488.40
90	70.14	0.001036	2.35953	2.36056	376.82	2117.70	2494.52
95	84.55	0.001040	1.98082	1.98186	397.86	2102.70	2500.56
100	101.3	0.001044	1.67185	1.67290	418.91	2087.58	2506.50
105	120.8	0.001047	1.41831	1.41936	440.00	2072.34	2512.34
110	143.3	0.001052	1.20909	1.21014	461.12	2056.96	2518.09
115	169.1	0.001056	1.03552	1.03658	482.28	2041.44	2523.72
120	198.5	0.001060	0.89080	0.89186	503.48	2025.76	2529.24
125	232.1	0.001065	0.76953	0.77059	524.72	2009.91	2534.63
130	270.1	0.001070	0.66744	0.66850	546.00	1993.90	2539.90
135	313.0	0.001075	0.58110	0.58217	567.34	1977.69	2545.03
140	361.3	0.001080	0.50777	0.50885	588.72	1961.30	2550.02
145	415.4	0.001085	0.44524	0.44632	610.16	1944.69	2554.86
150	475.9	0.001090	0.39169	0.39278	631.66	1927.87	2559.54
155	543.1	0.001096	0.34566	0.34676	653.23	1910.82	2564.04
160	617.8	0.001102	0.30596	0.30706	674.85	1893.52	2568.37
165	700.5	0.001108	0.27158	0.27269	696.55	1875.97	2572.51
170	791.7	0.001114	0.24171	0.24283	718.31	1858.14	2576.46
175	892.0	0.001121	0.21568	0.21680	740.16	1840.03	2580.19
180	1002.2	0.001127	0.19292	0.19405	762.08	1821.62	2583.70
185	1122.7	0.001134	0.17295	0.17409	784.08	1802.90	2586.98
190	1254.4	0.001141	0.15539	0.15654	806.17	1783.84	2590.01

TABLE B.1.1 (*continued*)
Saturated Water

Temp. (°C)	Press. (kPa)	ENTHALPY, kJ/kg			ENTROPY, kJ/kg-K		
		Sat. Liquid h_f	Evap. h_{fg}	Sat. Vapor h_g	Sat. Liquid s_f	Evap. s_{fg}	Sat. Vapor s_g
0.01	0.6113	0.00	2501.35	2501.35	0	9.1562	9.1562
5	0.8721	20.98	2489.57	2510.54	0.0761	8.9496	9.0257
10	1.2276	41.99	2477.75	2519.74	0.1510	8.7498	8.9007
15	1.705	62.98	2465.93	2528.91	0.2245	8.5569	8.7813
20	2.339	83.94	2454.12	2538.06	0.2966	8.3706	8.6671
25	3.169	104.87	2442.30	2547.17	0.3673	8.1905	8.5579
30	4.246	125.77	2430.48	2556.25	0.4369	8.0164	8.4533
35	5.628	146.66	2418.62	2565.28	0.5052	7.8478	8.3530
40	7.384	167.54	2406.72	2574.26	0.5724	7.6845	8.2569
45	9.593	188.42	2394.77	2583.19	0.6386	7.5261	8.1647
50	12.350	209.31	2382.75	2592.06	0.7037	7.3725	8.0762
55	15.758	230.20	2370.66	2600.86	0.7679	7.2234	7.9912
60	19.941	251.11	2358.48	2609.59	0.8311	7.0784	7.9095
65	25.03	272.03	2346.21	2618.24	0.8934	6.9375	7.8309
70	31.19	292.96	2333.85	2626.80	0.9548	6.8004	7.7552
75	38.58	313.91	2321.37	2635.28	1.0154	6.6670	7.6824
80	47.39	334.88	2308.77	2643.66	1.0752	6.5369	7.6121
85	57.83	355.88	2296.05	2651.93	1.1342	6.4102	7.5444
90	70.14	376.90	2283.19	2660.09	1.1924	6.2866	7.4790
95	84.55	397.94	2270.19	2668.13	1.2500	6.1659	7.4158
100	101.3	419.02	2257.03	2676.05	1.3068	6.0480	7.3548
105	120.8	440.13	2243.70	2683.83	1.3629	5.9328	7.2958
110	143.3	461.27	2230.20	2691.47	1.4184	5.8202	7.2386
115	169.1	482.46	2216.50	2698.96	1.4733	5.7100	7.1832
120	198.5	503.69	2202.61	2706.30	1.5275	5.6020	7.1295
125	232.1	524.96	2188.50	2713.46	1.5812	5.4962	7.0774
130	270.1	546.29	2174.16	2720.46	1.6343	5.3925	7.0269
135	313.0	567.67	2159.59	2727.26	1.6869	5.2907	6.9777
140	361.3	589.11	2144.75	2733.87	1.7390	5.1908	6.9298
145	415.4	610.61	2129.65	2740.26	1.7906	5.0926	6.8832
150	475.9	632.18	2114.26	2746.44	1.8417	4.9960	6.8378
155	543.1	653.82	2098.56	2752.39	1.8924	4.9010	6.7934
160	617.8	675.53	2082.55	2758.09	1.9426	4.8075	6.7501
165	700.5	697.32	2066.20	2763.53	1.9924	4.7153	6.7078
170	791.7	719.20	2049.50	2768.70	2.0418	4.6244	6.6663
175	892.0	741.16	2032.42	2773.58	2.0909	4.5347	6.6256
180	1002.2	763.21	2014.96	2778.16	2.1395	4.4461	6.5857
185	1122.7	785.36	1997.07	2782.43	2.1878	4.3586	6.5464
190	1254.4	807.61	1978.76	2786.37	2.2358	4.2720	6.5078

TABLE B.1.1 (*continued*)
Saturated Water

Temp. (°C)	Press. (kPa)	SPECIFIC VOLUME, m³/kg			INTERNAL ENERGY, kJ/kg		
		Sat. Liquid v_f	Evap. v_{fg}	Sat. Vapor v_g	Sat. Liquid u_f	Evap. u_{fg}	Sat. Vapor u_g
195	1397.8	0.001149	0.13990	0.14105	828.36	1764.43	2592.79
200	1553.8	0.001156	0.12620	0.12736	850.64	1744.66	2595.29
205	1723.0	0.001164	0.11405	0.11521	873.02	1724.49	2597.52
210	1906.3	0.001173	0.10324	0.10441	895.51	1703.93	2599.44
215	2104.2	0.001181	0.09361	0.09479	918.12	1682.94	2601.06
220	2317.8	0.001190	0.08500	0.08619	940.85	1661.49	2602.35
225	2547.7	0.001199	0.07729	0.07849	963.72	1639.58	2603.30
230	2794.9	0.001209	0.07037	0.07158	986.72	1617.17	2603.89
235	3060.1	0.001219	0.06415	0.06536	1009.88	1594.24	2604.11
240	3344.2	0.001229	0.05853	0.05976	1033.19	1570.75	2603.95
245	3648.2	0.001240	0.05346	0.05470	1056.69	1546.68	2603.37
250	3973.0	0.001251	0.04887	0.05013	1080.37	1522.00	2602.37
255	4319.5	0.001263	0.04471	0.04598	1104.26	1496.66	2600.93
260	4688.6	0.001276	0.04093	0.04220	1128.37	1470.64	2599.01
265	5081.3	0.001289	0.03748	0.03877	1152.72	1443.87	2596.60
270	5498.7	0.001302	0.03434	0.03564	1177.33	1416.33	2593.66
275	5941.8	0.001317	0.03147	0.03279	1202.23	1387.94	2590.17
280	6411.7	0.001332	0.02884	0.03017	1227.43	1358.66	2586.09
285	6909.4	0.001348	0.02642	0.02777	1252.98	1328.41	2581.38
290	7436.0	0.001366	0.02420	0.02557	1278.89	1297.11	2575.99
295	7992.8	0.001384	0.02216	0.02354	1305.21	1264.67	2569.87
300	8581.0	0.001404	0.02027	0.02167	1331.97	1230.99	2562.96
305	9201.8	0.001425	0.01852	0.01995	1359.22	1195.94	2555.16
310	9856.6	0.001447	0.01690	0.01835	1387.03	1159.37	2546.40
315	10547	0.001472	0.01539	0.01687	1415.44	1121.11	2536.55
320	11274	0.001499	0.01399	0.01549	1444.55	1080.93	2525.48
325	12040	0.001528	0.01267	0.01420	1474.44	1038.57	2513.01
330	12845	0.001561	0.01144	0.01300	1505.24	993.66	2498.91
335	13694	0.001597	0.01027	0.01186	1537.11	945.77	2482.88
340	14586	0.001638	0.00916	0.01080	1570.26	894.26	2464.53
345	15525	0.001685	0.00810	0.00978	1605.01	838.29	2443.30
350	16514	0.001740	0.00707	0.00881	1641.81	776.58	2418.39
355	17554	0.001807	0.00607	0.00787	1681.41	707.11	2388.52
360	18651	0.001892	0.00505	0.00694	1725.19	626.29	2351.47
365	19807	0.002011	0.00398	0.00599	1776.13	526.54	2302.67
370	21028	0.002213	0.00271	0.00493	1843.84	384.69	2228.53
374.1	22089	0.003155	0	0.00315	2029.58	0	2029.58

TABLE B.1.1 (*continued*)
Saturated Water

Temp. (°C)	Press. (kPa)	ENTHALPY, kJ/kg			ENTROPY, kJ/kg-K		
		Sat. Liquid h_f	Evap. h_{fg}	Sat. Vapor h_g	Sat. Liquid s_f	Evap. s_{fg}	Sat. Vapor s_g
195	1397.8	829.96	1959.99	2789.96	2.2835	4.1863	6.4697
200	1553.8	852.43	1940.75	2793.18	2.3308	4.1014	6.4322
205	1723.0	875.03	1921.00	2796.03	2.3779	4.0172	6.3951
210	1906.3	897.75	1900.73	2798.48	2.4247	3.9337	6.3584
215	2104.2	920.61	1879.91	2800.51	2.4713	3.8507	6.3221
220	2317.8	943.61	1858.51	2802.12	2.5177	3.7683	6.2860
225	2547.7	966.77	1836.50	2803.27	2.5639	3.6863	6.2502
230	2794.9	990.10	1813.85	2803.95	2.6099	3.6047	6.2146
235	3060.1	1013.61	1790.53	2804.13	2.6557	3.5233	6.1791
240	3344.2	1037.31	1766.50	2803.81	2.7015	3.4422	6.1436
245	3648.2	1061.21	1741.73	2802.95	2.7471	3.3612	6.1083
250	3973.0	1085.34	1716.18	2801.52	2.7927	3.2802	6.0729
255	4319.5	1109.72	1689.80	2799.51	2.8382	3.1992	6.0374
260	4688.6	1134.35	1662.54	2796.89	2.8837	3.1181	6.0018
265	5081.3	1159.27	1634.34	2793.61	2.9293	3.0368	5.9661
270	5498.7	1184.49	1605.16	2789.65	2.9750	2.9551	5.9301
275	5941.8	1210.05	1574.92	2784.97	3.0208	2.8730	5.8937
280	6411.7	1235.97	1543.55	2779.53	3.0667	2.7903	5.8570
285	6909.4	1262.29	1510.97	2773.27	3.1129	2.7069	5.8198
290	7436.0	1289.04	1477.08	2766.13	3.1593	2.6227	5.7821
295	7992.8	1316.27	1441.78	2758.05	3.2061	2.5375	5.7436
300	8581.0	1344.01	1404.93	2748.94	3.2533	2.4511	5.7044
305	9201.8	1372.33	1366.38	2738.72	3.3009	2.3633	5.6642
310	9856.6	1401.29	1325.97	2727.27	3.3492	2.2737	5.6229
315	10547	1430.97	1283.48	2714.44	3.3981	2.1821	5.5803
320	11274	1461.45	1238.64	2700.08	3.4479	2.0882	5.5361
325	12040	1492.84	1191.13	2683.97	3.4987	1.9913	5.4900
330	12845	1525.29	1140.56	2665.85	3.5506	1.8909	5.4416
335	13694	1558.98	1086.37	2645.35	3.6040	1.7863	5.3903
340	14586	1594.15	1027.86	2622.01	3.6593	1.6763	5.3356
345	15525	1631.17	964.02	2595.19	3.7169	1.5594	5.2763
350	16514	1670.54	893.38	2563.92	3.7776	1.4336	5.2111
355	17554	1713.13	813.59	2526.72	3.8427	1.2951	5.1378
360	18651	1760.48	720.52	2481.00	3.9146	1.1379	5.0525
365	19807	1815.96	605.44	2421.40	3.9983	0.9487	4.9470
370	21028	1890.37	441.75	2332.12	4.1104	0.6868	4.7972
374.1	22089	2099.26	0	2099.26	4.4297	0	4.4297

Table B.1.2
Saturated Water Pressure Entry

Press. (kPa)	Temp. (°C)	SPECIFIC VOLUME, m³/kg			INTERNAL ENERGY, kJ/kg		
		Sat. Liquid v_f	Evap. v_{fg}	Sat. Vapor v_g	Sat. Liquid u_f	Evap. u_{fg}	Sat. Vapor u_g
0.6113	0.01	0.001000	206.131	206.132	0	2375.3	2375.3
1	6.98	0.001000	129.20702	129.20802	29.29	2355.69	2384.98
1.5	13.03	0.001001	87.97913	87.98013	54.70	2338.63	2393.32
2	17.50	0.001001	67.00285	67.00385	73.47	2326.02	2399.48
2.5	21.08	0.001002	54.25285	54.25385	88.47	2315.93	2404.40
3	24.08	0.001003	45.66402	45.66502	101.03	2307.48	2408.51
4	28.96	0.001004	34.79915	34.80015	121.44	2293.73	2415.17
5	32.88	0.001005	28.19150	28.19251	137.79	2282.70	2420.49
7.5	40.29	0.001008	19.23674	19.23775	168.76	2261.74	2430.50
10	45.81	0.001010	14.67254	14.67355	191.79	2246.10	2437.89
15	53.97	0.001014	10.02117	10.02218	225.90	2222.83	2448.73
20	60.06	0.001017	7.64835	7.64937	251.35	2205.36	2456.71
25	64.97	0.001020	6.20322	6.20424	271.88	2191.21	2463.08
30	69.10	0.001022	5.22816	5.22918	289.18	2179.22	2468.40
40	75.87	0.001026	3.99243	3.99345	317.51	2159.49	2477.00
50	81.33	0.001030	3.23931	3.24034	340.42	2143.43	2483.85
75	91.77	0.001037	2.21607	2.21711	394.29	2112.39	2496.67
100	99.62	0.001043	1.69296	1.69400	417.33	2088.72	2506.06
125	105.99	0.001048	1.37385	1.37490	444.16	2069.32	2513.48
150	111.37	0.001053	1.15828	1.15933	466.92	2052.72	2519.64
175	116.06	0.001057	1.00257	1.00363	486.78	2038.12	2524.90
200	120.23	0.001061	0.88467	0.88573	504.47	2025.02	2529.49
225	124.00	0.001064	0.79219	0.79325	520.45	2013.10	2533.56
250	127.43	0.001067	0.71765	0.71871	535.08	2002.14	2537.21
275	130.60	0.001070	0.65624	0.65731	548.57	1991.95	2540.53
300	133.55	0.001073	0.60475	0.60582	561.13	1982.43	2543.55
325	136.30	0.001076	0.56093	0.56201	572.88	1973.46	2546.34
350	138.88	0.001079	0.52317	0.52425	583.93	1964.98	2548.92
375	141.32	0.001081	0.49029	0.49137	594.38	1956.93	2551.31
400	143.63	0.001084	0.46138	0.46246	604.29	1949.26	2553.55
450	147.93	0.001088	0.41289	0.41398	622.75	1934.87	2557.62
500	151.86	0.001093	0.37380	0.37489	639.66	1921.57	2561.23
550	155.48	0.001097	0.34159	0.34268	655.30	1909.17	2564.47
600	158.85	0.001101	0.31457	0.31567	669.88	1897.52	2567.40
650	162.01	0.001104	0.29158	0.29268	683.55	1886.51	2570.06
700	164.97	0.001108	0.27176	0.27286	696.43	1876.07	2572.49
750	167.77	0.001111	0.25449	0.25560	708.62	1866.11	2574.73
800	170.43	0.001115	0.23931	0.24043	720.20	1856.58	2576.79

TABLE B.1.2 (*Continued*)
Saturated Water Pressure Entry

Press. (kPa)	Temp. (°C)	ENTHALPY, kJ/kg			ENTROPY, kJ/kg-K		
		Sat. Liquid h_f	Evap. h_{fg}	Sat. Vapor h_g	Sat. Liquid s_f	Evap. s_{fg}	Sat. Vapor s_g
0.6113	0.01	0.00	2501.3	2501.3	0	9.1562	9.1562
1.0	6.98	29.29	2484.89	2514.18	0.1059	8.8697	8.9756
1.5	13.03	54.70	2470.59	2525.30	0.1956	8.6322	8.8278
2.0	17.50	73.47	2460.02	2533.49	0.2607	8.4629	8.7236
2.5	21.08	88.47	2451.56	2540.03	0.3120	8.3311	8.6431
3.0	24.08	101.03	2444.47	2545.50	0.3545	8.2231	8.5775
4.0	28.96	121.44	2432.93	2554.37	0.4226	8.0520	8.4746
5.0	32.88	137.79	2423.66	2561.45	0.4763	7.9187	8.3950
7.5	40.29	168.77	2406.02	2574.79	0.5763	7.6751	8.2514
10	45.81	191.81	2392.82	2584.63	0.6492	7.5010	8.1501
15	53.97	225.91	2373.14	2599.06	0.7548	7.2536	8.0084
20	60.06	251.38	2358.33	2609.70	0.8319	7.0766	7.9085
25	64.97	271.90	2346.29	2618.19	0.8930	6.9383	7.8313
30	69.10	289.21	2336.07	2625.28	0.9439	6.8247	7.7686
40	75.87	317.55	2319.19	2636.74	1.0258	6.6441	7.6700
50	81.33	340.47	2305.40	2645.87	1.0910	6.5029	7.5939
75	91.77	384.36	2278.59	2662.96	1.2129	6.2434	7.4563
100	99.62	417.44	2258.02	2675.46	1.3025	6.0568	7.3593
125	105.99	444.30	2241.05	2685.35	1.3739	5.9104	7.2843
150	111.37	467.08	2226.46	2693.54	1.4335	5.7897	7.2232
175	116.06	486.97	2213.57	2700.53	1.4848	5.6868	7.1717
200	120.23	504.68	2201.96	2706.63	1.5300	5.5970	7.1271
225	124.00	520.69	2191.35	2712.04	1.5705	5.5173	7.0878
250	127.43	535.34	2181.55	2716.89	1.6072	5.4455	7.0526
275	130.60	548.87	2172.42	2721.29	1.6407	5.3801	7.0208
300	133.55	561.45	2163.85	2725.30	1.6717	5.3201	6.9918
325	136.30	573.23	2155.76	2728.99	1.7005	5.2646	6.9651
350	138.88	584.31	2148.10	2732.40	1.7274	5.2130	6.9404
375	141.32	594.79	2140.79	2735.58	1.7527	5.1647	6.9174
400	143.63	604.73	2133.81	2738.53	1.7766	5.1193	6.8958
450	147.93	623.24	2120.67	2743.91	1.8206	5.0359	6.8565
500	151.86	640.21	2108.47	2748.67	1.8606	4.9606	6.8212
550	155.48	655.91	2097.04	2752.94	1.8972	4.8920	6.7892
600	158.85	670.54	2086.26	2756.80	1.9311	4.8289	6.7600
650	162.01	684.26	2076.04	2760.30	1.9627	4.7704	6.7330
700	164.97	697.20	2066.30	2763.50	1.9922	4.7158	6.7080
750	167.77	709.45	2056.98	2766.43	2.0199	4.6647	6.6846
800	170.43	721.10	2048.04	2769.13	2.0461	4.6166	6.6627

Table B.1.2 (*continued*)
Saturated Water Pressure Entry

Press. (kPa)	Temp. (°C)	Sat. Liquid v_f	Evap. v_{fg}	Sat. Vapor v_g	Sat. Liquid u_f	Evap. u_{fg}	Sat. Vapor u_g
		SPECIFIC VOLUME, m³/kg			INTERNAL ENERGY, kJ/kg		
850	172.96	0.001118	0.22586	0.22698	731.25	1847.45	2578.69
900	175.38	0.001121	0.21385	0.21497	741.81	1838.65	2580.46
950	177.69	0.001124	0.20306	0.20419	751.94	1830.17	2582.11
1000	179.91	0.001127	0.19332	0.19444	761.67	1821.97	2583.64
1100	184.09	0.001133	0.17639	0.17753	780.08	1806.32	2586.40
1200	187.99	0.001139	0.16220	0.16333	797.27	1791.55	2588.82
1300	191.64	0.001144	0.15011	0.15125	813.42	1777.53	2590.95
1400	195.07	0.001149	0.13969	0.14084	828.68	1764.15	2592.83
1500	198.32	0.001154	0.13062	0.13177	843.14	1751.3	2594.5
1750	205.76	0.001166	0.11232	0.11349	876.44	1721.39	2597.83
2000	212.42	0.001177	0.09845	0.09963	906.42	1693.84	2600.26
2250	218.45	0.001187	0.08756	0.08875	933.81	1668.18	2601.98
2500	223.99	0.001197	0.07878	0.07998	959.09	1644.04	2603.13
2750	229.12	0.001207	0.07154	0.07275	982.65	1621.16	2603.81
3000	233.90	0.001216	0.06546	0.06668	1004.76	1599.34	2604.10
3250	238.38	0.001226	0.06029	0.06152	1025.62	1578.43	2604.04
3500	242.60	0.001235	0.05583	0.05707	1045.41	1558.29	2603.70
4000	250.40	0.001252	0.04853	0.04978	1082.28	1519.99	2602.27
5000	263.99	0.001286	0.03815	0.03944	1147.78	1449.34	2597.12
6000	275.64	0.001319	0.03112	0.03244	1205.41	1384.27	2589.69
7000	285.88	0.001351	0.02602	0.02737	1257.51	1322.97	2580.48
8000	295.06	0.001384	0.02213	0.02352	1305.54	1264.25	2569.79
9000	303.40	0.001418	0.01907	0.02048	1350.47	1207.28	2557.75
10000	311.06	0.001452	0.01657	0.01803	1393.00	1151.40	2544.41
11000	318.15	0.001489	0.01450	0.01599	1433.68	1096.06	2529.74
12000	324.75	0.001527	0.01274	0.01426	1472.92	1040.76	2513.67
13000	330.93	0.001567	0.01121	0.01278	1511.09	984.99	2496.08
14000	336.75	0.001611	0.00987	0.01149	1548.53	928.23	2476.76
15000	342.24	0.001658	0.00868	0.01034	1585.58	869.85	2455.43
16000	347.43	0.001711	0.00760	0.00931	1622.63	809.07	2431.70
17000	352.37	0.001770	0.00659	0.00836	1660.16	744.80	2404.96
18000	357.06	0.001840	0.00565	0.00749	1698.86	675.42	2374.28
19000	361.54	0.001924	0.00473	0.00666	1739.87	598.18	2338.05
20000	365.81	0.002035	0.00380	0.00583	1785.47	507.58	2293.05
21000	369.89	0.002206	0.00275	0.00495	1841.97	388.74	2230.71
22000	373.80	0.002808	0.00072	0.00353	1973.16	108.24	2081.39
22089	374.14	0.003155	0	0.00315	2029.58	0	2029.58

TABLE B.1.2 (*Continued*)
Saturated Water Pressure Entry

Press. (kPa)	Temp. (°C)	ENTHALPY, kJ/kg			ENTROPY, kJ/kg-K		
		Sat. Liquid h_f	Evap. h_{fg}	Sat. Vapor h_g	Sat. Liquid s_f	Evap. s_{fg}	Sat. Vapor s_g
850	172.96	732.20	2039.43	2771.63	2.0709	4.5711	6.6421
900	175.38	742.82	2031.12	2773.94	2.0946	4.5280	6.6225
950	177.69	753.00	2023.08	2776.08	2.1171	4.4869	6.6040
1000	179.91	762.79	2015.29	2778.08	2.1386.	4.4478	6.5864
1100	184.09	781.32	2000.36	2781.68	2.1791	4.3744	6.5535
1200	187.99	798.64	1986.19	2784.82	2.2165	4.3067	6.5233
1300	191.64	814.91	1972.67	2787.58	2.2514	4.2438	6.4953
1400	195.07	830.29	1959.72	2790.00	2.2842	4.1850	6.4692
1500	198.32	844.87	1947.28	2792.15	2.3150	4.1298	6.4448
1750	205.76	878.48	1917.95	2796.43	2.3851	4.0044	6.3895
2000	212.42	908.77	1890.74	2799.51	2.4473	3.8935	6.3408
2250	218.45	936.48	1865.19	2801.67	2.5034	3.7938	6.2971
2500	223.99	962.09	1840.98	2803.07	2.5546	3.7028	6.2574
2750	229.12	985.97	1817.89	2803.86	2.6018	3.6190	6.2208
3000	233.90	1008.41	1795.73	2804.14	2.6456	3.5412	6.1869
3250	238.38	1029.60	1774.37	2803.97	2.6866	3.4685	6.1551
3500	242.60	1049.73	1753.70	2803.43	2.7252	3.4000	6.1252
4000	250.40	1087.29	1714.09	2801.38	2.7963	3.2737	6.0700
5000	263.99	1154.21	1640.12	2794.33	2.9201	3.0532	5.9733
6000	275.64	1213.32	1571.00	2784.33	3.0266	2.8625	5.8891
7000	285.88	1266.97	1505.10	2772.07	3.1210	2.6922	5.8132
8000	295.06	1316.61	1441.33	2757.94	3.2067	2.5365	5.7431
9000	303.40	1363.23	1378.88	2742.11	3.2857	2.3915	5.6771
10000	311.06	1407.53	1317.14	2724.67	3.3595	2.2545	5.6140
11000	318.15	1450.05	1255.55	2705.60	3.4294	2.1233	5.5527
12000	324.75	1491.24	1193.59	2684.83	3.4961	1.9962	5.4923
13000	330.93	1531.46	1130.76	2662.22	3.5604	1.8718	5.4323
14000	336.75	1571.08	1066.47	2637.55	3.6231	1.7485	5.3716
15000	342.24	1610.45	1000.04	2610.49	3.6847	1.6250	5.3097
16000	347.43	1650.00	930.59	2580.59	3.7460	1.4995	5.2454
17000	352.37	1690.25	856.90	2547.15	3.8078	1.3698	5.1776
18000	357.06	1731.97	777.13	2509.09	3.8713	1.2330	5.1044
19000	361.54	1776.43	688.11	2464.54	3.9387	1.0841	5.0227
20000	365.81	1826.18	583.56	2409.74	4.0137	0.9132	4.9269
21000	369.89	1888.30	446.42	2334.72	4.1073	0.6942	4.8015
22000	373.80	2034.92	124.04	2158.97	4.3307	0.1917	4.5224
22089	374.14	2099.26	0	2099.26	4.4297	0	4.4297

TABLE B.1.3
Superheated Vapor Water

Temp. (°C)	v (m³/kg)	u (kJ/kg)	h (kJ/kg)	s (kJ/kg-K)	v (m³/kg)	u (kJ/kg)	h (kJ/kg)	s (kJ/kg-K)
	P = 10 kPa (45.81)				P = 50 kPa (81.33)			
Sat.	14.67355	2437.89	2584.63	8.1501	3.24034	2483.85	2645.87	7.5939
50	14.86920	2443.87	2592.56	8.1749	—	—	—	—
100	17.19561	2515.50	2687.46	8.4479	3.41833	2511.61	2682.52	7.6947
150	19.51251	2587.86	2782.99	8.6881	3.88937	2585.61	2780.08	7.9400
200	21.82507	2661.27	2879.52	8.9037	4.35595	2659.85	2877.64	8.1579
250	24.13559	2735.95	2977.31	9.1002	4.82045	2734.97	2975.99	8.3555
300	26.44508	2812.06	3076.51	9.2812	5.28391	2811.33	3075.52	8.5372
400	31.06252	2968.89	3279.51	9.6076	6.20929	2968.43	3278.89	8.8641
500	35.67896	3132.26	3489.05	9.8977	7.13364	3131.94	3488.62	9.1545
600	40.29488	3302.45	3705.40	10.1608	8.05748	3302.22	3705.10	9.4177
700	44.91052	3479.63	3928.73	10.4028	8.98104	3479.45	3928.51	9.6599
800	49.52599	3663.84	4159.10	10.6281	9.90444	3663.70	4158.92	9.8852
900	54.14137	3855.03	4396.44	10.8395	10.82773	3854.91	4396.30	10.0967
1000	58.75669	4053.01	4640.58	11.0392	11.75097	4052.91	4640.46	10.2964
1100	63.37198	4257.47	4891.19	11.2287	12.67418	4257.37	4891.08	10.4858
1200	67.98724	4467.91	5147.78	11.4090	13.59737	4467.82	5147.69	10.6662
1300	72.60250	4683.68	5409.70	14.5810	14.52054	4683.58	5409.61	10.8382
	100 kPa (99.62)				200 kPa (120.23)			
Sat.	1.69400	2506.06	2675.46	7.3593	0.88573	2529.49	2706.63	7.1271
150	1.93636	2582.75	2776.38	7.6133	0.95964	2576.87	2768.80	7.2795
200	2.17226	2658.05	2875.27	7.8342	1.08034	2654.39	2870.46	7.5066
250	2.40604	2733.73	2974.33	8.0332	1.19880	2731.22	2970.98	7.7085
300	2.63876	2810.41	3074.28	8.2157	1.31616	2808.55	3071.79	7.8926
400	3.10263	2967.85	3278.11	8.5434	1.54930	2966.69	3276.55	8.2217
500	3.56547	3131.54	3488.09	8.8341	1.78139	3130.75	3487.03	8.5132
600	4.02781	3301.94	3704.72	9.0975	2.01297	3301.36	3703.96	8.7769
700	4.48986	3479.24	3928.23	9.3398	2.24426	3478.81	3927.66	9.0194
800	4.95174	3663.53	4158.71	9.5652	2.47539	3663.19	4158.27	9.2450
900	5.41353	3854.77	4396.12	9.7767	2.70643	3854.49	4395.77	9.4565
1000	5.87526	4052.78	4640.31	9.9764	2.93740	4052.53	4640.01	9.6563
1100	6.33696	4257.25	4890.95	10.1658	3.16834	4257.01	4890.68	9.8458
1200	6.79863	4467.70	5147.56	10.3462	3.39927	4467.46	5147.32	10.0262
1300	7.26030	4683.47	5409.49	10.5182	3.63018	4683.23	5409.26	10.1982
	300 kPa (133.55)				400 kPa (143.63)			
Sat.	0.60582	2543.55	2725.30	6.9918	0.46246	2553.55	2738.53	6.8958
150	0.63388	2570.79	2760.95	7.0778	0.47084	2564.48	2752.82	6.9299
200	0.71629	2650.65	2865.54	7.3115	0.53422	2646.83	2860.51	7.1706

TABLE B.1.3 (*continued*)
Superheated Vapor Water

Temp. (°C)	v (m³/kg)	u (kJ/kg)	h (kJ/kg)	s (kJ/kg-K)	v (m³/kg)	u (kJ/kg)	h (kJ/kg)	s (kJ/kg-K)
	300 kPa (133.55)				400 kPa (143.63)			
250	0.79636	2728.69	2967.59	7.5165	0.59512	2726.11	2964.16	7.3788
300	0.87529	2806.69	3069.28	7.7022	0.65484	2804.81	3066.75	7.5661
400	1.03151	2965.53	3274.98	8.0329	0.77262	2964.36	3273.41	7.8984
500	1.18669	3129.95	3485.96	8.3250	0.88934	3129.15	3484.89	8.1912
600	1.34136	3300.79	3703.20	8.5892	1.00555	3300.22	3702.44	8.4557
700	1.49573	3478.38	3927.10	8.8319	1.12147	3477.95	3926.53	8.6987
800	1.64994	3662.85	4157.83	9.0575	1.23722	3662.51	4157.40	8.9244
900	1.80406	3854.20	4395.42	9.2691	1.35288	3853.91	4395.06	9.1361
1000	1.95812	4052.27	4639.71	9.4689	1.46847	4052.02	4639.41	9.3360
1100	2.11214	4256.77	4890.41	9.6585	1.58404	4256.53	4890.15	9.5255
1200	2.26614	4467.23	5147.07	9.8389	1.69958	4466.99	5146.83	9.7059
1300	2.42013	4682.99	5409.03	10.0109	1.81511	4682.75	5408.80	9.8780
	500 kPa (151.86)				600 kPa (158.85)			
Sat.	0.37489	2561.23	2748.67	6.8212	0.31567	2567.40	2756.80	6.7600
200	0.42492	2642.91	2855.37	7.0592	0.35202	2638.91	2850.12	6.9665
250	0.47436	2723.50	2960.68	7.2708	0.39383	2720.86	2957.16	7.1816
300	0.52256	2802.91	3064.20	7.4598	0.43437	2801.00	3061.63	7.3723
350	0.57012	2882.59	3167.65	7.6328	0.47424	2881.12	3165.66	7.5463
400	0.61728	2963.19	3271.83	7.7937	0.51372	2962.02	3270.25	7.7078
500	0.71093	3128.35	3483.82	8.0872	0.59199	3127.55	3482.75	8.0020
600	0.80406	3299.64	3701.67	8.3521	0.66974	3299.07	3700.91	8.2673
700	0.89691	3477.52	3925.97	8.5952	0.74720	3477.08	3925.41	8.5107
800	0.98959	3662.17	4156.96	8.8211	0.82450	3661.83	4156.52	8.7367
900	1.08217	3853.63	4394.71	9.0329	0.90169	3853.34	4394.36	8.9485
1000	1.17469	4051.76	4639.11	9.2328	0.97883	4051.51	4638.81	9.1484
1100	1.26718	4256.29	4889.88	9.4224	1.05594	4256.05	4889.61	9.3381
1200	1.35964	4466.76	5146.58	9.6028	1.13302	4466.52	5146.34	9.5185
1300	1.45210	4682.52	5408.57	9.7749	1.21009	4682.28	5408.34	9.6906
	800 kPa (170.43)				1000 kPa (179.91)			
Sat.	0.24043	2576.79	2769.13	6.6627	0.19444	2583.64	2778.08	6.5864
200	0.26080	2630.61	2839.25	6.8158	0.20596	2621.90	2827.86	6.6939
250	0.29314	2715.46	2949.97	7.0384	0.23268	2709.91	2942.59	6.9246
300	0.32411	2797.14	3056.43	7.2327	0.25794	2793.21	3051.15	7.1228
350	0.35439	2878.16	3161.68	7.4088	0.28247	2875.18	3157.65	7.3010
400	0.38426	2959.66	3267.07	7.5715	0.30659	2957.29	3263.88	7.4650
500	0.44331	3125.95	3480.60	7.8672	0.35411	3124.34	3478.44	7.7621
600	0.50184	3297.91	3699.38	8.1332	0.40109	3296.76	3697.85	8.0289

TABLE B.1.3 (*continued*)
Superheated Vapor Water

Temp. (°C)	v (m³/kg)	u (kJ/kg)	h (kJ/kg)	s (kJ/kg-K)	v (m³/kg)	u (kJ/kg)	h (kJ/kg)	s (kJ/kg-K)
	800 kPa (170.43)				1000 kPa (179.91)			
700	0.56007	3476.22	3924.27	8.3770	0.44779	3475.35	3923.14	8.2731
800	0.61813	3661.14	4155.65	8.6033	0.49432	3660.46	4154.78	8.4996
900	0.67610	3852.77	4393.65	8.8153	0.54075	3852.19	4392.94	8.7118
1000	0.73401	4051.00	4638.20	9.0153	0.58712	4050.49	4637.60	8.9119
1100	0.79188	4255.57	4889.08	9.2049	0.63345	4255.09	4888.55	9.1016
1200	0.84974	4466.05	5145.85	9.3854	0.67977	4465.58	5145.36	9.2821
1300	0.90758	4681.81	5407.87	9.5575	0.72608	4681.33	5407.41	9.4542
	1200 kPa (187.99)				1400 kPa (195.07)			
Sat.	0.16333	2588.82	2784.82	6.5233	0.14084	2592.83	2790.00	6.4692
200	0.16930	2612.74	2815.90	6.5898	0.14302	2603.09	2803.32	6.4975
250	0.19235	2704.20	2935.01	6.8293	0.16350	2698.32	2927.22	6.7467
300	0.21382	2789.22	3045.80	7.0316	0.18228	2785.16	3040.35	6.9533
350	0.23452	2872.16	3153.59	7.2120	0.20026	2869.12	3149.49	7.1359
400	0.25480	2954.90	3260.66	7.3773	0.21780	2952.50	3257.42	7.3025
500	0.29463	3122.72	3476.28	7.6758	0.25215	3121.10	3474.11	7.6026
600	0.33393	3295.60	3696.32	7.9434	0.28596	3294.44	3694.78	7.8710
700	0.37294	3474.48	3922.01	8.1881	0.31947	3473.61	3920.87	8.1160
800	0.41177	3659.77	4153.90	8.4149	0.35281	3659.09	4153.03	8.3431
900	0.45051	3851.62	4392.23	8.6272	0.38606	3851.05	4391.53	8.5555
1000	0.48919	4049.98	4637.00	8.8274	0.41924	4049.47	4636.41	8.7558
1100	0.52783	4254.61	4888.02	9.0171	0.45239	4254.14	4887.49	8.9456
1200	0.56646	4465.12	5144.87	9.1977	0.48552	4464.65	5144.38	9.1262
1300	0.60507	4680.86	5406.95	9.3698	0.51864	4680.39	5406.49	9.2983
	1600 kPa (201.40)				1800 kPa (207.15)			
Sat.	0.12380	2595.95	2794.02	6.4217	0.11042	2598.38	2797.13	6.3793
250	0.14184	2692.26	2919.20	6.6732	0.12497	2686.02	2910.96	6.6066
300	0.15862	2781.03	3034.83	6.8844	0.14021	2776.83	3029.21	6.8226
350	0.17456	2866.05	3145.35	7.0693	0.15457	2862.95	3141.18	7.0099
400	0.19005	2950.09	3254.17	7.2373	0.16847	2947.66	3250.90	7.1793
500	0.22029	3119.47	3471.93	7.5389	0.19550	3117.84	3469.75	7.4824
600	0.24998	3293.27	3693.23	7.8080	0.22199	3292.10	3691.69	7.7523
700	0.27937	3472.74	3919.73	8.0535	0.24818	3471.87	3918.59	7.9983
800	0.30859	3658.40	4152.15	8.2808	0.27420	3657.71	4151.27	8.2258
900	0.33772	3850.47	4390.82	8.4934	0.30012	3849.90	4390.11	8.4386
1000	0.36678	4048.96	4635.81	8.6938	0.32598	4048.45	4635.21	8.6390
1100	0.39581	4253.66	4886.95	8.8837	0.35180	4253.18	4886.42	8.8290
1200	0.42482	4464.18	5143.89	9.0642	0.37761	4463.71	5143.40	9.0096
1300	0.45382	4679.92	5406.02	9.2364	0.40340	4679.44	5405.56	9.1817

TABLE B.1.3 (*continued*)
Superheated Vapor Water

Temp. (°C)	v (m³/kg)	u (kJ/kg)	h (kJ/kg)	s (kJ/kg-K)	v (m³/kg)	u (kJ/kg)	h (kJ/kg)	s (kJ/kg-K)
	2000 kPa (212.42)				2500 kPa (223.99)			
Sat.	0.09963	2600.26	2799.51	6.3408	0.07998	2603.13	2803.07	6.2574
250	0.11144	2679.58	2902.46	6.5452	0.08700	2662.55	2880.06	6.4084
300	0.12547	2772.56	3023.50	6.7663	0.09890	2761.56	3008.81	6.6437
350	0.13857	2859.81	3136.96	6.9562	0.10976	2851.84	3126.24	6.8402
400	0.15120	2045.21	3247.60	7.1270	0.12010	2939.03	3239.28	7.0147
450	0.16353	3030.41	3357.48	7.2844	0.13014	3025.43	3350.77	7.1745
500	0.17568	3116.20	3467.55	7.4316	0.13998	3112.08	3462.04	7.3233
600	0.19960	3290.93	3690.14	7.7023	0.15930	3287.99	3686.25	7.5960
700	0.22323	3470.99	3917.45	7.9487	0.17832	3468.80	3914.59	7.8435
800	0.24668	3657.03	4150.40	8.1766	0.19716	3655.30	4148.20	8.0720
900	0.27004	3849.33	4389.40	8.3895	0.21590	3847.89	4387.64	8.2853
1000	0.29333	4047.94	4634.61	8.5900	0.23458	4046.67	4633.12	8.4860
1100	0.31659	4252.71	4885.89	8.7800	0.25322	4251.52	4884.57	8.6761
1200	0.33984	4463.25	5142.92	8.9606	0.27185	4462.08	5141.70	8.8569
1300	0.36306	4678.97	5405.10	9.1328	0.29046	4677.80	5403.95	9.0291
	3000 kPa (233.90)				4000 kPa (250.40)			
Sat.	0.06668	2604.10	2804.14	6.1869	0.04978	2602.27	2801.38	6.0700
250	0.07058	2644.00	2855.75	6.2871	—	—	—	—
300	0.08114	2750.05	2993.48	6.5389	0.05884	2725.33	2960.68	6.3614
350	0.09053	2843.66	3115.25	6.7427	0.06645	2826.65	3092.43	6.5820
400	0.09936	2932.75	3230.82	6.9211	0.07341	2919.88	3213.51	6.7689
450	0.10787	3020.38	3344.00	7.0833	0.08003	3010.13	3330.23	6.9362
500	0.11619	3107.92	3456.48	7.2337	0.08643	3099.49	3445.21	7.0900
600	0.13243	3285.03	3682.34	7.5084	0.09885	3279.06	3674.44	7.3688
700	0.14838	3466.59	3911.72	7.7571	0.11095	3462.15	3905.94	7.6198
800	0.16414	3653.58	4146.00	7.9862	0.12287	3650.11	4141.59	7.8502
900	0.17980	3846.46	4385.87	8.1999	0.13469	3843.59	4382.34	8.0647
1000	0.19541	4045.40	4631.63	8.4009	0.14645	4042.87	4628.65	8.2661
1100	0.21098	4250.33	4883.26	8.5911	0.15817	4247.96	4880.63	8.4566
1200	0.22652	4460.92	5140.49	8.7719	0.16987	4458.60	5138.07	8.6376
1300	0.24206	4676.63	5402.81	8.9442	0.18156	4674.29	5400.52	8.8099

TABLE B.1.3 (*continued*)
Superheated Vapor Water

Temp. (°C)	v (m³/kg)	u (kJ/kg)	h (kJ/kg)	s (kJ/kg-K)	v (m³/kg)	u (kJ/kg)	h (kJ/kg)	s (kJ/kg-K)
	5000 kPa (263.99)				6000 kPa (275.64)			
Sat.	0.03944	2597.12	2794.33	5.9733	0.03244	2589.69	2784.33	5.8891
300	0.04532	2697.94	2924.53	6.2083	0.03616	2667.22	2884.19	6.0673
350	0.05194	2808.67	3068.39	6.4492	0.04223	2789.61	3042.97	6.3334
400	0.05781	2906.58	3195.64	6.6458	0.04739	2892.81	3177.17	6.5407
450	0.06330	2999.64	3316.15	6.8185	0.05214	2988.90	3301.76	6.7192
500	0.06857	3090.92	3433.76	6.9758	0.05665	3082.20	3422.12	6.8802
550	0.07368	3181.82	3550.23	7.1217	0.06101	3174.57	3540.62	7.0287
600	0.07869	3273.01	3666.47	7.2588	0.06525	3266.89	3658.40	7.1676
700	0.08849	3457.67	3900.13	7.5122	0.07352	3453.15	3894.28	7.4234
800	0.09811	3646.62	4137.17	7.7440	0.08160	3643.12	4132.74	7.6566
900	0.10762	3840.71	4378.82	7.9593	0.08958	3837.84	4375.29	7.8727
1000	0.11707	4040.35	4625.69	8.1612	0.09749	4037.83	4622.74	8.0751
1100	0.12648	4245.61	4878.02	8.3519	0.10536	4243.26	4875.42	8.2661
1200	0.13587	4456.30	5135.67	8.5330	0.11321	4454.00	5133.28	8.4473
1300	0.14526	4671.96	5398.24	8.7055	0.12106	4669.64	5395.97	8.6199
	8000 kPa (295.06)				10000 kPa (311.06)			
Sat.	0.02352	2569.79	2757.94	5.7431	0.01803	2544.41	2724.67	5.6140
300	0.02426	2590.93	2784.98	5.7905	—	—	—	—
350	0.02995	2747.67	2987.30	6.1300	0.02242	2699.16	2923.39	5.9442
400	0.03432	2863.75	3138.28	6.3633	0.02641	2832.38	3096.46	6.2119
450	0.03817	2966.66	3271.99	6.5550	0.02975	2943.32	3240.83	6.4189
500	0.04175	3064.30	3398.27	6.7239	0.03279	3045.77	3373.63	6.5965
550	0.04516	3159.76	3521.01	6.8778	0.03564	3144.54	3500.92	6.7561
600	0.04845	3254.43	3642.03	7.0205	0.03837	3241.68	3625.34	6.9028
700	0.05481	3444.00	3882.47	7.2812	0.04358	3434.72	3870.52	7.1687
800	0.06097	3636.08	4123.84	7.5173	0.04859	3628.97	4114.91	7.4077
900	0.06702	3832.08	4368.26	7.7350	0.05349	3826.32	4361.24	7.6272
1000	0.07301	4032.81	4616.87	7.9384	0.05832	4027.81	4611.04	7.8315
1100	0.07896	4238.60	4870.25	8.1299	0.06312	4233.97	4865.14	8.0236
1200	0.08489	4449.45	5128.54	8.3115	0.06789	4444.93	5123.84	8.2054
1300	0.09080	4665.02	5391.46	8.4842	0.07265	4660.44	5386.99	8.3783

TABLE B.1.3 (*continued*)
Superheated Vapor Water

Temp. (°C)	v (m³/kg)	u (kJ/kg)	h (kJ/kg)	s (kJ/kg-K)	v (m³/kg)	u (kJ/kg)	h (kJ/kg)	s (kJ/kg-K)
	15000 kPa (342.24)				20000 kPa (365.81)			
Sat.	0.01034	2455.43	2610.49	5.3097	0.00583	2293.05	2409.74	4.9269
350	0.01147	2520.36	2692.41	5.4420	—	—	—	—
400	0.01565	2740.70	2975.44	5.8810	0.00994	2619.22	2818.07	5.5539
450	0.01845	2879.47	3156.15	6.1403	0.01270	2806.16	3060.06	5.9016
500	0.02080	2996.52	3308.53	6.3442	0.01477	2942.82	3238.18	6.1400
550	0.02293	3104.71	3448.61	6.5198	0.01656	3062.34	3393.45	6.3347
600	0.02491	3208.64	3582.30	6.6775	0.01818	3174.00	3537.57	6.5048
650	0.02680	3310.37	3712.32	6.8223	0.01969	3281.46	3675.32	6.6582
700	0.02861	3410.94	3840.12	6.9572	0.02113	3386.46	3809.09	6.7993
800	0.03210	3610.99	4092.43	7.2040	0.02385	3592.73	4069.80	7.0544
900	0.03546	3811.89	4343.75	7.4279	0.02645	3797.44	4326.37	7.2830
1000	0.03875	4015.41	4596.63	7.6347	0.02897	4003.12	4582.45	7.4925
1100	0.04200	4222.55	4852.56	7.8282	0.03145	4211.30	4840.24	7.6874
1200	0.04523	4433.78	5112.27	8.0108	0.03391	4422.81	5100.96	7.8706
1300	0.04845	4649.12	5375.94	8.1839	0.03636	4637.95	5365.10	8.0441
	30000 kPa				40000 kPa			
375	0.001789	1737.75	1791.43	3.9303	0.001641	1677.09	1742.71	3.8289
400	0.002790	2067.34	2151.04	4.4728	0.001908	1854.52	1930.83	4.1134
425	0.005304	2455.06	2614.17	5.1503	0.002532	2096.83	2198.11	4.5028
450	0.006735	2619.30	2821.35	5.4423	0.003693	2365.07	2512.79	4.9459
500	0.008679	2820.67	3081.03	5.7904	0.005623	2678.36	2903.26	5.4699
550	0.010168	2970.31	3275.36	6.0342	0.006984	2869.69	3149.05	5.7784
600	0.011446	3100.53	3443.91	6.2330	0.008094	3022.61	3346.38	6.0113
650	0.012596	3221.04	3598.93	6.4057	0.009064	3158.04	3520.58	6.2054
700	0.013661	3335.84	3745.67	6.5606	0.009942	3283.63	3681.29	6.3750
800	0.015623	3555.60	4024.31	6.8332	0.011523	3517.89	3978.80	6.6662
900	0.017448	3768.48	4291.93	7.0717	0.012963	3739.42	4257.93	6.9150
1000	0.019196	3978.79	4554.68	7.2867	0.014324	3954.64	4527.59	7.1356
1100	0.020903	4189.18	4816.28	7.4845	0.015643	4167.38	4793.08	7.3364
1200	0.022589	4401.29	5078.97	7.6691	0.016940	4380.11	5057.72	7.5224
1300	0.024266	4615.96	5343.95	7.8432	0.018229	4594.28	5323.45	7.6969

Table B.1.4
Compressed Liquid Water

Temp. (°C)	v (m³/kg)	u (kJ/kg)	h (kJ/kg)	s (kJ/kg-K)	v (m³/kg)	u (kJ/kg)	h (kJ/kg)	s (kJ/kg-K)
	500 kPa (151.86)				2000 kPa (212.42)			
Sat.	0.001093	639.66	640.21	1.8606	0.001177	906.42	908.77	2.4473
0.01	0.000999	0.01	0.51	0.0000	0.000999	0.03	2.03	0.0001
20	0.001002	83.91	84.41	0.2965	0.001001	83.82	85.82	.2962
40	0.001008	167.47	167.98	0.5722	0.001007	167.29	169.30	.5716
60	0.001017	251.00	251.51	0.8308	0.001016	250.73	252.77	.8300
80	0.001029	334.73	335.24	1.0749	0.001028	334.38	336.44	1.0739
100	0.001043	418.80	419.32	1.3065	0.001043	418.36	420.45	1.3053
120	0.001060	503.37	503.90	1.5273	0.001059	502.84	504.96	1.5259
140	0.001080	588.66	589.20	1.7389	0.001079	588.02	590.18	1.7373
160	—	—	—	—	0.001101	674.14	676.34	1.9410
180	—	—	—	—	0.001127	761.46	763.71	2.1382
200	—	—	—	—	0.001156	850.30	852.61	2.3301
	5000 kPa (263.99)				10000 kPa (311.06)			
Sat	0.001286	1147.78	1154.21	2.9201	0.001452	1393.00	1407.53	3.3595
0	0.000998	0.03	5.02	0.0001	0.000995	0.10	10.05	0.0003
20	0.001000	83.64	88.64	0.2955	0.000997	83.35	93.32	0.2945
40	0.001006	166.93	171.95	0.5705	0.001003	166.33	176.36	0.5685
60	0.001015	250.21	255.28	0.8284	0.001013	249.34	259.47	0.8258
80	0.001027	333.69	338.83	1.0719	0.001025	332.56	342.81	1.0687
100	0.001041	417.50	422.71	1.3030	0.001039	416.09	426.48	1.2992
120	0.001058	501.79	507.07	1.5232	0.001055	500.07	510.61	1.5188
140	0.001077	586.74	592.13	1.7342	0.001074	584.67	595.40	1.7291
160	0.001099	672.61	678.10	1.9374	0.001195	670.11	681.07	1.9316
180	0.001124	759.62	765.24	2.1341	0.001120	756.63	767.83	2.1274
200	0.001153	848.08	853.85	2.3254	0.001148	844.49	855.97	2.3178
220	0.001187	938.43	944.36	2.5128	0.001181	934.07	945.88	2.5038
240	0.001226	1031.34	1037.47	2.6978	0.001219	1025.94	1038.13	2.6872
260	0.001275	1127.92	1134.30	2.8829	0.001265	1121.03	1133.68	2.8698
280					0.001322	1220.90	1234.11	3.0547
300					0.001397	1328.34	1342.31	3.2468

Temp. (°C)	v (m³/kg)	u (kJ/kg)	h (kJ/kg)	s (kJ/kg-K)	v (m³/kg)	u (kJ/kg)	h (kJ/kg)	s (kJ/kg-K)
	15000 kPa (342.24)				20000 kPa (365.81)			
Sat.	0.001658	1585.58	1610.45	3.6847	0.002035	1785.47	1826.18	4.0137
0	0.000993	0.15	15.04	0.0004	0.000990	0.20	20.00	0.0004
20	0.000995	83.05	97.97	0.2934	0.000993	82.75	102.61	0.2922
40	0.001001	165.73	180.75	0.5665	0.000999	165.15	185.14	0.5646
60	0.001011	248.49	263.65	0.8231	0.001008	247.66	267.82	0.8205
80	0.001022	331.46	346.79	1.0655	0.001020	330.38	350.78	1.0623
100	0.001036	414.72	430.26	1.2954	0.001034	413.37	434.04	1.2917
120	0.001052	498.39	514.17	1.5144	0.001050	496.75	517.74	1.5101
140	0.001071	582.64	598.70	1.7241	0.001068	580.67	602.03	1.7192
160	0.001092	667.69	684.07	1.9259	0.001089	665.34	687.11	1.9203
180	0.001116	753.74	770.48	2.1209	0.001112	750.94	773.18	2.1146
200	0.001143	841.04	858.18	2.3103	0.001139	837.70	860.47	2.3031
220	0.001175	929.89	947.52	2.4952	0.001169	925.89	949.27	2.4869
240	0.001211	1020.82	1038.99	2.6770	0.001205	1015.94	1040.04	2.6673
260	0.001255	1114.59	1133.41	2.8575	0.001246	1108.53	1133.45	2.8459
280	0.001308	1212.47	1232.09	3.0392	0.001297	1204.69	1230.62	3.0248
300	0.001377	1316.58	1337.23	3.2259	0.001360	1306.10	1333.29	3.2071
320	0.001472	1431.05	1453.13	3.4246	0.001444	1415.66	1444.53	3.3978
340	0.001631	1567.42	1591.88	3.6545	0.001568	1539.64	1571.01	3.6074
360					0.001823	1702.78	1739.23	3.8770
	30000 kPa				50000 kPa			
0	0.000986	0.25	29.82	0.0001	0.000977	0.20	49.03	−0.0014
20	0.000989	82.16	111.82	0.2898	0.000980	80.98	130.00	0.2847
40	0.000995	164.01	193.87	0.5606	0.000987	161.84	211.20	0.5526
60	0.001004	246.03	276.16	0.8153	0.000996	242.96	292.77	0.8051
80	0.001016	328.28	358.75	1.0561	0.001007	324.32	374.68	1.0439
100	0.001029	410.76	441.63	1.2844	0.001020	405.86	456.87	1.2703
120	0.001044	493.58	524.91	1.5017	0.001035	487.63	539.37	1.4857
140	0.001062	576.86	608.73	1.7097	0.001052	569.76	622.33	1.6915
160	0.001082	660.81	693.27	1.9095	0.001070	652.39	705.91	1.8890
180	0.001105	745.57	778.71	2.1024	0.001091	735.68	790.24	2.0793
200	0.001130	831.34	865.24	2.2892	0.001115	819.73	875.46	2.2634
220	0.001159	918.32	953.09	2.4710	0.001141	904.67	961.71	2.4419
240	0.001192	1006.84	1042.60	2.6489	0.001170	990.69	1049.20	2.6158
260	0.001230	1097.38	1134.29	2.8242	0.001203	1078.06	1138.23	2.7860
280	0.001275	1190.69	1228.96	2.9985	0.001242	1167.19	1229.26	2.9536
300	0.001330	1287.89	1327.80	3.1740	0.001286	1258.66	1322.95	3.1200
320	0.001400	1390.64	1432.63	3.3538	0.001339	1353.23	1420.17	3.2867
340	0.001492	1501.71	1546.47	3.5425	0.001403	1451.91	1522.07	3.4556
360	0.001627	1626.57	1675.36	3.7492	0.001484	1555.97	1630.16	3.6290
380	0.001869	1781.35	1837.43	4.0010	0.001588	1667.13	1746.54	3.8100

TABLE B.1.5
Saturated Solid–Saturated Vapor, Water

Temp. (°C)	Press. (kPa)	Specific Volume, m³/kg			Internal Energy, kJ/kg		
		Sat. Solid v_i	Evap. v_{ig}	Sat. Vapor v_g	Sat. Solid u_i	Evap. u_{ig}	Sat. Vapor u_g
0.01	0.6113	0.0010908	206.152	206.153	−333.40	2708.7	2375.3
0	0.6108	0.0010908	206.314	206.315	−333.42	2708.7	2375.3
−2	0.5177	0.0010905	241.662	241.663	−337.61	2710.2	2372.5
−4	0.4376	0.0010901	283.798	283.799	−341.78	2711.5	2369.8
−6	0.3689	0.0010898	334.138	334.139	−345.91	2712.9	2367.0
−8	0.3102	0.0010894	394.413	394.414	−350.02	2714.2	2364.2
−10	0.2601	0.0010891	466.756	466.757	−354.09	2715.5	2361.4
−12	0.2176	0.0010888	553.802	553.803	−358.14	2716.8	2358.7
−14	0.1815	0.0010884	658.824	658.824	−362.16	2718.0	2355.9
−16	0.1510	0.0010881	785.906	785.907	−366.14	2719.2	2353.1
−18	0.1252	0.0010878	940.182	940.183	−370.10	2720.4	2350.3
−20	0.10355	0.0010874	1128.112	1128.113	−374.03	2721.6	2347.5
−22	0.08535	0.0010871	1357.863	1357.864	−377.93	2722.7	2344.7
−24	0.07012	0.0010868	1639.752	1639.753	−381.80	2723.7	2342.0
−26	0.05741	0.0010864	1986.775	1986.776	−385.64	2724.8	2339.2
−28	0.04684	0.0010861	2415.200	2415.201	−389.45	2725.8	2336.4
−30	0.03810	0.0010858	2945.227	2945.228	−393.23	2726.8	2333.6
−32	0.03090	0.0010854	3601.822	3601.823	−396.98	2727.8	2330.8
−34	0.02499	0.0010851	4416.252	4416.253	−400.71	2728.7	2328.0
−36	0.02016	0.0010848	5430.115	5430.116	−404.40	2729.6	2325.2
−38	0.01618	0.0010844	6707.021	6707.022	−408.06	2730.5	2322.4
−40	0.01286	0.0010841	8366.395	8366.396	−411.70	2731.3	2319.6

TABLE B.1.5 (*continued*)
Saturated Solid-Saturated Vapor, Water

Temp. (°C)	Press. (kPa)	ENTHALPY, kJ/kg			ENTROPY, kJ/kg-K		
		Sat. Solid h_i	Evap. h_{ig}	Sat. Vapor h_g	Sat. Solid s_i	Evap. s_{ig}	Sat. Vapor s_g
0.01	0.6113	−333.40	2834.7	2501.3	−1.2210	10.3772	9.1562
0	0.6108	−333.42	2834.8	2501.3	−1.2211	10.3776	9.1565
−2	0.5177	−337.61	2835.3	2497.6	−1.2369	10.4562	9.2193
−4	0.4376	−341.78	2835.7	2494.0	−1.2526	10.5358	9.2832
−6	0.3689	−345.91	2836.2	2490.3	−1.2683	10.6165	9.3482
−8	0.3102	−350.02	2836.6	2486.6	−1.2839	10.6982	9.4143
−10	0.2601	−354.09	2837.0	2482.9	−1.2995	10.7809	9.4815
−12	0.2176	−358.14	2837.3	2479.2	−1.3150	10.8648	9.5498
−14	0.1815	−362.16	2837.6	2475.5	−1.3306	10.9498	9.6192
−16	0.1510	−366.14	2837.9	2471.8	−1.3461	11.0359	9.6898
−18	0.1252	−370.10	2838.2	2468.1	−1.3617	11.1233	9.7616
−20	0.10355	−374.03	2838.4	2464.3	−1.3772	11.2120	9.8348
−22	0.08535	−377.93	2838.6	2460.6	−1.3928	11.3020	9.9093
−24	0.07012	−381.80	2838.7	2456.9	−1.4083	11.3935	9.9852
−26	0.05741	−385.64	2838.9	2453.2	−1.4239	11.4864	10.0625
−28	0.04684	−389.45	2839.0	2449.5	−1.4394	11.5808	10.1413
−30	0.03810	−393.23	2839.0	2445.8	−1.4550	11.6765	10.2215
−32	0.03090	−396.98	2839.1	2442.1	−1.4705	11.7733	10.3028
−34	0.02499	−400.71	2839.1	2438.4	−1.4860	11.8713	10.3853
−36	0.02016	−404.40	2839.1	2434.7	−1.5014	11.9704	10.4690
−38	0.01618	−408.06	2839.0	2431.0	−1.5168	12.0714	10.5546
−40	0.01286	−411.70	2838.9	2427.2	−1.5321	12.1768	10.6447

Table B.2
Thermodynamic Properties of Ammonia

Table B.2.1
Saturated Ammonia

Temp. (°C)	Press. (kPa)	SPECIFIC VOLUME, m³/kg			INTERNAL ENERGY, kJ/kg		
		Sat. Liquid v_f	Evap. v_{fg}	Sat. Vapor v_g	Sat. Liquid u_f	Evap. u_{fg}	Sat. Vapor u_g
−50	40.9	0.001424	2.62557	2.62700	−43.82	1309.1	1265.2
−45	54.5	0.001437	2.00489	2.00632	−22.01	1293.5	1271.4
−40	71.7	0.001450	1.55111	1.55256	−0.10	1277.6	1277.4
−35	93.2	0.001463	1.21466	1.21613	21.93	1261.3	1283.3
−30	119.5	0.001476	0.96192	0.96339	44.08	1244.8	1288.9
−25	151.6	0.001490	0.76970	0.77119	66.36	1227.9	1294.3
−20	190.2	0.001504	0.62184	0.62334	88.76	1210.7	1299.5
−15	236.3	0.001519	0.50686	0.50838	111.30	1193.2	1304.5
−10	290.9	0.001534	0.41655	0.41808	133.96	1175.2	1309.2
−5	354.9	0.001550	0.34493	0.34648	156.76	1157.0	1313.7
0	429.6	0.001566	0.28763	0.28920	179.69	1138.3	1318.0
5	515.9	0.001583	0.24140	0.24299	202.77	1119.2	1322.0
10	615.2	0.001600	0.20381	0.20541	225.99	1099.7	1325.7
15	728.6	0.001619	0.17300	0.17462	249.36	1079.7	1329.1
20	857.5	0.001638	0.14758	0.14922	272.89	1059.3	1332.2
25	1003.2	0.001658	0.12647	0.12813	296.59	1038.4	1335.0
30	1167.0	0.001680	0.10881	0.11049	320.46	1016.9	1337.4
35	1350.4	0.001702	0.09397	0.09567	344.50	994.9	1339.4
40	1554.9	0.001725	0.08141	0.08313	368.74	972.2	1341.0
45	1782.0	0.001750	0.07073	0.07248	393.19	948.9	1342.1
50	2033.1	0.001777	0.06159	0.06337	417.87	924.8	1342.7
55	2310.1	0.001804	0.05375	0.05555	442.79	899.9	1342.7
60	2614.4	0.001834	0.04697	0.04880	467.99	874.2	1342.1
65	2947.8	0.001866	0.04109	0.04296	493.51	847.4	1340.9
70	3312.0	0.001900	0.03597	0.03787	519.39	819.5	1338.9
75	3709.0	0.001937	0.03148	0.03341	545.70	790.4	1336.1
80	4140.5	0.001978	0.02753	0.02951	572.50	759.9	1332.4
85	4608.6	0.002022	0.02404	0.02606	599.90	727.8	1327.7
90	5115.3	0.002071	0.02093	0.02300	627.99	693.7	1321.7
95	5662.9	0.002126	0.01815	0.02028	656.95	657.4	1314.4
100	6253.7	0.002188	0.01565	0.01784	686.96	618.4	1305.3
105	6890.4	0.002261	0.01337	0.01564	718.30	575.9	1294.2
110	7575.7	0.002347	0.01128	0.01363	751.37	529.1	1280.5
115	8313.3	0.002452	0.00933	0.01178	786.82	476.2	1263.1
120	9107.2	0.002589	0.00744	0.01003	825.77	414.5	1240.3
125	9963.5	0.002783	0.00554	0.00833	870.69	337.7	1208.4
130	10891.6	0.003122	0.00337	0.00649	929.29	226.9	1156.2
132.3	11333.2	0.004255	0	0.00426	1037.62	0	1037.6

Table B.2.1 (*continued*)
Saturated Ammonia

Temp. (°C)	Press. (kPa)	ENTHALPY, kJ/kg			ENTROPY, kJ/kg-K		
		Sat. Liquid h_f	Evap. h_{fg}	Sat. Vapor h_g	Sat. Liquid s_f	Evap. s_{fg}	Sat. Vapor s_g
−50	40.9	−43.76	1416.3	1372.6	−0.1916	6.3470	6.1554
−45	54.5	−21.94	1402.8	1380.8	−0.0950	6.1484	6.0534
−40	71.7	0	1388.8	1388.8	0	5.9567	5.9567
−35	93.2	22.06	1374.5	1396.5	0.0935	5.7715	5.8650
−30	119.5	44.26	1359.8	1404.0	0.1856	5.5922	5.7778
−25	151.6	66.58	1344.6	1411.2	0.2763	5.4185	5.6947
−20	190.2	89.05	1329.0	1418.0	0.3657	5.2498	5.6155
−15	236.3	111.66	1312.9	1424.6	0.4538	5.0859	5.5397
−10	290.9	134.41	1296.4	1430.8	0.5408	4.9265	5.4673
−5	354.9	157.31	1279.4	1436.7	0.6266	4.7711	5.3977
0	429.6	180.36	1261.8	1442.2	0.7114	4.6195	5.3309
5	515.9	203.58	1243.7	1447.3	0.7951	4.4715	5.2666
10	615.2	226.97	1225.1	1452.0	0.8779	4.3266	5.2045
15	728.6	250.54	1205.8	1456.3	0.9598	4.1846	5.1444
20	857.5	274.30	1185.9	1460.2	1.0408	4.0452	5.0860
25	1003.2	298.25	1165.2	1463.5	1.1210	3.9083	5.0293
30	1167.0	322.42	1143.9	1466.3	1.2005	3.7734	4.9738
35	1350.4	346.80	1121.8	1468.6	1.2792	3.6403	4.9196
40	1554.9	371.43	1098.8	1470.2	1.3574	3.5088	4.8662
45	1782.0	396.31	1074.9	1471.2	1.4350	3.3786	4.8136
50	2033.1	421.48	1050.0	1471.5	1.5121	3.2493	4.7614
55	2310.1	446.96	1024.1	1471.0	1.5888	3.1208	4.7095
60	2614.4	472.79	997.0	1469.7	1.6652	2.9925	4.6577
65	2947.8	499.01	968.5	1467.5	1.7415	2.8642	4.6057
70	3312.0	525.69	938.7	1464.4	1.8178	2.7354	4.3533
75	3709.0	552.88	907.2	1460.1	1.8943	2.6058	4.5001
80	4140.5	580.69	873.9	1454.6	1.9712	2.4746	4.4458
85	4608.6	609.21	838.6	1447.8	2.0488	2.3413	4.3901
90	5115.3	638.59	800.8	1439.4	2.1273	2.2051	4.3325
95	5662.9	668.99	760.2	1429.2	2.2073	2.0650	4.2723
100	6253.7	700.64	716.2	1416.9	2.2893	1.9195	4.2088
105	6890.4	733.87	668.1	1402.0	2.3740	1.7667	4.1407
110	7575.7	769.15	614.6	1383.7	2.4625	1.6040	4.0665
115	8313.3	807.21	553.8	1361.0	2.5566	1.4267	3.9833
120	9107.2	849.36	482.3	1331.7	2.6593	1.2268	3.8861
125	9963.5	898.42	393.0	1291.4	2.7775	0.9870	3.7645
130	10892	963.29	263.7	1227.0	2.9326	0.6540	3.5866
132.3	11333	1085.85	0	1085.9	3.2316	0	3.2316

Table B.2.2
Superheated Ammonia

Temp. (°C)	v (m³/kg)	u (kJ/kg)	h (kJ/kg)	s (kJ/kg-K)	v (m³/kg)	u (kJ/kg)	h (kJ/kg)	s (kJ/kg-K)
	50 kPa (−46.53)				100 kPa (−33.60)			
Sat.	2.1752	1269.6	1378.3	6.0839	1.1381	1284.9	1398.7	5.8401
−30	2.3448	1296.2	1413.4	6.2333	1.1573	1291.0	1406.7	5.8734
−20	2.4463	1312.3	1434.6	6.3187	1.2101	1307.8	1428.8	5.9626
−10	2.5471	1328.4	1455.7	6.4006	1.2621	1324.6	1450.8	6.0477
0	2.6474	1344.5	1476.9	6.4795	1.3136	1341.3	1472.6	6.1291
10	2.7472	1360.7	1498.1	6.5556	1.3647	1357.9	1494.4	6.2073
20	2.8466	1377.0	1519.3	6.6293	1.4153	1374.5	1516.1	6.2826
30	2.9458	1393.3	1540.6	6.7008	1.4657	1391.2	1537.7	6.3553
40	3.0447	1409.8	1562.0	6.7703	1.5158	1407.9	1559.5	6.4258
50	3.1435	1426.3	1583.5	6.8379	1.5658	1424.7	1581.2	6.4943
60	3.2421	1443.0	1605.1	6.9038	1.6156	1441.5	1603.1	6.5609
70	3.3406	1459.9	1626.9	6.9682	1.6653	1458.5	1625.1	6.6258
80	3.4390	1476.9	1648.8	7.0312	1.7148	1475.6	1647.1	6.6892
100	3.6355	1511.4	1693.2	7.1533	1.8137	1510.3	1691.7	6.8120
120	3.8318	1546.6	1738.2	7.2708	1.9124	1545.7	1736.9	6.9300
140	4.0280	1582.5	1783.9	7.3842	2.0109	1581.7	1782.8	7.0439
160	4.2240	1619.2	1830.4	7.4941	2.1093	1618.5	1829.4	7.1540
180	4.4199	1656.7	1877.7	7.6008	2.2075	1656.0	1876.8	7.2609
200	4.6157	1694.9	1925.7	7.7045	2.3057	1694.3	1924.9	7.3648
	150 kPa (−25.22)				200 kPa (−18.86)			
Sat.	0.7787	1294.1	1410.9	5.6983	0.5946	1300.6	1419.6	5.5979
−20	0.7977	1303.3	1422.9	5.7465	—	—	—	—
−10	0.8336	1320.7	1445.7	5.8349	0.6193	1316.7	1440.6	5.6791
0	0.8689	1337.9	1468.3	5.9189	0.6465	1334.5	1463.8	5.7659
10	0.9037	1355.0	1490.6	5.9992	0.6732	1352.1	1486.8	5.8484
20	0.9382	1372.0	1512.8	6.0761	0.6995	1369.5	1509.4	5.9270
30	0.9723	1389.0	1534.9	6.1502	0.7255	1386.8	1531.9	6.0025
40	1.0062	1406.0	1556.9	6.2217	0.7513	1404.0	1554.3	6.0751
50	1.0398	1423.0	1578.9	6.2910	0.7769	1421.3	1576.6	6.1453
60	1.0734	1440.0	1601.0	6.3583	0.8023	1438.5	1598.9	6.2133
70	1.1068	1457.2	1623.2	6.4238	0.8275	1455.8	1621.3	6.2794
80	1.1401	1474.4	1645.4	6.4877	0.8527	1473.1	1643.7	6.3437
100	1.2065	1509.3	1690.2	6.6112	0.9028	1508.2	1688.8	6.4679
120	1.2726	1544.8	1735.6	6.7297	0.9527	1543.8	1734.4	6.5869
140	1.3386	1580.9	1781.7	6.8439	1.0024	1580.1	1780.6	6.7015
160	1.4044	1617.8	1828.4	6.9544	1.0519	1617.0	1827.4	6.8123
180	1.4701	1655.4	1875.9	7.0615	1.1014	1654.7	1875.0	6.9196
200	1.5357	1693.7	1924.1	7.1656	1.1507	1693.2	1923.3	7.0239
220	1.6013	1732.9	1973.1	7.2670	1.2000	1732.4	1972.4	7.1255

TABLE B.2.2 (*continued*)
Superheated Ammonia

Temp. (°C)	v (m³/kg)	u (kJ/kg)	h (kJ/kg)	s (kJ/kg-K)	v (m³/kg)	u (kJ/kg)	h (kJ/kg)	s (kJ/kg-K)
	300 kPa (−9.24)				400 kPa (−1.89)			
Sat	0.40607	1309.9	1431.7	5.4565	0.30942	1316.4	1440.2	5.3559
0	0.42382	1327.5	1454.7	5.5420	0.31227	1320.2	1445.1	5.3741
10	0.44251	1346.1	1478.9	5.6290	0.32701	1339.9	1470.7	5.4663
20	0.46077	1364.4	1502.6	5.7113	0.34129	1359.1	1495.6	5.5525
30	0.47870	1382.3	1526.0	5.7896	0.35520	1377.7	1519.8	5.6338
40	0.49636	1400.1	1549.0	5.8645	0.36884	1396.1	1543.6	5.7111
50	0.51382	1417.8	1571.9	5.9365	0.38226	1414.2	1567.1	5.7850
60	0.53111	1435.4	1594.7	6.0060	0.39550	1432.2	1590.4	5.8560
70	0.54827	1453.0	1617.5	6.0732	0.40860	1450.1	1613.6	5.9244
80	0.56532	1470.6	1640.2	6.1385	0.42160	1468.0	1636.7	5.9907
100	0.59916	1506.1	1685.8	6.2642	0.44732	1503.9	1682.8	6.1179
120	0.63276	1542.0	1731.8	6.3842	0.47279	1540.1	1729.2	6.2390
140	0.66618	1578.5	1778.3	6.4996	0.49808	1576.8	1776.0	6.3552
160	0.69946	1615.6	1825.4	6.6109	0.52323	1614.1	1823.4	6.4671
180	0.73263	1653.4	1873.2	6.7188	0.54827	1652.1	1871.4	6.5755
200	0.76572	1692.0	1921.7	6.8235	0.57321	1690.8	1920.1	6.6806
220	0.79872	1731.3	1970.9	6.9254	0.59809	1730.3	1969.5	6.7828
240	0.83167	1771.4	2020.9	7.0247	0.62289	1770.5	2019.6	6.8825
260	0.86455	1812.2	2071.6	7.1217	0.64764	1811.4	2070.5	6.9797
	500 kPa (4.13)				600 kPa (9.28)			
Sat.	0.25035	1321.3	1446.5	5.2776	0.21038	1325.2	1451.4	5.2133
10	0.25757	1333.5	1462.3	5.3340	0.21115	1326.7	1453.4	5.2205
20	0.26949	1353.6	1488.3	5.4244	0.22154	1347.9	1480.8	5.3156
30	0.28103	1373.0	1513.5	5.5090	0.23152	1368.2	1507.1	5.4037
40	0.29227	1392.0	1538.1	5.5889	0.24118	1387.8	1532.5	5.4862
50	0.30328	1410.6	1562.2	5.6647	0.25059	1406.9	1557.3	5.5641
60	0.31410	1429.0	1586.1	5.7373	0.25981	1425.7	1581.6	5.6383
70	0.32478	1447.3	1609.6	5.8070	0.26888	1444.3	1605.7	5.7094
80	0.33535	1465.4	1633.1	5.8744	0.27783	1462.8	1629.5	5.7778
100	0.35621	1501.7	1679.8	6.0031	0.29545	1499.5	1676.8	5.9081
120	0.37681	1538.2	1726.6	6.1253	0.31281	1536.3	1724.0	6.0314
140	0.39722	1575.2	1773.8	6.2422	0.32997	1573.5	1771.5	6.1491
160	0.41748	1612.7	1821.4	6.3548	0.34699	1611.2	1819.4	6.2623
180	0.43764	1650.8	1869.6	6.4636	0.36389	1649.5	1867.8	6.3717
200	0.45771	1689.6	1918.5	6.5691	0.38071	1688.5	1916.9	6.4776
220	0.47770	1729.2	1968.1	6.6717	0.39745	1728.2	1966.6	6.5806
240	0.49763	1769.5	2018.3	6.7717	0.41412	1768.6	2017.0	6.6808
260	0.51749	1810.6	2069.3	6.8692	0.43073	1809.8	2068.2	6.7786

TABLE B.2.2 *(continued)*
Superheated Ammonia

Temp. (°C)	v (m³/kg)	u (kJ/kg)	h (kJ/kg)	s (kJ/kg-K)	v (m³/kg)	u (kJ/kg)	h (kJ/kg)	s (kJ/kg-K)
	800 kPa (17.85)				1000 kPa (24.90)			
Sat.	0.15958	1330.9	1458.6	5.1110	0.12852	1334.9	1463.4	5.0304
20	0.16138	1335.8	1464.9	5.1328	—	—	—	—
30	0.16947	1358.0	1493.5	5.2287	0.13206	1347.1	1479.1	5.0826
40	0.17720	1379.0	1520.8	5.3171	0.13868	1369.8	1508.5	5.1778
50	0.18465	1399.3	1547.0	5.3996	0.14499	1391.3	1536.3	5.2654
60	0.19189	1419.0	1572.5	5.4774	0.15106	1412.1	1563.1	5.3471
70	0.19896	1438.3	1597.5	5.5513	0.15695	1432.2	1589.1	5.4240
80	0.20590	1457.4	1622.1	5.6219	0.16270	1451.9	1614.6	5.4971
100	0.21949	1495.0	1670.6	5.7555	0.17389	1490.5	1664.3	5.6342
120	0.23280	1532.5	1718.7	5.8811	0.18477	1528.6	1713.4	5.7622
140	0.24590	1570.1	1766.9	6.0006	0.19545	1566.8	1762.2	5.8834
160	0.25886	1608.2	1815.3	6.1150	0.20597	1605.2	1811.2	5.9992
180	0.27170	1646.8	1864.2	6.2254	0.21638	1644.2	1860.5	6.1105
200	0.28445	1686.1	1913.6	6.3322	0.22669	1683.7	1910.4	6.2182
220	0.29712	1726.0	1963.7	6.4358	0.23693	1723.9	1960.8	6.3226
240	0.30973	1766.7	2014.5	6.5367	0.24710	1764.8	2011.9	6.4241
260	0.32228	1808.1	2065.9	6.6350	0.25720	1806.4	2063.6	6.5229
280	0.33477	1850.2	2118.0	6.7310	0.26726	1848.8	2116.0	6.6194
300	0.34722	1893.1	2170.9	6.8248	0.27726	1891.8	2169.1	6.7137
	1200 kPa (30.94)				1400 kPa (36.26)			
Sat.	0.10751	1337.8	1466.8	4.9635	0.09231	1339.8	1469.0	4.9060
40	0.11287	1360.0	1495.4	5.0564	0.09432	1349.5	1481.6	4.9463
50	0.11846	1383.0	1525.1	5.1497	0.09942	1374.2	1513.4	5.0462
60	0.12378	1404.8	1553.3	5.2357	0.10423	1397.2	1543.1	5.1370
70	0.12890	1425.8	1580.5	5.3159	0.10882	1419.2	1571.5	5.2209
80	0.13387	1446.2	1606.8	5.3916	0.11324	1440.3	1598.8	5.2994
100	0.14347	1485.8	1658.0	5.5325	0.12172	1481.0	1651.4	5.4443
120	0.15275	1524.7	1708.0	5.6631	0.12986	1520.7	1702.5	5.5775
140	0.16181	1563.3	1757.5	5.7860	0.13777	1559.9	1752.8	5.7023
160	0.17071	1602.2	1807.1	5.9031	0.14552	1599.2	1802.9	5.8208
180	0.17950	1641.5	1856.9	6.0156	0.15315	1638.8	1853.2	5.9343
200	0.18819	1681.3	1907.1	6.1241	0.16068	1678.9	1903.8	6.0437
220	0.19680	1721.8	1957.9	6.2292	0.16813	1719.6	1955.0	6.1495
240	0.20534	1762.9	2009.3	6.3313	0.17551	1761.0	2006.7	6.2523
260	0.21382	1804.7	2061.3	6.4308	0.18283	1803.0	2059.0	6.3523
280	0.22225	1847.3	2114.0	6.5278	0.19010	1845.8	2111.9	6.4498
300	0.23063	1890.6	2167.3	6.6225	0.19732	1889.3	2165.5	6.5450
320	0.23897	1934.6	2221.3	6.7151	0.20450	1933.5	2219.8	6.6380

Table B.2.2 (*continued*)
Superheated Ammonia

Temp. (°C)	v (m³/kg)	u (kJ/kg)	h (kJ/kg)	s (kJ/kg-K)	v (m³/kg)	u (kJ/kg)	h (kJ/kg)	s (kJ/kg-K)
	1600 kPa (41.03)				2000 kPa (49.37)			
Sat.	0.08079	1341.2	1470.5	4.8553	0.06444	1342.6	1471.5	4.7680
50	0.08506	1364.9	1501.0	4.9510	0.06471	1344.5	1473.9	4.7754
60	0.08951	1389.3	1532.5	5.0472	0.06875	1372.3	1509.8	4.8848
70	0.09372	1412.3	1562.3	5.1351	0.07246	1397.8	1542.7	4.9821
80	0.09774	1434.3	1590.6	5.2167	0.07595	1421.6	1573.5	5.0707
100	0.10539	1476.2	1644.8	5.3659	0.08248	1466.1	1631.1	5.2294
200	0.11268	1516.6	1696.9	5.5018	0.08861	1508.3	1685.5	5.3714
140	0.11974	1556.4	1748.0	5.6286	0.09447	1549.3	1738.2	5.5022
160	0.12662	1596.1	1798.7	5.7485	0.10016	1589.9	1790.2	5.6251
180	0.13339	1636.1	1849.5	5.8631	0.10571	1630.6	1842.0	5.7420
200	0.14005	1676.5	1900.5	5.9734	0.11116	1671.6	1893.9	5.8540
220	0.14663	1717.4	1952.0	6.0800	0.11652	1713.1	1946.1	5.9621
240	0.15314	1759.0	2004.1	6.1834	0.12182	1755.2	1998.8	6.0668
260	0.15959	1801.3	2056.7	6.2839	0.12705	1797.9	2052.0	6.1685
280	0.16599	1844.3	2109.9	6.3819	0.13224	1841.3	2105.8	6.2675
300	0.17234	1888.0	2163.7	6.4775	0.13737	1885.4	2160.1	6.3641
320	0.17865	1932.4	2218.2	6.5710	0.14246	1930.2	2215.1	6.4583
340	0.18492	1977.5	2273.4	6.6624	0.14751	1975.6	2270.7	6.5505
360	0.19115	2023.3	2329.1	6.7519	0.15253	2021.8	2326.8	6.6406
	5000 kPa (88.90)				10000 kPa (125.20)			
Sat.	0.02365	1323.2	1441.4	4.3454	0.00826	1206.8	1289.4	3.7587
100	0.02636	1369.7	1501.5	4.5091	—	—	—	—
120	0.03024	1435.1	1586.3	4.7306	—	—	—	—
140	0.03350	1489.8	1657.3	4.9068	0.01195	1341.8	1461.3	4.1839
160	0.03643	1539.5	1721.7	5.0591	0.01461	1432.2	1578.3	4.4610
180	0.03916	1586.9	1782.7	5.1968	0.01666	1500.6	1667.2	4.6617
200	0.04174	1633.1	1841.8	5.3245	0.01842	1560.3	1744.5	4.8287
220	0.04422	1678.9	1900.0	5.4450	0.02001	1615.8	1816.0	4.9767
240	0.04662	1724.8	1957.9	5.5600	0.02150	1669.2	1884.2	5.1123
260	0.04895	1770.9	2015.6	5.6704	0.02290	1721.6	1950.6	5.2392
280	0.05123	1817.4	2073.6	5.7771	0.02424	1773.6	2015.9	5.3596
300	0.05346	1864.5	2131.8	5.8805	0.02552	1825.5	2080.7	5.4746
320	0.05565	1912.1	2190.3	5.9809	0.02676	1877.6	2145.2	5.5852
340	0.05779	1960.3	2249.2	6.0786	0.02796	1930.0	2209.6	5.6921
360	0.05990	2009.1	2308.6	6.1738	0.02913	1982.8	2274.1	5.7955
380	0.06198	2058.5	2368.4	6.2668	0.03026	2036.1	2338.7	5.8960
400	0.06403	2108.4	2428.6	6.3576	0.03137	2089.8	2403.5	5.9937
420	0.06606	2159.0	2489.3	6.4464	0.03245	2143.9	2468.5	6.0888
440	0.06806	2210.1	2550.4	6.5334	0.03351	2198.5	2533.7	6.1815

TABLE B.3
Thermodynamic Properties of R-12

TABLE B.3.1
Saturated R-12

Temp. (°C)	Press. (kPa)	SPECIFIC VOLUME, m³/kg			INTERNAL ENERGY, kJ/kg		
		Sat. Liquid v_f	Evap. v_{fg}	Sat. Vapor v_g	Sat. Liquid u_f	Evap. u_{fg}	Sat. Vapor u_g
−90	2.8	0.000608	4.41494	4.41555	−43.29	177.20	133.91
−80	6.2	0.000617	2.13773	2.13835	−34.73	172.54	137.82
−70	12.3	0.000627	1.12665	1.12728	−26.14	167.94	141.81
−60	22.6	0.000637	0.63727	0.63791	−17.50	163.36	145.86
−50	39.1	0.000648	0.38246	0.38310	−8.80	158.76	149.95
−45	50.4	0.000654	0.30203	0.30268	−4.43	156.44	152.01
−40	64.2	0.000659	0.24125	0.24191	−0.04	154.11	154.07
−35	80.7	0.000666	0.19473	0.19540	4.37	151.77	156.13
−30	100.4	0.000672	0.15870	0.15937	8.79	149.40	158.19
−29.8	101.3	0.000672	0.15736	0.15803	8.98	149.30	158.28
−25	123.7	0.000679	0.13049	0.13117	13.24	147.01	160.25
−20	150.9	0.000685	0.10816	0.10885	17.71	144.59	162.31
−15	182.6	0.000693	0.09033	0.09102	22.20	142.15	164.35
−10	219.1	0.000700	0.07595	0.07665	26.72	139.67	166.39
−5	261.0	0.000708	0.06426	0.06496	31.26	137.16	168.42
0	308.6	0.000716	0.05467	0.05539	35.83	134.61	170.44
5	362.6	0.000724	0.04676	0.04749	40.43	132.01	172.44
10	423.3	0.000733	0.04018	0.04091	45.06	129.36	174.42
15	491.4	0.000743	0.03467	0.03541	49.73	126.65	176.38
20	567.3	0.000752	0.03003	0.03078	54.45	123.87	178.32
25	651.6	0.000763	0.02609	0.02685	59.21	121.03	180.23
30	744.9	0.000774	0.02273	0.02351	64.02	118.09	182.11
35	847.7	0.000786	0.01986	0.02064	68.88	115.06	183.95
40	960.7	0.000798	0.01737	0.01817	73.82	111.92	185.74
45	1084.3	0.000811	0.01522	0.01603	78.83	108.66	187.49
50	1219.3	0.000826	0.01334	0.01417	83.93	105.24	189.17
55	1366.3	0.000841	0.01170	0.01254	89.12	101.66	190.78
60	1525.9	0.000858	0.01025	0.01111	94.43	97.88	192.31
65	1698.8	0.000877	0.00897	0.00985	99.87	93.86	193.73
70	1885.8	0.000897	0.00783	0.00873	105.46	89.56	195.03
75	2087.5	0.000920	0.00680	0.00772	111.23	84.94	196.17
80	2304.6	0.000946	0.00588	0.00682	117.21	79.90	197.11
85	2538.0	0.000976	0.00503	0.00600	123.45	74.34	197.80
90	2788.5	0.001012	0.00425	0.00526	130.02	68.12	198.14
95	3056.9	0.001056	0.00351	0.00456	137.01	60.98	197.99
100	3344.1	0.001113	0.00279	0.00390	144.59	52.48	197.07
105	3650.9	0.001197	0.00205	0.00324	153.15	41.58	194.73
110	3978.5	0.001364	0.00110	0.00246	164.12	24.08	188.20
112.0	4116.8	0.001792	0	0.00179	176.06	0	176.06

TABLE B.3.1 (*continued*)
Saturated R-12

Temp. (°C)	Press. (kPa)	ENTHALPY, kJ/kg			ENTROPY, kJ/kg-K		
		Sat. Liquid h_f	Evap. h_{fg}	Sat. Vapor h_g	Sat. Liquid s_f	Evap. s_{fg}	Sat. Vapor s_g
−90	2.8	−43.28	189.75	146.46	−0.2086	1.0359	0.8273
−80	6.2	−34.72	185.74	151.02	−0.1631	0.9616	0.7984
−70	12.3	−26.13	181.76	155.64	−0.1198	0.8947	0.7749
−60	22.6	−17.49	177.77	160.29	−0.0783	0.8340	0.7557
−50	39.1	−8.78	173.73	164.95	−0.0384	0.7785	0.7401
−45	50.4	−4.40	171.68	167.28	−0.0190	0.7524	0.7334
−40	64.2	0	169.59	169.59	0	0.7274	0.7274
−35	80.7	4.42	167.48	171.90	0.0187	0.7032	0.7219
−30	100.4	8.86	165.34	174.20	0.0371	0.6799	0.7170
−29.8	101.3	9.05	165.24	174.29	0.0379	0.6790	0.7168
−25	123.7	13.33	163.15	176.48	0.0552	0.6574	0.7126
−20	150.9	17.82	160.92	178.74	0.0731	0.6356	0.7087
−15	182.6	22.33	158.64	180.97	0.0906	0.6145	0.7051
−10	219.1	26.87	156.31	183.19	0.1080	0.5940	0.7019
−5	261.0	31.45	153.93	185.37	0.1251	0.5740	0.6991
0	308.6	36.05	151.48	187.53	0.1420	0.5545	0.6965
5	362.6	40.69	148.96	189.65	0.1587	0.5355	0.6942
10	423.3	45.37	146.37	191.74	0.1752	0.5169	0.6921
15	491.4	50.10	143.68	193.78	0.1915	0.4986	0.6902
20	567.3	54.87	140.91	195.78	0.2078	0.4806	0.6884
25	651.6	59.70	138.03	197.73	0.2239	0.4629	0.6868
30	744.9	64.59	135.03	199.62	0.2399	0.4454	0.6853
35	847.7	69.55	131.90	201.45	0.2559	0.4280	0.6839
40	960.7	74.59	128.61	203.20	0.2718	0.4107	0.6825
45	1084.3	79.71	125.16	204.87	0.2877	0.3934	0.6811
50	1219.3	84.94	121.51	206.45	0.3037	0.3760	0.6797
55	1366.3	90.27	117.65	207.92	0.3197	0.3585	0.6782
60	1525.9	95.74	113.52	209.26	0.3358	0.3407	0.6765
65	1698.8	101.36	109.10	210.46	0.3521	0.3226	0.6747
70	1885.8	107.15	104.33	211.48	0.3686	0.3040	0.6726
75	2087.5	113.15	99.14	212.29	0.3854	0.2847	0.6702
80	2304.6	119.39	93.44	212.83	0.4027	0.2646	0.6672
85	2538.0	125.93	87.11	213.04	0.4204	0.2432	0.6636
90	2788.5	132.84	79.96	212.80	0.4389	0.2202	0.6590
95	3056.9	140.23	71.71	211.94	0.4583	0.1948	0.6531
100	3344.1	148.31	61.81	210.12	0.4793	0.1656	0.6449
105	3650.9	157.52	49.05	206.57	0.5028	0.1297	0.6325
110	3978.5	169.55	28.44	197.99	0.5333	0.0742	0.6076
112.0	4116.8	183.43	0	183.43	0.5689	0	0.5689

TABLE B.3.2
Superheated R-12

Temp. (°C)	v (m³/kg)	u (kJ/kg)	h (kJ/kg)	s (kJ/kg-K)	v (m³/kg)	u (kJ/kg)	h (kJ/kg)	s (kJ/kg-K)
	50 kPa (−45.18)				100 kPa (−33.10)			
Sat.	0.30515	151.94	167.19	0.7336	0.15999	158.15	174.15	0.7171
−30	0.32738	159.18	175.55	0.7691	0.16006	158.20	174.21	0.7174
−20	0.34186	164.08	181.17	0.7917	0.16770	163.22	179.99	0.7406
−10	0.35623	169.08	186.89	0.8139	0.17522	168.32	185.84	0.7633
0	0.37051	174.18	192.70	0.8356	0.18265	173.50	191.77	0.7854
10	0.38472	179.38	198.61	0.8568	0.18999	178.77	197.77	0.8070
20	0.39886	184.67	204.62	0.8776	0.19728	184.13	203.85	0.8281
30	0.41296	190.06	210.71	0.8981	0.20451	189.57	210.02	0.8488
40	0.42701	195.54	216.89	0.9181	0.21169	195.09	216.26	0.8691
50	0.44103	201.11	223.16	0.9378	0.21884	200.70	222.58	0.8889
60	0.45502	206.76	229.51	0.9572	0.22596	206.39	228.98	0.9084
70	0.46898	212.50	235.95	0.9762	0.23305	212.15	235.46	0.9276
80	0.48292	218.31	242.46	0.9949	0.24011	218.00	242.01	0.9464
90	0.49684	224.21	249.05	1.0133	0.24716	223.91	248.63	0.9649
100	0.51074	230.18	255.71	1.0314	0.25419	229.90	255.32	0.9831
110	0.52463	236.22	262.45	1.0493	0.26121	235.96	262.08	1.0009
120	0.53851	242.33	269.26	1.0668	0.26821	242.09	268.91	1.0185
	200 kPa (−12.53)				400 kPa (8.15)			
Sat.	0.08354	165.36	182.07	0.7035	0.04321	173.69	190.97	0.6928
0	0.08861	172.08	189.80	0.7325	—	—	—	—
10	0.09255	177.51	196.02	0.7548	0.04363	174.76	192.21	0.6972
20	0.09642	183.00	202.28	0.7766	0.04584	180.57	198.91	0.7204
30	0.10023	188.55	208.60	0.7978	0.04797	186.39	205.58	0.7428
40	0.10399	194.17	214.97	0.8184	0.05005	192.23	212.25	0.7645
50	0.10771	199.86	221.41	0.8387	0.05207	198.11	218.94	0.7855
60	0.11140	205.62	227.90	0.8585	0.05406	204.03	225.65	0.8060
70	0.11506	211.45	234.46	0.8779	0.05601	209.99	232.40	0.8259
80	0.11869	217.35	241.09	0.8969	0.05794	216.01	239.19	0.8454
90	0.12230	223.31	247.77	0.9156	0.05985	222.08	246.02	0.8645
100	0.12590	229.34	254.53	0.9339	0.06173	228.20	252.89	0.8831
110	0.12948	235.44	261.34	0.9519	0.06360	234.37	259.81	0.9015
120	0.13305	241.60	268.21	0.9696	0.06546	240.60	266.79	0.9194
130	0.13661	247.82	275.15	0.9870	0.06730	246.89	273.81	0.9370
140	0.14016	254.11	282.14	1.0042	0.06913	253.22	280.88	0.9544
150	0.14370	260.45	289.19	1.0210	0.07095	259.61	287.99	0.9714

TABLE B.3.2 (*continued*)
Superheated R-12

Temp. (°C)	v (m³/kg)	u (kJ/kg)	h (kJ/kg)	s (kJ/kg-K)	v (m³/kg)	u (kJ/kg)	h (kJ/kg)	s (kJ/kg-K)
	500 kPa (15.60)				1000 kPa (41.64)			
Sat.	0.03482	176.62	194.03	0.6899	0.01744	186.32	203.76	0.6820
30	0.03746	185.23	203.96	0.7235	—	—	—	—
40	0.03921	191.20	210.81	0.7457	—	—	—	—
50	0.04091	197.19	217.64	0.7672	0.01837	191.95	210.32	0.7026
60	0.04257	203.20	224.48	0.7881	0.01941	198.56	217.97	0.7259
70	0.04418	209.24	231.33	0.8083	0.02040	205.09	225.49	0.7481
80	0.04577	215.32	238.21	0.8281	0.02134	211.57	232.91	0.7695
90	0.04734	221.44	245.11	0.8473	0.02225	218.03	240.28	0.7900
100	0.04889	227.61	252.05	0.8662	0.02313	224.48	247.61	0.8100
110	0.05041	233.83	259.03	0.8847	0.02399	230.94	254.93	0.8293
120	0.05193	240.09	266.06	0.9028	0.02483	237.41	262.25	0.8482
130	0.05343	246.41	273.12	0.9205	0.02566	243.91	269.57	0.8665
140	0.05492	252.77	280.23	0.9379	0.02647	250.43	276.90	0.8845
150	0.05640	259.19	287.39	0.9550	0.02728	256.98	284.26	0.9021
160	0.05788	265.65	294.59	0.9718	0.02807	263.56	291.63	0.9193
170	0.05934	272.16	301.83	0.9884	0.02885	270.18	299.04	0.9362
180	0.06080	278.72	309.12	1.0046	0.02963	276.84	306.47	0.9528
	1500 kPa (59.22)				2000 kPa (72.88)			
Sat.	0.01132	192.08	209.06	0.6768	0.00813	195.70	211.97	0.6713
70	0.01226	200.05	218.44	0.7046	—	—	—	—
80	0.01305	207.16	226.73	0.7284	0.00870	201.61	219.02	0.6914
90	0.01377	214.11	234.77	0.7508	0.00941	209.41	228.23	0.7171
100	0.01446	220.95	242.65	0.7722	0.01003	216.87	236.94	0.7408
110	0.01512	227.73	250.41	0.7928	0.01061	224.11	245.34	0.7630
120	0.01575	234.47	258.10	0.8126	0.01116	231.21	253.53	0.7841
130	0.01636	241.20	265.74	0.8318	0.01168	238.22	261.58	0.8043
140	0.01696	247.91	273.35	0.8504	0.01217	245.18	269.53	0.8238
150	0.01754	254.63	280.94	0.8686	0.01265	252.10	277.41	0.8426
160	0.01811	261.36	288.52	0.8863	0.01312	259.00	285.24	0.8609
170	0.01867	268.10	296.11	0.9036	0.01357	265.90	293.04	0.8787
180	0.01922	274.87	303.70	0.9205	0.01401	272.79	300.82	0.8961
190	0.01977	281.65	311.31	0.9371	0.01445	279.69	308.59	0.9131
200	0.02031	288.47	318.93	0.9534	0.01488	286.61	316.36	0.9297
210	0.02084	295.31	326.58	0.9694	0.01530	293.54	324.14	0.9459
220	0.02137	302.19	334.24	0.9851	0.01572	300.49	331.92	0.9619

TABLE B.4
Thermodynamic Properties of R-22

TABLE B.4.1
Saturated R-22

Temp. (°C)	Press. (kPa)	SPECIFIC VOLUME, m³/kg			INTERNAL ENERGY, kJ/kg		
		Sat. Liquid v_f	Evap. v_{fg}	Sat. Vapor v_g	Sat. Liquid u_f	Evap. u_{fg}	Sat. Vapor u_g
−70	20.5	0.000670	0.94027	0.94094	−30.62	230.13	199.51
−65	28.0	0.000676	0.70480	0.70547	−25.68	227.21	201.54
−60	37.5	0.000682	0.53647	0.53715	−20.68	224.25	203.57
−55	49.5	0.000689	0.41414	0.41483	−15.62	221.21	205.59
−50	64.4	0.000695	0.32386	0.32456	−10.50	218.11	207.61
−45	82.7	0.000702	0.25629	0.25699	−5.32	214.94	209.62
−40.8	101.3	0.000708	0.21191	0.21261	−0.87	212.18	211.31
−40	104.9	0.000709	0.20504	0.20575	−0.07	211.68	211.60
−35	131.7	0.000717	0.16568	0.16640	5.23	208.34	213.57
−30	163.5	0.000725	0.13512	0.13584	10.61	204.91	215.52
−25	201.0	0.000733	0.11113	0.11186	16.04	201.39	217.44
−20	244.8	0.000741	0.09210	0.09284	21.55	197.78	219.32
−15	295.7	0.000750	0.07688	0.07763	27.11	194.07	221.18
−10	354.3	0.000759	0.06458	0.06534	32.74	190.25	222.99
−5	421.3	0.000768	0.05457	0.05534	38.44	186.33	224.77
0	497.6	0.000778	0.04636	0.04714	44.20	182.30	226.50
5	583.8	0.000789	0.03957	0.04036	50.03	178.15	228.17
10	680.7	0.000800	0.03391	0.03471	55.92	173.87	229.79
15	789.1	0.000812	0.02918	0.02999	61.88	169.47	231.35
20	909.9	0.000824	0.02518	0.02600	67.92	164.92	232.85
25	1043.9	0.000838	0.02179	0.02262	74.04	160.22	234.26
30	1191.9	0.000852	0.01889	0.01974	80.23	155.35	235.59
35	1354.8	0.000867	0.01640	0.01727	86.53	150.30	236.82
40	1533.5	0.000884	0.01425	0.01514	92.92	145.02	237.94
45	1729.0	0.000902	0.01238	0.01328	99.42	139.50	238.93
50	1942.3	0.000922	0.01075	0.01167	106.06	133.70	239.76
55	2174.4	0.000944	0.00931	0.01025	112.85	127.56	240.41
60	2426.6	0.000969	0.00803	0.00900	119.83	121.01	240.84
65	2699.9	0.000997	0.00689	0.00789	127.04	113.94	240.98
70	2995.9	0.001030	0.00586	0.00689	134.54	106.22	240.76
75	3316.1	0.001069	0.00491	0.00598	142.44	97.61	240.05
80	3662.3	0.001118	0.00403	0.00515	150.92	87.71	238.63
85	4036.8	0.001183	0.00317	0.00436	160.32	75.78	236.10
90	4442.5	0.001282	0.00228	0.00356	171.51	59.90	231.41
95	4883.5	0.001521	0.00103	0.00255	188.93	29.89	218.83
96.0	4969.0	0.001906	0	0.00191	203.07	0	203.07

TABLE B.4.1 (*continued*)
Saturated R-22

Temp. (°C)	Press. (kPa)	ENTHALPY, kJ/kg			ENTROPY, kJ/kg-K		
		Sat. Liquid h_f	Evap. h_{fg}	Sat. Vapor h_g	Sat. Liquid s_f	Evap. s_{fg}	Sat. Vapor s_g
−70	20.5	−30.61	249.43	218.82	−0.1401	1.2277	1.0876
−65	28.0	−25.66	246.93	221.27	−0.1161	1.1862	1.0701
−60	37.5	−20.65	244.35	223.70	−0.0924	1.1463	1.0540
−55	49.5	−15.59	241.70	226.12	−0.0689	1.1079	1.0390
−50	64.4	−10.46	238.96	228.51	−0.0457	1.0708	1.0251
−45	82.7	−5.26	236.13	230.87	−0.0227	1.0349	1.0122
−40.8	101.3	−0.80	233.65	232.85	−0.0034	1.0053	1.0019
−40	104.9	0	233.20	233.20	0	1.0002	1.0002
−35	131.7	5.33	230.16	235.48	0.0225	0.9664	0.9889
−30	163.5	10.73	227.00	237.73	0.0449	0.9335	0.9784
−25	201.0	16.19	223.73	239.92	0.0670	0.9015	0.9685
−20	244.8	21.73	220.33	242.06	0.0890	0.8703	0.9593
−15	295.7	27.33	216.80	244.13	0.1107	0.8398	0.9505
−10	354.3	33.01	213.13	246.14	0.1324	0.8099	0.9422
−5	421.3	38.76	209.32	248.09	0.1538	0.7806	0.9344
0	497.6	44.59	205.36	249.95	0.1751	0.7518	0.9269
5	583.8	50.49	201.25	251.73	0.1963	0.7235	0.9197
10	680.7	56.46	196.96	253.42	0.2173	0.6956	0.9129
15	789.1	62.52	192.49	255.02	0.2382	0.6680	0.9062
20	909.9	68.67	187.84	256.51	0.2590	0.6407	0.8997
25	1043.9	74.91	182.97	257.88	0.2797	0.6137	0.8934
30	1191.9	81.25	177.87	259.12	0.3004	0.5867	0.8871
35	1354.8	87.70	172.52	260.22	0.3210	0.5598	0.8809
40	1533.5	94.27	166.88	261.15	0.3417	0.5329	0.8746
45	1729.0	100.98	160.91	261.90	0.3624	0.5058	0.8682
50	1942.3	107.85	154.58	262.43	0.3832	0.4783	0.8615
55	2174.4	114.91	147.80	262.71	0.4042	0.4504	0.8546
60	2426.6	122.18	140.50	262.68	0.4255	0.4217	0.8472
65	2699.9	129.73	132.55	262.28	0.4472	0.3920	0.8391
70	2995.9	137.63	123.77	261.40	0.4695	0.3607	0.8302
75	3316.1	145.99	113.90	259.89	0.4927	0.3272	0.8198
80	3662.3	155.01	102.47	257.49	0.5173	0.2902	0.8075
85	4036.8	165.09	88.60	253.69	0.5445	0.2474	0.7918
90	4442.5	177.20	70.04	247.24	0.5767	0.1929	0.7695
95	4883.5	196.36	34.93	231.28	0.6273	0.0949	0.7222
96.0	4969.0	212.54	0	212.54	0.6708	0	0.6708

TABLE B.4.2
Superheated R-22

Temp. (°C)	v (m³/kg)	u (kJ/kg)	h (kJ/kg)	s (kJ/kg-K)	v (m³/kg)	u (kJ/kg)	h (kJ/kg)	s (kJ/kg-K)
	50 kPa (−54.80)				100 kPa (−41.03)			
Sat.	0.41077	205.67	226.21	1.0384	0.21525	211.19	232.72	1.0026
−40	0.44063	212.69	234.72	1.0762	0.21633	211.70	233.34	1.0052
−30	0.46064	217.57	240.60	1.1008	0.22675	216.68	239.36	1.0305
−20	0.48054	222.56	246.59	1.1250	0.23706	221.76	245.47	1.0551
−10	0.50036	227.66	252.68	1.1485	0.24728	226.94	251.67	1.0791
0	0.52010	232.87	258.87	1.1717	0.25742	232.21	257.96	1.1026
10	0.53977	238.19	265.18	1.1943	0.26748	237.60	264.35	1.1256
20	0.55939	243.62	271.59	1.2166	0.27750	243.08	270.83	1.1481
30	0.57897	249.17	278.12	1.2385	0.28747	248.67	277.42	1.1702
40	0.59851	254.82	284.74	1.2600	0.29739	254.36	284.10	1.1919
50	0.61801	260.58	291.48	1.2811	0.30729	260.16	290.89	1.2132
60	0.63749	266.44	298.32	1.3020	0.31715	266.06	297.77	1.2342
70	0.65694	272.42	305.26	1.3225	0.32699	272.06	304.76	1.2548
80	0.67636	278.50	312.31	1.3428	0.33680	278.16	311.84	1.2752
90	0.69577	284.68	319.47	1.3627	0.34660	284.37	319.03	1.2952
100	0.71516	290.96	326.72	1.3824	0.35637	290.67	326.31	1.3150
110	0.73454	297.34	334.07	1.4019	0.36614	297.07	333.69	1.3345
	150 kPa (−32.02)				200 kPa (−25.12)			
Sat.	0.14727	214.74	236.83	0.9826	0.11237	217.39	239.87	0.9688
−20	0.15585	220.94	244.32	1.0129	0.11520	220.10	243.14	0.9818
−10	0.16288	226.20	250.63	1.0373	0.12065	225.44	249.57	1.0068
0	0.16982	231.55	257.02	1.0612	0.12600	230.87	256.07	1.0310
10	0.17670	236.99	263.50	1.0844	0.13129	236.38	262.63	1.0546
20	0.18352	242.53	270.06	1.1072	0.13651	241.97	269.27	1.0776
30	0.19028	248.17	276.71	1.1295	0.14168	247.66	275.99	1.1002
40	0.19701	253.90	283.45	1.1514	0.14681	253.43	282.80	1.1222
50	0.20370	259.73	290.29	1.1729	0.15190	259.31	289.69	1.1439
60	0.21036	265.67	297.22	1.1940	0.15696	265.27	296.66	1.1652
70	0.21700	271.70	304.25	1.2148	0.16200	271.33	303.73	1.1861
80	0.22361	277.83	311.37	1.2353	0.16701	277.49	310.89	1.2066
90	0.23020	284.05	318.58	1.2554	0.17200	283.74	318.14	1.2269
100	0.23678	290.38	325.90	1.2753	0.17697	290.09	325.48	1.2468
110	0.24333	296.80	333.30	1.2948	0.18193	296.53	332.91	1.2665
120	0.24988	303.32	340.80	1.3142	0.18688	303.06	340.44	1.2858
130	0.25642	309.93	348.39	1.3332	0.19181	309.69	348.05	1.3050

TABLE B.4.2 (*continued*)
Superheated R-22

Temp. (°C)	v (m³/kg)	u (kJ/kg)	h (kJ/kg)	s (kJ/kg-K)	v (m³/kg)	u (kJ/kg)	h (kJ/kg)	s (kJ/kg-K)
	300 kPa (−14.61)				400 kPa (−6.52)			
Sat.	0.07657	221.32	244.29	0.9499	0.05817	224.23	247.50	0.9367
−10	0.07834	223.88	247.38	0.9617	—	—	—	—
0	0.08213	229.47	254.10	0.9868	0.06013	228.00	252.05	0.9536
10	0.08583	235.11	260.86	1.0111	0.06306	233.80	259.02	0.9787
20	0.08947	240.83	267.67	1.0347	0.06591	239.64	266.01	1.0029
30	0.09305	246.62	274.53	1.0577	0.06871	245.55	273.03	1.0265
40	0.09659	252.48	281.46	1.0802	0.07146	251.51	280.09	1.0494
50	0.10009	258.43	288.46	1.1022	0.07416	257.54	287.21	1.0717
60	0.10355	264.47	295.54	1.1238	0.07683	263.65	294.39	1.0936
70	0.10699	270.59	302.69	1.1449	0.07947	269.84	301.63	1.1150
80	0.11040	276.80	309.92	1.1657	0.08209	276.11	308.94	1.1361
90	0.11379	283.10	317.24	1.1861	0.08468	282.46	316.33	1.1567
100	0.11716	289.49	324.64	1.2062	0.08725	288.89	323.80	1.1770
110	0.12052	295.97	332.13	1.2260	0.08981	295.41	331.34	1.1969
120	0.12387	302.54	339.70	1.2455	0.09236	302.02	338.96	1.2165
130	0.12720	309.20	347.36	1.2648	0.09489	308.71	346.66	1.2359
140	0.13052	315.95	355.10	1.2837	0.09741	315.48	354.45	1.2550
	500 kPa (0.15)				600 kPa (5.88)			
Sat.	0.04692	226.55	250.00	0.9267	0.03929	228.46	252.04	0.9185
10	0.04936	232.43	257.11	0.9522	0.04018	231.00	255.11	0.9295
20	0.05175	238.42	264.30	0.9772	0.04228	237.15	262.52	0.9552
30	0.05408	244.44	271.48	1.0013	0.04431	243.30	269.89	0.9799
40	0.05636	250.51	278.69	1.0247	0.04628	249.48	277.25	1.0038
50	0.05859	256.63	285.93	1.0474	0.04820	255.70	284.62	1.0270
60	0.06079	262.82	293.22	1.0696	0.05008	261.97	292.02	1.0495
70	0.06295	269.08	300.55	1.0913	0.05193	268.30	299.46	1.0715
80	0.06509	275.40	307.95	1.1126	0.05375	274.69	306.94	1.0930
90	0.06721	281.81	315.41	1.1334	0.05555	281.14	314.48	1.1140
100	0.06930	288.29	322.94	1.1539	0.05733	287.67	322.07	1.1347
110	0.07138	294.85	330.54	1.1740	0.05909	294.27	329.73	1.1549
120	0.07345	301.49	338.21	1.1937	0.06084	300.95	337.46	1.1748
130	0.07550	308.21	345.96	1.2132	0.06258	307.71	345.26	1.1944
140	0.07755	315.02	353.79	1.2324	0.06430	314.54	353.12	1.2137
150	0.07958	321.90	361.69	1.2513	0.06601	321.46	361.07	1.2327
160	0.08160	328.87	369.67	1.2699	0.06772	328.45	369.08	1.2514

TABLE B.4.2 (*continued*)
Superheated R-22

Temp. (°C)	v (m³/kg)	u (kJ/kg)	h (kJ/kg)	s (kJ/kg-K)	v (m³/kg)	u (kJ/kg)	h (kJ/kg)	s (kJ/kg-K)
	800 kPa (15.47)				1000 kPa (23.42)			
Sat.	0.02958	231.50	255.16	0.9056	0.02364	233.82	257.46	0.8954
20	0.03037	234.44	258.74	0.9179	—	—	—	—
30	0.03203	240.91	266.53	0.9440	0.02460	238.31	262.91	0.9136
40	0.03363	247.34	274.24	0.9690	0.02599	245.05	271.04	0.9400
50	0.03518	253.77	281.91	0.9931	0.02732	251.72	279.05	0.9651
60	0.03667	260.21	289.55	1.0164	0.02860	258.37	286.97	0.9893
70	0.03814	266.69	297.20	1.0391	0.02984	265.02	294.86	1.0126
80	0.03957	273.21	304.87	1.0611	0.03104	271.69	302.73	1.0352
90	0.04097	279.79	312.57	1.0826	0.03221	278.39	310.60	1.0572
100	0.04236	286.42	320.30	1.1036	0.03336	285.12	318.49	1.0786
110	0.04373	293.11	328.09	1.1242	0.03449	291.91	326.41	1.0996
120	0.04508	299.86	335.93	1.1444	0.03561	298.75	334.36	1.1200
130	0.04641	306.69	343.82	1.1642	0.03671	305.65	342.36	1.1401
140	0.04774	313.59	351.78	1.1837	0.03780	312.61	350.41	1.1599
150	0.04905	320.56	359.80	1.2029	0.03887	319.64	358.51	1.1792
160	0.05036	327.60	367.89	1.2218	0.03994	326.74	366.68	1.1983
170	0.05166	334.72	376.04	1.2404	0.04100	333.90	374.90	1.2171
	1200 kPa (30.26)				1400 kPa (36.31)			
Sat.	0.01960	235.66	259.18	0.8868	0.01668	237.12	260.48	0.8792
40	0.02085	242.58	267.60	0.9141	0.01712	239.89	263.86	0.8901
50	0.02205	249.55	276.01	0.9405	0.01825	247.22	272.77	0.9181
60	0.02319	256.43	284.26	0.9657	0.01930	254.38	281.40	0.9444
70	0.02428	263.28	292.42	0.9898	0.02029	261.45	289.86	0.9694
80	0.02534	270.10	300.51	1.0131	0.02125	268.45	298.20	0.9934
90	0.02636	276.94	308.57	1.0356	0.02217	275.44	306.47	1.0165
100	0.02736	283.79	316.62	1.0574	0.02306	282.42	314.70	1.0388
110	0.02833	290.68	324.68	1.0788	0.02393	289.42	322.92	1.0606
120	0.02929	297.61	332.76	1.0996	0.02477	296.44	331.13	1.0817
130	0.03024	304.59	340.87	1.1199	0.02561	303.50	339.35	1.1024
140	0.03117	311.62	349.02	1.1399	0.02643	310.61	347.60	1.1226
150	0.03208	318.71	357.21	1.1595	0.02723	317.76	355.89	1.1424
160	0.03299	325.86	365.45	1.1787	0.02803	324.97	364.21	1.1618
170	0.03389	333.07	373.74	1.1977	0.02882	332.23	372.57	1.1809
180	0.03479	340.35	382.09	1.2163	0.02960	339.55	380.99	1.1997
190	0.03567	347.69	390.50	1.2346	0.03037	346.94	389.45	1.2182

TABLE B.4.2 (*continued*)
Superheated R-22

Temp. (°C)	v (m³/kg)	u (kJ/kg)	h (kJ/kg)	s (kJ/kg-K)	v (m³/kg)	u (kJ/kg)	h (kJ/kg)	s (kJ/kg-K)
	1600 kPa (41.75)				2000 kPa (51.28)			
Sat.	0.01446	238.30	261.43	0.8724	0.01129	239.95	262.53	0.8598
50	0.01535	244.70	269.26	0.8969	—	—	—	—
60	0.01635	252.20	278.36	0.9246	0.01213	247.29	271.56	0.8873
70	0.01728	259.52	287.17	0.9507	0.01301	255.29	281.31	0.9161
80	0.01817	266.73	295.80	0.9755	0.01381	263.02	290.64	0.9429
90	0.01901	273.88	304.30	0.9992	0.01456	270.57	299.70	0.9682
100	0.01983	281.00	312.73	1.0221	0.01528	278.02	308.57	0.9923
110	0.02061	288.12	321.10	1.0442	0.01596	285.40	317.32	1.0155
120	0.02138	295.25	329.46	1.0658	0.01662	292.75	325.99	1.0378
130	0.02213	302.39	337.81	1.0867	0.01726	300.09	334.61	1.0594
140	0.02287	309.57	346.16	1.1072	0.01788	307.44	343.20	1.0805
150	0.02359	316.79	354.54	1.1272	0.01849	314.80	351.78	1.1010
160	0.02430	324.06	362.95	1.1469	0.01909	322.19	360.37	1.1211
170	0.02501	331.37	371.39	1.1661	0.01967	329.62	368.97	1.1407
180	0.02570	338.74	379.87	1.1851	0.02025	337.09	377.60	1.1600
190	0.02639	346.17	388.40	1.2037	0.02082	344.61	386.25	1.1788
200	0.02707	353.66	396.97	1.2220	0.02138	352.17	394.94	1.1974
	3000 kPa (70.09)				4000 kPa (84.53)			
Sat.	0.00688	240.75	261.38	0.8300	0.00443	236.40	254.13	0.7935
80	0.00775	251.29	274.53	0.8678	—	—	—	—
90	0.00847	260.65	286.04	0.9000	0.00504	245.48	265.63	0.8254
100	0.00910	269.37	296.66	0.9288	0.00580	257.78	281.00	0.8672
110	0.00967	277.72	306.74	0.9555	0.00641	268.13	293.75	0.9009
120	0.01021	285.84	316.47	0.9805	0.00692	277.58	305.27	0.9306
130	0.01072	293.80	325.96	1.0044	0.00739	286.52	316.08	0.9578
140	0.01120	301.67	335.27	1.0272	0.00782	295.14	326.42	0.9831
150	0.01166	309.47	344.47	1.0492	0.00823	303.54	336.45	1.0071
160	0.01211	317.24	353.58	1.0705	0.00861	311.81	346.25	1.0300
170	0.01255	325.00	362.65	1.0912	0.00898	319.97	355.89	1.0520
180	0.01298	332.75	371.68	1.1113	0.00933	328.08	365.41	1.0732
190	0.01339	340.52	380.70	1.1310	0.00968	336.15	374.85	1.0939
200	0.01380	348.31	389.71	1.1502	0.01001	344.20	384.24	1.1139
210	0.01420	356.12	398.73	1.1691	0.01033	352.25	393.59	1.1335
220	0.01460	363.98	407.77	1.1876	0.01065	360.31	402.93	1.1526
230	0.01499	371.87	416.83	1.2058	0.01097	368.38	412.25	1.1713

TABLE B.5
Thermodynamic Properties of R-134a

TABLE B.5.1
Saturated R-134a

Temp. (°C)	Press. (kPa)	SPECIFIC VOLUME, m³/kg			INTERNAL ENERGY, kJ/kg		
		Sat. Liquid v_f	Evap. v_{fg}	Sat. Vapor v_g	Sat. Liquid u_f	Evap. u_{fg}	Sat. Vapor u_g
−70	8.3	0.000675	1.97207	1.97274	119.46	218.74	338.20
−65	11.7	0.000679	1.42915	1.42983	123.18	217.76	340.94
−60	16.3	0.000684	1.05199	1.05268	127.52	216.19	343.71
−55	22.2	0.000689	0.78609	0.78678	132.36	214.14	346.50
−50	29.9	0.000695	0.59587	0.59657	137.60	211.71	349.31
−45	39.6	0.000701	0.45783	0.45853	143.15	208.99	352.15
−40	51.8	0.000708	0.35625	0.35696	148.95	206.05	355.00
−35	66.8	0.000715	0.28051	0.28122	154.93	202.93	357.86
−30	85.1	0.000722	0.22330	0.22402	161.06	199.67	360.73
−26.3	101.3	0.000728	0.18947	0.19020	165.73	197.16	362.89
−25	107.2	0.000730	0.17957	0.18030	167.30	196.31	363.61
−20	133.7	0.000738	0.14576	0.14649	173.65	192.85	366.50
−15	165.0	0.000746	0.11932	0.12007	180.07	189.32	369.39
−10	201.7	0.000755	0.09845	0.09921	186.57	185.70	372.27
−5	244.5	0.000764	0.08181	0.08257	193.14	182.01	375.15
0	294.0	0.000773	0.06842	0.06919	199.77	178.24	378.01
5	350.9	0.000783	0.05755	0.05833	206.48	174.38	380.85
10	415.8	0.000794	0.04866	0.04945	213.25	170.42	383.67
15	489.5	0.000805	0.04133	0.04213	220.10	166.35	386.45
20	572.8	0.000817	0.03524	0.03606	227.03	162.16	389.19
25	666.3	0.000829	0.03015	0.03098	234.04	157.83	391.87
30	771.0	0.000843	0.02587	0.02671	241.14	153.34	394.48
35	887.6	0.000857	0.02224	0.02310	248.34	148.68	397.02
40	1017.0	0.000873	0.01915	0.02002	255.65	143.81	399.46
45	1160.2	0.000890	0.01650	0.01739	263.08	138.71	401.79
50	1318.1	0.000908	0.01422	0.01512	270.63	133.35	403.98
55	1491.6	0.000928	0.01224	0.01316	278.33	127.68	406.01
60	1681.8	0.000951	0.01051	0.01146	286.19	121.66	407.85
65	1889.9	0.000976	0.00899	0.00997	294.24	115.22	409.46
70	2117.0	0.001005	0.00765	0.00866	302.51	108.27	410.78
75	2364.4	0.001038	0.00645	0.00749	311.06	100.68	411.74
80	2633.6	0.001078	0.00537	0.00645	319.96	92.26	412.22
85	2926.2	0.001128	0.00437	0.00550	329.35	82.67	412.01
90	3244.5	0.001195	0.00341	0.00461	339.51	71.24	410.75
95	3591.5	0.001297	0.00243	0.00373	351.17	56.25	407.42
100	3973.2	0.001557	0.00108	0.00264	368.55	28.19	396.74
101.2	4064.0	0.001969	0	0.00197	382.97	0	382.97

TABLE B.5.1 (*continued*)
Saturated R-134a

Temp. (°C)	Press. (kPa)	ENTHALPY, kJ/kg			ENTROPY, kJ/kg-K		
		Sat. Liquid h_f	Evap. h_{fg}	Sat. Vapor h_g	Sat. Liquid s_f	Evap. s_{fg}	Sat. Vapor s_g
−70	8.3	119.47	235.15	354.62	0.6645	1.1575	1.8220
−65	11.7	123.18	234.55	357.73	0.6825	1.1268	1.8094
−60	16.3	127.53	233.33	360.86	0.7031	1.0947	1.7978
−55	22.2	132.37	231.63	364.00	0.7256	1.0618	1.7874
−50	29.9	137.62	229.54	367.16	0.7493	1.0286	1.7780
−45	39.6	143.18	227.14	370.32	0.7740	0.9956	1.7695
−40	51.8	148.98	224.50	373.48	0.7991	0.9629	1.7620
−35	66.8	154.98	221.67	376.64	0.8245	0.9308	1.7553
−30	85.1	161.12	218.68	379.80	0.8499	0.8994	1.7493
−26.3	101.3	165.80	216.36	382.16	0.8690	0.8763	1.7453
−25	107.2	167.38	215.57	382.95	0.8754	0.8687	1.7441
−20	133.7	173.74	212.34	386.08	0.9007	0.8388	1.7395
−15	165.0	180.19	209.00	389.20	0.9258	0.8096	1.7354
−10	201.7	186.72	205.56	392.28	0.9507	0.7812	1.7319
−5	244.5	193.32	202.02	395.34	0.9755	0.7534	1.7288
0	294.0	200.00	198.36	398.36	1.0000	0.7262	1.7262
5	350.9	206.75	194.57	401.32	1.0243	0.6995	1.7239
10	415.8	213.58	190.65	404.23	1.0485	0.6733	1.7218
15	489.5	220.49	186.58	407.07	1.0725	0.6475	1.7200
20	572.8	227.49	182.35	409.84	1.0963	0.6220	1.7183
25	666.3	234.59	177.92	412.51	1.1201	0.5967	1.7168
30	771.0	241.79	173.29	415.08	1.1437	0.5716	1.7153
35	887.6	249.10	168.42	417.52	1.1673	0.5465	1.7139
40	1017.0	256.54	163.28	419.82	1.1909	0.5214	1.7123
45	1160.2	264.11	157.85	421.96	1.2145	0.4962	1.7106
50	1318.1	271.83	152.08	423.91	1.2381	0.4706	1.7088
55	1491.6	279.72	145.93	425.65	1.2619	0.4447	1.7066
60	1681.8	287.79	139.33	427.13	1.2857	0.4182	1.7040
65	1889.9	296.09	132.21	428.30	1.3099	0.3910	1.7008
70	2117.0	304.64	124.47	429.11	1.3343	0.3627	1.6970
75	2364.4	313.51	115.94	429.45	1.3592	0.3330	1.6923
80	2633.6	322.79	106.40	429.19	1.3849	0.3013	1.6862
85	2926.2	332.65	95.45	428.10	1.4117	0.2665	1.6782
90	3244.5	343.38	82.31	425.70	1.4404	0.2267	1.6671
95	3591.5	355.83	64.98	420.81	1.4733	0.1765	1.6498
100	3973.2	374.74	32.47	407.21	1.5228	0.0870	1.6098
101.2	4064.0	390.98	0	390.98	1.5658	0	1.5658

TABLE B.5.2
Superheated R-134a

Temp. (°C)	v (m³/kg)	u (kJ/kg)	h (kJ/kg)	s (kJ/kg-K)	v (m³/kg)	u (kJ/kg)	h (kJ/kg)	s (kJ/kg-K)
		50 kPa (−40.67)				100 kPa (−26.54)		
Sat.	0.36889	354.61	373.06	1.7629	0.19257	362.73	381.98	1.7456
−20	0.40507	368.57	388.82	1.8279	0.19860	367.36	387.22	1.7665
−10	0.42222	375.53	396.64	1.8582	0.20765	374.51	395.27	1.7978
0	0.43921	382.63	404.59	1.8878	0.21652	381.76	403.41	1.8281
10	0.45608	389.90	412.70	1.9170	0.22527	389.14	411.67	1.8578
20	0.47287	397.32	420.96	1.9456	0.23392	396.66	420.05	1.8869
30	0.48958	404.90	429.38	1.9739	0.24250	404.31	428.56	1.9155
40	0.50623	412.64	437.96	2.0017	0.25101	412.12	437.22	1.9436
50	0.52284	420.55	446.70	2.0292	0.25948	420.08	446.03	1.9712
60	0.53941	428.63	455.60	2.0563	0.26791	428.20	454.99	1.9985
70	0.55595	436.86	464.66	2.0831	0.27631	436.47	464.10	2.0255
80	0.57247	445.26	473.88	2.1096	0.28468	444.89	473.36	2.0521
90	0.58896	453.82	483.26	2.1358	0.29302	453.47	482.78	2.0784
100	0.60544	462.53	492.81	2.1617	0.30135	462.21	492.35	2.1044
110	0.62190	471.41	502.50	2.1874	0.30967	471.11	502.07	2.1301
120	0.63835	480.44	512.36	2.2128	0.31797	480.16	511.95	2.1555
130	0.65479	489.63	522.37	2.2379	0.32626	489.36	521.98	2.1807
		150 kPa (−17.29)				200 kPa (−10.22)		
Sat.	0.13139	368.06	387.77	1.7372	0.10002	372.15	392.15	1.7320
−10	0.13602	373.44	393.84	1.7606	0.10013	372.31	392.34	1.7328
0	0.14222	380.85	402.19	1.7917	0.10501	379.91	400.91	1.7647
10	0.14828	388.36	410.60	1.8220	0.10974	387.55	409.50	1.7956
20	0.15424	395.98	419.11	1.8515	0.11436	395.27	418.15	1.8256
30	0.16011	403.71	427.73	1.8804	0.11889	403.10	426.87	1.8549
40	0.16592	411.59	436.47	1.9088	0.12335	411.04	435.71	1.8836
50	0.17168	419.60	445.35	1.9367	0.12776	419.11	444.66	1.9117
60	0.17740	427.76	454.37	1.9642	0.13213	427.31	453.74	1.9394
70	0.18308	436.06	463.53	1.9913	0.13646	435.65	462.95	1.9666
80	0.18874	444.52	472.83	2.0180	0.14076	444.14	472.30	1.9935
90	0.19437	453.13	482.28	2.0444	0.14504	452.78	481.79	2.0200
100	0.19999	461.89	491.89	2.0705	0.14930	461.56	491.42	2.0461
110	0.20559	470.80	501.64	2.0963	0.15355	470.50	501.21	2.0720
120	0.21117	479.87	511.54	2.1218	0.15777	479.58	511.13	2.0976
130	0.21675	489.08	521.60	2.1470	0.16199	488.81	521.21	2.1229
140	0.22231	498.45	531.80	2.1720	0.16620	498.19	531.43	2.1479

Table B.5.2 (*continued*)
Superheated R-134a

Temp. (°C)	v (m³/kg)	u (kJ/kg)	h (kJ/kg)	s (kJ/kg-K)	v (m³/kg)	u (kJ/kg)	h (kJ/kg)	s (kJ/kg-K)
	300 kPa (0.56)				400 kPa (8.84)			
Sat.	0.06787	378.33	398.69	1.7259	0.05136	383.02	403.56	1.7223
10	0.07111	385.84	407.17	1.7564	0.05168	383.98	404.65	1.7261
20	0.07441	393.80	416.12	1.7874	0.05436	392.22	413.97	1.7584
30	0.07762	401.81	425.10	1.8175	0.05693	400.45	423.22	1.7895
40	0.08075	409.90	434.12	1.8468	0.05940	408.70	432.46	1.8195
50	0.08382	418.09	443.23	1.8755	0.06181	417.03	441.75	1.8487
60	0.08684	426.39	452.44	1.9035	0.06417	425.44	451.10	1.8772
70	0.08982	434.82	461.76	1.9311	0.06648	433.95	460.55	1.9051
80	0.09277	443.37	471.21	1.9582	0.06877	442.58	470.09	1.9325
90	0.09570	452.07	480.78	1.9850	0.07102	451.34	479.75	1.9595
100	0.09861	460.90	490.48	2.0113	0.07325	460.22	489.52	1.9860
110	0.10150	469.87	500.32	2.0373	0.07547	469.24	499.43	2.0122
120	0.10437	478.99	510.30	2.0631	0.07767	478.40	509.46	2.0381
130	0.10723	488.26	520.43	2.0885	0.07985	487.69	519.63	2.0636
140	0.11008	497.66	530.69	2.1136	0.08202	497.13	529.94	2.0889
150	0.11292	507.22	541.09	2.1385	0.08418	506.71	540.38	2.1139
160	0.11575	516.91	551.64	2.1631	0.08634	516.43	550.97	2.1386
	500 kPa (15.66)				600 kPa (21.52)			
Sat.	0.04126	386.82	407.45	1.7198	0.03442	390.01	410.66	1.7179
20	0.04226	390.52	411.65	1.7342	—	—	—	—
30	0.04446	398.99	421.22	1.7663	0.03609	397.44	419.09	1.7461
40	0.04656	407.44	430.72	1.7971	0.03796	406.11	428.88	1.7779
50	0.04858	415.91	440.20	1.8270	0.03974	414.75	438.59	1.8084
60	0.05055	424.44	449.72	1.8560	0.04145	423.41	448.28	1.8379
70	0.05247	433.06	459.29	1.8843	0.04311	432.13	457.99	1.8666
80	0.05435	441.77	468.94	1.9120	0.04473	440.93	467.76	1.8947
90	0.05620	450.59	478.69	1.9392	0.04632	449.82	477.61	1.9222
100	0.05804	459.53	488.55	1.9660	0.04788	458.82	487.55	1.9492
110	0.05985	468.60	498.52	1.9924	0.04943	467.94	497.59	1.9758
120	0.06164	477.79	508.61	2.0184	0.05095	477.18	507.75	2.0019
130	0.06342	487.13	518.83	2.0440	0.05246	486.55	518.03	2.0277
140	0.06518	496.59	529.19	2.0694	0.05396	496.05	528.43	2.0532
150	0.06694	506.20	539.67	2.0945	0.05544	505.69	538.95	2.0784
160	0.06869	515.95	550.29	2.1193	0.05692	515.46	549.61	2.1033
170	0.07043	525.83	561.04	2.1438	0.05839	525.36	560.40	2.1279

Table B.5.2 (*continued*)
Superheated R-134a

Temp. (°C)	v (m³/kg)	u (kJ/kg)	h (kJ/kg)	s (kJ/kg-K)	v (m³/kg)	u (kJ/kg)	h (kJ/kg)	s (kJ/kg-K)
	800 kPa (31.30)				1000 kPa (39.37)			
Sat.	0.02571	395.15	415.72	1.7150	0.02038	399.16	419.54	1.7125
40	0.02711	403.17	424.86	1.7446	0.02047	399.78	420.25	1.7148
50	0.02861	412.23	435.11	1.7768	0.02185	409.39	431.24	1.7494
60	0.03002	421.20	445.22	1.8076	0.02311	418.78	441.89	1.7818
70	0.03137	430.17	455.27	1.8373	0.02429	428.05	452.34	1.8127
80	0.03268	439.17	465.31	1.8662	0.02542	437.29	462.70	1.8425
90	0.03394	448.22	475.38	1.8943	0.02650	446.53	473.03	1.8713
100	0.03518	457.35	485.50	1.9218	0.02754	455.82	483.36	1.8994
110	0.03639	466.58	495.70	1.9487	0.02856	465.18	493.74	1.9268
120	0.03758	475.92	505.99	1.9753	0.02956	474.62	504.17	1.9537
130	0.03876	485.37	516.38	2.0014	0.03053	484.16	514.69	1.9801
140	0.03992	494.94	526.88	2.0271	0.03150	493.81	525.30	2.0061
150	0.04107	504.64	537.50	2.0525	0.03244	503.57	536.02	2.0318
160	0.04221	514.46	548.23	2.0775	0.03338	513.46	546.84	2.0570
170	0.04334	524.42	559.09	2.1023	0.03431	523.46	557.77	2.0820
180	0.04446	534.51	570.08	2.1268	0.03523	533.60	568.83	2.1067
	1200 kPa (46.31)				1400 kPa (52.42)			
Sat.	0.01676	402.37	422.49	1.7102	0.01414	404.98	424.78	1.7077
50	0.01724	406.15	426.84	1.7237	—	—	—	—
60	0.01844	416.08	438.21	1.7584	0.01503	413.03	434.08	1.7360
70	0.01953	425.74	449.18	1.7908	0.01608	423.20	445.72	1.7704
80	0.02055	435.27	459.92	1.8217	0.01704	433.09	456.94	1.8026
90	0.02151	444.74	470.55	1.8514	0.01793	442.83	467.93	1.8333
100	0.02244	454.20	481.13	1.8801	0.01878	452.50	478.79	1.8628
110	0.02333	463.71	491.70	1.9081	0.01958	462.17	489.59	1.8914
120	0.02420	473.27	502.31	1.9354	0.02036	471.87	500.38	1.9192
130	0.02504	482.91	512.97	1.9621	0.02112	481.63	511.19	1.9463
140	0.02587	492.65	523.70	1.9884	0.02186	491.46	522.05	1.9730
150	0.02669	502.48	534.51	2.0143	0.02258	501.37	532.98	1.9991
160	0.02750	512.43	545.43	2.0398	0.02329	511.39	543.99	2.0248
170	0.02829	522.50	556.44	2.0649	0.02399	521.51	555.10	2.0502
180	0.02907	532.68	567.57	2.0898	0.02468	531.75	566.30	2.0752

TABLE B.5.2 (*continued*)
Superheated R-134a

Temp. (°C)	v (m³/kg)	u (kJ/kg)	h (kJ/kg)	s (kJ/kg-K)	v (m³/kg)	u (kJ/kg)	h (kJ/kg)	s (kJ/kg-K)
	1600 kPa (57.90)				2000 kPa (67.48)			
Sat.	0.01215	407.11	426.54	1.7051	0.00930	410.15	428.75	1.6991
60	0.01239	409.49	429.32	1.7135	—	—	—	—
70	0.01345	420.37	441.89	1.7507	0.00958	413.37	432.53	1.7101
80	0.01438	430.72	453.72	1.7847	0.01055	425.20	446.30	1.7497
90	0.01522	440.79	465.15	1.8166	0.01137	436.20	458.95	1.7850
100	0.01601	450.71	476.33	1.8469	0.01211	446.78	471.00	1.8177
110	0.01676	460.57	487.39	1.8762	0.01279	457.12	482.69	1.8487
120	0.01748	470.42	498.39	1.9045	0.01342	467.34	494.19	1.8783
130	0.01817	480.30	509.37	1.9321	0.01403	477.51	505.57	1.9069
140	0.01884	490.23	520.38	1.9591	0.01461	487.68	516.90	1.9346
150	0.01949	500.24	531.43	1.9855	0.01517	497.89	528.22	1.9617
160	0.02013	510.33	542.54	2.0115	0.01571	508.15	539.57	1.9882
170	0.02076	520.52	553.73	2.0370	0.01624	518.48	550.96	2.0142
180	0.02138	530.81	565.02	2.0622	0.01676	528.89	562.42	2.0398
	3000 kPa (86.20)				4000 kPa (100.33)			
Sat.	0.00528	411.83	427.67	1.6759	0.00252	394.86	404.94	1.6036
90	0.00575	418.93	436.19	1.6995	—	—	—	—
100	0.00665	433.77	453.73	1.7472	—	—	—	—
110	0.00734	446.48	468.50	1.7862	0.00428	429.74	446.84	1.7148
120	0.00792	458.27	482.04	1.8211	0.00500	445.97	465.99	1.7642
130	0.00845	469.58	494.91	1.8535	0.00556	459.63	481.87	1.8040
140	0.00893	480.61	507.39	1.8840	0.00603	472.19	496.29	1.8394
150	0.00937	491.49	519.62	1.9133	0.00644	484.15	509.92	1.8720
160	0.00980	502.30	531.70	1.9415	0.00683	495.77	523.07	1.9027
170	0.01021	513.09	543.71	1.9689	0.00718	507.19	535.92	1.9320
180	0.01060	523.89	555.69	1.9956	0.00752	518.51	548.57	1.9603
	6000 kPa				10000 kPa			
90	0.001059	328.34	334.70	1.4081	0.000991	320.72	330.62	1.3856
100	0.001150	346.71	353.61	1.4595	0.001040	336.45	346.85	1.4297
110	0.001307	368.06	375.90	1.5184	0.001100	352.74	363.73	1.4744
120	0.001698	396.59	406.78	1.5979	0.001175	369.69	381.44	1.5200
130	0.002396	426.81	441.18	1.6843	0.001272	387.44	400.16	1.5670
140	0.002985	448.34	466.25	1.7458	0.001400	405.97	419.98	1.6155
150	0.003439	465.19	485.82	1.7926	0.001564	424.99	440.63	1.6649
160	0.003814	479.89	502.77	1.8322	0.001758	443.77	461.34	1.7133
170	0.004141	493.45	518.30	1.8676	0.001965	461.65	481.30	1.7589
180	0.004435	506.35	532.96	1.9004	0.002172	478.40	500.12	1.8009

TABLE B.6
Thermodynamic Properties of Nitrogen

TABLE B.6.1
Saturated Nitrogen

Temp. (K)	Press. (kPa)	SPECIFIC VOLUME, m³/kg			INTERNAL ENERGY, kJ/kg		
		Sat. Liquid v_f	Evap. v_{fg}	Sat. Vapor v_g	Sat. Liquid u_f	Evap. u_{fg}	Sat. Vapor u_g
63.1	12.5	0.001150	1.48074	1.48189	−150.92	196.86	45.94
65	17.4	0.001160	1.09231	1.09347	−147.19	194.37	47.17
70	38.6	0.001191	0.52513	0.52632	−137.13	187.54	50.40
75	76.1	0.001223	0.28052	0.28174	−127.04	180.47	53.43
77.3	101.3	0.001240	0.21515	0.21639	−122.27	177.04	54.76
80	137.0	0.001259	0.16249	0.16375	−116.86	173.06	56.20
85	229.1	0.001299	0.10018	0.10148	−106.55	165.20	58.65
90	360.8	0.001343	0.06477	0.06611	−96.06	156.76	60.70
95	541.1	0.001393	0.04337	0.04476	−85.35	147.60	62.25
100	779.2	0.001452	0.02975	0.03120	−74.33	137.50	63.17
105	1084.6	0.001522	0.02066	0.02218	−62.89	126.18	63.29
110	1467.6	0.001610	0.01434	0.01595	−50.81	113.11	62.31
115	1939.3	0.001729	0.00971	0.01144	−37.66	97.36	59.70
120	2513.0	0.001915	0.00608	0.00799	−22.42	76.63	54.21
125	3208.0	0.002355	0.00254	0.00490	−0.83	40.73	39.90
126.2	3397.8	0.003194	0	0.00319	18.94	0	18.94

Temp. (K)	Press. (kPa)	ENTHALPY, kJ/kg			ENTROPY, kJ/kg-K		
		Sat. Liquid h_f	Evap. h_{fg}	Sat. Vapor h_g	Sat. Liquid s_f	Evap. s_{fg}	Sat. Vapor s_g
63.1	12.5	−150.91	215.39	64.48	2.4234	3.4109	5.8343
65	17.4	−147.17	213.38	66.21	2.4816	3.2828	5.7645
70	38.6	−137.09	207.79	70.70	2.6307	2.9684	5.5991
75	76.1	−126.95	201.82	74.87	2.7700	2.6909	5.4609
77.3	101.3	−122.15	198.84	76.69	2.8326	2.5707	5.4033
80	137.0	−116.69	195.32	78.63	2.9014	2.4415	5.3429
85	229.1	−106.25	188.15	81.90	3.0266	2.2135	5.2401
90	360.8	−95.58	180.13	84.55	3.1466	2.0015	5.1480
95	541.1	−84.59	171.07	86.47	3.2627	1.8007	5.0634
100	779.2	−73.20	160.68	87.48	3.3761	1.6068	4.9829
105	1084.6	−61.24	148.59	87.35	3.4883	1.4151	4.9034
110	1467.6	−48.45	134.15	85.71	3.6017	1.2196	4.8213
115	1939.3	−34.31	116.19	81.88	3.7204	1.0104	4.7307
120	2513.0	−17.61	91.91	74.30	3.8536	0.7659	4.6195
125	3208.0	6.73	48.88	55.60	4.0399	0.3910	4.4309
126.2	3397.8	29.79	0	29.79	4.2193	0	4.2193

TABLE B.6.2
Superheated Nitrogen

Temp. (K)	v (m³/kg)	u (kJ/kg)	h (kJ/kg)	s (kJ/kg-K)	v (m³/kg)	u (kJ/kg)	h (kJ/kg)	s (kJ/kg-K)
	100 kPa (77.24 K)				200 kPa (83.62 K)			
Sat.	0.21903	54.70	76.61	5.4059	0.11520	58.01	81.05	5.2673
100	0.29103	72.84	101.94	5.6944	0.14252	71.73	100.24	5.4775
120	0.35208	87.94	123.15	5.8878	0.17397	87.14	121.93	5.6753
140	0.41253	102.95	144.20	6.0501	0.20476	102.33	143.28	5.8399
160	0.47263	117.91	165.17	6.1901	0.23519	117.40	164.44	5.9812
180	0.53254	132.83	186.09	6.3132	0.26542	132.41	185.49	6.1052
200	0.59231	147.74	206.97	6.4232	0.29551	147.37	206.48	6.2157
220	0.65199	162.63	227.83	6.5227	0.32552	162.31	227.41	6.3155
240	0.71161	177.51	248.67	6.6133	0.35546	177.23	248.32	6.4064
260	0.77118	192.39	269.51	6.6967	0.38535	192.14	269.21	6.4900
280	0.83072	207.26	290.33	6.7739	0.41520	207.04	290.08	6.5674
300	0.89023	222.14	311.16	6.8457	0.44503	221.93	310.94	6.6393
350	1.03891	259.35	363.24	7.0063	0.51952	259.18	363.09	6.8001
400	1.18752	296.66	415.41	7.1456	0.59392	296.52	415.31	6.9396
450	1.33607	334.16	467.77	7.2690	0.66827	334.04	467.70	7.0630
500	1.48458	371.95	520.41	7.3799	0.74258	371.85	520.37	7.1740
600	1.78154	448.79	626.94	7.5741	0.89114	448.71	626.94	7.3682
700	2.07845	527.74	735.58	7.7415	1.03965	527.68	735.61	7.5357
800	2.37532	609.07	846.60	7.8897	1.18812	609.02	846.64	7.6839
900	2.67217	692.79	960.01	8.0232	1.33657	692.75	960.07	7.8175
1000	2.96900	778.78	1075.68	8.1451	1.48501	778.74	1075.75	7.9393

Temp. (K)	v (m³/kg)	u (kJ/kg)	h (kJ/kg)	s (kJ/kg-K)	v (m³/kg)	u (kJ/kg)	h (kJ/kg)	s (kJ/kg-K)
		400 kPa (91.22 K)				600 kPa (96.37 K)		
Sat	0.05992	61.13	85.10	5.1268	0.04046	62.57	86.85	5.0411
100	0.06806	69.30	96.52	5.2466	0.04299	66.41	92.20	5.0957
120	0.08486	85.48	119.42	5.4556	0.05510	83.73	116.79	5.3204
140	0.10085	101.06	141.40	5.6250	0.06620	99.75	139.47	5.4953
160	0.11647	116.38	162.96	5.7690	0.07689	115.34	161.47	5.6422
180	0.13186	131.55	184.30	5.8947	0.08734	130.69	183.10	5.7696
200	0.14712	146.64	205.49	6.0063	0.09766	145.91	204.50	5.8823
220	0.16228	161.68	226.59	6.1069	0.10788	161.04	225.76	5.9837
240	0.17738	176.67	247.62	6.1984	0.11803	176.11	246.92	6.0757
260	0.19243	191.64	268.61	6.2824	0.12813	191.13	268.01	6.1601
280	0.20745	206.58	289.56	6.3600	0.13820	206.13	289.05	6.2381
300	0.22244	221.52	310.50	6.4322	0.14824	221.11	310.06	6.3105
350	0.25982	258.85	362.78	6.5934	0.17326	258.52	362.48	6.4722
400	0.29712	296.25	415.10	6.7331	0.19819	295.97	414.89	6.6121
450	0.33437	333.81	467.56	6.8567	0.22308	333.57	467.42	6.7359
500	0.37159	371.65	520.28	6.9678	0.24792	371.45	520.20	6.8471
600	0.44595	448.55	626.93	7.1622	0.29755	448.40	626.93	7.0416
700	0.52025	527.55	735.65	7.3298	0.34712	527.43	735.70	7.2093
800	0.59453	608.92	846.73	7.4781	0.39666	608.82	846.82	7.3576
900	0.66878	692.67	960.19	7.6117	0.44618	692.59	960.30	7.4912
1000	0.74302	778.68	1075.89	7.7335	0.49568	778.61	1076.02	7.6131
		800 kPa (100.38 K)				1000 kPa (103.73 K)		
Sat.	0.03038	63.21	87.52	4.9768	0.02416	63.35	87.51	4.9237
120	0.04017	81.88	114.02	5.2191	0.03117	79.91	111.08	5.1357
140	0.04886	98.41	137.50	5.4002	0.03845	97.02	135.47	5.3239
160	0.05710	114.28	159.95	5.5501	0.04522	113.20	158.42	5.4772
180	0.06509	129.82	181.89	5.6793	0.05173	128.94	180.67	5.6082
200	0.07293	145.17	203.51	5.7933	0.05809	144.43	202.52	5.7234
220	0.08067	160.40	224.94	5.8954	0.06436	159.76	224.11	5.8263
240	0.08835	175.54	246.23	5.9880	0.07055	174.98	245.53	5.9194
260	0.09599	190.63	267.42	6.0728	0.07670	190.13	266.83	6.0047
280	0.10358	205.68	288.54	6.1511	0.08281	205.23	288.04	6.0833
300	0.11115	220.70	309.62	6.2238	0.08889	220.29	309.18	6.1562
350	0.12998	258.19	362.17	6.3858	0.10401	257.86	361.87	6.3187
400	0.14873	295.69	414.68	6.5260	0.11905	295.42	414.47	6.4591
500	0.18609	371.25	520.12	6.7613	0.14899	371.04	520.04	6.6947
600	0.22335	448.24	626.93	6.9560	0.17883	448.09	626.92	6.8895
700	0.26056	527.31	735.76	7.1237	0.20862	527.19	735.81	7.0573
800	0.29773	608.73	846.91	7.2721	0.23837	608.63	847.00	7.2057
900	0.33488	692.52	960.42	7.4058	0.26810	692.44	960.54	7.3394
1000	0.37202	778.55	1076.16	7.5277	0.29782	778.49	1076.30	7.4614

TABLE B.6.2 (*continued*)
Superheated Nitrogen

Temp. (K)	v (m³/kg)	u (kJ/kg)	h (kJ/kg)	s (kJ/kg-K)	v (m³/kg)	u (kJ/kg)	h (kJ/kg)	s (kJ/kg-K)
	1500 kPa (110.38 K)				2000 kPa (115.58 K)			
Sat.	0.01555	62.17	85.51	4.8148	0.01100	59.25	81.25	4.7193
120	0.01899	74.26	102.75	4.9650	0.01260	66.90	92.10	4.8116
140	0.02452	93.36	130.15	5.1767	0.01752	89.37	124.40	5.0618
160	0.02937	110.44	154.50	5.3394	0.02144	107.55	150.43	5.2358
180	0.03393	126.71	177.60	5.4755	0.02503	124.42	174.48	5.3775
200	0.03832	142.56	200.03	5.5937	0.02844	140.66	197.53	5.4989
220	0.04260	158.14	222.05	5.6987	0.03174	156.52	219.99	5.6060
240	0.04682	173.57	243.80	5.7933	0.03496	172.15	242.08	5.7021
260	0.05099	188.87	265.36	5.8796	0.03814	187.62	263.90	5.7894
280	0.05512	204.10	286.78	5.9590	0.04128	202.97	285.53	5.8696
300	0.05922	219.27	308.10	6.0325	0.04440	218.24	307.03	5.9438
350	0.06940	257.03	361.13	6.1960	0.05209	256.21	360.39	6.1083
400	0.07949	294.73	413.96	6.3371	0.05971	294.05	413.47	6.2500
450	0.08953	332.53	466.82	6.4616	0.06727	331.95	466.49	6.3750
500	0.09953	370.54	519.84	6.5733	0.07480	370.05	519.65	6.4870
600	0.11948	447.71	626.92	6.7685	0.08980	447.33	626.93	6.6825
700	0.13937	526.89	735.94	6.9365	0.10474	526.59	736.07	6.8507
800	0.15923	608.39	847.22	7.0851	0.11965	608.14	847.45	6.9994
900	0.17906	692.24	960.83	7.2189	0.13454	692.04	961.13	7.1333
1000	0.19889	778.32	1076.65	7.3409	0.14942	778.16	1077.01	7.2553
	3000 kPa (123.61 K)				10000 kPa			
Sat.	0.00582	46.03	63.47	4.5032	—	—	—	—
140	0.01038	79.98	111.13	4.8706	0.00200	0.84	20.87	4.0373
160	0.01350	101.35	141.85	5.0763	0.00291	47.44	76.52	4.4088
180	0.01614	119.68	168.09	5.2310	0.00402	82.44	122.65	4.6813
200	0.01857	136.78	192.49	5.3596	0.00501	108.21	158.35	4.8697
220	0.02088	153.24	215.88	5.4711	0.00590	129.86	188.88	5.0153
240	0.02312	169.30	238.66	5.5702	0.00672	149.42	216.64	5.1362
260	0.02531	185.10	261.02	5.6597	0.00749	167.77	242.72	5.2406
280	0.02746	200.72	283.09	5.7414	0.00824	185.34	267.69	5.3331
300	0.02958	216.21	304.94	5.8168	0.00895	202.38	291.90	5.4167
350	0.03480	254.57	358.96	5.9834	0.01067	243.57	350.26	5.5967
400	0.03993	292.70	412.50	6.1264	0.01232	283.59	406.79	5.7477
500	0.05008	369.06	519.29	6.3647	0.01551	362.42	517.48	5.9948
600	0.06013	446.57	626.95	6.5609	0.01861	441.47	627.58	6.1955
700	0.07012	525.99	736.35	6.7295	0.02167	521.96	738.65	6.3667
800	0.08008	607.67	847.92	6.8785	0.02470	604.42	851.43	6.5172
900	0.09003	691.65	961.73	7.0125	0.02771	689.02	966.15	6.6523
1000	0.09996	777.85	1077.72	7.1347	0.03072	775.68	1082.84	6.7753

TABLE B.7
Thermodynamic Properties of Methane

TABLE B.7.1
Saturated Methane

Temp. (K)	P (kPa)	SPECIFIC VOLUME			INTERNAL ENERGY		
		v_f	v_{fg}	v_g	u_f	u_{fg}	u_g
90.7	11.7	0.002215	3.97941	3.98163	−358.10	496.59	138.49
95	19.8	0.002243	2.44845	2.45069	−343.79	488.62	144.83
100	34.4	0.002278	1.47657	1.47885	−326.90	478.96	152.06
105	56.4	0.002315	0.93780	0.94012	−309.79	468.89	159.11
110	88.2	0.002353	0.62208	0.62443	−292.50	458.41	165.91
111.7	101.3	0.002367	0.54760	0.54997	−286.74	454.85	168.10
115	132.3	0.002395	0.42800	0.43040	−275.05	447.48	172.42
120	191.6	0.002439	0.30367	0.30610	−257.45	436.02	178.57
125	269.0	0.002486	0.22108	0.22357	−239.66	423.97	184.32
130	367.6	0.002537	0.16448	0.16701	−221.65	411.25	189.60
135	490.7	0.002592	0.12458	0.12717	−203.40	397.77	194.37
140	641.6	0.002653	0.09575	0.09841	−184.86	383.42	198.56
145	823.7	0.002719	0.07445	0.07717	−165.97	368.06	202.09
150	1040.5	0.002794	0.05839	0.06118	−146.65	351.53	204.88
155	1295.6	0.002877	0.04605	0.04892	−126.82	333.61	206.79
160	1592.8	0.002974	0.03638	0.03936	−106.35	314.01	207.66
165	1935.9	0.003086	0.02868	0.03177	−85.06	292.30	207.24
170	2329.3	0.003222	0.02241	0.02563	−62.67	267.81	205.14
175	2777.6	0.003393	0.01718	0.02058	−38.75	239.47	200.72
180	3286.4	0.003623	0.01266	0.01629	−12.43	205.16	192.73
185	3863.2	0.003977	0.00846	0.01243	18.47	159.49	177.96
190	4520.5	0.004968	0.00300	0.00797	69.10	67.010	136.11
190.6	4599.2	0.006148	0	0.00615	101.46	0	101.46

TABLE B.7.1 (*continued*)
Saturated Methane

Temp. (k)	P (kPa)	ENTHALPY			ENTROPY		
		h_f	h_{fg}	h_g	s_f	s_{fg}	s_g
90.7	11.7	−358.07	543.12	185.05	4.2264	5.9891	10.2155
95	19.8	−343.75	537.18	193.43	4.3805	5.6545	10.035
100	34.4	−326.83	529.77	202.94	4.5538	5.2977	9.8514
105	56.4	−309.66	521.82	212.16	4.7208	4.9697	9.6905
110	88.2	−292.29	513.29	221.00	4.8817	4.6663	9.5480
111.7	101.3	−286.50	510.33	223.83	4.9336	4.5706	9.5042
115	132.3	−274.74	504.12	229.38	5.0368	4.3836	9.4205
120	191.6	−256.98	494.20	237.23	5.1867	4.1184	9.3051
125	269.0	−238.99	483.44	244.45	5.3321	3.8675	9.1996
130	367.6	−220.72	471.72	251.00	5.4734	3.6286	9.1020
135	490.7	−202.13	458.90	256.77	5.6113	3.3993	9.0106
140	641.6	−183.16	444.85	261.69	5.7464	3.1775	8.9239
145	823.7	−163.73	429.38	265.66	5.8794	2.9613	8.8406
150	1040.5	−143.74	412.29	268.54	6.0108	2.7486	8.7594
155	1295.6	−123.09	393.27	270.18	6.1415	2.5372	8.6787
160	1592.8	−101.61	371.96	270.35	6.2724	2.3248	8.5971
165	1935.9	−79.08	347.82	268.74	6.4046	2.1080	8.5126
170	2329.3	−55.17	320.02	264.85	6.5399	1.8824	8.4224
175	2777.6	−29.33	287.20	257.87	6.6811	1.6411	8.3223
180	3286.4	−0.53	246.77	246.25	6.8333	1.3710	8.2043
185	3863.2	33.83	192.16	226.00	7.0095	1.0387	8.0483
190	4520.5	91.56	80.58	172.14	7.3015	0.4241	7.7256
190.6	4599.2	129.74	0	129.74	7.4999	0	7.4999

Table B.7.2
Superheated Methane

Temp. (K)	v (m³/kg)	u (kJ/kg)	h (kJ/kg)	s (kJ/kg-K)	v (m³/kg)	u (kJ/kg)	h (kJ/kg)	s (kJ/kg-K)
	100 kPa (111.50 K)				200 kPa (120.61 K)			
Sat.	0.55665	167.90	223.56	9.5084	0.29422	179.30	238.14	9.2918
125	0.63126	190.21	253.33	9.7606	0.30695	186.80	248.19	9.3736
150	0.76586	230.18	306.77	10.1504	0.37700	227.91	303.31	9.7759
175	0.89840	269.72	359.56	10.4759	0.44486	268.05	357.02	10.1071
200	1.02994	309.20	412.19	10.7570	0.51165	307.88	410.21	10.3912
225	1.16092	348.90	464.99	11.0058	0.57786	347.81	463.38	10.6417
250	1.29154	389.12	518.27	11.2303	0.64370	388.19	516.93	10.8674
275	1.42193	430.17	572.36	11.4365	0.70931	429.36	571.22	11.0743
300	1.55215	472.36	627.58	11.6286	0.77475	471.65	626.60	11.2670
325	1.68225	516.00	684.23	11.8100	0.84008	515.37	683.38	11.4488
350	1.81226	561.34	742.57	11.9829	0.90530	560.77	741.83	11.6220
375	1.94220	608.58	802.80	12.1491	0.97046	608.07	802.16	11.7885
400	2.07209	657.89	865.10	12.3099	1.03557	657.41	864.53	11.9495
425	2.20193	709.36	929.55	12.4661	1.10062	708.92	929.05	12.1059

TABLE B.7.2 (*continued*)
Superheated Methane

Temp. (K)	v (m³/kg)	u (kJ/kg)	h (kJ/kg)	s (kJ/kg-K)	v (m³/kg)	u (kJ/kg)	h (kJ/kg)	s (kJ/kg-K)
	400 kPa (131.42 K)				600 kPa (138.72 K)			
Sat.	0.15427	191.01	252.72	9.0754	0.10496	197.54	260.51	8.9458
150	0.18233	223.16	296.09	9.3843	0.11717	218.08	288.38	9.1390
175	0.21799	264.61	351.81	9.7280	0.14227	261.03	346.39	9.4970
200	0.25246	305.19	406.18	10.0185	0.16603	302.44	402.06	9.7944
225	0.28631	345.61	460.13	10.2726	0.18911	343.37	456.84	10.0525
250	0.31978	386.32	514.23	10.5007	0.21180	384.44	511.52	10.2830
275	0.35301	427.74	568.94	10.7092	0.23424	426.11	566.66	10.4931
300	0.38606	470.23	624.65	10.9031	0.25650	468.80	622.69	10.6882
325	0.41899	514.10	681.69	11.0857	0.27863	512.82	680.00	10.8716
350	0.45183	559.63	740.36	11.2595	0.30067	558.48	738.88	11.0461
375	0.48460	607.03	800.87	11.4265	0.32264	605.99	799.57	11.2136
400	0.51731	656.47	863.39	11.5879	0.34456	655.52	862.25	11.3754
425	0.54997	708.05	928.04	11.7446	0.36643	707.18	927.04	11.5324
450	0.58260	761.85	994.89	11.8974	0.38826	761.05	994.00	11.6855
475	0.61520	817.89	1063.97	12.0468	0.41006	817.15	1063.18	11.8351
500	0.64778	876.18	1135.29	12.1931	0.43184	875.48	1134.59	11.9816
525	0.68033	936.67	1208.81	12.3366	0.45360	936.03	1208.18	12.1252
	800 kPa (144.40 K)				1000 kPa (149.13 K)			
Sat.	0.07941	201.70	265.23	8.8505	0.06367	204.45	268.12	8.7735
150	0.08434	212.53	280.00	8.9509	0.06434	206.28	270.62	8.7902
175	0.10433	257.30	340.76	9.3260	0.08149	253.38	334.87	9.1871
200	0.12278	299.62	397.85	9.6310	0.09681	296.73	393.53	9.5006
225	0.14050	341.10	453.50	9.8932	0.11132	338.79	450.11	9.7672
250	0.15781	382.53	508.78	10.1262	0.12541	380.61	506.01	10.0028
275	0.17485	424.47	564.35	10.3381	0.13922	422.82	562.04	10.2164
300	0.19172	467.36	620.73	10.5343	0.15285	465.91	618.76	10.4138
325	0.20845	511.55	678.31	10.7186	0.16635	510.26	676.61	10.5990
350	0.22510	557.33	737.41	10.8938	0.17976	556.18	735.94	10.7748
375	0.24167	604.95	798.28	11.0617	0.19309	603.91	797.00	10.9433
400	0.25818	654.57	861.12	11.2239	0.20636	653.62	859.98	11.1059
425	0.27465	706.31	926.03	11.3813	0.21959	705.44	925.03	11.2636
450	0.29109	760.24	993.11	11.5346	0.23279	759.44	992.23	11.4172
475	0.30749	816.40	1062.40	11.6845	0.24595	815.66	1061.61	11.5672
500	0.32387	874.79	1133.89	11.8311	0.25909	874.10	1133.19	11.7141
525	0.34023	935.38	1207.56	11.9749	0.27221	934.73	1206.95	11.8580
550	0.35657	998.14	1283.45	12.1161	0.28531	997.53	1282.84	11.9992

TABLE B.7.2 (*continued*)
Superheated Methane

Temp. (K)	v (m³/kg)	u (kJ/kg)	h (kJ/kg)	s (kJ/kg-K)	v (m³/kg)	u (kJ/kg)	h (kJ/kg)	s (kJ/kg-K)
	1500 kPa (158.52 K)				2000 kPa (165.86 K)			
Sat.	0.04196	207.53	270.47	8.6215	0.03062	207.01	268.25	8.4975
175	0.05078	242.64	318.81	8.9121	0.03504	229.90	299.97	8.6839
200	0.06209	289.13	382.26	9.2514	0.04463	280.91	370.17	9.0596
225	0.07239	332.85	441.44	9.5303	0.05289	326.64	432.43	9.3532
250	0.08220	375.70	499.00	9.7730	0.06059	370.67	491.84	9.6036
275	0.09171	418.65	556.21	9.9911	0.06796	414.40	550.31	9.8266
300	0.10103	462.27	613.82	10.1916	0.07513	458.59	608.85	10.0303
325	0.11022	507.04	672.37	10.3790	0.08216	503.80	668.12	10.2200
350	0.11931	553.30	732.26	10.5565	0.08909	550.40	728.58	10.3992
375	0.12832	601.30	793.78	10.7263	0.09594	598.69	790.57	10.5703
400	0.13728	651.24	857.16	10.8899	0.10274	648.87	854.34	10.7349
425	0.14619	703.26	922.54	11.0484	0.10949	701.08	920.06	10.8942
450	0.15506	757.43	990.02	11.2027	0.11620	755.43	987.84	11.0491
475	0.16391	813.80	1059.66	11.3532	0.12289	811.94	1057.72	11.2003
500	0.17273	872.37	1131.46	11.5005	0.12955	870.64	1129.74	11.3480
525	0.18152	933.12	1205.41	11.6448	0.13619	931.51	1203.88	11.4927
550	0.19031	996.02	1281.48	11.7864	0.14281	994.51	1280.13	11.6346
	4000 kPa (186.10 K)				8000 kPa			
Sat.	0.01160	172.96	219.34	8.0035				
200	0.01763	237.70	308.23	8.4675	0.00412	55.58	88.54	7.2069
225	0.02347	298.52	392.39	8.8653	0.00846	217.30	284.98	8.1344
250	0.02814	349.08	461.63	9.1574	0.01198	298.05	393.92	8.5954
275	0.03235	396.67	526.07	9.4031	0.01469	357.88	475.39	8.9064
300	0.03631	443.48	588.73	9.6212	0.01705	411.71	548.15	9.1598
325	0.04011	490.62	651.07	9.8208	0.01924	463.52	617.40	9.3815
350	0.04381	538.70	713.93	10.0071	0.02130	515.02	685.39	9.5831
375	0.04742	588.18	777.86	10.1835	0.02328	567.12	753.34	9.7706
400	0.05097	639.34	843.24	10.3523	0.02520	620.38	821.95	9.9477
425	0.05448	692.38	910.31	10.5149	0.02707	675.14	891.71	10.1169
450	0.05795	747.43	979.23	10.6725	0.02891	731.63	962.92	10.2796
475	0.06139	804.55	1050.12	10.8258	0.03072	789.99	1035.75	10.4372
500	0.06481	863.78	1123.01	10.9753	0.03251	850.28	1110.34	10.5902
525	0.06820	925.11	1197.93	11.1215	0.03428	912.54	1186.74	10.7393
550	0.07158	988.53	1274.86	11.2646	0.03603	976.77	1264.99	10.8849
575	0.07495	1053.98	1353.77	11.4049	0.03776	1042.96	1345.07	11.0272

APPENDIX C
IDEAL-GAS SPECIFIC HEAT

Three types of energy storage or possession were identified in Section 2.6, of which two, translation and intramolecular energy, are associated with the individual molecules. These comprise the ideal-gas model, with the third type, the system intermolecular potential energy, then accounting for the behavior of real (nonideal-gas) substances. This appendix deals with the ideal-gas contributions. Since these contribute to the energy, and therefore also the enthalpy, they also contribute to the specific heat of each gas. The different possibilities can be grouped according to the intramolecular energy contributions as follows:

MONATOMIC GASES (INERT GASES AR, HE, NE, XE, KR, ALSO N, O, H, CL, F, . . .)

$$h = h_{\text{translation}} + h_{\text{electronic}} = h_t + h_e$$

$$\frac{dh}{dT} = \frac{dh_t}{dT} + \frac{dh_e}{dT}, \qquad C_{P0} = C_{P0t} + C_{P0e} = \frac{5}{2}R + f_e(T)$$

where the electronic contribution, $f_e(T)$, is usually small, except at very high T (common exceptions are O, Cl, F).

DIATOMIC AND LINEAR POLYATOMIC GASES (N_2, O_2, CO, OH, . . . , CO_2, N_2O, . . .)

In addition to translational and electronic contributions to specific heat, these also have molecular rotation (about the center of mass of the molecule) and also $(3a - 5)$ independent modes of molecular vibration of the a atoms in the molecule relative to one another, such that

$$C_{P0} = C_{P0t} + C_{P0r} + C_{P0v} + C_{P0e} = \frac{5}{2}R + R + f_v(T) + f_e(T)$$

where the vibrational contribution is

$$f_v(T) = R \sum_{i=1}^{3a-5} [x_i^2 e^{x_i}/(e^{x_i} - 1)^2], \qquad x_i = \frac{\theta_i}{T}$$

and the electronic contribution, $f_e(T)$, is usually small, except at very high T (common exceptions are O_2, NO, OH).

EXAMPLE C.1	N_2, $3a - 5 = 1$ vibrational mode, with $\theta_i = 3392$ K.

N_2, $3a - 5 = 1$ vibrational mode, with $\theta_i = 3392$ K.

At $T = 300$ K, $C_{P0} = 0.742 + 0.2968 + 0.0005 + \approx 0 = 1.0393$ kJ/kg K.

At $T = 1000$ K, $C_{P0} = 0.742 + 0.2968 + 0.123 + \approx 0 = 1.1618$ kJ/kg K.

(an increase of 11.8% from 300 K).

EXAMPLE C.2 CO_2, $3a - 5 = 4$ vibrational modes, with $\theta_i = 960$ K, 960 K, 1993 K, 3380 K

At $T = 300$ K, $C_{P0} = 0.4723 + 0.1889 + 0.1826 + \approx 0 = 0.8438$ kJ/kg K.

At $T = 1000$ K, $C_{P0} = 0.4723 + 0.1889 + 0.5659 + \approx 0 = 1.2271$ kJ/kg K.

(an increase of 45.4% from 300 K).

NONLINEAR POLYATOMIC MOLECULES (H_2O, NH_3, CH_4, C_2H_6, ...)

Contributions to specific heat are similar to those for linear molecules, except that the rotational contribution is larger and there are $(3a - 6)$ independent vibrational modes, such that

$$C_{P0} = C_{P0t} + C_{P0r} + C_{P0v} + C_{P0e} = \frac{5}{2}R + \frac{3}{2}R + f_v(T) + f_e(T)$$

where the vibrational contribution is

$$f_v(T) = R\sum_{i=1}^{3a-6}[x_i^2 e^{x_i}/(e^{x_i} - 1)^2], \qquad x_i = \frac{\theta_i}{T}$$

and $f_e(T)$ is usually small, except at very high temperatures.

EXAMPLE C.3 CH_4, $3a - 6 = 9$ vibrational modes, with $\theta_i = 4196$ K, 2207 K (two modes), 1879 K (three), 4343 K (three)

At $T = 300$ K, $C_{P0} = 1.2958 + 0.7774 + 0.1527 + \approx 0 = 2.2259$ kJ/kg K.

At $T = 1000$ K, $C_{P0} = 1.2958 + 0.7774 + 2.4022 + \approx 0 = 4.4754$ kJ/kg K.

(an increase of 101.1% from 300 K).

APPENDIX D
EQUATIONS OF STATE

Some of the most used pressure-explicit equations of state can be shown in a form with two parameters. This form is known as a cubic equation of state and contains as a special case the ideal-gas law:

$$P = \frac{RT}{v - b} - \frac{a}{v^2 + cbv + db^2}$$

where (a, b) are parameters and (c, d) define the model as shown in the following table with the acentric factor (w) and

$$b = b_0 RT_c/P_c \qquad \text{and} \qquad a = a_0 R^2 T_c^2/P_c$$

The acentric factor is defined by the saturation pressure at a reduced temperature $T_r = 0.7$

$$\omega = -\frac{\ln P_r^{\text{sat}} \text{ at } T_r = 0.7}{\ln 10} - 1$$

TABLE D.1
Equations of State

Model	c	d	b_0	a_0
Ideal gas	0	0	0	0
van der Waals	0	0	1/8	27/64
Redlich–Kwong	1	0	0.08664	$0.42748\, T_r^{-1/2}$
Soave	1	0	0.08664	$0.42748\, [1 + f(1 - T_r^{1/2})]^2$
Peng–Robinson	2	−1	0.0778	$0.45724\, [1 + f(1 - T_r^{1/2})]^2$

$$f = 0.48 + 1.574\omega - 0.176\omega^2 \qquad \text{for Soave}$$
$$f = 0.37464 + 1.54226\omega - 0.26992\omega^2 \qquad \text{for Peng–Robinson}$$

Appendix D

725

TABLE D.2
*Empirical Constants for Benedict–Webb–Rubin Equation**

Gas	Formula	A_0	B_0	$C_0 \times 10^{-6}$	a	b	$c \times 10^{-6}$	$\alpha \times 10^3$	$\gamma \times 10^2$
Methane	CH_4	1.85500	0.042600	0.022570	0.49400	0.00038004	0.002545	0.124359	0.600
Ethylene	C_2H_4	3.33958	0.0556833	0.131140	0.25900	0.008600	0.021120	0.17800	0.923
Ethane	C_2H_6	4.15556	0.0627724	0.179592	0.34516	0.011122	0.032767	0.243389	1.180
Propylene	C_3H_6	6.11220	0.0850647	0.439182	0.774056	0.0187059	0.102611	0.455696	1.829
Propane	C_3H_8	6.872	0.097313	0.508256	0.94770	0.022500	0.12900	0.607175	2.200
n-Butane	C_4H_{10}	10.0847	0.124361	0.992830	1.88231	0.0399983	0.316400	1.10132	3.400
n-Pentane	C_5H_{12}	12.1794	0.156751	2.12121	4.07480	0.066812	0.82417	1.81000	4.750
n-Hexane	C_6H_{14}	14.4373	0.177813	3.31935	7.11671	0.109131	1.51276	2.81086	6.66849
n-Heptane	C_7H_{16}	17.5206	0.199005	4.74574	10.36475	0.151954	2.47000	4.35611	9.000
Nitrogen	N_2	1.19250	0.04580	0.0058891	0.01490	0.00198154	0.000548064	0.291545	0.750
Oxygen	O_2	1.49880	0.046524	0.0038617	-0.040507	-0.000027963	-0.00020376	0.008641	0.359
Ammonia	NH_3	3.78928	0.0516461	0.178567	0.10354	0.000719561	0.000157536	0.00465189	1.980
Carbon dioxide	CO_2	2.67340	0.045628	0.11333	0.051689	0.0030819	0.0070672	0.11271	0.494

*Units: atmospheres, liters, moles, K, gas constant $R = 0.08206$.

TABLE D.3
The Lee–Kesler Equation of State

The Lee–Kesler generalized equation of state is

$$Z = \frac{P_r v_r'}{T_r} = 1 + \frac{B}{v_r'} + \frac{C}{v_r'^2} + \frac{D}{v_r'^5} + \frac{c_4}{T_r^3 v_r'^2} + \left(\beta + \frac{\gamma}{v_r'^2}\right) \exp\left(-\frac{\gamma}{v_r'^2}\right)$$

$$B = b_1 - \frac{b_2}{T_r} - \frac{b_3}{T_r^2} - \frac{b_4}{T_r^3}$$

$$C = c_1 - \frac{c_2}{T_r} + \frac{c_3}{T_r^3}$$

$$D = d_1 + \frac{d_2}{T_r}$$

in which

$$T_r = \frac{T}{T_c}, \quad P_r = \frac{P}{P_c}, \quad v_r' = \frac{v}{RT_c/P_c}$$

The set of constants is as follows:

Constant	Simple Fluids	Constant	Simple Fluids
b_1	0.118 119 3	c_3	0.0
b_2	0.265 728	c_4	0.042 724
b_3	0.154 790	$d_1 \times 10^4$	0.155 488
b_4	0.030 323	$d_2 \times 10^4$	0.623 689
c_1	0.023 674 4	β	0.653 92
c_2	0.018 698 4	γ	0.060 167

TABLE D.4
Saturated Liquid–Vapor Compressibilities, Lee–Kesler Simple Fluid

T_r	0.40	0.50	0.60	0.70	0.80	0.85	0.90	0.95	1
P_r sat	2.7E-4	4.6E-3	0.028	0.099	0.252	0.373	0.532	0.737	1
Z_f	6.5E-5	9.5E-4	0.0052	0.017	0.042	0.062	0.090	0.132	0.29
Z_g	0.999	0.988	0.957	0.897	0.807	0.747	0.673	0.569	0.29

TABLE D.5
Acentric Factor for Some Substances

Substance		ω	Substance		ω
Ammonia	NH_3	0.25	Water	H_2O	0.344
Argon	Ar	0.001	*n*-Butane	C_4H_{10}	0.199
Bromine	Br_2	0.108	Ethane	C_2H_6	0.099
Helium	He	−0.365	Methane	CH_4	0.011
Neon	Ne	−0.029	R-32		0.277
Nitrogen	N_2	0.039	R-125		0.305

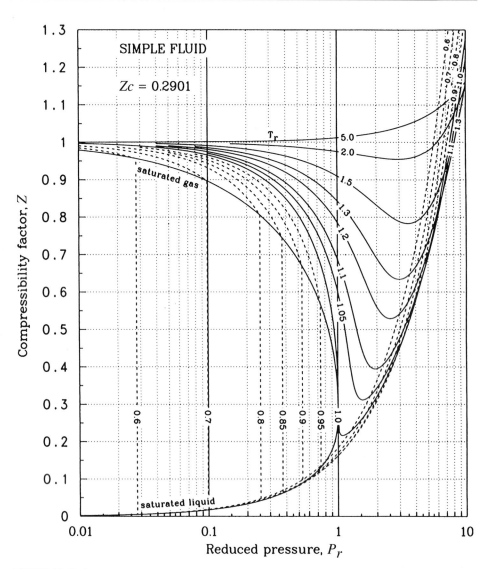

FIGURE D.1 Lee–Kesler Simple Fluid Compressibility Factor.

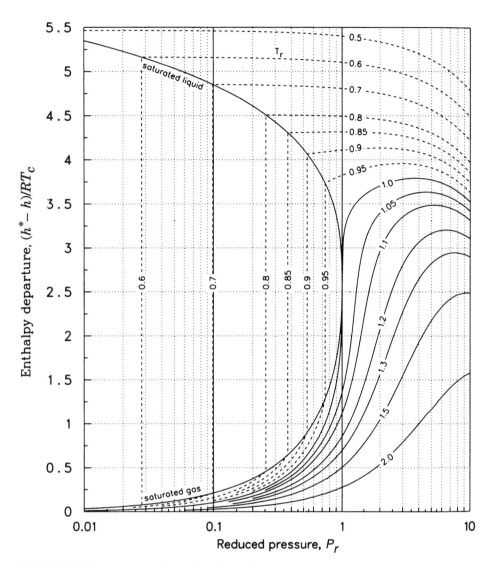

FIGURE D.2 Lee–Kesler Simple Fluid Enthalpy Departure.

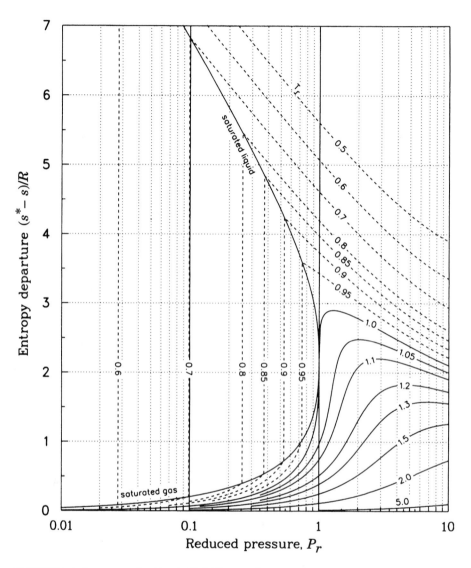

FIGURE D.3 Lee–Kesler Simple Fluid Entropy Departure.

APPENDIX E
FIGURES

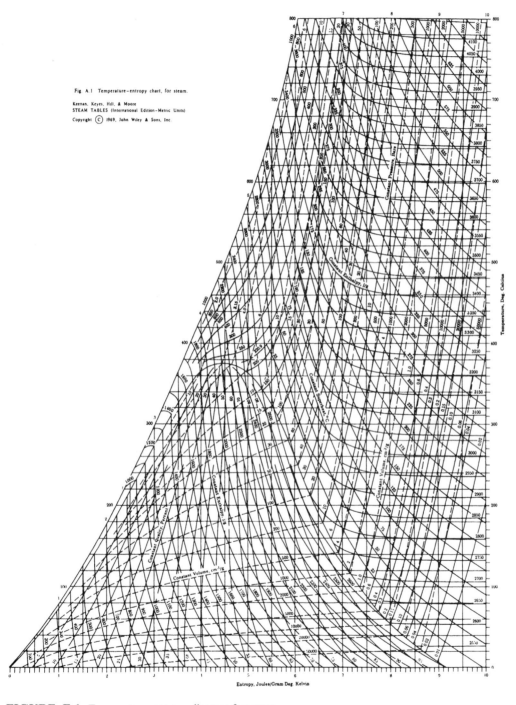

FIGURE E.1 Temperature–entropy diagram for water.

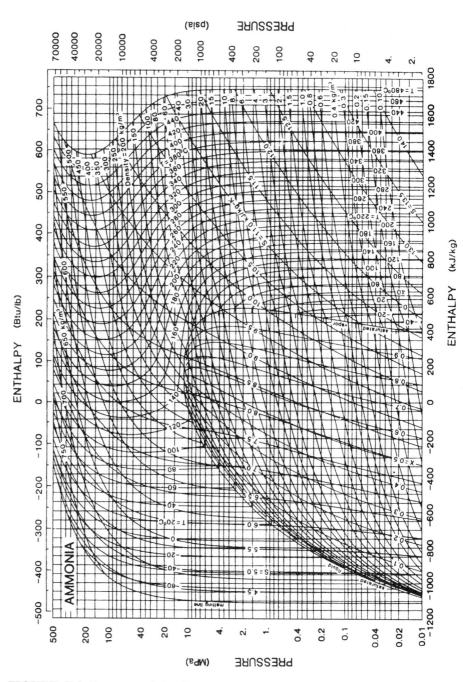

FIGURE E.2 Pressure–enthalpy diagram for ammonia.

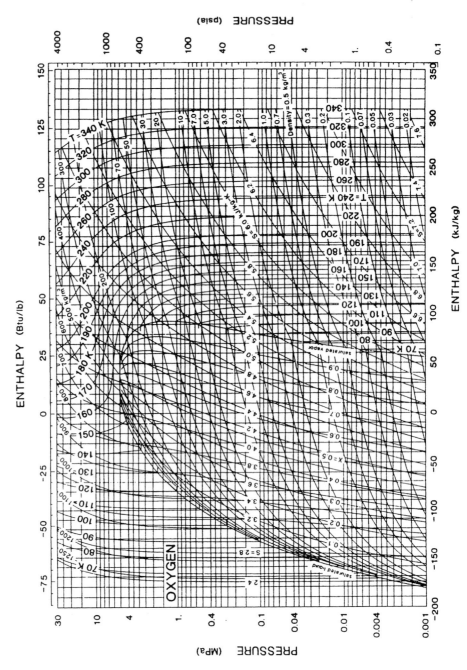

FIGURE E.3 Pressure–enthalpy diagram for oxygen.

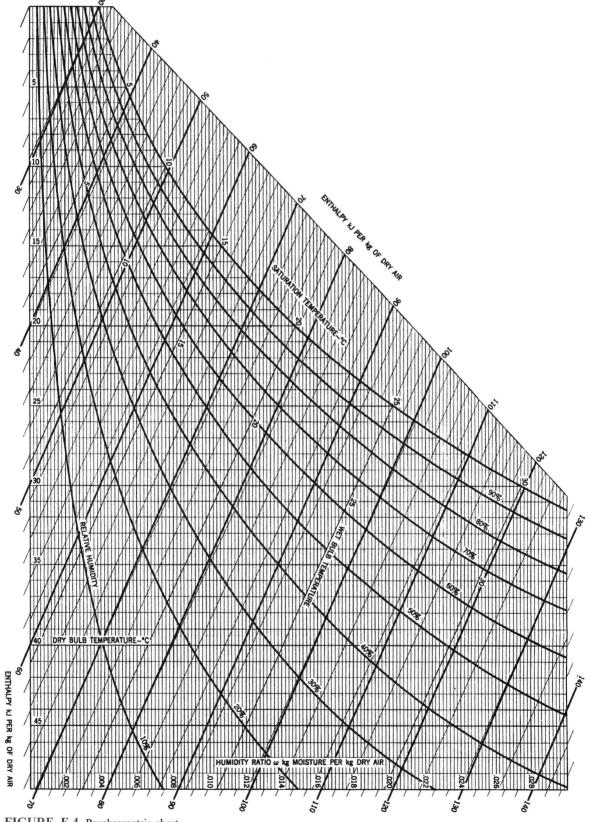

FIGURE E.4 Psychrometric chart.

APPENDIX F
ENGLISH UNIT TABLES

TABLE F.1
Critical Constants (English Units)

Substance	Formula	Molec. Weight	Temp. (R)	Pressure (lbf/in.^{2})	Volume (ft^3/lbm)
Ammonia	NH_3	17.031	729.9	1646	0.0682
Argon	Ar	39.948	271.4	706	0.0300
Bromine	Br_2	159.808	1058.4	1494	0.0127
Carbon dioxide	CO_2	44.010	547.4	1070	0.0342
Carbon monoxide	CO	28.010	239.2	508	0.0533
Chlorine	Cl_2	70.906	750.4	1157	0.0280
Fluorine	F_2	37.997	259.7	757	0.0279
Helium	He	4.003	9.34	32.9	0.2300
Hydrogen (normal)	H_2	2.016	59.76	188.6	0.5170
Krypton	Kr	83.800	376.9	798	0.0174
Neon	Ne	20.183	79.92	400	0.0330
Nitric oxide	NO	30.006	324.0	940	0.0308
Nitrogen	N_2	28.013	227.2	492	0.0514
Nitrogen dioxide	NO_2	46.006	775.8	1465	0.0584
Nitrous oxide	N_2O	44.013	557.3	1050	0.0354
Oxygen	O_2	31.999	278.3	731	0.0367
Sulfur dioxide	SO_2	64.063	775.4	1143	0.0306
Water	H_2O	18.015	1165.1	3208	0.0508
Xenon	Xe	131.300	521.5	847	0.0144
Acetylene	C_2H_2	26.038	554.9	891	0.0693
Benzene	C_6H_6	78.114	1012.0	709	0.0531
n-Butane	C_4H_{10}	58.124	765.4	551	0.0703
Chlorodifluoroethane (142b)	CH_3CClF_2	100.495	738.5	616	0.0368
Chlorodifluoromethane (22)	$CHClF_2$	86.469	664.7	721	0.0307
Dichlorodifluoroethane (141)	CH_3CCl_2F	116.950	866.7	658	0.0345
Dichlorotrifluoroethane (123)	$CHCl_2CF_3$	152.930	822.4	532	0.0291
Difluoroethane (152a)	CHF_2CH_3	66.050	695.5	656	0.0435
Difluoromethane (32)	CH_2F_2	52.024	632.3	838	0.0378
Ethane	C_2H_6	30.070	549.7	708	0.0790
Ethyl alcohol	C_2H_5OH	46.069	925.0	891	0.0581
Ethylene	C_2H_4	28.054	508.3	731	0.0744
n-Heptane	C_7H_{16}	100.205	972.5	397	0.0691
n-Hexane	C_6H_{14}	86.178	913.5	437	0.0688
Methane	CH_4	16.043	342.7	667	0.0990
Methyl alcohol	CH_3OH	32.042	922.7	1173	0.0590
n-Octane	C_8H_{18}	114.232	1023.8	361	0.0690
Pentafluoroethane (125)	CHF_2CF_3	120.022	610.6	525	0.0282
n-Pentane	C_5H_{12}	72.151	845.5	489	0.0675
Propane	C_3H_8	44.094	665.6	616	0.0964
Propene	C_3H_6	42.081	656.8	667	0.0689
Tetrafluoroethane (134a)	CF_3CH_2F	102.030	673.6	589	0.0311

TABLE F.2
Properties of Selected Solids at 77 F

Substance	ρ (lbm/ft³)	C_p (Btu/lbm R)
Asphalt	132.3	0.225
Brick, common	112.4	0.20
Carbon, diamond	202.9	0.122
Carbon, graphite	125–156	0.146
Coal	75–95	0.305
Concrete	137	0.21
Glass, plate	156	0.191
Glass, wool	12.5	0.158
Granite	172	0.212
Ice (32°F)	57.2	0.487
Paper	43.7	0.287
Plexiglas	73.7	0.344
Polystyrene	57.4	0.549
Polyvinyl chloride	86.1	0.229
Rubber, soft	68.7	0.399
Sand, dry	93.6	0.191
Salt, rock	130–156	0.2196
Silicon	145.5	0.167
Snow, firm	35	0.501
Wood, hard (oak)	44.9	0.301
Wood, soft (pine)	31.8	0.33
Wool	6.24	0.411
Metals		
Aluminum, duralumin	170	0.215
Brass, 60-40	524	0.0898
Copper, commercial	518	0.100
Gold	1205	0.03082
Iron, cast	454	0.100
Iron, 304 St Steel	488	0.110
Lead	708	0.031
Magnesium, 2% Mn	111	0.239
Nickel, 10% Cr	541	0.1066
Silver, 99.9% Ag	657	0.0564
Sodium	60.6	0.288
Tin	456	0.0525
Tungsten	1205	0.032
Zinc	446	0.0927

TABLE F.3
Properties of Some Liquids at 77 F

Substance	ρ (lbm/ft³)	C_p (Btu/lbm R)
Ammonia	37.7	1.151
Benzene	54.9	0.41
Butane	34.7	0.60
CCl_4	98.9	0.20
CO_2	42.5	0.69
Ethanol	48.9	0.59
Gasoline	46.8	0.50
Glycerine	78.7	0.58
Kerosene	50.9	0.48
Methanol	49.1	0.61
n-octane	43.2	0.53
Oil, engine	55.2	0.46
Oil, light	57	0.43
Propane	31.8	0.61
R-12	81.8	0.232
R-22	74.3	0.30
R-32	60	0.463
R-125	74.4	0.337
R-134a	75.3	0.34
Water	62.2	1.00
Liquid Metals		
Bismuth, Bi	627	0.033
Lead, Pb	665	0.038
Mercury, Hg	848	0.033
NaK (56/44)	55.4	0.27
Potassium, K	51.7	0.193
Sodium, Na	58	0.33
Tin, Sn	434	0.057
Zinc, Zn	410	0.12

TABLE F.4

Properties of Various Ideal Gases at 77 F, 1 atm* (English Units)

Gas	Chemical Formula	Molecular Mass	R (ft-lbf/lbm R)	$\rho \times 10^3$ (lbm/ft^3)	C_{p0} (Btu/lbm R)	C_{v0} (Btu/lbm R)	k C_{p0}/C_{v0}
Steam	H_2O	18.015	85.76	1.442	0.447	0.337	1.327
Acetylene	C_2H_2	26.038	59.34	65.55	0.406	0.330	1.231
Air	—	28.97	53.34	72.98	0.240	0.171	1.400
Ammonia	NH_3	17.031	90.72	43.325	0.509	0.392	1.297
Argon	Ar	39.948	38.68	100.7	0.124	0.0745	1.667
Butane	C_4H_{10}	58.124	26.58	150.3	0.410	0.376	1.091
Carbon dioxide	CO_2	44.01	35.10	110.8	0.201	0.156	1.289
Carbon monoxide	CO	28.01	55.16	70.5	0.249	0.178	1.399
Ethane	C_2H_6	30.07	51.38	76.29	0.422	0.356	1.186
Ethanol	C_2H_5OH	46.069	33.54	117.6	0.341	0.298	1.145
Ethylene	C_2H_4	28.054	55.07	71.04	0.370	0.299	1.237
Helium	He	4.003	386.0	10.08	1.240	0.744	1.667
Hydrogen	H_2	2.016	766.5	5.075	3.394	2.409	1.409
Methane	CH_4	16.043	96.35	40.52	0.538	0.415	1.299
Methanol	CH_3OH	32.042	48.22	81.78	0.336	0.274	1.227
Neon	Ne	20.183	76.55	50.81	0.246	0.148	1.667
Nitric oxide	NO	30.006	51.50	75.54	0.237	0.171	1.387
Nitrogen	N_2	28.013	55.15	70.61	0.249	0.178	1.400
Nitrous oxide	N_2O	44.013	35.10	110.8	0.210	0.165	1.274
n-octane	C_8H_{18}	114.23	13.53	5.74	0.409	0.391	1.044
Oxygen	O_2	31.999	48.28	80.66	0.220	0.158	1.393
Propane	C_3H_8	44.094	35.04	112.9	0.401	0.356	1.126
R-12	CCl_2F_2	120.914	12.78	310.9	0.147	0.131	1.126
R-22	$CHClF_2$	86.469	17.87	221.0	0.157	0.134	1.171
R-32	CF_2H_2	52.024	29.70	132.6	0.196	0.158	1.242
R-125	CHF_2CF_3	120.022	12.87	307.0	0.189	0.172	1.097
R-134a	CF_3CH_2F	102.03	15.15	262.2	0.203	0.184	1.106
Sulfur dioxide	SO_2	64.059	24.12	163.4	0.149	0.118	1.263
Sulfur trioxide	SO_3	80.053	19.30	204.3	0.152	0.127	1.196

*Or saturation pressure if it is less than 1 atm.

TABLE F.5
Ideal-Gas Properties of Air (English Units)0, Standard Entropy at 1 atm = 101.325 kPa = 14.696 lbf/in.²

T (R)	u (Btu/lbm)	h (Btu/lbm)	s_T^0 (Btu/lbm R)	T (R)	u (Btu/lbm)	h (Btu/lbm)	s_T^0 (Btu/lbm R)
400	68.212	95.634	1.56788	1950	357.243	490.928	1.96404
440	75.047	105.212	1.59071	2000	367.642	504.755	1.97104
480	81.887	114.794	1.61155	2050	378.096	518.636	1.97790
520	88.733	124.383	1.63074	2100	388.602	532.570	1.98461
536.67	91.589	128.381	1.63831	2150	399.158	546.554	1.99119
540	92.160	129.180	1.63979	2200	409.764	560.588	1.99765
560	95.589	133.980	1.64852	2300	431.114	588.793	2.01018
600	102.457	143.590	1.66510	2400	452.640	617.175	2.02226
640	109.340	153.216	1.68063	2500	474.330	645.721	2.03391
680	116.242	162.860	1.69524	2600	496.175	674.421	2.04517
720	123.167	172.528	1.70906	2700	518.165	703.267	2.05606
760	130.118	182.221	1.72216	2800	540.286	732.244	2.06659
800	137.099	191.944	1.73463	2900	562.532	761.345	2.07681
840	144.114	201.701	1.74653	3000	584.895	790.564	2.08671
880	151.165	211.494	1.75791	3100	607.369	819.894	2.09633
920	158.255	221.327	1.76884	3200	629.948	849.328	2.10567
960	165.388	231.202	1.77935	3300	652.625	878.861	2.11476
1000	172.564	241.121	1.78947	3400	675.396	908.488	2.12361
1040	179.787	251.086	1.79924	3500	698.257	938.204	2.13222
1080	187.058	261.099	1.80868	3600	721.203	968.005	2.14062
1120	194.378	271.161	1.81783	3700	744.230	997.888	2.14880
1160	201.748	281.273	1.82670	3800	767.334	1027.848	2.15679
1200	209.168	291.436	1.83532	3900	790.513	1057.882	2.16459
1240	216.640	301.650	1.84369	4000	813.763	1087.988	2.17221
1280	224.163	311.915	1.85184	4100	837.081	1118.162	2.17967
1320	231.737	322.231	1.85977	4200	860.466	1148.402	2.18695
1360	239.362	332.598	1.86751	4300	883.913	1178.705	2.19408
1400	247.037	343.016	1.87506	4400	907.422	1209.069	2.20106
1440	254.762	353.483	1.88243	4500	930.989	1239.492	2.20790
1480	262.537	364.000	1.88964	4600	954.613	1269.972	2.21460
1520	270.359	374.565	1.89668	4700	978.292	1300.506	2.22117
1560	278.230	385.177	1.90357	4800	1002.023	1331.093	2.22761
1600	286.146	395.837	1.91032	4900	1025.806	1361.732	2.23392
1650	296.106	409.224	1.91856	5000	1049.638	1392.419	2.24012
1700	306.136	422.681	1.92659	5100	1073.518	1423.155	2.24621
1750	316.232	436.205	1.93444	5200	1097.444	1453.936	2.25219
1800	326.393	449.794	1.94209	5300	1121.414	1484.762	2.25806
1850	336.616	463.445	1.94957	5400	1145.428	1515.632	2.26383
1900	346.901	477.158	1.95689				

TABLE F.6
Ideal-Gas Properties of Various Substances (English Units), Entropies at 1 atm Pressure

	NITROGEN, DIATOMIC (N_2) $\overline{h}^0_{f,537} = 0$ Btu/lb mol $M = 28.013$		NITROGEN, MONATOMIC (N) $\overline{h}^0_{f,537} = 203\,216$ Btu/lb mol $M = 14.007$	
T R	$\overline{h} - \overline{h}^0_{537}$ Btu/lb mol	$\overline{s}^0_T$ Btu/lbmol R	$\overline{h} - \overline{h}^0_{537}$ Btu/lb mol	$\overline{s}^0_T$ Btu/lbmol R
0	−3727	0	−2664	0
200	−2341	38.877	−1671	31.689
400	−950	43.695	−679	35.130
537	0	45.739	0	36.589
600	441	46.515	314	37.143
800	1837	48.524	1307	38.571
1000	3251	50.100	2300	39.679
1200	4693	51.414	3293	40.584
1400	6169	52.552	4286	41.349
1600	7681	53.561	5279	42.012
1800	9227	54.472	6272	42.597
2000	10804	55.302	7265	43.120
2200	12407	56.066	8258	43.593
2400	14034	56.774	9251	44.025
2600	15681	57.433	10244	44.423
2800	17345	58.049	11237	44.791
3000	19025	58.629	12230	45.133
3200	20717	59.175	13223	45.454
3400	22421	59.691	14216	45.755
3600	24135	60.181	15209	46.038
3800	25857	60.647	16202	46.307
4000	27587	61.090	17195	46.562
4200	29324	61.514	18189	46.804
4400	31068	61.920	19183	47.035
4600	32817	62.308	20178	47.256
4800	34571	62.682	21174	47.468
5000	36330	63.041	22171	47.672
5500	40745	63.882	24670	48.148
6000	45182	64.654	27186	48.586
6500	49638	65.368	29724	48.992
7000	54109	66.030	32294	49.373
7500	58595	66.649	34903	49.733
8000	63093	67.230	37559	50.076
8500	67603	67.777	40270	50.405
9000	72125	68.294	43040	50.721
9500	96658	68.784	45875	51.028
10000	81203	69.250	48777	51.325

TABLE F.6 (*continued*)
Ideal-Gas Properties of Various Substances (English Units), Entropies at 1 atm Pressure

	OXYGEN, DIATOMIC (O₂) $\overline{h}^0_{f,537} = 0$ Btu/lb mol $M = 31.999$		OXYGEN, MONATOMIC (O) $\overline{h}^0_{f,537} = 107\ 124$ Btu/lb mol $M = 16.00$	
T **R**	$\overline{h} - \overline{h}^0_{537}$ **Btu/lb mol**	$\overline{s}^0_T$ **Btu/lbmol R**	$\overline{h} - \overline{h}^0_{537}$ **Btu/lb mol**	$\overline{s}^0_T$ **Btu/lbmol R**
0	−3733	0	−2891	0
200	−2345	42.100	−1829	33.041
400	−955	46.920	−724	36.884
537	0	48.973	0	38.442
600	446	49.758	330	39.023
800	1881	51.819	1358	40.503
1000	3366	53.475	2374	41.636
1200	4903	54.876	3383	42.556
1400	6487	56.096	4387	43.330
1600	8108	57.179	5389	43.999
1800	9761	58.152	6389	44.588
2000	11438	59.035	7387	45.114
2200	13136	59.844	8385	45.589
2400	14852	60.591	9381	46.023
2600	16584	61.284	10378	46.422
2800	18329	61.930	11373	46.791
3000	20088	62.537	12369	47.134
3200	21860	63.109	13364	47.455
3400	23644	63.650	14359	47.757
3600	25441	64.163	15354	48.041
3800	27250	64.652	16349	48.310
4000	29071	65.119	17344	48.565
4200	30904	65.566	18339	48.808
4400	32748	65.995	19334	49.039
4600	34605	66.408	20330	49.261
4800	36472	66.805	21327	49.473
5000	38350	67.189	22325	49.677
5500	43091	68.092	24823	50.153
6000	47894	68.928	27329	50.589
6500	52751	69.705	29847	50.992
7000	57657	70.433	32378	51.367
7500	62608	71.116	34924	51.718
8000	67600	71.760	37485	52.049
8500	72633	72.370	40063	52.362
9000	77708	72.950	42658	52.658
9500	82828	73.504	45270	52.941
10000	87997	74.034	47897	53.210

TABLE F.6 (*continued*)
Ideal-Gas Properties of Various Substances (English Units), Entropies at 1 atm Pressure

	CARBON DIOXIDE (CO_2) $\overline{h}^0_{f,537} = -169\,184$ Btu/lb mol $M = 44.01$		CARBON MONOXIDE (CO) $\overline{h}^0_{f,537} = -47\,518$ Btu/lb mol $M = 28.01$	
T R	$\overline{h} - \overline{h}^0_{537}$ Btu/lb mol	$\overline{s}^0_T$ Btu/lbmol R	$\overline{h} - \overline{h}^0_{537}$ Btu/lb mol	$\overline{s}^0_T$ Btu/lbmol R
0	−4026	0	−3728	0
200	−2636	43.466	−2343	40.319
400	−1153	48.565	−951	45.137
537	0	51.038	0	47.182
600	573	52.047	441	47.959
800	2525	54.848	1842	49.974
1000	4655	57.222	3266	51.562
1200	6927	59.291	4723	52.891
1400	9315	61.131	6220	54.044
1600	11798	62.788	7754	55.068
1800	14358	64.295	9323	55.992
2000	16982	65.677	10923	56.835
2200	19659	66.952	12549	57.609
2400	22380	68.136	14197	58.326
2600	25138	69.239	15864	58.993
2800	27926	70.273	17547	59.616
3000	30741	71.244	19243	60.201
3200	33579	72.160	20951	60.752
3400	36437	73.026	22669	61.273
3600	39312	73.847	24395	61.767
3800	42202	74.629	26128	62.236
4000	45105	75.373	27869	62.683
4200	48021	76.084	29614	63.108
4400	50948	76.765	31366	63.515
4600	53885	77.418	33122	63.905
4800	56830	78.045	34883	64.280
5000	59784	78.648	36650	64.641
5500	67202	80.062	41089	65.487
6000	74660	81.360	45548	66.263
6500	82155	82.560	50023	66.979
7000	89682	83.675	54514	67.645
7500	97239	84.718	59020	68.267
8000	104823	85.697	63539	68.850
8500	112434	86.620	68069	69.399
9000	120071	87.493	72610	69.918
9500	127734	88.321	77161	70.410
10000	135426	89.110	81721	70.878

TABLE F.6 (*continued*)
Ideal-Gas Properties of Various Substances (English Units), Entropies at 1 atm Pressure

	WATER (H_2O) $\bar{h}_{f,537}^0 = -103\,966$ Btu/lb mol $M = 18.015$		HYDROXYL (OH) $\bar{h}_{f,537}^0 = 16\,761$ Btu/lb mol $M = 17.007$	
T R	$\bar{h}-\bar{h}_{537}^0$ Btu/lb mol	$\bar{s}_T^0$ Btu/lbmol R	$\bar{h}-\bar{h}_{537}^0$ Btu/lb mol	$\bar{s}_T^0$ Btu/lbmol R
0	−4528	0	−3943	0
200	−2686	37.209	−2484	36.521
400	−1092	42.728	−986	41.729
537	0	45.076	0	43.852
600	509	45.973	452	44.649
800	2142	48.320	1870	46.689
1000	3824	50.197	3280	48.263
1200	5566	51.784	4692	49.549
1400	7371	53.174	6112	50.643
1600	9241	54.422	7547	51.601
1800	11178	55.563	9001	52.457
2000	13183	56.619	10477	53.235
2200	15254	57.605	11978	53.950
2400	17388	58.533	13504	54.614
2600	19582	59.411	15054	55.235
2800	21832	60.245	16627	55.817
3000	24132	61.038	18220	56.367
3200	26479	61.796	19834	56.887
3400	28867	62.520	21466	57.382
3600	31293	63.213	23114	57.853
3800	33756	63.878	24777	58.303
4000	36251	64.518	26455	58.733
4200	38774	65.134	28145	59.145
4400	41325	65.727	29849	59.542
4600	43899	66.299	31563	59.922
4800	46496	66.852	33287	60.289
5000	49114	67.386	35021	60.643
5500	55739	68.649	39393	61.477
6000	62463	69.819	43812	62.246
6500	69270	70.908	48272	62.959
7000	76146	71.927	52767	63.626
7500	83081	72.884	57294	64.250
8000	90069	73.786	61851	64.838
8500	97101	74.639	66434	65.394
9000	104176	75.448	71043	65.921
9500	111289	76.217	75677	66.422
10000	118440	76.950	80335	66.900

TABLE F.6 *(continued)*
Ideal-Gas Properties of Various Substances (English Units), Entropies at 1 atm Pressure

	Hydrogen (H_2) $\overline{h}^0_{f,537} = 0$ Btu/lb mol $M = 2.016$		Hydrogen, Monatomic (H) $\overline{h}^0_{f,537} = 93\ 723$ Btu/lb mol $M = 1.008$	
T R	$\overline{h} - \overline{h}^0_{537}$ Btu/lb mol	$\overline{s}^0_T$ Btu/lbmol R	$\overline{h} - \overline{h}^0_{537}$ Btu/lb mol	$\overline{s}^0_T$ Btu/lbmol R
0	−3640	0	−2664	0
200	−2224	24.703	−1672	22.473
400	−927	29.193	−679	25.914
537	0	31.186	0	27.373
600	438	31.957	314	27.927
800	1831	33.960	1307	29.355
1000	3225	35.519	2300	30.463
1200	4622	36.797	3293	31.368
1400	6029	37.883	4286	32.134
1600	7448	38.831	5279	32.797
1800	8884	39.676	6272	33.381
2000	10337	40.441	7265	33.905
2200	11812	41.143	8258	34.378
2400	13309	41.794	9251	34.810
2600	14829	42.401	10244	35.207
2800	16372	42.973	11237	35.575
3000	17938	43.512	12230	35.917
3200	19525	44.024	13223	36.238
3400	21133	44.512	14215	36.539
3600	22761	44.977	15208	36.823
3800	24407	45.422	16201	37.091
4000	26071	45.849	17194	37.346
4200	27752	46.260	18187	37.588
4400	29449	46.655	19180	37.819
4600	31161	47.035	20173	38.040
4800	32887	47.403	21166	38.251
5000	34627	47.758	22159	38.454
5500	39032	48.598	24641	38.927
6000	43513	49.378	27124	39.359
6500	48062	50.105	29606	39.756
7000	52678	50.789	32088	40.124
7500	57356	51.434	34571	40.467
8000	62094	52.045	37053	40.787
8500	66889	52.627	39535	41.088
9000	71738	53.182	42018	41.372
9500	76638	53.712	44500	41.640
10000	81581	54.220	46982	41.895

TABLE F.6 (*continued*)
Ideal-Gas Properties of Various Substances (English Units), Entropies at 1 atm Pressure

	NITRIC OXIDE (NO) $\bar{h}^0_{f,537} = 38\ 818$ Btu/lb mol $M = 30.006$		NITROGEN DIOXIDE (NO_2) $\bar{h}^0_{f,537} = 14\ 230$ Btu/lb mol $M = 46.005$	
T **R**	$\bar{h} - \bar{h}^0_{537}$ **Btu/lb mol**	$\bar{s}^0_T$ **Btu/lbmol R**	$\bar{h} - \bar{h}^0_{537}$ **Btu/lb mol**	$\bar{s}^0_T$ **Btu/lbmol R**
0	−3952	0	−4379	0
200	−2457	43.066	−2791	49.193
400	−979	48.207	−1172	54.789
537	0	50.313	0	57.305
600	451	51.107	567	58.304
800	1881	53.163	2469	61.034
1000	3338	54.788	4532	63.333
1200	4834	56.152	6733	65.337
1400	6372	57.337	9044	67.118
1600	7948	58.389	11442	68.718
1800	9557	59.336	13905	70.168
2000	11193	60.198	16421	71.493
2200	12853	60.989	18978	72.712
2400	14532	61.719	21567	73.838
2600	16228	62.397	24182	74.885
2800	17937	63.031	26819	75.861
3000	19657	63.624	29473	76.777
3200	21388	64.183	32142	77.638
3400	23128	64.710	34823	78.451
3600	24875	65.209	37515	79.220
3800	26629	65.684	40215	79.950
4000	28389	66.135	42923	80.645
4200	30154	66.565	45637	81.307
4400	31924	66.977	48358	81.940
4600	33698	67.371	51083	82.545
4800	35476	67.750	53813	83.126
5000	37258	68.113	56546	83.684
5500	41726	68.965	63395	84.990
6000	46212	69.746	70260	86.184
6500	50714	70.467	77138	87.285
7000	55229	71.136	84026	88.306
7500	59756	71.760	90923	89.258
8000	64294	72.346	97826	90.149
8500	68842	72.898	104735	90.986
9000	73401	73.419	111648	91.777
9500	77968	73.913	118565	92.525
10000	82544	74.382	125485	93.235

Table F.7
Thermodynamic Properties of Water

Table F.7.1
Saturated Water

Temp. (F)	Press. (psia)	SPECIFIC VOLUME, ft³/lbm			INTERNAL ENERGY, Btu/lbm		
		Sat. Liquid v_f	Evap. v_{fg}	Sat. Vapor v_g	Sat. Liquid u_f	Evap. u_{fg}	Sat. Vapor u_g
32	0.0887	0.01602	3301.6545	3301.6705	0	1021.21	1021.21
35	0.100	0.01602	2947.5021	2947.5181	2.99	1019.20	1022.19
40	0.122	0.01602	2445.0713	2445.0873	8.01	1015.84	1023.85
45	0.147	0.01602	2036.9527	2036.9687	13.03	1012.47	1025.50
50	0.178	0.01602	1703.9867	1704.0027	18.05	1009.10	1027.15
60	0.256	0.01603	1206.7283	1206.7443	28.08	1002.36	1030.44
70	0.363	0.01605	867.5791	867.5952	38.09	995.64	1033.72
80	0.507	0.01607	632.6739	632.6900	48.08	988.91	1036.99
90	0.699	0.01610	467.5865	467.6026	58.06	982.18	1040.24
100	0.950	0.01613	349.9602	349.9764	68.04	975.43	1043.47
110	1.276	0.01617	265.0548	265.0709	78.01	968.67	1046.68
120	1.695	0.01620	203.0105	203.0267	87.99	961.88	1049.87
130	2.225	0.01625	157.1419	157.1582	97.96	955.07	1053.03
140	2.892	0.01629	122.8567	122.8730	107.95	948.21	1056.16
150	3.722	0.01634	96.9611	96.9774	117.94	941.32	1059.26
160	4.745	0.01639	77.2079	77.2243	127.94	934.39	1062.32
170	5.997	0.01645	61.9983	62.0148	137.94	927.41	1065.35
180	7.515	0.01651	50.1826	50.1991	147.96	920.38	1068.34
190	9.344	0.01657	40.9255	40.9421	157.99	913.29	1071.29
200	11.530	0.01663	33.6146	33.6312	168.03	906.15	1074.18
210	14.126	0.01670	27.7964	27.8131	178.09	898.95	1077.04
212.0	14.696	0.01672	26.7864	26.8032	180.09	897.51	1077.60
220	17.189	0.01677	23.1325	23.1492	188.16	891.68	1079.84
230	20.781	0.01685	19.3677	19.3846	198.25	884.33	1082.58
240	24.968	0.01692	16.3088	16.3257	208.36	876.91	1085.27
250	29.823	0.01700	13.8077	13.8247	218.48	869.41	1087.90
260	35.422	0.01708	11.7503	11.7674	228.64	861.82	1090.46
270	41.848	0.01717	10.0483	10.0655	238.81	854.14	1092.95
280	49.189	0.01726	8.6325	8.6498	249.02	846.35	1095.37
290	57.535	0.01735	7.4486	7.4660	259.25	838.46	1097.71
300	66.985	1.01745	6.4537	6.4712	269.51	830.45	1099.96
310	77.641	0.01755	5.6136	5.6312	279.80	822.32	1102.13
320	89.609	0.01765	4.9010	4.9186	290.13	814.07	1104.20
330	103.00	0.01776	4.2938	4.3115	300.50	805.68	1106.17
340	117.94	0.01787	3.7742	3.7921	310.90	797.14	1108.04
350	134.54	0.01799	3.3279	3.3459	321.35	788.45	1109.80

TABLE F.7.1 (*continued*)
Saturated Water

Temp. (F)	Press. (psia)	ENTHALPY, Btu/lbm			ENTROPY, Btu/lbm R		
		Sat. Liquid h_f	Evap. h_{fg}	Sat. Vapor h_g	Sat. Liquid s_f	Evap. s_{fg}	Sat. Vapor s_g
32	0.0887	0	1075.38	1075.39	0	2.1869	2.1869
35	0.100	2.99	1073.71	1076.70	0.0061	2.1703	2.1764
40	0.122	8.01	1070.89	1078.90	0.0162	2.1430	2.1591
45	0.147	13.03	1068.06	1081.10	0.0262	2.1161	2.1423
50	0.178	18.05	1065.24	1083.29	0.0361	2.0898	2.1259
60	0.256	28.08	1059.59	1087.67	0.0555	2.0388	2.0943
70	0.363	38.09	1053.95	1092.04	0.0746	1.9896	2.0642
80	0.507	48.08	1048.31	1096.39	0.0933	1.9423	2.0356
90	0.699	58.06	1042.65	1100.72	0.1116	1.8966	2.0083
100	0.950	68.04	1036.98	1105.02	0.1296	1.8526	1.9822
110	1.276	78.01	1031.28	1109.29	0.1473	1.8101	1.9574
120	1.695	87.99	1025.55	1113.54	0.1646	1.7690	1.9336
130	2.225	97.97	1019.78	1117.75	0.1817	1.7292	1.9109
140	2.892	107.96	1013.96	1121.92	0.1985	1.6907	1.8892
150	3.722	117.95	1008.10	1126.05	0.2150	1.6533	1.8683
160	4.745	127.95	1002.18	1130.14	0.2313	1.6171	1.8484
170	5.997	137.96	996.21	1134.17	0.2473	1.5819	1.8292
180	7.515	147.98	990.17	1138.15	0.2631	1.5478	1.8109
190	9.344	158.02	984.06	1142.08	0.2786	1.5146	1.7932
200	11.530	168.07	977.87	1145.94	0.2940	1.4822	1.7762
210	14.126	178.13	971.61	1149.74	0.3091	1.4507	1.7599
212.0	14.696	180.13	970.35	1150.49	0.3121	1.4446	1.7567
220	17.189	188.21	965.26	1153.47	0.3240	1.4201	1.7441
230	20.781	198.31	958.81	1157.12	0.3388	1.3901	1.7289
240	24.968	208.43	952.27	1160.70	0.3533	1.3609	1.7142
250	29.823	218.58	945.61	1164.19	0.3677	1.3324	1.7001
260	35.422	228.75	938.84	1167.59	0.3819	1.3044	1.6864
270	41.848	238.95	931.95	1170.90	0.3960	1.2771	1.6731
280	49.189	249.17	924.93	1174.10	0.4098	1.2504	1.6602
290	57.535	259.43	917.76	1177.19	0.4236	1.2241	1.6477
300	66.985	269.73	910.45	1180.18	0.4372	1.1984	1.6356
310	77.641	280.06	902.98	1183.03	0.4507	1.1731	1.6238
320	89.609	290.43	895.34	1185.76	0.4640	1.1483	1.6122
330	103.00	300.84	887.52	1188.36	0.4772	1.1238	1.6010
340	117.94	311.29	879.51	1190.80	0.4903	1.0997	1.5900
350	134.54	321.80	871.30	1193.10	0.5033	1.0760	1.5793

TABLE F.7.1 (*continued*)
Saturated Water

Temp. (F)	Press. (psia)	SPECIFIC VOLUME, ft³/lbm			INTERNAL ENERGY, Btu/lbm		
		Sat. Liquid v_f	Evap. v_{fg}	Sat. Vapor v_g	Sat. Liquid u_f	Evap. u_{fg}	Sat. Vapor u_g
360	152.93	0.01811	2.9430	2.9611	331.83	779.60	1111.43
370	173.24	0.01823	2.6098	2.6280	342.37	770.57	1112.94
380	195.61	0.01836	2.3203	2.3387	352.95	761.37	1114.31
390	220.17	0.01850	2.0680	2.0865	363.58	751.97	1115.55
400	347.08	0.01864	1.8474	1.8660	374.26	742.37	1116.63
410	276.48	0.01878	1.6537	1.6725	385.00	732.56	1117.56
420	308.52	0.01894	1.4833	1.5023	395.80	722.52	1118.32
430	343.37	0.01909	1.3329	1.3520	406.67	712.24	1118.91
440	381.18	0.01926	1.1998	1.2191	417.61	701.71	1119.32
450	422.13	0.01943	1.0816	1.1011	428.63	690.90	1119.53
460	466.38	0.01961	0.9764	0.9961	439.73	679.82	1119.55
470	514.11	0.01980	0.8826	0.9024	450.92	668.43	1119.35
480	565.50	0.02000	0.7986	0.8186	462.21	656.72	1118.93
490	620.74	0.02021	0.7233	0.7435	473.60	644.67	1118.28
500	680.02	0.02043	0.6556	0.6761	485.11	632.26	1117.37
510	743.53	0.02066	0.5946	0.6153	496.75	619.46	1116.21
520	811.48	0.02091	0.5395	0.5604	508.53	606.23	1114.76
530	884.07	0.02117	0.4896	0.5108	520.46	592.56	1113.02
540	961.51	0.02145	0.4443	0.4658	532.56	578.39	1110.95
550	1044.02	0.02175	0.4031	0.4249	544.85	563.69	1108.54
560	1131.85	0.02207	0.3656	0.3876	557.35	548.42	1105.76
570	1225.21	0.02241	0.3312	0.3536	570.07	532.50	1102.56
580	1324.37	0.02278	0.2997	0.3225	583.05	515.87	1098.91
590	1429.58	0.02318	0.2707	0.2939	596.31	498.44	1094.76
600	1541.13	0.02362	0.2440	0.2676	609.91	480.11	1090.02
610	1659.32	0.02411	0.2193	0.2434	623.87	460.76	1084.63
620	1784.48	0.02465	0.1963	0.2209	638.26	440.20	1078.46
630	1916.96	0.02525	0.1747	0.2000	653.17	418.22	1071.38
640	2057.17	0.02593	0.1545	0.1804	668.68	394.52	1063.20
650	2205.54	0.02673	0.1353	0.1620	684.96	368.66	1053.63
660	2362.59	0.02766	0.1169	0.1446	702.24	340.02	1042.26
670	2528.88	0.02882	0.0990	0.1278	720.91	307.52	1028.43
680	2705.09	0.03031	0.0809	0.1112	741.70	269.26	1010.95
690	2891.99	0.03248	0.0618	0.0943	766.34	220.82	987.16
700	3090.47	0.03665	0.0377	0.0743	801.66	145.92	947.57
705.4	3203.79	0.05053	0	0.0505	872.56	0	872.56

TABLE F.7.1 (*continued*)
Saturated Water

Temp. (F)	Press. (psia)	ENTHALPY, Btu/lbm			ENTROPY, Btu/lbm R		
		Sat. Liquid h_f	Evap. h_{fg}	Sat. Vapor h_g	Sat. Liquid s_f	Evap. s_{fg}	Sat. Vapor s_g
360	152.93	332.35	862.88	1195.23	0.5162	1.0526	1.5688
370	173.24	342.95	854.24	1197.19	0.5289	1.0295	1.5584
380	195.61	353.61	845.36	1198.97	0.5416	1.0067	1.5483
390	220.17	364.33	836.23	1200.56	0.5542	0.9841	1.5383
400	247.08	375.11	826.84	1201.95	0.5667	0.9617	1.5284
410	276.48	385.96	817.17	1203.13	0.5791	0.9395	1.5187
420	308.52	396.89	807.20	1204.09	0.5915	0.9175	1.5090
430	343.37	407.89	796.93	1204.82	0.6038	0.8957	1.4995
440	381.18	418.97	786.34	1205.31	0.6160	0.8740	1.4900
450	422.13	430.15	775.40	1205.54	0.6282	0.8523	1.4805
460	466.38	441.42	764.09	1205.51	0.6404	0.8308	1.4711
470	514.11	452.80	752.40	1205.20	0.6525	0.8093	1.4618
480	565.50	464.30	740.30	1204.60	0.6646	0.7878	1.4524
490	620.74	475.92	727.76	1203.68	0.6767	0.7663	1.4430
500	680.02	487.68	714.76	1202.44	0.6888	0.7447	1.4335
510	743.53	499.59	701.27	1200.86	0.7009	0.7232	1.4240
520	811.48	511.67	687.25	1198.92	0.7130	0.7015	1.4144
530	884.07	523.93	672.66	1196.58	0.7251	0.6796	1.4048
540	961.51	536.38	657.45	1193.83	0.7374	0.6576	1.3950
550	1044.02	549.05	641.58	1190.63	0.7496	0.6354	1.3850
560	1131.85	561.97	624.98	1186.95	0.7620	0.6129	1.3749
570	1225.21	575.15	607.59	1182.74	0.7745	0.5901	1.3646
580	1324.37	588.63	589.32	1177.95	0.7871	0.5668	1.3539
590	1429.58	602.45	570.06	1172.51	0.7999	0.5431	1.3430
600	1541.13	616.64	549.71	1166.35	0.8129	0.5187	1.3317
610	1659.32	631.27	528.08	1159.36	0.8262	0.4937	1.3199
620	1784.48	646.40	505.00	1151.41	0.8397	0.4677	1.3075
630	1916.96	662.12	480.21	1142.33	0.8537	0.4407	1.2943
640	2057.17	678.55	453.33	1131.89	0.8681	0.4122	1.2803
650	2205.54	695.87	423.89	1119.76	0.8831	0.3820	1.2651
660	2362.59	714.34	391.13	1105.47	0.8990	0.3493	1.2483
670	2528.88	734.39	353.83	1088.23	0.9160	0.3132	1.2292
680	2705.09	756.87	309.77	1066.64	0.9350	0.2718	1.2068
690	2891.99	783.72	253.88	1037.60	0.9575	0.2208	1.1783
700	3090.47	822.61	167.47	990.09	0.9901	0.1444	1.1345
705.4	3203.79	902.52	0	902.52	1.0580	0	1.0580

TABLE F.7.2
Superheated Vapor Water

Temp. (F)	v (ft³/lbm)	u (Btu/lbm)	h (Btu/lbm)	s (Btu/lbm R)	v (ft³/lbm)	u (Btu/lbm)	h (Btu/lbm)	s (Btu/lbm R)
	1 psia (101.70)				5 psia (162.20)			
Sat.	333.58	1044.02	1105.75	1.9779	73.531	1062.99	1131.03	1.8441
200	392.51	1077.49	1150.12	2.0507	78.147	1076.25	1148.55	1.8715
240	416.42	1091.22	1168.28	2.0775	83.001	1090.25	1167.05	1.8987
280	440.32	1105.02	1186.50	2.1028	87.831	1104.27	1185.53	1.9244
320	464.19	1118.92	1204.82	2.1269	92.645	1118.32	1204.04	1.9487
360	488.05	1132.92	1223.23	2.1499	97.447	1132.42	1222.59	1.9719
400	511.91	1147.02	1241.75	2.1720	102.24	1146.61	1241.21	1.9941
440	535.76	1161.23	1260.37	2.1932	107.03	1160.89	1259.92	2.0154
500	571.53	1182.77	1288.53	2.2235	114.21	1182.50	1288.17	2.0458
600	631.13	1219.30	1336.09	2.2706	126.15	1219.10	1335.82	2.0930
700	690.72	1256.65	1384.47	2.3142	138.08	1256.50	1384.26	2.1367
800	750.30	1294.86	1433.70	2.3549	150.01	1294.73	1433.53	2.1774
900	809.88	1333.94	1483.81	2.3932	161.94	1333.84	1483.68	2.2157
1000	869.45	1373.93	1534.82	2.4294	173.86	1373.85	1534.71	2.2520
1100	929.03	1414.83	1586.75	2.4638	185.78	1414.77	1586.66	2.2864
1200	988.60	1456.67	1639.61	2.4967	197.70	1456.61	1639.53	2.3192
1300	1048.17	1499.43	1693.40	2.5281	209.62	1499.38	1693.33	2.3507
1400	1107.74	1543.13	1748.12	2.5584	221.53	1543.09	1748.06	2.3809
	10 psia (193.19)				14.696 psia (211.99)			
Sat.	38.424	1072.21	1143.32	1.7877	26.803	1077.60	1150.49	1.7567
200	38.848	1074.67	1146.56	1.7927	—	—	—	—
240	41.320	1089.03	1165.50	1.8205	27.999	1087.87	1164.02	1.7764
280	43.768	1103.31	1184.31	1.8467	29.687	1102.40	1183.14	1.8030
320	46.200	1117.56	1203.05	1.8713	31.359	1116.83	1202.11	1.8280
360	48.620	1131.81	1221.78	1.8948	33.018	1131.22	1221.01	1.8516
400	51.032	1146.10	1240.53	1.9171	34.668	1145.62	1239.90	1.8741
440	53.438	1160.46	1259.34	1.9385	36.313	1160.05	1258.80	1.8956
500	57.039	1182.16	1287.71	1.9690	38.772	1181.83	1287.27	1.9262
600	63.027	1218.85	1335.48	2.0164	42.857	1218.61	1335.16	1.9737
700	69.006	1256.30	1384.00	2.0601	46.932	1256.12	1383.75	2.0175
800	74.978	1294.58	1433.32	2.1009	51.001	1294.43	1433.13	2.0584
900	80.946	1333.72	1483.51	2.1392	55.066	1333.60	1483.35	2.0967
1000	86.912	1373.74	1534.57	2.1755	59.128	1373.65	1534.44	2.1330
1100	92.875	1414.68	1586.54	2.2099	63.188	1414.60	1586.44	2.1674
1200	98.837	1456.53	1639.43	2.2428	67.247	1456.47	1639.34	2.2003
1300	104.798	1499.32	1693.25	2.2743	71.304	1499.26	1693.17	2.2318
1400	110.759	1543.03	1747.99	2.3045	75.361	1542.98	1747.92	2.2620
1500	116.718	1587.67	1803.66	2.3337	79.417	1587.63	1803.60	2.2912
1600	122.678	1633.24	1860.25	2.3618	83.473	1633.20	1860.20	2.3194

TABLE F.7.2 (*continued*)
Superheated Vapor Water

Temp. (F)	v (ft³/lbm)	u (Btu/lbm)	h (Btu/lbm)	s (Btu/lbm R)	v (ft³/lbm)	u (Btu/lbm)	h (Btu/lbm)	s (Btu/lbm R)
	20 psia (227.96)				40 psia (267.26)			
Sat.	20.091	1082.02	1156.38	1.7320	10.501	1092.27	1170.00	1.6767
240	20.475	1086.54	1162.32	1.7405	—	—	—	—
280	21.734	1101.36	1181.80	1.7676	10.711	1097.31	1176.59	1.6857
320	22.976	1116.01	1201.04	1.7929	11.360	1112.81	1196.90	1.7124
360	24.206	1130.55	1220.14	1.8168	11.996	1127.98	1216.77	1.7373
400	25.427	1145.06	1239.17	1.8395	12.623	1142.95	1236.38	1.7606
440	26.642	1159.59	1258.19	1.8611	13.243	1157.82	1255.84	1.7827
500	28.456	1181.46	1286.78	1.8919	14.164	1180.06	1284.91	1.8140
600	31.466	1218.35	1334.80	1.9395	15.685	1217.33	1333.43	1.8621
700	34.466	1255.91	1383.47	1.9834	17.196	1255.14	1382.42	1.9063
800	37.460	1294.27	1432.91	2.0243	18.701	1293.65	1432.08	1.9474
900	40.450	1333.47	1483.17	2.0626	20.202	1332.96	1482.50	1.9859
1000	43.437	1373.54	1534.30	2.0989	21.700	1373.12	1533.74	2.0222
1100	46.422	1414.51	1586.32	2.1334	23.196	1414.16	1585.86	2.0568
1200	49.406	1456.39	1639.24	2.1663	24.690	1456.09	1638.85	2.0897
1300	52.389	1499.19	1693.08	2.1978	26.184	1498.94	1692.75	2.1212
1400	55.371	1542.92	1747.85	2.2280	27.677	1542.70	1747.56	2.1515
1500	58.352	1587.58	1803.54	2.2572	29.169	1587.38	1803.29	2.1807
1600	61.333	1633.15	1860.14	2.2854	30.660	1632.97	1859.92	2.2089
	60 psia (292.73)				80 psia (312.06)			
Sat.	7.177	1098.33	1178.02	1.6444	5.474	1102.56	1183.61	1.6214
320	7.485	1109.46	1192.56	1.6633	5.544	1105.95	1188.02	1.6270
360	7.924	1125.31	1213.29	1.6893	5.886	1122.53	1209.67	1.6541
400	8.353	1140.77	1233.52	1.7134	6.217	1138.53	1230.56	1.6790
440	8.775	1156.01	1253.44	1.7360	6.541	1154.15	1250.98	1.7022
500	9.399	1178.64	1283.00	1.7678	7.017	1177.19	1281.07	1.7346
600	10.425	1216.31	1332.06	1.8165	7.794	1215.28	1330.66	1.7838
700	11.440	1254.35	1381.37	1.8609	8.561	1253.57	1380.31	1.8285
800	12.448	1293.03	1431.24	1.9022	9.322	1292.41	1430.40	1.8700
900	13.452	1332.46	1481.82	1.9408	10.078	1331.95	1481.14	1.9087
1000	14.454	1372.71	1533.19	1.9773	10.831	1372.29	1532.63	1.9453
1100	15.454	1413.81	1585.39	2.0119	11.583	1413.46	1584.93	1.9799
1200	16.452	1455.80	1638.46	2.0448	12.333	1455.51	1638.08	2.0129
1300	17.449	1498.69	1692.42	2.0764	13.082	1498.43	1692.09	2.0445
1400	18.445	1542.48	1747.28	2.1067	13.830	1542.26	1746.99	2.0749
1500	19.441	1587.18	1803.04	2.1359	14.577	1586.99	1802.79	2.1041
1600	20.436	1632.79	1859.70	2.1641	15.324	1632.62	1859.48	2.1323
1800	22.426	1726.69	1975.69	2.2178	16.818	1726.54	1975.50	2.1861
2000	24.415	1824.02	2095.10	2.2685	18.310	1823.88	2094.94	2.2367

TABLE F.7.2 (*continued*)
Superheated Vapor Water

Temp. (F)	v (ft³/lbm)	u (Btu/lbm)	h (Btu/lbm)	s (Btu/lbm R)	v (ft³/lbm)	u (Btu/lbm)	h (Btu/lbm)	s (Btu/lbm R)
	100 psia (327.85)				150 psia (358.47)			
Sat.	4.4340	1105.76	1187.81	1.6034	3.0163	1111.19	1194.91	1.5704
350	4.5917	1115.39	1200.36	1.6191	—	—	—	—
400	4.9344	1136.21	1227.53	1.6517	3.2212	1130.10	1219.51	1.5997
450	5.2646	1156.20	1253.62	1.6812	3.4547	1151.47	1247.36	1.6312
500	5.5866	1175.72	1279.10	1.7085	3.6789	1171.93	1274.04	1.6598
550	5.9032	1195.02	1304.25	1.7340	3.8970	1191.88	1300.05	1.6862
600	6.2160	1214.23	1329.26	1.7582	4.1110	1211.58	1325.69	1.7110
700	6.8340	1252.78	1379.24	1.8033	4.5309	1250.78	1376.55	1.7568
800	7.4455	1291.78	1429.56	1.8449	4.9441	1290.21	1427.44	1.7989
900	8.0528	1331.45	1480.47	1.8838	5.3529	1330.18	1478.76	1.8381
1000	8.6574	1371.87	1532.08	1.9204	5.7590	1370.83	1530.68	1.8750
1100	9.2599	1413.12	1584.47	1.9551	6.1630	1412.24	1583.31	1.9098
1200	9.8610	1455.21	1637.69	1.9882	6.5655	1454.47	1636.71	1.9430
1300	10.4610	1498.18	1691.76	2.0198	6.9670	1497.55	1690.93	1.9747
1400	11.0602	1542.04	1746.71	2.0502	7.3677	1541.49	1745.99	2.0052
1500	11.6588	1586.79	1802.54	2.0794	7.7677	1586.30	1801.91	2.0345
1600	12.2570	1632.44	1859.25	2.1076	8.1673	1632.00	1858.70	2.0627
1800	13.4525	1726.38	1975.32	2.1614	8.9657	1726.00	1974.86	2.1165
2000	14.6472	1823.74	2094.78	2.2120	9.7633	1823.38	2094.38	2.1672
	200 psia (381.86)				300 psia (417.42)			
Sat.	2.2892	1114.55	1199.28	1.5464	1.5441	1118.14	1203.86	1.5115
400	2.3609	1123.45	1210.83	1.5600	—	—	—	—
450	2.5477	1146.44	1240.73	1.5938	1.6361	1135.37	1226.20	1.5365
500	2.7238	1167.96	1268.77	1.6238	1.7662	1159.47	1257.52	1.5701
550	2.8932	1188.65	1295.72	1.6512	1.8878	1181.85	1286.65	1.5997
600	3.0580	1208.87	1322.05	1.6767	2.0041	1203.24	1314.50	1.6266
700	3.3792	1248.76	1373.82	1.7234	2.2269	1244.63	1368.26	1.6751
800	3.6932	1288.62	1425.31	1.7659	2.4421	1285.41	1420.99	1.7187
900	4.0029	1328.90	1477.04	1.8055	2.6528	1326.31	1473.58	1.7589
1000	4.3097	1369.77	1529.28	1.8425	2.8604	1367.65	1526.45	1.7964
1100	4.6145	1411.36	1582.15	1.8776	3.0660	1409.60	1579.80	1.8317
1200	4.9178	1453.73	1635.74	1.9109	3.2700	1452.24	1633.77	1.8653
1300	5.2200	1496.91	1690.10	1.9427	3.4730	1495.63	1688.43	1.8972
1400	5.5214	1540.93	1745.28	1.9732	3.6751	1539.82	1743.84	1.9279
1500	5.8222	1585.81	1801.29	2.0025	3.8767	1584.82	1800.03	1.9573
1600	6.1225	1631.55	1858.15	2.0308	4.0777	1630.66	1857.04	1.9857
1800	6.7223	1725.62	1974.41	2.0847	4.4790	1724.85	1973.50	2.0396
2000	7.3214	1823.02	2093.99	2.1354	4.8794	1822.32	2093.20	2.0904

TABLE F.7.2 (*continued*)
Superheated Vapor Water

Temp. (F)	v (ft³/lbm)	u (Btu/lbm)	h (Btu/lbm)	s (Btu/lbm R)	v (ft³/lbm)	u (Btu/lbm)	h (Btu/lbm)	s (Btu/lbm R)
	400 psia (444.69)				600 psia (486.33)			
Sat.	1.1619	1119.44	1205.45	1.4856	0.7702	1118.54	1204.06	1.4464
450	1.1745	1122.63	1209.57	1.4901	—	—	—	—
500	1.2843	1150.11	1245.17	1.5282	0.7947	1127.97	1216.21	1.4592
550	1.3834	1174.56	1276.95	1.5605	0.8749	1158.23	1255.36	1.4990
600	1.4760	1197.33	1306.58	1.5892	0.9456	1184.50	1289.49	1.5320
700	1.6503	1240.38	1362.54	1.6396	1.0728	1231.51	1350.62	1.5871
800	1.8163	1282.14	1416.59	1.6844	1.1900	1275.42	1407.55	1.6343
900	1.9776	1323.69	1470.07	1.7252	1.3021	1318.36	1462.92	1.6766
1000	2.1357	1365.51	1523.59	1.7632	1.4108	1361.15	1517.79	1.7155
1100	2.2917	1407.81	1577.44	1.7989	1.5173	1404.20	1572.66	1.7519
1200	2.4462	1450.73	1631.79	1.8327	1.6222	1447.68	1627.80	1.7861
1300	2.5995	1494.34	1686.76	1.8648	1.7260	1491.74	1683.38	1.8186
1400	2.7520	1538.70	1742.40	1.8956	1.8289	1536.44	1739.51	1.8497
1500	2.9039	1583.83	1798.78	1.9251	1.9312	1581.84	1796.26	1.8794
1600	3.0553	1629.77	1855.93	1.9535	2.0330	1627.98	1853.71	1.9080
1700	3.2064	1676.52	1913.86	1.9810	2.1345	1674.88	1911.87	1.9355
1800	3.3573	1724.08	1972.59	2.0076	2.2357	1722.55	1970.78	1.9622
2000	3.6585	1821.61	2092.41	2.0584	2.4375	1820.20	2090.84	2.0131
	800 psia (518.36)				1000 psia (544.74)			
Sat.	0.5691	1115.02	1199.26	1.4160	0.4459	1109.86	1192.37	1.3903
550	0.6154	1138.83	1229.93	1.4469	0.4534	1114.77	1198.67	1.3965
600	0.6776	1170.10	1270.41	1.4861	0.5140	1153.66	1248.76	1.4450
650	0.7324	1197.22	1305.64	1.5186	0.5637	1184.74	1289.06	1.4822
700	0.7829	1222.08	1337.98	1.5471	0.6080	1212.03	1324.54	1.5135
800	0.8764	1268.45	1398.19	1.5969	0.6878	1261.21	1388.49	1.5664
900	0.9640	1312.88	1455.60	1.6408	0.7610	1307.26	1448.08	1.6120
1000	1.0482	1356.71	1511.88	1.6807	0.8305	1352.17	1505.86	1.6530
1100	1.1300	1400.52	1567.81	1.7178	0.8976	1396.77	1562.88	1.6908
1200	1.2102	1444.60	1623.76	1.7525	0.9630	1441.46	1619.67	1.7260
1300	1.2892	1489.11	1679.97	1.7854	1.0272	1486.45	1676.53	1.7593
1400	1.3674	1534.17	1736.59	1.8167	1.0905	1531.88	1733.67	1.7909
1500	1.4448	1579.85	1793.74	1.8467	1.1531	1577.84	1791.21	1.8210
1600	1.5218	1626.19	1851.49	1.8754	1.2152	1624.40	1849.27	1.8499
1700	1.5985	1673.25	1909.89	1.9031	1.2769	1671.61	1907.91	1.8777
1800	1.6749	1721.03	1968.98	1.9298	1.3384	1719.51	1967.18	1.9046
2000	1.8271	1818.80	2089.28	1.9808	1.4608	1817.41	2087.74	1.9557

TABLE F.7.2 *(continued)*
Superheated Vapor Water

Temp. (F)	v (ft³/lbm)	u (Btu/lbm)	h (Btu/lbm)	s (Btu/lbm R)	v (ft³/lbm)	u (Btu/lbm)	h (Btu/lbm)	s (Btu/lbm R)
	1500 psia (596.38)				2000 psia (635.99)			
Sat.	0.2769	1091.81	1168.67	1.3358	0.1881	1066.63	1136.25	1.2861
650	0.3329	1146.95	1239.34	1.4012	0.2057	1091.06	1167.18	1.3141
700	0.3716	1183.44	1286.60	1.4429	0.2487	1147.74	1239.79	1.3782
750	0.4049	1214.13	1326.52	1.4766	0.2803	1187.32	1291.07	1.4216
800	0.4350	1241.79	1362.53	1.5058	0.3071	1220.13	1333.80	1.4562
850	0.4631	1267.69	1396.23	1.5321	0.3312	1249.46	1372.03	1.4860
900	0.4897	1292.53	1428.46	1.5562	0.3534	1276.78	1407.58	1.5126
1000	0.5400	1340.43	1490.32	1.6001	0.3945	1328.10	1474.09	1.5598
1100	0.5876	1387.16	1550.26	1.6398	0.4325	1377.17	1537.23	1.6017
1200	0.6334	1433.45	1609.25	1.6765	0.4685	1425.19	1598.58	1.6398
1300	0.6778	1479.68	1667.82	1.7108	0.5031	1472.74	1658.95	1.6751
1400	0.7213	1526.06	1726.28	1.7431	0.5368	1520.15	1718.81	1.7082
1500	0.7641	1572.77	1784.86	1.7738	0.5697	1567.64	1778.48	1.7395
1600	0.8064	1619.90	1843.72	1.8301	0.6020	1615.37	1838.18	1.7692
1700	0.8482	1667.53	1902.98	1.8312	0.6340	1663.45	1898.08	1.7976
1800	0.8899	1715.73	1962.73	1.8582	0.6656	1711.97	1958.32	1.8248
1900	0.9313	1764.53	2023.03	1.8843	0.6971	1760.99	2018.99	1.8511
2000	0.9725	1813.97	2083.91	1.9096	0.7284	1810.56	2080.15	1.8765
	4000 psia				8000 psia			
650	0.02447	657.71	675.82	0.8574	0.02239	627.01	660.16	0.8278
700	0.02867	742.13	763.35	0.9345	0.02418	688.59	724.39	0.8844
750	0.06332	960.69	1007.56	1.1395	0.02671	755.67	795.21	0.9441
800	0.10523	1095.04	1172.93	1.2740	0.03061	830.67	875.99	1.0095
850	0.12833	1156.47	1251.46	1.3352	0.03706	915.81	970.67	1.0832
900	0.14623	1201.47	1309.71	1.3789	0.04657	1003.68	1072.63	1.1596
950	0.16152	1239.20	1358.75	1.4143	0.05721	1079.59	1164.28	1.2259
1000	0.17520	1272.94	1402.62	1.4449	0.06722	1141.04	1240.55	1.2791
1100	0.19954	1333.90	1481.60	1.4973	0.08445	1236.84	1361.85	1.3595
1200	0.22129	1390.11	1553.91	1.5423	0.09892	1314.18	1460.62	1.4210
1300	0.24137	1443.72	1622.38	1.5823	0.11161	1382.27	1547.50	1.4718
1400	0.26029	1495.73	1688.39	1.6188	0.12309	1444.85	1627.08	1.5158
1500	0.27837	1546.73	1752.78	1.6525	0.13372	1503.78	1701.74	1.5549
1600	0.29586	1597.12	1816.11	1.6841	0.14373	1560.12	1772.89	1.5904
1700	0.31291	1647.17	1878.79	1.7138	0.15328	1614.58	1841.49	1.6229
1800	0.32964	1697.11	1941.11	1.7420	0.16251	1667.69	1908.27	1.6531
1900	0.34616	1747.10	2003.32	1.7689	0.17151	1719.85	1973.75	1.6815
2000	0.36251	1797.27	2065.60	1.7948	0.18034	1771.38	2038.36	1.7083

TABLE F.7.3
Compressed Liquid Water

Temp. (F)	v (ft³/lbm)	u (Btu/lbm)	h (Btu/lbm)	s (Btu/lbm R)	v (ft³/lbm)	u (Btu/lbm)	h (Btu/lbm)	s (Btu/lbm R)
	500 psia (467.12)				1000 psia (544.74)			
Sat.	0.01975	447.69	449.51	0.6490	0.02159	538.37	542.36	0.74318
32	0.01599	0.00	1.48	0.0000	0.01597	0.02	2.98	0.0000
50	0.01599	18.02	19.50	0.0360	0.01599	17.98	20.94	0.0359
75	0.0160	42.98	44.46	0.0838	0.0160	42.87	45.83	0.0836
100	0.0161	67.87	69.36	0.1293	0.0161	67.70	70.67	0.1290
125	0.0162	92.75	94.24	0.1728	0.0162	92.52	95.51	0.1724
150	0.0163	117.66	119.17	0.2146	0.0163	117.37	120.39	0.2141
175	0.0165	142.62	144.14	0.2547	0.0164	142.28	145.32	0.2542
200	0.0166	167.64	168.18	0.2934	0.0166	167.25	170.32	0.2928
225	0.0168	192.76	194.31	0.3308	0.0168	192.30	195.40	0.3301
250	0.0170	217.99	219.56	0.3670	0.0169	217.46	220.60	0.3663
275	0.0172	243.36	244.95	0.4022	0.0171	242.77	245.94	0.4014
300	0.0174	268.91	270.52	0.4364	0.0174	268.24	271.45	0.4355
325	0.0177	294.68	296.32	0.4698	0.0176	293.91	297.17	0.4688
350	0.0180	320.70	322.36	0.5025	0.0179	319.83	323.14	0.5014
375	0.0183	347.01	348.70	0.5345	0.0182	346.02	349.39	0.5333
400	0.0186	373.68	375.40	0.5660	0.0185	372.55	375.98	0.5647
425	0.0190	400.77	402.52	0.5971	0.0189	399.47	402.97	0.5957
450	0.0194	428.39	430.19	0.6280	0.0193	426.89	430.47	0.6263
	2000 psia (635.99)				8000 psia			
Sat.	0.02565	662.38	671.87	0.8622	—	—	—	—
50	0.01592	17.91	23.80	0.0357	0.01563	17.38	40.52	0.0342
75	0.0160	42.66	48.57	0.0832	0.0157	41.42	64.65	0.0804
100	0.0160	67.36	73.30	0.1284	0.01577	65.49	88.83	0.1246
125	0.0161	92.07	98.04	0.1716	0.01586	89.62	113.10	0.1670
150	0.0162	116.82	122.84	0.2132	0.01597	113.81	137.45	0.2078
175	0.0164	141.62	147.68	0.2531	0.01610	138.04	161.87	0.2471
200	0.0165	166.48	172.60	0.2916	0.01623	162.31	186.34	0.2849
225	0.0167	191.42	197.59	0.3288	0.01639	186.61	210.87	0.3214
250	0.0169	216.45	222.69	0.3648	0.01655	210.97	235.47	0.3567
275	0.0171	241.61	247.93	0.3998	0.01675	235.39	260.16	0.3909
300	0.0173	266.92	273.33	0.4337	0.01693	259.91	284.97	0.4241
325	0.0176	292.42	298.92	0.4669	0.01714	284.53	309.91	0.4564
350	0.0178	318.14	324.74	0.4993	0.01737	309.29	335.01	0.4878
400	0.0184	370.38	377.20	0.5621	0.01788	359.26	385.73	0.5486
450	0.0192	424.03	431.13	0.6231	0.01848	409.94	437.30	0.6069
500	0.0201	479.84	487.29	0.6832	0.01918	461.56	489.95	0.6633
600	0.0233	605.37	613.99	0.8086	0.02106	569.36	600.53	0.7728

TABLE F.7.4

Saturated Solid–Saturated Vapor, Water (English Units)

| Temp. (F) | Press. (lbf/in.²) | SPECIFIC VOLUME, ft³/lbm | | INTERNAL ENERGY, Btu/lbm | | |
		Sat. Solid v_i	Sat. Vapor $v_g \times 10^{-3}$	Sat. Solid u_i	Evap. u_{ig}	Sat. Vapor u_g
32.02	0.08866	0.017473	3.302	−143.34	1164.5	1021.2
32	0.08859	0.01747	3.305	−143.35	1164.5	1021.2
30	0.08083	0.01747	3.607	−144.35	1164.9	1020.5
25	0.06406	0.01746	4.505	−146.84	1165.7	1018.9
20	0.05051	0.01745	5.655	−149.31	1166.5	1017.2
15	0.03963	0.01745	7.133	−151.75	1167.3	1015.6
10	0.03093	0.01744	9.043	−154.16	1168.1	1013.9
5	0.02402	0.01743	11.522	−156.56	1168.8	1012.2
0	0.01855	0.01742	14.761	−158.93	1169.5	1010.6
−5	0.01424	0.01742	19.019	−161.27	1170.2	1008.9
−10	0.01086	0.01741	24.657	−163.59	1170.8	1007.3
−15	0.00823	0.01740	32.169	−165.89	1171.5	1005.6
−20	0.00620	0.01740	42.238	−168.16	1172.1	1003.9
−25	0.00464	0.01739	55.782	−170.40	1172.7	1002.3
−30	0.00346	0.01738	74.046	−172.63	1173.2	1000.6
−35	0.00256	0.01737	98.890	−174.82	1173.8	998.9
−40	0.00187	0.01737	134.017	−177.00	1174.3	997.3

TABLE F.7.4 (*continued*)
Saturated Solid–Saturated Vapor, Water (English Units)

Temp. (F)	Press. (lbf/in.²)	ENTHALPY, Btu/lbm			ENTROPY, Btu/lbm R		
		Sat. Solid h_i	Evap. h_{ig}	Sat. Vapor h_g	Sat. Solid s_i	Evap. s_{ig}	Sat. Vapor s_g
32.02	0.08866	−143.34	1218.7	1075.4	−0.2916	2.4786	2.1869
32	0.08859	−143.35	1218.7	1075.4	−0.2917	2.4787	2.1870
30	0.08083	−144.35	1218.8	1074.5	−0.2938	2.4891	2.1953
25	0.06406	−146.84	1219.1	1072.3	−0.2990	2.5154	2.2164
20	0.05051	−149.31	1219.4	1070.1	−0.3042	2.5422	2.2380
15	0.03963	−151.75	1219.6	1067.9	−0.3093	2.5695	2.2601
10	0.03093	−154.16	1219.8	1065.7	−0.3145	2.5973	2.2827
5	0.02402	−156.56	1220.0	1063.5	−0.3197	2.6256	2.3059
0	0.01855	−158.93	1220.2	1061.2	−0.3248	2.6544	2.3296
−5	0.01424	−161.27	1220.3	1059.0	−0.3300	2.6839	2.3539
−10	0.01086	−163.59	1220.4	1056.8	−0.3351	2.7140	2.3788
−15	0.00823	−165.89	1220.5	1054.6	−0.3403	2.7447	2.4044
−20	0.00620	−168.16	1220.5	1052.4	−0.3455	2.7761	2.4307
−25	0.00464	−170.40	1220.6	1050.2	−0.3506	2.8081	2.4575
−30	0.00346	−172.63	1220.6	1048.0	−0.3557	2.8406	2.4849
−35	0.00256	−174.82	1220.6	1045.7	−0.3608	2.8737	2.5129
−40	0.00187	−177.00	1220.5	1043.5	−0.3659	2.9084	2.5425

TABLE F.8
Thermodynamic Properties of Ammonia

TABLE F.8.1
Saturated Ammonia

Temp. F T	Press. psia P	SPECIFIC VOLUME, ft³/lbm			INTERNAL ENERGY, Btu/lbm		
		Sat. Liquid v_f	Evap. v_{fg}	Sat. Vapor v_g	Sat. Liquid u_f	Evap. u_{fg}	Sat. Vapor u_g
−60	5.547	0.02277	44.7397	44.7625	−20.92	564.27	543.36
−50	7.663	0.02299	33.0702	33.0932	−10.51	556.84	546.33
−40	10.404	0.02322	24.8464	24.8696	−0.04	549.25	549.20
−30	13.898	0.02345	18.9490	18.9724	10.48	541.50	551.98
−28.0	14.696	0.02350	17.9833	18.0068	12.59	539.93	552.52
−20	18.289	0.02369	14.6510	14.6747	21.07	533.57	554.64
−10	23.737	0.02394	11.4714	11.4953	31.73	252.47	557.20
0	30.415	0.02420	9.0861	9.1103	42.46	517.18	559.64
10	38.508	0.02446	7.2734	7.2979	53.26	508.71	561.96
20	48.218	0.02474	5.8792	5.9039	64.12	500.04	564.16
30	59.756	0.02502	4.7945	4.8195	75.06	491.17	566.23
40	73.346	0.02532	3.9418	3.9671	86.07	482.09	568.15
50	89.226	0.02564	3.2647	3.2903	97.16	472.78	569.94
60	107.641	0.02597	2.7221	2.7481	108.33	463.24	571.56
70	128.849	0.02631	2.2835	2.3098	119.58	453.44	573.02
80	153.116	0.02668	1.9260	1.9526	130.92	443.37	574.30
90	180.721	0.02706	1.6323	1.6594	142.36	433.01	573.37
100	211.949	0.02747	1.3894	1.4168	153.89	422.34	576.23
110	247.098	0.02790	1.1870	1.2149	165.53	411.32	576.85
120	286.473	0.02836	1.0172	1.0456	177.28	399.92	577.20
130	330.392	0.02885	0.8740	0.9028	189.17	388.10	577.27
140	379.181	0.02938	0.7524	0.7818	201.20	375.82	577.02
150	433.181	0.02995	0.6485	0.6785	213.40	363.01	576.41
160	492.742	0.03057	0.5593	0.5899	225.80	349.61	575.41
170	558.231	0.03124	0.4822	0.5135	238.42	335.53	573.95
180	630.029	0.03199	0.4153	0.4472	251.33	320.66	571.99
190	708.538	0.03281	0.3567	0.3895	264.58	304.87	569.45
200	794.183	0.03375	0.3051	0.3388	278.24	287.96	566.20
210	887.424	0.03482	0.2592	0.2941	292.43	269.70	562.13
220	988.761	0.03608	0.2181	0.2542	307.28	249.72	557.00
230	1098.766	0.03759	0.1807	0.2183	323.03	227.47	550.50
240	1218.113	0.03950	0.1460	0.1855	340.05	202.02	542.06
250	1347.668	0.04206	0.1126	0.1547	359.03	171.57	530.60
260	1488.694	0.04599	0.0781	0.1241	381.74	131.74	513.48
270.1	1643.742	0.06816	0	0.0682	446.09	0	446.09

TABLE F.8.1 (*continued*)
Saturated Ammonia

Temp. F T	Press. psia P	ENTHALPY, Btu/lbm			ENTROPY, Btu/lbm R		
		Sat. Liquid h_f	Evap. h_{fg}	Sat. Vapor h_g	Sat. Liquid s_f	Evap. s_{fg}	Sat. Vapor s_g
−60	5.547	−20.89	610.19	589.30	−0.0510	1.5267	1.4758
−50	7.663	−10.48	603.73	593.26	−0.0252	1.4737	1.4485
−40	10.404	0	597.08	597.08	0	1.4227	1.4227
−30	13.898	10.54	590.23	600.77	0.0248	1.3737	1.3985
−28.0	14.696	12.65	588.84	601.49	0.0297	1.3641	1.3938
−20	18.289	21.15	583.15	604.31	0.0492	1.3263	1.3755
−10	23.737	31.84	575.85	607.69	0.0731	1.2806	1.3538
0	30.415	42.60	568.32	610.92	0.0967	1.2364	1.3331
10	38.508	53.43	560.54	613.97	0.1200	1.1935	1.3134
20	48.218	64.34	552.50	616.84	0.1429	1.1518	1.2947
30	59.756	75.33	544.18	619.52	0.1654	1.1113	1.2768
40	73.346	86.41	535.59	622.00	0.1877	1.0719	1.2596
50	89.226	97.58	526.68	624.26	0.2097	1.0334	1.2431
60	107.641	108.84	517.46	626.30	0.2314	0.9957	1.2271
70	128.849	120.21	507.89	628.09	0.2529	0.9589	1.2117
80	153.116	131.68	497.94	629.62	0.2741	0.9227	1.1968
90	180.721	143.26	487.60	630.86	0.2951	0.8871	1.1822
100	211.949	154.97	476.83	631.80	0.3159	0.8520	1.1679
110	247.098	166.80	465.59	632.40	0.3366	0.8173	1.1539
120	286.473	178.79	453.84	632.63	0.3571	0.7829	1.1400
130	330.392	190.93	441.54	632.47	0.3774	0.7488	1.1262
140	379.181	203.26	428.61	631.87	0.3977	0.7147	1.1125
150	433.181	215.80	415.00	630.80	0.4180	0.6807	1.0987
160	492.742	228.58	400.61	629.19	0.4382	0.6465	1.0847
170	558.231	241.65	385.35	627.00	0.4586	0.6120	1.0705
180	630.029	255.06	369.08	624.14	0.4790	0.5770	1.0560
190	708.538	268.88	351.63	620.51	0.4997	0.5412	1.0410
200	794.183	283.20	332.80	616.00	0.5208	0.5045	1.0253
210	887.424	298.14	312.27	610.42	0.5424	0.4663	1.0087
220	988.761	313.88	289.63	603.51	0.5647	0.4261	0.9909
230	1098.766	330.67	264.21	594.89	0.5882	0.3831	0.9713
240	1218.113	348.95	234.93	583.87	0.6132	0.3358	0.9490
250	1347.668	369.52	199.65	569.17	0.6410	0.2813	0.9224
260	1488.694	394.41	153.25	547.66	0.6743	0.2129	0.8872
270.1	1643.742	466.83	0	466.83	0.7718	0	0.7718

TABLE F.8.2
Superheated Ammonia

Temp. F	v ft³/lbm	h Btu/lbm	s Btu/lbm R	v ft³/lbm	h Btu/lbm	s Btu/lbm R	v ft³/lbm	h Btu/lbm	s Btu/lbm R
	5 psia (−63.09)			10 psia (−41.33)			15 psia (−27.27)		
Sat.	49.32002	588.05	1.4846	25.80648	596.58	1.4261	17.66533	601.75	1.3921
−40	52.3487	599.56	1.5128	25.8962	597.27	1.4277	—	—	—
−20	54.9506	609.53	1.5360	27.2401	607.60	1.4518	17.9999	605.63	1.4010
0	57.5366	619.51	1.5582	28.5674	617.88	1.4746	18.9086	616.22	1.4245
20	60.1099	629.50	1.5795	29.8814	628.12	1.4964	19.8036	626.72	1.4469
40	62.6732	639.52	1.5999	31.1852	638.34	1.5173	20.6880	637.15	1.4682
60	65.2288	649.57	1.6197	32.4809	648.56	1.5374	21.5641	647.54	1.4886
80	67.7782	659.67	1.6387	33.7703	658.80	1.5567	22.4338	657.91	1.5082
100	70.3228	669.84	1.6572	35.0549	669.07	1.5754	23.2985	668.29	1.5271
120	72.8637	680.06	1.6752	36.3356	679.38	1.5935	24.1593	678.70	1.5453
140	75.4015	690.36	1.6926	37.6133	689.75	1.6111	25.0170	689.14	1.5630
160	77.9370	700.74	1.7097	38.8886	700.19	1.6282	25.8723	699.64	1.5803
180	80.4706	711.20	1.7263	40.1620	710.70	1.6449	26.7256	710.21	1.5970
200	83.0026	721.75	1.7425	41.4338	721.30	1.6612	27.5774	720.84	1.6134
220	85.5334	732.39	1.7584	42.7043	731.98	1.6771	28.4278	731.56	1.6294
240	88.0631	743.13	1.7740	43.9737	742.74	1.6928	29.2772	742.36	1.6451
260	90.5918	753.96	1.7892	45.2422	753.61	1.7081	30.1256	753.24	1.6604
280	93.1199	764.90	1.8042	46.5100	764.56	1.7231	30.9733	764.23	1.6755
	20 psia (−16.63)			25 psia (−7.95)			30 psia (−0.57)		
Sat.	13.49628	605.47	1.3680	10.95013	608.37	1.3494	9.22850	610.74	1.3342
0	14.0774	614.54	1.3881	11.1771	612.82	1.3592	9.2423	611.06	1.3349
20	14.7635	625.30	1.4111	11.7383	623.86	1.3827	9.7206	622.39	1.3591
40	15.4385	635.94	1.4328	12.2881	634.72	1.4049	10.1872	633.49	1.3817
60	16.1051	646.51	1.4535	12.8291	645.46	1.4260	10.6447	644.41	1.4032
80	16.7651	657.02	1.4734	13.3634	656.12	1.4461	11.0954	655.21	1.4236
100	17.4200	667.51	1.4925	13.8926	666.73	1.4654	11.5407	665.93	1.4431
120	18.0709	678.01	1.5109	14.4176	677.32	1.4840	11.9820	676.62	1.4618
140	18.7187	688.53	1.5287	14.9395	687.91	1.5020	12.4200	687.29	1.4799
160	19.3640	699.09	1.5461	15.4589	698.54	1.5194	12.8554	697.98	1.4975
180	20.0073	709.71	1.5629	15.9763	709.20	1.5363	13.2888	708.70	1.5145
200	20.6491	720.39	1.5794	16.4920	719.93	1.5528	13.7206	719.47	1.5311
220	21.2895	731.14	1.5954	17.0065	730.72	1.5689	14.1511	730.29	1.5472
240	21.9288	741.97	1.6111	17.5198	741.58	1.5847	14.5804	741.19	1.5630
260	22.5673	752.88	1.6265	18.0322	752.52	1.6001	15.0088	752.16	1.5785
280	23.2049	763.89	1.6416	18.5439	763.55	1.6152	15.4365	763.21	1.5936
300	23.8419	774.99	1.6564	19.0548	774.67	1.6301	15.8634	774.36	1.6085
320	24.4783	786.18	1.6709	19.5652	785.89	1.6446	16.2898	785.59	1.6231

TABLE F.8.2 (*continued*)
Superheated Ammonia

Temp. F	v ft³/lbm	h Btu/lbm	s Btu/lbm R	v ft³/lbm	h Btu/lbm	s Btu/lbm R	v ft³/lbm	h Btu/lbm	s Btu/lbm R
	35 psia (5.89)			40 psia (11.66)			50 psia (21.66)		
Sat.	7.98414	612.73	1.3214	7.04135	614.45	1.3103	5.70491	617.30	1.2917
20	8.2786	620.90	1.3387	7.1964	619.39	1.3206	—	—	—
40	8.6860	632.23	1.3618	7.5596	630.96	1.3443	5.9814	628.37	1.3142
60	9.0841	643.34	1.3836	7.9132	642.26	1.3665	6.2731	640.07	1.3372
80	9.4751	654.29	1.4043	8.2596	653.37	1.3874	6.5573	651.49	1.3588
100	9.8606	665.14	1.4240	8.6004	664.33	1.4074	6.8356	662.70	1.3792
120	10.2420	675.92	1.4430	8.9370	675.21	1.4265	7.1096	673.79	1.3986
140	10.6202	686.67	1.4612	9.2702	686.04	1.4449	7.3800	684.78	1.4173
160	10.9957	697.42	1.4788	9.6008	696.86	1.4626	7.6478	695.73	1.4352
180	11.3692	708.19	1.4959	9.9294	707.69	1.4798	7.9135	706.67	1.4526
200	11.7410	719.01	1.5126	10.2562	718.54	1.4965	8.1775	717.61	1.4695
220	12.1115	729.87	1.5288	10.5817	729.44	1.5128	8.4400	728.59	1.4859
240	12.4808	740.80	1.5447	10.9061	741.40	1.5287	8.7014	739.62	1.5018
260	12.8493	751.80	1.5602	11.2296	751.43	1.5442	8.9619	750.70	1.5175
280	13.2169	762.88	1.5753	11.5522	762.54	1.5594	9.2216	761.86	1.5327
300	13.5838	774.04	1.5902	11.8741	773.72	1.5744	9.4805	773.09	1.5477
320	13.9502	785.29	1.6049	12.1955	785.00	1.5890	9.7389	784.40	1.5624
340	14.3160	796.64	1.6192	12.5163	796.36	1.6034	9.9967	795.80	1.5769
	60 psia (30.19)			70 psia (37.68)			80 psia (44.38)		
Sat.	4.80091	619.57	1.2764	4.14732	621.44	1.2635	3.65200	623.02	1.2523
40	4.9277	625.69	1.2888	4.1738	622.94	1.2665	—	—	—
60	5.1787	637.82	1.3126	4.3961	635.52	1.2912	3.8083	633.16	1.2721
80	5.4217	649.57	1.3348	4.6099	647.62	1.3140	4.0005	645.63	1.2956
100	5.6586	661.05	1.3557	4.8174	659.37	1.3354	4.1861	657.66	1.3175
120	5.8909	672.34	1.3755	5.0201	670.88	1.3556	4.3667	669.39	1.3381
140	6.1197	683.50	1.3944	5.2191	682.21	1.3749	4.5435	680.90	1.3577
160	6.3456	694.59	1.4126	5.4153	693.44	1.3933	4.7174	692.27	1.3763
180	6.5694	705.64	1.4302	5.6093	704.60	1.4110	4.8890	703.55	1.3942
200	6.7915	716.68	1.4472	5.8014	715.73	1.4281	5.0588	714.79	1.4115
220	7.0121	727.73	1.4637	5.9921	726.87	1.4448	5.2270	726.00	1.4283
240	7.2316	738.83	1.4798	6.1816	738.03	1.4610	5.3941	737.23	1.4446
260	7.4501	749.97	1.4955	6.3702	749.23	1.4767	5.5602	748.50	1.4604
280	7.6678	761.17	1.5108	6.5579	760.49	1.4922	5.7254	759.80	1.4759
300	7.8848	772.45	1.5259	6.7449	771.81	1.5073	5.8900	771.17	1.4911
320	8.1011	783.80	1.5406	6.9313	783.21	1.5221	6.0538	782.61	1.5059
340	8.3169	795.24	1.5551	7.1171	794.68	1.5366	6.2172	794.12	1.5205
360	8.5323	806.77	1.5693	7.3025	806.24	1.5509	6.3801	805.71	1.5348

TABLE F.8.2 (*continued*)
Superheated Ammonia

Temp. F	v ft³/lbm	h Btu/lbm	s Btu/lbm R	v ft³/lbm	h Btu/lbm	s Btu/lbm R	v ft³/lbm	h Btu/lbm	s Btu/lbm R
	90 psia (50.45)			100 psia (56.02)			125 psia (68.28)		
Sat.	3.26324	624.36	1.2423	2.94969	625.52	1.2334	2.37866	627.80	1.2143
60	3.35063	630.74	1.2547	2.9831	628.25	1.2387	—	—	—
80	3.5260	643.59	1.2790	3.1459	641.51	1.2637	2.4597	636.11	1.2299
100	3.6947	655.92	1.3014	3.3013	654.16	1.2867	2.5917	649.59	1.2544
120	3.8583	667.88	1.3224	3.4513	666.36	1.3082	2.7177	662.44	1.2770
140	4.0179	679.58	1.3423	3.5972	678.24	1.3283	2.8392	674.83	1.2980
160	4.1745	691.10	1.3612	3.7400	689.91	1.3475	2.9574	686.90	1.3178
180	4.3287	702.50	1.3793	3.8804	701.44	1.3658	3.0730	698.74	1.3366
200	4.4811	713.83	1.3967	4.0188	712.87	1.3834	3.1865	710.44	1.3546
220	4.6319	725.13	1.4136	4.1558	724.25	1.4004	3.2985	722.04	1.3720
240	4.7816	736.43	1.4300	4.2915	735.63	1.4169	3.4091	733.59	1.3887
260	4.9302	747.75	1.4459	4.4261	747.01	1.4329	3.5187	745.13	1.4050
280	5.0779	759.11	1.4615	4.5599	758.42	1.4485	3.6274	756.68	1.4208
300	5.2250	770.53	1.4767	4.6930	769.88	1.4638	3.7353	768.27	1.4362
320	5.3714	782.01	1.4916	4.8254	781.40	1.4788	3.8426	779.89	1.4514
340	5.5173	793.56	1.5063	4.9573	792.99	1.4935	3.9493	791.58	1.4662
360	5.6626	805.18	1.5206	5.0887	804.66	1.5079	4.0555	803.33	1.4807
380	5.8076	816.90	1.5348	5.2196	816.40	1.5220	4.1613	815.15	1.4949
	150 psia (78.79)			175 psia (88.03)			200 psia (96.31)		
Sat.	1.99226	629.45	1.1986	1.71282	630.64	1.1850	1.50102	631.49	1.1731
80	1.9997	630.36	1.2003	—	—	—	—	—	—
100	2.1170	644.81	1.2265	1.7762	639.77	1.2015	1.5190	634.45	1.1785
120	2.2275	658.37	1.2504	1.8762	654.13	1.2267	1.6117	649.71	1.2052
140	2.3331	671.31	1.2723	1.9708	667.67	1.2497	1.6984	663.90	1.2293
160	2.4351	683.80	1.2928	2.0614	680.62	1.2710	1.7807	677.36	1.2514
180	2.5343	695.99	1.3122	2.1491	693.17	1.2909	1.8598	690.30	1.2719
200	2.6313	707.96	1.3306	2.2345	705.44	1.3098	1.9365	702.87	1.2913
220	2.7267	719.79	1.3483	2.3181	717.51	1.3278	2.0114	715.20	1.3097
240	2.8207	731.54	1.3653	2.4002	729.46	1.3451	2.0847	727.35	1.3273
260	2.9136	743.24	1.3818	2.4813	741.33	1.3619	2.1569	739.39	1.3443
280	3.0056	754.93	1.3978	2.5613	753.16	1.3781	2.2280	751.38	1.3607
300	3.0968	766.63	1.4134	2.6406	764.99	1.3939	2.2984	763.33	1.3767
320	3.1873	778.37	1.4287	2.7192	776.84	1.4092	2.3680	775.30	1.3922
340	3.2772	790.15	1.4436	2.7972	788.72	1.4243	2.4370	787.28	1.4074
360	3.3667	801.99	1.4582	2.8746	800.65	1.4390	2.5056	799.30	1.4223
380	3.4557	813.90	1.4726	2.9516	812.64	1.4535	2.5736	811.38	1.4368
400	3.5442	825.88	1.4867	3.0282	824.70	1.4677	2.6412	823.51	1.4511

TABLE F.8.2 *(continued)*
Superheated Ammonia

Temp. F	v ft³/lbm	h Btu/lbm	s Btu/lbm R	v ft³/lbm	h Btu/lbm	s Btu/lbm R	v ft³/lbm	h Btu/lbm	s Btu/lbm R
	250 psia (110.78)			300 psia (123.20)			350 psia (134.14)		
Sat.	1.20063	632.43	1.1528	0.99733	632.63	1.1356	0.85027	632.28	1.1205
120	1.2384	640.21	1.1663	—	—	—	—	—	—
140	1.3150	655.95	1.1930	1.0568	647.32	1.1605	0.8696	637.87	1.1299
160	1.3863	670.53	1.2170	1.1217	663.27	1.1866	0.9309	655.48	1.1588
180	1.4539	684.34	1.2389	1.1821	678.07	1.2101	0.9868	671.46	1.1842
200	1.5188	697.59	1.2593	1.2394	692.08	1.2317	1.0391	686.34	1.2071
220	1.5815	710.45	1.2785	1.2943	705.55	1.2518	1.0886	700.47	1.2282
240	1.6426	723.05	1.2968	1.3474	718.63	1.2708	1.1362	714.08	1.2479
260	1.7024	735.46	1.3142	1.3991	731.44	1.2888	1.1822	727.32	1.2666
280	1.7612	747.76	1.3311	1.4497	744.07	1.3062	1.2270	740.31	1.2844
300	1.8191	759.98	1.3474	1.4994	756.58	1.3228	1.2708	753.12	1.3015
320	1.8762	772.18	1.3633	1.5482	769.02	1.3390	1.3138	765.82	1.3180
340	1.9328	784.37	1.3787	1.5965	781.43	1.3547	1.3561	778.46	1.3340
360	1.9887	796.59	1.3938	1.6441	793.84	1.3701	1.3979	791.07	1.3496
380	2.0442	808.83	1.4085	1.6913	806.27	1.3850	1.4391	803.67	1.3648
400	2.0993	821.13	1.4230	1.7380	818.72	1.3997	1.4798	816.30	1.3796
420	2.1540	833.48	1.4372	1.7843	831.23	1.4141	1.5202	828.95	1.3942
440	2.2083	845.90	1.4512	1.8302	843.78	1.4282	1.5602	841.65	1.4085
	400 psia (143.97)			600 psia (175.93)			800 psia (200.65)		
Sat.	0.73876	631.50	1.1070	0.47311	625.39	1.0620	0.33575	615.67	1.0242
160	0.7860	647.06	1.1324	—	—	—	—	—	—
180	0.8392	664.44	1.1601	0.4834	630.48	1.0700	—	—	—
200	0.8880	680.32	1.1845	0.5287	652.67	1.1041	—	—	—
220	0.9338	695.21	1.2067	0.5680	671.78	1.1327	0.3769	642.62	1.0645
240	0.9773	709.40	1.2273	0.6035	689.03	1.1577	0.4115	665.08	1.0971
260	1.0192	723.10	1.2466	0.6366	705.06	1.1803	0.4419	684.62	1.1246
280	1.0597	736.47	1.2650	0.6678	720.26	1.2011	0.4694	702.36	1.1489
300	1.0992	749.60	1.2825	0.6976	734.88	1.2206	0.4951	718.93	1.1710
320	1.1379	762.58	1.2993	0.7264	749.09	1.2391	0.5193	734.69	1.1915
340	1.1758	775.45	1.3156	0.7542	763.02	1.2567	0.5425	749.89	1.2108
360	1.2131	788.27	1.3315	0.7814	776.75	1.2737	0.5648	764.68	1.2290
380	1.2499	801.06	1.3469	0.8079	790.34	1.2901	0.5864	779.19	1.2465
400	1.2862	813.85	1.3619	0.8340	803.86	1.3060	0.6074	793.50	1.2634
420	1.3221	826.66	1.3767	0.8595	817.32	1.3215	0.6279	807.68	1.2797
440	1.3576	839.51	1.3911	0.8847	830.76	1.3366	0.6480	821.76	1.2955
460	1.3928	852.39	1.4053	0.9095	844.21	1.3514	0.6677	835.80	1.3109
480	1.4277	865.34	1.4192	0.9340	857.67	1.3658	0.6871	849.80	1.3260

TABLE F.9
Thermodynamic Properties of R-22

TABLE F.9.1
Saturated R-22

Temp. F T	Press. psia P	SPECIFIC VOLUME, ft³/lbm			INTERNAL ENERGY, Btu/lbm		
		Sat. Liquid v_f	Evap. v_{fg}	Sat. Vapor v_g	Sat. Liquid u_f	Evap. u_{fg}	Sat. Vapor u_g
−100	2.398	0.01066	18.4219	18.4326	−14.57	99.76	85.19
−90	3.423	0.01077	13.2243	13.2351	−12.22	98.38	86.16
−80	4.782	0.01088	9.6840	9.6949	−9.85	96.98	87.13
−70	6.552	0.01099	7.2208	7.2318	−7.44	95.54	88.10
−60	8.818	0.01111	5.4733	5.4844	−5.01	94.07	89.06
−50	11.674	0.01124	4.2111	4.2224	−2.54	92.56	90.02
−41.4	14.696	0.01135	3.3944	3.4058	−0.37	91.22	90.85
−40	15.222	0.01136	3.2844	3.2957	−0.03	91.01	90.97
−30	19.573	0.01150	2.5934	2.6049	2.51	89.41	91.91
−20	24.845	0.01163	2.0709	2.0826	5.08	87.76	92.84
−10	31.162	0.01178	1.6707	1.6825	7.68	86.07	93.75
0	38.657	0.01193	1.3603	1.3723	10.32	84.33	94.65
10	47.464	0.01209	1.1170	1.1290	13.00	82.53	95.53
20	57.727	0.01226	0.9241	0.9363	15.71	80.67	96.38
30	69.591	0.01243	0.7697	0.7821	18.45	78.76	97.21
40	83.206	0.01262	0.6449	0.6575	21.23	76.79	98.02
50	98.727	0.01282	0.5432	0.5561	24.04	74.75	98.79
60	116.312	0.01303	0.4597	0.4727	26.89	72.65	99.54
70	136.123	0.01325	0.3905	0.4037	29.78	70.46	100.24
80	158.326	0.01349	0.3327	0.3462	32.71	68.19	100.91
90	183.094	0.01375	0.2841	0.2979	35.69	65.83	101.53
100	210.604	0.01404	0.2430	0.2570	38.72	63.37	102.09
110	241.042	0.01435	0.2079	0.2222	41.81	60.78	102.59
120	274.604	0.01469	0.1777	0.1924	44.96	58.05	103.01
130	311.496	0.01508	0.1515	0.1666	48.19	55.14	103.33
140	351.944	0.01552	0.1287	0.1442	51.52	52.02	103.54
150	396.194	0.01603	0.1085	0.1245	54.97	48.63	103.60
160	444.525	0.01663	0.0904	0.1070	58.58	44.88	103.46
170	497.259	0.01737	0.0739	0.0913	62.42	40.62	103.04
180	554.783	0.01833	0.0585	0.0768	66.62	35.57	102.18
190	617.590	0.01973	0.0431	0.0628	71.46	29.10	100.55
200	686.356	0.02244	0.0250	0.0474	78.01	18.81	96.83
204.8	720.698	0.03053	0	0.0305	87.30	0	87.30

TABLE F.9.1 (*continued*)
Saturated R-22

Temp. F T	Press. psia P	ENTHALPY, Btu/lbm			ENTROPY, Btu/lbm R		
		Sat. Liquid h_f	Evap. h_{fg}	Sat. Vapor h_g	Sat. Liquid s_f	Evap. s_{fg}	Sat. Vapor s_g
−100	2.398	−14.56	107.94	93.37	−0.0373	0.3001	0.2627
−90	3.423	−12.22	106.76	94.54	−0.0309	0.2888	0.2579
−80	4.782	−9.84	105.55	95.71	−0.0246	0.2780	0.2534
−70	6.552	−7.43	104.30	96.87	−0.0183	0.2676	0.2493
−60	8.818	−4.99	103.00	98.01	−0.0121	0.2577	0.2456
−50	11.674	−2.51	101.66	99.14	−0.0060	0.2481	0.2421
−41.4	14.696	−0.34	100.45	100.11	−0.0008	0.2401	0.2393
−40	15.222	0	100.26	100.26	0	0.2389	0.2389
−30	19.573	2.55	98.80	101.35	0.0060	0.2299	0.2359
−20	24.845	5.13	97.28	102.42	0.0119	0.2213	0.2332
−10	31.162	7.75	95.70	103.46	0.0178	0.2128	0.2306
0	38.657	10.41	94.06	104.47	0.0236	0.2046	0.2282
10	47.464	13.10	92.34	105.44	0.0293	0.1966	0.2259
20	57.727	15.84	90.55	106.38	0.0350	0.1888	0.2238
30	69.591	18.61	88.67	107.28	0.0407	0.1811	0.2218
40	83.206	21.42	86.72	108.14	0.0463	0.1735	0.2199
50	98.727	24.27	84.68	108.95	0.0519	0.1661	0.2180
60	116.312	27.17	82.54	109.71	0.0574	0.1588	0.2163
70	136.123	30.12	80.30	110.41	0.0630	0.1516	0.2146
80	158.326	33.11	77.94	111.05	0.0685	0.1444	0.2129
90	183.094	36.16	75.46	111.62	0.0739	0.1373	0.2112
100	210.604	39.27	72.84	112.11	0.0794	0.1301	0.2096
110	241.042	42.45	70.05	112.50	0.0849	0.1230	0.2079
120	274.604	45.71	67.08	112.78	0.0904	0.1157	0.2061
130	311.496	49.06	63.88	112.94	0.0960	0.1083	0.2043
140	351.944	52.53	60.40	112.93	0.1016	0.1007	0.2023
150	396.194	56.14	56.58	112.73	0.1074	0.0928	0.2002
160	444.525	59.95	52.32	112.26	0.1133	0.0844	0.1978
170	497.259	64.02	47.42	111.44	0.1196	0.0753	0.1949
180	554.783	68.50	41.57	110.07	0.1263	0.0650	0.1913
190	617.590	73.71	34.02	107.73	0.1341	0.0524	0.1865
200	686.356	80.86	21.99	102.85	0.1446	0.0333	0.1779
204.8	720.698	91.38	0	91.38	0.1602	0	0.1602

TABLE F.9.2
Superheated R-22

Temp. F	v ft³/lbm	h Btu/lbm	s Btu/lbm R	v ft³/lbm	h Btu/lbm	s Btu/lbm R	v ft³/lbm	h Btu/lbm	s Btu/lbm R
	5 psia (−78.62)			10 psia (−55.59)			15 psia (−40.57)		
Sat.	9.30117	95.87	0.2528	4.87779	98.52	0.2440	3.34121	100.19	0.2391
−40	10.2935	101.09	0.2659	5.0838	100.69	0.2493	3.3463	100.28	0.2393
−20	10.8034	103.89	0.2724	5.3460	103.53	0.2559	3.5261	103.16	0.2460
0	11.3114	106.73	0.2787	5.6060	106.41	0.2623	3.7037	106.09	0.2525
20	11.8177	109.64	0.2849	5.8643	109.36	0.2686	3.8794	109.06	0.2588
40	12.3227	112.61	0.2910	6.1212	112.35	0.2747	4.0537	112.09	0.2650
60	12.8265	115.64	0.2969	6.3769	115.40	0.2807	4.2268	115.17	0.2710
80	13.3293	118.72	0.3027	6.6316	118.51	0.2865	4.3989	118.30	0.2769
100	13.8313	121.87	0.3085	6.8855	121.67	0.2923	4.5701	121.48	0.2827
120	14.3327	125.07	0.3141	7.1387	124.89	0.2979	4.7406	124.72	0.2884
140	14.8335	128.33	0.3196	7.3913	128.17	0.3035	4.9105	128.00	0.2940
160	15.3337	131.64	0.3250	7.6434	131.49	0.3089	5.0799	131.34	0.2995
180	15.8336	135.01	0.3304	7.8951	134.87	0.3143	5.2489	134.74	0.3049
200	16.3331	138.44	0.3357	8.1464	138.31	0.3196	5.4174	138.18	0.3102
220	16.8323	141.92	0.3409	8.3974	141.80	0.3248	5.5857	141.68	0.3154
240	17.3312	145.45	0.3460	8.6481	145.34	0.3300	5.7537	145.23	0.3205
260	17.8298	149.03	0.3510	8.8986	148.93	0.3350	5.9215	148.83	0.3256
280	18.3283	152.67	0.3560	9.1489	152.57	0.3400	6.0890	152.48	0.3306
	20 psia (−29.12)			25 psia (−19.73)			30 psia (−11.71)		
Sat.	2.55270	101.44	0.2357	2.07040	102.44	0.2331	1.74388	103.28	0.2310
0	2.7521	105.76	0.2454	2.1808	105.42	0.2397	1.7997	105.08	0.2350
20	2.8867	108.77	0.2518	2.2908	108.47	0.2462	1.8933	108.16	0.2415
40	3.0198	111.83	0.2580	2.3992	111.56	0.2525	1.9853	111.29	0.2479
60	3.1516	114.93	0.2641	2.5063	114.69	0.2586	2.0760	114.45	0.2541
80	3.2823	118.08	0.2701	2.6123	117.86	0.2646	2.1655	117.64	0.2602
100	3.4122	121.28	0.2759	2.7175	121.09	0.2705	2.2542	120.89	0.2661
120	3.5414	124.54	0.2816	2.8219	124.36	0.2762	2.3421	124.18	0.2718
140	3.6700	127.84	0.2872	2.9257	127.68	0.2819	2.4294	127.51	0.2775
160	3.7981	131.19	0.2927	3.0289	131.04	0.2874	2.5162	130.89	0.2830
180	3.9257	134.60	0.2981	3.1318	134.46	0.2928	2.6025	134.32	0.2885
200	4.0529	138.05	0.3034	3.2342	137.93	0.2982	2.6884	137.80	0.2938
220	4.1799	141.56	0.3087	3.3363	141.44	0.3034	2.7739	141.32	0.2991
240	4.3065	145.12	0.3138	3.4382	145.01	0.3086	2.8592	144.90	0.3043
260	4.4329	148.73	0.3189	3.5397	148.62	0.3137	2.9443	148.52	0.3094
280	4.5591	152.38	0.3239	3.6411	152.28	0.3187	3.0291	152.19	0.3144
300	4.6851	156.09	0.3288	3.7423	155.99	0.3236	3.1138	155.90	0.3194
320	4.8109	159.84	0.3337	3.8434	159.75	0.3285	3.1983	159.67	0.3243

TABLE F.9.2 (*continued*)
Superheated R-22

Temp. F	v ft³/lbm	h Btu/lbm	s Btu/lbm R	v ft³/lbm	h Btu/lbm	s Btu/lbm R	v ft³/lbm	h Btu/lbm	s Btu/lbm R
		40 psia (1.63)			50 psia (12.61)			60 psia (22.03)	
Sat.	1.32853	104.63	0.2278	1.07436	105.69	0.2253	0.90223	106.57	0.2234
20	1.3959	107.54	0.2340	1.0968	106.90	0.2279	—	—	—
40	1.4676	110.73	0.2405	1.1564	110.16	0.2346	0.9486	109.58	0.2295
60	1.5378	113.95	0.2468	1.2145	113.44	0.2410	0.9987	112.92	0.2361
80	1.6068	117.20	0.2530	1.2714	116.74	0.2472	1.0475	116.28	0.2424
100	1.6749	120.48	0.2589	1.3272	120.08	0.2533	1.0952	119.66	0.2486
120	1.7423	123.81	0.2648	1.3822	123.44	0.2592	1.1420	123.06	0.2545
140	1.8090	127.18	0.2705	1.4366	126.84	0.2649	1.1882	126.50	0.2604
160	1.8751	130.59	0.2761	1.4903	130.28	0.2706	1.2338	129.96	0.2660
180	1.9407	134.04	0.2816	1.5436	133.75	0.2761	1.2788	133.47	0.2716
200	2.0060	137.54	0.2869	1.5965	137.28	0.2815	1.3235	137.01	0.2771
220	2.0709	141.08	0.2922	1.6491	140.84	0.2869	1.3678	140.60	0.2824
240	2.1356	144.67	0.2974	1.7013	144.45	0.2921	1.4118	144.22	0.2877
260	2.2000	148.31	0.3026	1.7533	148.10	0.2972	1.4556	147.89	0.2928
280	2.2641	151.99	0.3076	1.8051	151.80	0.3023	1.4991	151.60	0.2979
300	2.3281	155.72	0.3126	1.8567	155.54	0.3073	1.5424	155.35	0.3029
320	2.3919	159.49	0.3175	1.9081	159.32	0.3122	1.5856	159.15	0.3079
340	2.4556	163.31	0.3223	1.9594	163.15	0.3171	1.6286	162.99	0.3127
		70 psia (30.32)			80 psia (37.76)			100 psia (50.77)	
Sat.	0.77766	107.31	0.2217	0.68319	107.95	0.2203	0.54908	109.01	0.2179
40	0.7998	108.97	0.2251	0.6878	108.35	0.2211	—	—	—
60	0.8443	112.39	0.2318	0.7282	111.84	0.2279	0.5650	110.70	0.2212
80	0.8874	115.81	0.2382	0.7671	115.32	0.2345	0.5982	114.32	0.2280
100	0.9293	119.23	0.2445	0.8048	118.80	0.2408	0.6300	117.91	0.2345
120	0.9704	122.68	0.2505	0.8415	122.29	0.2470	0.6608	121.49	0.2408
140	1.0107	126.15	0.2564	0.8775	125.80	0.2529	0.6908	125.08	0.2469
160	1.0504	129.65	0.2621	0.9129	129.33	0.2587	0.7201	128.67	0.2528
180	1.0896	133.18	0.2677	0.9477	132.89	0.2643	0.7488	132.29	0.2586
200	1.1284	136.75	0.2732	0.9821	136.48	0.2699	0.7771	135.93	0.2642
220	1.1669	140.35	0.2786	1.0161	140.10	0.2753	0.8050	139.60	0.2696
240	1.2050	143.99	0.2839	1.0498	143.76	0.2806	0.8326	143.30	0.2750
260	1.2428	147.68	0.2891	1.0833	147.46	0.2858	0.8599	147.03	0.2803
280	1.2805	151.40	0.2942	1.1165	151.20	0.2909	0.8869	150.80	0.2854
300	1.3179	155.17	0.2992	1.1495	154.98	0.2960	0.9137	154.61	0.2905
320	1.3552	158.97	0.3042	1.1823	158.80	0.3009	0.9404	158.45	0.2955
340	1.3923	162.82	0.3090	1.2150	162.66	0.3058	0.9669	162.33	0.3004
360	1.4292	166.72	0.3138	1.2476	166.56	0.3106	0.9932	166.25	0.3052

TABLE F.9.2 (*continued*)
Superheated R-22

Temp. F	v ft³/lbm	h Btu/lbm	s Btu/lbm R	v ft³/lbm	h Btu/lbm	s Btu/lbm R	v ft³/lbm	h Btu/lbm	s Btu/lbm R
	125 psia (64.53)			150 psia (76.38)			175 psia (86.85)		
Sat.	0.43988	110.04	0.2155	0.36587	110.83	0.2135	0.31224	111.45	0.2117
80	0.4622	112.99	0.2210	0.3705	111.56	0.2148	—	—	—
100	0.4896	116.74	0.2279	0.3953	115.50	0.2220	0.3273	114.18	0.2167
120	0.5158	120.46	0.2344	0.4187	119.37	0.2288	0.3488	118.22	0.2238
140	0.5411	124.15	0.2406	0.4410	123.18	0.2353	0.3691	122.17	0.2305
160	0.5656	127.84	0.2467	0.4624	126.97	0.2415	0.3884	126.07	0.2369
180	0.5896	131.53	0.2526	0.4832	130.74	0.2475	0.4070	129.94	0.2430
200	0.6130	135.23	0.2583	0.5034	134.52	0.2533	0.4250	133.79	0.2489
220	0.6360	138.96	0.2638	0.5232	138.31	0.2589	0.4426	137.64	0.2547
240	0.6587	142.71	0.2693	0.5426	142.11	0.2645	0.4597	141.49	0.2603
260	0.6810	146.48	0.2746	0.5618	145.93	0.2698	0.4765	145.36	0.2657
280	0.7032	150.29	0.2798	0.5806	149.78	0.2751	0.4930	149.25	0.2711
300	0.7251	154.13	0.2849	0.5993	153.65	0.2803	0.5094	153.16	0.2763
320	0.7468	158.00	0.2900	0.6177	157.55	0.2854	0.5255	157.10	0.2814
340	0.7683	161.91	0.2949	0.6360	161.49	0.2903	0.5414	161.06	0.2864
360	0.7898	165.85	0.2998	0.6541	165.46	0.2952	0.5572	165.06	0.2913
380	0.8111	169.83	0.3046	0.6721	169.46	0.3001	0.5728	169.08	0.2962
400	0.8322	173.85	0.3093	0.6900	173.50	0.3048	0.5884	173.14	0.3010
	200 psia (96.27)			250 psia (112.76)			300 psia (126.98)		
Sat.	0.27150	111.93	0.2102	0.21352	112.59	0.2074	0.17400	112.90	0.2049
100	0.2755	112.75	0.2116	—	—	—	—	—	—
120	0.2959	117.00	0.2191	0.2204	114.30	0.2104	—	—	—
140	0.3149	121.11	0.2261	0.2379	118.82	0.2180	0.1852	116.20	0.2104
160	0.3327	125.13	0.2327	0.2540	123.14	0.2251	0.2006	120.94	0.2182
180	0.3497	129.10	0.2390	0.2690	127.34	0.2318	0.2146	125.44	0.2253
200	0.3661	133.03	0.2450	0.2833	131.46	0.2381	0.2276	129.78	0.2320
220	0.3820	136.95	0.2509	0.2969	135.53	0.2442	0.2398	134.04	0.2384
240	0.3974	140.87	0.2566	0.3100	139.58	0.2501	0.2515	138.22	0.2445
260	0.4125	144.79	0.2621	0.3228	143.60	0.2558	0.2628	142.38	0.2503
280	0.4273	148.72	0.2675	0.3352	147.63	0.2613	0.2737	146.50	0.2560
300	0.4419	152.67	0.2727	0.3474	151.66	0.2667	0.2843	150.62	0.2615
320	0.4563	156.64	0.2779	0.3593	155.70	0.2719	0.2947	154.74	0.2668
340	0.4705	160.63	0.2830	0.3711	159.76	0.2770	0.3048	158.86	0.2720
360	0.4845	164.65	0.2879	0.3827	163.83	0.2821	0.3148	163.00	0.2771
380	0.4984	168.70	0.2928	0.3942	167.93	0.2870	0.3247	167.15	0.2821
400	0.5122	172.78	0.2976	0.4055	172.05	0.2919	0.3344	171.31	0.2870
420	0.5259	176.89	0.3023	0.4167	176.20	0.2966	0.3440	175.51	0.2919

TABLE F.9.2 (*continued*)
Superheated R-22

Temp. F	v ft³/lbm	h Btu/lbm	s Btu/lbm R	v ft³/lbm	h Btu/lbm	s Btu/lbm R	v ft³/lbm	h Btu/lbm	s Btu/lbm R
	400 psia (150.82)			500 psia (170.50)			600 psia (187.29)		
Sat.	0.12297	112.70	0.2000	0.09053	111.38	0.1947	0.06663	108.51	0.1880
160	0.1305	115.52	0.2046	—	—	—	—	—	—
180	0.1446	121.01	0.2133	0.0987	115.06	0.2005	—	—	—
200	0.1567	126.02	0.2210	0.1122	121.43	0.2103	0.0791	115.12	0.1981
220	0.1677	130.76	0.2281	0.1232	126.95	0.2186	0.0919	122.32	0.2089
240	0.1778	135.31	0.2347	0.1328	132.05	0.2260	0.1020	128.29	0.2175
260	0.1874	139.76	0.2410	0.1416	136.89	0.2328	0.1106	133.70	0.2252
280	0.1965	144.13	0.2470	0.1498	141.57	0.2392	0.1184	138.79	0.2321
300	0.2052	148.45	0.2527	0.1576	146.14	0.2453	0.1256	143.67	0.2386
320	0.2137	152.74	0.2583	0.1650	150.63	0.2512	0.1323	148.41	0.2448
340	0.2219	157.01	0.2637	0.1721	155.08	0.2568	0.1387	153.05	0.2507
360	0.2299	161.27	0.2690	0.1789	159.49	0.2622	0.1449	157.63	0.2563
380	0.2378	165.54	0.2741	0.1856	163.88	0.2675	0.1508	162.16	0.2618
400	0.2455	169.81	0.2791	0.1921	168.26	0.2727	0.1566	166.66	0.2671
420	0.2530	174.09	0.2841	0.1985	172.63	0.2777	0.1622	171.15	0.2722
440	0.2605	178.38	0.2889	0.2048	177.01	0.2826	0.1677	175.62	0.2773
460	0.2679	182.69	0.2936	0.2109	181.40	0.2875	0.1730	180.09	0.2822
480	0.2752	187.02	0.2983	0.2170	185.80	0.2922	0.1783	184.57	0.2870
	700 psia (201.88)			800 psia			900 psia		
Sat.	0.04365	101.02	0.1750	—	—	—	—	—	—
220	0.0671	116.05	0.1975	0.0422	104.39	0.1788	0.0305	95.43	0.1648
240	0.0788	123.79	0.2087	0.0600	118.02	0.1986	0.0434	109.91	0.1857
260	0.0879	130.09	0.2176	0.0702	125.88	0.2097	0.0557	120.85	0.2011
280	0.0956	135.73	0.2253	0.0782	132.34	0.2186	0.0643	128.53	0.2117
300	0.1025	141.01	0.2324	0.0851	138.13	0.2263	0.0713	135.01	0.2203
320	0.1089	146.05	0.2389	0.0913	143.54	0.2333	0.0774	140.87	0.2279
340	0.1149	150.93	0.2451	0.0970	148.70	0.2399	0.0830	146.36	0.2349
360	0.1206	155.69	0.2510	0.1023	153.69	0.2460	0.0881	151.60	0.2413
380	0.1260	160.39	0.2566	0.1074	158.56	0.2519	0.0929	156.67	0.2475
400	0.1312	165.02	0.2621	0.1122	163.34	0.2575	0.0975	161.62	0.2533
420	0.1363	169.62	0.2674	0.1169	168.07	0.2630	0.1018	166.49	0.2589
440	0.1412	174.20	0.2725	0.1214	172.76	0.2682	0.1060	171.29	0.2643
460	0.1460	178.76	0.2775	0.1258	177.41	0.2734	0.1101	176.05	0.2695
480	0.1507	183.31	0.2824	0.1301	182.05	0.2783	0.1141	180.77	0.2746
500	0.1553	187.87	0.2872	0.1343	186.67	0.2832	0.1179	185.47	0.2795
520	0.1599	192.42	0.2919	0.1384	191.30	0.2880	0.1217	190.17	0.2844

Table F.10
Thermodynamic Properties of R-134a

Table F.10.1
Saturated R-134a

Temp. F	Press. (psia)	SPECIFIC VOLUME, ft³/lbm			INTERNAL ENERGY, Btu/lbm		
		Sat. Liquid v_f	Evap. v_{fg}	Sat. Vapor v_g	Sat. Liquid u_f	Evap. u_{fg}	Sat. Vapor u_g
−100	0.951	0.01077	39.5032	39.5139	50.47	94.15	144.62
−90	1.410	0.01083	27.3236	27.3345	52.03	93.89	145.92
−80	2.047	0.01091	19.2731	19.2840	53.96	93.27	147.24
−70	2.913	0.01101	13.8538	13.8648	56.19	92.38	148.57
−60	4.067	0.01111	10.1389	10.1501	58.64	91.26	149.91
−50	5.575	0.01122	7.5468	7.5580	61.27	89.99	151.26
−40	7.511	0.01134	5.7066	5.7179	64.04	88.58	152.62
−30	9.959	0.01146	4.3785	4.3900	66.90	87.09	153.99
−20	13.009	0.01159	3.4049	3.4165	69.83	85.53	155.36
−15.3	14.696	0.01166	3.0350	3.0466	71.25	84.76	156.02
−10	16.760	0.01173	2.6805	2.6922	72.83	83.91	156.74
0	21.315	0.01187	2.1340	2.1458	75.88	82.24	158.12
10	26.787	0.01202	1.7162	1.7282	78.96	80.53	159.50
20	33.294	0.01218	1.3928	1.4050	82.09	78.78	160.87
30	40.962	0.01235	1.1398	1.1521	85.25	76.99	162.24
40	49.922	0.01253	0.9395	0.9520	88.45	75.16	163.60
50	60.311	0.01271	0.7794	0.7921	91.68	73.27	164.95
60	72.271	0.01291	0.6503	0.6632	94.95	71.32	166.28
70	85.954	0.01313	0.5451	0.5582	98.27	69.31	167.58
80	101.515	0.01335	0.4588	0.4721	101.63	67.22	168.85
90	119.115	0.01360	0.3873	0.4009	105.04	65.04	170.09
100	138.926	0.01387	0.3278	0.3416	108.51	62.77	171.28
110	161.122	0.01416	0.2777	0.2919	112.03	60.38	172.41
120	185.890	0.01448	0.2354	0.2499	115.62	57.85	173.48
130	213.425	0.01483	0.1993	0.2142	119.29	55.17	174.46
140	243.932	0.01523	0.1684	0.1836	123.04	52.30	175.34
150	277.630	0.01568	0.1415	0.1572	126.89	49.21	176.11
160	314.758	0.01620	0.1181	0.1343	130.86	45.85	176.71
170	355.578	0.01683	0.0974	0.1142	134.99	42.12	177.11
180	400.392	0.01760	0.0787	0.0963	139.32	37.91	177.23
190	449.572	0.01862	0.0614	0.0801	143.97	32.94	176.90
200	503.624	0.02013	0.0444	0.0645	149.19	26.59	175.79
210	563.438	0.02334	0.0238	0.0471	156.18	16.17	172.34
214.1	589.953	0.03153	0	0.0315	164.65	0	164.65

TABLE F.10.1 (*continued*)
Saturated R-134a

Temp. (F)	Press. (psia)	ENTHALPY, Btu/lbm			ENTROPY, Btu/lbm R		
		Sat. Liquid h_f	Evap. h_{fg}	Sat. Vapor h_g	Sat. Liquid s_f	Evap. s_{fg}	Sat. Vapor s_g
−100	0.951	50.47	101.10	151.57	0.1563	0.2811	0.4373
−90	1.410	52.04	101.02	153.05	0.1605	0.2733	0.4338
−80	2.047	53.97	100.58	154.54	0.1657	0.2649	0.4306
−70	2.913	56.19	99.85	156.04	0.1715	0.2562	0.4277
−60	4.067	58.65	98.90	157.55	0.1777	0.2474	0.4251
−50	5.575	61.29	97.77	159.06	0.1842	0.2387	0.4229
−40	7.511	64.05	96.52	160.57	0.1909	0.2300	0.4208
−30	9.959	66.92	95.16	162.08	0.1976	0.2215	0.4191
−20	13.009	69.86	93.72	163.59	0.2044	0.2132	0.4175
−15.3	14.696	71.28	93.02	164.30	0.2076	0.2093	0.4169
−10	16.760	72.87	92.22	165.09	0.2111	0.2051	0.4162
0	21.315	75.92	90.66	166.58	0.2178	0.1972	0.4150
10	26.787	79.02	89.04	168.06	0.2244	0.1896	0.4140
20	33.294	82.16	87.36	169.53	0.2310	0.1821	0.4132
30	40.962	85.34	85.63	170.98	0.2375	0.1749	0.4124
40	49.922	88.56	83.83	172.40	0.2440	0.1678	0.4118
50	60.311	91.82	81.97	173.79	0.2504	0.1608	0.4112
60	72.271	95.13	80.02	175.14	0.2568	0.1540	0.4108
70	85.954	98.48	77.98	176.46	0.2631	0.1472	0.4103
80	101.515	101.88	75.84	177.72	0.2694	0.1405	0.4099
90	119.115	105.34	73.58	178.92	0.2757	0.1339	0.4095
100	138.926	108.86	71.19	180.06	0.2819	0.1272	0.4091
110	161.122	112.46	68.66	181.11	0.2882	0.1205	0.4087
120	185.890	116.12	65.95	182.07	0.2945	0.1138	0.4082
130	213.425	119.88	63.04	182.92	0.3008	0.1069	0.4077
140	243.932	123.73	59.90	183.63	0.3071	0.0999	0.4070
150	277.630	127.70	56.49	184.18	0.3135	0.0926	0.4061
160	314.758	131.81	52.73	184.53	0.3200	0.0851	0.4051
170	355.578	136.09	48.53	184.63	0.3267	0.0771	0.4037
180	400.392	140.62	43.74	184.36	0.3336	0.0684	0.4020
190	449.572	145.52	38.05	183.56	0.3409	0.0586	0.3995
200	503.624	151.07	30.73	181.80	0.3491	0.0466	0.3957
210	563.438	158.61	18.65	177.26	0.3601	0.0278	0.3879
214.1	589.953	168.09	0	168.09	0.3740	0	0.3740

TABLE F.10.2
Superheated R-134a

Temp. (F)	v (ft³/lbm)	u (Btu/lbm)	h (Btu/lbm)	s (Btu/lbm R)	v (ft³/lbm)	u (Btu/lbm)	h (Btu/lbm)	s (Btu/lbm R)
	5 psia (−53.51)				15 psia (−14.44)			
Sat.	8.3676	150.78	158.53	0.4236	2.9885	156.13	164.42	0.4168
−20	9.1149	156.03	164.47	0.4377	—	—	—	—
0	9.5533	159.27	168.11	0.4458	3.1033	158.58	167.19	0.4229
20	9.9881	162.58	171.83	0.4537	3.2586	162.01	171.06	0.4311
40	10.4202	165.99	175.63	0.4615	3.4109	165.51	174.97	0.4391
60	10.8502	169.48	179.52	0.4691	3.5610	169.07	178.95	0.4469
80	11.2786	173.06	183.50	0.4766	3.7093	172.70	183.00	0.4545
100	11.7059	176.73	187.56	0.4840	3.8563	176.41	187.12	0.4620
120	12.1322	180.49	191.71	0.4913	4.0024	180.20	191.31	0.4694
140	12.5578	184.33	195.95	0.4985	4.1476	184.08	195.59	0.4767
160	12.9828	188.27	200.28	0.5056	4.2922	188.03	199.95	0.4838
180	13.4073	192.29	204.69	0.5126	4.4364	192.07	204.39	0.4909
200	13.8314	196.39	209.19	0.5195	4.5801	196.19	208.91	0.4978
220	14.2551	200.58	213.77	0.5263	4.7234	200.40	213.51	0.5047
240	14.6786	204.86	218.44	0.5331	4.8665	204.68	218.19	0.5115
260	15.1019	209.21	223.19	0.5398	5.0093	209.05	222.96	0.5182
280	15.5250	213.65	228.02	0.5464	5.1519	213.50	227.80	0.5248
300	15.9478	218.17	232.93	0.5530	5.2943	218.03	232.72	0.5314
320	16.3706	222.78	237.92	0.5595	5.4365	222.64	237.73	0.5379
	30 psia (15.15)				40 psia (28.83)			
Sat.	1.5517	160.21	168.82	0.4136	1.1787	162.08	170.81	0.4125
20	1.5725	161.09	169.82	0.4157	—	—	—	—
40	1.6559	164.73	173.93	0.4240	1.2157	164.18	173.18	0.4173
60	1.7367	168.41	178.05	0.4321	1.2796	167.95	177.42	0.4256
80	1.8155	172.14	182.21	0.4400	1.3413	171.74	181.67	0.4336
100	1.8929	175.92	186.43	0.4477	1.4015	175.57	185.95	0.4414
120	1.9691	179.77	190.70	0.4552	1.4604	179.46	190.27	0.4490
140	2.0445	183.68	195.03	0.4625	1.5184	183.41	194.65	0.4565
160	2.1192	187.68	199.44	0.4697	1.5757	187.43	199.09	0.4637
180	2.1933	191.74	203.92	0.4769	1.6324	191.52	203.60	0.4709
200	2.2670	195.89	208.48	0.4839	1.6886	195.69	208.18	0.4780
220	2.3403	200.12	213.11	0.4908	1.7444	199.93	212.84	0.4849
240	2.4133	204.42	217.82	0.4976	1.7999	204.24	217.57	0.4918
260	2.4860	208.80	222.61	0.5044	1.8552	208.64	222.37	0.4985
280	2.5585	213.27	227.47	0.5110	1.9102	213.11	227.25	0.5052
300	2.6309	217.81	232.41	0.5176	1.9650	217.66	232.20	0.5118
320	2.7030	222.42	237.43	0.5241	2.0196	222.28	237.23	0.5184
340	2.7750	227.12	242.53	0.5306	2.0741	226.99	242.34	0.5248
360	2.8469	231.89	247.70	0.5370	2.1285	231.76	247.52	0.5312

TABLE F.10.2 (*continued*)
Superheated R-134a

Temp. (F)	v (ft³/lbm)	u (Btu/lbm)	h (Btu/lbm)	s (Btu/lbm R)	v (ft³/lbm)	u (Btu/lbm)	h (Btu/lbm)	s (Btu/lbm R)
	60 psia (49.72)				80 psia (65.81)			
Sat.	0.7961	164.91	173.75	0.4113	0.5996	167.04	175.91	0.4105
60	0.8204	166.95	176.06	0.4157	—	—	—	—
80	0.8657	170.89	180.51	0.4241	0.6262	169.97	179.24	0.4168
100	0.9091	174.85	184.94	0.4322	0.6617	174.06	183.86	0.4252
120	0.9510	178.82	189.38	0.4400	0.6954	178.15	188.44	0.4332
140	0.9918	182.85	193.86	0.4476	0.7279	182.25	193.03	0.4410
160	1.0318	186.92	198.38	0.4550	0.7595	186.39	197.64	0.4485
180	1.0712	191.06	202.95	0.4623	0.7903	190.58	202.28	0.4559
200	1.1100	195.26	207.59	0.4694	0.8205	194.83	206.98	0.4632
220	1.1484	199.54	212.29	0.4764	0.8503	199.14	211.72	0.4702
240	1.1865	203.88	217.05	0.4833	0.8796	203.51	216.53	0.4772
260	1.2243	208.30	221.89	2.4902	0.9087	207.95	221.41	0.4841
280	1.2618	212.79	226.80	0.4969	0.9375	212.47	226.34	0.4909
300	1.2991	217.36	231.78	0.5035	0.9661	217.05	231.35	0.4975
320	1.3362	222.00	236.83	0.1501	0.9945	221.71	236.43	0.5041
340	1.3732	226.71	241.96	0.5166	1.0227	226.44	241.58	0.5107
360	1.4100	231.51	247.16	0.5230	1.0508	231.24	246.80	0.5171
380	1.4468	236.37	252.43	0.5294	1.0788	236.12	252.09	0.5235
400	1.4834	241.31	257.78	0.5357	1.1066	241.07	257.46	0.5298
	100 psia (79.08)				125 psia (93.09)			
Sat.	0.4794	168.74	177.61	0.4100	0.3814	170.46	179.28	0.4094
80	0.4809	168.93	177.83	0.4104	—	—	—	—
100	0.5122	173.20	182.68	0.4192	0.3910	172.01	181.06	0.4126
120	0.5414	177.42	187.44	0.4276	0.4171	176.43	186.08	0.4214
140	0.5691	181.62	192.15	0.4356	0.4413	180.77	190.98	0.4297
160	0.5957	185.84	196.86	0.4433	0.4642	185.10	195.84	0.4377
180	0.6215	190.08	201.58	0.4508	0.4861	189.43	200.68	0.4454
200	0.6466	194.38	206.34	0.4581	0.5073	193.79	205.52	0.4529
220	0.6712	198.72	211.15	0.4653	0.5278	198.19	210.40	0.4601
240	0.6954	203.13	216.00	0.4723	0.5480	202.64	215.32	0.4673
260	0.7193	207.60	220.91	0.4792	0.5677	207.15	220.28	0.4743
280	0.7429	212.14	225.88	0.4861	0.5872	211.72	225.30	0.4811
300	0.7663	216.74	230.92	0.4928	0.6064	216.35	230.38	0.4879
320	0.7895	221.42	236.03	0.4994	0.6254	221.05	235.51	0.4946
340	0.8125	226.16	241.20	0.5060	0.6442	225.81	240.71	0.5012
360	0.8353	230.98	246.44	0.5124	0.6629	230.65	245.98	0.5077
380	0.8580	235.87	251.75	0.5188	0.6814	235.56	251.32	0.5141
400	0.8806	240.83	257.13	0.5252	0.6998	240.53	256.72	0.5205

TABLE F.10.2 (*continued*)
Superheated R-134a

Temp. (F)	v (ft³/lbm)	u (Btu/lbm)	h (Btu/lbm)	s (Btu/lbm R)	v (ft³/lbm)	u (Btu/lbm)	h (Btu/lbm)	s (Btu/lbm R)
	150 psia (105.13)				200 psia (125.25)			
Sat.	0.3150	171.87	180.61	0.4089	0.2304	174.00	182.53	0.4080
120	0.3332	175.33	184.57	0.4159	—	—	—	—
140	0.3554	179.85	189.72	0.4246	0.2459	177.72	186.82	0.4152
160	0.3761	184.31	194.75	0.4328	0.2645	182.54	192.33	0.4242
180	0.3955	188.74	199.72	0.4407	0.2814	187.23	197.64	0.4327
200	0.4141	193.18	204.67	0.4484	0.2971	191.86	202.85	0.4407
220	0.4321	197.64	209.63	0.4558	0.3120	196.46	208.01	0.4484
240	0.4496	202.14	214.62	0.4630	0.3262	201.08	213.15	0.4559
260	0.4666	206.69	219.64	0.4701	0.3400	205.72	218.31	0.4631
280	0.4833	211.29	224.70	0.4770	0.3534	210.40	223.48	0.4702
300	0.4998	215.95	229.82	0.4838	0.3664	215.13	228.69	0.4772
320	0.5160	220.67	235.00	0.4906	0.3792	219.91	233.94	0.4840
340	0.5320	225.46	240.23	0.4972	0.3918	224.74	239.24	0.4907
360	0.5479	230.32	245.52	0.5037	0.4042	229.64	244.60	0.4973
380	0.5636	235.24	250.88	0.5102	0.4165	234.60	250.01	0.5038
400	0.5792	240.23	256.31	0.5166	0.4286	239.62	255.48	0.5103
	250 psia (141.87)				300 psia (156.14)			
Sat.	0.1783	175.50	183.75	0.4068	0.1428	176.50	184.43	0.4055
160	0.1955	180.42	189.46	0.4162	0.1467	177.70	185.84	0.4078
180	0.2117	185.49	195.28	0.4255	0.1637	183.44	192.53	0.4184
200	0.2261	190.38	200.84	0.4340	0.1779	188.71	198.59	0.4278
220	0.2394	195.18	206.26	0.4421	0.1905	193.77	204.35	0.4364
240	0.2519	199.94	211.60	0.4498	0.2020	198.72	209.93	0.4445
260	0.2638	204.70	216.90	0.4573	0.2128	203.62	215.43	0.4522
280	0.2752	209.47	222.21	0.4646	0.2230	208.50	220.88	0.4597
300	0.2863	214.27	227.52	0.4717	0.2328	213.39	226.31	0.4669
320	0.2971	219.12	232.86	0.4786	0.2423	218.30	231.75	0.4740
340	0.3076	224.01	238.24	0.4854	0.2515	223.25	237.21	0.4809
360	0.3180	228.95	243.66	0.4921	0.2605	228.24	242.70	0.4877
380	0.3282	233.95	249.13	0.4987	0.2693	233.29	248.24	0.4944
400	0.3382	239.01	254.65	0.5052	0.2779	238.38	253.81	0.5009

TABLE F.10.2 (*continued*)
Superheated R-134a

Temp. (F)	v (ft³/lbm)	u (Btu/lbm)	h (Btu/lbm)	s (Btu/lbm R)	v (ft³/lbm)	u (Btu/lbm)	h (Btu/lbm)	s (Btu/lbm R)
	400 psia (179.92)				500 psia (199.36)			
Sat.	0.0965	177.23	184.37	0.4020	0.0655	175.90	181.96	0.3960
180	0.0966	177.26	184.41	0.4020	—	—	—	—
200	0.1146	184.44	192.92	0.4152	0.0666	176.38	182.54	0.3969
220	0.1277	190.41	199.86	0.4255	0.0867	185.78	193.80	0.4137
240	0.1386	195.92	206.19	0.4347	0.0990	192.46	201.62	0.4251
260	0.1484	201.21	212.20	0.4432	0.1089	198.40	208.47	0.4347
280	0.1573	206.38	218.03	0.4512	0.1174	204.00	214.86	0.4435
300	0.1657	211.49	223.76	0.4588	0.1252	209.41	220.99	0.4517
320	0.1737	216.58	229.44	0.4662	0.1323	214.74	226.98	0.4594
340	0.1813	221.68	235.09	0.4733	0.1390	220.01	232.87	0.4669
360	0.1886	226.79	240.75	0.4803	0.1454	225.27	238.73	0.4741
380	0.1957	231.93	246.42	0.4872	0.1516	230.53	244.56	0.4812
400	0.2027	237.12	252.12	0.4939	0.1575	235.82	250.39	0.4880
	750 psia				1000 psia			
180	0.01640	136.22	138.49	0.3285	0.01593	134.77	137.71	0.3262
200	0.01786	144.85	147.32	0.3421	0.01700	142.70	145.84	0.3387
220	0.02069	155.27	158.14	0.3583	0.01851	151.26	154.69	0.3519
240	0.03426	173.83	178.58	0.3879	0.02102	160.95	164.84	0.3666
260	0.05166	187.78	194.95	0.4110	0.02603	172.59	177.40	0.3843
280	0.06206	196.16	204.77	0.4244	0.03411	184.70	191.01	0.4029
300	0.06997	203.08	212.79	0.4351	0.04208	194.58	202.37	0.4181
320	0.07662	209.37	220.00	0.4445	0.04875	202.67	211.69	0.4302
340	0.08250	215.33	226.78	0.4531	0.05441	209.79	219.86	0.4406
360	0.08786	221.11	233.30	0.4611	0.05938	216.36	227.35	0.4498
380	0.09284	226.78	239.66	0.4688	0.06385	222.61	234.43	0.4583
400	0.09753	232.39	245.92	0.4762	0.06797	228.67	241.25	0.4664

Table F.11

Enthalpy of Formation and Absolute Entropy of Various Substances at 77 F, 1 atm Pressure

Substance	Formula	M	State	$\bar{h}_f^0$ Btu/lbmol	$\bar{s}_f^0$ Btu/lbmol R
Water	H_2O	18.015	gas	−103 966	45.076
Water	H_2O	18.015	liq	−122 885	16.707
Hydrogen peroxide	H_2O_2	34.015	gas	−58 515	55.623
Ozone	O_3	47.998	gas	+61 339	57.042
Carbon (graphite)	C	12.011	solid	0	1.371
Carbon monoxide	CO	28.011	gas	−47 518	47.182
Carbon dioxide	CO_2	44.010	gas	−169 184	51.038
Methane	CH_4	16.043	gas	−32 190	44.459
Acetylene	C_2H_2	26.038	gas	+97 477	47.972
Ethene	C_2H_4	28.054	gas	+22.557	52.360
Ethane	C_2H_6	30.070	gas	−36 432	54.812
Propene	C_3H_6	42.081	gas	+8 783	63.761
Propane	C_3H_8	44.094	gas	−44 669	64.442
Butane	C_4H_{10}	58.124	gas	−54 256	73.215
Pentane	C_5H_{12}	72.151	gas	−62 984	83.318
Benzene	C_6H_6	78.114	gas	+35 675	64.358
Hexane	C_6H_{14}	86.178	gas	−71 926	92.641
Heptane	C_7H_{16}	100.205	gas	−80 782	102.153
n-Octane	C_8H_{18}	114.232	gas	−89 682	111.399
n-Octane	C_8H_{18}	114.232	liq	−107 526	86.122
Methanol	CH_3OH	32.042	gas	−86 543	57.227
Ethanol	C_2H_5OH	46.069	gas	−101 032	67.434
Ammonia	NH_3	17.031	gas	−19 656	45.969
T-T-Diesel	$C_{14.4}H_{24.9}$	198.06	liq	−74 807	125.609
Sulfur	S	32.06	solid	0	7.656
Sulfur dioxide	SO_2	64.059	gas	−127 619	59.258
Sulfur trioxide	SO_3	80.058	gas	−170 148	61.302
Nitrogen oxide	N_2O	44.013	gas	+35 275	52.510
Nitromethane	CH_3NO_2	61.04	liq	−48 624	41.034

ANSWERS TO SELECTED PROBLEMS

2.27 19.613 N; 0.102 kg

2.30 0.37%

2.33 4800 N; 3.82 s

2.36 2300 N

2.39 1.28 kg/m³

2.42 0.5 m³/kg; 16 m³/kmol

2.45 198 kPa

2.48 24.38 m/s²

2.51 a. 149.9 kPa

    b. 162.8 kPa

    c. 145.6 kPa

2.54 250 kPa; 99.6 kPa

2.60 199 kPa; 125.5 kPa

2.63 17.66 m

2.66 3.476 m

2.69 8.72 m

2.72 25.84 kPa

2.75 106.4 kPa

2.78 0.18%

2.81 23.94 kPa

2.84 370 kPa

2.93E 1165 lbf; 3.42 s

2.96E 15 ft³/lbm; 480 ft³/lbmol

2.99E 16.86 lbf/in.²

2.102E 141.3 lbf/in.²

2.105E 0.28 in.

3.21 7.38 MPa; 472 kg/m³; 6.14 MPa; 275 kg/m³

3.24 a. 1234 K

    b. 1356 K

3.36 a. 0.2319 m³/kg

    b. 0.000 733 m³/kg

    c. 0.090 58 m³/kg

    d. 0.4922 m³/kg

    e. 1.4153 m³/kg

3.39 a. 0.012 28 m³/kg

    b. 0.044 93 m³/kg

    c. 0.005 563 m³/kg

    d. 0.013 58 m³/kg

3.42 23.8 MPa

3.48 120.2°C; 0.05 m

3.51 18.9 kPa

3.54 771 kPa; 1318 kPa, 0.4678

3.57 0.948

3.60 1.318 MPa, 93.295 kg

3.63 900 kPa, 1.6263 m³

3.69 2152 kPa

3.72 0.603 kg

3.75 2.75%; 9.4%; 25.3%

3.78 204.4 kPa

3.81 4.5%; 1.4%

3.84 356 kg/m³

3.87 925 kPa, 0.007 m³/kg

3.90 a. 120°C; 0.516

    b. 117 K; 0.959

3.93 a. 71.4°C

    b. 858 kPa, 0.666

3.96 a. 99.9 kg

    b. 149 kPa

3.99 1450 kPa

3.102 1554 kPa, 0.118

3.105 641°C

3.111 a. 372.5 K

    b. 53 mm

3.114 60%; 1%

3.117 a. 6.7°C, 0.1729 m³/kg

     b. 1.9878 m³/kg

     c. 336.8 kPa

3.120 0.9578 m³/kg

3.138E a. 1.752 ft³/lbm

     b. 0.011 78 ft³/lbm

     c. 1.0669 ft³/lbm

     d. 21.564 ft³/lbm

3.141E 70.36 lbm

3.144E 0.111 lbm

3.147E 1966.88 lbm

3.150E 0.021 54 lbm

3.153E 217 lbf/in.²

3.156E −8.1 F, 1.761 ft³/lbm

4.21 9807 J

4.24 0.000 833 m³; 0.0833 m, 0.0278 m

4.27 303.5 m

4.30 500 N; 0.05 m, 0.083 33 m; 25 J

4.33 40 kJ

4.36 −128.7 kJ

4.39 0.2375 kJ

4.42 a. 404°C

     d. 163.35 kJ

4.45 0.0427 kJ

4.48 −80.4 kJ

4.51 1.969; −51.8 kJ/kg

4.54 117.5 kJ

4.57 43.2 kJ/kg

4.60 −49.4 kJ

4.63 0 kJ; −20.22 kJ

4.66 −38.64 kJ/kg

4.69 583.2 kJ

4.72 0.318 m, 17.47 J

4.75 5.89 × 10⁻⁵ J

4.84 186 W

4.87 98.8 km/h

4.90 1500 W

4.93 1 kW

4.96 15.8°C

4.99 480 m²

4.102 45°C

4.105 725°C

4.108 396 kPa; −26.7 kJ

4.111 −13.4 kJ

4.114 143.6°C; 0.4625 m³; 145 kJ

4.123E 154.2 Btu; 12.85 Btu

4.126E 0.030 873 ft³; 0.3087 ft, 0.1029 ft

4.129E −10.49 Btu

4.132E 3.33 Btu

4.135E 44.42 Btu

4.138E 2.94 ft lbf

4.141E 17 325 Btu/h

5.21 31 kJ

5.24 463 kJ, 117.7 kJ

5.27 9.9 m/s; −5.49 m

5.30 a. 1435 kJ/kg, 0.2645 m³/kg

     b. 1374.5 kJ/kg, 1.4153 m³/kg

     c. 1200 kPa, 1383 kJ/kg

5.33 a. 68.1°C, 0.043 87 m³/kg, 208.1 kJ/kg

     b. 680.7 kPa, 0.0289 m³/kg, 219.7 kJ/kg, 0.8287

     c. 1017 kPa, 0.0177 m³/kg, 382.1 kJ/kg, 0.8788

5.36 133.6°C, 0 kJ, −1148 kJ

5.39 −263.3 kJ

5.42 −23.5 kJ

5.45 877.8 kJ, 0 kJ

5.48 287.7 m³

5.51 179.9°C, 1684 kJ

5.54 0.0327 m³; 0.23 m³; 605 kJ

5.57 803°C, 587 kJ

5.60 25°C

5.63 25 510 kJ

5.66 −8395 kJ

5.69 a. 0.9314 kg, 0.5798 m³

     b. 85.24 kJ

     c. 588 kJ

5.72 191.3 kJ

5.75 41.82 MJ

5.78 65.9°C

5.81 90°C

5.84 a. 397.2 kJ/kg

     b. 490.1 kJ/kg

     c. 485.3 kJ/kg

5.87 1. 1237.7 kJ/kg

     b. 1839 kJ/kg

c. 1767.8 kJ/kg

d. 1764.2 kJ/kg

5.90 188.3 kPa, 498.4 K

5.93 220.7 kJ

5.96 1160 K; 71.0 kJ

5.99 0.03 m^3; 1913 kPa; −83.5 kJ; −319.9 kJ

5.102 b. 1000 kPa, 1200 K

d. 32.3 kJ, 212 kJ

5.105 −0.192 kJ; −0.072 kJ

5.108 668 K; 121.4 kJ/kg, 60.5 kJ/kg

5.111 1491 kPa; 41.5 kJ; 1024.9 kJ

5.114 27.3 kJ

5.117 4.2 min.

5.120 15.08 h

5.123 8°C

5.126 −3115 kJ, −32 276 kJ

5.129 51.2°C, 33.2°C, 0.1 m^3, 359.8 kJ/kg

5.132 361 kPa; 2080 kJ; 60 kJ

5.135 a. 0.0132 m^3

b. 8.48 kJ

c. 129.8 kJ

d. 43.3 min.

5.138 0 kJ; −15.7 kJ

5.147E 22.28 Btu

5.150E a. 225 lbf/in.^{2}, 0.2104 ft^3/lbm

b. 628.4 Btu/lbm, 6.457 ft^3/lbm

c. 55 lbf/in.^{2}, 0.999 ft^3/lbm, 185.2 Btu/lbm

5.153E −223 Btu

5.156E 12 218 ft^3

5.159E a. 0.249

b. 662 Btu

5.162E 163 lbf/in.^{2}, 633.5 Btu/lbm, 0 Btu/lbm, 312.9 Btu/lbm

5.165E 337.4 Btu

5.168E −575.5 Btu

5.171E 74.1 Btu/lbm, 288.3 Btu/lbm; 0 Btu/lbm, −214.2 Btu/lbm

5.174E 122.1 lbf/in.^{2}, 825.9 R, −0.0139 Btu

5.177E 16.3 Btu, 34.6 Btu

5.180E 32.9 s

5.183E 4452 Btu

6.24 60 min.

6.27 6.36 m/s

6.30 381.9 m/s

6.33 581.8 m/s

6.36 123.9 kPa; 319.7 K

6.39 22.9°C; 215.7 kPa

6.42 −20°C; 3679%

6.45 0.0579; 240 m/s

6.48 482.3 kJ/kg; 964.6 kW

6.51 664 m^3/s

6.54 −9.9 kW

6.57 3.333 kg/s

6.60 a. 1.538

b. −140 kJ/kg; 73.8 kJ/kg

6.63 1374.5 kJ/kg

6.66 0.866 kW; 0.99 kW

6.69 0.106 kg/s

6.72 0.98 kW

6.75 29.4 m/s; 2.31 kg/s

6.78 91.565 MW

6.81 131.2 m/s; 1056 kW

6.84 69 kg/s

6.87 1.815 kg/s

6.90 426.3 K

6.93 119.6°C; 3.0 m^3/s

6.96 0.964 kg/s

6.99 49 m/s; 24 800 kW

6.102 a. 2673.9 kJ/kg, 0.9755

b. 22.489 MW

c. 18.394 MW

d. 0.26

6.105 123 075 kg/h

6.108 16.96 kg; −468.9 kJ

6.111 520°C; 0.342 m^3

6.114 −379 636 kJ

6.117 8.90 kg; 25.459 MJ

6.120 0.903 m

6.123 270 390 kJ

6.126 8.743 kg/s

6.129 a. −118.77 MW

b. −127.8 kW

6.132 773.7 K

6.135 0.52°C

6.138 1.79 m^2; 4.39 m^2

6.144E 3.0 ft/s

6.147E 1.453 Btu/s; 4.255 Btu/s; 20.64 ft^3/s^2

6.150E 17.72 lbf/in.^{2}; 569.7 R

6.153E 7.57 lbm/h

6.156E 0.432 hp

6.159E 0.264 lbm/s

6.162E 61 lbm/s

6.165E 1.904 lbm/s

6.168E 4.75×10^7 Btu/h; 2.291×10^8 Btu/h

6.171E 2.667 ft^3; 15.8 Btu

6.174E 47 479 Btu

7.18 0.225

7.21 2.164

7.24 1.53 g/s; 42.9 kW

7.27 0.1875

7.30 26.8 kJ; 35.8 s

7.42 7.07

7.45 0.051

7.48 2.56 kW

7.54 3 J; 0.000 33

7.57 5886.7 kJ; 0.0142

7.60 0.731

7.63 9.76 kW

7.75 38.4°C

7.78 378.4 kJ/kg; 126.1 kJ/kg; 0.667

7.81 27.2°C; 10.4°C

7.84 0.687

7.87 153.1 kW

7.99 23.06 Btu; 32.4 s

7.102E 85 100 Btu

7.108E 33.8 Btu/s

7.111E 0.58 Btu

7.114E 505 680 Btu; 0.28; 0

7.117E 1591 Btu

8.27 a. 2573.8 kJ/kg; 65.0°C; 0.981

    b. 682°C; 0.7139 kJ/kg K

    c. 0.7325 kJ/kg K

    d. 0.045 06 m^3/kg; 1.3022 kJ/kg K

    e. 1356.7 kJ/kg

8.33 1940 kJ/kg; 0.46 kJ

8.39 363.75 kJ; 396.97 kJ

8.42 7.11 kJ; 59.65 kJ

8.45 2000 kPa; 471.2 kJ

8.48 30°C; −31.6 kJ/kg

8.51 −108.4 kJ

8.54 172°C; −132 kJ

8.57 0.385 m^3

8.60 4910 kJ; 1290.3 kJ

8.63 312.2°C; 0.225 kJ/K

8.66 a. 3.662 kJ/K; 3.950 kJ/K; 12.929 kJ/K

8.69 −66.2 kJ; −824.1 kJ; 0.716 kJ/K

8.72 0.395 kJ/kg K

8.75 334.6 kJ/kg; 1.0091 kJ/kg K; 334.4 kJ/kg;
     1.0086 kJ/kg K

8.78 2.57 kJ/K

8.81 81 946 kJ

8.84 0.202 kJ/K

8.87 660.8 kJ; 0.661 kJ/K

8.90 1.92 cm^3; 0.145 kJ/K

8.93 b. 11.48 cm^3; 405.3 K; 8.44 cm^3; 298 K

8.96 48.6 cm; 0.2935 kJ

8.99 0.315 kJ/K

8.102 6.52 kg

8.105 300 kPa; 400 K; 0.52 kJ/K

8.111 1.81 kJ; −0.96 kJ

8.114 1.538; −182.1 kJ; 147.3 kJ; 0.0936 kJ/K

8.117 −191.3 kJ/kg; −47.9 kJ/kg;
     0.0374 kJ/kg

8.120 0.1 kW/K; 0.1 kW/K

8.123 0.05 W/K; 0.11 W/K; 0.168 W/K

8.129 3.33 kJ; 30.43 kJ; 9.04 kJ

8.132 12.18 kJ/K

8.135 a. 2.54 kJ

8.138 0.768 kJ/K

8.150E b. 0.214; 0.9326

     c. 7.995

8.153E 100 Btu; 0

8.156E 2802.7 Btu; 543.3 Btu

8.159E 0.0645 Btu/R

8.162E 0.0956 Btu/lbm R

8.165E 0.1277 Btu/R

8.168E 12.392 atm; 0.1614 ft^3

8.171E 10.53 lbm

8.174E 69.4 lbf/in.^{2}; −235.6 Btu/lbm;
     −377 Btu/lbm

8.177E 1.305; 0.854 ft^3; −23.35 Btu;
 −5.56 Btu; 0.002 26 Btu/R

8.180E 0.000 878 Btu/R s; 0.002 59 Btu/R s

9.21 349.7°C; 953.9 kJ/kg

9.24 −4.1 kW

9.27 259.1 K; 93.6 kPa

9.30 149.6 kPa; 322.9 K

9.33 69.29 kW; 69.29 kW

9.36 167.2 kW; 661.2 kW

9.39 a. 989.3 kJ/kg
 b. 510 kPa

9.42 a. 1569 kJ/kg; 0.8433
 b. −20.1 kJ/kg; 187.6 kJ/kg
 c. 0.428

9.45 744.4 K

9.48 0; 187.2 kJ/kg; 0.1626 kJ/kg K

9.51 0.017 kW/K

9.54 47.29 kg/min; 8.94 kJ/K

9.60 0.952 kg/s; 4.05 kg/s; 0.852 kW/K

9.63 0.506 kJ/kg K

9.66 13 186 kJ; 12.37 kJ/K

9.69 1.076 × 10^6; 680 K; −2.322 × 10^8 kJ

9.72 0.361 kg; 332.9°C; 0.186 kJ/K

9.75 0.466 kJ/K

9.78 132.2 kPa; 500 kW

9.81 14.08 m/s; 10.1 m

9.84 −4.0 kW

9.87 17.6°C; 100.67 kPa

9.90 365.8 kW

9.93 −51.3 kJ/kg; 135.8 kJ/kg; 0.117 kJ/kg K

9.96 0.854; 0.1492 kJ/kg K

9.99 49.7 kW

9.102 a. 1333.7 kJ/kg; 0.943
 b. 23.7 kJ/kg; 191.2 kJ/kg
 c. 0.362

9.105 951 K; 0.1268 kJ/kg K

9.108 0.92; 0.028 kJ/kg K

9.111 −180.3 kJ/kg; 55.9°C

9.114 −2.77 kW

9.117 281°C; 0.724 kW/K

9.120 1.867; 0.9639

9.123 0.157 m^3; 0.3469 kJ/K

9.126 a. 0.0952
 b. −0.6 kJ/kg; 328.1 kJ/kg
 c. 0.3088 kJ/kg K

9.129 12.022 kg; 362.3 kJ/kg; 4140 kPa; −539.2 kJ; 4.423 kJ/K

9.132 a. 353.3 K
 b. 35.9 kW; 19.2 kW
 c. 0.0163 kW/K

9.144E 21.6 lbf/in.^{2}; 579.3 R

9.147E 59.59 Btu/s; 59.59 Btu/s

9.150E 12.2 Btu/R s

9.153E 917.8 Btu/s; 2.033 Btu/R s

9.156E 2.132 × 10^6 lbm; 1221 R;
 182.8 lbf/in.^{2}; −2.158 × 10^8 Btu

9.159E −3.53 hp

9.162E 454.6 Btu/s; 94.8 Btu/s

9.165E 63.13 lbf/in.^{2}; 1349.2 R

9.168E 55 lbf/in.^{2}; 137.3 F

9.171E 1.003 × 10^6 Btu; 77.2 Btu/R s

10.21 40 kW

10.24 −48.2 kJ/kg

10.27 −0.504 kW

10.30 1483.9 kJ/kg; 1636.8 kJ/kg

10.33 120.3 kJ

10.36 419.9 kW

10.39 1269 kW

10.42 7444 kW

10.45 46.3°C; 19.8 kJ/kg

10.48 10.0 kg/s; 0.057

10.51 6.237 kJ/kg

10.54 1787.5 kJ/kg; 218.5 kJ/kg; 1.51 kJ/kg; 21.61 kJ/kg

10.57 271.96 kJ/kg

10.60 −36.1 kJ/kg

10.63 a. 500 W
 b. 244 W
 c. 0

10.66 300.7 K; −44.0 kJ

10.69 0.92

10.72 1.007 kg/s; 0.77

10.75 0.918; 0.871

10.78 55.3 kJ/kg; 0.91

10.81 0.403

10.84 377.5 kJ/kg

10.87 10 kW, 5 kW, 5 kW; 6.25 kW, 5 kW, 0.194 kW

10.90 0.94

10.93 96.33 kJ

10.96 6.6 kJ; 271.4 kJ

10.99 702.4 kJ/kg; 764.8 kJ/kg; 0.319; 0.479

10.102 1.4 MPa; 1085.8 kJ; 1147.6 kJ

10.105 44.9°C; 125.4 kPa

10.114E −5.43 Btu/lbm; −19.3 Btu/lbm

10.117E 540.3 R; 19 078 Btu

10.120E 31.68 Btu/s; 5467 Btu/s

10.123E 580.3 R; 8.7 Btu/lbm

10.126E 1.136 Btu/lbm

10.129E a. 500 W

       b. 225 W

       c. 0

10.132E 171.3 Btu

10.135E 21.3 Btu/lbm; 0.946

10.138E 0.335; 0.70

10.141E 157.2 Btu; 213.2 Btu

10.144E 2102 ft/s; 0.95

11.21 0.323

11.27 0.103

11.30 15.43 kW

11.33 0.308

11.36 6487 kW; 14 578 kW

11.39 0.3574; 0.059; 0.3587, 0.0913

11.42 0.1661; 1.0 kJ/kg; 4.5 kJ/kg

11.45 0.357; 896.2 kJ/kg

11.48 7.257 kg/s

11.51 0.191; 49.3 kW

11.54 0.31; 0.803

11.57 0.274

11.60 0.289; 0.117 kJ/kg K

11.63 40.3°C; 29.185 MW; 11.606 MW

11.66 7424 kW

11.69 3036 kW; 7320 kW; 0.484

11.72 1596.5 K; 26.66 kg/s

11.75 375 kPa; 442.2 kJ/kg; 0.339 kg/s;
      958.8 K; 0.687

11.78 253 kJ/kg; 271 kJ/kg

11.81 −163.3 kJ/kg; −133.2 kJ/kg;
      −133.2 kJ/kg

11.84 360.4 kPa; 425.7 kJ/kg; 0.352 kg/s; 975.2 K; 0.678

11.87 544.8 kPa; 1011 m/s

11.90 530 kPa; 1229.8 K; 830 K; 957 m/s

11.93 2502 K; 6338 kPa

11.96 2677 K; 1458 kJ/kg; 1165 K

11.99 7.67; 262 kJ/kg; 4883 kPa

11.102 7946 kPa; 1303.6 kJ/kg; 1055 kPa

11.105 7351 kPa; 2660 K; 0.578

11.108 6298 kPa; 550.5 kJ/kg; 0.653

11.111 20.92; 894.8 kPa

11.114 0.458

11.117 900 K; 429.9 kJ/kg; 15.6

11.120 3.198; 3.172

11.123 0.0397 kg/s

11.126 0.0403 kg/s; 5.21 kW; 0.26 kg/s

11.129 15.55 kW

11.132 0.0917 kW/K; 0.04 kW/K

11.135 1.015 kJ

11.138 0.369

11.141 26.584 kg/s; 133.6°C

11.144 1.843; 0.44

11.147 0.5657

11.150 0.765

11.153 a. 6600 kW; 125 kW

       b. 22 579 kg/s; 23 432 kg/s

       c. 0.033

11.156 0.368; 1033.4 kJ/kg

11.159 304.9 kPa; 323.2 kJ/kg; 0.464 kg/s; 886.8 K; 0.451

11.162 276 kPa; 529.6 kJ/kg; 0.536

11.165 a. 1.433

       b. 1.032

11.168E 0.277

11.171E 0.104

11.174E 0.45; 0.7515

11.177E 12.656 lbm/s

11.180E 0.284; 0.0153 Btu/lbm R

11.183E 2813 R; 63.985 lbm/s

11.186E 61.9 Btu/lbm; 275 Btu/lbm; 0.775

11.189E 1033 lbf/in.^{2}; 5789 R; 0.541;
       188 lbf/in.^{2}

11.192E 1033 lbf/in.^{2}; 5789 R; 0.488;
       169.5 lbf/in.^{2}

11.195E 12.24; 0.5843; 139.8 lbf/in.^{2}

11.198E 0.458

11.201E 1620 R; 198.0 Btu/lbm; 18.1

11.204E 3.206

11.207E 206.3 Btu/lbm; 528.9 Btu/lbm; 0.61

11.210E 1.94 Btu/lbm; 67.76 Btu/lbm;
50.58 Btu/lbm

11.213E 0.357; 421.2 Btu/lbm

11.216E 242.7 Btu/lbm; 449.5 Btu/lbm; 0.427

12.21 0.5427, 0.2094, 0.2479; 0.3221 kJ/kg K; 5.065 m^3

12.24 0.381, 0.180, 0.439; 0.096 45 kJ/kg K;
0.8298 kJ/kg K; 0.7334 kJ/kg K

12.27 0.2513 kJ/kg K; 1.005 m^3

12.30 72.586; 0.114 54 kJ/kg K; 1.1655

12.33 1.675 m^3; 372.8 kJ

12.36 334.5 K; 305.7 kPa

12.39 1096 kW

12.42 1247 kW

12.45 34.575 kJ/kmol; 679 K

12.48 2.863 kJ; 20.16 kJ

12.51 279 kPa; 418.6 K; −231 kJ

12.54 320.4 K; −46.16 kJ/kg

12.57 13 236 kPa; 0.933; −1595.7 kJ;
−1988 kJ; −6.7656 kJ/K

12.60 304.7 K; 0.1789 kJ/kg K

12.63 301.2 K

12.66 2831 kW

12.69 616 K; −0.339 kW/K

12.72 697.7 kPa; 3647.2 kJ; 5.4 kJ/K

12.75 0.56

12.78 0.387; 15.15 kW

12.81 0.007 73; 0.0155 kg/s; 29.1°C

12.84 −3.055 kW

12.87 −5.15 kJ; 0.0123 kg; 155.9 kPa

12.90 0.0189; 0.0108; 48.6 kJ/kg

12.93 0.019; 0.0099 kg/kg air; 35 kJ/kg air

12.96 a. 25°C

b. 0.71; 0.0143

12.99 33.33 kg/s; 0.323 kg/s

12.102 18.4 kg/h; −4.21 kW

12.105 46°C, 12%; 1.165 kW

12.108 a. 25°C

b. 0.008, 21.4°C

c. 71%; 0.016

12.111 17%, 16 kJ/kg; 100%, −15 kJ/kg

12.114 a. 20.6°C, 23.5°C

b. 0.0086, 20.2°C

c. 37%, 27.2°C

d. 80%, 8.2°C

12.117 54.936 kW; 38.087 kW

12.120 361.3 K; −2.4 kJ

12.123 0.3857 kJ/K

12.126 811.7 kg/s; 0.0163

12.129 0.001 05 kg/s; 4.537 kW; 25.0 kJ/kg; 42°C; 0.30

12.132 −3.48 kJ; 0.201 kJ/K

12.144E 72.586; 21.285 ft lbf/lbm R; 1.1656

12.147E 1123.7 Btu/s

12.150E 38 lbf/in.^{2}; 565.1 R

12.153E 1184 Btu/s

12.156E 0.836 Btu/R

12.159E 7.8 F; −1.49 Btu

12.162E −4.18 Btu; 0.0227 lbm; 20.76 lbf/in.^{2}

12.165E 1.235 Btu/s; −0.78 Btu/s

12.168E 0.123 lbm/min; 0.04 lbm/min; 94 F; 9%

13.21 0.0098 kPa

13.27 4.247 kPa

13.30 −150.6 kW

13.33 3.83 × 10^{-2} Pa

13.45 2.44 kJ

13.48 5.23 × 10^{-9} 1/kPa

13.51 1100 m/s; −66.7 J/kg

13.66 0.9743 kJ/kmol K

13.69 0.354; 13.51 kg

13.72 1.063 MPa; 0.002 38 kg; 0.753 kJ

13.75 −62.1 kJ/kg; −379 kJ/kg

13.78 −0.4703 R

13.81 5.0°C; 4.7°C

13.84 0.471 kg; 36.0 kJ; 101.2 kJ

13.87 55.4 kJ

13.90 −1644 kJ

13.93 5.04

13.96 a. 0.044 m^3; 0.040 33 m^3

13.99 0.88; 0.638

13.102 −8309 kW

13.105 a. −7.71 kJ; −7.71 kJ

b. −10.85 kJ; −7.81 kJ

13.108 −981.4 kJ

13.111 62.57 kW

13.114 575 kPa

13.117 a. 254 K

   b. 469 806 kJ

   c. 259.1 K; −451 523 kJ

13.120E 0.2912 lbf/in.^{2}; 132 ft^3/lbm

13.123E 27.4 F

13.126E 3848 ft/s

13.129E −123.9 Btu/lbmol

13.132E 817.4 R; 98.8 Btu

13.135E −78.4 Btu/lbm; −202.1 Btu/lbm

13.138E −42.1 Btu/lbm; −149.1 Btu/lbm

13.141E −704 Btu

13.144E −892 Btu

14.21 8.943 kg/kg; 14.28 kmol/kmol

14.24 0.80; 125%

14.27 824.1 kg; 23.765 kmol; 32.778 kmol

14.30 145%

14.33 9.444 kg/kg

14.36 0.718 kmol/kmol gas

14.39 43.2°C; 0.0639 kg/kg fuel

14.42 2854 K

14.45 −1 234 557 kJ/kmol

14.48 −255 816 kJ/kmol

14.51 −372 738 kJ/kmol

14.54 1.1123; −1 182 480 kJ

14.57 232 009 kJ/kmol

14.60 20 986 kJ/kg

14.63 −627 058 kJ/kmol

14.66 1668 kJ/m^3

14.69 13 101 kJ/kg fuel; 13 101 kJ/kg fuel; 1216 K

14.72 72.6°C; 2524.5 K

14.75 1773 K

14.78 2048 K

14.81 a. 2909 K

   b. 7400 K

14.84 5712 K

14.87 148.6%; 287 MJ/kmol fuel

14.90 2150 kJ/K kmol fuel

14.93 a. 5.757

   b. 1414.3 kJ/K

14.96 238.3 kPa; −1.613 × 10^6 kJ; 4070 K

14.99 2039 K

14.102 410%; 30.5°C

14.105 1.109 V

14.108 0.013 26; −149 931 kJ

14.111 0.328; 0.414

14.114 238%

14.120 a. 2011 kPa; 666.4 K

   b. 2907 K; 8772 kPa; 512.62 kJ/K

   c. 152 860 kJ

14.126E 125.8 lbf/in.^{2}; −194 945 Btu

14.129E −369 746 Btu/lbmol;
   −337 570 Btu/lbmol

14.132E 6.91; 9.689

14.135E 79.85 Btu/ft^3

14.138E 3628 R

14.141E 3510 R

14.144E a. 5.07

   b. 307.99 Btu/R

14.147E 34.9 Btu/s; −67.45 Btu/s

14.150E 0.353; 0.419

15.21 34.36 MPa

15.24 29.682 MPa

15.27 exp(−12.8407)

15.30 2980 K

15.33 50.3% H

15.36 1444 K

15.39 1108 kPa; 93.7% O$_2$; 97 681 kJ

15.42 0.0237

15.45 exp(−2.1665); exp(−2.4716)

15.48 exp(−8.293)

15.51 69.1% NH$_3$, 21.8% N$_2$, 9.1% H$_2$

15.54 176 811 kJ

15.57 30.87% CH$_4$; 30.87% H$_2$O, 28.7% H$_2$, 9.56% CO

15.60 66.21% H$_2$O, 32.42% C$_2$H$_4$,
   1.37% C$_2$H$_5$OH; 41 330 kJ

15.63 a. 6.202

   b. 85.44% NH$_3$, 10.92% H$_2$, 3.64% N$_2$

15.69 75.904% N$_2$, 10.065% O$_2$, 7.764% H$_2$O, 6.211% CO$_2$, 0.055% NO, 0.001% NO$_2$

15.72 66.11% H$_2$O, 12.91% H$_2$, 9.9% OH, 5.68% H, 5.4% O$_2$

15.75 0.009 67

15.81 16.92 N_2, 3.3538 H_2O, 2.6462 CO_2, 0.6462 H_2, 0.3538 CO

15.84 1.96

15.87 214 306 k/kmol air

15.90E exp(−185.85); exp(+5.127)

15.93E 0.1015

15.96E 75 360 Btu

15.99E a. 6.826

b. 85.91% NH_3, 10.57% H_2, 3.52% N_2

15.102E 66.24% H_2O, 12.88% H_2, 9.87% OH, 5.63% H, 5.38% O_2

15.105E 73.7%; 0.128%

INDEX